ELECTROMAGNETICS

McGraw-Hill Series in Electrical and Computer Engineering

Senior Consulting Editor

Stephen W. Director, ***University of Michigan, Ann Arbor***

Circuits and Systems
Communications and Signal Processing
Computer Engineering
Control Theory
Electromagnetics
Electronics and VLSI Circuits
Introductory
Power
Radar and Antennas

Previous Consulting Editors

Ronald N. Bracewell, Colin Cherry, James F. Gibbons, Willis W. Harman, Hubert Heffner, Edward W. Herold, John G. Linvill, Simon Ramo, Ronald A. Rohrer, Anthony E. Siegman, Charles Susskind, Frederick E. Terman, John G. Truxal, Ernst Weber, and John R. Whinnery

Electromagnetics

Consulting Editor

Stephen W. Director, ***University of Michigan, Ann Arbor***

Goodman: *Introduction to Fourier Optics*
Hayt: *Engineering Electromagnetics*
Kraus and Fleisch: *Electromagnetics with Applications*
Nasar: *Solved Problems in Electromagnetics*
Paul and Nasar: *Introduction to Electromagnetic Fields*
Plonus: *Applied Electromagnetics*

ELECTROMAGNETICS
with Applications

Fifth Edition

JOHN D. KRAUS

McDougal Professor of Electrical
Engineering and Astronomy, Emeritus
Ohio State University

DANIEL A. FLEISCH

Chief Scientist, Aeroflex-Lintek Corp.
Visiting Assistant Professor, Wittenberg University

with a chapter on "Electromagnetic Effects in High-Speed Digital Systems"
by **SAMUEL H. RUSS**
Assistant Professor of Electrical Engineering
Mississippi State University

Boston Burr Ridge, IL Dubuque, IA Madison, WI
New York San Francisco St. Louis
Bangkok Bogotá Caracas Lisbon London Madrid
Mexico City Milan New Delhi Seoul Singapore Sydney Taipei Toronto

WCB/McGraw-Hill

A Division of The ***McGraw-Hill*** *Companies*

ELECTROMAGNETICS WITH APPLICATIONS

This book is printed on acid-free paper.

2 3 4 5 6 7 8 9 0 DOC/DOC 3 2 1 0

ISBN 0-07-289969-7

Vice president and editorial director: *Kevin T. Kane*
Publisher: *Thomas Casson*
Executive editor: *Elizabeth A. Jones*
Editorial coordinator: *Emily J. Gray*
Marketing manager: *John T. Wannemacher*
Project manager: *Kimberly Schau*
Senior production supervisor: *Madelyn S. Underwood*
Designer: *Kiera Cunningham*
Supplement coordinator: *Carol Loreth*
Compositor: *Publication Services*
Typeface: *10/12 Times Roman*
Printer: *R. R. Donnelley & Sons Company*

Front cover: Traveling waves on a lossy transmission line (incident wave to left, reflected wave to right).

Back cover: Maxwell's equations in differential time-harmonic form.

Library of Congress Cataloging-in-Publication Data

Kraus, John Daniel
Electromagnetics : with applications / John D. Kraus and Daniel A. Fleisch : with a chapter on "Electromagnetic effects in high speed digital systems" by Samuel H. Russ. – 5th ed.
P. cm. – (McGraw-Hill series in electrical and computer engineering)
Includes bibliographical references and index.
ISBN 0-07-289969-7
1. Electromagnetic theory. I. Fleisch, Daniel A. II. Title. III. Series.
QC661.K72 1999
537–dc21 98-34935

http://www.mhhe.com

*To **Michael Faraday,** who performed the pioneering experiments,*

*To **James Clerk Maxwell,** who used them to formulate the all-encompassing theory of electromagnetics, and*

*To **Heinrich Hertz,** whose experiments validated the theory and gave us wireless.*

ABOUT THE AUTHORS

John Kraus received his B.S., M.S., and Ph.D. degrees from the University of Michigan. For several years he did research with the University's 100-ton cyclotron, then the world's most powerful particle accelerator. During World War II he was with the U.S. Navy and Harvard University's Radio Research Laboratory. Joining the Ohio State University Electrical Engineering faculty in 1946, he was also a member of the faculties of Physics and Astronomy.

Dr. Kraus has published hundreds of scientific and technical articles. He also is the author of several widely translated books: *Antennas* (1950), *Electromagnetics* (1953), *Radio Astronomy* (1966), and *Big Ear* (1976), all now in newer editions. He was elected a member of the National Academy of Engineering in 1972. He has received numerous medals and awards, including the Sullivant gold medal of Ohio State and the Edison and Heinrich Hertz gold medals of the Institute of Electrical and Electronic Engineers.

Daniel Fleisch received his B.S. degree from Georgetown University and his M.S. and Ph.D. degrees from Rice University. For several years he did upper atmosphere research at the Arecibo, Puerto Rico, and Jicamarca, Peru, ionospheric observatories. Later, he did design work for a company producing microelectronic equipment.

In 1993 Dr. Fleisch joined Aeroflex-Lintek of Powell, Ohio, as Chief Scientist. Concurrently, he is on the faculty of Wittenberg University.

CONTENTS

PREFACE

With the explosive growth of wireless technology and with less time available in curricula, this book is structured to get to practical applications early. The first five chapters form the basic core with the last five providing supplementary material that can be used as desired. The book is designed for a one-semester or two-quarter course with a variety of assignment options. A few assignment schedules are given on page xix.

The book has 11 chapters:

Chapter 1. Introduction. Explains the language of electromagnetics and gives an introduction to vector analysis.

Chapter 2. Electric and Magnetic Fields. Provides a concise but thorough introduction to the fundamental concepts culminating in Maxwell's equations.

Chapter 3. Transmission Lines. Explains them from both circuit and field viewpoints with early reference to the ubiquitous microstrip line. Discusses matching and the propagation of both continuous signals and short pulses.

Chapter 4. Wave Propagation, Attenuation, Polarization, Reflection, Refraction and Diffraction. Waves in space and their interaction with media are discussed with analogies to wave behavior on transmission lines.

Chapter 5. Antennas, Radiation, and Wireless Systems. Basic antenna concepts are followed by array theory and the design of over two dozen types of antennas with many applications to wireless systems. The many topics include pulsed and doppler radar, global position systems, and passive remote sensing.

Chapter 6. Electrodynamics. Gives a concise treatment of particle dynamics and the operation of cathode-ray-tubes, motors and generators both mechanical and Hall.

Chapter 7. Dielectric and Magnetic Materials. Provides further information about dielectric and magnetic materials with sections on dielectric polarization, magnetization, and hysteresis.

Chapter 8. Waveguides, Resonators, and Fiber Optics. Covers wave propagation in rectangular and circular guides, along dielectric sheets and fibers and waves trapped inside a resonator.

Chapter 9. Bioelectromagnetics. A short but important chapter on the electromagnetics of living systems, radiation hazards, and related environmental issues.

Chapter 10. Electromagnetic Effects in High-Speed Digital Systems. Another short but important chapter with insights into the role of electromagnetics in the design and proper operation of high-speed digital systems. Includes a discussion of electromagnetic interference (EMI), electromagnetic compatibility (EMC), government regulations and testing procedures.

Chapter 11. Numerical Methods. An important chapter on repetitive Laplace or the finite difference method, the integral equation and the moment method, finite element, and other methods. These are accompanied by computer programs in Appendix C.

Appendixes. Include tables of units, material constants, mathematical formulas, computer programs, and other useful relations.

The format of this edition is modular. Explanatory material is typically followed by a worked example and problems with answers. At the end of each chapter are many additional problems and also a few instructive projects that may be assigned with or without extra class credit.

It would not have been possible to prepare this edition without the dedicated assistance of Dr. Erich Pacht, who has been involved in all aspects of the editorial process. Dr. Jerry Ehman has provided great assistance in proofing both text and problems. We have also benefited from the comments and suggestions of many others who have read all or parts of the manuscript. These include: Richard Mallozzi, University of California, Berkeley; Professor Benedikt Munk, Ohio State University; and the McGraw-Hill reviewers Professors Jim Akers, Mississippi State University; Brian Austin, Liverpool University; Keith Carver, University of Massachusetts; Elbadawy Elsbarawy, Arizona State University; Haralambos Kritikos, University of Pennsylvania; Raymond Luebbers, Pennsylvania State University; Irene Peden, University of Washington; and Andrew Peterson, Georgia Institute of Technology. The McGraw-Hill editors and supervisors Lynn Cox, Nina Kreiden, Betsy Jones, Emily Gray, and Kimberly Schau were most helpful.

Finally, John Kraus thanks his wife, Alice, for her patience, encouragement, and dedication through all the years of work it has taken, and Daniel Fleisch thanks Jill for her constant support.

Supplementary material, new examples, and projects are available on the book's Web site: **www.elmag5.com.**

A word about *wireless.* After Heinrich Hertz first demonstrated radiation from antennas, it was called *wireless* (German: *drahtlos*, French: *sans fils*). And wireless it was until broadcasting began about 1920 and the word *radio* was introduced. Now *wireless* is back to describe the many systems that operate without wires as distinguished from *radio* which to most persons now implies AM or FM.

Suggested Assignments

8 pages per class day (avg.)

Chapter	One Semester Days/Chapter	One Semester Optional	One Semester Optional	Two Quarters Days/Chapter (First quarter)	Two Quarters Days/Chapter (Second quarter)	Second quarter Optional	Second quarter Optional
1	3			3			
2	9			9			
3	5			5			
	Exam 1			Exam 1			
4	9			9			
5	12			Basics → 2	10	10	10
	Exam 1			Exam 1			
	↓			30 (First quarter)			
6	↓				3	3	3
7	↓	2			5	6	5
	↓				Exam 1	1	1
8	↓	2			4	5	3
9	2				2	2	1
10	2				2	2	1
11			4		2		5
	Exam 1	1	1		Exam 1	1	1
	45	45	45 days		30	30	30 days

CHAPTER 1

INTRODUCTION

1-1 ELECTROMAGNETICS: ITS IMPORTANCE

Electromagnetics is important because it provides a real-world, three-dimensional understanding of electricity and magnetism.

By circuit theory the battery of Fig. 1-1 applies a voltage V sending a current I through the wires to the load. This is a simplistic view. The energy is conveyed from the battery to the load almost entirely by electromagnetic fields external to the wires, the wires acting as guides for the energy as in the figure. An electric field extends between the wires and a magnetic field surrounds them. With alternating currents

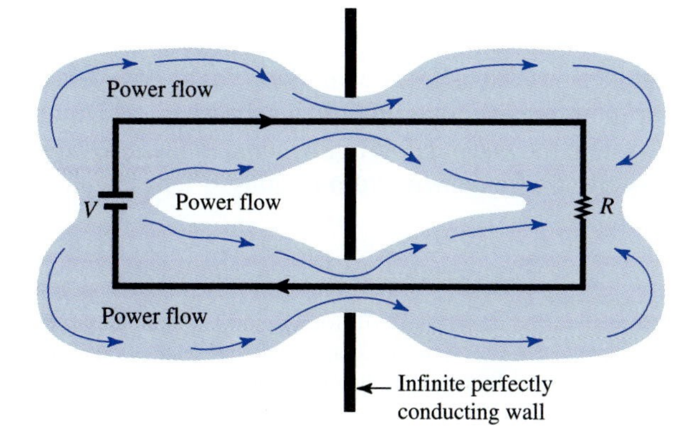

FIGURE 1-1
Simple circuit showing power flow from generator to load R through holes in a perfectly conducting wall.

some energy is radiated into space. At high enough frequencies nearly all may be radiated, the circuit acting as an antenna.

The vast grids of the earth's 50- and 60-Hz power transmission lines not only supply the world's electrical needs but their "hum" is the most powerful continuous electromagnetic signal sent out from our planet.

Many electromagnetic devices are inadvertently coupled to other systems. Video display units (VDUs) of computers and television sets may emit sufficient unintentional radiation that it can be picked up, decoded and the screen displays reproduced at distances of a kilometer. Any device that radiates is coupled, in principle, to the entire universe. By reciprocity, the universe is coupled to it.

This is the real world and field theory is essential to understand it. The purpose of *Electromagnetics,* fifth edition, is to provide insights and answers.

1-2 DIMENSIONS AND UNITS

A ***dimension*** defines some physical characteristic. For example, length, mass, time, velocity, and force are dimensions. The dimensions of length, mass, time, electric current, temperature, and luminous intensity are considered as the *fundamental dimensions* since other dimensions can be defined in terms of these six. This choice is arbitrary but convenient. Let the letters L, M, T, I, $\mathcal{T}$, and $\mathcal{I}$ represent the dimensions of length, mass, time, electric current, temperature, and luminous intensity. Other dimensions are then secondary dimensions. For example, area is a secondary dimension which can be expressed in terms of the fundamental dimension of length squared (L^2). As other examples, the fundamental dimensions of velocity are L/T and of force are ML/T^2.

A ***unit*** is a standard or reference by which a dimension can be expressed numerically. Thus, the meter is a unit in terms of which the dimension of length can be expressed, and the kilogram is a unit in terms of which the dimension of mass can be expressed. For example, the length (dimension) of a steel rod might be 2 meters, and its mass (dimension), 5 kilograms.

1-3 FUNDAMENTAL AND SECONDARY UNITS

The units for the fundamental dimensions are called the *fundamental* or *base units.* In this book the *metric system* or more precisely the International System of Units, abbreviated SI, is used.† In this system the *meter, kilogram, second, ampere, kelvin,* and *candela* are the base units for the six fundamental dimensions of length, mass,

†The International System of Units is the modernized version of the metric system. The abbreviation SI is from the French name *Système Internationale d'Unités.* For the complete official description of the system see U.S. National Bureau of Standards (now National Institute of Standards and Technology) *Spec. Pub.* 330, 1971.

TABLE 1-1
Frequencies and wavelengths of the electromagnetic spectrum from almost dc to gamma-rays

Band	Frequency, $f = c/\lambda$			Wavelength, $\lambda = c/f$		Relevant dimensions
Radio	1 Hz —	Hz	1	300	Mm	
			10	30		
			100	3		— Earth diameter
					— 10^6 m	
	10^3 Hz —	kHz	1	300	km	
			10	30		
			100	3		— Mt. Everest
					— 10^3 m	
	10^6 Hz —	MHz	1	300	m	
			10	30		— Redwood tree
			100	3		
					— 1 m	— Man
	10^9 Hz —	GHz	1	300	mm	— Hydrogen line
			10	30		O_2 line
			100	3		Molecular lines
					— 10^{-3} m	Sand grain
Infrared	10^{12} Hz —	THz	1	300	μm	
			10	30		
			100	3		— Bacterium
Visible					— 10^{-6} m	
UV	10^{15} Hz —	PHz	1	300	nm	
			10	30		— Virus
			100	3		
X-ray					— 10^{-9} m	— Atomic spacing
	10^{18} Hz —	EHz	1	300	pm	— Atom
			10	30		
			100	3		
					— 10^{-12} m	
Gamma ray	10^{21} Hz —		1	300	fm	
			10	30		
			100	3		
					— 10^{-15} m	— Atomic nucleus

Radio-frequency band names†

Name	Frequency	Principal use
ELF†	3–30 Hz	
SLF	30–300 Hz	Power grids
ULF	300–3000 Hz	
VLF	3–30 kHz	Submarines
LF	30–300 kHz	Beacons
MF	300–3000 kHz	AM broadcast
HF	3–30 MHz	Shortwave broadcast
VHF	30–300 MHz	FM, TV
UHF	300–3000 MHz	TV, LAN, cellular, GPS
SHF	3–30 GHz	Radar, GSO satellites, data
EHF	30–300 GHz	Radar, automotive, data

Microwave bands

"Old"	"New"	Frequency
L	D	1–2 GHz
S	E, F	2–4 GHz
C	G, H	4–8 GHz
X	I, J	8–12 GHz
Ku	J	12–18 GHz
K	J	18–26 GHz
Ka	K	26–40 GHz

†ELF = Extremely Low Frequency, SLF = Super-Low Frequency, VLF = Very Low Frequency, MF = Medium Frequency, HF = High Frequency, UHF = Ultrahigh Frequency, etc.

TABLE 1-2
Across the universe at the speed of light or radio waves (300,000 kilometers per second)

	Light or radio travel time†
Across atomic nucleus	1 microattosecond = 10^{-24} second
Across a virus	100 attoseconds = 100×10^{-18} second
Across a bacterium	1 femtosecond = 10^{-15} second
1 meter	3 nanoseconds = 3×10^{-9} second
New York to Los Angeles	13 milliseconds = 0.013 second
New York to London	18 milliseconds = 0.018 second
New York to Tokyo	35 milliseconds = 0.035 second
Around the earth	133 milliseconds = 0.133 second
Earth to geostationary relay satellite and back	250 milliseconds = 0.25 second
Earth to moon	1.3 seconds
Earth to Sun	500 seconds = 8.3 minutes
Earth to Venus	2 minutes min. to 14 minutes max.
Earth to Mars	3 minutes min. to 21 minutes max.
Earth to Jupiter	45 minutes average
Earth to Pluto	5.5 hours average
Diameter of solar system	11 hours
Sun to nearest star	4 years
Sun to center of our galaxy	30,000 years
Diameter of our galaxy	100,000 years
Distance to Andromeda galaxy	2 million years = 2 megayears
Distance to Cygnus A	1 billion years = 1000 megayears = 1 gigayear
Distance to quasar OH471	14 billion years = 14 gigayears
Distance to the "edge" of the known universe	15 billion years = 15 gigayears = 15×10^9 years

†Light or radio travel time is a measure of distance.

Problem 1-3-1. Light-year distance. What is the distance equivalent of one light-year?
Ans. $60 \times 60 \times 24 \times 365$ (s) $\times 3 \times 10^8$ (m/s) $= 9.46 \times 10^{15}$ m.

time, electric current, temperature, and luminous intensity. The definitions for these fundamental units are:

Meter (m). Equal to the path length traveled by light in vacuum in a time $t = 1/299{,}792{,}458$ second.

Kilogram (kg). Equal to mass of international prototype kilogram, a cylinder of platinum-iridium alloy kept at Sèvres, France. This standard kilogram is the only artifact among the SI base units.

Second (s). Equal to duration of 9,192,631,770 periods of radiation corresponding to the transition between two hyperfine levels of the ground state of cesium 133. The second was formerly defined as 1/86,400 part of a mean solar day. The earth's rotation rate is gradually slowing down, but the atomic (cesium 133) transition is much more constant and is now the standard. The two standards differ by about 1 second per year. Atomic clocks are accurate to about 1 microsecond per year. Distant fast rotating (1000 rps) pulsars may soon replace atomic clocks as a still better standard (nanoseconds per year accuracy).

Ampere (A). Equal to the electric current flowing in each of two infinitely long parallel wires in vacuum separated by 1 meter which produces a force of 200 nanonewtons per meter of length (200 nN m^{-1} = 2×10^{-7} N m^{-1}).

Kelvin (K). Temperature equal to 1/273.16 of the triple point of water (or triple point of water equals 273.16 kelvins). Note that the degree sign (°) is not used with kelvins. Thus, the boiling temperature of water = 100°C = 373 K.

Candela (cd). Luminous intensity equal to that of 1/600,000 square meter of a perfect radiator at the temperature of freezing platinum at a pressure of 1 standard atmosphere.

The units for other dimensions are called *secondary* or *derived* units and are based on these fundamental units (see Appendix A).

The material in this book deals almost exclusively with the four fundamental dimensions *length, mass, time,* and *electric current* (dimensional symbols L, M, T, and I). The four fundamental units for these dimensions are the basis of what was formerly called the *meter-kilogram-second-ampere* (mksa) *system,* now a subsystem of the SI.

The complete SI involves not only units but also other recommendations, one of which is that multiples and submultiples of the SI units be stated in steps of 10^3 or 10^{-3}. Thus, the kilometer (1 km = 10^3 m) and the millimeter (1 mm = 10^{-3} m) are preferred units of length. For example, the proper SI designation for the width of motion-picture film is 35 mm, not 3.5 cm. For a list of the preferred units see Table 1-3 on page 6. This table also gives the pronunciation, abbreviation, and derivation of these units.

In this book *rationalized* SI units are used. The rationalized system has the advantage that the factor 4π does not appear in Maxwell's equations, although it does appear in certain other relations. A complete table of units in this system is given in Appendix A-1. The table lists dimensions, or quantities, alphabetically under each of the following headings: Fundamental, Mechanical, Electrical, and Magnetic. For each quantity the mathematical symbol (as used in equations), description, SI unit and abbreviation, equivalent units, and fundamental dimensions are listed.

It is a good idea to refer to the table as each new quantity and unit is discussed to become familiar with its fundamental dimensions.

1-4 HOW TO READ THE SYMBOLS AND NOTATION

In this book *quantities,* or *dimensions,* which are scalars, like charge Q, mass M, or resistance R, are always in italics. Quantities which may be vectors *or* scalars are boldface as vectors and italics as scalars, e.g., electric field $\mathbf{E}$ (vector) or E (scalar). Unit vectors are always boldface with a hat (circumflex) over the letter, e.g., $\hat{\mathbf{x}}$ or $\hat{\mathbf{r}}$.†

Units are in roman type, i.e., *not* italic; for example, H for henry, s for second, or A for ampere. The abbreviation for a unit is capitalized if the unit is derived from

†In longhand notation a vector may be indicated by a bar over the letter and hat (^) over the unit vector. Also, no distinction is usually made between quantities (italics) and units (roman). However, it can be done by placing a bar under the letter to indicate italics or by writing the letter with a distinct slant.

TABLE 1-3
Metric prefixes

Numerical value	Prefix	Pronunciation	Symbol	As used in this book (U.S. meaning)	Meaning in other countries
1 000 000 000 000 000 000 = 10^{18}	exa	(ex a)	E	one quintillion	trillion
1 000 000 000 000 000 = 10^{15}	peta	(pet a)	P	one quadrillion	thousand billion
1 000 000 000 000 = 10^{12}	tera	(tare a)	T	one trillion	billion
1 000 000 000 = 10^{9}	giga	(jig a)	G	one billion	milliard
1 000 000 = 10^{6}	mega	(meg a)	M	one million	
1 000 = 10^{3}	kilo	(key lo)	k	one thousand	
100 = 10^{2}	hecto	(hek toe)	h	one hundred	
10 = 10	deka	(dek a)	da	ten	
0.1 = 10^{-1}	deci	(dec i)	d	one tenth	
0.01 = 10^{-2}	centi	(cent i)	c	one hundredth	
0.001 = 10^{-3}	milli	(mill i)	m	one thousandth	
0.000 001 = 10^{-6}	micro	(my kro)	μ	one millionth	
0.000 000 001 = 10^{-9}	nano	(nan o)	n	one billionth	milliardth
0.000 000 000 001 = 10^{-12}	pico	(pee ko)	p	one trillionth	billionth
0.000 000 000 000 001 = 10^{-15}	femto	(fem toe)	f	one quadrillionth	thousand billionth
0.000 000 000 000 000 001 = 10^{-18}	atto	(at o)	a	one quintillionth	trillionth

Examples: 1 kilometer (1 km) = 1000 meters (m)
1 micrometer (1 μm) = one millionth of a meter
1 kilowatt (1 kW) = 1000 watts (W)
1 milliwatt (1 mW) = one thousandth of a watt

Note that whereas billion and trillion (and billionth and trillionth) may be ambiguous, giga, tera, nano, and pico are not.

a proper name; otherwise it is lowercase (small letter). Thus, we have C for coulomb but m for meter. Note that when the unit is written out, it is always lowercase even though derived from a proper name. *Prefixes* for units are also roman, like n in nC for nanocoulomb or M in MW for megawatt.

Example 1-1. Notation.

$$\mathbf{D} = \hat{\mathbf{x}}\, 200 \text{ pC m}^{-2}$$

means that the electric flux density **D** is a vector in the positive x direction with a magnitude of 200 picocoulombs per square meter ($= 2 \times 10^{-10}$ coulombs per square meter).

Example 1-2. Notation.

$$V = 10 \text{ V}$$

means that the voltage V (a scalar) equals 10 volts. Distinguish carefully between V (italics) for voltage, V (roman) for volts, **v** (lowercase, boldface) for velocity (a vector), and v (lowercase, italics) for volume.

Example 1-3. Notation.

$$S = 4 \text{ W m}^{-2} \text{ Hz}^{-1}$$

means that the flux density S (a scalar) equals 4 watts per square meter per hertz. This can also be written $S = 4 \text{ W/m}^2 \text{ Hz}$ or $4 \text{ W/(m}^2 \text{ Hz)}$, but the form $\text{W m}^{-2} \text{ Hz}^{-1}$ is more direct and less ambiguous. However the slash form (W/m^2 Hz) is easier to write and is used in the problem statements.

Note that for conciseness, prefixes are used where appropriate instead of exponents. Thus, the velocity of light is given as $c = 300 \text{ Mm s}^{-1}$ (300 megameters per second) and *not* $3 \times 10^8 \text{ m s}^{-1}$. However, in solving a problem the exponential form ($3 \times 10^8 \text{ m s}^{-1}$) is used.

Problem 1-4-1. Lead weight. What are the dimensions and units for a lead weight? *Ans.* Volume, m^3; and mass, kg.

Note: The number 1-4-1 indicates that the problem is in Chap. 1, Sec. 4 and is the first problem in that section.

Problem 1-4-2. Velocity change. What are the dimensions and units for a rate of change of velocity? *Ans.* Acceleration, m s^{-2}.

Problem 1-4-3. Power bill. What are the dimensions and units of the quantity you pay for on your electric power bill? *Ans.* Electric energy in kilowatt-hours, kW-hr $= 3.6 \times 10^6$ J.

The metric prefixes are in steps of 10^{-3} or 10^3 and go from *atto* (10^{-18}) to *exa* (10^{18}), a ratio or range of 10^{36}. These are adequate for most purposes. Outside this range the exponential form is used. Thus, there are 10^{79} atoms in the universe.

The modernized metric (SI) units and the conventions used herein combine to give a concise, exact, and unambiguous notation, and if one is attentive to the details, it will be seen to possess both elegance and beauty.

1-5 EQUATION AND PROBLEM NUMBERING

Important equations and those referred to in the text are numbered consecutively beginning with each section. When reference is made to an equation in a different section, its number is preceded by the chapter and section number. Thus, (2-8-4) refers to Chap. 2, Sec. 8, Eq. (4). A reference to this same equation within Sec. 8 of Chap. 2 would read simply (4). Note that the chapter number and name, and section number and name are printed at the top of each left-hand and right-hand page, respectively.

Problems are numbered according to the section of the book which is relevant. Thus, a problem numbered 1-5-2 is the second problem involving the subject matter of Sec. 1-5.

Example 1-4. Dimensional balance. Is this hypothetical formula balanced?

$$\frac{M}{L} = DA$$

where M = mass
L = length
D = density (mass per unit volume)
A = area

Solution. The dimensional symbols for the left side are M/L, the same as those used. The dimensional symbols for the right side are

$$\frac{M}{L^3}L^2 = \frac{M}{L}$$

Therefore, both sides of this equation have the dimensions of mass per length, and the equation is balanced dimensionally. *Ans.*

This is not a guarantee that the equation is correct; i.e., it is not a *sufficient* condition for correctness. It is, however, a *necessary* condition for correctness, and it is frequently helpful to analyze equations in this way to determine whether or not they are dimensionally balanced.

Such *dimensional analysis* is also useful for determining the dimensions of a quantity.

Example 1-5. Dimensional symbols. Find the dimensions of force.

Solution. From Newton's second law,

$$\text{Force} = \text{mass} \times \text{acceleration}$$

Since acceleration has the dimensions of length per time squared (see Appendix A, Table A-1), the dimensions of force are

$$\frac{\text{Mass} \times \text{length}}{\text{time}^2} \quad \text{or in dimensional symbols} \quad \text{force} = \frac{ML}{T^2} \quad \textit{Ans.}$$

Problem 1-5-1. Dimensional balance. Are the following equations balanced dimensionally? (*a*) Force, $F = MLT^{-2}$; (*b*) length, $L = vT$ where v = velocity; (*c*) Energy $E = Mv^2$; and (*d*) power, $P = ML^2/T^2$. *Ans.* All yes except (*d*).

Problem 1-5-2. Dimensional symbols and units. What are (*a*) the dimensional symbols for momentum (mass × velocity = force × time) and (*b*) the units?
Ans. $M(L/T) = (ML/T^2)T$; (*b*) kg m/s = N s.

1-6 VECTOR ANALYSIS

Introduction

Vector analysis is a concise language or mathematical shorthand which greatly facilitates the analysis of electric and magnetic fields. In this section the basic ideas of vector analysis are developed in a very concise manner. Students who have had little or no vector analysis can study this section before or concurrently with the text. Students who have had a course in vector analysis may wish to use this section for review or reference.

After defining a scalar and a vector, we proceed to a discussion of vector addition and subtraction and the multiplication and division of a vector by a scalar. Then we consider the resolution of a vector into components. This is followed by the dot (or scalar) product of two vectors and its application to line and surface integrals. The cross (or vector) product of two vectors is then discussed. A brief introduction is given to coordinate systems: rectangular, cylindrical, and spherical.

The vector concepts of gradient, divergence, and curl are developed in Chap. 2.

Scalars and Vectors

A ***scalar*** quantity has magnitude only. It is a quantity that can be described by a simple number. Thus, temperature, mass, volume, and energy are scalars.

A ***vector*** quantity has both magnitude and direction. For example, force and velocity are vectors. To distinguish vectors from scalars, the letter representing the vector is printed (in books) in boldface or heavy type, as **F** or **v**. In handwritten notation a vector is designated by a bar over the letter, as $\overline{\mathrm{F}}$ or $\overline{\mathrm{v}}$. A *unit vector* is printed in boldface with a hat or cap (ˆ) over the letter, as $\hat{\mathbf{x}}$, $\hat{\mathbf{y}}$, $\hat{\mathbf{z}}$, $\hat{\mathbf{r}}$, and $\hat{\mathbf{n}}$, while in hand notation, only the hat is used, as $\hat{\mathrm{x}}$, $\hat{\mathrm{y}}$, $\hat{\mathrm{z}}$, $\hat{\mathrm{r}}$, and $\hat{\mathrm{n}}$.

Although the air temperature in a room is a scalar function of position (having magnitude only), the *rate of change of T with distance x* is a vector $\hat{\mathbf{x}}\,(dT/dx)$ giving the magnitude of the component in the x-direction. In Chap. 2 we call dT/dx the x component of the gradient of T.

Vector Addition and Subtraction

As shown in Fig. 1-2, a vector may be represented graphically by a line with an arrow. The line has an origin and an endpoint (at the arrow). The orientation of the line and the arrow indicates the direction of the vector. (The arrow is required to avoid a 180° ambiguity in direction.) If the origin and endpoint of the vector (Fig. 1-2) are reversed so that the arrow is at the left but the orientation of the line is unchanged, it is said that the *sense* of the vector is changed (direction changed by 180°).

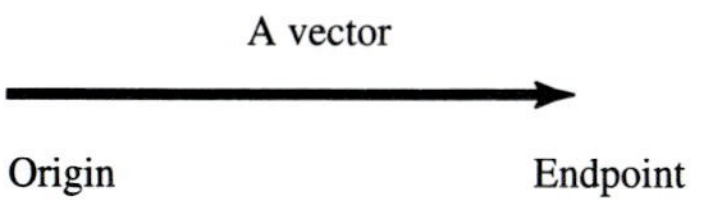

A vector

Origin Endpoint

FIGURE 1-2
A vector.

Example 1-6. Addition of two vectors; commutative law. Add the two vectors **A** and **B** shown in Fig. 1-3*a*.

Solution. Place the origin of **B** coincident with the endpoint of **A**. Both vectors are understood to *apply to the same point in space* (x, y, z), but for convenience in graphical addition either may be moved, provided the magnitude and direction are unaltered. The *sum* or *resultant* of the two vectors is then the new vector **C** which extends from the origin of **A** to the end of **B**, as in Fig. 1-3*a*. In symbols this is expressed

$$\mathbf{A} + \mathbf{B} = \mathbf{C} \quad \textit{Ans.} \tag{1}$$

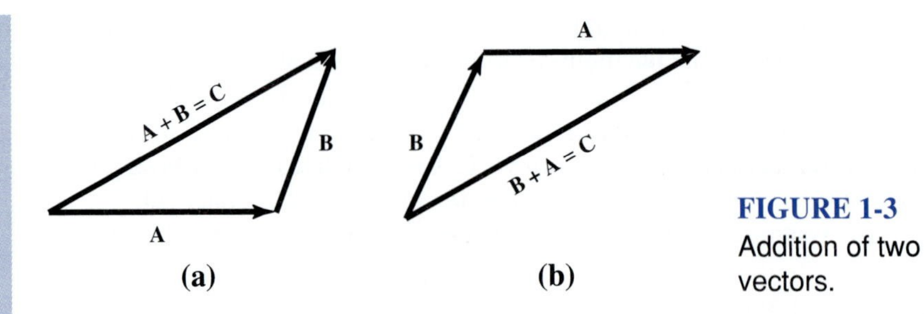

FIGURE 1-3
Addition of two vectors.

The same resultant **C** is also obtained if the origin of **A** is placed coincident with the end of **B** as in Fig. 1-3*b*. Thus we may write

$$\mathbf{B} + \mathbf{A} = \mathbf{C}$$

or

$$\mathbf{A} + \mathbf{B} = \mathbf{B} + \mathbf{A} \tag{2}$$

From (2) we deduce that the sum of two vectors is independent of the order in which they are added, that is, vector addition obeys the ***commutative law.***

Example 1-7. Addition of three vectors; associative law. Add the three vectors **A**, **B**, and **C** shown in Fig. 1-4*a*.

Solution. First obtain the sum of **A** and **B** as done in Example 1-6. Then add this resultant to **C**. This procedure is indicated by parentheses; thus (**A** + **B**) + **C**. *Ans.* The same total or resultant vector is obtained by first adding **B** to **C** and then adding this sum to **A** as in Fig. 1-4*b*. This procedure is expressed by **A** + (**B** + **C**). It follows that

$$(\mathbf{A} + \mathbf{B}) + \mathbf{C} = \mathbf{A} + (\mathbf{B} + \mathbf{C}) \tag{3}$$

From (3) we deduce that the sum of three vectors is independent of the way in which **B** is associated with **A** or **C**; that is, vector addition obeys the ***associative law.***

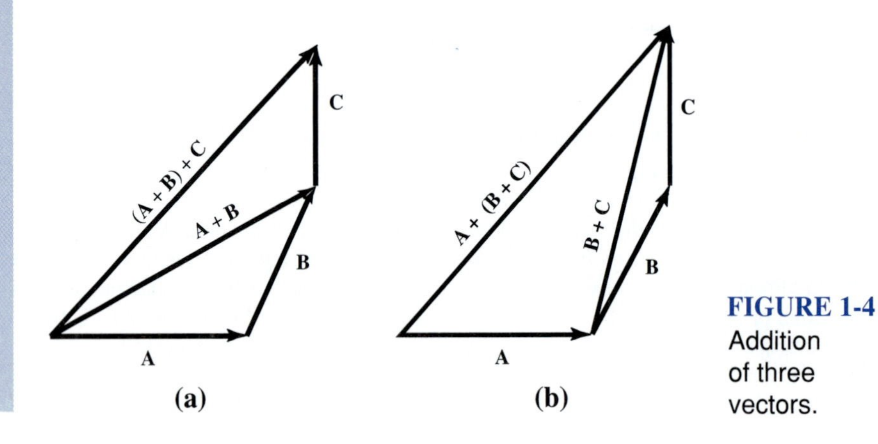

FIGURE 1-4
Addition of three vectors.

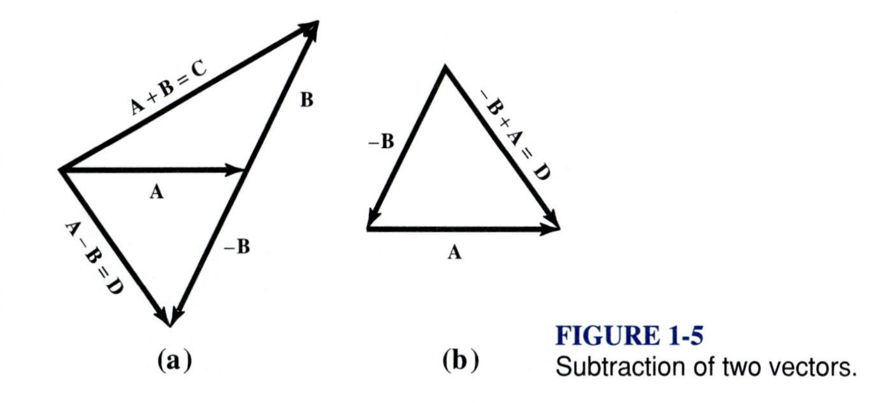

FIGURE 1-5
Subtraction of two vectors.

Example 1-8. Subtraction of two vectors. Referring to Fig. 1-3, subtract vector **B** from vector **A**.

Solution. Add **B** with sense reversed to **A** obtaining their difference **D** as shown in Fig. 1-5*a*. In symbols

$$\mathbf{A} + (-\mathbf{B}) = \mathbf{D} \qquad \text{or} \qquad \mathbf{A} - \mathbf{B} = \mathbf{D} \quad \textit{Ans.}$$

From the commutative law it also follows that

$$-\mathbf{B} + \mathbf{A} = \mathbf{D} \qquad \text{as in Fig. 1-5}b.$$

It is important to keep in mind that the vectors **A**, **B**, **C**, and **D** *all refer to the same point in space.* The displacements in Figs. 1-3, 1-4, and 1-5 are only for the convenience of illustrating graphically the addition or subtraction.

Multiplication and Division of a Vector by a Scalar

A vector may be multiplied or divided by a scalar. The magnitude of the vector changes (provided the scalar is other than unity), but its direction remains unaltered. Thus, the vector **A** multiplied by the scalar number 3 is a vector 3 times as long as **A** but in the same direction, as in Fig. 1-6. Thus, length is equivalent to magnitude.

Newton's second law provides an example of a vector multiplied by a scalar. Thus,

$$m\mathbf{a} = \mathbf{F}$$

where m = a scalar = mass of object, kg
$\mathbf{a}$ = a vector = acceleration of object, m s^{-2}
$\mathbf{F}$ = a vector = force on object, N

3A

FIGURE 1-6
Multiplication of a vector by a scalar.

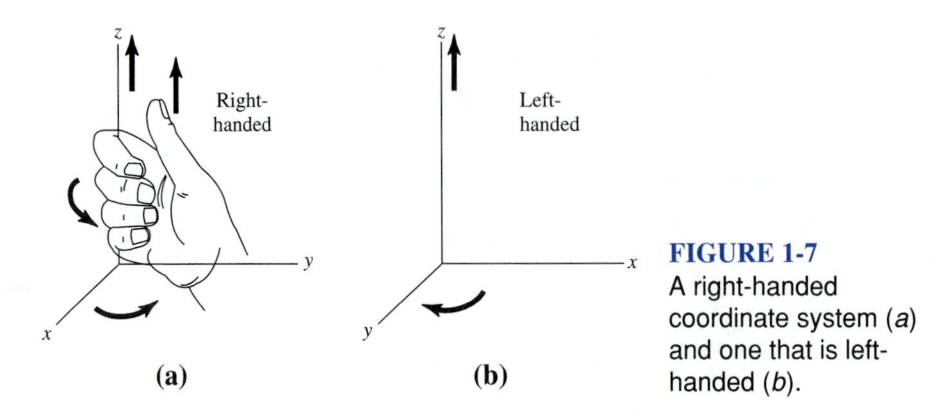

FIGURE 1-7
A right-handed coordinate system (*a*) and one that is left-handed (*b*).

Rectangular Coordinates and the Resolution of a Vector into Components

A rectangular or cartesian coordinate system has three mutually perpendicular axes, called the x, y, and z axes as illustrated in Fig. 1-7. The system may be either *right-handed* or *left-handed.* For a right-handed system, rotation from the positive x axis toward the positive y axis (in the direction of the fingers of the right hand), as in Fig. 1-7*a*, has the thumb pointing in the direction of the positive z axis. Thus, the system in Fig. 1-7*a* is right-handed, but the one in Fig. 1-7*b* is left-handed. It is customary to use right-handed systems, and these are used in this book.

A vector **A** at the origin of the rectangular coordinate system, as in Fig. 1-8, may be resolved into three component vectors, each parallel to one of the coordinate axes. Thus,

$$\mathbf{A} = \mathbf{A}_x + \mathbf{A}_y + \mathbf{A}_z \tag{4}$$

Each of the component vectors may in turn be expressed as the product of a scalar magnitude and a *unit vector,* i.e., a vector of unit length, in the direction of the coordinate axis. A unit vector is a dimensionless quantity of unit magnitude. Thus,

$$\mathbf{A} = \hat{\mathbf{x}}A_x + \hat{\mathbf{y}}A_y + \hat{\mathbf{z}}A_z = \frac{\mathbf{A}_x}{|\mathbf{A}_x|}|\mathbf{A}_x| + \frac{\mathbf{A}_y}{|\mathbf{A}_y|}|\mathbf{A}_y| + \frac{\mathbf{A}_z}{|\mathbf{A}_z|}|\mathbf{A}_z| \tag{5a}$$

The vertical bars signify *magnitude of.* Thus $|\mathbf{A}|$ is the scalar magnitude A of the vector **A** as given by $|\mathbf{A}| = A$. The meanings of the other symbols are as follows:

$\hat{\mathbf{x}}$ = unit vector in x direction $= \dfrac{\mathbf{A}_x}{|\mathbf{A}_x|}$ $\qquad A_x = x$ component of **A** (scalar magnitude of **A** in x direction) $= |\mathbf{A}_x|$

$\hat{\mathbf{y}}$ = unit vector in y direction $= \dfrac{\mathbf{A}_y}{|\mathbf{A}_y|}$ $\qquad A_y = y$ component of $\mathbf{A} = |\mathbf{A}_y|$

$\hat{\mathbf{z}}$ = unit vector in z direction $= \dfrac{\mathbf{A}_z}{|\mathbf{A}_z|}$ $\qquad A_z = z$ component of $\mathbf{A} = |\mathbf{A}_z|$

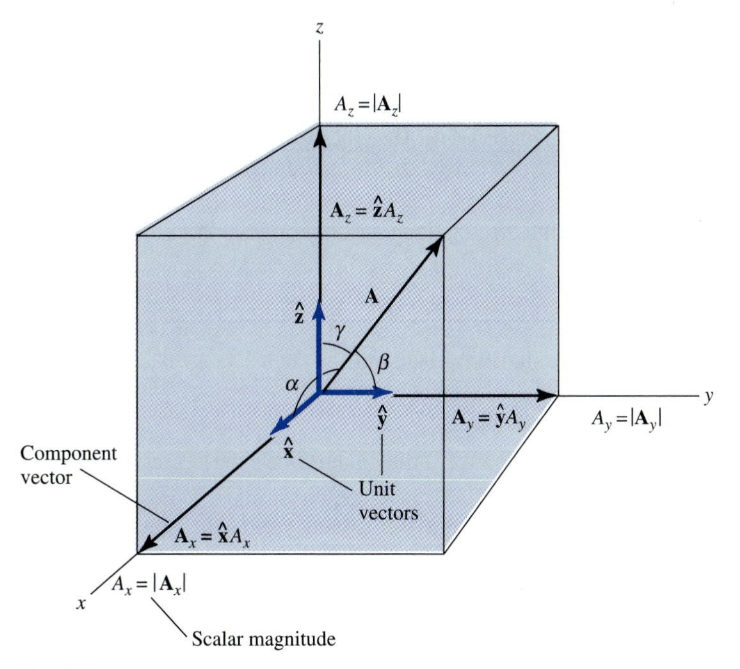

FIGURE 1-8
Resolution of a vector into component vectors and into unit vectors multiplied by scalar magnitudes.

From the theorem of Pythagoras, the scalar magnitude of **A**, written $|\mathbf{A}|$ or simply A, is given by

$$|\mathbf{A}| = A = \sqrt{A_x^2 + A_y^2 + A_z^2} \tag{5b}$$

Problem 1-6-1. Addition of two vectors. Add vector $\mathbf{A} = \hat{\mathbf{x}}6 + \hat{\mathbf{y}}3$ and vector $\mathbf{B} = \hat{\mathbf{x}}4 - \hat{\mathbf{y}}7$. What are (*a*) the magnitude of the resulting vector **C** and (*b*) its angle with respect to the x-axis? *Ans.* (*a*) $|\mathbf{C}| = 10.8$; (*b*) $\alpha = -21.8°$.

Problem 1-6-2. Addition of three vectors. Add $\mathbf{A} = -\hat{\mathbf{x}}8 + \hat{\mathbf{y}}12$, $\mathbf{B} = -\hat{\mathbf{x}}5 + \hat{\mathbf{y}}15$, and $\mathrm{C} = \hat{\mathbf{x}}7 - \hat{\mathbf{y}}9$. What are (*a*) the magnitude of the resulting vector **D** and (*b*) its angle with respect to the $+x$-axis? *Ans.* $|\mathbf{D}| = 19.0$, (*b*) $\alpha = 108.4°$.

Problem 1-6-3. Subtraction of two vectors. Subtract $\mathbf{A} = \hat{\mathbf{y}}12 - \hat{\mathbf{z}}7$ from $\mathrm{B} = \hat{\mathbf{x}}4 + \hat{\mathbf{y}}10$. Find: (*a*) the magnitude of the resultant vector **C** and (*b*) its angle γ with the z-axis. *Ans.* (*a*) $|\mathbf{C}| = 8.31$, (*b*) $\gamma = 32.6°$.

Problem 1-6-4. Magnitude and angles of a vector. Referring to Fig. 1-8, if the vector $\mathbf{A} = \hat{\mathbf{x}}10 + \hat{\mathbf{y}}10 + \hat{\mathbf{z}}10$, find (*a*) magnitude of **A** and (*b*) angles α, β, and γ. *Ans.* (*a*) $|\mathbf{A}| = 17.3$, (*b*) $\alpha = \beta = \gamma = 54.7°$.

The Scalar or Dot Product of Two Vectors

The ***scalar product*** of two vectors is defined as the product of their magnitudes multiplied by the cosine of the angle between the vectors. Thus, the scalar product of **A** and **B** is written

$$\mathbf{A} \cdot \mathbf{B} = |\mathbf{A}||\mathbf{B}| \cos\theta = AB\cos\theta \tag{6}$$

where θ is the angle between **A** and **B**. The dot (•) between **A** and **B** indicates the scalar product. Hence it is also called the ***dot product.*** The dot product of two vectors yields a scalar.

It is to be noted that the scalar or dot product obeys the commutative law; that is,

$$\mathbf{A} \cdot \mathbf{B} = \mathbf{B} \cdot \mathbf{A}$$

If **A** and **B** are perpendicular, $\theta = 90°$, and

$$\mathbf{A} \cdot \mathbf{B} = 0 \tag{7}$$

Conversely, we may deduce from (7) that **A** and **B** are perpendicular provided neither is zero.

The scalar product of a vector with itself yields the square of its magnitude. Thus,

$$\mathbf{A} \cdot \mathbf{A} = |\mathbf{A}||\mathbf{A}| \cos 0° = A^2 \tag{8}$$

The scalar product of a vector and a unit vector yields the component of the vector in the unit vector direction. Thus, we have

$$\begin{aligned}
\hat{\mathbf{x}} \cdot \mathbf{A} &= |\hat{\mathbf{x}}||\mathbf{A}| \cos\alpha = A_x \\
\hat{\mathbf{y}} \cdot \mathbf{A} &= |\hat{\mathbf{y}}||\mathbf{A}| \cos\beta = A_y \\
\hat{\mathbf{z}} \cdot \mathbf{A} &= |\hat{\mathbf{z}}||\mathbf{A}| \cos\gamma = A_z
\end{aligned} \tag{9}$$

where $\cos\alpha$, $\cos\beta$, and $\cos\gamma$ are the ***direction cosines*** with α the angle between **A** and the x axis, β the angle between **A** and the y axis, and γ the angle between **A** and the z axis (Fig. 1-8). By the Pythagorean theorem, the sum of the squares of the direction cosines is unity. We consider direction cosines in greater detail in Sec. 1-7.

The scalar products of the rectangular coordinate unit vectors $\hat{\mathbf{x}}$, $\hat{\mathbf{y}}$, and $\hat{\mathbf{z}}$ are as follows:

$$\hat{\mathbf{x}} \cdot \hat{\mathbf{y}} = \hat{\mathbf{y}} \cdot \hat{\mathbf{z}} = \hat{\mathbf{z}} \cdot \hat{\mathbf{x}} = 0 \tag{10}$$

but

$$\hat{\mathbf{x}} \cdot \hat{\mathbf{x}} = \hat{\mathbf{y}} \cdot \hat{\mathbf{y}} = \hat{\mathbf{z}} \cdot \hat{\mathbf{z}} = 1$$

In other words, the scalar or dot product of one coordinate unit vector with a different one is always zero, while the scalar product of a unit vector with itself is always unity (for orthogonal coordinate systems).

In rectangular coordinates the scalar product of two vectors is

$$\begin{aligned}
\mathbf{A} \cdot \mathbf{B} &= (\hat{\mathbf{x}}A_x + \hat{\mathbf{y}}A_y + \hat{\mathbf{z}}A_z) \cdot (\hat{\mathbf{x}}B_x + \hat{\mathbf{y}}B_y + \hat{\mathbf{z}}B_z) \\
&= \hat{\mathbf{x}} \cdot \hat{\mathbf{x}}A_xB_x + \hat{\mathbf{x}} \cdot \hat{\mathbf{y}}A_xB_y + \hat{\mathbf{x}} \cdot \hat{\mathbf{z}}A_xB_z \\
&\quad + \hat{\mathbf{y}} \cdot \hat{\mathbf{x}}A_yB_x + \hat{\mathbf{y}} \cdot \hat{\mathbf{y}}A_yB_y + \hat{\mathbf{y}} \cdot \hat{\mathbf{z}}A_yB_z \\
&\quad + \hat{\mathbf{z}} \cdot \hat{\mathbf{x}}A_zB_x + \hat{\mathbf{z}} \cdot \hat{\mathbf{y}}A_zB_y + \hat{\mathbf{z}} \cdot \hat{\mathbf{z}}A_zB_z
\end{aligned} \tag{11}$$

Applying the unit vector relations of (10), (11) reduces to

$$\mathbf{A} \cdot \mathbf{B} = A_x B_x + A_y B_y + A_z B_z \tag{12}$$

Example 1-9. Dot product. Vector $\mathbf{A} = \hat{\mathbf{y}}3 + \hat{\mathbf{z}}2$ and vector $\mathbf{B} = \hat{\mathbf{x}}5 + \hat{\mathbf{y}}8$ extend from the origin. Find $\mathbf{A} \cdot \mathbf{B}$.

Solution. From (12) $(\hat{\mathbf{y}}3 + \hat{\mathbf{z}}2) \cdot (\hat{\mathbf{x}}5 + \hat{\mathbf{y}}8) = \hat{\mathbf{y}} \cdot \hat{\mathbf{y}}24 = 24$ *Ans.*

Problem 1-6-5. Dot product. Vector $\mathbf{A} = -\hat{\mathbf{x}}7 + \hat{\mathbf{y}}12 + \hat{\mathbf{z}}3$, vector $\mathbf{B} = \hat{\mathbf{x}}4 - \hat{\mathbf{y}}2 + \hat{\mathbf{z}}16$. Find $\mathbf{A} \cdot \mathbf{B}$. *Ans.* -4.

The Line Integral

An important application of the scalar (or dot) product involves the line integral. Suppose we move along a curved path from point P_1 to point P_2 in a radial force field $\mathbf{F}$, with force $\mathbf{F}$ acting on an object in the r (radial) direction (Fig. 1-9). At any point P, the product of a path length dL and the component of $\mathbf{F}$ parallel to it is given (Fig. 1-9) by

$$F \cos \theta \, dL = F_L \, dL \tag{13}$$

where F_L = component of $\mathbf{F}$ in direction of path and θ = angle between positive directions of path and $\mathbf{F}$. We note also from Fig. 1-9 that the component of dL in the r (and F) direction is given by

$$dr = \cos \theta \, dL \tag{14}$$

Using vector notation (dot product), we have

$$\mathbf{F} \cdot d\mathbf{L} = F \cos \theta \, dL = F_L \, dL = F \, dr \tag{15}$$

where $d\mathbf{L}$ = vector incremental length (magnitude dL in direction of path).

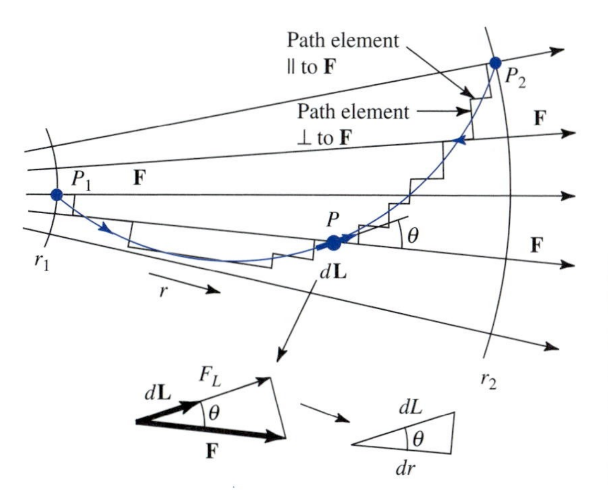

FIGURE 1-9
Line integral from P_1 to P_2 along a curved path in a uniform force field **F**. A segmented path parallel and perpendicular to **F** is also shown.

The product of a force F and a distance dr represents an incremental amount of work dW done by the force $\mathbf{F}$ in moving an object a distance $\cos\theta\, dL = dr$. Thus,

$$dW = \mathbf{F} \cdot d\mathbf{L} = F \cos\theta\, dL \tag{16}$$

If the path is broken up into segments parallel and perpendicular to $\mathbf{F}$ (Fig. 1-9), we note from (16) that contributions to the work occur only for the segments parallel to $\mathbf{F}$ ($\theta = 0$) with no work for the segments perpendicular to $\mathbf{F}$ ($\theta = 90°$). Summing the contributions of the segments parallel to $\mathbf{F}$, we obtain the total work W between the endpoints of the path. For finite length segments dL, this value is approximate. As $dL \to 0$, it becomes exact as given by

$$W = \int_{P_1 \text{(start)}}^{P_2 \text{(end)}} \mathbf{F} \cdot d\mathbf{L} \tag{17}$$

where P_1 = starting point (lower limit of integration)
P_2 = endpoint (upper limit of integration)

The formulation of (17) is called a ***line integral*** giving the total work W done by $\mathbf{F}$ on an object (equals energy imparted to the object) moved over the path from P_1 to P_2.

For path segments formulated as in (16), we observe the sign of $\cos\theta$. In the integral form of (17), the endpoints or limits of integration determine the correct sign of the result.

For the path in Fig. 1-9, (17) becomes,

$$W = \int_{r_1 \text{(start)}}^{r_2 \text{(end)}} F \cos\theta\, dL = \int_{r_1}^{r_2} F\, dr \tag{18}$$

Reversing the path (going from P_2 to P_1, or r_2 to r_1), we reverse the limits of integration so

$$W = \int_{r_2 \text{(start)}}^{r_1 \text{(end)}} F\, dr = -\int_{r_1}^{r_2} F\, dr \tag{19}$$

In (19), W is the work done in moving *against* $\mathbf{F}$ and is the negative of (18).

To illustrate the line integral, let us consider two examples.

Example 1-10. Work in linear field. A force field $\mathbf{F}$ is in the x direction and increases linearly with distance x. Thus, $\mathbf{F} = \hat{\mathbf{x}}x$. Find the work done by the force $\mathbf{F}$ in moving an object from a point $x = 1$ to a point $x = 2$.

Solution. $\mathbf{F} \cdot d\mathbf{L} = x\, dx$, so

$$W = \int \mathbf{F} \cdot d\mathbf{L} = \int_1^2 x\, dx = \frac{1}{2}x^2\Big|_1^2 = \frac{3}{2} \quad \textit{Ans.}$$

If F is in newtons and x is in meters, the work is in joules. (*End of Example.*)

Example 1-11. Work in radial field. A radial force **F** decreases with distance as given by $\mathbf{F} = \hat{\mathbf{r}}r^{-2}$ (Fig. 1-10). This inverse square variation with radial distance is the same as for the electric field **E** around a point charge Q as discussed in Chap. 2. Find the work done in moving from a point at $r = \sqrt{2}$ to a point at $r = 2\sqrt{2}$ by a direct (radial) path and one following rectangular coordinates (see Sec. 1-7).

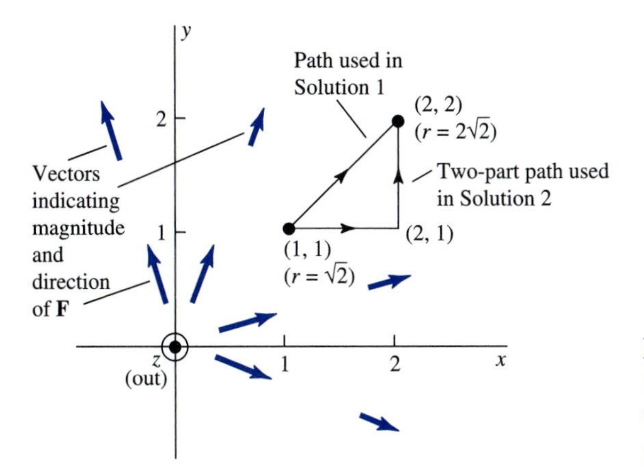

FIGURE 1-10 Line integral paths from point (1, 1) to (2, 2).

Solution 1.

$$\mathbf{F} \cdot d\mathbf{L} = r^{-2}\,dr$$

so
$$W = \int \mathbf{F} \cdot d\mathbf{L} = \int_{\sqrt{2}}^{2\sqrt{2}} r^{-2}\,dr = -\frac{1}{r}\bigg|_{\sqrt{2}}^{2\sqrt{2}} = \frac{1}{2\sqrt{2}} \qquad \textit{Ans.}$$

Solution 2. Instead of integrating along the direct (radial) path from $r = \sqrt{2}$ (at $x, y = 1, 1$) to $r = 2\sqrt{2}$ (at $x, y = 2, 2$) (Fig. 1-10), let us follow a rectangular coordinate path from $x, y = 1, 1$ to $x, y = 2, 1$ (y constant) and then from $x, y = 2, 1$ to $x, y = 2, 2$ (x constant). Thus, for the constant y path we have

$$(\mathbf{F} \cdot d\mathbf{L})_y = \frac{\hat{\mathbf{r}}}{x^2 + y^2} \cdot \hat{\mathbf{x}}\,dx$$

where $\hat{\mathbf{r}} \cdot \hat{\mathbf{x}} = \cos 45° = x/\sqrt{x^2 + y^2}$ and

$$(\mathbf{F} \cdot d\mathbf{L})_y = \frac{x}{(x^2 + y^2)\sqrt{x^2 + y^2}}$$

Since $x = y$ and $dx = dy$, the x-constant and y-constant terms are equal. Thus, the total work is twice the work for the y-constant path, or

$$W = \int \mathbf{F} \cdot d\mathbf{L} = 2\int_1^2 \frac{1}{(x^2 + x^2)} \frac{x}{\sqrt{x^2 + x^2}}\,dx$$

$$= 2\int_1^2 \frac{x}{(2x^2)^{3/2}}\,dx = \frac{2}{\sqrt{8}}\int_1^2 \frac{dx}{x^2} = -\frac{1}{\sqrt{2}}\frac{1}{x}\bigg|_1^2 = \frac{1}{2\sqrt{2}} \qquad \textit{Ans.}$$

the same as for Solution 1.

For a vector like **F**, *the line integral depends only on the endpoints* so we could follow *any* path from $x, y = 1, 1$ to $x, y = 2, 2$. Further, if we integrate **F** around a *closed path,* starting say at $x, y = 1, 1$ and ending back at $x, y = 1, 1$, the result is zero. Thus,

$$\oint \mathbf{F} \cdot d\mathbf{L} = 0 \tag{20}$$

where $\oint$ indicates integration around a closed path. Any field for which the line integral around a closed path is zero is called a ***conservative*** or ***lamellar field.*** Not all fields are lamellar.

In Chap. 2 we discuss the electric field **E** (or force per unit charge) and the work we do moving a charge *against* this field. To make this work positive to match an increase in electric potential energy, a negative sign is introduced, so (17) becomes $W = -\int \mathbf{E} \cdot d\mathbf{L}$.

Problem 1-6-6. Line integral for work. Find the work required to move a 5-kg mass from $x = 0, y = 0$ to $x = 8, y = 7$ m against a force $\mathbf{F} = \hat{\mathbf{x}}x^2$ N. *Ans.* 171 J.

Problem 1-6-7. Line integral for satellite to orbit. Find the work done in lifting a 6-tonne ($m_s = 6 \times 10^3$ kg) satellite to GSO (geostationary orbit) height = 37,000 km. The force of gravity is

$$\mathbf{F}_g = \hat{\mathbf{r}} G \frac{m_s m_e}{r^2} \quad \text{(N)}$$

where m_s = mass of satellite, kg
m_e = mass of earth = 6×10^{24} kg
r = distance of satellite from center of earth, m
G = gravitational constant = 6.67×10^{-11} N m^2 kg^{-2}

The circumference of the earth = 40,000 km. *Ans.* 89 MW-hr or over 100 million horsepower for 1 hour.

The Surface Integral

Suppose water is flowing at a uniform rate of B liters per minute per square meter (l min^{-1} m^{-2}) through a square loop of area A as in Fig. 1-11*a*. The *flow* or *flux of water* through the loop depends on three things: **B** (the rate and direction of the flow), the area A of the loop, and the loop's angle with respect to **B**. If we define the area as a vector of magnitude A and direction perpendicular to its surface (Fig. 1-11*a*), we can express the flow or flux ψ of water as a scalar or dot product. Thus,

$$\psi = BA\cos\theta = \mathbf{B} \cdot \mathbf{A} \qquad \text{(l/min)} \tag{21}$$

where $\mathbf{A} = \hat{\mathbf{n}}A$
$\hat{\mathbf{n}}$ = unit vector perpendicular to loop surface
A = scalar magnitude of the loop area

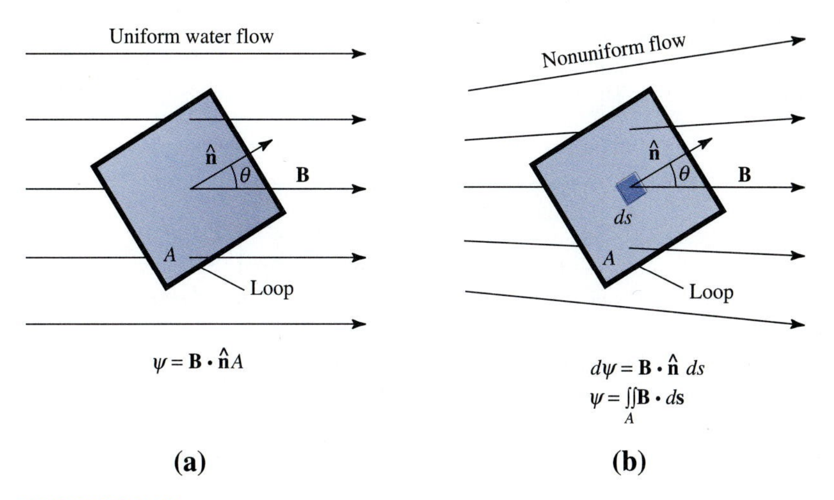

FIGURE 1-11
(*a*) Uniform water flow **B** through a loop of area *A* with unit normal $\hat{\mathbf{n}}$ at angle θ with respect to the flow direction of **B**. (*b*) Nonuniform flow case. Incremental flux $d\psi$ of vector **B** through surface element *ds* is given by $d\psi = \mathbf{B} \cdot \hat{\mathbf{n}}\, ds$.

If the flow is nonuniform (**B** a function of position), we need to calculate the incremental flux of water $d\psi$ through a surface area ds at a point (Fig. 1-11*b*) as given by

$$d\psi = \mathbf{B} \cdot \hat{\mathbf{n}}\, ds = \mathbf{B} \cdot d\mathbf{s} \tag{22}$$

where $\hat{\mathbf{n}}$ = unit normal to surface area ds
ds = scalar magnitude of surface area
$d\mathbf{s}$ = vector magnitude and direction of the surface area = $\hat{\mathbf{n}}\, ds$

Then we add up or integrate the contributions at all points across the surface of the loop obtaining the total flow or flux ψ of water. Thus,

$$\psi = \iint_{\substack{\text{Area}\\ A}} \mathbf{B} \cdot \hat{\mathbf{n}}\, ds = \iint_{\substack{\text{Area}\\ A}} \mathbf{B} \cdot d\mathbf{s} \qquad \text{(l/min)} \tag{23}$$

Equation (23) is called a ***surface integral.*** The integral signs in (23) imply that ds is an infinitesimal area. This leads to an exact value of ψ. If the area ds is small but finite, we use a summation sign $\sum$ obtaining an approximate value of ψ. As $ds \to 0$, the result becomes exact.

Example 1-12. Water flux. Water flowing in the x direction (Fig. 1-12) has a rate of flow as a function of y and z given by $B_x = 3yz$ liters min^{-1} m^{-2}. Find the total flow (or flux) of water through the rectangular area with corners (0, 0, 0), (0, 3, 0), (0, 0, 2), and (0, 3, 2) m.

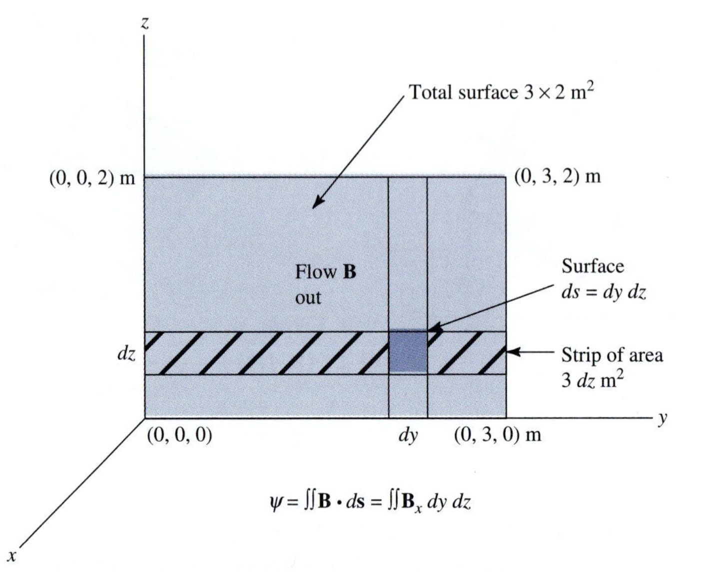

FIGURE 1-12
Surface over which water flow is integrated to obtain ψ in total liters per minute.

Solution. The flow rate $B_x = 3yz$ liters min^{-1} m^{-2} indicates that the flow rate is zero at the origin (0, 0, 0) and a maximum at the corner (0, 3, 2). Integrating first with respect to y (z constant), we get the flow through the strip of area 3 m (in the y direction) by width dz (shaded in Fig. 1-12). Then integrating with respect to z we sum the flow through all dz strips from $z = 0$ to $z = 2$ m for the total flow. Thus,

$$\psi = \iint_{\text{Surface}} \mathbf{B} \cdot d\mathbf{s} = \iint_{\text{Surface}} B_x \, dy \, dz = 3 \int_{z=0}^{z=2} \int_{y=0}^{y=3} yz \, dy \, dz$$

$$= 3 \frac{y^2}{2}\bigg|_0^3 \int_{z=0}^{z=2} z \, dz = 3 \times \frac{9}{2} \int_{z=0}^{z=2} z \, dz = 3 \times \frac{9}{2} \frac{z^2}{2}\bigg|_0^2 = 27 \text{ liters min}^{-1} \quad \textit{Ans.}$$

Note that if B_x had been a constant (= 3 liters min^{-1} m^{-2}) and not a function of y and z,

$$\psi = B_x \iint dy \, dz = 3 \times 6 = 18 \text{ liters min}^{-1} \quad \textit{Ans.}$$

Problem 1-6-8. Surface integral. The air flowing through a square opening in the y-z plane from $0 \leq y \leq 2$ m and $0 \leq z \leq 2$ m has a velocity $\mathbf{v} = \hat{\mathbf{x}} 3yz^2$ m/s. Find the total flow. *Ans.* 16 m^3/s.

Problem 1-6-9. Surface integral. Find the gas flowing through the shaded area of Fig. P1-6-9 at a radius $r = 3$ m if the gas velocity $\mathbf{v} = \hat{\mathbf{r}} 7\theta / r^3$ m/s. *Ans.* 0.48 m^3/s.

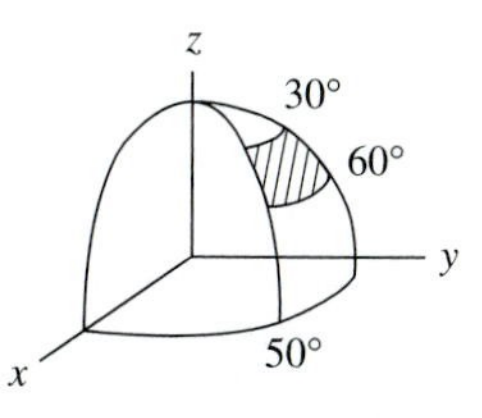

FIGURE P1-6-9

Problem 1-6-10. Surface integral for jet engine. The air-gas exhaust velocity from a jet engine varies linearly from a maximum of 300 m/s at the center of the circular exhaust opening to zero at the edges. If the exhaust diameter is 1.6 m, find the exhaust flow. *Ans.* 201 m^3/s.

The Volume Integral

Extending the surface integral concept to a volume, consider the following example.

Example 1-13. Mass. If the mass density ρ is a function of position x, y, z, as given by $\rho = 5xyz$ kg m^{-3}, find the total mass in a volume bounded by the planes $x = 0$, $x = 2$, $y = 0$, $y = 4$, $z = 0$, $z = 6$ m (Fig. 1-13).

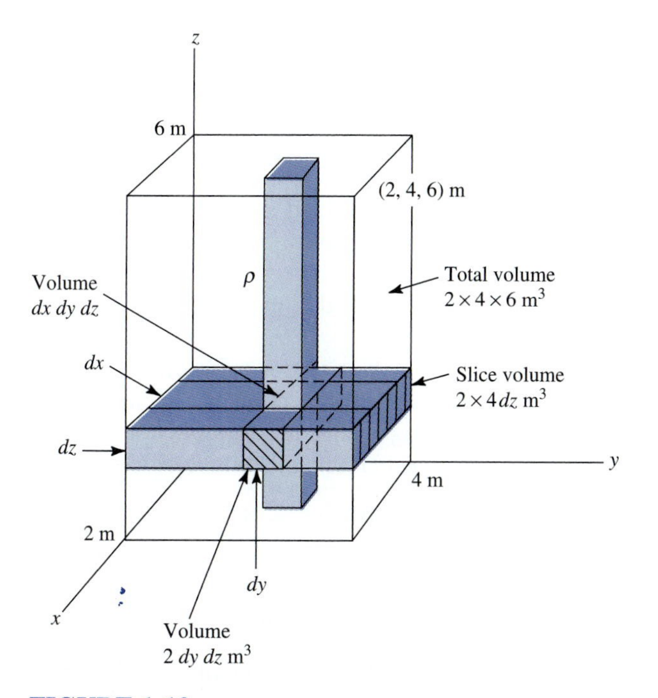

FIGURE 1-13
Volume throughout which mass density ρ is integrated to obtain total mass of the volume.

Solution. The mass density $\rho = 5xyz$ kg m^{-3} indicates that the density is zero at the origin (0, 0, 0) and a maximum at the corner (2, 4, 6). Integrating first with respect to x (y and z constant), we get the mass in the volume of 2 m length (in the x direction) and of cross section $dy\,dz$ (Fig. 1-13). Then integrating with respect to y, we sum the masses in all these volumes in a slice 2 × 4 m (in xy plane) by thickness dz. Finally, integrating with respect to z, we sum the mass in all the slices from $z = 0$ to $z = 6$ m for the total mass M. Thus,

$$M = \iiint \rho\,dv = 5\iiint xyz\,dx\,dy\,dz = 5\int_{z=0}^{z=6}\int_{y=0}^{y=4}\int_{x=0}^{x=2} xyz\,dx\,dy\,dz$$

$$= 5\frac{x^2}{2}\Big|_0^2 \int_{z=0}^{z=6}\int_{y=0}^{y=4} yz\,dy\,dz = 5 \times \frac{4}{2}\frac{y^2}{2}\Big|_0^4 \int_{z=0}^{z=6} z\,dz$$

$$= 5 \times \frac{4}{2} \times \frac{16}{2}\frac{z^2}{2}\Big|_0^6 = 5 \times \frac{4}{2} \times \frac{16}{2} \times \frac{36}{2} = 1440 \text{ kg} \quad \textit{Ans.}$$

Note that if ρ had been a constant (= 5 kg m^{-3}) and not a function of x, y, and z,

$$M = 5\iiint dx\,dy\,dz = 5 \times 2 \times 4 \times 6 = 240 \text{ kg} \quad \textit{Ans.}$$

Problem 1-6-11. Volume integral for spherical mass. Find the mass in a spherically symmetrical volume extending from $2 \le r \le 4$ m, if the mass density $\rho = 6/r^2$ kg/m^3. *Ans.* 151 kg.

Problem 1-6-12. Volume integral for barrel of soil. A cylindrical container 1 m in diameter and 3 m tall contains soil whose density decreases with height z as given by $\rho = 10^{(9-z)/3}$ kg/m^3. Find the mass of soil in the barrel. *Ans.* 921 kg.

Vector or Cross-Product of Two Vectors

The *vector product* of two vectors is defined as a third vector of magnitude equal to the product of the vector magnitudes multiplied by the sine of the angle between them. The direction of the third or resultant vector is perpendicular to the plane containing the first two and in such a sense that the three vectors form a right-handed set.

Referring to Fig. 1-14*a*, let **A** be a vector coincident with the x axis and **B** a vector in the x-y plane. To obtain the vector product of **A** and **B**, we turn **A** into **B** and proceed in the direction of a right-handed screw obtaining the vector **C** in the z direction (see Figure 1-7*a* on page 12). The magnitude of **C** is given by

$$|\mathbf{C}| = |\mathbf{A}||\mathbf{B}|\sin\theta = AB\sin\theta \tag{24}$$

where θ is the angle between **A** and **B**. In vector notation the relation is expressed

$$\mathbf{A} \times \mathbf{B} = \mathbf{C} = \hat{\mathbf{n}}AB\sin\theta \tag{25}$$

where $\hat{\mathbf{n}}$ is a unit vector normal to the plane containing **A** and **B**. The cross (×) between **A** and **B** indicates the vector product. Hence, it is also called the *cross-product.* The vector product does *not* obey the commutative law. Thus, $\mathbf{A} \times \mathbf{B} = \mathbf{C}$ but $\mathbf{B} \times \mathbf{A} = -\mathbf{C}$.

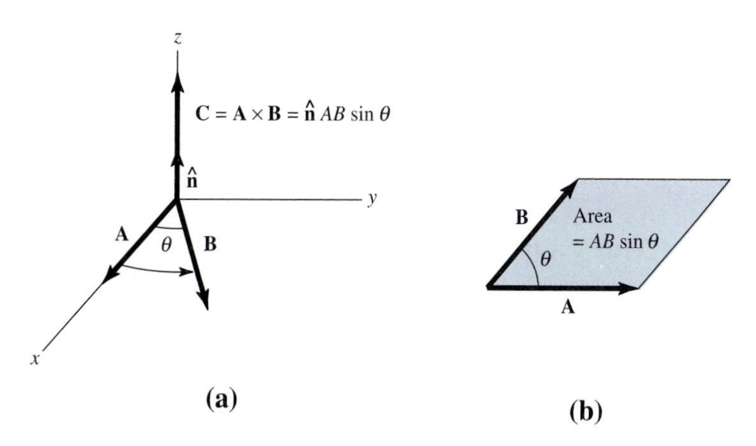

FIGURE 1-14
(*a*) Cross-product of vectors **A** and **B** is a third vector **C** normal to the plane containing **A** and **B** with direction in a right-hand sense. Thus, turning **A** into **B** (in the *x*-*y* plane), **C** is in the direction of the *z* axis with magnitude $AB \sin \theta$. (*b*) Magnitude of cross-product of **A** and **B** ($= AB \sin \theta$) is equal to the area of parallelogram shown.

From Fig. 1-14*b* we note that the magnitude of **C** ($= AB \sin \theta$) is given by the area of the parallelogram having **A** and **B** as two adjacent sides.

The vector product also conveniently expresses the right-handed relation of the rectangular coordinate unit vectors $\hat{\mathbf{x}}$, $\hat{\mathbf{y}}$, $\hat{\mathbf{z}}$. Thus,

$$\begin{aligned} \hat{\mathbf{x}} \times \hat{\mathbf{y}} &= \hat{\mathbf{z}} \\ \hat{\mathbf{y}} \times \hat{\mathbf{z}} &= \hat{\mathbf{x}} \\ \hat{\mathbf{z}} \times \hat{\mathbf{x}} &= \hat{\mathbf{y}} \end{aligned} \tag{26}$$

We also note that

$$\hat{\mathbf{x}} \times \hat{\mathbf{x}} = \hat{\mathbf{y}} \times \hat{\mathbf{y}} = \hat{\mathbf{z}} \times \hat{\mathbf{z}} = 0 \tag{27}$$

In rectangular coordinates the vector product of two vectors is

$$\begin{aligned} \mathbf{A} \times \mathbf{B} &= (\hat{\mathbf{x}}A_x + \hat{\mathbf{y}}A_y + \hat{\mathbf{z}}A_z) \times (\hat{\mathbf{x}}B_x + \hat{\mathbf{y}}B_y + \hat{\mathbf{z}}B_z) \\ &= \hat{\mathbf{x}} \times \hat{\mathbf{x}}A_xB_x + \hat{\mathbf{x}} \times \hat{\mathbf{y}}A_xB_y + \hat{\mathbf{x}} \times \hat{\mathbf{z}}A_xB_z \\ &\quad + \hat{\mathbf{y}} \times \hat{\mathbf{x}}A_yB_x + \hat{\mathbf{y}} \times \hat{\mathbf{y}}A_yB_y + \hat{\mathbf{y}} \times \hat{\mathbf{z}}A_yB_z \\ &\quad + \hat{\mathbf{z}} \times \hat{\mathbf{x}}A_zB_x + \hat{\mathbf{z}} \times \hat{\mathbf{y}}A_zB_y + \hat{\mathbf{z}} \times \hat{\mathbf{z}}A_zB_z \end{aligned} \tag{28}$$

Applying the unit vector relations of (26) and (27) reduces (28) to

$$\mathbf{A} \times \mathbf{B} = \hat{\mathbf{x}}(A_yB_z - A_zB_y) + \hat{\mathbf{y}}(A_zB_x - A_xB_z) + \hat{\mathbf{z}}(A_xB_y - A_yB_x) \tag{29}$$

This may be expressed concisely in determinant form as follows:

$$\mathbf{A} \times \mathbf{B} = \begin{vmatrix} \hat{\mathbf{x}} & \hat{\mathbf{y}} & \hat{\mathbf{z}} \\ A_x & A_y & A_z \\ B_x & B_y & B_z \end{vmatrix} \tag{30}$$

Example 1-14. Cross-product. Vector $\mathbf{A} = \hat{\mathbf{x}}8 + \hat{\mathbf{y}}3 - \hat{\mathbf{z}}10$ and vector $\mathbf{B} = -\hat{\mathbf{x}}15 + \hat{\mathbf{y}}6 + \hat{\mathbf{z}}17$. Find $\mathbf{A} \times \mathbf{B}$.

Solution. From (29),

$$\begin{aligned} \mathbf{A} \times \mathbf{B} &= \hat{\mathbf{x}}(3 \times 17 + 10 \times 6) + \hat{\mathbf{y}}(10 \times 15 - 8 \times 17) + \hat{\mathbf{z}}(8 \times 6 + 3 \times 15) \\ &= \hat{\mathbf{x}}111 + \hat{\mathbf{y}}14 + \hat{\mathbf{z}}93 \quad \textit{Ans.} \end{aligned}$$

Problem 1-6-13. Cross-product. Vector $\mathbf{A} = \hat{\mathbf{y}}20 - \hat{\mathbf{z}}5$, vector $\mathbf{B} = -\hat{\mathbf{x}}6 + \hat{\mathbf{y}}14$. Find $\mathbf{A} \times \mathbf{B}$. *Ans.* $\hat{\mathbf{x}}70 + \hat{\mathbf{y}}30 + \hat{\mathbf{z}}120$.

1-7 INTRODUCTION TO COORDINATE SYSTEMS

The three most common coordinate systems are rectangular (coordinates x, y, z), cylindrical (coordinates r, ϕ, z), and spherical (coordinates r, θ, ϕ) as shown in Figure 1-15.

In *rectangular coordinates* a point P is specified by x, y, z, where these values are all measured from the origin (Fig. 1-15*a*). A vector at the point P is specified in terms of three mutually perpendicular components with unit vectors $\hat{\mathbf{x}}$, $\hat{\mathbf{y}}$, and $\hat{\mathbf{z}}$. The unit vectors $\hat{\mathbf{x}}$, $\hat{\mathbf{y}}$, and $\hat{\mathbf{z}}$ form a right-handed set; that is, turning $\hat{\mathbf{x}}$ into $\hat{\mathbf{y}}$ like a right-handed screw, we move in the $\hat{\mathbf{z}}$ direction.

In *cylindrical coordinates* a point P is specified by r, ϕ, z, where ϕ is measured from the x axis (or x-z plane) (Fig. 1-15*b*). A vector at the point P is specified in terms of three mutually perpendicular components with unit vectors $\hat{\mathbf{r}}$ perpendicular to the cylinder of radius r, $\hat{\boldsymbol{\phi}}$ perpendicular to the plane through the z axis at angle ϕ, and $\hat{\mathbf{z}}$ perpendicular to the x-y plane at distance z. The unit vectors $\hat{\mathbf{r}}$, $\hat{\boldsymbol{\phi}}$, $\hat{\mathbf{z}}$ form a right-handed set.

In *spherical coordinates* a point P is specified by r, θ, ϕ, where r is measured from the origin, θ is measured from the z axis, and ϕ is measured from the x axis (or x-z plane) (Fig. 1-15*c*). With z axis up, as in Fig. 1-15*c*, θ is sometimes called the *zenith* angle and ϕ the *azimuth* angle. A vector at the point P is specified in terms of three mutually perpendicular components with unit vectors $\hat{\mathbf{r}}$ perpendicular to the sphere of radius r, $\hat{\boldsymbol{\theta}}$ perpendicular to the cone of angle θ, and $\hat{\boldsymbol{\phi}}$ perpendicular to the plane through the z axis at angle ϕ. The unit vectors $\hat{\mathbf{r}}$, $\hat{\boldsymbol{\theta}}$, $\hat{\boldsymbol{\phi}}$ form a right-handed set.

An infinitesimal length in the *rectangular* system is given by

$$dL = \sqrt{dx^2 + dy^2 + dz^2} \tag{1}$$

and an infinitesimal volume by

$$dv = dx\,dy\,dz \tag{2}$$

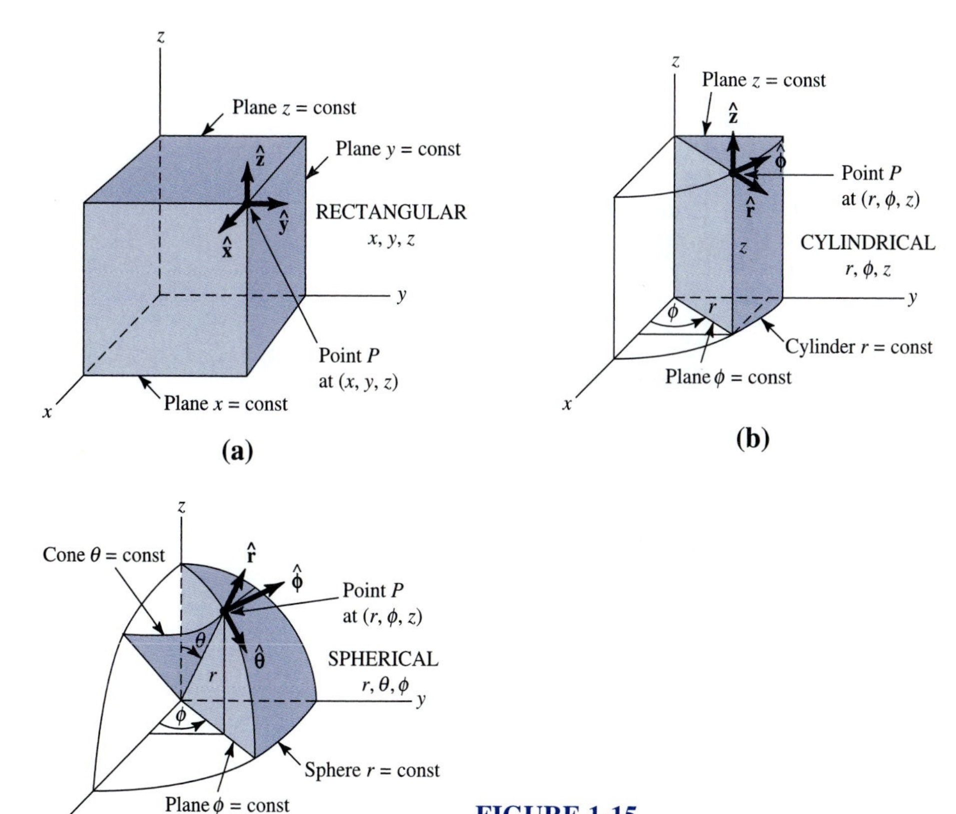

FIGURE 1-15
Coordinate systems. (*a*) Rectangular. (*b*) Cylindrical. (*c*) Spherical.

In the *cylindrical* system the corresponding quantities are

$$dL = \sqrt{dr^2 + r^2\,d\phi^2 + dz^2} \tag{3}$$

and

$$dv = dr\,r\,d\phi\,dz \tag{4}$$

In the *spherical* system we have

$$dL = \sqrt{dr^2 + r^2\,d\theta^2 + r^2\sin^2\theta\,d\phi^2} \tag{5}$$

and

$$dv = dr\,r\,d\theta\,r\sin\theta\,d\phi \tag{6}$$

Referring to Fig. 1-16, the projection x of the scalar distance r on the x axis is given by $r\cos\alpha$ where α is the angle between r and the x axis. The projection of r on the y axis is given by $r\cos\beta$, and the projection on the z axis by $r\cos\gamma$. Note that $\gamma = \theta$ so $\cos\gamma = \cos\theta$.

The quantities $\cos\alpha$, $\cos\beta$, and $\cos\gamma$ are called the *direction cosines*. From the theorem of Pythagoras,

$$\cos^2\alpha + \cos^2\beta + \cos^2\gamma = 1 \tag{7}$$

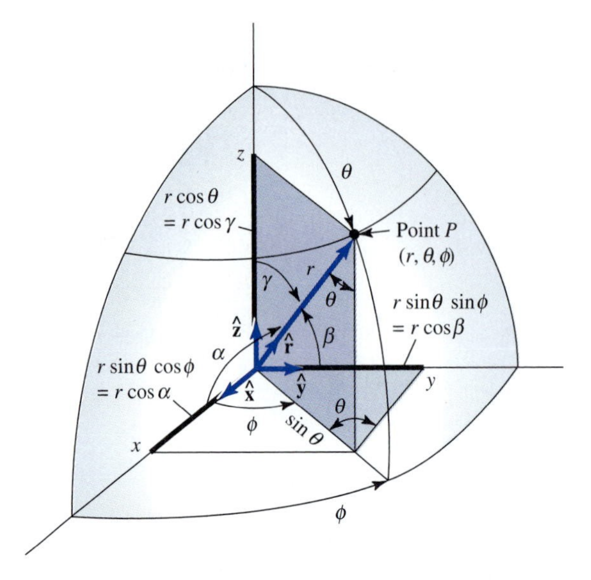

FIGURE 1-16
Scalar distance r to point P at (r, θ, ϕ) resolved into rectangular (x, y, z) components. Rectangular $(\hat{\mathbf{x}}, \hat{\mathbf{y}}, \hat{\mathbf{z}})$ unit vectors are also shown.

From Fig. 1-16, the scalar distance r of a spherical coordinate system transforms into rectangular coordinate distance

$$x = r\cos\alpha = r\sin\theta\cos\phi \tag{8}$$

$$y = r\cos\beta = r\sin\theta\sin\phi \tag{9}$$

$$z = r\cos\gamma = r\cos\theta \tag{10}$$

from which

$$\left.\begin{aligned} \cos\alpha &= \sin\theta\cos\phi \quad (11)\\ \cos\beta &= \sin\theta\sin\phi \quad (12)\\ \cos\gamma &= \cos\theta \quad (13)\end{aligned}\right\}\ \text{direction cosines}$$

As the converse of (8), (9), and (10), the spherical coordinate values (r, θ, ϕ) may be expressed in terms of rectangular coordinate distances as follows:

$$r = \sqrt{x^2 + y^2 + z^2} \qquad r \geq 0 \tag{14}$$

$$\theta = \cos^{-1}\frac{z}{\sqrt{x^2 + y^2 + z^2}} \qquad (0 \leq \theta \leq \pi) \tag{15}$$

$$\phi = \tan^{-1}\frac{y}{x} \tag{16}$$

From these and similar coordinate transformations of spherical to rectangular and rectangular to spherical coordinates, we may express a vector $\mathbf{A}$ at some point P with spherical components A_r, A_θ, and A_ϕ as the rectangular components A_x, A_y, and A_z, where

$$A_x = A_r \sin\theta \cos\phi + A_\theta \cos\theta \cos\phi - A_\phi \sin\phi \tag{17}$$

$$A_y = A_r \sin\theta \sin\phi + A_\theta \cos\theta \sin\phi + A_\phi \cos\phi \tag{18}$$

$$A_z = A_r \cos\theta - A_\theta \sin\theta \tag{19}$$

We note also from Fig. 1-15 that the dot products of the unit vector $\hat{\mathbf{r}}$ with the rectangular units vectors $\hat{\mathbf{x}}$, $\hat{\mathbf{y}}$, and $\hat{\mathbf{z}}$ are equal to the *direction cosines* as given by

$$\hat{\mathbf{x}} \cdot \hat{\mathbf{r}} = \cos\alpha = \sin\theta \cos\phi \tag{20}$$

$$\hat{\mathbf{y}} \cdot \hat{\mathbf{r}} = \cos\beta = \sin\theta \sin\phi \tag{21}$$

$$\hat{\mathbf{z}} \cdot \hat{\mathbf{r}} = \cos\gamma = \cos\theta \tag{22}$$

These and other dot product combinations are listed in Table 1-4.

In addition to rectangular, cylindrical, and spherical coordinate systems, there are many other systems such as the elliptical, spheroidal (both prolate and oblate), and paraboloidal systems. Although the number of possible systems is infinite, all of them can be treated in terms of a *generalized curvilinear coordinate system.* However, we will not need to deal with these systems in this book, the rectangular, cylindrical, and spherical systems being sufficient for our requirements.

TABLE 1-4
Dot products of unit vectors in three coordinate systems

		Rectangular			Cylindrical			Spherical		
	•	$\hat{\mathbf{x}}$	$\hat{\mathbf{y}}$	$\hat{\mathbf{z}}$	$\hat{\mathbf{r}}$	$\hat{\boldsymbol{\phi}}$	$\hat{\mathbf{z}}$	$\hat{\mathbf{r}}$	$\hat{\boldsymbol{\theta}}$	$\hat{\boldsymbol{\phi}}$
Rectangular	$\hat{\mathbf{x}}$	1	0	0	$\cos\phi$	$-\sin\phi$	0	$\sin\theta\cos\phi$	$\cos\theta\cos\phi$	$-\sin\phi$
	$\hat{\mathbf{y}}$	0	1	0	$\sin\phi$	$\cos\phi$	0	$\sin\theta\sin\phi$	$\cos\theta\sin\phi$	$\cos\phi$
	$\hat{\mathbf{z}}$	0	0	1	0	0	1	$\cos\theta$	$-\sin\theta$	0
Cylindrical	$\hat{\mathbf{r}}$	$\cos\phi$	$\sin\phi$	0	1	0	0	$\sin\theta$	$\cos\theta$	0
	$\hat{\boldsymbol{\phi}}$	$-\sin\phi$	$\cos\phi$	0	0	1	0	0	0	1
	$\hat{\mathbf{z}}$	0	0	1	0	0	1	$\cos\theta$	$-\sin\theta$	0
Spherical	$\hat{\mathbf{r}}$	$\sin\theta\cos\phi$	$\sin\theta\sin\phi$	$\cos\theta$	$\sin\theta$	0	$\cos\theta$	1	0	0
	$\hat{\boldsymbol{\theta}}$	$\cos\theta\cos\phi$	$\cos\theta\sin\phi$	$-\sin\theta$	$\cos\theta$	0	$-\sin\theta$	0	1	0
	$\hat{\boldsymbol{\phi}}$	$-\sin\phi$	$\cos\phi$	0	0	1	0	0	0	1

Note that the unit vectors $\hat{\mathbf{r}}$ in the cylindrical and spherical systems are *not* the same. For example,

Spherical	*Cylindrical*
$\hat{\mathbf{r}} \cdot \hat{\mathbf{x}} = \sin\theta\cos\phi$	$\hat{\mathbf{r}} \cdot \hat{\mathbf{x}} = \cos\phi$
$\hat{\mathbf{r}} \cdot \hat{\mathbf{y}} = \sin\theta\sin\phi$	$\hat{\mathbf{r}} \cdot \hat{\mathbf{y}} = \sin\phi$
$\hat{\mathbf{r}} \cdot \hat{\mathbf{z}} = \cos\theta$	$\hat{\mathbf{r}} \cdot \hat{\mathbf{z}} = 0$

The fundamental parameters of the rectangular, cylindrical, and spherical coordinate systems are summarized in Table 1-5 on page 29.

Table 1-6, also on page 29, gives the unit vector dot products in rectangular coordinates for both rectangular-cylindrical and rectangular-spherical coordinates.

Table 1-7 on page 30 summarizes the transformations between the three systems.

Example 1-15. Coordinate transformation. Vectors $\mathbf{A} = \hat{\mathbf{r}}7 + \hat{\boldsymbol{\theta}}5 + \hat{\boldsymbol{\phi}}3$ and $\mathbf{B} = \hat{\mathbf{r}}2 + \hat{\boldsymbol{\theta}}3 + \hat{\boldsymbol{\phi}}4$ are situated at a point $(r, \theta, \phi) = (3, 45°, 45°)$. Find (*a*) $\mathbf{A} + \mathbf{B}$; (*b*) $\mathbf{A} \cdot \mathbf{B}$; and (*c*) $\mathbf{A} \times \mathbf{B}$.

Solution. From Table 1-7 on page 30,

$$\begin{aligned} A_x &= A_r \sin\theta\cos\phi + A_\theta\cos\theta\cos\phi - A_\phi\sin\phi \\ A_y &= A_r\sin\theta\sin\phi + A_\theta\cos\theta\sin\phi + A_\phi\cos\phi \\ A_z &= A_r\cos\theta - A_\theta\sin\theta \end{aligned}$$

$$\begin{aligned} A_x &= 7\sin 45°\cos 45° + 5\cos 45°\cos 45° - 3\sin 45° = 3.879 \\ A_y &= 7\sin 45°\sin 45° + 5\cos 45°\sin 45° + 3\cos 45° = 8.121 \\ A_z &= 7\cos 45° - 5\sin 45° = 1.414 \end{aligned}$$

Similarly,

$$\begin{aligned} B_x &= 2\sin 45°\cos 45° + 3\cos 45°\cos 45° - 4\sin 45° = -0.328 \\ B_y &= 2\sin 45°\sin 45° + 3\cos 45°\sin 45° + 4\cos 45° = 5.328 \\ B_z &= 2\cos 45° - 3\sin 45° = -0.707 \end{aligned}$$

$$\begin{aligned} (a)\ \hat{\mathbf{A}} + \hat{\mathbf{B}} &= \hat{\mathbf{x}}(A_x + B_x) + \hat{\mathbf{y}}(A_y + B_y) + \hat{\mathbf{z}}(A_z + B_z) \\ &= \hat{\mathbf{x}}3.55 + \hat{\mathbf{y}}13.45 + \hat{\mathbf{z}}0.707 \quad \textit{Ans.} \end{aligned}$$

$$\begin{aligned} (b)\ \mathbf{A}\cdot\mathbf{B} &= A_xB_x + A_yB_y + A_zB_z = A_rB_r + A_\theta B_\theta + A_\phi B_\phi \\ &= 7\times 2 + 5\times 3 + 3\times 4 = 14 + 15 + 12 = 41 \quad \textit{Ans.} \end{aligned}$$

$$(c)\ \mathbf{A}\times\mathbf{B} = \begin{vmatrix} \hat{\mathbf{x}} & \hat{\mathbf{y}} & \hat{\mathbf{z}} \\ 3.879 & 8.121 & 1.414 \\ -0.328 & 5.328 & -0.707 \end{vmatrix} = -\hat{\mathbf{x}}13.27 + \hat{\mathbf{y}}2.28 + \hat{\mathbf{z}}23.33 \quad \textit{Ans.}$$

Problem 1-7-1. Coordinate transformation. If $\mathbf{A} = \hat{\mathbf{x}}2 + \hat{\mathbf{y}}0 + \hat{\mathbf{z}}2$ and $\mathbf{B} = \hat{\mathbf{r}}2$ at $\theta = 45°$, $\phi = 0°$, find the angle between **A** and **B**. *Ans.* 0°.

Problem 1-7-2. Coordinate transformation. Find $\mathbf{A}\cdot\mathbf{B}$, where $\mathbf{A} = \hat{\mathbf{x}}3 + \hat{\mathbf{y}}2 + \hat{\mathbf{z}}$ and $\mathbf{B} = \hat{\mathbf{r}}$ at $\theta = 45°$, $\phi = 30°$. *Ans.* 3.25.

Problem 1-7-3. Coordinate transformation. Vectors $\mathbf{A} = \hat{\mathbf{r}}9$ at $\theta = 75°$, $\phi = 60°$ and $\mathbf{B} = \hat{\mathbf{x}}8 + \hat{\mathbf{y}}5 - \hat{\mathbf{z}}6$ extend from the origin. Find (*a*) $\mathbf{A}\cdot\mathbf{B}$, (*b*) $\mathbf{A}\times\mathbf{B}$, and (*c*) the angle between **A** and **B**. *Ans.* (*a*) 58.5; (*b*) $-\hat{\mathbf{x}}56.8 + \hat{\mathbf{y}}44.7 - \hat{\mathbf{z}}38.5$; (*c*) 54.5°.

TABLE 1-5

Parameters of rectangular, cylindrical, and spherical coordinate systems

Coordinate system	Coordinates	Range	Unit vectors	Length elements	Coordinate surfaces	
Rectangular (Fig. 1-15*a*)	x	$-\infty$ to $+\infty$	$\hat{\mathbf{x}}$	dx	Plane	x = constant
	y	$-\infty$ to $+\infty$	$\hat{\mathbf{y}}$	dy	Plane	y = constant
	z	$-\infty$ to $+\infty$	$\hat{\mathbf{z}}$	dz	Plane	z = constant
Cylindrical (Fig. 1-15*b*)	r	0 to ∞	$\hat{\mathbf{r}}$	dr	Cylinder	r = constant
	ϕ	0 to 2π	$\hat{\boldsymbol{\phi}}$	$r\,d\phi$	Plane	ϕ = constant
	z	$-\infty$ to $+\infty$	$\hat{\mathbf{z}}$	dz	Plane	z = constant
Spherical (Fig. 1-15*c*)	r	0 to ∞	$\hat{\mathbf{r}}$	dr	Cylinder	r = constant
	θ	0 to π	$\hat{\boldsymbol{\theta}}$	$r\,d\theta$	Cone	θ = constant
	ϕ	0 to 2π	$\hat{\boldsymbol{\phi}}$	$r\sin\theta\,d\phi$	Plane	ϕ = constant

TABLE 1-6

Unit vector dot products for rectangular-cylindrical and rectangular-spherical coordinates

Rectangular-cylindrical product in rectangular coordinates				Rectangular-spherical product in rectangular coordinates			
•	$\hat{\mathbf{x}}$	$\hat{\mathbf{y}}$	$\hat{\mathbf{z}}$	•	$\hat{\mathbf{x}}$	$\hat{\mathbf{y}}$	$\hat{\mathbf{z}}$
$\hat{\mathbf{r}}$	$\frac{x}{\sqrt{x^2+y^2}}$	$\frac{y}{\sqrt{x^2+y^2}}$	0	$\hat{\mathbf{r}}$	$\frac{x}{\sqrt{x^2+y^2+z^2}}$	$\frac{y}{\sqrt{x^2+y^2+z^2}}$	$\frac{z}{\sqrt{x^2+y^2+z^2}}$
$\hat{\boldsymbol{\phi}}$	$\frac{-y}{\sqrt{x^2+y^2}}$	$\frac{x}{\sqrt{x^2+y^2}}$	0	$\hat{\boldsymbol{\theta}}$	$\frac{xz}{\sqrt{x^2+y^2}\sqrt{x^2+y^2+z^2}}$	$\frac{yz}{\sqrt{x^2+y^2}\sqrt{x^2+y^2+z^2}}$	$-\frac{\sqrt{x^2+y^2}}{\sqrt{x^2+y^2+z^2}}$
$\hat{\mathbf{z}}$	0	0	1	$\hat{\boldsymbol{\phi}}$	$-\frac{y}{\sqrt{x^2+y^2}}$	$\frac{x}{\sqrt{x^2+y^2}}$	0
Example: $\hat{\boldsymbol{\phi}}\cdot\hat{\mathbf{y}} = \cos\phi = \frac{x}{\sqrt{x^2+y^2}}$				Example: $\hat{\mathbf{x}}\cdot\hat{\mathbf{r}} = \sin\theta\cos\phi = \frac{x}{\sqrt{x^2+y^2+z^2}}$			

TABLE 1-7
Coordinate transformations

Rectangular to cylindrical

$$A_r = A_x \frac{x}{\sqrt{x^2 + y^2}} + A_y \frac{x}{\sqrt{x^2 + y^2}}$$

$$A_\phi = -A_x \frac{y}{\sqrt{x^2 + y^2}} + A_y \frac{x}{\sqrt{x^2 + y^2}}$$

$$A_z = A_z$$

Rectangular to spherical

$$A_r = A_x \frac{x}{\sqrt{x^2 + y^2 + z^2}} + A_y \frac{y}{\sqrt{x^2 + y^2 + z^2}} + A_z \frac{z}{\sqrt{x^2 + y^2 + z^2}}$$

$$A_\theta = A_x \frac{xz}{\sqrt{x^2 + y^2}\sqrt{x^2 + y^2 + z^2}} + A_y \frac{yz}{\sqrt{x^2 + y^2}\sqrt{x^2 + y^2 + z^2}} - A_z \frac{\sqrt{x^2 + y^2}}{\sqrt{x^2 + y^2 + z^2}}$$

$$A_\phi = -A_x \frac{y}{\sqrt{x^2 + y^2}} + A_y \frac{x}{\sqrt{x^2 + y^2}}$$

Cylindrical to rectangular

$$A_x = A_r \cos\phi - A_\phi \sin\phi$$

$$A_y = A_r \sin\phi - A_\phi \cos\phi$$

$$A_z = A_z$$

Spherical to rectangular

$$A_x = A_r \sin\theta\cos\phi + A_\theta \cos\theta\cos\phi - A_\phi \sin\phi$$

$$A_y = A_r \sin\theta\sin\phi + A_\theta \cos\theta\sin\phi + A_\phi \cos\phi$$

$$A_z = A_r \cos\theta - A_\theta \sin\theta$$

PROBLEMS

Note: The problem numbers indicate chapter and section. Thus, Problem 1-4-4 is the 4th problem of Sec. 4, Chap. 1. Since 1-4-4 is the first problem listed below in this end-of-chapter set, it tells you also that three problems related to material in Sec. 4 have already been given with the text of Sec. 4.

Tips: The following suggestions will often help in solving the problems in this and other chapters:

1. Make a sketch whenever possible. This helps you visualize the situation and often provides a clue about the best coordinate system to use.
2. Be aware of the octant or octants occupied by the points, surfaces, and volumes of the problem. Thus, a point with positive x, positive y, and positive z coordinates should have θ between 0° and 90° and ϕ between 0° and 90°.
3. In most cases, converting from cylindrical or spherical to rectangular coordinates will simplify problems involving angles between vectors, distances between arrow tips, and dot or cross-products.

4. When determining unit normal vectors, check that the magnitude of your result equals 1. If it doesn't, you've made an error.

Answers are given in Appendix E to problems followed by the @ symbol.

1-4-4. Unit reciprocals. What is the reciprocal of (*a*) femtojoule, (*b*) terawatt, and (*c*) kilometer? @

1-5-3. Are the following equations balanced dimensionally? @

$$F = MLT^{-2} \qquad \text{where } F = \text{force}$$

$$P = MLT^{-3} \qquad \text{where } P = \text{power}$$

$$L = vT \qquad \text{where } v = \text{velocity}$$

$$w = ML^{-1}T^{-2} \qquad \text{where } w = \text{energy density}$$

$$E = Mv^2 \qquad \text{where } E = \text{energy}$$

1-6-14. Vectors $\mathbf{A} = \hat{\mathbf{y}}6 + \hat{\mathbf{z}}$ and $\mathbf{B} = \hat{\mathbf{x}}2 + \hat{\mathbf{y}}4$ extend from the origin. Find (*a*) $\mathbf{A} \cdot \mathbf{B}$, (*b*) $\mathbf{A} \times \mathbf{B}$, and (*c*) the angle between **A** and **B**. @

1-6-15. What is the work done in moving against a force $\mathbf{F} = \hat{\mathbf{r}}\, 5/r^2$ (N) from $r = 10$ m to $r = 1$ m?

1-6-16. What is the distance along a spherical surface between points at $\theta = 45°$, $\phi = 10°$ and $\theta = -20°$, $\phi = 100°$, if the radius of the sphere is 20 cm? @

1-6-17. If $\mathbf{A} = \hat{\mathbf{x}}5 + \hat{\mathbf{y}}12 - \hat{\mathbf{z}}13$ and $\mathbf{B} = \hat{\mathbf{x}}5 + \hat{\mathbf{y}}12 + \hat{\mathbf{z}}13$, find the angle between **A** and **B**.

1-6-18. (*a*) Find the area of a sphere of 2 meter radius between the angles $20° \geq \theta \geq 45°$ and $100° \geq \phi \geq 150°$. (*b*) What is the solid angle subtended? @

1-6-19. Integrate the vector $\mathbf{B} = \hat{\mathbf{x}}x^2y^3z^{1/4}$ over the plane square surface bounded by the points $(x, y, z) = (3, 2, 0)\,(-3, 2, 0)\,(3, 2, 4)\,(-3, 2, 4)$.

1-6-20. The difference in electric potential between two points is equal to the work done per unit charge in transporting a quantity of charge between the points. Find the change in electrical potential in going from the point $x = 0$ m to the point $x = 10$ m along the x axis in the presence of the electric field $\mathbf{E} = \hat{\mathbf{x}}3x^2$ V/m. @

1-6-21. A scalar function $V = x^2y^2 + z^{1/2}$. The rate of change of this function with position is a vector given by

$$-\mathbf{E} = \hat{\mathbf{x}}\frac{\partial V}{\partial x} + \hat{\mathbf{y}}\frac{\partial V}{\partial y} + \hat{\mathbf{z}}\frac{\partial V}{\partial z}$$

At the point (4, 1, 2), find (*a*) the magnitude of **E** and (*b*) the unit vector in the direction of **E**.

1-6-22. Vectors $\mathbf{A} = \hat{\mathbf{x}}6 + \hat{\mathbf{y}}2 + \hat{\mathbf{z}}5$ and $\mathbf{B} = \hat{\mathbf{x}}2 + \hat{\mathbf{y}}4 + \hat{\mathbf{z}}7$ extend out from the origin. Find (*a*) the angle between **A** and **B**, (*b*) the distance between their tips, (*c*) the unit vectors normal to the plane containing **A** and **B**, (*d*) the area of the parallelogram of which **A** and **B** are adjacent sides, and (*e*) the unit vectors parallel to the bisector of the angle between the two vectors. @

1-6-23. Vector $\mathbf{A} = \hat{\mathbf{x}}4 + \hat{\mathbf{y}}2 + \hat{\mathbf{z}}8$. Find a vector **B** such that $\mathbf{A} \cdot \mathbf{B} = 12$.

1-6-24. Vectors extend from the origin to points $(-4, 4, 2)$ and $(5, -2, -3)$. Find (*a*) the angle between the vectors, (*b*) the distance between their tips, (*c*) the unit vector normal to the plane containing them, (*d*) the area of the parallelogram of which the vectors form adjacent sides, and (*e*) the unit vectors parallel to the bisector of the angle between the two vectors. @

1-6-25. Vectors $\mathbf{A} = \hat{\mathbf{x}}3 - \hat{\mathbf{y}}2 + \hat{\mathbf{z}}6$ and $\mathbf{B} = \hat{\mathbf{x}} - \hat{\mathbf{y}}7 - \hat{\mathbf{z}}5$ extend out from the origin. Find (*a*) the angle between **A** and **B**, (*b*) the distance between the tips of the vectors, (*c*) the unit vectors normal to the plane containing **A** and **B**, (*d*) the area of the parallelogram of which **A** and **B** are adjacent sides, and (*e*) the unit vector parallel to the bisector of the angle between the two vectors.

1-6-26. Three vectors extend from the origin to the points $(x, y, z) = (1, 3, 2), (3, -2, 1)$, and $(-2, 2, 3)$. Find (*a*) the length of the perimeter of the triangle defined by the three points, (*b*) the area of the triangle, and (*c*) the angle between the surface of the triangle and the z axis. @

1-6-27. Vectors $\mathbf{A} = \hat{\mathbf{x}}3 + \hat{\mathbf{y}}2 + \hat{\mathbf{z}}6$ and $\mathbf{B} = \hat{\mathbf{x}} + \hat{\mathbf{y}}4 + \hat{\mathbf{z}}2$ are situated at point (x, y, z). Find (*a*) $\mathbf{A} + \mathbf{B}$, (*b*) $\mathbf{A} \cdot \mathbf{B}$, (*c*) the angle between **A** and **B**, (*d*) $\mathbf{A} \times \mathbf{B}$, (*e*) unit normals to the plane containing **A** and **B**, and (*f*) the area of the parallelogram of which **A** and **B** are adjacent sides.

1-6-28. Find the values of **B** such that the angle between the vectors $\mathbf{A} = \hat{\mathbf{x}} + \hat{\mathbf{y}}5 + \hat{\mathbf{z}}6$ and $\mathbf{B} = -\hat{\mathbf{x}}2 - \hat{\mathbf{y}}4 + \hat{\mathbf{z}}B_z$ is (*a*) 45°, (*b*) 90°, and (*c*) 135°. @

1-6-29. Vectors $\mathbf{A} = \hat{\mathbf{x}}4 + \hat{\mathbf{y}}6 + \hat{\mathbf{z}}10$ and $\mathbf{B} = \hat{\mathbf{x}}7 + \hat{\mathbf{y}}12 + \hat{\mathbf{z}}8$ are situated at point (x, y, z). Find (*a*) $\mathbf{A} + \mathbf{B}$, (*b*) $\mathbf{A} \cdot \mathbf{B}$, (*c*) the angle between **A** and **B**, (*d*) $\mathbf{A} \times \mathbf{B}$, (*e*) the unit normals to the plane containing **A** and **B**, (*f*) the area of the parallelogram of which **A** and **B** are adjacent sides.

1-6-30. If the vector function of position $\mathbf{F} = \hat{\mathbf{x}}x + \hat{\mathbf{y}}y + \hat{\mathbf{z}}z$, find (*a*) the line integral from the point (2, 1, 4) to the point (5, 9, 7) and (*b*) the flux of **F** through the surface at $x = 5$ from $y = 0$ to $y = 4$ and $z = 0$ to $z = 5$. @

1-7-4. A point in rectangular coordinates is at (2, 9, 3). Express its position in (*a*) cylindrical coordinates and (*b*) spherical coordinates.

1-7-5. A point in rectangular coordinates is at $(-2, 4, -1)$. Express its position in (*a*) cylindrical coordinates and (*b*) spherical coordinates. @

1-7-6. A vector $\mathbf{A} = \hat{\mathbf{x}}2 + \hat{\mathbf{y}}4 + \hat{\mathbf{z}}3$ is at the point $(x, y, z) = (1, 5, 2)$. Express **A** in (*a*) cylindrical coordinates and (*b*) spherical coordinates.

1-7-7. Two points have coordinates $(r, \theta, \phi) = (3, 60°, 30°)$ and $(r, \theta, \phi) = (3, 30°, 120°)$. Find (*a*) the straight-line distance between the two points, (*b*) the angle between the two lines extending from the origin to the two points, (*c*) the distance between the two points measured along the spherical surface for which $r = 3$, and (*d*) the area contained between the two lines and the circle of radius $r = 3$. @

1-7-8. A point in cylindrical coordinates is at (4, 45°, 2). Express its position in (*a*) rectangular coordinates and (*b*) spherical coordinates.

1-7-9. A point in spherical coordinates is at (5, 60°, 150°). Express its position in (*a*) rectangular coordinates and (*b*) cylindrical coordinates. @

1-7-10. A vector $\hat{\mathbf{x}}5 + \hat{\mathbf{y}} + \hat{\mathbf{z}}3$ is located at the point $(x, y, z) = (2, 5, 1)$. Express the vector in (*a*) cylindrical coordinates and (*b*) spherical coordinates.

1-7-11. Vectors $\mathbf{A} = \hat{\mathbf{r}}4 + \hat{\boldsymbol{\phi}}3 + \hat{\mathbf{z}}5$ and $\mathbf{B} = \hat{\mathbf{r}}7 + \hat{\boldsymbol{\phi}}2 + \hat{\mathbf{z}}$ are situated at point $(r, \phi, z) = (6, 45°, 2)$. Find (*a*) $\mathbf{A} + \mathbf{B}$, (*b*) $\mathbf{A} \cdot \mathbf{B}$, and (*c*) $\mathbf{A} \times \mathbf{B}$. @

1-7-12. Vectors $\mathbf{A} = \hat{\mathbf{r}}5 + \hat{\boldsymbol{\theta}}2 + \hat{\boldsymbol{\phi}}$ and $\mathbf{B} = \hat{\mathbf{r}} + \hat{\boldsymbol{\theta}}4 + \hat{\boldsymbol{\phi}}2$ are situated at point $(r, \theta, \phi) = (5, 60°, 90°)$. Find (*a*) $\mathbf{A} + \mathbf{B}$, (*b*) $\mathbf{A} \cdot \mathbf{B}$, and (*c*) $\mathbf{A} \times \mathbf{B}$.

1-7-13. A vector $\mathbf{A} = \hat{\mathbf{x}}4 + \hat{\mathbf{y}}5 + \hat{\mathbf{z}}6$ is at the point $(x, y, z) = (2, 3, 5)$. Find (*a*) cylindrical coordinates of the point, (*b*) the r component of **A** (normal to the cylinder of radius r), (*c*) the ϕ component of **A** (tangent to the cylinder of radius r), and (*d*) the

z component of **A** (tangent to the cylinder of radius r and parallel to the z axis) (Fig. 1-15*b*). @

1-7-14. A vector $\mathbf{A} = \hat{\mathbf{x}}4 + \hat{\mathbf{y}}6 + \hat{\mathbf{z}}9$ is at the point $(x, y, z) = (5, 2, 4)$. Find (*a*) the spherical coordinates of the point, (*b*) the r component of **A** (normal to the sphere of radius r), (*c*) the θ component of **A** (tangent to the sphere of radius r), and (*d*) the ϕ component of **A** (tangent to the cone of angle θ) (Fig. 1-15*c*).

1-7-15. Vector **A** extends from the origin to the point $(x, y, z) = (2, 6, 4)$ and vector **B** from the origin to the point $(r, \phi, z) = (7, 55°, 3)$. Find (*a*) **A** + **B** (in rectangular coordinates), (*b*) **A** • **B**, (*c*) the angle between **A** and **B**, (*d*) **A** × **B**, (*e*) the unit normals to the plane containing **A** and **B** (in rectangular coordinates), and (*f*) the area of the parallelogram of which **A** and **B** are adjacent sides. @

1-7-16. Vector **A** extends from the origin to the point $(x, y, z) = (5, 3, 2)$ and vector **B** from the origin to the point $(r, \theta, \phi) = (7, 30°, 70°)$. Find (*a*) **A** + **B**, (*b*) **A** • **B**, (*c*) the angle between **A** and **B**, (*d*) **A** × **B**, (*e*) unit normals to the plane containing **A** and **B**, and (*f*) the area of the parallelogram of which **A** and **B** are adjacent sides.

1-7-17. Vectors $\mathbf{A} = \hat{\mathbf{r}}2 + \hat{\boldsymbol{\phi}}5 + \hat{\mathbf{z}}3$ and $\mathbf{B} = \hat{\mathbf{r}}5 + \hat{\boldsymbol{\theta}}2 - \hat{\boldsymbol{\phi}}$ are situated at the point $(x, y, z) = (7, 2, 3)$. Find (*a*) **A** + **B**, (*b*) **A** • **B**, (*c*) the angle between **A** and **B**, (*d*) **A** × **B**, (*e*) unit normals to the plane containing **A** and **B**, and (*f*) the area of the parallelogram of which **A** and **B** are adjacent sides. @

1-7-18. A vector $\mathbf{A} = \hat{\boldsymbol{\theta}}2$ is located at the point $r = 5$, $\theta = 30°$, $\phi = 10°$. Express **A** in terms of its rectangular components.

1-7-19. A vector $\mathbf{A} = \hat{\mathbf{r}}4 + \hat{\boldsymbol{\theta}} + \hat{\boldsymbol{\phi}}6$ is located at the point $r = 5$, $\theta = 45°$, $\phi = 45°$. Express **A** in terms of its rectangular components. @

1-7-20. Find the line integral of the cylindrically symmetrical vector function $\mathbf{F} = \hat{\mathbf{r}}r^{-3/2}$ from the point $(r, \phi, z) = (4, 25°, 2)$ to the point $(r, \phi, z) = (1, 10°, 4)$.

1-7-21. Find the line integral of the spherically vector function $\mathbf{F} = \hat{\mathbf{r}}r^{-2}$ from the point $(r, \theta, \phi) = (3, 50°, 20°)$ to the point $(r, \theta, \phi) = (2, 75°, 80°)$. @

1-7-22. Find the flux of the cylindrically symmetrical vector function $\mathbf{F} = \hat{\mathbf{r}}r^{-3}$ through the surface at $r = 5$ between $\phi = 0°$ and $\phi = 45°$ and between $z = 2$ and $z = 6$.

1-7-23. Find the flux of the spherically symmetrical vector function $\mathbf{F} = \hat{\mathbf{r}}r^{-3}$ through the surface at $r = 4$ between $\theta = 45°$ and $\theta = 125°$ and between $\phi = 20°$ and $\phi = 180°$. @

1-7-24. If vector **A** extends from point (5, 25°, 55°) to point (8, 5°, 75°), find (*a*) $|\mathbf{A}|$, (*b*) A_r, A_θ, A_ϕ, and (*c*) the angles between **A** and the x, y, and z axis.

1-7-25. The surfaces of a volume are defined by $r = 1$ and $r = 7$, $\theta = 10°$ and $\theta = 80°$, and $\phi = 20°$ and $\phi = 180°$. Find (*a*) the total surface area and (*b*) the volume enclosed. @

1-7-26. A surface is defined by $r = 7$, $\theta = 15°$ to $80°$, and $\phi = 45°$ to $135°$. Find its area.

1-7-27. A surface for which $r = 8$ has three sides extending between the points (8, 45°, 4), (8, 90°, 3), and (8, 80°, 7). Find the area of the surface. The sides are the shortest distances between the points as measured along the $r = 8$ surface. @

1-7-28. Integrate the function $5x^2y^4z^5$ throughout the plane-sided volume with corners at $(x, y, z) = (5, -3, 2)\ (7, -3, 2)\ (5, 3, 2)\ (7, 3, 2)\ (5, -3, 6)\ (7, -3, 6)\ (5, 3, 6)\ (7, 3, 6)$.

CHAPTER 2

ELECTRIC AND MAGNETIC FIELDS

2-1 INTRODUCTION

This chapter is an introduction to electric and magnetic fields, forming a bridge from circuit theory to the transmission lines, waves, and wireless systems discussed in Chaps. 3, 4 and 5.

Circuit theory treats resistors, capacitors, and inductors as two-terminal devices connected by wires. Field theory deals with the space *both inside and outside* these devices, providing a three-dimensional, real-world understanding of how they work. As pointed out in Sec. 1-1, most of the energy flows *outside* the conductors of a transmission line rather than inside them as usually postulated by circuit theory.

If the spacing of the conductors of a two-wire transmission line is increased sufficiently, most of the energy leaves the conductors and is radiated into space, forming a radiating system or antenna. Going from circuits to fields to transmission lines to antennas, we follow a logical sequence that gets us quickly to topics of great practical importance.

2-2 ELECTRIC FIELDS

Electric Charge Q and Electric Field E

POINT CHARGES. The basic electrical quantity is charge Q. An isolated charge is surrounded by an electric field that exerts a force on all other charges. Thus, a charge Q_2 at a distance r from a charge Q_1, as in Fig. 2-1*a*, experiences a force given by Coulomb's law as

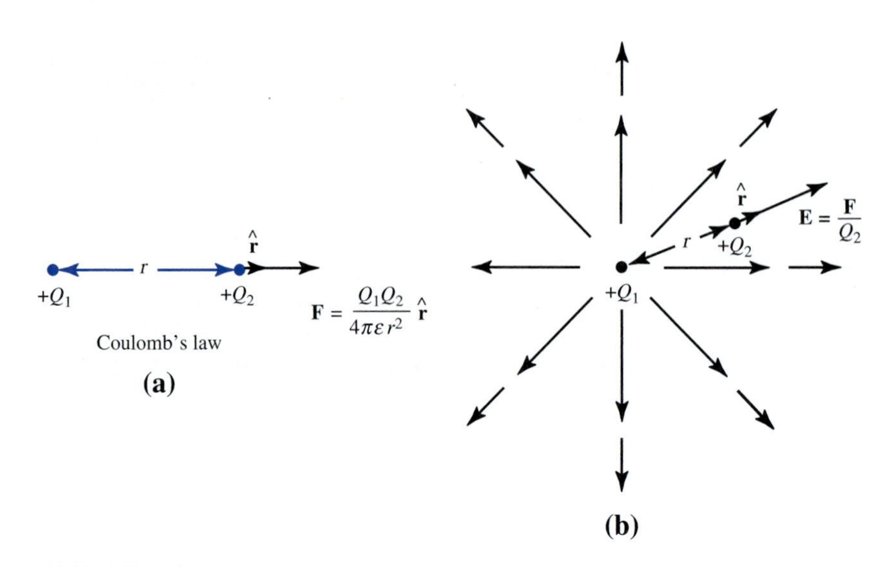

FIGURE 2-1
(*a*) Two positive charges separated by distance r. The force of repulsion **F** is given by Coulomb's law. (*b*) The ***force per charge*** on Q_2 equals the ***electric field*** **E**. Using Q_2 as a probe or test charge around Q_1 shows that **E** is radial and decreases with distance as $1/r^2$.

$$\mathbf{F} = F\hat{\mathbf{r}} = \frac{Q_1 Q_2}{4\pi\varepsilon r^2}\hat{\mathbf{r}} \qquad \text{(N)} \qquad \textit{Coulomb's law} \tag{1}$$

where Q_1 = charge 1, C
Q_2 = charge 2, C
r = radial distance from charge 1 to charge 2, m
$\hat{\mathbf{r}}$ = unit vector in radial direction, dimensionless
ε = permittivity (or dielectric constant), F m^{-1}

The ***permittivity*** ε is conveniently expressed as the product of the permittivity ε_0 of vacuum multiplied by a dimensionless number ε_r called the *relative permittivity*. Thus,

$$\varepsilon = \varepsilon_r \varepsilon_0$$

where ε_r = relative permittivity and ε_0 = permittivity of vacuum = 8.85×10^{-12} farads/meter, F m^{-1}. For a dielectric material $\varepsilon_r > 1$. For polystyrene $\varepsilon_r = 2.7$, so its permittivity is given by

$$\varepsilon = \varepsilon_r \varepsilon_0 = 2.7 \times 8.85 \times 10^{-12}\ \text{F m}^{-1} = 23.9\ \text{pF m}^{-1}$$

For air at atmospheric pressure $\varepsilon_r = 1.0006$. This differs so little from vacuum ($\varepsilon_r = 1$ exactly) that for most practical situations $\varepsilon_0 = 8.85 \times 10^{-12}$ F m^{-1} is taken as the permittivity of both air and vacuum.

Dividing (1) by Q_2 gives a ***force per unit charge*** which is defined as the ***electric field*** **E**. Its units are newtons per coulomb or volts per meter. Thus,

$$\mathbf{E} = \frac{\mathbf{F}}{Q_2} = \frac{Q_1}{4\pi\varepsilon_0 r^2}\hat{\mathbf{r}} \qquad (\mathrm{N\ C^{-1}\ or\ V\ m^{-1}}) \tag{2}$$

By moving Q_2 around Q_1 as a probe or test charge, the electric field **E** around Q_1 may be explored. As shown in Fig. 2-1*b*, the field is radial and decreases with distance as $1/r^2$. This variation is suggested by the length of the arrows.

Example 2-1. Field of point charge. (*a*) If the electric field E (Fig. 2-1) is 100 V m^{-1} at a distance of 2 m from a point charge Q, find Q. $\varepsilon_r = 1$. (*b*) Is **E** radially in or outward?

Solution

(*a*)

$$Q = 4\pi\varepsilon_0 r^2 E = 4\pi 8.85 \times 10^{-12} \times 2^2 \times 100$$
$$= 44.5 \times 10^{-9}\ \mathrm{C} \qquad \text{or } 44.5\ \mathrm{nC} \quad \textit{Ans.} \quad (a)$$

(*b*) Outward if Q is positive, inward if Q is negative. *Ans.* (*b*)

Problem 2-2-1. Force on charge. The electric field $E = 11.2$ V/m at a distance r from a point charge $Q = 125$ fC. $\varepsilon_r = 1$. Find: (*a*) r and (*b*) force on point charge of 50 fC at a distance r. *Ans.* (*a*) 1 cm; (*b*) 560 fN.

Example 2-2. Vector addition of four fields. Four 3-pC charges are at the corners of a 1-m square. The two charges an the left side of the square are positive. The two charges on the right side are negative. Find the field E at the center of the square. $\varepsilon_r = 1$.

Solution. Draw a sketch (Fig. 2-2). The sketch solves the problem geometrically. Thus, $E = 4E_0 \cos\theta$.

The numerical evaluation gives

$$E = 4\frac{Q}{4\pi\varepsilon_0 r^2}\cos\theta = 4\frac{3 \times 10^{-12}}{4\pi 8.85 \times 10^{-12} 0.707^2} 0.707$$
$$= 0.153\ \mathrm{V\ m^{-1}} = 153\ \mathrm{mV\ m^{-1}} \quad \textit{Ans.}$$

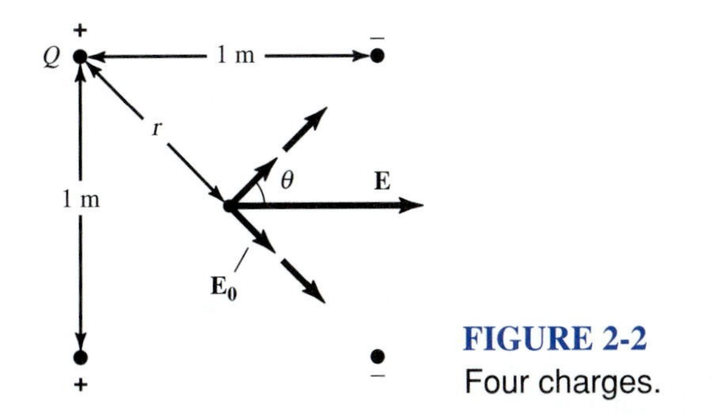

FIGURE 2-2
Four charges.

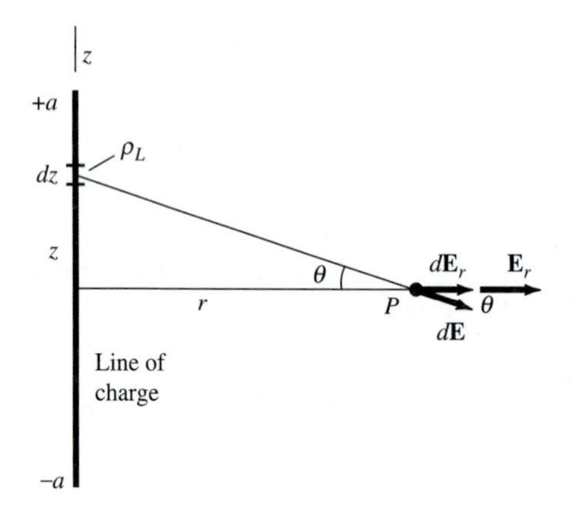

FIGURE 2-3
Line of charge of length $2a$ produces a radial field $\mathbf{E}_r$ at P that is the integral of the components $d\mathbf{E}$ from all line segments dz.

Problem 2-2-2. Field between two charges. A 7-pC positive charge is situated 60 cm from a 3-pC negative charge. Find the field E half-way between the charges. $\varepsilon_r = 1$. *Ans.* 1.0 V/m.

LINE CHARGES. In Fig. 2-3 a line charge of length $2a$ has a linear charge density ρ_L (C m^{-1}). The elemental length dz of the line has a charge $\rho_L\,dz$ producing an electric field dE at the point P given by

$$dE = \frac{\rho_L}{4\pi\varepsilon_0}\frac{dz}{(r^2 + z^2)} \tag{3}$$

where ρ_L = charge per unit length, C m^{-1}. The field component in the direction r (perpendicular to the line) is

$$dE_r = dE\cos\theta = dE\frac{r}{\sqrt{(r^2 + z^2)}} \tag{4}$$

From symmetry the positive and negative z components add to zero and the radial field at P from the line is then the integral of dE_r over the length $2a$ of the line, or

$$E_r = \frac{\rho_L r}{4\pi\varepsilon_0}\int_{-a}^{+a}\frac{dz}{(r^2 + z^2)^{3/2}} = \frac{\rho_L}{2\pi\varepsilon_0 r\sqrt{\left(\frac{r}{a}\right)^2 + 1}} \tag{5}$$

For a long line ($a \gg r$), (5) reduces to

$$E_r = \frac{\rho_L}{2\pi\varepsilon_0 r} \tag{6}$$

Example 2-3. Field under dc transmission line. Two long parallel conductors of a dc transmission line separated by 2 m have charges of $\rho_L = 5\ \mu$C m^{-1} of opposite sign. Both lines are 8 m above ground. What is the magnitude of the electric field 4 m directly below one of the wires? $\varepsilon_r = 1$.

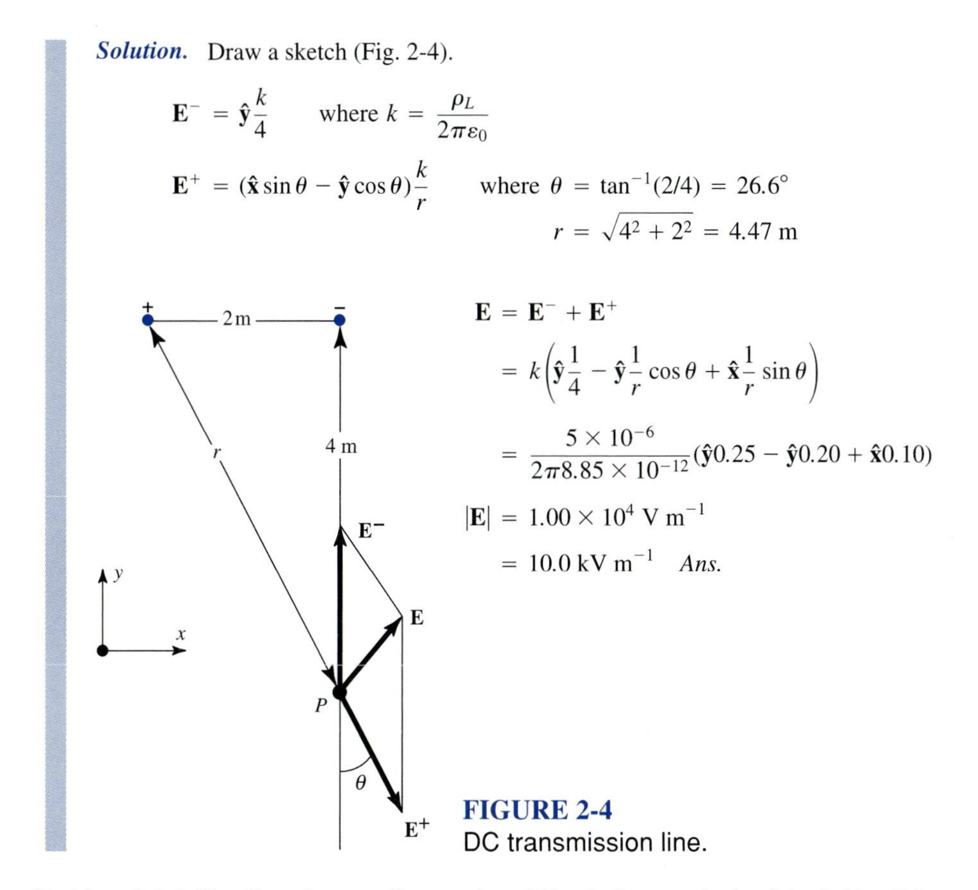

Solution. Draw a sketch (Fig. 2-4).

$$\mathbf{E}^- = \hat{\mathbf{y}}\frac{k}{4} \qquad \text{where } k = \frac{\rho_L}{2\pi\varepsilon_0}$$

$$\mathbf{E}^+ = (\hat{\mathbf{x}}\sin\theta - \hat{\mathbf{y}}\cos\theta)\frac{k}{r} \qquad \text{where } \theta = \tan^{-1}(2/4) = 26.6^\circ$$

$$r = \sqrt{4^2 + 2^2} = 4.47 \text{ m}$$

$$\mathbf{E} = \mathbf{E}^- + \mathbf{E}^+$$

$$= k\left(\hat{\mathbf{y}}\frac{1}{4} - \hat{\mathbf{y}}\frac{1}{r}\cos\theta + \hat{\mathbf{x}}\frac{1}{r}\sin\theta\right)$$

$$= \frac{5 \times 10^{-6}}{2\pi 8.85 \times 10^{-12}}(\hat{\mathbf{y}}0.25 - \hat{\mathbf{y}}0.20 + \hat{\mathbf{x}}0.10)$$

$$|\mathbf{E}| = 1.00 \times 10^4 \text{ V m}^{-1}$$

$$= 10.0 \text{ kV m}^{-1} \quad \textit{Ans.}$$

FIGURE 2-4
DC transmission line.

Problem 2-2-3. Two line charges of same sign. What is the magnitude of the field at P in Example 2-3 (Fig. 2-4) if both lines are of the same sign? *Ans.* 41.4 kV/m.

2-3 ELECTRIC POTENTIAL V AND ITS GRADIENT E

In (2-2-2), the electric field E at a point is defined as the force F on a unit charge $+Q$ at the point. Thus,

$$E = \frac{F}{Q} = \frac{\text{force}}{\text{charge}} = \text{electric field} \qquad \text{N C}^{-1} \tag{1}$$

By moving the charge $+Q$ against a field E between the two points a and b in Fig. 2-5a, work is done. Thus,

$$E\Delta x = \frac{\text{force}}{\text{charge}} \times \text{distance} = \frac{\text{work}}{\text{charge}} \qquad \left(\frac{\text{N}}{\text{C}}\text{ m} = \frac{\text{J}}{\text{C}}\right) \tag{2}$$

This work per charge is the *electric potential difference* ΔV between the points a and b. The units are joules per coulomb or volts. Thus,

$$\Delta V = E\,\Delta x \qquad \left(\frac{\text{J}}{\text{C}} = \text{V}\right) \tag{3}$$

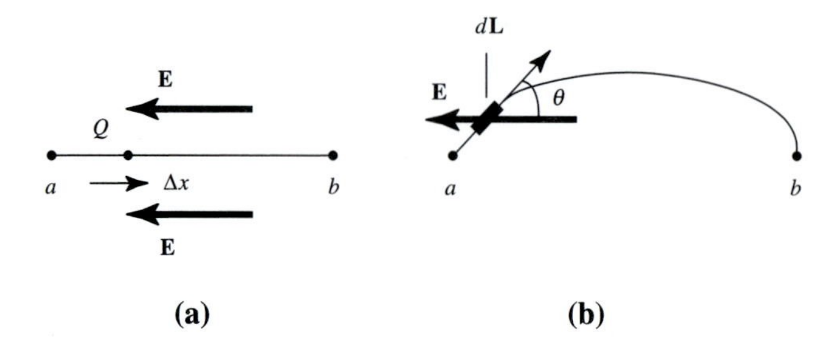

FIGURE 2-5
Work in moving a charge $+Q$ against a field.

or

$$\text{Electric field } E = \frac{\Delta V}{\Delta x} \qquad (\text{V m}^{-1}) \tag{4}$$

In vector notation

$$\mathbf{E} = \hat{\mathbf{x}}E = -\hat{\mathbf{x}}\frac{\Delta V}{\Delta x} \qquad (\text{V m}^{-1}) \tag{5}$$

The negative sign in (5) reminds us that moving against the electric field results in positive work. Integrating **E** between a and b by some other path, as in Fig. 2-5b, the voltage difference V is given by the line integral of **E** between a and b.

$$V = -\int_a^b \mathbf{E} \cdot d\mathbf{L} = -\int_a^b E\cos\theta \, dL \qquad (\text{V}) \tag{6}$$

Expanding (5) to the general case where **E** also has y and z components gives

$$\mathbf{E} = -\left(\hat{\mathbf{x}}\frac{\partial}{\partial x} + \hat{\mathbf{y}}\frac{\partial}{\partial y} + \hat{\mathbf{z}}\frac{\partial}{\partial z}\right)V \qquad (\text{V m}^{-1}) \tag{7}$$

E in (7) is called the ***gradient of V.*** Abbreviating *gradient* to *grad* or the operator *del* (∇), (7) may be written as

$$\mathbf{E} = -\text{grad } V = -\nabla V \qquad (\text{V m}^{-1}) \tag{8}$$

where ∇ = vector operator called *del*

$$\nabla = \hat{\mathbf{x}}\frac{\partial}{\partial x} + \hat{\mathbf{y}}\frac{\partial}{\partial y} + \hat{\mathbf{z}}\frac{\partial}{\partial z} \text{ (in rectangular coordinates)}$$

Expressions for gradient in other coordinate systems are given in Appendix A, Table A-2.

The gradient is in the direction of the *maximum rate of change* of the potential V and in a direction *opposite* to that of the field.

From (2-2-2), the field of a point charge Q is given by

$$\mathbf{E} = \frac{Q}{4\pi\varepsilon_0 r^2}\hat{\mathbf{r}} \qquad (\text{V m}^{-1}) \tag{9}$$

Example 2-4. Gradient of V in uniform field. In Fig. 2-6, the equipotential lines change by 2 V m^{-1} in the x direction and by 1 V m^{-1} in the y direction. The potential does not change in the z direction. Find the electric field **E**.

Solution. **E** is equal to the gradient of V as given by (7) or (8). Thus,

$$\mathbf{E} = -\nabla V = -\left(\hat{\mathbf{x}}\frac{\partial V}{\partial x} + \hat{\mathbf{y}}\frac{\partial V}{\partial y}\right) = (\hat{\mathbf{x}}2 + \hat{\mathbf{y}}1) = 2.24\angle 26.6^\circ \quad \text{V m}^{-1} \quad \textit{Ans.}$$

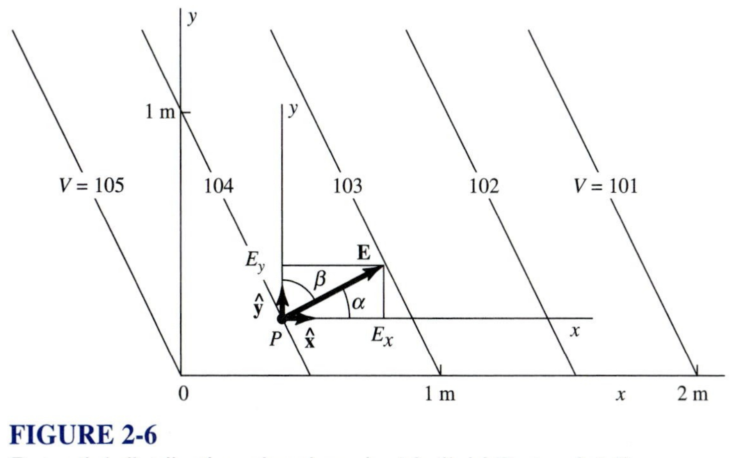

FIGURE 2-6
Potential distribution showing electric field **E** at point P.

Problem 2-3-1. Gradient in nonuniform field. If the potential as a function of position is given by $V = 2x^2 + 3y + 4z^{1/2}$ V, find the magnitude of the electric field at the point $x = y = z = 0.5$ m. *Ans.* 4.58 V/m.

The field is radial and extends to infinity and $\varepsilon = \varepsilon_0$. The work to move a unit charge from infinity to a radial distance r_1 is then

$$V = -\int_\infty^{r_1} \mathbf{E} \cdot d\mathbf{r} = -\frac{Q}{4\pi\varepsilon_0}\int_\infty^{r_1} \frac{dr}{r^2} = \frac{Q}{4\pi\varepsilon_0}\left(\frac{1}{r_1} - \frac{1}{\infty}\right) = \frac{Q}{4\pi\varepsilon_0 r_1} \tag{10}$$

The potential in (10) is called the ***absolute potential,*** that is, the potential at r_1 referred to zero potential at infinity. Potential is always a difference in voltage between two points. If one point is at infinity, the potential is then an absolute potential.

Example 2-5. Work and potential for a point charge. (*a*) What is the absolute potential at a point P which is 2 m from a point charge $Q = +5$ μC? (*b*) What is the work required to move a +8 nC charge from infinity to P? $\varepsilon_r = 1$.

Solution. From (10)

$$V = \frac{5 \times 10^{-6}}{4\pi 8.85 \times 10^{-12} \times 2} = 22.5 \text{ kV} \quad \textit{Ans.} \quad (a)$$

Since 1 kV = 1 kJ C^{-1},

$$\text{Work} = 22.5\ (\text{kJ C}^{-1}) \times 8\ (\text{nC}) = 22.5 \times 10^3 \times 8 \times 10^{-9} = 180 \times 10^{-6}\ \text{J}$$
$$= 180\ \mu\text{J} \quad \textit{Ans.} \quad (b)$$

Problem 2-3-2. Absolute potential from sphere. A conducting sphere of 5-cm radius has a surface charge density of 1 pC/m^2. If the medium $\varepsilon_r = 2$, find the absolute potential at a distance of 3 m from the center of the sphere. *Ans.* 47.1 μV.

Figure 2-7 shows the radial field lines and the circular equipotentials around a point charge. In three dimensions the equipotentials are spherical surfaces with Q at the center. An *equipotential surface or line* is a contour along which a charge moves with *zero work.* The *maximum amount of work* per unit distance is performed moving normal or perpendicular to an equipotential surface in the *direction of the electric field,* as shown in Fig. 2-7*a*.

The work to transport a charge around any *closed path* in a static field is zero since the path starts and ends at the same point. Thus,

$$\oint \mathbf{E} \cdot d\mathbf{L} = 0 \tag{11}$$

To summarize:

1. *The line integral of the field* **E** *gives the potential V between two points.*
2. *The line integral of* **E** *from infinity to a point gives the absolute potential of the point.*
3. *The gradient of V gives the field* **E** *at the point.*

Example 2-6. Potentials for a point charge. Referring to Fig. 2-7, find the potential difference between point a at radius $r_1 = 400$ mm and point b at radius $r_2 = 100$ mm from a point charge $Q = 222$ pC. $\varepsilon_r = 1$.

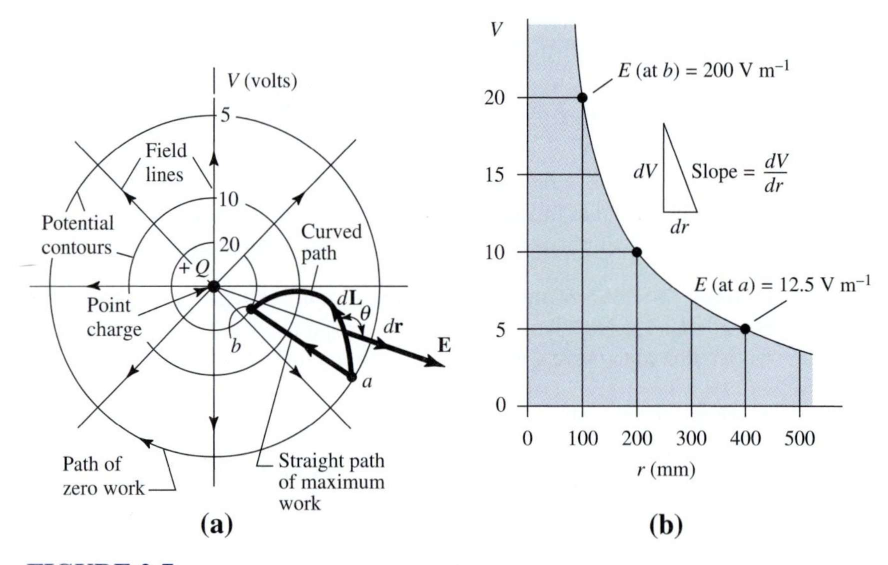

FIGURE 2-7
(*a*) Field lines and equipotentials around a point charge. (*b*) Absolute potential versus radial distance from the charge.

Solution. From (2-2-2) for E and putting $dr = \cos\theta\, dL$ in (6) where dr is an infinitesimal element of radial distance, we have

$$\Delta V = \frac{-Q}{4\pi\varepsilon_0}\int_a^b \frac{dr}{r^2} = \frac{Q}{4\pi\varepsilon_0}\left(\frac{1}{b} - \frac{1}{a}\right)$$

$$= 9\times 10^9 \times 222 \times 10^{-12}\left(\frac{1}{0.1} - \frac{1}{0.4}\right) = 2\left(\frac{1}{0.1} - \frac{1}{0.4}\right)$$

$$= 20 - 5 = 15 \text{ V} \quad \textit{Ans.}$$

The result is identical by curved or straight path.

Example 2-7. Absolute potential and gradient for a point charge. (*a*) Find the absolute potential at a and b in Fig. 2-7. (*b*) Find **E** as the gradient of V at a and b.

Solution. The absolute potential of the point b at $r = 100$ mm is obtained by removing the point a to infinity so that

$$V\,(\text{at } b) = \frac{Q}{4\pi\varepsilon_0}\left(\frac{1}{b} - \frac{1}{a}\right) = 2\left(\frac{1}{0.1} - \frac{1}{\infty}\right) = 20 \text{ V} \quad \textit{Ans.} \quad (a)$$

For the absolute potential of the point a at $r = 400$ mm, we have

$$V\,(\text{at } a) = 2\left(\frac{1}{0.4} - \frac{1}{\infty}\right) = 5 \text{ V} \quad \textit{Ans.} \quad (a)$$

E is the gradient (or slope) of V as given by

$$|\mathbf{E}| = -\frac{dV}{dr} = -\frac{Q}{4\pi\varepsilon_0}\frac{\partial}{\partial r}r^{-1} = \frac{Q}{4\pi\varepsilon_0 r^2}$$

Compare with (2-2-2).

At (*b*)

$$E = \frac{Q}{4\pi\varepsilon_0 0.1^2} = 2/0.1^2 = 200 \text{ V m}^{-1} \quad \textit{Ans.} \quad (b)$$

At (*a*)

$$E = 2/0.4^2 = 12.5 \text{ V m}^{-1} \quad \textit{Ans.} \quad (b)$$

Problem 2-3-3. Charged ring. A circular ring 2.4 m in diameter has a charge density of 8 μC/m. Find V and **E** at the center of the ring. *Ans.* $V = 452$ kV, $\mathbf{E} = 0$.

Problem 2-3-4. Line charge. Find the electric field of a long line charge at a radial distance where the potential is 24 V higher than at a radial distance $r_1 = 3$ m where $E = 4$ V/m. *Ans.* 29.5 V/m.

Variation of E and *V* for point and line charges

	E	***V***
Point charge	$1/r^2$	$1/r$
Infinite line charge	$1/r$	$\ln \frac{r_2}{r_1}$

Superposition of Potential

Since the electric scalar potential due to a single point charge is a linear function of the value of its charge, it follows that the potentials of more than one point charge are linearly superposable by scalar (algebraic) addition. As a generalization, this may be stated as the *principle of superposition* applied to electric potential† as follows: ***The total electric potential at a point is the algebraic sum of the individual potentials at the point.*** Thus, for the three point charges Q_1, Q_2, and Q_3 in Fig. 2-7.1, the total electric potential (work per unit charge) at the point P is given by

$$V_P = \frac{1}{4\pi\varepsilon_0}\left(\frac{Q_1}{r_1} + \frac{Q_2}{r_2} + \frac{Q_3}{r_3}\right) \quad \text{(V)} \qquad \textbf{\textit{Point charges}} \tag{12}$$

where r_1 = distance from Q_1 to P
r_2 = distance from Q_2 to P
r_3 = distance from Q_3 to P

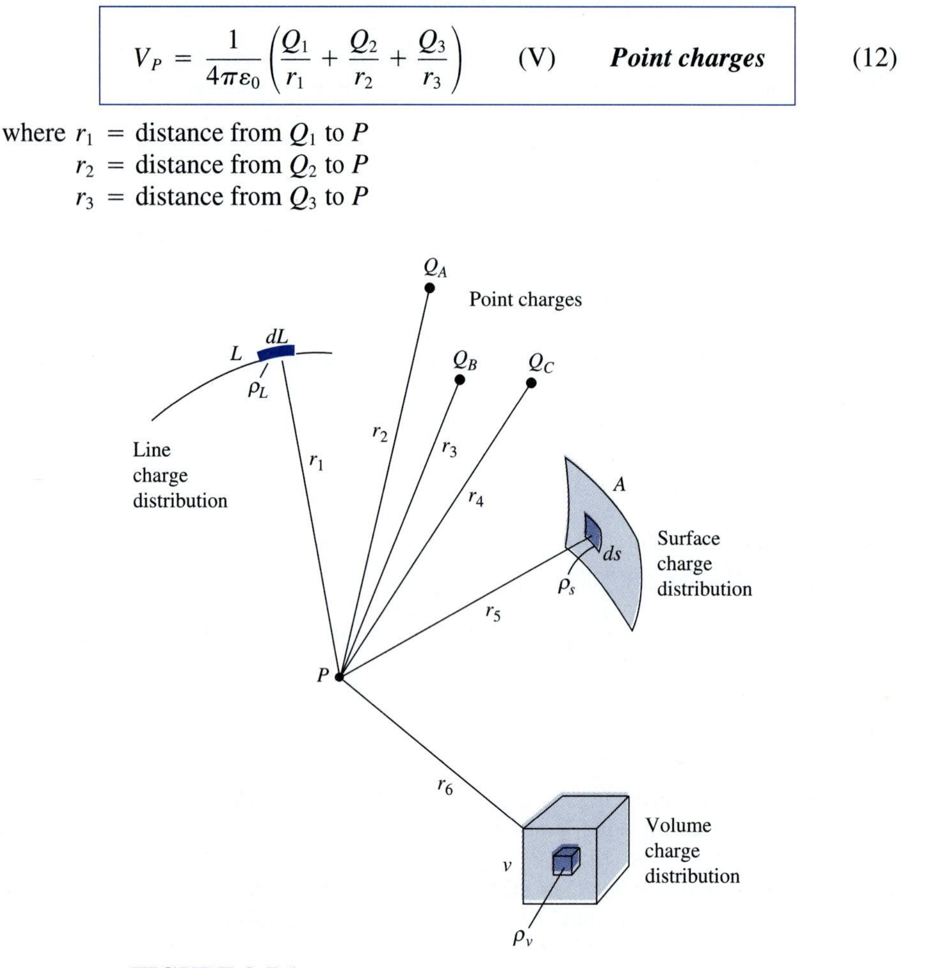

FIGURE 2-7.1
Potential V at point P is the sum of the potentials of the point charges, line charge, surface charge, and volume charge.

† Although *electric scalar potential* is implied, the word *scalar* is usually omitted for brevity.

This can also be expressed with a summation sign. Thus,

$$V_P = \frac{1}{4\pi\varepsilon_0}\sum_{n=1}^{3}\frac{Q_n}{r_n} \tag{13}$$

For the linear charge distribution in Fig. 2-7.1, the potential at P is given by

$$V_L = \frac{1}{4\pi\varepsilon_0}\int \frac{\rho_L}{r}\,dL \qquad \text{(V)} \qquad \textbf{\textit{Line charge}} \tag{14}$$

where ρ_L = linear charge density, C m^{-1}, and dL = element of length of line, m. The integration is carried out over the entire line of charge.

For the surface charge distribution in Fig. 2-7.1, the potential at P is given by

$$V_s = \frac{1}{4\pi\varepsilon_0}\iint \frac{\rho_s}{r}\,ds \qquad \text{(V)} \qquad \textbf{\textit{Surface charge}} \tag{15}$$

where ρ_s = surface charge density, C m^{-2}
ds = element of surface, m^2

The integration is carried out over the entire surface of charge.

For the volume charge distribution in Fig. 2-7.1, the potential at P is

$$V_v = \frac{1}{4\pi\varepsilon_0}\iiint \frac{\rho_v}{r}\,dv \qquad \text{(V)} \qquad \textbf{\textit{Volume charge}} \tag{16}$$

where ρ_v = (volume) charge density, C m^{-3}
dv = element of volume, m^3

The integration is carried out over the volume containing charge.

If the point charges, the line charge distribution, the surface charge distribution, and the volume charge distribution of Fig. 2-7.1 are all present simultaneously, the total electric potential at the point P due to all these distributions is, by the superposition principle, the algebraic sum of the individual component potentials. Thus, in general,

$$V = V_P + V_L + V_s + V_v \qquad \text{or} \tag{17}$$

$$V = \frac{1}{4\pi\varepsilon_0}\left(\sum_{1}^{N}\frac{Q_n}{r_n} + \int\frac{\rho_L}{r}\,dL + \iint\frac{\rho_s}{r}\,ds + \iiint\frac{\rho_v}{r}\,dv\right) \qquad \text{(V)} \qquad \textbf{\textit{Superposition of all potentials}} \tag{18}$$

Example 2-7.1. Superposition of potential from line charge. (a) Find the potential V at P, 1 m from a 2-m straight-line charge shown solid in Fig. 2-7.2. $\rho_L = 10^{-10}$ C m^{-1}. Medium is air.

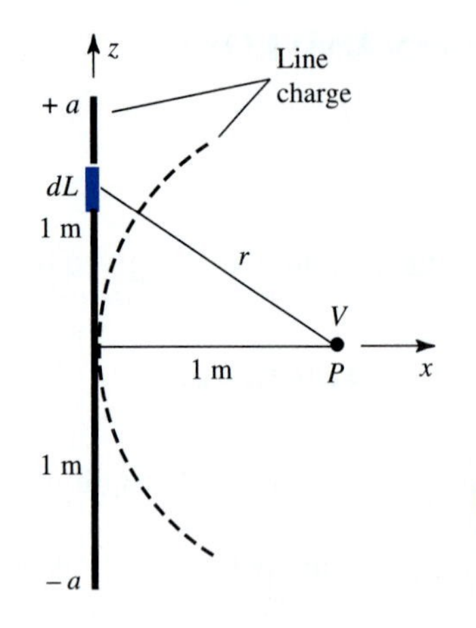

FIGURE 2-7.2
Potential at V from straight 2-m line charge (solid) and circular 2-m line charge (dashed).

Solution. (*a*) The geometry is identical to that in Fig. 2-3 for calculating the field E at a point. Finding the potential is simpler, involving scalar instead of vector quantities. Thus, from (14)

$$V = \frac{\rho_L}{4\pi\varepsilon_0}\int_{-a}^{+a}\frac{1}{\sqrt{z^2+x^2}}\,dz = \frac{\rho_L}{4\pi\varepsilon_0}\ln\left(z+\sqrt{z^2+x^2}\right)\Big|_{-a}^{+a}$$

$$= \frac{\rho_L}{4\pi\varepsilon_0}2\ln\left(1+\sqrt{1+1}\right) = \frac{10^{-10}\times 1.76}{4\pi\times 8.85\times 10^{-12}}$$

$$= 1.59\ \text{V} \quad \textit{Ans.} \quad (a)$$

(*b*) Find the potential V at P from the 2-m curved-line charge shown dashed in Fig. 2-7.2 at a constant radius of 1 m from point P.

Solution. (*b*)

$$V = \frac{10^{-10}\times 2}{4\pi\times 8.85\times 10^{-12}} = 1.80\ \text{V} \quad \textit{Ans.} \quad (b)$$

This potential is the same as for a point charge $= 2\times 10^{-10}$ C at a distance of 1 m. The potential at P due to the straight line charge is $1.59/1.80 = 0.88$ as much or 12% less because of the greater distance to the ends of the straight line charge as compared to the circular line which is at a constant distance.

Problem 2-3-5. Superposition of potential. Referring to Fig. 2-7.1, $r_1 = 0.55$ m, $r_2 = 0.65$ m, $r_3 = 0.55$ m, $r_4 = 0.60$ m, $r_5 = 0.50$ m, $r_6 = 0.50$ m, $L = 0.35$ m, $A = 0.03\ \text{m}^2$, $v = 0.001\ \text{m}^3$, $\rho_L = 10^{-10}\ \text{C m}^{-1}$, $\rho_s = 10^{-9}\ \text{C m}^{-2}$, $\rho_v = 10^{-8}\ \text{C m}^{-3}$, $Q_A = 2\times 10^{-11}$ C, $Q_B = 8\times 10^{-11}$ C, $Q_C = 3\times 10^{-11}$ C. Medium is air. Using (18), find V at P. The line charge is at a constant distance r_1 and the surface charge at a constant distance r_5. The radius r_6 assumes all of the volume charge at one point as would be obtained by a triple or volume integral. *Ans.* 3.34 V.

2-4 ELECTRIC FIELD STREAMLINES AND EQUIPOTENTIAL CONTOURS; ORTHOGONALITY

A field line indicates the direction of the force on a positive test charge introduced into the field. If the test charge is released, it accelerates in the direction of a field line.

In a uniform field the **E** lines are parallel and the equipotential lines form a parallel orthogonal set of lines as in Fig. 2-8*a*. Actually the equipotentials are plane surfaces perpendicular to **E** and for a fixed voltage increment ΔV (= 10 V in the figure) are spaced uniformly.

In a nonuniform field the **E** lines diverge in going from a stronger to a weaker field region, as in Fig. 2-8*b*. Furthermore, for a fixed voltage increment, such as 10 V, the equipotential surfaces become more widely spaced in the weaker field region. The uniform and nonuniform fields are shown in three dimensions in Figs. 2-8*c* and *d*.

Electric field lines and equipotentials are everywhere orthogonal (at right angles).

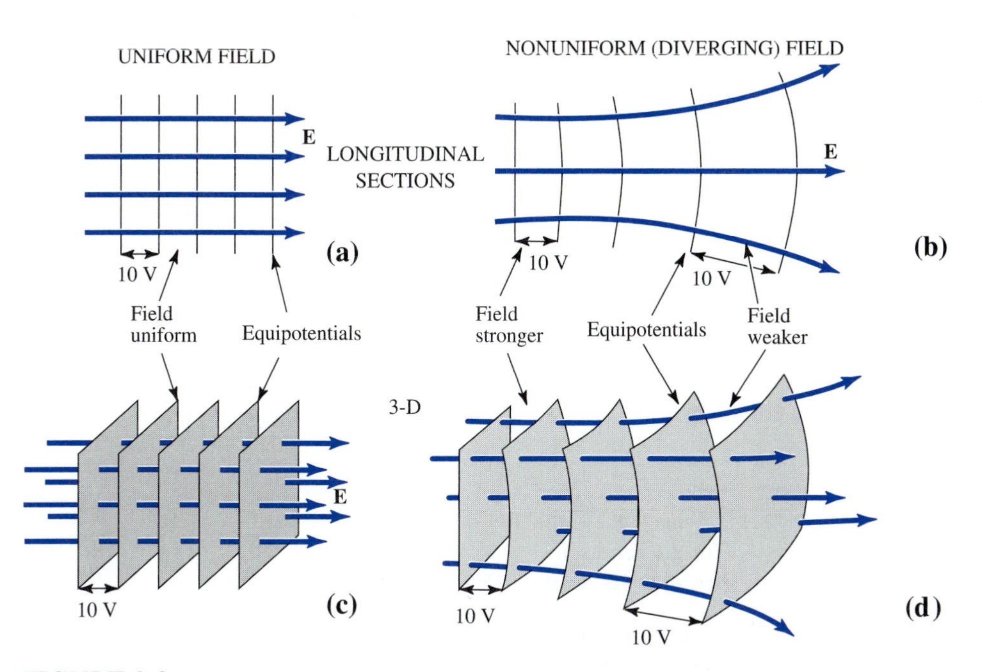

FIGURE 2-8
(*a*) Uniform electric field. (*b*) Nonuniform electric field. Equipotential surfaces are planar in (*a*) and curved in (*b*) as seen in three dimensions in (*c*) and (*d*). For a given potential difference or increment (10 V), the equipotential surfaces are equally spaced in a uniform field (*c*), but in the nonuniform field (*d*) the spacing increases as the field becomes weaker.

2-5 MULTICONDUCTOR TRANSMISSION LINES

Line Charges

Transmission lines are used to transmit power, radio, television, and data. They are a vital part of our technological society. As an example, two long parallel wires spaced $2s$ apart as in Fig. 2-9 form a two-wire transmission line. The fields derived here for the static electric case, as with a line operating at dc, are also the same for the traveling waves of high frequencies (line assumed uniform, lossless, and with spacing a small fraction of a wavelength).

The potential at P at any point (x, y) is the *scalar sum* of the potentials due to the two lines. Taking the origin of the coordinates in Fig. 2-9 as the reference for potential, the potential difference between P and the origin for the positive line is, by integration of (2-2-6),

$$V^+ = \frac{\rho_L}{2\pi\varepsilon}\int_{r_2}^{s}\frac{dr}{r} = \frac{\rho_L}{2\pi\varepsilon}\ln\frac{s}{r_2} \qquad \text{(V)} \tag{1}$$

Similarly for the negatively charged line,

$$V^- = -\frac{\rho_L}{2\pi\varepsilon}\ln\frac{s}{r_1} \qquad \text{(V)} \tag{2}$$

The total potential difference V between P and the origin is the algebraic sum of the individual potentials of (1) and (2), or

$$V = V^+ + V^- = \frac{\rho_L}{2\pi\varepsilon}\ln\frac{r_1}{r_2} \tag{3}$$

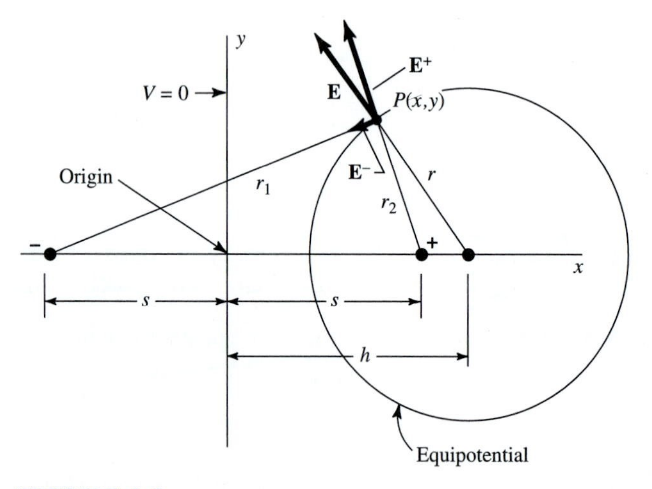

FIGURE 2-9

Cross section of two-wire transmission line. The total potential V at P with respect to the origin is the algebraic sum of the individual potentials. The total field at the point P is the vector sum of the fields of the two wires.

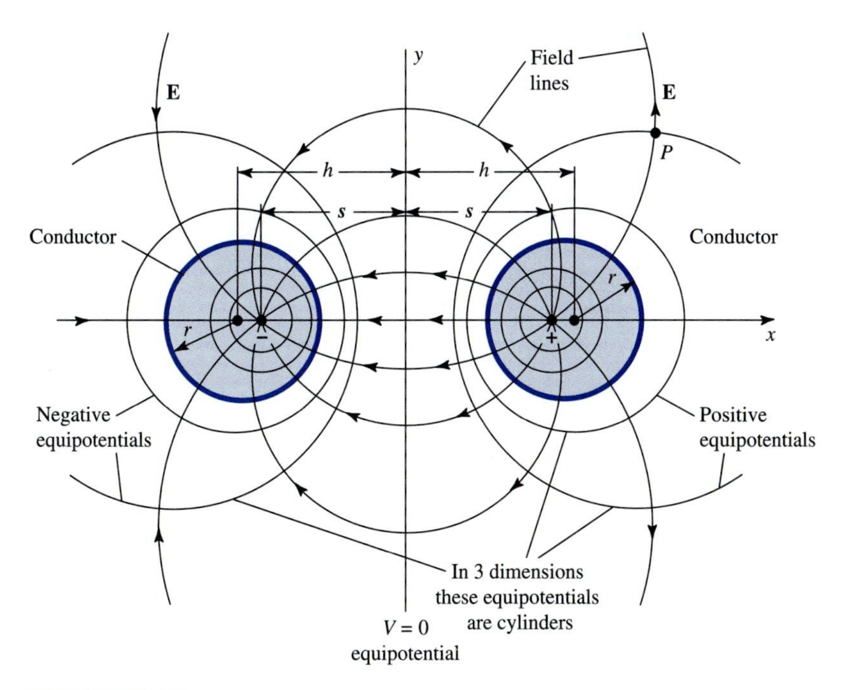

FIGURE 2-10
Field lines and equipotentials around a two-conductor transmission line. The conductors have a radius r with centers spaced $2h$. The conductors of radius r coincide with equipotentials of the two-wire line of spacing $2s$.

The total field at $P(x, y)$ is the vector sum of the fields of the two wires, or

$$\mathbf{E} = \mathbf{E}^{+} + \mathbf{E}^{-} \tag{4}$$

where $E^{+} = \dfrac{\rho_L}{2\pi\varepsilon r_2}$ and $E^{-} = \dfrac{\rho_L}{2\pi\varepsilon r_1}$.

For V equal to a constant, (3) is the equation of an equipotential line. Equipotentials and field lines for the two-wire transmission line are shown in Fig. 2-10.

Example 2-8. Electric field under high-voltage power line. A 765-kV rms, 3-phase, 60-Hz transmission line has conductors spaced 16 m. Their height is 12 m above ground as in Fig. 2-11. Each conductor is a bundle of smaller conductors. It has an effective diameter of 0.6 m. A fluorescent lamp bulb held 2 m above ground at point P under an outside conductor, as in the figure, lights to full brilliance. No wires are connected to the bulb. What is the magnitude of the rms electric field at P?

Solution. From (2-3-6) and (2-2-6) the potential difference between conductors 1 and 2 is

$$V = 765\text{ kV} = -\int \mathbf{E}\cdot d\mathbf{r} = -\frac{\rho_L}{2\pi\varepsilon_0}\int_{0.3\text{ m}}^{15.7\text{ m}}\frac{dr}{r} = \frac{\rho_L}{2\pi\varepsilon_0}\ln\Big|_{0.3}^{15.7} = \frac{\rho_L}{2\pi\varepsilon_0}3.95$$

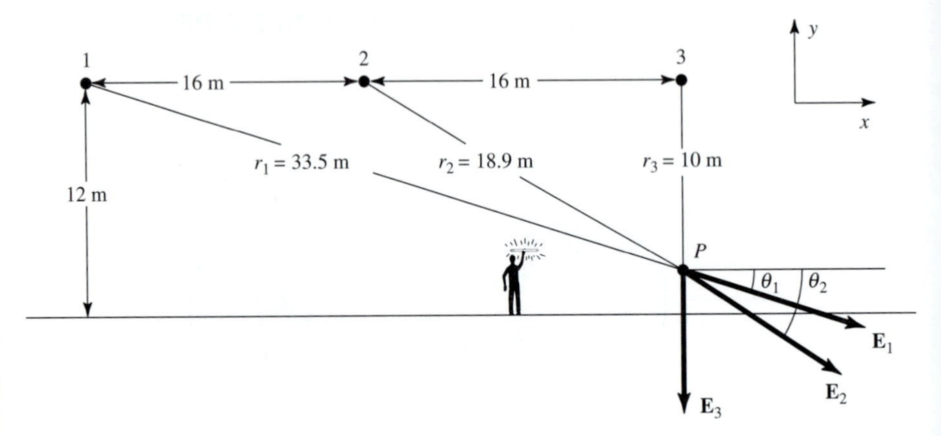

FIGURE 2-11
Man with fluorescent bulb under power line.

and

$$\rho_L = \frac{2\pi\varepsilon_0 V}{3.95} = 10.76\ \mu\text{C m}^{-1} = \text{charge on conductor}$$

Therefore, the field at P from conductor 3 is

$$\begin{aligned}
\mathbf{E}_3 &= -\hat{\mathbf{y}}\frac{\rho_L}{2\pi\varepsilon_0 r_3} = -\hat{\mathbf{y}}\frac{10.76\times10^{-6}}{2\pi 8.85\times10^{-12}\times 10} \\
&= -\hat{\mathbf{y}}19.3\times10^3\ \text{V m}^{-1} \\
\mathbf{E}_2 &= 10.2(-\hat{\mathbf{y}}\sin 32^\circ + \hat{\mathbf{x}}\cos 32^\circ)\times10^3 \\
&= (-\hat{\mathbf{y}}5.40 + \hat{\mathbf{x}}8.66)\times10^3\ \text{V m}^{-1} \\
\mathbf{E}_1 &= 5.76(-\hat{\mathbf{y}}\sin 17.4^\circ + \hat{\mathbf{x}}\cos 17.4^\circ)\times10^3 \\
&= (-\hat{\mathbf{y}}172 + \hat{\mathbf{x}}5.50)\times10^3\ \text{V m}^{-1} \\
E\ (\text{total max.}) &= [-\hat{\mathbf{y}}(19.3\angle 0^\circ + 5.40\angle 120^\circ + 1.72\angle -120^\circ) \\
&\quad + \hat{\mathbf{x}}(8.66\angle 120^\circ + 5.50\angle -120^\circ)]\times10^3\ \text{V m}^{-1} \\
&= \sqrt{16.0^2 + 7.62^2}\times10^3 = 23.6\times10^3\ \text{V/m} \\
&= 23.6\ \text{kV m}^{-1} \quad \textit{Ans.}
\end{aligned}$$

In standard operation, voltage is applied between the ends of a fluorescent bulb. The field under the power line excites the bulb to brilliance without any wires. And the bulb glows equally well whether held horizontally or vertically.

Problem 2-5-1. Power line field. Find the field at a point 8 m below the center wire of Example 2-8. *Ans.* 15 kV/m max. The x and y field components are nearly in phase in Example 2-8 but not in this problem.

Problem 2-5-2. Two-conductor transmission line. At point P (x = 43 cm, y = 32 cm) in Fig. 2-10, $|\mathbf{E}|$ = 450 V/m. If s = 25 cm, h = 29 cm, and r = 12.5 cm, find the potential difference ΔV of the two conductors. *Ans.* ΔV = 1.13 kV independent of permittivity of medium.

2-6 ELECTRIC FLUX AND ELECTRIC FLUX DENSITY (OR DISPLACEMENT): GAUSS'S LAW

Surface Charge, Uniform Case

The *electric flux* ψ through a surface area is the integral of the normal component of the electric field (times ε) over the area. Thus

$$\psi = \iint \varepsilon \mathbf{E} \cdot d\mathbf{s} \tag{1}$$

In the (assumed) uniform field **E** between the two plates in Fig. 2-12 the integral reduces to a simple scalar product. Thus, the flux over the small area a is

$$\psi = \varepsilon E a \qquad \text{(C)} \tag{2}$$

and the total flux between the plates is

$$\psi = \varepsilon E A \qquad \text{(C)} \tag{2.1}$$

where ε = permittivity of medium, F m^{-1}. Dividing the electric flux of (2.1) by the area A gives the *flux density* D or charge per unit area. Thus,

$$D = \frac{\psi}{A} = \varepsilon E \qquad (\text{C m}^{-2}) \tag{2.2}$$

$$\mathbf{D} = \varepsilon \mathbf{E} = \varepsilon_r \varepsilon_0 \mathbf{E} \qquad (\text{C m}^{-2}) \tag{2.3}$$

where $\mathbf{D}$ = flux density, C m^{-2}
$\mathbf{E}$ = electric field, V m^{-1}
ε = permittivity, F m^{-1}
ε_0 = permittivity of air or vacuum
= 8.85×10^{-12}, F m^{-1}
$\varepsilon_r = \varepsilon/\varepsilon_0$ = relative permittivity, dimensionless

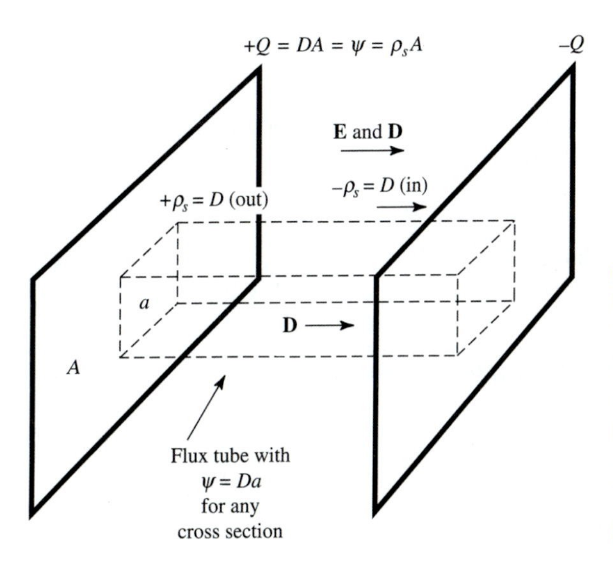

FIGURE 2-12
Uniform flux density case. Two parallel conducting plates with charges $+Q$ and $-Q$, respectively. Between the plates there is a uniform flux density **D**. The total flux between the plates is $DA = \rho_s A = Q$.

The greater the permittivity, the greater the flux density **D** for a given electric field **E**. The significance of the name becomes obvious if we say that a larger permittivity "permits" a greater flux density.

The flux density **D** is also called the *electric displacement*, hence, the symbol **D**. In an *isotropic* medium (properties independent of direction) **D** and **E** are in the same direction and ε is a scalar quantity. In an *anisotropic* medium **E** and **D** may be in different directions and ε is a tensor. Only isotropic media are considered in this book.

Assuming a uniform field between the plates (fringing effects neglected), the total flux between the plates of Fig. 2-12 is

$$\psi = DA \qquad \text{(C)} \tag{3}$$

where A = total plate area, m^2, and the flux through any cross-section of the flux tube is

$$\psi = Da \qquad \text{(C)} \tag{4}$$

where a = area of flux tube, m^2.

For this uniform case the charge is distributed uniformly over the plates so that the total charge Q is equal to the *charge density* ρ_s on the plates times the area A, or

$$Q = \rho_s A \qquad \text{(C)} \tag{5}$$

The flux ψ in (3) is equal to the charge Q. Thus, comparing (3) and (5), we note that the charge density ρ_s is equal to the magnitude of **D**, or

$$\rho_s = D \qquad (\text{C m}^{-2}) \tag{5.1}$$

and the total flux between the plates is

$$\psi = DA = \rho_s A = Q \qquad \text{(C)} \tag{5.2}$$

Surface Charge, Nonuniform Case

Figure 2-13 shows two point charges $+Q$ and $-Q$ connected by *flux tubes* along lines of electric field. Each tube contains a constant amount of *flux*

$$\psi = DA \qquad \text{(C)} \tag{6}$$

Near each charge, where D is large, A is small, and halfway between, where D is weaker, A is larger so that DA = constant along each tube. Figure 2-13 shows only three tubes of the many that fill all the space between the charges.

Adding the flux through N tubes filling all space, we may write

$$\sum_{n=1}^{n=N} DA = Q \qquad \text{(C)} \tag{7}$$

where N = total number of tubes of flux.

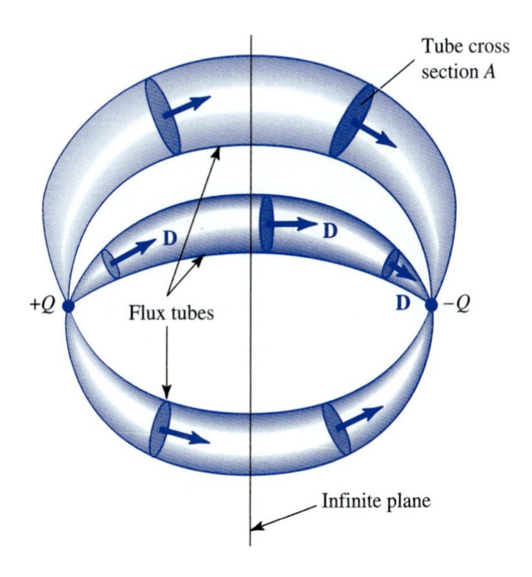

FIGURE 2-13
Nonuniform flux density case of two point charges. Flux tubes, each with a constant amount of flux, $\psi = DA$, join two equal point charges of opposite sign. The tubes follow the streamlines of the electric field **E**. For clarity only 3 of the N flux tubes filling all space are shown.

Instead of *summing* discrete tubes of flux to obtain the total charge, let us *integrate* the flux density **D** over a spherical surface surrounding a point charge Q. Figure 2-14 shows one octant or 1/8 of the spherical surface enclosing Q. The elemental electric flux $d\psi$ through an elemental surface element ds is given by

$$d\psi = \mathbf{D} \cdot d\mathbf{s} \tag{8}$$

Integrating **D** over the spherical surface, we have

$$\psi = \oint_s \mathbf{D} \cdot d\mathbf{s} = Q \qquad \text{(C)} \tag{9}$$

where $\oint_s = \oiint$, double or surface integral over a closed surface.

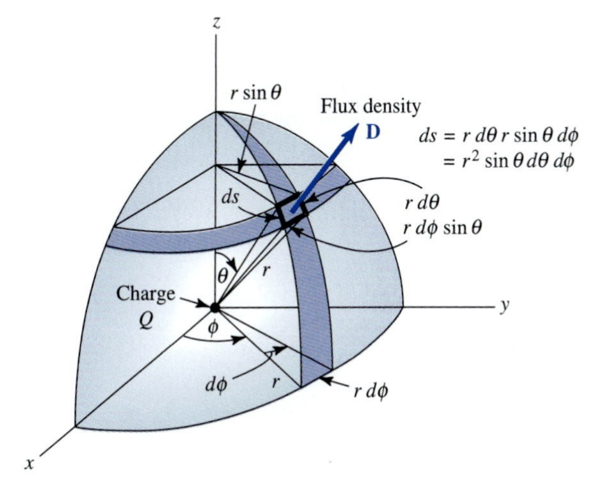

FIGURE 2-14
Point charge Q surrounded by sphere of radius r and surface area $4\pi r^2$.

From (2-2-2) the electric field **E** from a point charge Q is equal to $Q/4\pi\varepsilon r^2$. Since $\mathbf{D} = \varepsilon\mathbf{E}$ and $d\Omega = \sin\theta\, d\theta\, d\phi$, we have on expanding (9)

$$\psi = \frac{Q}{4\pi}\iint d\Omega = \frac{Q}{4\pi}\int_0^{2\pi}\int_0^{\pi}\sin\theta\, d\theta\, d\phi$$

$$= \frac{Q}{4\pi}[-\cos\theta]_0^{\pi}\int_0^{2\pi} d\phi = \frac{Q}{4\pi}\times 2\times 2\pi = Q \tag{10}$$

Thus, the total electric flux over the sphere (obtained by integrating the normal component of the flux density **D** *over the sphere) is equal to the charge Q enclosed by the sphere.*

Example 2-9. Electric flux through part of sphere. A sphere of 2-m radius has a point charge of 8 nC at its center. Find the electric flux passing through that part of the sphere between ±60° latitude and ±20° longitude. Note that latitudes are measured north and south of the equator but θ (Fig. 2-14) is measured from the north pole.

Solution. From (10)

$$\psi = \frac{Q}{4\pi}\int_{-\pi/9}^{\pi/9}\int_{30^\circ}^{150^\circ}\sin\theta\, d\theta\, d\phi = \frac{8\times 2\pi\times 10^{-9}}{4\pi\times 9}\left(-\cos\theta\Big|_{30^\circ}^{150^\circ}\right)$$

$$= \frac{4}{9}\times 1.73\times 10^{-9}$$

$$= 0.769\times 10^{-9}\text{ C} = 769\text{ pC} \quad \textit{Ans.}$$

Problem 2-6-1. Flux from line charge. A uniform line charge of $\rho_L = 6$ nC/m is situated coincident with the x axis. Find the electric flux per unit length of line passing through a plane strip extending in the x direction with edges at $y = 1, z = 0$ and $y = 1, z = 5$ m. *Ans.* 1.31 nC/m.

Volume Charge and Gauss's Law

A cube of volume charge density ρ (C m^{-3}) and volume v is shown in Fig. 2-15. Integrating the flux density **D** over all six faces we obtain the total electric flux from the cube as a surface integral,

$$\psi = \int_0^{y_1}\int_0^{z_1} 2D_x\, dy\, dz + \int_0^{x_1}\int_0^{z_1} 2D_y\, dx\, dz + \int_0^{x_1}\int_0^{y_1} 2D_z\, dx\, dy = Q$$

The total charge is also given by the volume integral of the volume charge density ρ throughout the cube or

$$\psi = \oint_v \rho\, dv = \rho\int_0^{x_1}\int_0^{y_1}\int_0^{z_1} dx\, dy\, dz$$

$$= \rho x_1 y_1 z_1 = \rho v$$

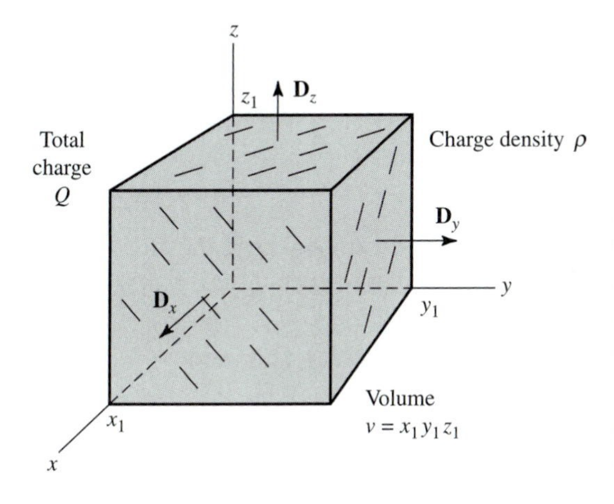

FIGURE 2-15
Cube of total charge Q, volume charge density ρ and volume $v = xyz$. Integrating the flux density **D** over all six faces yields the total charge Q. Integrating the charge density ρ over the volume also equals the charge Q.

Combining surface and volume integrals we obtain *Gauss's law* for electric fields, formulated by Karl Friedrich Gauss in 1813. It states that the integral of the flux density **D** over any closed surface equals the charge enclosed.

$$\oint_s \mathbf{D} \cdot d\mathbf{s} = \oint_v \rho \, dv = Q \qquad \textbf{\textit{Gauss's law}} \tag{11}$$

where $\oint_s$ = double, or surface, integral over closed surface

$\oint_v$ = triple, or volume, integral throughout region enclosed

Any *closed* surface, real or imaginary, may be called a ***Gaussian surface.*** Thus, the double or surface integral with the circle ($\oint_s$) implies a Gaussian or closed surface.

Gauss's law is a basic theorem of electrostatics. It is a necessary consequence of the inverse-square law (Coulomb's law). Thus, if **D** for a point charge did not vary as $1/r^2$, the total flux over a surface enclosing it would not equal the charge.

Example 2-10. Charge in sphere. A sphere of radius $r_1 = 30$ cm has a charge density variation with radius given by $\rho_o r/r_1$ where $\rho_o = 200$ pC m^{-3}. Find the total charge of the sphere.

Solution. In the spherical volume of thickness dr

$$dQ = \rho_o(r/r_1)4\pi r^2 \, dr$$

and

$$Q = \frac{4\pi\rho_o}{r_1}\int_0^{r_1} r^3 \, dr = \frac{4\pi\rho_o}{4r_1} r_1^4 \, dr$$
$$= 200\pi(10^{-12})(0.3^3)$$
$$= 17 \times 10^{-12} \text{ C} = 17 \text{ pC} \quad \textit{Ans.}$$

Problem 2-6-2. Charge in spherical shell. A spherical shell of outer radius 1.5 m and inner radius 0.5 m has a charge density $\rho = 5$ nC/m^3. Find the charge in the shell. *Ans.* 68.1 nC.

2-7 DIVERGENCE

According to Gauss's law, the integral of the normal component of the electric flux density **D** over a closed surface yields the electric charge Q enclosed. Thus,

$$\oiint \mathbf{D} \cdot d\mathbf{s} = \iiint \rho \, dv = Q \qquad \text{(C)} \tag{1}$$

For incremental charges ΔQ and volumes Δv,

$$\oiint D_n \, ds = \rho \, \Delta v = \Delta Q \qquad \text{(C)} \tag{2}$$

where D_n = electric flux density normal to surface.

If the charge density is not uniform throughout Δv, we have in the limit as the volume Δv shrinks to zero that

$$\lim_{\Delta v \to 0} \frac{\iint D_n \, ds}{\Delta v} = \text{div}\,\mathbf{D} = \rho \qquad (\text{C m}^{-3}) \tag{3}$$

where div **D** = *divergence of D* $= \rho$.

The divergence of **D** *yields the electric charge density* ρ *at a point.* Div **D** has a value wherever charge is present.

Consider the cubical volume of Fig. 2-16 with sides Δx, Δy, Δz. If the electric flux density has only an x component D_x with $D_y = D_z = 0$, we have from Gauss's law that

$$\iint \mathbf{D} \cdot d\mathbf{s} = \left(D_x + \frac{\partial D_x}{\partial x}\Delta x\right)\Delta y \, \Delta z - D_x \, \Delta y \, \Delta z = Q = \rho \, \Delta v \tag{4}$$

or

$$\iint \mathbf{D} \cdot d\mathbf{s} = \frac{\partial D_x}{\partial x}\Delta x \, \Delta y \, \Delta z = \rho \, \Delta v \qquad \text{(C)} \tag{5}$$

Taking the limit of (5) as Δv shrinks to zero, we have

$$\text{div}\,\mathbf{D} = \lim_{\Delta v \to 0} \frac{\iint \mathbf{D} \cdot d\mathbf{s}}{\Delta v} = \frac{\partial D_x}{\partial x} = \rho \quad (\text{C m}^{-3}) \tag{6}$$

where div **D** = divergence of **D** $= \rho$.

In general, when D_y and D_z are not zero,

$$\boxed{\text{div}\,\mathbf{D} = \frac{\partial D_x}{\partial x} + \frac{\partial D_y}{\partial y} + \frac{\partial D_z}{\partial z} = \rho \quad (\text{C m}^{-3}) \quad \textit{Divergence of } \mathbf{D}} \tag{7}$$

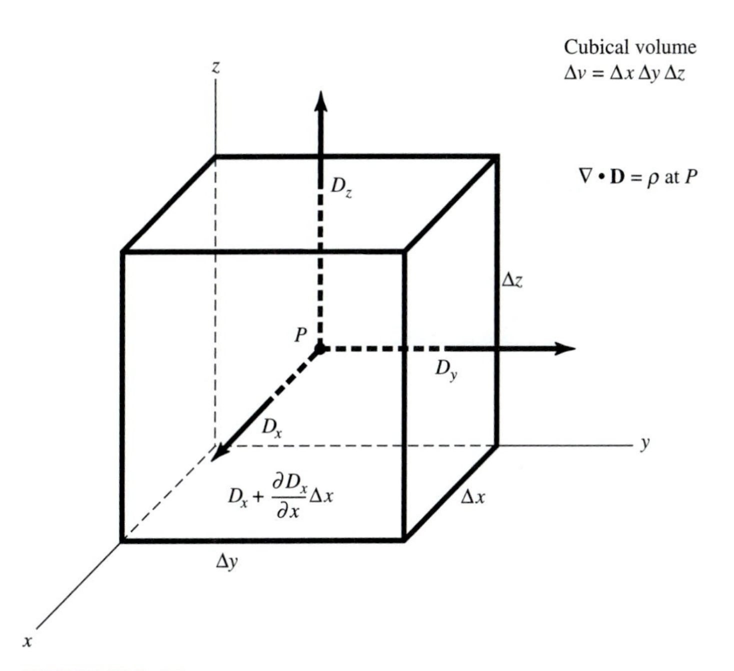

FIGURE 2-16
Small cubical volume Δv $(= \Delta x\, \Delta y\, \Delta z)$ with P at the center. Integrating **D** over the surface of the volume yields the charge enclosed $(= \rho\, \Delta v)$. Dividing by Δv yields (in the limit) the divergence of **D** or the charge density ρ at P as Δv shrinks to zero around P.

In vector notation, the divergence of **D** is expressed as the scalar or dot product of the del operator ∇ and **D**. Thus,

$$\operatorname{div} \mathbf{D} = \nabla \cdot \mathbf{D} = \underbrace{\left(\hat{x}\frac{\partial}{\partial x} + \hat{y}\frac{\partial}{\partial y} + \hat{z}\frac{\partial}{\partial z}\right)}_{\nabla} \cdot \underbrace{(\hat{x}D_x + \hat{y}D_y + \hat{z}D_z)}_{\mathbf{D}} = \rho \quad (\text{C m}^{-3}) \quad (8)$$

Taking the dot product of (8) yields (7). Note that

$$\hat{\mathbf{x}} \cdot \hat{\mathbf{x}} = \hat{\mathbf{y}} \cdot \hat{\mathbf{y}} = \hat{\mathbf{z}} \cdot \hat{\mathbf{z}} = 1 \quad \text{but} \quad \hat{\mathbf{x}} \cdot \hat{\mathbf{y}} = \hat{\mathbf{x}} \cdot \hat{\mathbf{z}} = \hat{\mathbf{y}} \cdot \hat{\mathbf{x}} = \hat{\mathbf{y}} \cdot \hat{\mathbf{z}} = \hat{\mathbf{z}} \cdot \hat{\mathbf{x}} = \hat{\mathbf{z}} \cdot \hat{\mathbf{y}} = 0$$

Example 2-11. Divergence of air velocity. As a simple example of divergence, consider that a long hollow cylinder is filled with air under pressure. If the cover over one end of the cylinder is removed quickly, the air rushes out. It is apparent that the velocity of the air will be greatest near the open end of the cylinder, as suggested by the arrows representing the velocity vector **v** in Fig. 2-17. Suppose that the flow of air is free from turbulence, so that **v** has only an x component. Let us also assume that the velocity **v** in the cylinder is independent of y but is directly proportional to x, as indicated by

$$|\mathbf{v}| = v_x = Kx$$

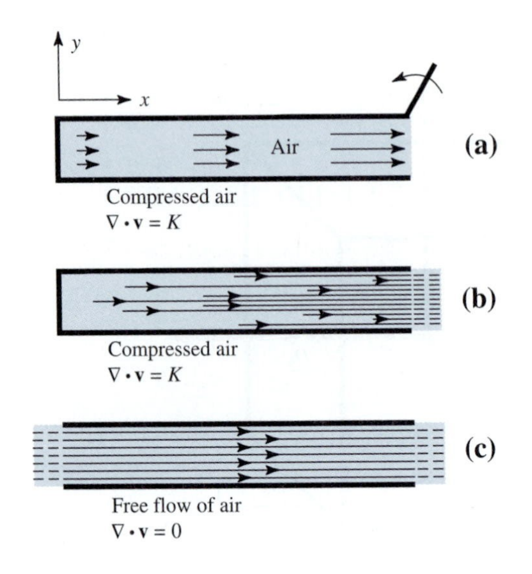

FIGURE 2-17
When air under pressure in a tube is released, the velocity of the air rushing out has divergence, (*a*) and (*b*). When air flows with uniform velocity through a tube open at both ends, as in (*c*), the divergence of the velocity is zero.

where K is a constant of proportionality. The question is: What is the divergence of **v** in the cylinder?

Solution. Apply the divergence operator to **v**:

$$\nabla \cdot \mathbf{v} = \frac{dv_x}{dx} = K$$

Hence, the divergence of **v** is equal to the constant K which equals the ***rate of change of the air velocity with respect to x.***

Problem 2-7-1. Flux density and charge from divergence. If $\nabla \cdot \mathbf{D} = dD_x/dx = 40$ pC/m^3, find: (*a*) D at $x = 5$ cm and (*b*) Q in the volume $0 \le x \le 10$, $5 \le y \le 15$, $0 \le z \le 12$ cm. *Ans.* (*a*) $D = D_x = 2$ pC/m^2; (*b*) $Q = 48$ fC.

Poisson's and Laplace's Equations

From (2-6-2.3)

$$\nabla \cdot \mathbf{D} = \nabla \cdot \varepsilon \mathbf{E} = \rho \tag{9}$$

and since from (2-3-8) $\mathbf{E} = -\nabla V$,

$$\nabla \cdot (-\varepsilon \nabla V) = -\varepsilon \nabla^2 V = \rho \tag{10}$$

Thus,

$$\boxed{\nabla^2 V = -\frac{\rho}{\varepsilon} \quad \textit{Poisson's equation}} \tag{11}$$

In charge-free space regions, (11) reduces to

$$\boxed{\nabla^2 V = 0 \quad \textit{Laplace's equation}} \tag{12}$$

2-8 BOUNDARY CONDITIONS; DIELECTRIC MEDIA

The line integral of the static electric field **E** around a closed path is zero. Thus, for $\Delta y \to 0$ in Fig. 2-18*a,*

$$\oint \mathbf{E} \bullet d\mathbf{L} = E_{t1}\Delta x - E_{t2}\Delta x = 0$$

or

$$\boxed{E_{t1} = E_{t2} \quad \textit{Tangential E}} \tag{1}$$

From Gauss's law, the surface integral of **D** over a closed surface equals the charge enclosed. Thus for $\Delta y \to 0$ in Fig. 2-18*b,*

$$\oiint \mathbf{D} \bullet d\mathbf{s} = D_{n1}A - D_{n2}A = \rho_s A$$

or

$$\boxed{D_{n1} - D_{n2} = \rho_s \quad \textit{Normal D}} \tag{2}$$

Thus, at the boundary of two media, such as air and polystyrene, the component of the electric field parallel to the interface (tangential **E**) is continuous. If the boundary is free of electric charge, the normal component of **D** is continuous.

Under static conditions, $\mathbf{E} = 0$ in a conductor. Therefore, the tangential field $E_t = 0$ and **E** is normal to a conducting sheet, as in Fig. 2-19. This is also true for time-changing fields if the sheet is *perfectly* conducting.

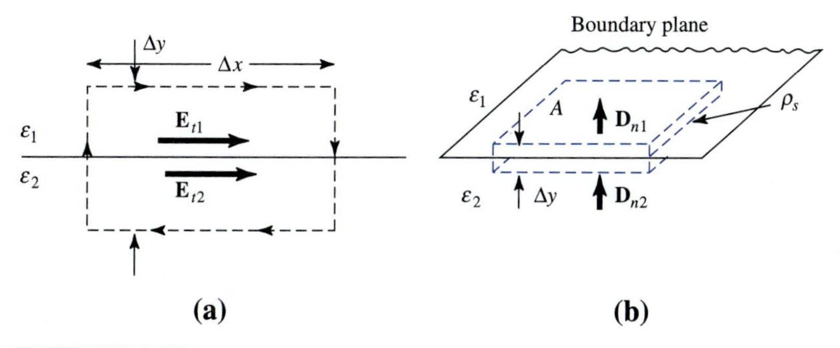

FIGURE 2-18
(*a*) Tangential **E** is continuous at the boundary. (*b*) Normal **D** is continuous at the boundary if the boundary is free of charge.

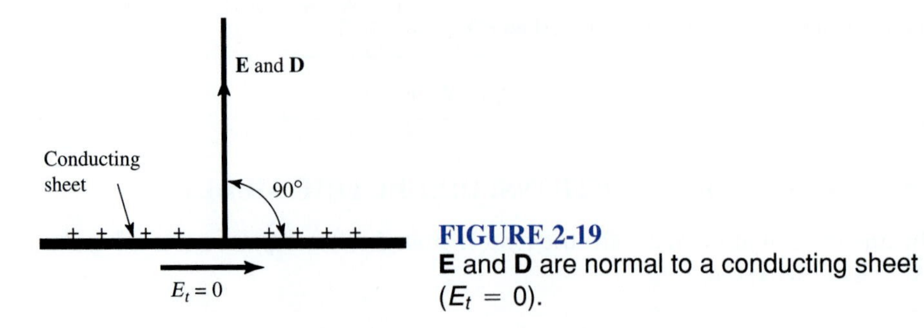

FIGURE 2-19
E and **D** are normal to a conducting sheet ($E_t = 0$).

Example 2-12. Fields at charge-free boundary at oblique incidence. Referring to Fig. 2-20, $d = 2s(\alpha_1 = 26.6°)$ and $E_1 = 10\ \text{V m}^{-1}$ at P_1. Find: (*a*) α_2, (*b*) E_2 at P_1, and (*c*) D_2 at P_1.

Solution
(*a*) P_1:

$$E_{t1} = E_1 \sin\alpha_1 \text{ and } E_{t2} = E_2 \sin\alpha_2$$

From (1),

$$E_1 \sin\alpha_1 = E_2 \sin\alpha_2$$

Also

$$D_{n1} = D_1 \cos\alpha_1 \text{ and } D_{n2} = D_2 \cos\alpha_2$$

From (2),

$$D_1 \cos\alpha_1 = D_2 \cos\alpha_2$$

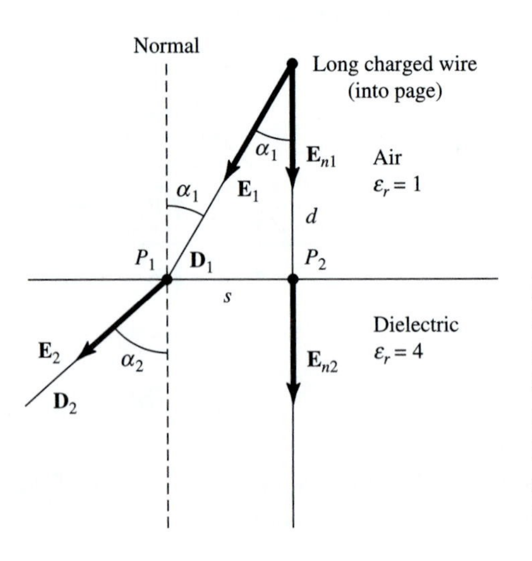

FIGURE 2-20
Long wire in air directly above point P_2 at the boundary of a dielectric medium with fields $\mathbf{E}_1$ and $\mathbf{E}_2$ at oblique incidence and $\mathbf{E}_{n1}$ and $\mathbf{E}_{n2}$ at normal incidence.

Therefore,

$$\frac{E_1}{D_1}\tan\alpha_1 = \frac{E_2}{D_2}\tan\alpha_2$$

and since $D_1 = \varepsilon_0 E_1$ and $D_2 = 4\varepsilon_0 E_2$

$$\tan\alpha_2 = 4\tan\alpha_1 = 4 \times \tfrac{1}{2} = 2 \text{ and } \alpha_2 = 63.4^\circ \quad \textit{Ans.} \quad (a)$$

$$(b)\ E_2(\text{at } P_1) = E_1\frac{\sin\alpha_1}{\sin\alpha_2} = \tfrac{1}{2}E_1 = 5\ \text{V m}^{-1} \quad \textit{Ans.} \quad (b)$$

$$(c)\ D_2(\text{at } P_1) = D_1\frac{\cos\alpha_1}{\cos\alpha_2} = 2D_1 = 2\varepsilon_0 E_1$$

$$= 2 \times 8.85 \times 10^{-12} \times 10$$

$$= 177\ \text{pC m}^{-2} \quad \textit{Ans.} \quad (c)$$

Example 2-13. Fields at a boundary at normal incidence. For same conditions as in above example (see Fig. 2-20). Find (*a*) charge density on wire, (*b*) D_2 at P_2, and (*c*) E_{n2} at P_2. $d = 1$ m, $s = 0.5$ m.

Solution

(*a*) $\rho_L = 2\pi\varepsilon_0 r E_1 = 2\pi 8.85 \times 10^{-12} \times 1.12 \times 10 = 623\ \text{pC m}^{-1}$ *Ans.* (*a*).

(*b*) $D_2(\text{at } P_2) = D_1(\text{at } P_2) = \varepsilon_0 E_{n1}$ (at P_2)

$= 8.85 \times 10^{-12} \times 11.2 = 99\ \text{pC m}^{-2}$ *Ans.* (*b*).

(*c*) $E_{n2}(\text{at } P_2) = D_2(\text{at } P_2)/4\varepsilon_0 = 11.2/4 = 2.8\ \text{V m}^{-1}$ *Ans.* (*c*).

Problem 2-8-1. Line charge. Find the charge density on the long line of Fig. 2-20 if E_2 at P_1 is 15 V/m and $\alpha_2 = 45^\circ$. $d = 1$ m, $s = 0.5$ m. *Ans.* 1.48 nC/m

2-9 CAPACITORS AND CAPACITANCE; CAPACITOR CELLS

The two flat conducting plates of Fig. 2-21*a* form a simple *capacitor.* Its *capacitance* C is given by

$$C = \frac{Q}{V} \quad \left(\frac{\text{coulombs}}{\text{volts}} \quad \text{or} \quad \text{farads, F}\right) \tag{1}$$

The greater the charge Q for a given applied voltage V, the greater the capacitance C. Assuming that the charge Q is uniformly distributed and the field between the plates is uniform,

$$Q = \rho_s A = DA = \varepsilon EA \quad \text{(C)} \tag{2}$$

where A = plate area, m^2
ρ_s = surface charge density, C m^{-2}
D = electric flux density, C m^{-2}
E = electric field, V m^{-1}
ε = permittivity of medium, F m^{-1}

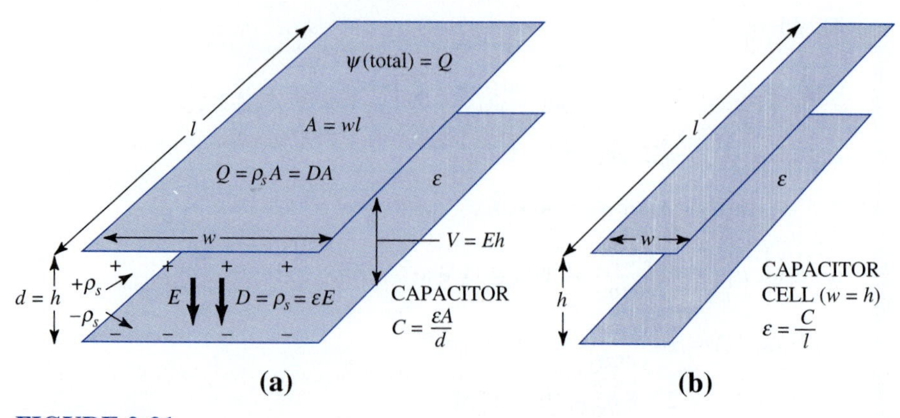

FIGURE 2-21
(*a*) Parallel-plate capacitor. The total flux ψ between the plates equals the total charge Q. (*b*) Parallel-plate capacitor with $w = h$ is a capacitor cell with a capacitance per unit length equal to the permittivity of the cell medium ($C/l = \varepsilon$).

Since $V = Eh = Ed$, the capacitance C is given by

$$C = \frac{Q}{V} = \frac{DA}{Ed} = \frac{\varepsilon EA}{Ed} = \frac{\varepsilon A}{d} \qquad \text{(F)} \tag{3}$$

or in general,

$$C = \frac{\varepsilon_r \varepsilon_0 A}{d} = \frac{\varepsilon A}{d} \qquad \text{(F)} \tag{4}$$

where ε_r = relative permittivity of medium = $\varepsilon/\varepsilon_0$ (dimensionless).

Thus, *the capacitance is determined by the capacitor's geometry*: the plate area A, the plate spacing d and the permittivity ε of the medium between the plates. It is independent of the voltage applied. However, if too high a voltage is applied, the medium may break down and arcing may occur. Hence, capacitors are usually labeled not only in farads or a subunit of farads but also in *maximum working voltage.*

The plate area $A = wl$, so

$$C = \frac{\varepsilon\, wl}{h} \qquad \text{(F)} \tag{5}$$

When $w = h$ the capacitor becomes a ***capacitor cell,*** as in Fig. 2-21*b* and

$$\frac{C}{l} = \varepsilon \qquad (\text{F m}^{-1}) \tag{6}$$

Thus, ***the capacitance per unit length of a capacitor cell equals the permittivity of the medium.***

Example 2-14. Sandwich capacitor. Referring to Fig. 2-22*a*, the capacitor is a sandwich of two dielectric media of the same thickness (d = 1 cm). Plate area

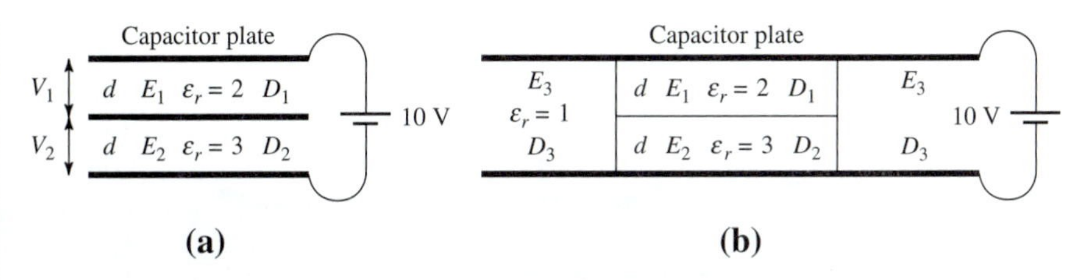

FIGURE 2-22
(*a*) Sandwich capacitor. (*b*) Same but with larger capacitor plates.

$A = 100$ cm^2. Neglect the field outside the capacitor, called *fringing field.* Find: (*a*) E_1, E_2, V_1, V_2, D_1 and D_2; (*b*) capacitance.

Solution

(*a*) $D_1 = D_2$ so

$$2\varepsilon_0 E_1 = 3\varepsilon_0 E_2 \qquad \text{or} \qquad E_1 = \frac{3}{2}E_2$$

Also $V_1 + V_2 = 10$ V or $E_1 d + E_2 d = 10$ V. Thus,

$$E_1 = 600 \text{ V m}^{-1},\ E_2 = 400 \text{ V m}^{-1},\ V_1 = 6 \text{ V},\ V_2 = 4 \text{ V},$$
$$D_1 = D_2 = 2\varepsilon_0 E_1 = 10.6 \text{ nC m}^{-2} \quad \textit{Ans.} \quad (a)$$

(*b*) $C = DA/V = 10.6$ pF *Ans.* (*b*)

Example 2-15. Referring to Fig. 2-22*b*, the capacitor is a sandwich as in Fig. 2-22*a* but the capacitor plates are larger ($A = 200$ cm^2). Neglect fringing. $d = 1$ cm. Find: (*a*) E_1, E_2, E_3; (*b*) capacitance.

Solution

(*a*) $E_3 = 10\text{V}/0.2 \text{ m} = 500 \text{ V m}^{-1}$, $E_1 = 600 \text{ V m}^{-1}$, $E_2 = 400 \text{ V m}^{-1}$ *Ans.* (*a*)
(*b*) $C = 15.0$ pF *Ans.* (*b*)

Problem 2-9-1. One-joule capacitor. What is the capacitance of a capacitor storing 1 J with 500 V applied? *Ans.* 8 μF.

Problem 2-9-2. Parallel-plate capacitor. A parallel-plate capacitor has 4×6 cm plates spaced 3 mm. Find the capacitance (*a*) if the medium between the plates is air ($\varepsilon_r = 1$) and (*b*) if $\varepsilon_r = 12$. (*c*) Find the charge per unit area on the capacitor plates if a potential of 100 V is applied. Neglect fringing. *Ans.* (*a*) 7.08 pF, (*b*) 85.0 pF, (*c*) 295 nC m^{-2} and 3.54 μC m^{-2}.

Example 2-15.1. Process control in a plastics factory. Plastics of relative permittivity 2 and 5 are mixed in a plastics factory to produce 4-mm-thick sheets of relative permittivity 3.2±0.1 at 1 GHz. A 100-cm^2 pad rides on the extruded sheet as shown in Fig. 2-23. (*a*) Assuming a homogeneous mix and uniform thickness, what change in current does a change in relative permittivity from 3.2 to 3.3 produce? $V = 25$ V rms. Take $\sigma = 0$. (*b*) What thickness tolerance is permissible to keep changes in current due to thickness variations an order of magnitude (factor of 10) less?

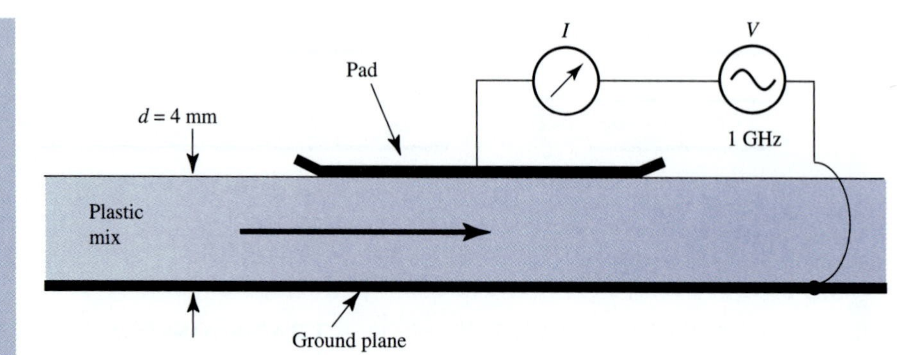

FIGURE 2-23
Permittivity measurement of moving plastic.

Solution. Pad capacitance is

$$C = \varepsilon_r \varepsilon_0 \frac{A}{d} = \frac{3.2 \times 8.885 \times 10^{-12} \times 100 \times 10^{-4}}{4 \times 10^{-3}} = 7.08 \times 10^{-11}\,\text{F}$$

$$I = V\omega C = 25 \times 2\pi 10^9 \times 7.08 \times 10^{-11} = 11.1\,\text{A}$$

$$\Delta I = \left(\frac{3.3 - 3.2}{3.2}\right) 11.1 = 0.347\,\text{A} = 347\,\text{mA} \quad Ans. \quad (a)$$

$$\frac{\Delta d}{4} \times 11.1 = 0.0347 \qquad \Delta d = \frac{0.0347 \times 4}{11.1} = 0.0125\,\text{mm} = \pm 12.5\,\mu\text{m} \quad Ans. \quad (b)$$

Problem 2-9-3. Measurement of permittivity. Two 20 × 20 cm capacitor plates are separated by 2 mm and connected to a 100 V dc source. A 1-mm-thick sheet of dielectric is inserted between the plates giving a 1/2-s pulse of 30 nA average current. What is the permittivity of the dielectric sheet? *Ans.* $\varepsilon_r = 12$.

Capacitor Energy and Energy Density

A capacitor stores an incremental energy given by the product of the voltage applied and the incremental charge. Thus,

$$dW = V dq = \frac{1}{C} q\,dq \qquad \text{(joules, J)} \tag{7}$$

or a *total energy* (Fig. 2-21*a*),

$$W = \frac{1}{C}\int_0^Q q\,dq = \frac{1}{2}\frac{Q^2}{C} = \frac{1}{2}CV^2$$
$$= \frac{1}{2}QV = \frac{1}{2}\varepsilon EAEh = \frac{1}{2}\varepsilon E^2 Ah \qquad \text{(J)} \tag{8}$$

where Ah = capacitor volume, m^3. Dividing by this volume, we obtain the ***energy density*** (assumed uniform) in the capacitor as

$$w = \frac{1}{2}\varepsilon E^2 \quad (\text{J m}^{-3}) \tag{9}$$

Problem 2-9-4. Thunderbolt. (*a*) How many 100-μF capacitors charged to 6000 V are required to store as much energy as a fully charged 12V, 100 Ah automobile storage battery? *Ans.* 1200. If all were connected in series they could be discharged instantaneously to produce a 14.4-million-volt thunderbolt. Attempts to discharge the battery instantaneously would cause it to explode. Thus, capacitors are better for quick release of energy. (*b*) If the capacitor energy is released in 1/2 s, what is the peak power of the thunderbolt? *Ans.* 17.8×10^6W. (*c*) Where did the energy go? *Ans.* Into acoustic, thermal, and electromagnetic radiation.

2-10 TWIN-STRIP AND MICROSTRIP TRANSMISSION LINES

The electric field of a transmission line of two flat-strip conductors is shown in cross section in Fig. 2-24.1. The field map is divided into 20 capacitor cells in parallel ($N_p = 20$) and 8 cells in series ($N_s = 8$). Each cell is square ($w = h$).

The lines of electric field **E** are vertical and the equipotential surfaces ($V =$ constant) are horizontal. The strip width $w' = 20$ mm and the strip spacing $h' = 8$ mm. Thus, the capacitance per unit length is given for air ($\varepsilon = \varepsilon_0$) by

$$C/l = \varepsilon_0 \frac{\text{strip width}}{\text{strip spacing}} = \varepsilon_0 \frac{N_p}{N_s} = \varepsilon_0 \frac{20}{8}$$

$$= 8.85 \times 10^{-12}\ \text{F m}^{-1} \times 2.5 = 22\ \text{pF m}^{-1} \tag{1}$$

For 16 V applied between the strips, the electric field is

$$E = \frac{V}{\text{spacing}} = \frac{16\,\text{V}}{0.008\,\text{m}} = 2000\ \text{V m}^{-1}$$

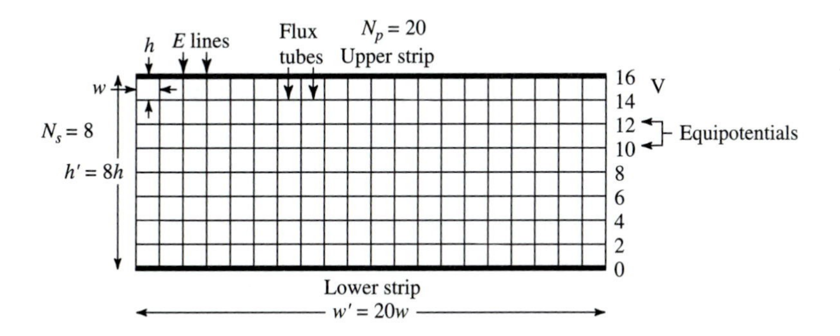

FIGURE 2-24.1
Field map of twin-strip transmission line divided into 20 × 8 = 160 field cells. Fringing of the field is ignored.

between the strips. The flux density between the strips is

$$D = \varepsilon_0 E = 8.85 \times 10^{-12}\ \mathrm{F\,m^{-1}} \times 2000\ \mathrm{V\,m^{-1}} = 17.7\ \mathrm{nC\,m^{-2}} \tag{2}$$

This is also the surface charge density ρ_s on the strips. Thus,

$$\rho_s = D = 17.7\ \mathrm{nC\,m^{-2}}$$

The total charge Q on the plates equals the surface charge density times the strip area $A(= w'l)$ or

$$Q = \rho_s A = \rho_s w'l \tag{3}$$

The charge per unit length of line is

$$Q/l = \rho_s w' = 17.7\ \mathrm{nC\,m^{-2}} \times 20\ \mathrm{mm} = 354\ \mathrm{pC\,m^{-1}} \tag{4}$$

It is a simplifying idealization to consider that the electric field is all contained in the rectangular space between the strips. The actual field distribution, shown in Fig. 2-24.2, has fringing field extending infinitely far outside. This field map is divided into cells that are either true squares or *curvilinear squares* but in all cases the cell width w equals the cell height h ($w = h$).

Methods for making field maps such as that of Fig. 2-24.2 are discussed in Appendix B.

To evaluate the actual field distribution shown in Fig. 2-24.2, the width and spacing of the strips no longer suffices. We need to use the information provided by the map. Thus, there are 30 cells in parallel ($N_p = 30$) and 8 in series ($N_s = 8$) so the total capacitance per unit length of the line is

$$C/l = \varepsilon_0 \frac{N_p}{N_s} = \varepsilon_0 \frac{30}{8} = 33\ \mathrm{pF\,m^{-1}} \tag{5}$$

which is *50 percent more* than calculated above for the case when the fringing field is ignored as in Fig. 2-24.1.

Fields and flux densities are weaker at cells which are larger, and stronger at cells which are smaller. From cell size we can evaluate the variation of the field quantities everywhere.

By placing a conducting sheet in the symmetry plane of Fig. 2-24.2 we obtain a single strip above a ground plane, a combination called a *microstrip transmission line* as shown in Fig. 2-25. The microstrip line has been chosen for exhibit because there are more microstrip lines in existence than any other kind. For example, they are used by millions as feed lines for patch antenna arrays and by the billions in computer and other electronic circuitry.

Usually the strip rests on a dielectric sheet or substrate but for simplicity the medium here is air ($\varepsilon_r = 1$).

Example 2-16. Microstrip transmission line. For an applied potential of 8 V to the line of Fig. 2-25, find (*a*) the electric field E at P_1, P_2, and P_3, (*b*) the flux density D at the same three points, and (*c*) the fraction of power transmitted in the fringing field region. (*Example continues on page 68.*)

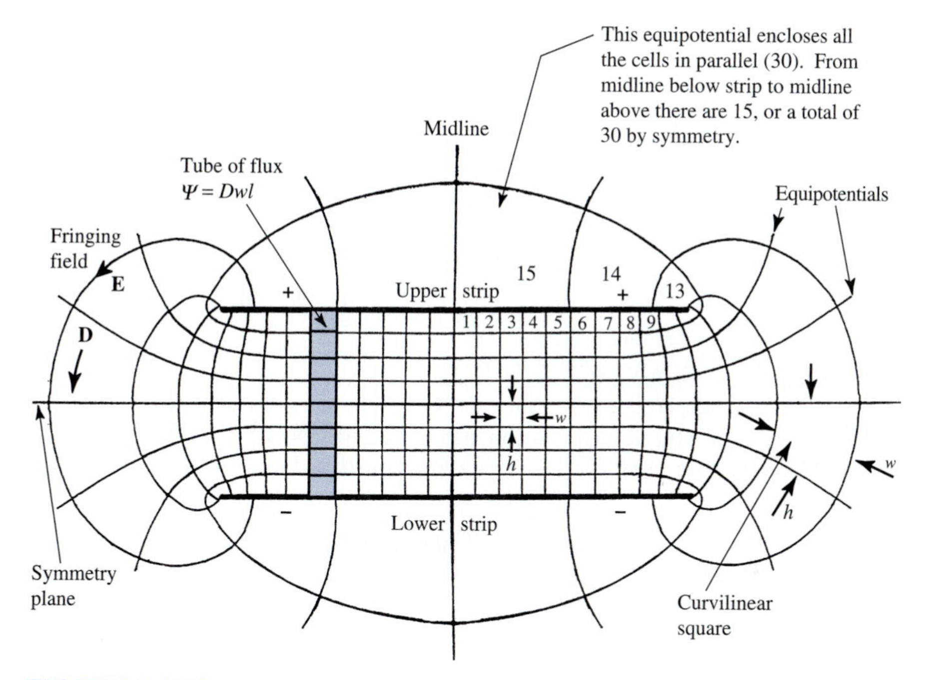

FIGURE 2-24.2
Same twin-strip transmission line as in Fig. 2-24.1 but showing the actual field distribution. Even though the fringing field extends to infinity, and some cells are outside the figure, the map informs us that there are 30 cells in parallel for a total of $30 \times 8 = 240$ (no more, no less) compared to 160 cells for Fig. 2-24.1. The 80 cell difference $(240 - 160)$ is in the friniging field which is ignored in FIg. 2-24.1. The actual capacitance of the line is, therefore, $240/160 = 1.50$ or 50 *percent more* than in Fig. 2-24.1.

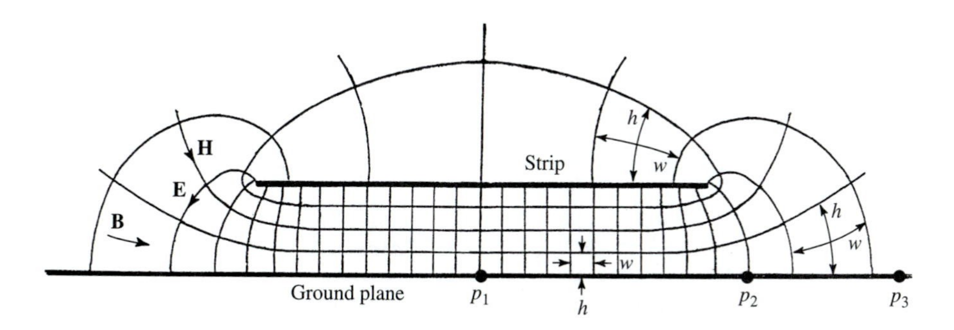

FIGURE 2-25
Transverse or cross-section view of the electric field around a microstrip transmission line with the map divided into field cells. The strip width is 20 mm and the spacing is 4 mm. **B** and **H** are the magnetic flux density and magnetic field, respectively, as discussed beginning in Sec. 2-12.

Example 2-16. (*Continued from page 66.*)

Solution

(*a*) E at $P_1 = V/\text{spacing} = 8\text{ V}/4\text{ mm} = 2000\text{ V m}^{-1}$ *Ans.*

$$E \text{ at } P_2 = 2000\text{ V m}^{-1}(h \text{ at } P_1/h \text{ at } P_2) = 2000 \times 0.41 = 824\text{ V m}^{-1} \quad \textit{Ans.}$$

$$E \text{ at } P_3 = 2000\text{ V m}^{-1}(h \text{ at } P_1/h \text{ at } P_3) = 2000 \times 0.18 = 360\text{ V m}^{-1} \quad \textit{Ans.}$$

(*b*) D at $P_1 = \varepsilon_0 E$ at $P_1 = 17.7\text{ nC m}^{-2}$ *Ans.*

$$D \text{ at } P_2 = 7.3\text{ nC m}^{-2} \quad \textit{Ans.}$$

$$D \text{ at } P_3 = 3.2\text{ nC m}^{-2} \quad \textit{Ans.}$$

(*c*) Since the potential difference ΔV across any cell is a constant, the power P per cell is a constant ($P \propto \Delta V^2$). There are $30 \times 4 = 120$ cells including fringing but only $20 \times 4 = 80$ cells with no fringing or $120 - 80 = 40$ fewer cells. Therefore, $40/120 = 0.33$ or 33 percent of the power is transmitted in the fringing field region. *Ans.*

Problem 2-10-1. Charge density on microstrip line. What is the charge density ($\rho_s = D$) (*a*) at center of strip of line of Fig. 2-25 and (*b*) at the midpoint on the ground plane? *Ans.* (*a*) 17.7 nC m^{-2} under strip, 3.0 nC m^{-2} above the strip; (*b*) 17.7 nC m^{-2}.

2-11 ELECTRIC CURRENTS

Electric charge in motion constitutes an *electric current*, and any current-carrying medium may be called a *conductor*. In metallic conductors the charge is carried by electrons. In plasmas or gaseous conductors the charge is carried by (*negative*) *electrons* and *positive ions* (electron-deficient atoms or molecules). A plasma contains equal numbers of positive and negative charges (total charge equals zero). The earth's ionosphere of positive ions and free electrons is a plasma. In liquid conductors (electrolytes) the charge is carried by ions, both positive and negative. In semiconductors the charge is carried by *electrons* and *holes*, the holes behaving like positive charges.

The fields of steady electric currents are discussed in this section. These fields are constant with time and, hence, are static fields.

Electric Current and Current Density

A test charge e introduced into an electric field $\mathbf{E}$ as in Fig. 2-26 experiences a force $\mathbf{F}$ given by

$$\boxed{\mathbf{F} = e\mathbf{E} \quad (\text{N}) \quad \textit{Force}} \tag{1}$$

If the charge is free to move, it will receive an acceleration which, from Newton's second law ($\mathbf{F} = m\mathbf{a}$), is

$$\boxed{\mathbf{a} = \frac{\mathbf{F}}{m} \quad (\text{m s}^{-2}) \quad \textit{Acceleration}} \tag{2}$$

where m is the mass of the charged particle in kilograms.

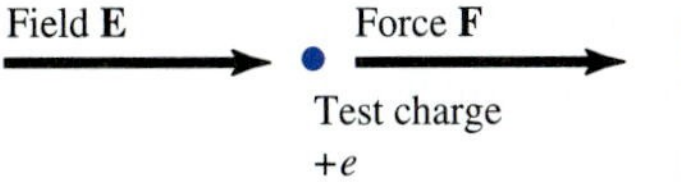

FIGURE 2-26
Test charge *e* in electric field **E** experiences a force **F**.

In the absence of restraints, the particle's velocity $v\,(= at)$ will increase indefinitely with the time t provided the electric field **E** is constant. However, in gaseous, liquid, or solid conductors the particle collides repeatedly with other particles, losing some of its energy and causing random changes in its direction of motion. If **E** is constant and the medium is homogeneous, the net effect of the collisions is to restrain the charged particle to a constant average velocity called the *drift velocity* $\mathbf{v}_d$. This drift velocity has the same direction as the electric field and is related to it by a constant called the *mobility* μ_m. Thus,

$$\mathbf{v}_d = \mu_m \mathbf{E} \quad (\text{m s}^{-1}) \quad \textit{Drift velocity} \tag{3}$$

where μ_m = mobility, $\text{m}^2\ \text{V}^{-1}\ \text{s}^{-1}$, and $\mathbf{E}$ = electric field intensity, V m^{-1}. Good conductors have high mobility.

If a medium of uniform cross section A, as in Fig. 2-27, contains many free-to-move charged particles of volume density ρ, then these moving charges will form a *current* **I** in coulombs per second passing a given reference point as given by

$$\mathbf{I} = \mathbf{v}_d \rho A \quad (\text{A}) \quad \textit{Current} \tag{4}$$

where $\mathbf{I}$ = current, C s^{-1} or A
$\mathbf{v}_d$ = drift velocity, m s^{-1}
ρ = charge density, C m^{-3}
A = area of conducting medium, m^2

The unit for current (coulombs per second) is the *ampere*.

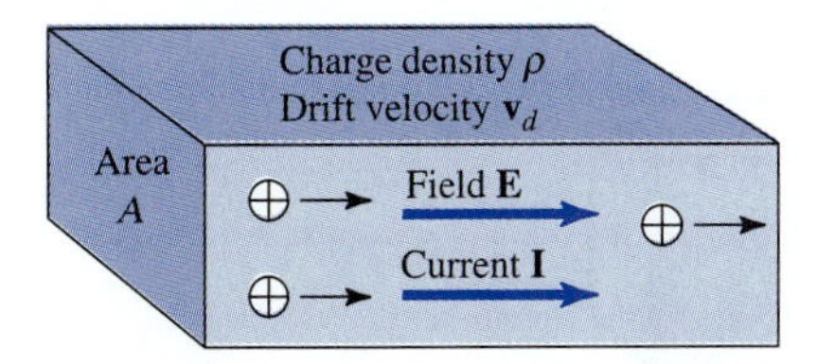

FIGURE 2-27
Medium of cross-sectional area *A* containing free-to-move charged particles has a current **I** in the presence of an electric field **E**. The current consists of charged particles of volume density ρ drifting with a velocity $\mathbf{v}_d$. Charges may be both positive and negative. Positive charges move in direction of field, negative charges in opposite direction but both add to the total current. For simplicity only positive charges are shown here.

According to (4), the current is proportional to the drift velocity, the charge density, and the area of the current-carrying medium or conductor. Dividing (4) by the area A, we obtain the current per unit area or the *current density* **J.** Thus,

$$\mathbf{J} = \frac{\mathbf{I}}{A} = \mathbf{v}_d \rho \quad (\mathrm{A\ m^{-2}}) \quad \textit{\textbf{Current density}} \tag{5}$$

Current density has the dimensions of current per area and is expressed in amperes per square meter (A m^{-2}).

Resistance and Conductance; Resistivity and Conductivity

Resistance R and conductance G apply to resistors or resistive devices.

Resistance R is expressed in *ohms*, Ω.

$$R = \frac{V}{I} \quad (\Omega)$$

where V = voltage across resistor, V, and I = current through resistor, A.

Conductance is the reciprocal of resistance, $G = 1/R$, and is expressed in *mhos* (℧) or in *siemens*. Mho is ohm spelled backwards, and ℧ is an upside-down omega (Ω). See Fig. 2-28*a*.

Resistivity, S, and conductivity, σ, are volumetric quantities which describe how well or poorly a material conducts electricity.

Resistivity S is expressed in *ohm-meters*, Ω m.

$$S = \frac{RA}{w} \qquad \text{(ohm-meters, } \Omega \text{ m)} \tag{6}$$

where R = resistance of block between end faces, Ω
w = length of block, m
A = area of the end faces of block, m^2

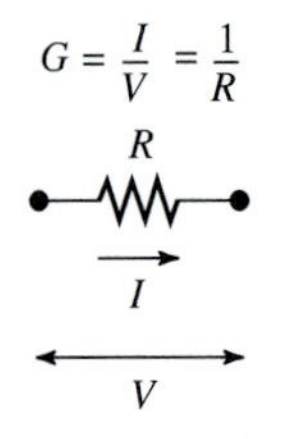

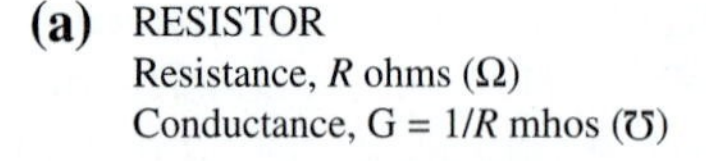
(a) RESISTOR
Resistance, R ohms (Ω)
Conductance, G = $1/R$ mhos (℧)

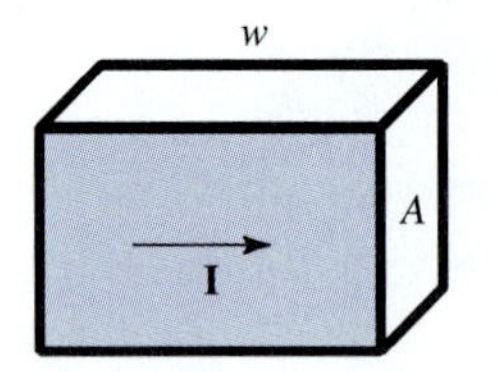

(b) RECTANGULAR BLOCK OF CONDUCTING MATERIAL
Resistivity, $S = RA/w$ ohm-meters (Ωm)
Conductivity, $\sigma = 1/S = w/RA$ mhos/meter (℧m^{-1})

FIGURE 2-28
(*a*) Resistor. (*b*) Block of conducting material.

See Fig. 2-28*b*. If the material has a resistivity $S = 1\ \Omega$ m, then a 1-m cube of the material has a resistance $R = 1\ \Omega$ and a 1-cm cube a resistance $R = 100\ \Omega$.

Conductivity, σ, is the reciprocal of resistivity, $\sigma = 1/S$, and is expressed in *mhos/meter,*

$$\sigma = \frac{1}{S} = \frac{w}{RA} \quad \text{(mhos per meter, ℧ m}^{-1}\text{)} \tag{7}$$

Example 2-17. Measurement of conductivity. (*a*) Calculate the conductivity of a 4-mm-diameter, 5-m-long wire if its measured resistance is 12 mΩ. (*b*) What is the material?

Solution

(*a*) $\sigma = \dfrac{w}{RA} = \dfrac{5}{12 \times 10^{-3}\pi(2 \times 10^{-3})^2} = 3.3 \times 10^7$ ℧ m^{-1} *Ans.*

(*b*) Aluminum alloy. Between aluminum and duralumin. See Table 2-1. *Ans.*

Problem 2-11-1. Resistance of copper wire. What is the resistance of a 2-mm-diameter copper wire 30 m long? σ(copper) $= 5.7 \times 10^7$ ℧ m^{-1}. *Ans.* 0.17 Ω.

Ohm's Law

In 1826 Georg Simon Ohm experimentally determined the relations between the voltage V over a length of conductor and the current I through the conductor in terms of a parameter characteristic of the conductor (see Fig. 2-29). This parameter, called the resistance R, is defined as the ratio of the voltage V to the current I. Thus,

$$R = \frac{V}{I} \quad \text{or} \quad \boxed{V = IR \qquad \textbf{\textit{Ohm's law}}} \tag{8}$$

Ohm's law states that the potential difference or voltage V between the ends of a conductor is equal to the product of its resistance R and the current I.

It is assumed that R is independent of I; that is, R is a constant. Conversely, such a resistance is said to obey Ohm's law. There exist circuit elements, however, such as rectifiers, whose resistance is not a constant. Such elements are said to be *nonlinear,* and a V versus I diagram is required to display their behavior. Nevertheless, the resistance of the nonlinear element is still defined by $R = V/I$, but R is not independent of I.

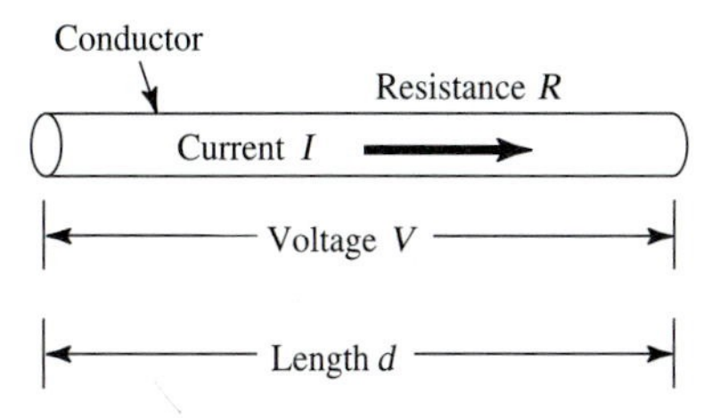

FIGURE 2-29
A voltage V across a conductor of length d produces a current I. The ratio of V to I is a measure of the resistance R of the conductor.

OHM'S LAW AT A POINT AND CURRENT DENSITY. From Ohm's law (8), the current through the conductor of Fig. 2-30 is given by

$$I = \frac{V}{R} \quad \text{(A)} \tag{9}$$

where V = potential across ends of block = Ew, V; and R = resistance of block, Ω.

The current I is equal to the current density J (A m^{-2}) times the cross-sectional area A of the block. Thus,

$$I = JA \quad \text{or} \quad J = \frac{I}{A}$$

Since

$$R = \frac{w}{\sigma A} \quad \text{and} \quad V = Ew \tag{10}$$

$$J = \frac{Ew\sigma A}{wA} \qquad \text{(A m}^{-2}\text{)} \tag{11}$$

or

$$\boxed{J = \sigma E \quad \textit{Ohm's law at a point}} \tag{12}$$

Uniform current density in the block is assumed. To achieve this, highly conducting (ideally perfectly conducting, $\sigma = \infty$) end plates are placed against the ends of the block as in Fig. 2-30.

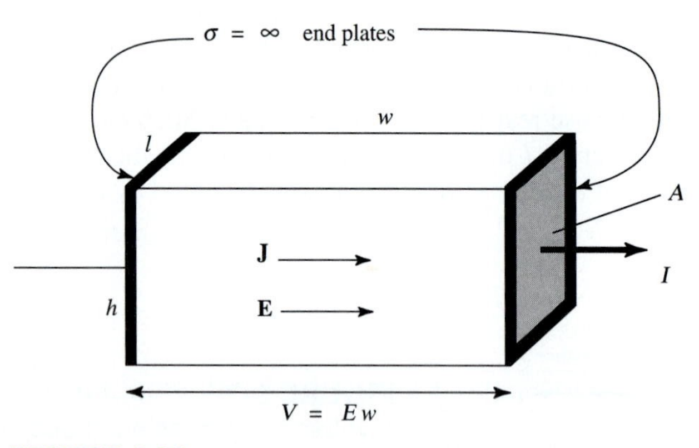

FIGURE 2-30
Conducting block with uniform current flow.

Power and Joule's Law

The input power P to a device is given by

$$P = VI \quad \text{(watts, W)} \tag{13}$$

where V = voltage applied, V
I = current, A

From Ohm's law ($V = IR$), (13) becomes

$$P = I^2R \quad \text{(W)} \tag{14}$$

This power may produce useful work or heat.

In a time T, the energy consumed by the device is given by Joule's law,

$$W = PT = I^2RT \quad \textit{Joule's law} \tag{15}$$

where W = energy, J = W s
P = power, W
I = current, A
R = resistance, Ω
T = time, s

It is assumed in (15) that P is constant over the time T. If not, I^2R is integrated over the time interval T. Thus,

$$W = R\int_0^T I^2\,dt \tag{16}$$

Example 2-18. Power in resistor with variable current. A resistor $R = 2\Omega$ absorbs 800 J of energy in 2 s from a current $I = I_0 \sin\omega t$, where $\omega = 2\pi f$ and $f = 60$ Hz. (*a*) What is the value of the peak current I_0? (*b*) Repeat for $f = 1$ kHz.

Solution. From (16),

$$I_0 = \sqrt{\frac{W}{R\int_0^2 \sin^2 \omega t\,dt}} = 20\text{ A} \quad \textit{Ans.} \quad (a) \text{ and } (b)$$

Problem 2-11-2. 50-Ω resistor heat energy. A 50-Ω resistor is connected to a 120-V (rms) 60-Hz line. What heat energy is produced in one hour? *Ans.* 288 W hr or 1.04 MJ.

Dielectrics, Conductors, and Semiconductors Compared—An Overview

In a ***dielectric,*** or ***insulator,*** charges are bound and not free to migrate so that the conductivity is ideally zero. Although a field applied to an insulator may produce no migration of charge, it can produce a polarization of the insulator or dielectric

(Sec. 4-6), i.e., a displacement of the electrons with respect to their equilibrium positions.

In ***plasmas, gases,*** and ***liquids,*** charges of both signs are usually present and free to migrate. Assuming that all negative particles are of the same kind and all positive particles (or ions) are of the same kind, the conductivity will have two terms, as follows:

$$\sigma = \rho_- \mu_- + \rho_+ \mu_+ \qquad (\mho\ \text{m}^{-1}) \tag{17}$$

where ρ_- = density of negatively charged particles, C m^{-3}
μ_- = mobility of negatively charged particles, m^2 V^{-1} s^{-1}
ρ_+ = density of positively charged particles (or ions), C m^{-3}
μ_+ = mobility of positively charged particles (or ions), m^2 V^{-1} s^{-1}

The first term represents the contribution to the conductivity from negatively charged particles moving opposite to the field **E** while the second term represents the contribution from positively charged particles moving with **E.**

In ordinary (metallic) conductors, such as copper and aluminum, the highest-energy electrons are readily detached from their atoms by an applied electric field and are free to migrate. The atoms, however, remain fixed in the conductor's lattice, so only the electrons have mobility. Therefore, the conductivity for conductors has only one term as given by

$$\sigma = \rho_e \mu_e \qquad (\mho\ \text{m}^{-1}) \tag{18}$$

where ρ_e = density of electrons, C m^{-3}
μ_e = mobility of the electrons, m^2 V^{-1} s^{-1}

In ***semiconductors*** the normal conduction by electrons is supplemented by another charge carrier called a *hole,* which represents a vacant space in the lattice structure of the semiconductor. These vacant spaces left by electrons can migrate from atom to atom in a semiconductor so that the hole tends to behave like a positively charged ion but its mobility is more like that of an electron. Thus, the conductivity of a semiconductor is given by

$$\sigma = \rho_e \mu_e + \rho_h \mu_h \qquad (\mho\ \text{m}^{-1}) \tag{19}$$

where ρ_e = charge density of electrons, C m^{-3}
μ_e = mobility of electrons, m^2 V^{-1} s^{-1}
ρ_h = charge density of holes, C m^{-3}
μ_h = mobility of holes, m^2 V^{-1} s^{-1}

Germanium and silicon are typical semiconductors. In their intrinsic or pure form, the electrons and vacancies (or electron-hole pairs) have only a short lifetime, disappearing as electrons and holes recombine. But new pairs are continually being formed, and so some are always present. Increasing the temperature accelerates pair formation, increasing the density of electrons and holes and, hence, the conductivity. This behavior with temperature is opposite to that for ordinary conductors, which decrease in conductivity (increase in resistance) with temperature.

If small amounts of certain impurities are added to the semiconductor, either during crystal growth or by diffusion, the carrier density and conductivity can be greatly increased. Impurities such as phosphorus provide more electrons and are called *donors,* forming *n-type* semiconductors in which electrons constitute most of the carriers, holes being in the minority. Impurities such as boron introduce more holes and are called *acceptors,* forming *p-type* semiconductors with holes predominating. The procedure of introducing impurities is called *doping.* Donor concentrations of less than one part per million can increase the conductivity by a factor of nearly a million. Thus, although the normal behavior of pure, or intrinsic, semiconductors continues in the presence of impurities, it is a minor effect in comparison with the extra electrons in *n*-type and extra holes in *p*-type semiconductors.

The boundary between *n*-type and *p*-type regions of a single semiconductor crystal forms a junction region utilized in diodes and transistors.

Table 2-1 lists conductivities of a wide range of materials from the best insulators (quartz), through poor insulators, poor conductors, and a semiconductor to the best ordinary conductors (copper and silver) and lastly to superconductors. The

TABLE 2-1
Conductivities†

	Material	Conductivity, $\mho$ m^{-1}
Insulators	Quartz, fused	$\sim 10^{-17}$
	Ceresin wax	$\sim 10^{-17}$
	Polystyrene	$\sim 10^{-16}$
	Sulfur	$\sim 10^{-15}$
	Mica	$\sim 10^{-15}$
	Paraffin	$\sim 10^{-15}$
	Rubber, hard	$\sim 10^{-15}$
	Porcelain	$\sim 10^{-14}$
	Glass	$\sim 10^{-12}$
	Bakelite	$\sim 10^{-9}$
	Distilled water	$\sim 10^{-4}$
Poor insulators	Dry, sandy soil	$\sim 10^{-3}$
	Marshy soil	$\sim 10^{-2}$
	Fresh water	$\sim 10^{-2}$
	Animal fat‡	4×10^{-2}
Poor conductors	Animal muscle ($\perp$ to fiber)‡	0.08
	Animal, body (ave)‡	0.2
	Animal muscle ($\parallel$ to fiber)‡	0.35
	Animal blood	0.7
	Germanium (semiconductor)	~ 2
Conductors	Seawater	~ 4
	Ferrite	10^2
	Tellurium	$\sim 5 \times 10^2$
Conductors	Silicon	10^3
	Carbon	$\sim 3 \times 10^4$
	Graphite	$\sim 10^5$
	Cast iron	$\sim 10^6$
	Mercury	10^6
	Nichrome	10^6
	Stainless steel	10^6
	Constantan	2×10^6
	Silicon steel	2×10^6
	German silver	3×10^6
	Lead	5×10^6
	Tin	9×10^6
	Phosphor bronze	10^7
	Brass	1×10^7
	Zinc	1.7×10^7
	Tungsten	1.8×10^7
	Duralumin	3×10^7
	Aluminum, hard-drawn	3.5×10^7
	Gold	4.1×10^7
	Copper	5.7×10^7
	Silver	6.1×10^7
Superconductors	Hg (at $<$4.1 K)	∞
	Nb (at $<$9.2 K)	∞
	Nb_3(Al-Ge) (at $<$21 K)	∞
	$YBa_2Cu_3O_7$ (at $<$80 K)	∞

†DC or low frequencies. At 20°C except where noted.
‡Typical of humans.

range in conductivities from quartz to silver is 25 orders of magnitude (10^{25}), and then the range goes to infinity for superconductors.

Example 2-19. Electron velocity. If 10 V is applied between the ends of a 1-mm-diameter copper wire 100 m long, find (*a*) current, (*b*) field, and (*c*) electron velocity. $T = 290$ K. Take electron mobility $\mu_e = 0.004$ m^2 V^{-1} s^{-1} and charge density $\rho = 1.4 \times 10^{10}$ C m^{-3}.

Solution. From (10) and Table 2-1, the resistance of the wire is given by

$$R = \frac{w}{\sigma a} = \frac{100}{5.7 \times 10^7 \times (\pi/4) \times 10^{-6}} = 2.2\ \Omega$$

Therefore, from Ohm's law, the current

$$I = \frac{V}{R} = \frac{10}{2.2} = 4.5\text{ A} \quad \textit{Ans.} \quad (a)$$

and field

$$E = \frac{V}{L} = \frac{10}{100} = 0.1\text{ V m}^{-1} \quad \textit{Ans.} \quad (b)$$

and from (3) and Table 2-1, the electron velocity

$$v_d = E\mu_e = 0.1 \times 0.004\text{ m s}^{-1} = 0.4\text{ mm s}^{-1} \quad \textit{Ans.} \quad (c)$$

Alternatively,

$$v_d = \frac{J}{\rho} = \frac{\sigma E}{\rho} = \frac{5.7 \times 10^7 \times 0.1}{1.4 \times 10^{10}} = 4 \times 10^{-4}\text{ m s}^{-1} = 0.4\text{ mm s}^{-1} \quad \textit{Ans.} \quad (c)$$

Although the electron's (average) velocity is only a fraction of a millimeter per second, a charge (or voltage change) may propagate along the wire at close to the speed of light. The situation is analogous to a sound impulse in air which propagates at 300 m s^{-1} while the air molecules move only a small amount.

Problem 2-11-3. Electron excursion distance. If the voltage applied in Example 2-19 is 10 V rms at 60 Hz, find the electron excursion distance (back and forth) on the wire. *Ans.* 1.1×10^{-6}m (rms), or about 100 atomic spaces.

Conductor Cells

Assuming uniform current flow through the block of Fig. 2-31*a* and dividing it into square conductor cells as in (*b*), the conductance per unit depth (*d*) is equal to the conductivity of the block material. Thus,

$$G/d = \frac{\sigma h}{w} = \sigma \tag{20}$$

or

$$G/d = \sigma \quad (\mho\text{ m}^{-1})$$

The potential difference V (cell) across each cell is the same and the current through each cell is the same. Thus,

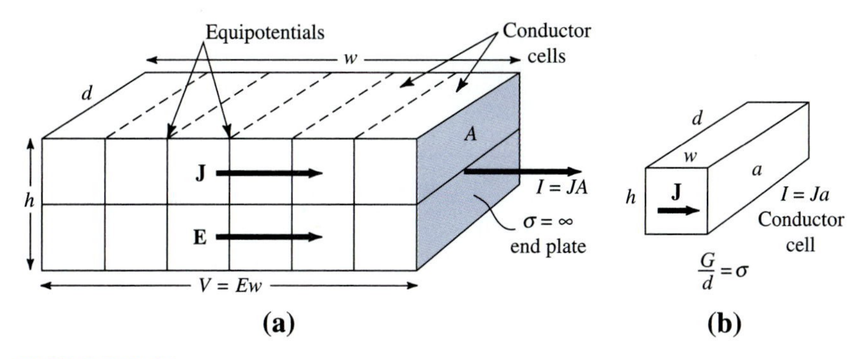

FIGURE 2-31
(*a*) Conducting block with uniform current flow divided into conductor cells. (*b*) Single conductor cell with conductance per unit length equal to the conductivity of the medium ($G/d = \sigma$).

$$V\,(\text{cell}) = IR = JaR = \frac{Jaw}{\sigma dh} = \frac{Jaw}{\sigma a} = \frac{Jw}{\sigma} = \frac{\sigma Ew}{\sigma} = Ew \quad (\text{V}) \tag{21}$$

$$I\,(\text{cell}) = Ja \quad (\text{A})$$

Each cell ($w = h$) has a resistance

$$R\,(\text{cell}) = \frac{w}{\sigma hd} = \frac{1}{\sigma d} \quad (\Omega) \tag{22}$$

Under steady conditions, as much current must flow into as away from a junction point (Kirchhoff's current law) and the integral of the current density **J** over a closed surface must be zero. Thus,

$$\oint_s \mathbf{J} \cdot d\mathbf{s} = 0$$

or, in differential form, the divergence of **J** is zero. Thus,

$$\nabla \cdot \mathbf{J} = 0 \quad \textit{Kirchhoff's current law} \tag{23}$$

To summarize:

$V = IR$	Ohm's law
$\mathbf{J} = \sigma\mathbf{E}$	Ohm's law at a point
$\Sigma V = I\Sigma R$	Kirchhoff's voltage law
$\nabla \cdot \mathbf{J} = 0$	Kirchhoff's current law
$\mathbf{D} = \varepsilon\mathbf{E}$	Electric flux density
$\nabla \cdot \mathbf{D} = \rho$	Gauss's law at a point
$C/l = \varepsilon$	Capacitance per unit length of a capacitor cell
$G/l = \sigma$	Conductance per unit length of a conductor cell

---

Example 2-20. Conducting block with and without saw cuts. See Fig. 2-32. (*a*) Find the resistance of the block without the saw cuts. (*b*) Find the resistance with saw cuts.

Solution

$$(a)\ R = \frac{1}{G} = \frac{w}{\sigma h l} = \frac{1}{\sigma(0.4)(0.6)} = \frac{4.17}{\sigma}\ \Omega \quad Ans.$$

$$(b)\ R = R(\text{cell})\frac{N_s}{N_p}$$

where N_s = number of cells in series
N_p = number of cells in parallel

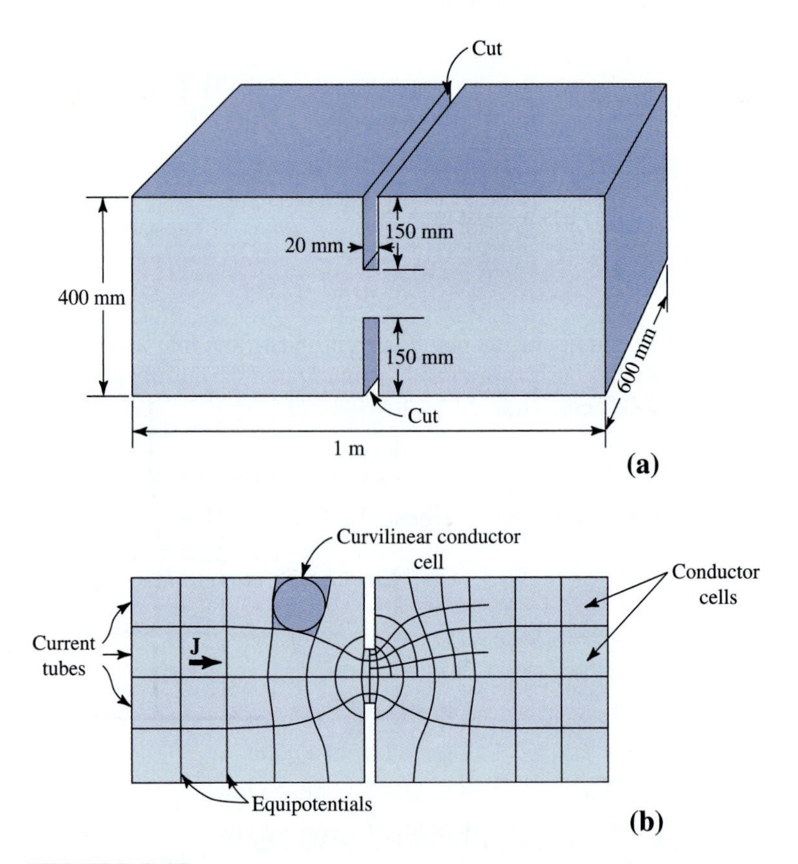

FIGURE 2-32
(*a*) Conducting bar with saw cuts. (*b*) Current map divided into conductor cells. Resistance of bar equals ratio of cells in series to cells in parallel multiplied by the resistance of each cell.

Thus,

$$R = \left(\frac{1}{\sigma l}\right)\frac{N_s}{N_p} = \left(\frac{1}{\sigma l}\right)\frac{13}{4} = \frac{13}{2.4\sigma} = \frac{5.42}{\sigma}\ \Omega \quad Ans.$$

This is 1.30 or 30 percent more than without the cuts.

Problem 2-11-4. Resistance three ways. A block with $\sigma = 3$ ℧/m has the field map and dimensions shown in Fig. P2-11-4. Find the resistance of the block for current flow in three directions: (*a*) end to end (left to right); (*b*) top to bottom; and (*c*) front to back. *Ans.* (*a*) 12.8 Ω; (*b*) 1.79 Ω; (*c*) 2.65 Ω.

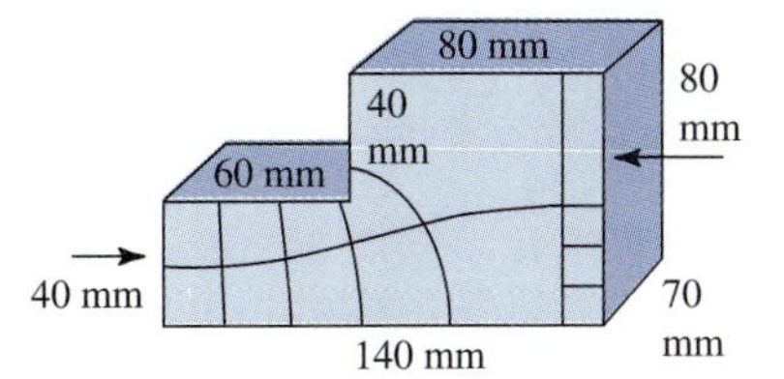

FIGURE P2-11-4
Block with step for Problem 2-11-4.

Problem 2-11-5. Block resistance with uneven current flow. Find the resistance of the block of Fig. P2-11-5 which is like the one in Fig. 2-32 but sawed in half and the two halves reassembled with the ends turned around and joined back-to-back. Contact is made with perfectly conducting strips attached to the center part of each end as shown. *Ans.* $5.42/\sigma$ Ω.

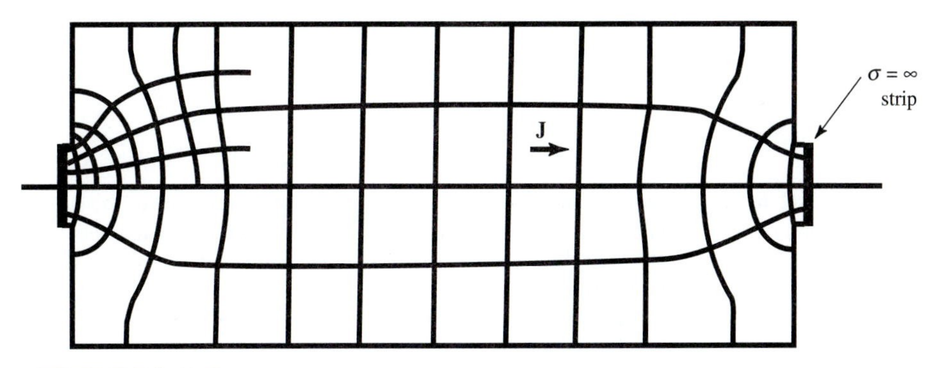

FIGURE P2-11-5

Boundary Conditions; Conducting Media

At the boundary of two conductors of different conductivities (σ_1 and σ_2), the normal components of the current density are continuous. Thus (see Fig. 2-33),

$$J_{n1} = J_{n2} \tag{24}$$

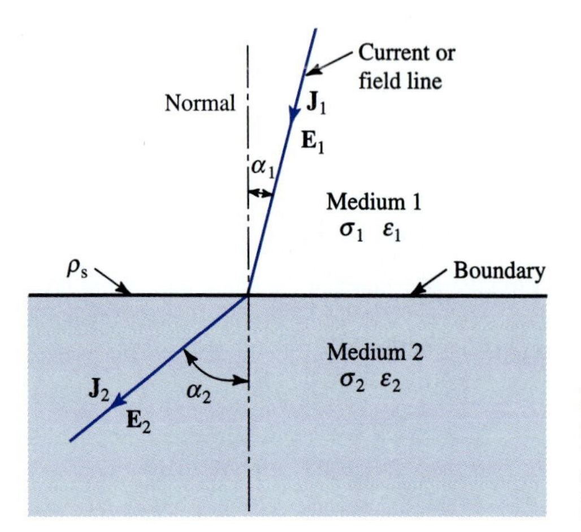

FIGURE 2-33
Boundary between two different conducting media showing change in direction of current or field line.

We also have

$$E_{t1} = E_{t2} \tag{25}$$

or

$$\frac{J_{t1}}{\sigma_1} = \frac{J_{t2}}{\sigma_2} \tag{26}$$

Dividing (26) by (24) gives

$$\frac{J_{t1}}{\sigma_1 J_{n1}} = \frac{J_{t2}}{\sigma_2 J_{n2}}$$

or

$$\frac{\tan\alpha_1}{\tan\alpha_2} = \frac{\sigma_1}{\sigma_2} \tag{27}$$

where α_1 and α_2 are shown in Fig. 2-33.

Problem 2-11-6. Carbon-copper boundary. Referring to Fig. 2-33, let medium 1 be carbon and medium 2 copper. If α_1 for J in the carbon is 10°, find α_2 for J in copper. σ_1 (carbon) $= 3\times10^4$ ℧ m^{-1}, σ_1 (copper) $= 5.7\times10^7$ ℧ m^{-1}. *Ans.* $\alpha_2 = 89.8°$, almost parallel to the interface.

Potential and emf

In the simple series circuit of Fig. 2-34*a*, the potential decreases across the resistor but increases across the battery as in Fig. 2-34*b*. The battery has an *electromotive force* (emf) voltage given by

$$\mathcal{V}_b = \int_c^d \mathbf{E}_b \cdot d\mathbf{L} \qquad \text{(V)}$$

where $\mathbf{E}_b$ = electric field in battery due to chemical action, V m^{-1}.

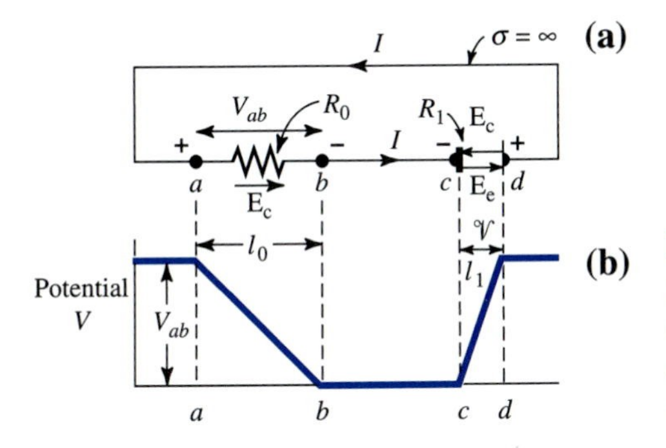

FIGURE 2-34
(*a*) Series circuit of battery and external resistance. (*b*) Graph showing variation of potential around circuit.

The symbol 𝒱 (script *V*) is used for *emf* to distinguish it from the scalar *potential V*. Both are expressed in volts so either may be referred to as a *voltage* if no distinction is made between potential and emf.

As discussed in Sec. 2-12, emf voltages are also produced by a conductor moving across a magnetic field, as in an electric generator.

2-12 MAGNETIC FIELDS OF ELECTRIC CURRENTS

A wire with a current *I* is surrounded by a region in which forces act on a magnetic or compass needle as suggested in Fig. 2-35*a*. Exploring the field with a compass, we find that the magnetic field **H** forms closed loops around the wire as in Fig. 2-35*b*. The direction of the field is given by the right-hand rule (Fig. 2-36).

Referring to Fig. 2-37, the magnetic field $d\mathbf{H}$ from a short section $d\mathbf{L}$ of a current-carrying wire is given by the Biot-Savart law as

$$d\mathbf{H} = \frac{I\,d\mathbf{L}\sin\theta}{4\pi r^2} \qquad (\mathrm{A\ m^{-1}}) \qquad \textit{Biot-Savart law} \tag{1}$$

Using the Biot-Savart law, the field of a long wire and of a loop are derived in the following two examples.

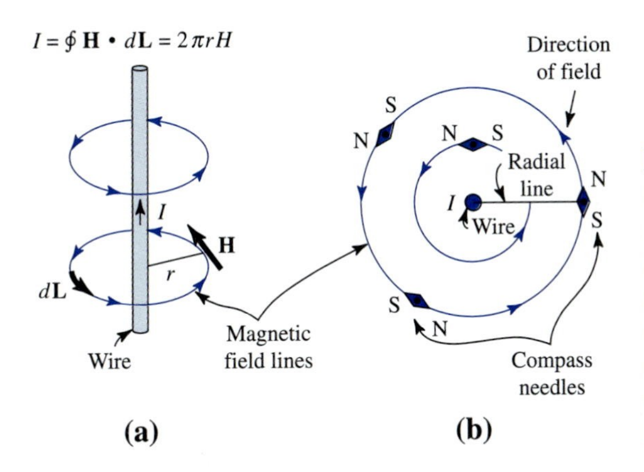

FIGURE 2-35
(*a*) Magnetic field **H** around a current-carrying wire.
(*b*) Cross section perpendicular to the wire with the current flowing out of the page. The magnetic field aligns the compass needles parallel to the field.

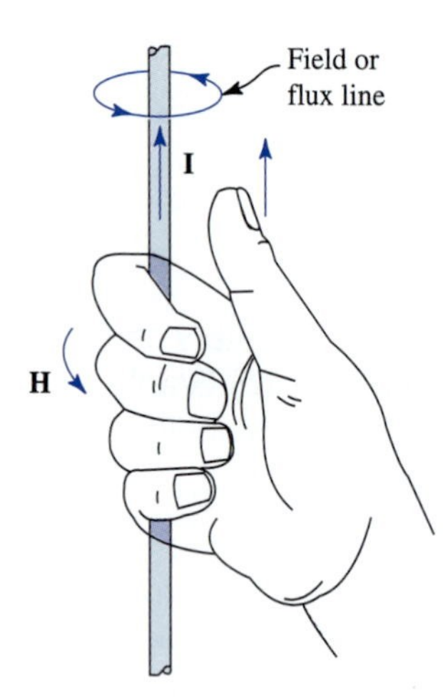

FIGURE 2-36
Right-hand rule relating direction of magnetic field **H** (fingers) to direction of current **I** (thumb).

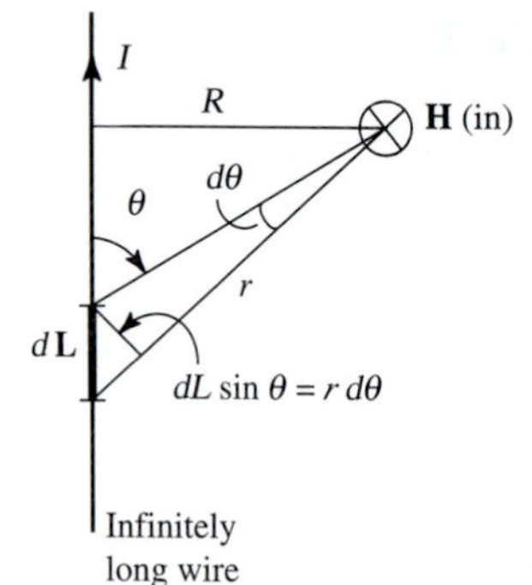

FIGURE 2-37
Construction for finding **H** near a long straight wire. The symbol ⊗ (tail of an arrow) indicates a direction into the page. The symbol ⊙ (head of an arrow) indicates a direction out of the page.

Example 2-21. Magnetic field of a long wire. Find the magnetic field H as a function of distance from a long current-carrying wire.

Solution. Referring to Fig. 2-37, $dL\sin\theta = r\,d\theta$ and $R = r\sin\theta$. Assuming the wire is infinitely long, we have from (1),

$$H = \frac{I}{4\pi R}\int_0^{\pi}\sin\theta\,d\theta = \frac{I}{2\pi R} \quad \text{A m}^{-1} \quad \textit{Ans.} \tag{2}$$

Note that the units for the magnetic field H are amperes per meter (A m^{-1}). Recall that the units for the electric field E are volts per meter (V m^{-1}).

Example 2-21.1. Magnetic field of loop. Find the magnetic field H for three cases: (*a*) at the center of the loop, (*b*) as a function of distance along the axis of the loop, and (*c*) at a large distance from the loop.

Solution

(*a*) Referring to Fig. 2-38, $\theta = 90°$, so from (1)

$$H = \frac{I}{4\pi R}\int_0^{2\pi}d\phi = \frac{I}{2R} \quad \text{A m}^{-1} \quad \textit{Ans.}\ (a) \tag{3}$$

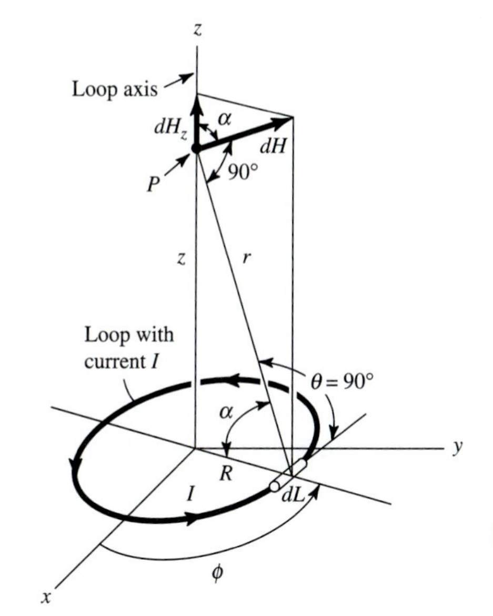

FIGURE 2-38
Construction for finding H on the axis of a loop (z direction).

(*b*) At point P on the z-axis, $\theta = 90°$, $dH_z = dH\cos\alpha = dH\,R/r$, $r = \sqrt{R^2 + z^2}$, and $dL = R\,d\phi$, so from (1),

$$H_z = \frac{IR^2}{4\pi(R^2 + z^2)^{3/2}} \int_0^{2\pi} d\phi = \frac{IR^2}{2(R^2 + z^2)^{3/2}} \quad \textit{Ans.} \quad (b) \tag{4}$$

At the center of the loop $z = 0$, and $H_z = I/2R$ as in (*a*).
(*c*) At large distances from the loop, $z \gg R$ and

$$H_z = \frac{IR^2}{2z^3} \quad \textit{Ans.} \quad (c) \tag{5}$$

Thus, the magnetic field decreases inversely as the cube of the axial distance z.

Problem 2-12-1. H from wire. A long straight wire carries a current $I = 10$ A. At what distance is the magnetic field $H = 1\ \mathrm{A\ m^{-1}}$? *Ans.* 1.59 m.

Problem 2-12-2. H from loop. (*a*) How much current must flow in a loop of radius 0.5 m to produce a magnetic field $H = 1$ mA/m? (*b*) What is the magnetic field H at a distance of 2 m from the loop along its axis? *Ans.* (*a*) 1 mA, (*b*) 14 μA/m.

For a long wire, we have from (2)

$$I = 2\pi r H \qquad \text{(A)} \tag{6}$$

as shown in Fig. 2-35.

More generally, I is the line integral of **H** around any path enclosing the wire, or

$$I = \oint \mathbf{H} \cdot d\mathbf{L} \qquad \text{(A)} \tag{7}$$

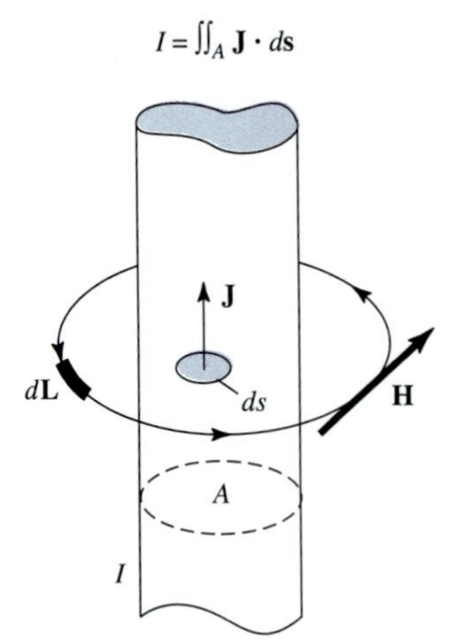

FIGURE 2-39
Magnetic field **H** around a current-carrying conductor of area *A* and current density **J** (A m^{-2}).

For a conductor of cross-section A (Fig. 2-39) and uniform current density J (amperes per square meter),

$$I = JA \qquad \text{(A)}$$

For the more general case, where the current density may be nonuniform, we have

$$I = \oint \mathbf{H} \cdot d\mathbf{L} = \int_s \mathbf{J} \cdot d\mathbf{s} \qquad \text{(A)} \qquad \textit{Ampere's law} \tag{8}$$

where $d\mathbf{s}$ = surface element, m^2
$d\mathbf{L}$ = line element, m
$\oint$ = line integral around a closed path
$\int_s$ = integral over surface enclosed by path
$\mathbf{J}$ = current density, A m^{-2}

Equation (8) is Ampere's law.

Example 2-22. Magnetic field addition for two-wire transmission line of Fig. 2-40*a*. Find the total field H at point P_1 if $d = 80$ cm, $\alpha = 30°$, and $|I_1| = |I_2| = 100$ A.

Solution

$$H = 2H_1 \cos\alpha = \frac{2I\cos^2\alpha}{\pi d} = 59.7 \text{ A m}^{-1} \quad \textit{Ans.}$$

Problem 2-12-3. H between wires. Find the magnetic field H at point P_2 halfway between the wires of Fig. 2-40*b* with d, I_1, and I_2 as in Example 2-22. *Ans.* 79.6 A/m.

Problem 2-12-4. H between strips. If $I = 100$ A, $w = 10$ cm and $h = 1$ cm for the strip line of Figs. 2-40*c* and 2-41, what is the magnetic field H at point P_3 between the strips? *Ans.* 1000 A/m.

If the wires of Fig. 2-40*a* and *b* are replaced by wide conducting strips, the field appears as in Fig. 2-40*c*. Whereas the magnetic field **H** is nonuniform between the wires, it is *essentially uniform between the two strips* with **H** a constant between them.

Neglecting edge effects or fringing,

$$I = Hw \qquad \text{(A)}$$

where w = strip width (m)

or

$$H = \frac{I}{w} \qquad (\text{A m}^{-1})$$

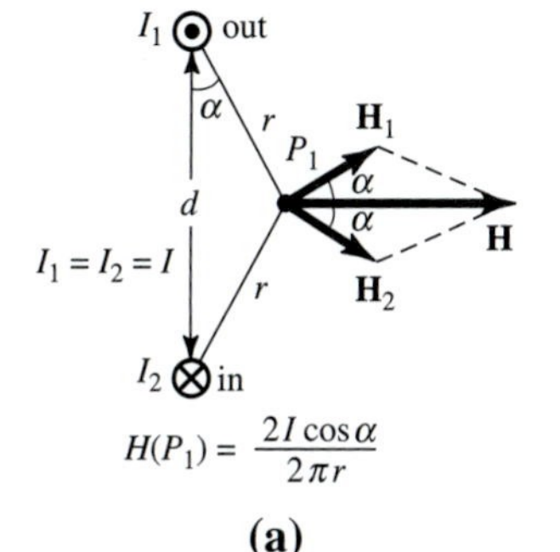

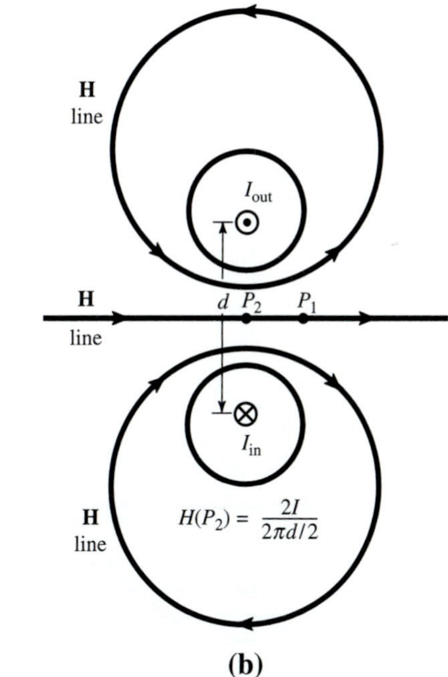

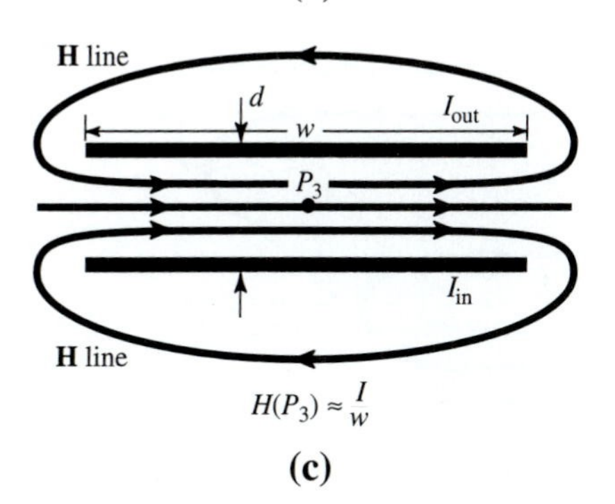

FIGURE 2-40
Magnetic field **H** from two-wire transmission line (*a* and *b*) and from two-strip line (*c*) running perpendicular to page. Vector addition of field from each wire is shown in (*a*). Resultant **H** field lines around wires are shown in (*b*).

Magnetic Flux ψ_m and Magnetic Flux Density B; Gauss's Law

The magnetic flux through a surface area is the integral of the normal component of the magnetic field times μ over the area. Thus,

$$\psi_m = \iint \mu \mathbf{H} \cdot d\mathbf{s} \qquad \text{(webers, Wb)} \tag{8.1}$$

where μ = permeability of medium, henrys/meter (H m^{-1}).

In the (assumed) uniform field **H** between the two strips of the transmission line of Fig. 2-41, the integral reduces to a simple scalar product. Thus the magnetic flux over the area A (= hl) is

$$\psi_m = \mu H h l = \mu H A \qquad \text{(Wb)}$$

Dividing the magnetic flux by the area A gives the *magnetic flux density B* or flux per unit area. Thus,

$$B = \frac{\psi_m}{A} = \mu H \qquad (\text{Wb m}^{-2} \text{ or teslas, T}) \tag{8.2}$$

In general, the magnetic flux through any surface is given by the surface integral of **B** over the surface, or

$$\psi_m = \int_s \mathbf{B} \cdot d\mathbf{s} \qquad \text{(Wb)} \tag{9}$$

The magnetic flux density **B** has the same direction as **H** in isotropic media with a magnitude $\mu\mathbf{H}$, or

$$\mathbf{B} = \mu \mathbf{H} = \mu_r \mu_0 \mathbf{H} \qquad (\text{Wb m}^{-2} \text{ or T}) \tag{10}$$

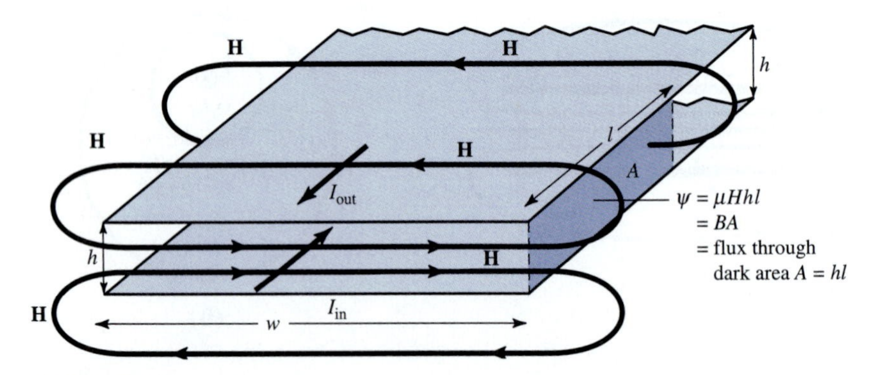

FIGURE 2-41
Twin-strip transmission line with magnetic field H (= I/w) essentially uniform between the strips. This is a three-dimensional view of the strip line shown in cross section in Fig. 2-40*c*.

where $\mathbf{B}$ = magnetic flux density, Wb m^{-2}
$\mathbf{H}$ = magnetic field, A m^{-1}
μ = permeability of medium, H m^{-1}
μ_0 = permeability of air or vacuum = $4\pi 10^{-7}$ H m^{-1}
$\mu_r = \mu/\mu_0$ = relative permeability (= 1 for air)

Since magnetic field lines are closed loops, it follows that as many lines leave as enter a volume or that the integral of **B** over a *closed surface* is zero. Thus,

$$\oint_s \mathbf{B} \cdot d\mathbf{s} = 0 \qquad \textbf{\textit{Gauss's law for magnetic fields}} \qquad (11)$$

Example 2-23. Magnetic flux between wires. Find the magnetic flux ψ_m between the wires of Fig. 2-40*b* for a 5-m length of line. $d = 0.8$ m, $I_1 = I_2 = 100$ A, wire radius $r = 10$ mm, $\mu_r = 1$.

Solution

$$\psi_m = 5\int_{0.01}^{0.79} 2\mu_0 \mathbf{H} \cdot d\mathbf{r} \qquad \text{where } H = \frac{I}{2\pi r}$$

so

$$\psi_m = \frac{5 \times 2\mu_0 I}{2\pi}\int_{0.01}^{0.79}\frac{dr}{r} = \frac{5 \times 2 \times 4\pi 10^{-7} 100 \times 4.37}{2\pi}$$

$$= 8.74 \times 10^{-4} \text{ Wb} = 874\ \mu\text{Wb} \quad \textit{Ans.}$$

Problem 2-12-5. Magnetic flux between strips. What is the magnetic flux between the strips of Fig. 2-40*c*? $I = 110$ A, $w = 0.1$ m, $h = 10$ mm. *Ans.* 62.8 μWb.

Lorentz Force or Motor Equation

A wire perpendicular to the page with current flowing inward has a magnetic field as shown in Fig. 2-42*a*. In the presence of a uniform magnetic field of flux density **B** to the right, the field above the wire is reinforced and is weakened below the wire resulting in a downward force on the wire as in Fig. 2-42*b*. This is the *Lorentz* or *motor force* as given (for a uniform field) by

$$F = IBL \qquad \text{(newtons, N)} \qquad (12)$$

where I = current, A
B = flux density, Wb m^{-2} or T
L = length of wire (into page), m

More generally, we have in vector notation,

$$\mathbf{F} = (\mathbf{I} \times \mathbf{B})L \qquad \text{(N)} \qquad \textbf{\textit{Lorentz force or motor equation}} \qquad (13)$$

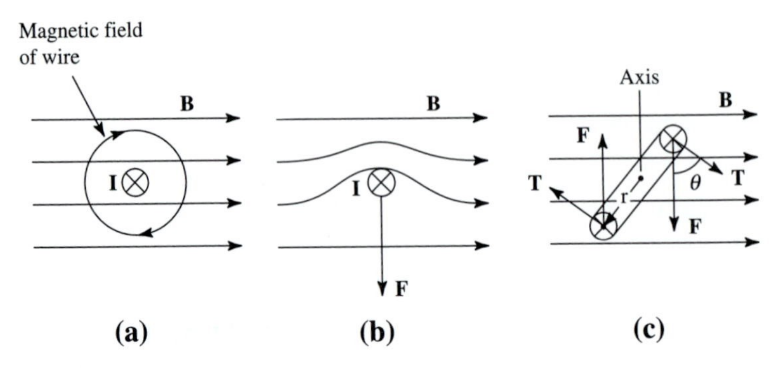

FIGURE 2-42
(*a*) Wire perpendicular to page in uniform magnetic field **B**. With current **I** flowing in, the magnetic field of the wire is clockwise reinforcing the field above and opposing (weakening) it below as suggested in (*b*). This results in a downward force **F** as shown. (*c*) Rotor which experiences turning torque **T**.

Equation (13) is the *Lorentz force or motor equation.* A conductor with current **I** in a magnetic flux density field **B** experiences a force **F** as shown in Fig. 2-42*b*.

In a motor, two conductors are mounted on a rotor with axis, as in Fig. 2-42*c*. The rotor experiences a turning torque

$$T = 2Fr\cos\theta = 2IBLr\cos\theta \qquad \text{(N m)} \tag{13.1}$$

Example 2-24. Rotary torque. What is the maximum torque of a rotor (Fig. 2-42*c*) with $I = 50$ A, $r = 6$ cm, $L = 12$ cm, $B = 0.1$ T?

Solution. From (13.1) $T = 2 \times 50 \times 0.1 \times 0.12 \times 0.06 = 0.072$ N m $= 72$ mN m *Ans.*

Problem 2-12-6. Average torque. What is the average torque on the rotor of Example 2-24? *Ans.* 46 mN m.

Dimensionally (13) is

$$\text{Force} = \text{current} \times \text{magnetic flux density} \times \text{length} \tag{13.2}$$

Since current = charge/time, we have, on rearranging (13.2) and multiplying both sides by a distance,

$$\frac{\text{Force}}{\text{charge}} \times \text{distance} = \frac{\text{distance}}{\text{time}} \times \text{magnetic flux density} \times \text{length} \tag{13.3}$$

or

$$\text{Electric field} \times \text{distance} = \text{velocity} \times \text{magnetic flux density} \times \text{length}$$

These are the dimensions of (14), which is a transformation of (13) to a *generator equation.*

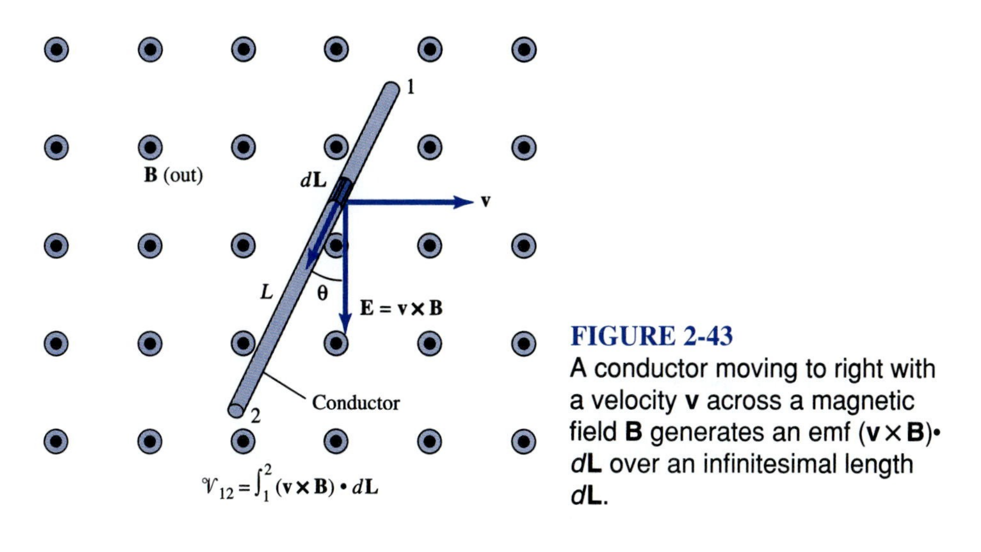

FIGURE 2-43
A conductor moving to right with a velocity **v** across a magnetic field **B** generates an emf $(\mathbf{v} \times \mathbf{B}) \cdot d\mathbf{L}$ over an infinitesimal length $d\mathbf{L}$.

$$\mathcal{V} = \int_0^L \mathbf{E} \cdot d\mathbf{L} = \int_0^L (\mathbf{v} \times \mathbf{B}) \cdot d\mathbf{L} \quad \text{(V)} \qquad \textit{Generator equation} \tag{14}$$

where, referring to Fig. 2-43,

$\mathcal{V}$ = emf between points 1 and 2 on conductor, m
$\mathbf{v}$ = velocity of conductor, m s^{-1}
$\mathbf{B}$ = magnetic flux density, Wb m^{-2} or T
$\mathbf{L}$ = conductor length, m

Thus, moving a conductor through a magnetic flux density **B** with velocity **v** produces an electric field along the conductor. Integrating **E** over the length **L** gives the voltage generated over that length.

Example 2-25. Generator voltage. Referring to Fig. 2-43, $L = 0.4$ m, $v = 3$ m s^{-1}, $B = 2$ mT, $\theta = 20°$. Find V.

Solution. $V = |\mathbf{v} \times \mathbf{B}| \cos\theta\, L = 3 \times 2 \times 10^{-3} \times 0.940 \times 0.4 = 2.26 \times 10^{-3}$ $= 2.26$ mV *Ans.*

Problem 2-12-7. Conductor angle. Referring to Fig. 2-43, $V = 13.5$ V, $L = 15$ m, $v = 1.5$ m s^{-1}, $B = 1$ T. Find θ. *Ans.* 53°.

Motors and generators are discussed in more detail in Chap. 6.

Inductance, Inductors, Energy, and Energy Density

Whereas capacitors are useful for storing electrical energy, solenoids or inductors are useful for storing magnetic energy.

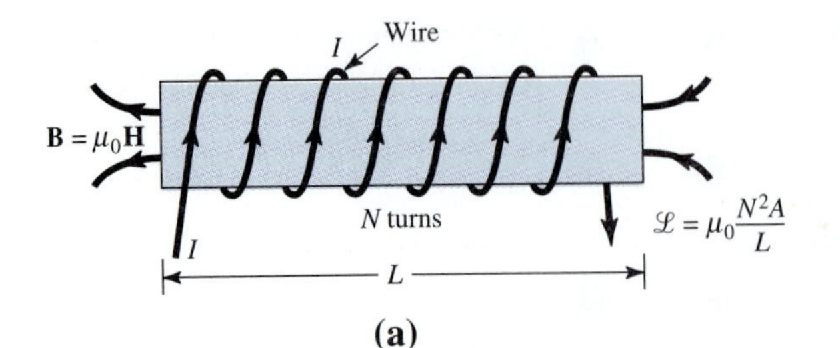

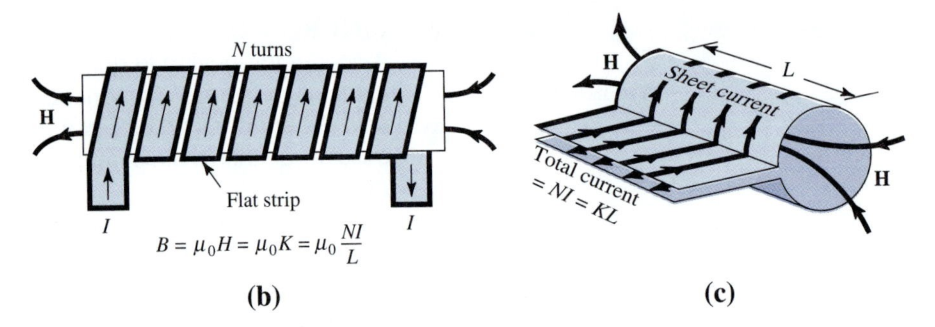

FIGURE 2-44
(*a*) Solenoid inductor of wire. (*b*) Solenoid of flat strips is equivalent to a full sheet conductor, as in (*c*), with total current *NI*.

The magnetic field H inside the coil of wire or solenoid of Fig. 2-44*a* is given by

$$H = \frac{NI}{L} \qquad (\text{A m}^{-1}) \tag{15}$$

where N = number of turns, dimensionless
I = solenoid current, A
L = length of solenoid, m

If the turns are flat strips close together as in Fig. 2-44*b*, the solenoid becomes equivalent to a continuous current sheet, as in Fig. 2-44*c*, with sheet current density,

$$K = \frac{NI}{L} \qquad (\text{A m}^{-1}) \tag{16}$$

or

$$NI = KL \qquad (\text{A}) \tag{17}$$

where K = sheet current density, A m^{-1}. Thus, inside the solenoid,

$$H = K \qquad (\text{A m}^{-1}) \tag{18}$$

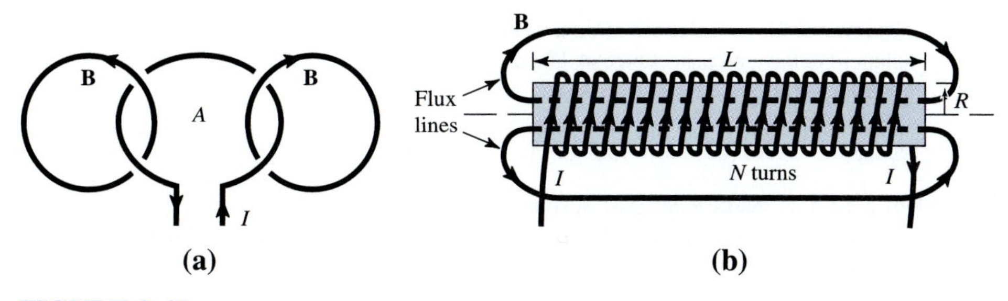

FIGURE 2-45
(a) The inductance $\mathcal{L}$ of a single-turn loop is equal to the ratio of the magnetic flux ψ_m ($= \iint \mathbf{B} \cdot d\mathbf{s}$) through the loop area A to the current I or $\mathcal{L} = \psi_m/I$.
(b) Inductance $\mathcal{L}$ of long solenoid of N turns is equal to the ratio of the magnetic flux linkage BAN to the current I or $\mathcal{L} = BAN/I$.

and the magnetic flux density (for air) is given by

$$B = \mu_0 H = \mu_0 K = \mu_0 \frac{NI}{L} \qquad (\text{Wb m}^{-2} \text{ or T}) \qquad (19)$$

The *inductance* $\mathcal{L}$ of the single-turn loop inductor of Fig. 2-45a is equal to the ratio of the magnetic flux ψ_m through the loop (or flux linkage Λ) to the loop current I, or†

$$\mathcal{L} = \frac{\psi_m}{I} = \frac{\iint \mathbf{B} \cdot d\mathbf{s}}{I} \qquad (\text{henries, H}) \qquad (20)$$

For the solenoid inductor of N turns, of Fig. 2-45b, the inductance is N times the magnetic flux/current ratio or,

$$\mathcal{L} = \frac{\psi_m N}{I} = \frac{BAN}{I} = \frac{\Lambda}{I} \qquad (\text{H}) \qquad (21)$$

where ψ_m = flux linkage of one turn, Wb m^{-2} or T
$\Lambda = BAN$ = total flux linkage of solenoid, Wb m^{-2} or T

We also have from (19) that

$$\mathcal{L} = \frac{BAN}{I} = \frac{\mu NIAN}{LI} = \frac{\mu N^2 A}{L} \qquad (22)$$

Thus, the inductance of the solenoid is determined by its geometry, that is, its area A ($= \pi R^2$), length L, the number of turns N and the medium μ. It is assumed that the length of the solenoid is much greater than its diameter.

†The analogous electric field relation is that the capacitance of a capacitor is equal to the ratio of its electric charge Q to the applied voltage V, or $C = Q/V$.

Example 2-25.1. Solenoid turns. How many turns are required for a 30-cm-long solenoid to have an inductance of 10 mH? The solenoid diameter is 4 cm. The medium is air ($\mu_r = 1$).

Solution. From (22) $N^2 = \mathcal{L}L/\mu_0 A$; $N = 1380$ *Ans.*

Problem 2-12-8. H inside solenoid. A solenoid has dimensions $L = 1.2$ m, $N = 750$ turns, diameter $= 10$ cm, and current $I = 1.75$ A. $\mu_r = 5$. Find the field inside the solenoid. *Ans.* 1.09 kA/m.

Inductor Energy and Energy Density

An inductor stores an incremental energy

$$dW = VI\,dt \qquad \text{(J)} \tag{23}$$

where V = voltage across inductor, V; and I = current through inductor, A.

From circuit theory,

$$V = \mathcal{L}\frac{dI}{dt} \qquad \text{(V)} \tag{24}$$

so the total energy

$$W = \mathcal{L}\int_0^I I\,dI = \frac{1}{2}\mathcal{L}I^2 = \frac{1}{2}\Lambda I = \frac{1}{2}\frac{\Lambda^2}{\mathcal{L}}$$

$$= \frac{1}{2}BANI = \frac{1}{2}BAKL = \frac{1}{2}BHAL \qquad \text{(J)} \tag{25}$$

where AL = volume of long solenoid, m^3. Dividing by this volume, we find the energy density (assumed uniform) in the solenoid is

$$w = \frac{1}{2}\mu H^2 \qquad (\text{J m}^{-3}) \tag{25.1}$$

Inductor Cells

Figure 2-46*a* shows a twin-strip transmission line divided into six inductor cells. From (20) the inductance of the twin-strip line is

$$\mathcal{L} = \frac{\psi_m}{I} \tag{26}$$

where $\psi_m = BA = Bhl = \mu Hhl$

$I = wH$

$\mu = \mu_r\mu_0$ = permeability of the medium

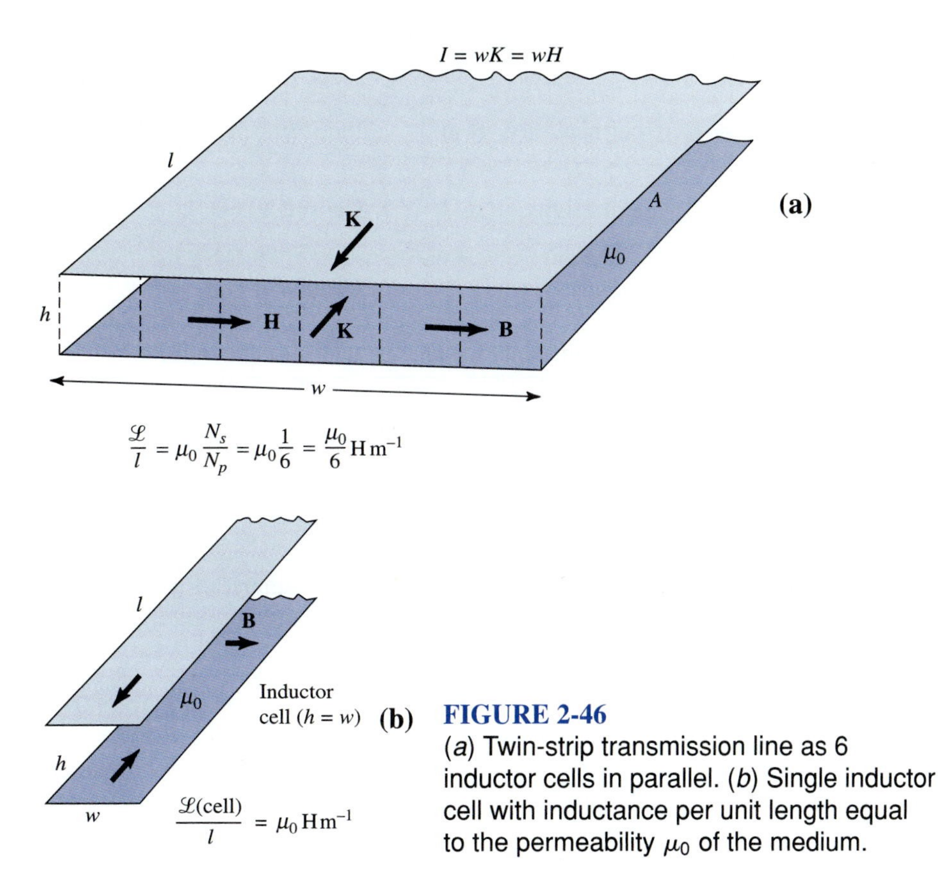

FIGURE 2-46
(*a*) Twin-strip transmission line as 6 inductor cells in parallel. (*b*) Single inductor cell with inductance per unit length equal to the permeability μ_0 of the medium.

For air ($\mu_r = 1$) the inductance $\mathscr{L}$ (cell) of the square ($w = h$) conductor cell at (*b*) is

$$\mathscr{L} = \frac{\mu_0 H h l}{w H} = \mu_0 l \tag{27}$$

or the inductance per unit length is

$$\frac{\mathscr{L}}{l} = \mu_0 \qquad (\mathrm{H\ m^{-1}}) \tag{28}$$

where $\mu_0 = 4\pi \times 10^{-7}$ H m^{-1}. Thus, the inductance per unit length of the inductor cell is equal to the permeability of the medium.

In summary:

$$\frac{\mathcal{L}}{l} = \mu = \text{permeability, H m}^{-1} \qquad \text{Inductor cell}$$

$$\frac{C}{l} = \varepsilon = \text{permittivity, F m}^{-1} \qquad \text{Capacitor cell}$$

$$\frac{G}{l} = \sigma = \text{conductivity, } \mho \text{ m}^{-1} \qquad \text{Conductor cell}$$

where μ, ε, and σ are the three constants which characterize a medium.

The twin line of Fig. 2-46*a* consists of several inductor cells in parallel. The inductance per unit length of the twin line is therefore

$$\frac{\mathcal{L}}{l} = \mu_0 \frac{N_s}{N_p} \qquad (\text{H m}^{-1}) \tag{29}$$

where N_s = number of inductor cells in series and N_p = number of inductor cells in parallel. Thus, for the twin line of Fig. 2-46*a* the inductance per unit length is

$$\frac{\mathcal{L}}{l} = \mu_0 \frac{N_s}{N_p} = \frac{\mu_0}{6} = (4\pi/6)10^{-7} = 209 \text{ nH m}^{-1}$$

The field map of a microstrip transmission line is shown in Fig. 2-47. Electric field *E* lines extend between the strip and the ground plane. Magnetic field *H* lines are at right angles and form closed loops.

The characteristic impedance of a transmission line from (3-2-11) is

$$Z = \sqrt{\frac{\mathcal{L}}{C}} = \sqrt{\frac{\mathcal{L}/l}{C/l}} \qquad (\Omega) \tag{30}$$

where l = unit length of line, m.

From (29) and (2-9-6), we have

$$Z = \sqrt{\frac{\mathcal{L}/l}{C/l}} = \sqrt{\frac{\mu \dfrac{N_s}{N_p}}{\varepsilon \dfrac{N_p}{N_s}}} = \sqrt{\frac{\mu_r \mu_0}{\varepsilon_r \varepsilon_0}} \frac{N_s}{N_p} = \sqrt{\frac{\mu_r}{\varepsilon_r}} \sqrt{\frac{4\pi \times 10^{-7} \text{ H m}^{-1}}{8.85 \times 10^{-12} \text{ F m}^{-1}}} \frac{N_s}{N_p}$$

$$= \sqrt{\frac{\mu_r}{\varepsilon_r}} 377 \frac{N_s}{N_p} \qquad (\Omega) \tag{31}$$

For air, $\mu_r = \varepsilon_r = 1$ and for one cell $N_s = N_p = 1$, so the impedance of a transmission line cell is

$$Z = 377 \ \Omega \tag{32}$$

and the impedance of the microstrip transmission line of Fig. 2-48 is

$$Z = 377 \frac{N_s}{N_p} \qquad (\Omega) \tag{33}$$

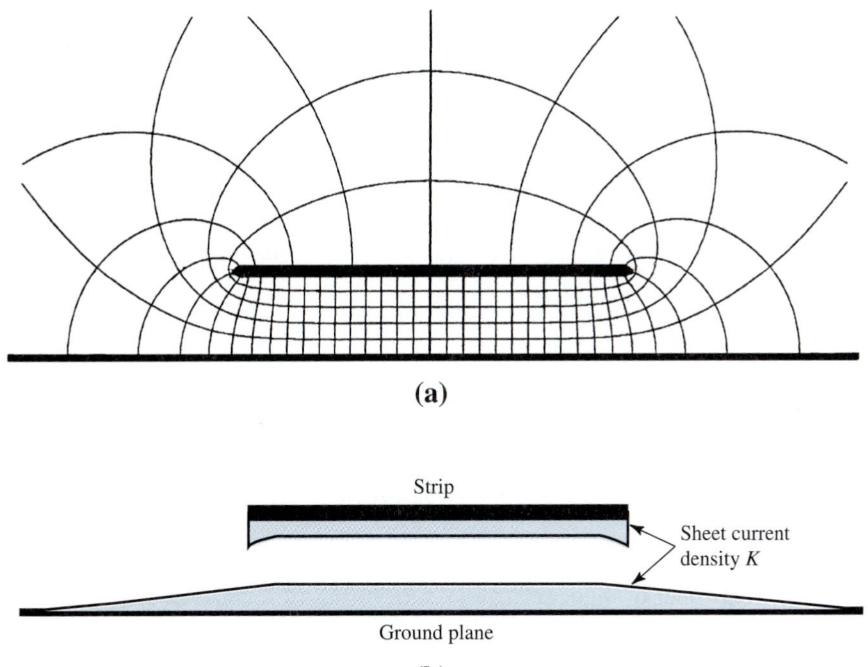

FIGURE 2-47
(a) Cross section of microstrip transmission line with field map in curvilinear squares. These provide information about the line impedance and the field and current densities. Map by Nicolas B. Piller, ETH, Zürich. (b) Shading suggests magnitude of sheet current density K on top of ground plane and under the strip. It tapers off gradually to either side on the ground plane but peaks up at the edges of the strip. K on the strip and ground plane equals the magnitude of $\mathbf{H}$ just adjacent to the conductors. However, $\mathbf{K}$ is perpendicular to the page while $\mathbf{H}$ is parallel to the page.

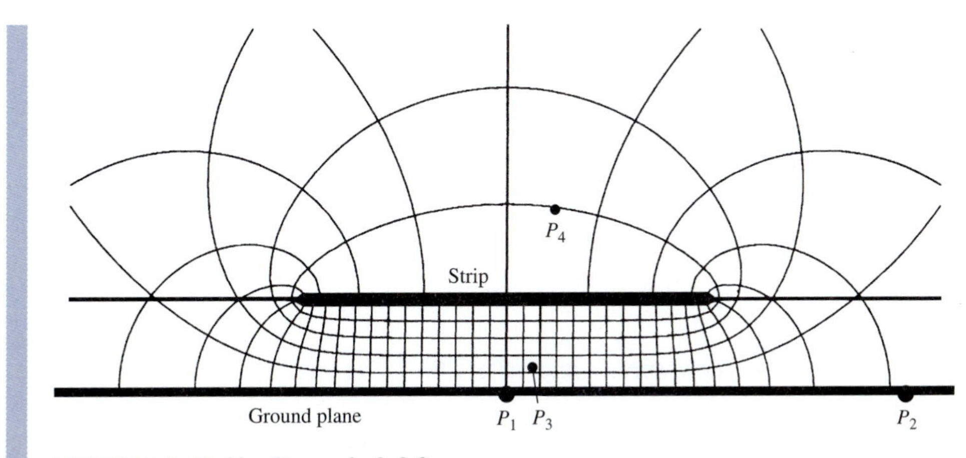

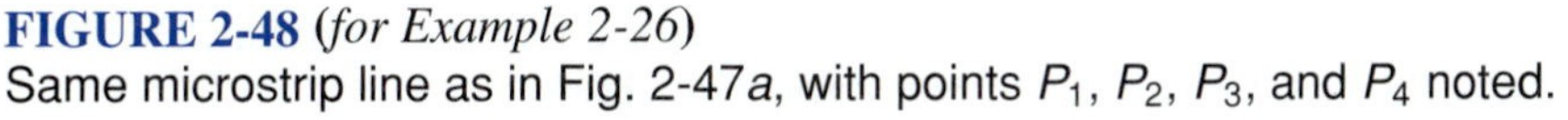

FIGURE 2-48 (*for Example 2-26*)
Same microstrip line as in Fig. 2-47a, with points P_1, P_2, P_3, and P_4 noted.

Example 2-26. Microstrip line parameters. If the voltage between the strip and ground plane of the line in Fig. 2-48 is 15 V, find (*a*) characteristic impedance of line, (*b*) $H = K$ at P_1, and (*c*) E at P_2.

Solution. From (33),

$$Z = 377\frac{N_s}{N_p} = 377\frac{5}{38} = 49.6\ \Omega \quad \textit{Ans.} \quad (a)$$

where N_s = cells in series, N_p = cells in parallel.
E (at P_1) = 15/5 = 3 V m^{-1},

$$H = E/Z = 3/50 = 0.06\ \text{A m}^{-1} = K \quad \textit{Ans.} \quad (b)$$

$$E\ (\text{at } P_2) = 3 \times 0.156 = 0.47\ \text{V m}^{-1} \quad \textit{Ans.} \quad (c)$$

0.156 is the inverse of the ratio of the cell sizes at P_2 and P_1.

Problem 2-12-9. Microstrip line. For the same line and conditions as in Example 2-26, find E at (*a*) P_3 and (*b*) P_4. *Ans.* (*a*) 3 V/m, (*b*) 0.52 V/m.

Thus, E (at P_3) = E (at P_1) while E (at P_2) and E (at P_4) is much less, or about 17 percent as much.

2-13 CHANGING MAGNETIC FIELDS, INDUCTION, AND FARADAY'S LAW

A steady electric current I produces a steady magnetic field **H** as given by Ampere's law. But a steady magnetic field will not produce an electric current. However, *a changing magnetic field* will. Thus, a changing magnetic flux ψ_m through a closed loop, as in Fig. 2-49, produces an emf or voltage $\mathcal{V}$ at the terminals as given by

$$\mathcal{V} = -\frac{d\psi_m}{dt} \qquad \text{(V)} \tag{1}$$

where the voltage is the integral of the electric field **E** around the loop. For a uniform magnetic field $\psi_m = BA$, where A = area of the loop. More generally, we have

$$\mathcal{V} = \oint \mathbf{E} \cdot d\mathbf{L} = -\iint \frac{\partial \mathbf{B}}{\partial t} \cdot d\mathbf{s} \qquad \text{(V)} \qquad \textit{Faraday's law} \tag{2}$$

where $\oint \mathbf{E} \cdot d\mathbf{L}$ = line integral of **E** around loop, V

$\iint \frac{\partial \mathbf{B}}{\partial t} \cdot d\mathbf{s}$ = surface integral of $\partial\mathbf{B}/\partial t$ over loop area A, V

Thus, the changing magnetic field produces a changing electric field **E** which adds up around the loop to a changing voltage at the loop terminals. On closing the terminals, a time-varying current flows in the loop.

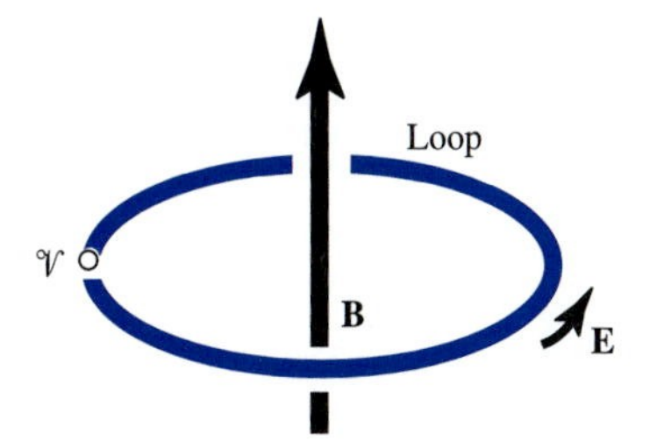

FIGURE 2-49
Open-circuit loop with voltage at its terminals due to a ***change*** in magnetic flux through the loop.

Equation (2) (*Faraday's law*) states that the line integral of the electric field around a stationary loop equals the (negative) surface integral of the time rate of change of the magnetic flux density **B** integrated over the loop area.

The total flux through a circuit is equal to the integral of the normal component of the flux density **B** over the surface bounded by the circuit. That is, the total magnetic flux is given by

$$\psi_m = \iint \mathbf{B} \cdot d\mathbf{s} = \int_S \mathbf{B} \cdot d\mathbf{s} \tag{3}$$

The surface over which the integration is carried out is the surface bounded by the periphery of the circuit, as in Fig. 2-50 shown on next page. Equation (3) applies to a closed single-conductor circuit of any number of turns or loops. It is important to note that *any closed circuit with any number of turns forms the boundary of a single surface* (see Fig. 2-50) and $\iint \mathbf{B} \cdot d\mathbf{s}$ over this surface yields the total flux. Thus integrating **B** over the surface in Fig. 2-50 yields all the flux. Lines of flux passing through the surface only once are integrated once, but those linking all four turns are integrated four times since they pass through the surface four times. Substituting (3) in (1) leads to†

$$\mathcal{V} = -\frac{d}{dt}\int_S \mathbf{B} \cdot d\mathbf{s} \tag{4}$$

where $\mathcal{V}$ = induced emf, V
$\mathbf{B}$ = flux density, T
$d\mathbf{s}$ = surface element, m^2
t = time, s

When the loop or closed circuit is stationary or fixed, (4) reduces to

$$\mathcal{V} = -\int_S \frac{\partial \mathbf{B}}{\partial t} \cdot d\mathbf{s} \tag{5}$$

This form of Faraday's law gives the induced emf due specifically to a time rate of change of **B** for a loop or circuit that is fixed with respect to the observer. This is sometimes called the ***transformer induction equation.***

†Recall from Sec. 1-6 that the symbol $\int_S$ indicates a double or surface integral ($\iint$) over a surface s.

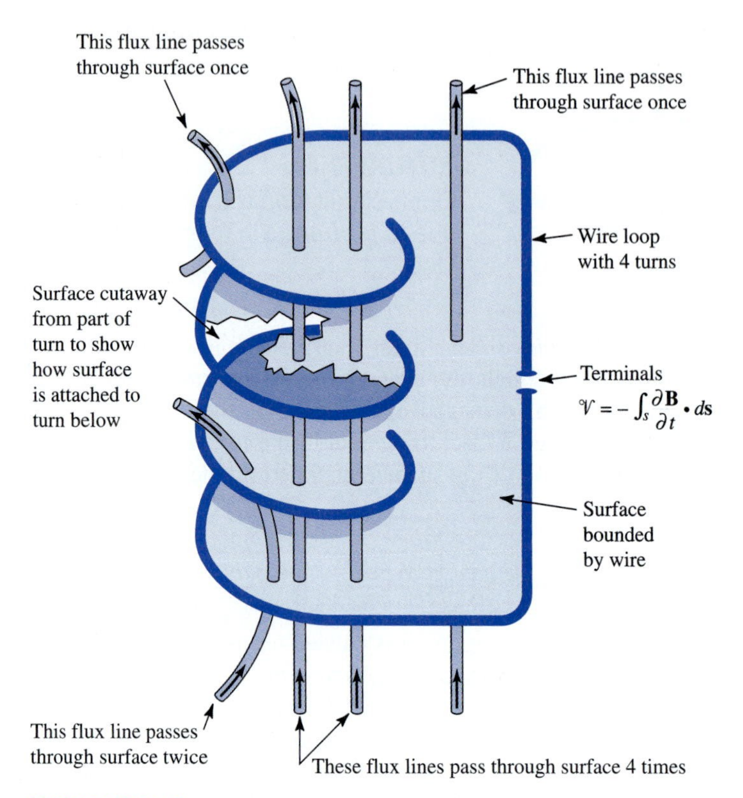

FIGURE 2-50
Circuit with four-turn coil. The wire forms the boundary of a single continuous surface. Part of the surface of one turn has been cut away to show how the surface is bounded by the turn below.

The more general relation is given by (2) as

$$\mathcal{V} = \oint \mathbf{E} \cdot d\mathbf{L} = -\int_s \frac{\partial \mathbf{B}}{\partial t} \cdot d\mathbf{s} \qquad \text{(V)} \qquad \textbf{\textit{Faraday's law}} \qquad (6)$$

According to (2) the total voltage or emf $\mathcal{V}$ inducted in the circuit is equal to the line integral of the electric field around a fixed closed circuit. It is also equal to the normal component of the time rate of decrease of the flux density **B** integrated over a surface bounded by the circuit.

When a wire cuts across a magnetic field **B**, an electric field **E** is generated along the wire. When the wire forms a closed circuit, the total emf induced by the ***motional induction*** is given by the *generator equation* (2-12-14) as

$$\mathcal{V} = \oint \mathbf{E} \cdot d\mathbf{L} = \oint (\mathbf{v} \times \mathbf{B}) \cdot d\mathbf{L} \qquad \text{(V)} \qquad (7)$$

and the integration is completely around the circuit.

When both kinds of changes are occurring simultaneously (**B** changing with time and circuit in motion), the total emf induced is equal to the sum of the emfs as given by

$$\mathcal{V} = \underbrace{\oint_L (\mathbf{v} \times \mathbf{B}) \cdot d\mathbf{L}}_{\text{Motion}} - \underbrace{\int_S \frac{\partial \mathbf{B}}{\partial t} \cdot d\mathbf{s}}_{\text{Time change}} \tag{8}$$

The first term of the right-hand member gives the emf induced by the motion, while the second term gives the emf induced by the time change of **B**. The line integral in the first term is taken around the entire circuit of length L, while the surface integral in the second term is taken over the entire surface S bounded by the circuit.

Equation (8) is a *general relation and gives the correct value of total induced emf in all cases.* For the special case of motion only, $\partial \mathbf{B}/\partial t = 0$, and (8) reduces to

$$\mathcal{V} = \oint (\mathbf{v} \times \mathbf{B}) \cdot d\mathbf{L} \tag{9}$$

For the special case of time change of flux density only, the velocity $\mathbf{v} = 0$ and (8) reduces to

$$\mathcal{V} = -\int_S \frac{\partial \mathbf{B}}{\partial t} \cdot d\mathbf{s} \tag{10}$$

In summary:

(I) $\mathcal{V} = \oint (\mathbf{v} \times \mathbf{B}) \cdot d\mathbf{L} - \int_S \frac{\partial \mathbf{B}}{\partial t} \cdot d\mathbf{s}$ ***General case***

(II) $\mathcal{V} = \oint (\mathbf{v} \times \mathbf{B}) \cdot d\mathbf{L}$ ***Motion only (motional induction)***

(III) $\mathcal{V} = -\int_S \frac{\partial \mathbf{B}}{\partial t} \cdot d\mathbf{s}$ ***B change only (transformer induction)***

2-14 EXAMPLES OF INDUCTION

Motion and Time-Changing Induction

Examples help to make theoretical concepts more meaningful. Here are seven different situations in which we use the above induction relations to determine the total emf inducted in a closed circuit. The general relation (I) gives the correct result in all cases. In some cases (motion only or time change only), (II) or (III) alone is sufficient.

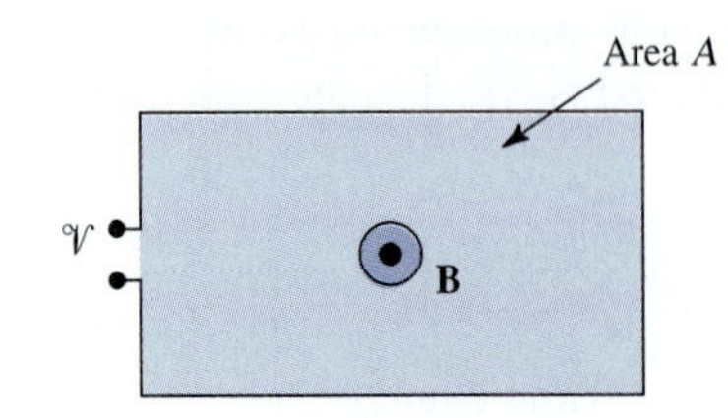

FIGURE 2-51
Fixed loop of area A with **B** changing.

Example 2-27. Loop: B change with no motion. Consider the fixed rectangular loop of area A shown in Fig. 2-51. The flux density **B** is normal to the plane of the loop (outward in Fig. 2-51) and is uniform over the area of the loop. However, the magnitude of **B** varies harmonically with respect to time as given by

$$B = B_0 \cos \omega t \tag{1}$$

where B_0 = maximum amplitude of **B**, T
ω = radian frequency ($= 2\pi f$, where f = frequency), rad s^{-1}
t = time, s

Find the total emf induced in the loop.

Solution. This is a case of B change only, there being no motion. Hence, from (III), the total emf induced in the loop is

$$\mathcal{V} = -\int_S \frac{\partial \mathbf{B}}{\partial t} \cdot d\mathbf{s} = A\omega B_0 \sin \omega t \qquad \text{(V)} \quad \textit{Ans.} \tag{2}$$

This emf $\mathcal{V}$ appears at the terminals of the loop (Fig. 2-51).

Problem 2-14-1. Induction in loop. (*a*) A 5-turn loop with 0.5-m^2 area situated in air has a uniform magnetic field normal to the plane of the loop. If the flux density changes 8 mT/s, what is the emf appearing at the terminals of the loop? (*b*) If the emf at the loop terminals is 150 mV, what is the rate of change of the magnetic field? *Ans.* (*a*) 20 mV; (*b*) 60 mT/s.

Example 2-28. Loop: Motion only, no B change. Consider the rectangular loop shown in Fig. 2-52. The width L of the loop is constant, but its length x is increased uniformly with time by moving the sliding conductor at a uniform velocity **v**. The flux density **B** is everywhere the same (normal to the plane of the loop) and is constant with respect to time. Find the total emf induced in the loop.

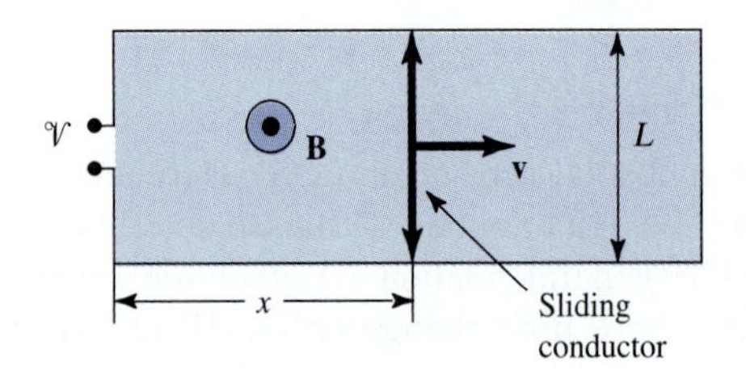

FIGURE 2-52
Sliding conductor for increasing loop area with **B** constant and **B** changing (Examples 2-28 and 2-29).

Solution. This is a pure case of motion only, the flux density **B** being constant. Hence, from (II),

$$\mathcal{V} = \oint (\mathbf{v} \times \mathbf{B}) \cdot d\mathbf{L} = vBL \qquad \text{(V)} \quad \textit{Ans.} \tag{3}$$

The entire emf in this case is induced in the moving conductor of length L.

Problem 2-14-2. Balloon loop. A conducting loop is painted around the equator of a spherical rubber balloon. A magnetic field $B = 0.2 \cos 4t$ T is applied perpendicular to the plane of the equator. The balloon is contracting inwardly with a radial velocity v. When the balloon radius is $r = 0.5$ m, the rms voltage induced in the loop is 500 mV. Find the velocity v at this instant. *Ans.* 0.52 m/s.

Example 2-29. Loop: Motion and B change. Consider the same loop with sliding conductor (Fig. 2-52) discussed in the preceding example. The flux density **B** is normal to the plane of the loop and is uniform everywhere. The sliding conductor moves with a uniform velocity **v**. These conditions are the same as in the preceding example. However, in this case let the magnitude of the flux density **B** vary harmonically with time as given by

$$B = B_0 \cos \omega t \tag{4}$$

Find the total emf induced in the loop.

Solution. This is a case involving both motion and time-changing **B**. The emf $\mathcal{V}_m$ due to the motion is given, from (II), by

$$\mathcal{V}_m = \oint (\mathbf{v} \times \mathbf{B}) \cdot d\mathbf{L} = vBL = vLB_0 \cos \omega t \tag{5}$$

The emf $\mathcal{V}_t$ due to a time-changing B is, from (III),

$$\mathcal{V}_t = -\int_S \frac{\partial \mathbf{B}}{\partial t} \cdot d\mathbf{s} = \omega x L B_0 \sin \omega t \tag{6}$$

According to (I) the total emf $\mathcal{V}$ is the sum of the emfs of (5) and (6), or

$$\begin{aligned} \mathcal{V} &= \mathcal{V}_m + \mathcal{V}_t = \oint (\mathbf{v} \times \mathbf{B}) \cdot d\mathbf{L} - \int_S \frac{\partial \mathbf{B}}{\partial t} \cdot d\mathbf{s} \\ &= vB_0 L \cos \omega t + \omega x B_0 L \sin \omega t \\ &= B_0 L \sqrt{v^2 + (\omega x)^2} \sin(\omega t + \delta) \qquad \text{(V)} \quad \textit{Ans.} \end{aligned} \tag{7}$$

where $\delta = \tan^{-1}(v/\omega x)$
$x =$ instantaneous length of loop

Problem 2-14-3. Expanding circular loop. A circular conducting rubber loop expands at a uniform velocity v. There is a uniform magnetic field perpendicular to the loop given by $B = B_0 t$, where $t =$ time. The radius of the loop is given by $r = vt$. (*a*) Find the emf induced in the loop (in symbols). (*b*) Evaluate this result if $v = 2$ m/s, $t = 9$ s, and $B_0 = 2$ T/s. *Ans.* (*a*) $B_0\pi r^2$ or $3B_0\pi r^3$; (*b*) 2.04 kV or 6.12 kV. (The two results are for positive and negative signs of B, respectively.)

Example 2-30. Moving strip: No B change. The circuit for a rectangular loop of width L and length x_1 is completed by sliding contacts through a thin conducting strip, as suggested in Fig. 2-53. The loop is stationary, but the strip moves longitudinally with a uniform velocity **v**. The magnetic flux density **B** is normal to the strip and the plane of the loop. It is constant with respect to time and is uniform everywhere. The width of the loop is L, the same as for the strip, although for clarity the loop is shown with a slightly greater width in Fig. 2-53. Find the total emf induced in the circuit.

Solution. This is another case of motion only. Therefore from (II) the total emf is given by

$$\mathcal{V}_m = \oint (\mathbf{v} \times \mathbf{B}) \cdot d\mathbf{L} = vBL \qquad \text{(V)} \quad \textit{Ans.} \tag{8}$$

The entire emf in this case is induced in the moving strip and appears at the terminals. A variation of the arrangement (Fig. 2-53) is provided by the Faraday disk generator (Problem 2-14-7).

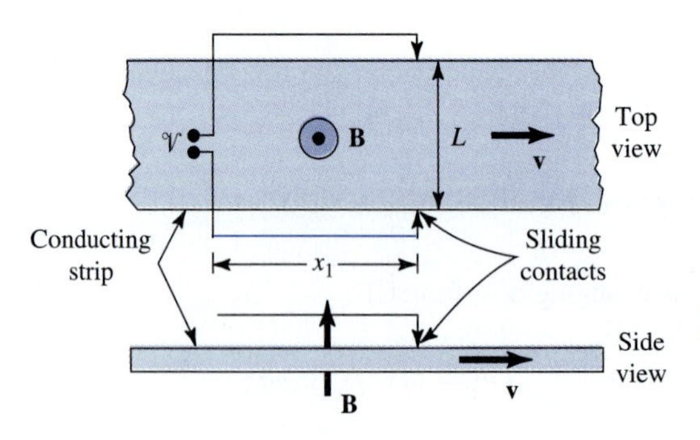

FIGURE 2-53
Fixed loop with sliding strip with **B** constant.

Example 2-31. Moving strip with B change. Consider now the same loop and strip as in the preceding example (Fig. 2-53) but let the magnitude of the flux density vary harmonically with time as given by

$$B = B_0 \cos \omega t \tag{9}$$

Find the total emf induced in the circuit.

Solution. This case involves both motion and a time-changing **B**. From (II) the emf $\mathcal{V}_m$ due to the motion is

$$\mathcal{V}_m = \oint (\mathbf{v} \times \mathbf{B}) \cdot d\mathbf{L} = vBL = vB_0L \cos \omega t \tag{10}$$

From (III) the emf $\mathcal{V}_t$ due to a time-changing **B** is†

$$\mathcal{V}_t = -\int_S \frac{\partial \mathbf{B}}{\partial t} \cdot d\mathbf{s} = \omega x_1 B_0 L \sin \omega t \qquad (11)$$

According to (I), the total emf $\mathcal{V}$ is the sum of (10) and (11), or

$$\begin{aligned} \mathcal{V} &= \mathcal{V}_m + \mathcal{V}_t = v B_0 L \cos \omega t + \omega x_1 B_0 L \sin \omega t \\ &= B_0 L \sqrt{v^2 + (\omega x_1)^2} \sin(\omega t + \delta) \qquad \text{(V)} \end{aligned} \qquad (12)$$

where $\delta = \tan^{-1}(v/\omega x_1)$.

Problem 2-14-4. Expanding square loop. Four straight conductors form a square with a magnetic field B perpendicular to the square. If all conductors move outward with the same velocity v while contacting each other at the corners, find V (rms) induced in the square loop at the instant when its area is 2 m^2, v = 4 m/s, and $B = \cos(2\pi f t)$ T, where f = 2 kHz. *Ans.* 17.8 kV.

Example 2-32. Rotating loop: No B change (ac generator). Consider next a rotating rectangular loop in a steady magnetic field as in Fig. 2-54*a*. The loop rotates with a uniform angular velocity ω rad s^{-1}. This arrangement represents a simple ac generator, the induced emf appearing at terminals connected to the slip rings. If the radius of the loop is R and its length L, find the total emf induced.

Solution. Since this is a case of motion only, the total emf can be obtained from (II). Referring to Fig. 2-54*b*, it is given by

$$\mathcal{V} = \oint (\mathbf{v} \times \mathbf{B}) \cdot d\mathbf{L} = 2vBL \sin \theta \qquad (13)$$

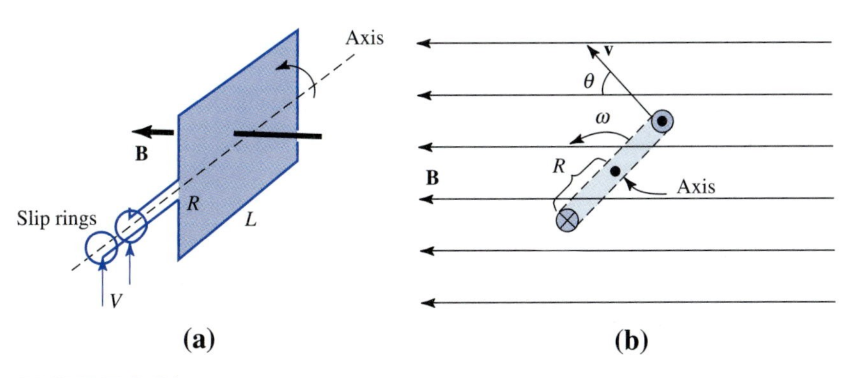

FIGURE 2-54
Rotating loop ac generator with **B** constant. (*a*) Perspective view and (*b*) cross section perpendicular to axis.

†At low frequencies the effect of eddy currents in the strip can be neglected. The effect of eddy currents will be even less if the strip is very thin and its conductivity is poor.

Since $\theta = \omega t$, we have

$$\mathcal{V} = 2\omega RLB \sin \omega t \qquad (14)$$

The factor 2 is necessary because there are two conductors of length L moving through the field, the emfs adding. Since $2RL = A$, the area of the loop, (14) reduces to

$$\mathcal{V} = \omega BA \sin \omega t \qquad \text{(V)} \qquad (15)$$

Problem 2-14-5. Earth inductor. (*a*) How many turns are required for a circular loop of 100-mm radius to develop a peak emf of 10 mV if the loop rotates 30 r/s (revolutions per second) in the earth's magnetic field? Take $B = 60\ \mu\text{T}$. (*b*) How must the axis of rotation be oriented to reduce the voltage to zero and determine the direction of the field? This arrangement constitutes an *earth* inductor, useful for measuring both the *magnitude and direction* of the earth's field. *Ans.* (*a*) 400 turns, (*b*) axis parallel to **B**.

Example 2-33. Rotating loop with B change. Consider finally the same rotating loop as in the preceding example with the modification that B varies with time as given by $B = B_0 \sin \omega t$ (same ω as in Example 2-32). Note that when $t = 0$, $B = 0$ and $\theta = 0$ (Fig. 2-54*b*). Find the total emf induced.

Solution. This case involves both motion and a time-changing B. From (II) the emf $\mathcal{V}_m$ due to the motion is

$$\mathcal{V}_m = 2\omega RLB_0 \sin^2 \omega t = \omega RLB_0 - \omega RLB_0 \cos 2\omega t \qquad (16)$$

From (III) the emf $\mathcal{V}_t$ due to a time-changing B is

$$\mathcal{V}_t = -2\omega RLB_0 \cos^2 \omega t = -\omega RLB_0 - \omega RLB_0 \cos 2\omega t \qquad (17)$$

From (I) the total emf $\mathcal{V}$ is given by the sum of (16) and (17), or

$$\mathcal{V} = \mathcal{V}_m + \mathcal{V}_t = -2\omega RLB_0 \cos 2\omega t = -\omega B_0 A \cos 2\omega t \qquad \text{(V)} \qquad (18)$$

The emf in this example is at twice the rotation, or magnetic field, frequency. It is to be noted that the emf calculated from either (II) or (III) alone contains a dc component. In adding the emfs of (II) and (III), the dc components cancel, yielding the correct total emf given by (18). The dc component is a mathematical artifice which disappears with the complete equation.

Problem 2-14-6. Generator. If $R = 2$ cm, $L = 6$ cm, and $B_0 = 3\ \text{Wb/m}^2$ in Fig. 2-54 and $\omega = 120\pi\ \text{s}^{-1}$, what is the rms voltage at the slip ring terminals? *Ans.* $V = 0.96$ V.

Problem 2-14-7. Faraday disk generator. (*a*) A thin copper disk 200 mm in diameter is situated with its plane normal to a constant, uniform magnetic field $B = 600$ mT. If the disk rotates 30 r/s, find the emf developed at the terminals connected to brushes as shown in Fig. 2-55. One brush contacts the periphery of the disk and the other contacts the axle or shaft. This arrangement is called a *Faraday disk generator.* (*b*) If the magnetic field varies with time, as given by $B = B_0 \sin \omega t$, where

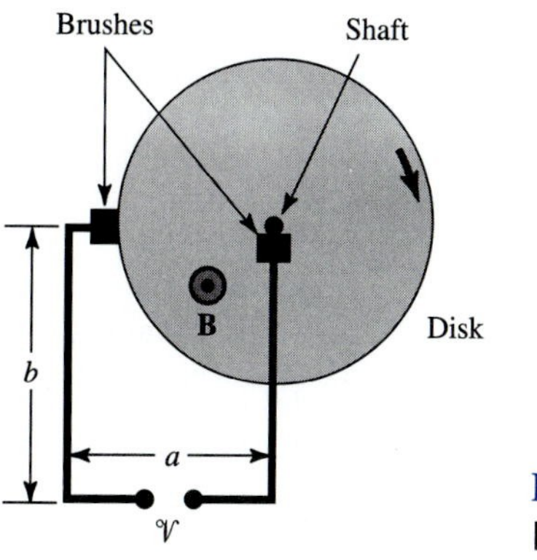

FIGURE 2-55
Faraday disk generator.

$B_0 = 600$ mT and $\omega = 2\pi \times 5$ rad/s, find the emf developed at the terminals. *Ans.* (*a*) 565 mV, (*b*) $0.57\sin(10\pi t) - 18.9ab\cos(10\pi t)$ V.

Coupling, Cross Talk, and Mutual Inductance

Interference and cross talk are the result of undesired signal coupling between circuits. The parameter of interest is the ratio of the voltage generated in one circuit by the rate-of-change of current in another. This ratio is the *mutual inductance M*. Thus,

$$M = \frac{V_1}{dI_2/dt} \qquad \text{(H)} \tag{19}$$

where V_1 = voltage induced in circuit 1, V
I_2 = current in circuit 2, A

For alternating current, $I = I_0 e^{j\omega t}$ and (19) becomes

$$M = \frac{V_1}{j\omega I_2} \qquad \text{and} \qquad V_1 = j\omega I_2 M \tag{20}$$

or

$$\frac{V_1}{I_2} = j\omega M = \text{mutual impedance} \qquad (\Omega) \tag{21}$$

The relation is also reciprocal, so

$$V_2 = j\omega I_1 M \qquad \text{and} \qquad \frac{V_2}{I_1} = j\omega M \qquad (\Omega) \tag{22}$$

The following example compares the mutual inductance of different circuit configurations.

Example 2-34. Mutual inductance of transmission lines. Fig. 2-56 shows two transmission lines in cross section in two configurations. In (*a*) the lines are side by side in the same plane. In (*b*) they are stacked side by side with spacing equal to the center-to-center spacing ($2d$) of the configuration in (*a*). $\mu_r = 1$. Find the mutual inductance per unit length of line for configurations (*a*) and (*b*).

Solution. The mutual inductance of the lines is equal to the normal component of the magnetic flux ψ_m from line 1 passing through the area A ($d \times$ length l into page) of line 2 divided by the current of line 1. Thus, the mutual inductance M per unit length of line l is given by

$$\frac{M}{l} = \frac{\psi_m \text{ (from line 1 through line 2)}}{I \text{ (line 1)}} = \frac{\int_{r_1}^{r_2} B\,dr}{I_1}$$

where B = magnetic flux density from line 1 at line 2 and r = distance between lines.

Solution. From the geometry,

$$r_1 = 1.5d \qquad r_2 = 2.5d \qquad r_3 = \sqrt{0.5^2 + 2^2}d$$

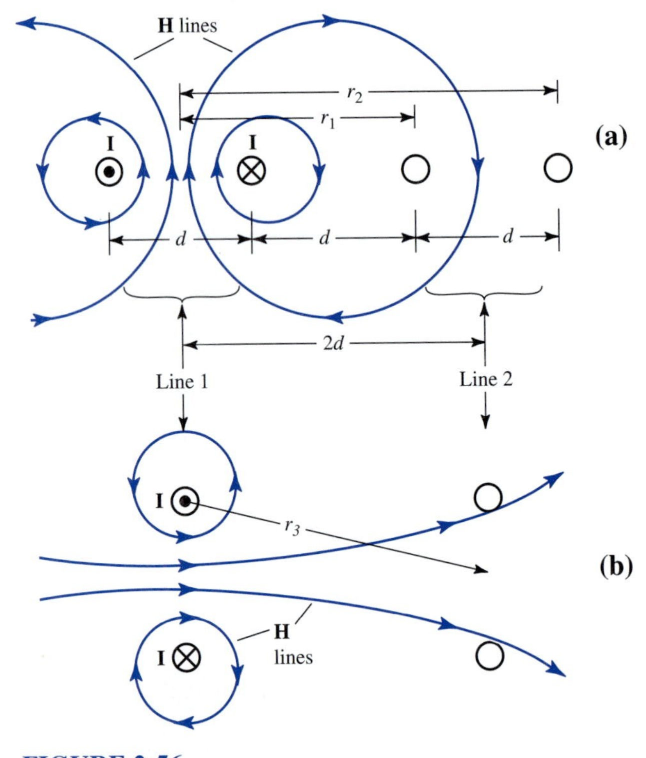

FIGURE 2-56
Transmission line configurations.

As a good approximation, the flux density from line 1 at the midpoint of line 2 is

$$B\text{ (from line 1)} = \frac{\mu I}{2\pi}\left(\frac{1}{r_1} - \frac{1}{r_2}\right)$$

so

$$\frac{M}{l} = \frac{\mu I}{2\pi d}\left(\frac{1}{1.5} - \frac{1}{2.5}\right)\frac{d}{I}$$

$$= \frac{4\pi 10^{-7}}{2\pi} \times 0.267 = 53.4\text{ nH m}^{-1} \quad \textit{Ans.} \quad (a)$$

Note that the mutual inductance depends only on the geometry.

$$\frac{M}{l} = \frac{\mu I}{2\pi}\left[\frac{2}{r_3}\sin\left(\tan^{-1}\frac{1}{4}\right)\right]\frac{d}{I} = 47.1\text{ nH m}^{-1} \quad \textit{Ans.} \quad (b)$$

or only about 11 percent less than for (*a*).

Problem 2-14-8. Mutual inductance of transmission lines. Find the mutual inductance per unit length for the two configurations of Fig. 2-57. *Ans.* (*a*) $M/l = 0$, (*b*) $M/l = 0$.

Problem 2-14-9. Induced voltage on telephone lines. Find the voltage per unit length induced in line 2 for telephone line configurations (*a*) and (*b*) of Fig. 2-56

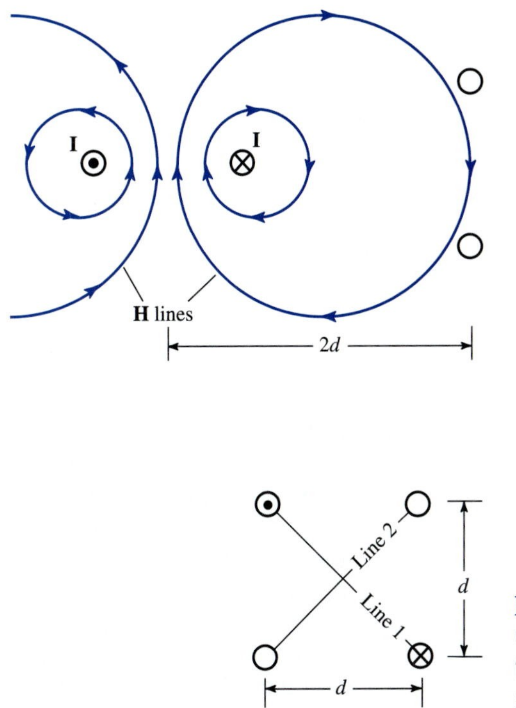

FIGURE 2-57
Configurations for pairs of telephone lines which result in zero cross talk.

and Fig. 2-57 if $I_1 = 5$ A, $f = 6$ kHz, and $d = 1$ cm. *Ans.* Fig. 2-56: (*a*) 7.14 mV/m, (*b*) 6.3 mV/m. Fig. 2-57: (*a*) 0, (*b*) 0.
Conclusion: To eliminate cross talk use the configurations of Fig. 2-57.

2-15 CURL

According to Ampere's law, integrating the magnetic field **H** around a path enclosing the conductor of Fig. 2-58 gives the *total current I* in the conductor. Thus,

$$\oint \mathbf{H} \cdot d\mathbf{L} = I \qquad \text{(A)} \tag{1}$$

Integrating over the sides of the incremental area $\Delta s = \Delta y \, \Delta z$ yields the current ΔI passing through Δs, or

$$\oint \mathbf{H} \cdot d\mathbf{L} = \Delta I_x \qquad \text{(A)} \tag{2}$$

Dividing by Δs we have

$$\frac{\oint \mathbf{H} \cdot d\mathbf{L}}{\Delta s} = \frac{\Delta I_x}{\Delta s} \qquad (\text{A m}^{-2}) \tag{3}$$

In the limit as Δs shrinks to zero around the point P, we obtain the *curl* of **H** which is equal to the current density **J** at the point P, or

$$\lim_{\Delta s \to 0} \frac{\oint \mathbf{H} \cdot d\mathbf{L}}{\Delta s} = \text{curl}_x \, \mathbf{H} = J_x \qquad (\text{A m}^{-2}) \tag{4}$$

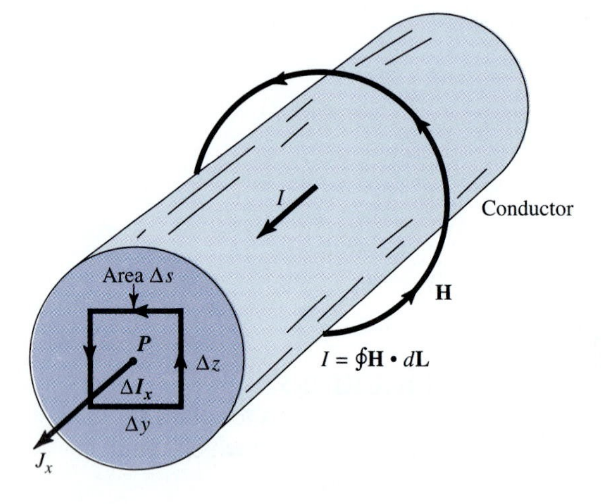

FIGURE 2-58
Integrating the magnetic field **H** around the conductor gives the total conductor current I. Integrating **H** around the sides of the area Δs yields the current ΔI_x through the area. Dividing by Δs yields (in the limit as Δs shrinks to zero around P) the curl of **H** or the current density J_x at P.

$\text{Curl}_x\ \mathbf{H}$ is the component of the curl in the x direction (perpendicular to Δs) and in the same direction as J_x.

In the most general situation of a conductor with current components flowing in the x, y, and z directions, the complete expression for curl **H** (in rectangular coordinates) is†

$$\text{curl}\ \mathbf{H} = \hat{\mathbf{x}}\left[\frac{\partial H_z}{\partial y} - \frac{\partial H_y}{\partial z}\right] + \hat{\mathbf{y}}\left[\frac{\partial H_x}{\partial z} - \frac{\partial H_z}{\partial x}\right] + \hat{\mathbf{z}}\left[\frac{\partial H_y}{\partial x} - \frac{\partial H_x}{\partial y}\right]$$
$$= \hat{\mathbf{x}}J_x + \hat{\mathbf{y}}J_y + \hat{\mathbf{z}}J_z = \mathbf{J} \tag{5}$$

or

$$\text{curl}\ \mathbf{H} = \mathbf{J} \qquad (\text{A m}^{-2}) \tag{6}$$

Curl **H** gives the current density **J** at a point. Curl **H** has a value wherever current is present.

Div **D** gives the charge density ρ at a point. Div **D** has a value wherever charge is present.

Curl **H** is conveniently expressed in vector notation as the cross product of the operator del (∇) and **H**, that is,

$$\text{curl}\ \mathbf{H} = \nabla \times \mathbf{H} = \underbrace{\left(\hat{\mathbf{x}}\frac{\partial}{\partial x} + \hat{\mathbf{y}}\frac{\partial}{\partial y} + \hat{\mathbf{z}}\frac{\partial}{\partial z}\right)}_{\nabla} \times \underbrace{(\hat{\mathbf{x}}H_x + \hat{\mathbf{y}}H_y + \hat{\mathbf{z}}H_z)}_{\mathbf{H}} = \mathbf{J} \tag{7}$$

or

$$\boxed{\nabla \times \mathbf{H} = \mathbf{J} \qquad (\text{A m}^{-2}) \qquad \textit{Curl of } \mathbf{H}} \tag{8}$$

From Faraday's law, the integral of the electric field **E** around a loop of incremental area Δs enclosing a time-changing magnetic flux density **B** equals the voltage V inducted in the loop, or

$$V = \oint \mathbf{E} \cdot d\mathbf{L} = -\iint_{\Delta s} \frac{\partial \mathbf{B}}{\partial t} \cdot d\mathbf{s} = -\frac{\partial \mathbf{B}}{\partial t}\Delta s \tag{9}$$

Dividing by Δs and taking the limit as Δs shrinks to zero around a point yields the curl of **E** at the point, or

$$\text{curl}\ \mathbf{E} = \lim_{\Delta s \to 0} \frac{\oint \mathbf{E} \cdot d\mathbf{L}}{\Delta s} = -\frac{\partial \mathbf{B}}{\partial t} \tag{10}$$

†For curl in cylindrical and spherical coordinates see Appendix A.

In vector notation,

$$\nabla \times \mathbf{E} = -\frac{\partial \mathbf{B}}{\partial t} \qquad \textit{Curl of } \mathbf{E} \tag{11}$$

From (9) and (11), we have

$$\oint_c \mathbf{E} \cdot d\mathbf{L} = \iint_s (\nabla \times \mathbf{E}) \cdot d\mathbf{s} \qquad \textit{Stokes's theorem} \tag{12}$$

Stokes's theorem states that the line integral of a vector function over a closed contour C equals the integral of the curl of that vector function over any surface having C as its boundary.

Example 2-35. Curl of water velocity. A rectangular trough carries water in the x direction. A section of the trough is shown in Fig. 2-59*a*, the vertical direction coinciding with the z axis. The width of the trough is b. Find the curl of the velocity $\mathbf{v}$ of the water for two assumed conditions.

(*a*) The velocity is everywhere uniform and equal to a constant, i.e.,

$$\mathbf{v} = \hat{\mathbf{x}}K \tag{13}$$

where $\hat{\mathbf{x}}$ = unit vector in positive x direction, dimensionless
K = a constant, m s^{-1}

A top view of the trough is shown in Fig. 2-59*b* with the positive x direction downward. The fact that the velocity $\mathbf{v}$ is constant is suggested by the arrows of uniform length and also by the graph v_x as a function of y in Fig. 2-59*c*.

Solution (*a*) Equation (13) can be reexpressed as

$$\mathbf{v} = \hat{\mathbf{x}}v_x \qquad (\text{m s}^{-1}) \tag{14}$$

where v_x is the component of velocity in x direction. Thus $v_x = K$. The curl of $\mathbf{v}$ has two terms involving v_x, namely $\partial v_x/\partial z$ and $\partial v_x/\partial y$. Since v_x is a constant, both terms are zero, and hence $\nabla \times \mathbf{v} = 0$ everywhere in the trough (see Fig. 2-59*d*).

(*b*) The velocity varies from zero at the edges of the trough to a maximum at the center, the quantitative variation being given by

$$\mathbf{v} = \hat{\mathbf{x}}K \sin \frac{\pi y}{b} \tag{15}$$

where K = a constant, m s^{-1}
b = width of trough, m

The sinusoidal variation of $\mathbf{v}$ is suggested by the length of the arrows in the top view of the trough in Fig. 2-59*e* and also by the graph of v_x as a function of y in Fig. 2-59*f*.

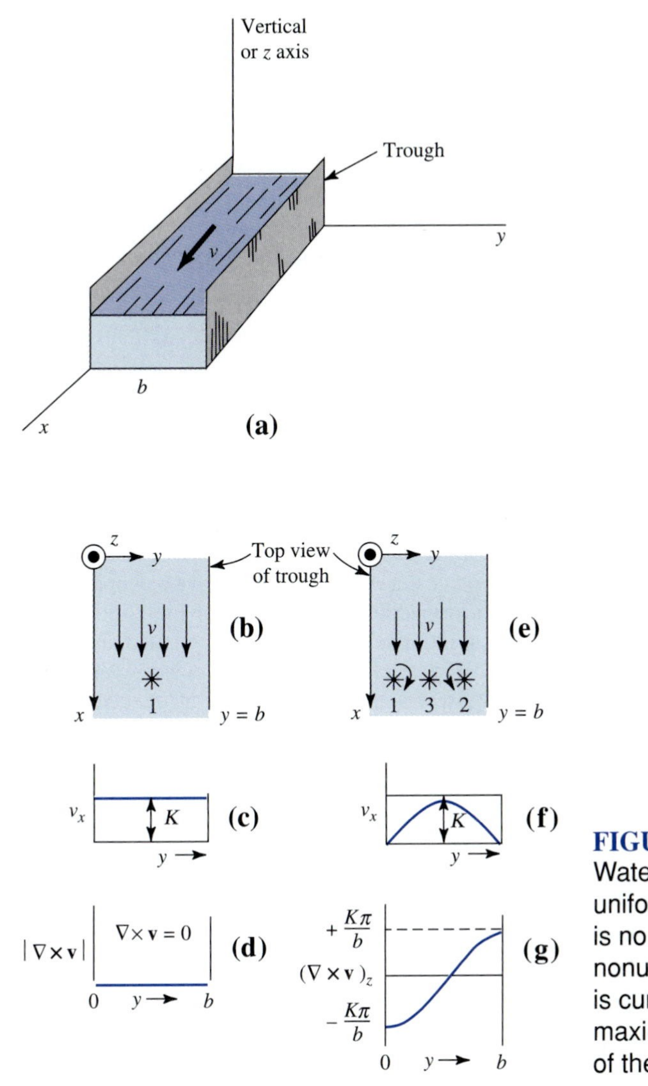

FIGURE 2-59
Water trough. With uniform flow there is no curl, but with nonuniform flow there is curl, which is a maximum at the edges of the trough.

Solution (*b*) From (15)

$$v_x = K \sin \frac{\pi y}{b} \qquad (\text{m s}^{-1}) \tag{16}$$

Since v_x is not a function of z, the derivative $\partial v_x / \partial z = 0$. However, v_x is a function of y so that

$$\frac{\partial v_x}{\partial y} = \frac{K\pi}{b} \cos \frac{\pi y}{b} \tag{17}$$

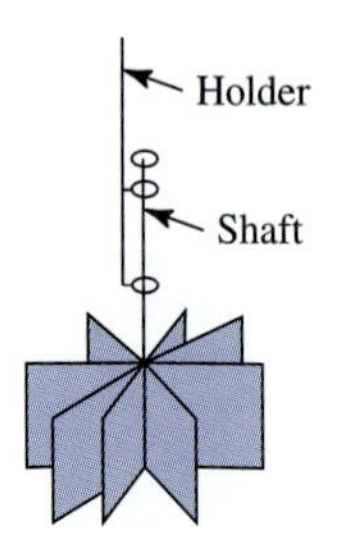

FIGURE 2-60
Paddle wheel for measuring curl.

and we have for the curl of **v**

$$\nabla \times \mathbf{v} = -\hat{\mathbf{z}}\frac{K\pi}{b}\cos\frac{\pi y}{b} \qquad (18)$$

where $\hat{\mathbf{z}}$ is the unit vector in the positive z direction. Thus at the left of the center of the trough the curl of **v** is in the negative z direction (downward in Fig. 2-59*a*) while to the right of the center it is in the positive z direction. The variation of the curl of **v** across the trough is presented graphically in Fig. 2-59*g*.

A physical interpretation of the curl of **v** in the above example can be obtained with the aid of the curlmeter, or paddle-wheel, device of Fig. 2-60. If this device is inserted with its shaft vertical into the trough with the assumed sinusoidal variation for the velocity of the water, it spins clockwise when it is at the left of the center of the trough and counterclockwise when it is at the right of the center of the trough, corresponding to negative and positive values of curl. At the center of the trough the curl meter does not rotate since the forces on the paddles are balanced. This corresponds to the curl of **v** being zero. The rate of rotation of the paddle-wheel shaft is proportional to the curl of **v** at the point where it is inserted. Thus, it rotates fastest near the edges of the trough. At any point the rate of rotation is also a maximum with the shaft vertical (rather than inclined to the vertical), indicating that $\nabla \times \mathbf{v}$ is in the z direction. It is assumed that the paddle wheel is small enough to avoid affecting the flow appreciably and to indicate closely the conditions at a point.

If the curlmeter with shaft vertical is inserted in water with uniform velocity, it will not rotate (curl **v** equals zero).

Problem 2-15-1. Curl. Find **J** if (*a*) $\mathbf{H} = \hat{\mathbf{x}}3+\hat{\mathbf{y}}7y+\hat{\mathbf{z}}2x$ A/m; (*b*) $\mathbf{H} = \hat{\mathbf{r}}6r+\hat{\boldsymbol{\phi}}2r+\hat{\mathbf{z}}5$ A/m. *Ans.* (*a*) $-\hat{\mathbf{y}}2$; (*b*) $\hat{\mathbf{z}}4$ A/m^2.

Problem 2-15-2. Curl. If $\mathbf{F} = \hat{\mathbf{x}}y + \hat{\mathbf{y}}x$, find $\nabla \times \mathbf{F}$. *Ans.* 0.

Problem 2-15-3. Curl. If $\mathbf{F} = \hat{\mathbf{x}}x^2 + \hat{\mathbf{y}}2yz - \hat{\mathbf{z}}x^2$, find $\nabla \times \mathbf{F}$ and the path of $\nabla \times \mathbf{F}$. *Ans.* $-\hat{\mathbf{x}}2y + \hat{\mathbf{y}}2x$, path is circular.

Problem 2-15-4. Curl. If $\mathbf{F} = \hat{\mathbf{x}}2x + \hat{\mathbf{y}}4xy^2z$, find (*a*) $\nabla \times \mathbf{F}$ and (*b*) $\nabla \times \nabla \times \mathbf{F}$. *Ans.* (*a*) $-\hat{\mathbf{x}}4xy^2 + \hat{\mathbf{z}}4y^2z$, (*b*) $\hat{\mathbf{x}}8yz + \hat{\mathbf{z}}8xy$.

2-16 MAXWELL'S EQUATIONS

In 1873 Professor James Clerk Maxwell of Cambridge University, England, assembled the laws of Ampere, Faraday, and Gauss (for electric and magnetic fields) into a set of four equations we call *Maxwell's equations.* Maxwell unified electromagnetic

$$I_t = \oint \mathbf{H} \cdot d\mathbf{L} = \iint_a \mathbf{J} \cdot ds + \iint_A \frac{\partial \mathbf{D}}{\partial t} \cdot ds$$

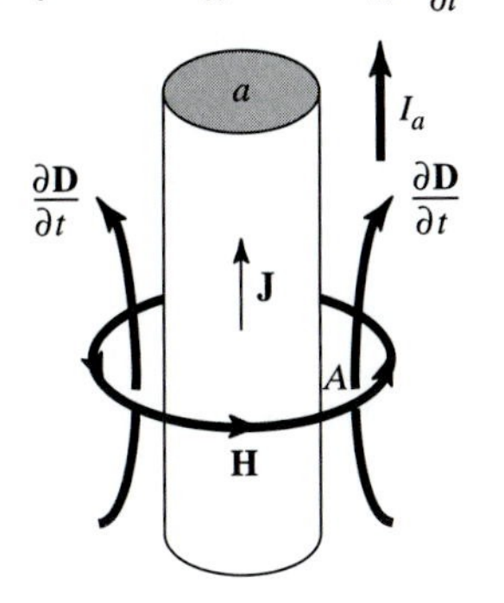

FIGURE 2-61
The integral of **H** ($= \oint \mathbf{H} \cdot d\mathbf{L}$) around area A yields the wire current I_a ($= \iint_a \mathbf{J} \cdot d\mathbf{s}$) plus the time-changing displacement current within A which includes the wire area a. The two components add up to a total current I_t. Note that the two components are in time-phase quadrature.

theory. He also added another term to Ampere's law to include a time-changing displacement current density ($\partial \mathbf{D}/\partial t$). Note that ($\partial \mathbf{D}/\partial t$) has the dimensions of current density (A m^{-2}) the same as J. See Fig. 2-61.

These equations, as we have derived them in both integral and differential forms, are listed below. *The integral equations have line, surface, and volume integrals. The differential equations involve divergence and curl and apply at a point.*

Law	Integral Form	Differential Form
Ampere	$\oint \mathbf{H} \cdot d\mathbf{L} = \int_s \left(\mathbf{J} + \frac{\partial \mathbf{D}}{\partial t}\right) \cdot ds = I_{\text{total}}$	$\nabla \times \mathbf{H} = \mathbf{J} + \frac{\partial \mathbf{D}}{\partial t}$
Faraday	$\oint \mathbf{E} \cdot d\mathbf{L} = -\int_s \frac{\partial \mathbf{B}}{\partial t} \cdot ds = V$	$\nabla \times \mathbf{E} = -\frac{\partial \mathbf{B}}{\partial t}$
Gauss for electric fields	$\oint_s \mathbf{D} \cdot ds = \int_v \rho \, dv = Q$	$\nabla \cdot \mathbf{D} = \rho$
Gauss for magnetic fields	$\oint_s \mathbf{B} \cdot ds = 0$	$\nabla \cdot \mathbf{B} = 0$

For harmonic variation, the phasor forms of Maxwell's integral and differential equations are

Integral	Differential
$\oint \mathbf{H} \cdot d\mathbf{L} = (\sigma + j\omega\varepsilon) \int_s \mathbf{E} \cdot ds$	$\nabla \times \mathbf{H} = (\sigma + j\omega\varepsilon)\mathbf{E}$
$\oint \mathbf{E} \cdot d\mathbf{L} = -j\omega\mu \int_s \mathbf{H} \cdot ds$	$\nabla \times \mathbf{E} = -j\omega\mu\mathbf{H}$
$\oint_s \mathbf{D} \cdot ds = \int_v \rho \, dv$	$\nabla \cdot \mathbf{D} = \rho$
$\oint_s \mathbf{B} \cdot ds = 0$	$\nabla \cdot \mathbf{B} = 0$

Constitutive relations: $\mathbf{D} = \varepsilon\mathbf{E}$, $\mathbf{B} = \mu\mathbf{H}$, $\mathbf{J} = \sigma\mathbf{E}$

TABLE 2-2
Electric and magnetic field equations

Electric fields	
Voltage and field	$V = \oint \mathbf{E} \cdot d\mathbf{L}$
Coulomb's force law	$F = \dfrac{Q_1 Q_2}{4\pi\varepsilon_0 r^2}$
Gauss's law	$\oiint \mathbf{D} \cdot d\mathbf{s} = \iiint \rho\, dv = Q$
Constitutive relation	$\mathbf{D} = \varepsilon_r \varepsilon_0 \mathbf{E}$ where $\varepsilon_r = \varepsilon/\varepsilon_0$
Capacitance	$C = \dfrac{Q}{V} = \dfrac{\varepsilon_0 A}{d}$
Capacitor energy	$W = \dfrac{1}{2}\dfrac{Q^2}{C} = \dfrac{1}{2}CV^2 = \dfrac{1}{2}QV$
Energy density	$w = \dfrac{1}{2}\varepsilon E^2$
Magnetic fields	
Ampere's law	$I = \oint \mathbf{H} \cdot d\mathbf{L} = \iint \mathbf{J} \cdot d\mathbf{s}$
Lorentz motor law	$\mathbf{F} = (\mathbf{I} \times \mathbf{B})L$
Force between wires of two-wire transmission line	$F = \dfrac{\mu_0 I_1 I_2}{2\pi d} L$
Gauss's law	$\oiint \mathbf{B} \cdot d\mathbf{s} = 0$
Faraday's law	$V = \oint \mathbf{E} \cdot d\mathbf{L} = -\iint \dfrac{\partial \mathbf{B}}{\partial t} \cdot d\mathbf{s}$
Constitutive relation	$\mathbf{B} = \mu_r \mu_0 \mathbf{H}$ where $\mu_r = \mu/\mu_0$
Inductance	$\mathcal{L} = \dfrac{\Lambda}{I} = \dfrac{\mu_0 N^2 A}{l}$
Inductor energy	$W = \dfrac{1}{2}\mathcal{L}I^2 = \dfrac{1}{2}\Lambda I = \dfrac{1}{2}\dfrac{\Lambda^2}{\mathcal{L}}$
Energy density	$w = \dfrac{1}{2}\mu H^2$

PROBLEMS

Answers are given in Appendix E to problems followed by the @ symbol.

2-2-4. Force of one point charge. A point charge of 7 nC is located at the point $(x, y, z) = (-1, 3, 2)$ m. Find the force **F** on another point charge of -3 nC at the point $(5, 1, 4)$ m. Express **F** in terms of its rectangular components and as a magnitude and unit vector. (See Appendix B, Part 2.) @

2-2-5. Two point charges. A charge of 130 nC is situated at the point $(r, \theta, \phi) = (7$ m, $45°, 10°)$. Find the force on a charge of 12 nC at $(x, y, z) = (-1, 4, 6)$ m. Express the force (*a*) in rectangular components, (*b*) in spherical components, and (*c*) as a magnitude and spherical-coordinate angles (θ and ϕ). (See Appendix B, Part 2.)

2-2-6. Field of two point charges. Rectangular coordinates. A charge of 30 nC is situated at the point $(x, y, z) = (0, 2, 0)$ m and a charge of -30 nC at the point $(0, -2, 0)$ m. Find **E** at the points: (*a*) $(4, 2, 4)$ m, (*b*) $(4, 0, 4)$ m, and (*c*) $(4, -2, 4)$ m. (See Appendix B, Part 2.) @

2-2-7. Field of two point charges. Cylindrical coordinates. A charge of 5 nC is situated at the point $(r, \phi, z) = (2$ m, $115°, 4$ m$)$ and another charge of 5 nC at the point $(2$ m, $-25°, 4$ m$)$. Find **E** at the origin. (See Appendix B, Part 2.)

2-2-8. Field of two point charges. Spherical coordinates. A charge of -7 nC is located at the point $(r, \theta, \phi) = (1$ m, $75°, 10°)$ and another charge of 3 nC at $(5$ m, $30°, 90°)$. Find **E** at the point $(2$ m, $45°, 45°)$. (See Appendix B, Part 2.) @

2-2-9. Forces between charges. Levitation. A small dielectric ball of mass 11 g slides freely on a vertical nonconducting string below a fixed, small dielectric ball with a charge of 11 nC. If the free-to-move ball has a charge of -1 nC, how far below the fixed ball will it "float"?

2-2-10. E from charged wire. A long straight wire carries a charge of 125 nC/m. Find **E** at a distance of 3 m from the wire. @

2-2-11. Dipole field. A charge $+Q$ is situated on the $+z$ axis at a distance $d/2$ from the origin. A charge of $-Q$ is situated on the $-z$ axis at a distance $d/2$ from the origin. The two charges of equal magnitude but opposite sign separated by distance d constitute an *electric dipole* of dipole moment Qd. Find the electric field **E** of the dipole at a point on the z axis for which $z \gg d$.

2-3-6. Absolute potential from point charge. At what distance is the absolute potential due to a 120-nC point charge equal to 1 kV (*a*) in vacuum and (*b*) in a medium with $\varepsilon_r = 4$? @

2-3-7. Gradient. Rectangular coordinates. If the electric scalar potential $V = 2x^2y + 4z$, find the electric field **E**.

2-3-8. Gradient. Cylindrical coordinates. If $V = \phi z^{1/2}/r$, find **E**. @

2-3-9. Gradient. Spherical coordinates. If $V = r^3\theta - 3\phi$, find **E**.

2-3-10. Gradient. If $V = 3e^{-2x}$, find **E**. @

2-3-11. Potential from electric field. If $\mathbf{E} = \hat{\mathbf{r}}(5r)$ V/m, find the difference in potential at points 1 m and 5 m from the origin.

2-3-12. Potential of two point charges. A point charge of -4 nC is situated at the origin and another point charge of 6 nC is located at the point $(3, 2, 1)$ m. Find V at the point $(1, 1, 1)$. @

2-3-13. Line charge. A uniform line of charge 5 m long with total charge 10 nC lies along the z axis with its center point 3 m from the origin. At a point on the y axis 5 m from the origin, find (*a*) V and (*b*) **E**. Repeat for a point charge of 10 nC on the z axis 3 m from the origin in place of the line charge, and compare the new values of V and **E** with those for the line charge.

2-3-14. Line charge. Find the electric field of a long line charge at a radial distance where the potential is 24 V higher than at a radial distance $r_1 = 3$ m where $|\mathbf{E}| = 4$ V/m. @

2-3-15. *V* and E of potential function. The potential function $V = 2x + y + 5z + 4$ V describes a plane equipotential surface for which $V = 4$ V at the origin. Find V and **E** at the origin, and show that V is a function of position but **E** is not (field is uniform).

2-5-3. Charged lines. A long two-wire transmission line has equal and opposite charges on the wires of 8 μC/m. The wires are spaced 2.4 m apart. Find V and **E** at a point midway between the wires. @

2-6-3. Line charge flux density. If $|\mathbf{D}| = 4$ nC m^{-2} at a distance of 30 cm from a long charged line immersed in oil ($\varepsilon_r = 2.2$), find (*a*) the line charge ρ_L, (*b*) the electric flux through a 2-m-long concentric cylinder of 10-cm radius, and (*c*) through a 2-m-long concentric cylinder of 70-cm radius. Repeat for air ($\varepsilon_r = 1$) as the medium.

2-6-4. Line charge density. Find the charge density on a long conductor if the flux density $|\mathbf{D}| = 26.6$ pC m^{-2} at a distance of 0.1 m. @

2-6-5. Charge inside spherical surface. Prove that with a charge Q *uniformly distributed* over the surface of a sphere of radius r, the field inside is zero. (*Hint:* Apply Gauss's law to a spherical surface of radius $r - \Delta r$.)

2-7-2. Divergence. If the field of a region of space is given by $\mathbf{E} = \hat{\mathbf{z}}2\cos\phi$, is this region free of charge? @

2-7-3. Charge density. Rectangular coordinates. If $\mathbf{D} = \hat{\mathbf{x}}2x^2 + \hat{\mathbf{y}}6y^2 + \hat{\mathbf{z}}z$ C/m^2, find the charge density at the point (2, 2, 5) m.

2-7-4. Charge density. Cylindrical coordinates. If $\mathbf{D} = \hat{\mathbf{r}}r^{-2} + \hat{\boldsymbol{\phi}}\sin\phi + \hat{\mathbf{z}}z^{-1/3}$ C/m^2, find the charge density at the point (1 m, 15°, 3 m). @

2-7-5. Charge density. Spherical coordinates. If $\mathbf{D} = \hat{\mathbf{r}}r^{-1} + \hat{\boldsymbol{\theta}}\cos\theta + \hat{\boldsymbol{\phi}}\cos\theta\sin\phi$ C/m^2, find the charge density at the point (5 m, 25°, 15°).

2-7-6. Charged cube. Find the total charge in a cube defined by the six planes for which $0 \le x \le 3$, $0 \le y \le 3$, $2 \le z \le 5$ if $\mathbf{D} = \hat{\mathbf{x}}x^2 + \hat{\mathbf{y}}2y + \hat{\mathbf{z}}3z^3$ nC/m^2. Solve by (*a*) integrating $\rho = \nabla \cdot \mathbf{D}$ throughout the volume of the cube and (*b*) integrating **D** over the surface of the cube. @

2-8-2. D and E at boundary. The y-z plane is the boundary between two dielectrics of relative permittivities $\varepsilon_r = 3$ and $\varepsilon_r = 9$. For negative values of x, $\mathbf{E} = \hat{\mathbf{x}}5 + \hat{\mathbf{y}}$ V/m. (*a*) Find the magnitude and direction of **D** for $x > 0$, and (*b*) draw a field line across the boundary. @

2-9-5. Parallel-plate capacitor. A parallel-plate capacitor has 2 × 3-cm plates spaced by 5 mm. Find the capacitance (*a*) if the medium between the plates is air ($\varepsilon_r = 1$) and (*b*) if $\varepsilon_r = 12$. Neglect fringing. @

2-9-6. Coaxial line. Two concentric conducting cylinders form a useful current-carrying arrangement called a *coaxial transmission line.* Show that the capacitance per unit length of a coaxial line is

$$\frac{C}{l} = \frac{2\pi\varepsilon}{\ln(b/a)}$$

where a and b are the inner and outer conductor diameters and ε is the permittivity of the material between the conductors. (*Hint:* The potential difference between concentric cylindrical conductors is

$$\frac{\rho_l}{2\pi\varepsilon} \ln(b/a)$$

where ρ_l is the charge per unit length on the inner conductor.)

2-9-7. Capacitor energy. An isolated parallel-plate capacitor with spacing d has a charge Q. Find the change in energy stored if the plate spacing is (a) halved and (b) doubled. (c) Account for energy changes and why (a) and (b) are not equal. @

2-9-8. Energy storage. If a 200-A-h, 12-V automobile battery can store $VIA = 12 \times 200 \times 3600 = 8.6$ MJ of energy, how many 1000-V, 1-μF capacitors are required to store the same amount of energy?

2-9-9. Thundercloud electrostatic energy. Electrostatically, a typical thundercloud may be represented by a capacitor model with horizontal plates 10 km^2 in area separated by a vertical distance of 1 km. The upper plate has a positive charge of 200 C and the lower plate an equal negative charge. Find (a) the electrostatic energy stored in the cloud, (b) the potential difference V between the top and bottom of the cloud, and (c) the average electric field $|\mathbf{E}|$ in the cloud. @

2-11-7. Electrostatic launcher. A golf ball with mass of 30 grams is given an electric charge of 1 C. What is the strength of the electric field needed to accelerate this golf ball to a velocity of 40 m/s in 0.1 second?

2-11-8. Wire resistance. Two wires, one made of copper and one of silver, have the same diameter and same resistance. Which wire is longer, and by how much? @

2-11-9. Wire material. When 12 V is applied to a 1-m length of 0.81-mm wire, a current $I = 25$ A is measured. What is the wire made of?

2-11-10. Electron and hole velocities. If the field $\mathbf{E}$ in a pure germanium semiconductor is the same as in the wire of Example 2-19, find the electron and hole drift velocities. Use $T = 290$ K. @

2-11-11. Conductivity measurement. Continuing Problem 2-9-3, two 20×20 cm capacitor plates are separated by 2 mm and connected by a 100-V dc source. A 1-mm-thick sheet of dielectric ($\varepsilon_r = 12$) is inserted between the plates. The two capacitor plates are now pressed tightly together on the dielectric sheet and the battery is disconnected. If the voltage falls to 10 V in 75 min, what is the conductivity of the sheet? Ignore the current drain of the voltmeter. @

2-11-12. Block resistance with uneven current flow. Find the resistance of a block like the one in Fig. 2-32 but sawed in half and reassembled back-to-back. Contact is made with perfectly conducting strips attached to each end as shown. Neglect thickness of saw cut. @

2-11-13. Current continuity equation. Kirchhoff's current law (2-11-23) is a special case of the general "continuity equation," which states that the divergence of the current density equals the negative time rate of change of the net charge density:

$$\nabla \cdot \mathbf{J} = -\frac{\partial \rho}{\partial t}$$

If a charge density ρ_0 is temporarily induced within a material of conductivity σ and permittivity ε, find an expression for the net charge density as a function of time.

2-11-14. Conductor boundary. At the plane boundary between two conductors, the tangential electric field is 5 V/m in medium 1, and the tangential current density in medium 2 is 23 A/m^2. Find the conductivity of medium 2. @

2-11-15. Battery emf. Find the emf of a battery in which chemical reactions produce an electric field $|\mathbf{E}| = 75x^{1/2}$ V/m over 10 cm.

2-12-10. Current element. A 1-cm current element with 5 A is coincident with the x axis. Find the magnitude of the magnetic flux density $|\mathbf{B}|$ at a radial distance of 5 m in the direction for which the angle θ with respect to the x axis equals (*a*) 0°, (*b*) 45°, and (*c*) 90°. @

2-12-11. Two current elements. Two current elements 15 mm long are situated with their centers at the origin. One is coincident with the y axis and carries 1 A in the $-y$ direction while the other is coincident with the x axis and carries 3 A in the $+x$ direction. Find the magnetic flux density $|\mathbf{B}|$ at points (*a*) (1, 4, 2) and (*b*) (3, 2, 1) m.

2-12-12. Magnetic field in cylinder. A conducting cylinder of radius r_0 carries a current I uniformly throughout its volume. Find an expression for the magnetic field $\mathbf{H}$ at a distance r from the axis of the cylinder (*a*) for $r < r_0$, and (*b*) for $r > r_0$. @

2-12-13. Magnetic flux. If $\mathbf{B} = \hat{\mathbf{x}}2xy + \hat{\mathbf{y}}3 - \hat{\mathbf{z}}2zy$ T, find the magnetic flux ψ_m through the surface of a volume enclosed by the six planes at $x = 0$, $x = 3$, $y = 2$, $y = 6$, $z = 1$, $z = 5$ m.

2-12-14. Moving wire. What is the emf induced when a 50-cm wire aligned with the x axis moves in the plane $z = 3$ m with a velocity $\mathbf{v} = \hat{\mathbf{x}}2 + \hat{\mathbf{y}}4$ m/s in a magnetic field with flux density $\mathbf{B} = \hat{\mathbf{x}}3x^2z + \hat{\mathbf{y}}6 - \hat{\mathbf{z}}3xz^2$ Wb/m^2? @

2-12-15. B in solenoid. Find an expression for the flux density at the center of a solenoid with a length L not much greater than its radius R. (*Hint:* Integrate expression (2-12-4) over the length of the solenoid and use NI/L as the current.)

2-12-16. Toroid. A long solenoid bent around on itself forms a toroid. Show that the inductance of a toroid of radius R with coil radius r is given by

$$\mathcal{L} = \mu \frac{N^2 r^2}{2R}$$

where N is the number of turns and μ is the permeability of the medium inside the coil.

2-12-17. Solenoid energy. What is the energy stored in the solenoid of Problem 2-12-8?

2-15-5. Curl. Rectangular coordinates. If $\mathbf{F} = \hat{\mathbf{x}}z + \hat{\mathbf{y}}x + \hat{\mathbf{z}}x$, find $\nabla \times \mathbf{F}$. @

2-15-6. Curl. Cylindrical coordinates. If $\mathbf{F} = \hat{\mathbf{r}}3r + \hat{\boldsymbol{\phi}}r + \hat{\mathbf{z}}2$, find $\nabla \times \mathbf{F}$. @

2-15-7. Curl. Spherical coordinates. If $\mathbf{F} = \hat{\mathbf{r}}(-2r) + \hat{\boldsymbol{\theta}}5r + \hat{\boldsymbol{\phi}} \sin\theta$, find $\nabla \times \mathbf{F}$. @

CHAPTER 3

TRANSMISSION LINES

3-1 INTRODUCTION

Transmission lines are the interconnections that convey electromagnetic energy from one point to another. The energy may be for light, heat, mechanical work, or information—speech, music, pictures, data.

Figure 3-1, on the next page, shows a few examples of transmission lines. They include two-wire, coaxial, strip, microstrip, waveguide, and optical fiber lines. In a broad sense, a wireless link is a transmission line. And we should also include the most ubiquitous transmission line of all, the axon nerve line, which is both noiseless and lossless.

In this chapter, transmission lines are explained from both circuit and field viewpoints. The wave equation is developed and wave propagation on lines is discussed.

3-2 CIRCUIT THEORY

From the circuit point of view, a transmission line has two terminals into which energy is fed and two terminals from which energy is received. Thus, as in Fig. 3-2, a transmission line can be regarded as a four-terminal network with

Series resistance R, Ω m^{-1}
Series inductance $\mathcal{L}$, H m^{-1}
Shunt conductance G, ℧ m^{-1}
Shunt capacitance C, F m^{-1}

Note that these symbols are per unit length.

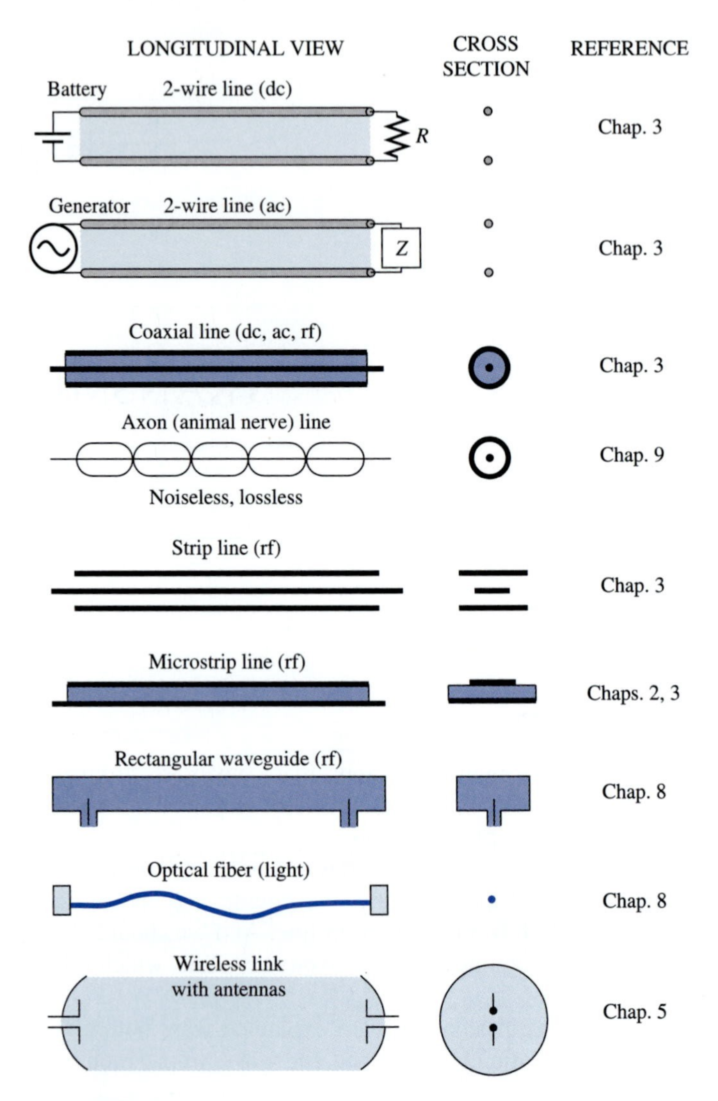

FIGURE 3-1
A few examples of transmission lines shown in longitudinal and cross-section views.

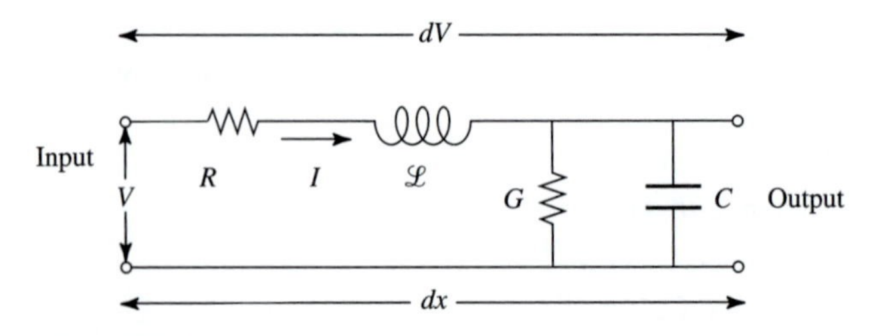

FIGURE 3-2
Transmission line as a four-terminal network.

For a lossless line $R = G = 0$ and the change in voltage dV (or change in current dI) per distance dx is given by

$$\frac{dV}{dx} = \mathcal{L}\frac{dI}{dt} \qquad (\text{V m}^{-1}) \tag{1}$$

and

$$\frac{dI}{dx} = C\frac{dV}{dt} \qquad (\text{A m}^{-1}) \tag{2}$$

Differentiating (1) with respect to distance x and (2) with respect to time t, we obtain

$$\boxed{\frac{d^2V}{dt^2} = \frac{1}{\mathcal{L}C}\frac{d^2V}{dx^2} \qquad \textit{Transmission line wave equation}} \tag{3}$$

Dimensionally, (3) is

$$\frac{V}{\text{time}^2} = \frac{1}{\mathcal{L}C}\frac{V}{\text{distance}^2}$$

or

$$\frac{1}{\sqrt{\mathcal{L}C}} = \text{velocity} = v \qquad (\text{m s}^{-1}) \tag{4}$$

where v is the velocity with which electromagnetic energy, as a wave or signal, propagates along a lossless line.

Referring to Fig. 3-3, on the next page, we have in the more general case for sinusoidal variation of V and I with R and C not zero, a *series impedance*

$$Z = R + j\omega\mathcal{L} = R + jX \quad (\Omega\ \text{m}^{-1})$$

and a *shunt admittance*

$$Y = G + j\omega C = G + jB \quad (\mho\ \text{m}^{-1})$$

where R, $\mathcal{L}$, G, and C are as given above and

$$\omega = 2\pi f,\ \text{angular frequency, rad s}^{-1}$$

$$f = 1/T = \text{frequency, Hz}$$

$$T = \text{period, s}$$

$$X = \omega\mathcal{L} = \text{series reactance, } \Omega\ \text{m}^{-1}$$

$$B = \omega C = \text{shunt susceptance, } \mho\ \text{m}^{-1}$$

Thus, (1) becomes

$$\frac{dV}{dx} = IZ \tag{5a}$$

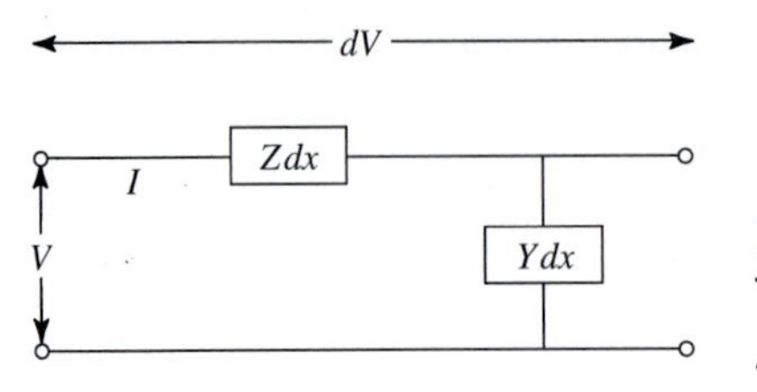

FIGURE 3-3
Transmission line with series impedance Z and shunt admittance Y.

and (2) becomes

$$\frac{dI}{dx} = VY \tag{5b}$$

Differentiating (5*a*) and (5*b*) with respect to the distance *x*, we have for a uniform line (Z and Y constant)

$$\frac{d^2V}{dx^2} - ZYV = 0 \tag{6}$$

$$\frac{d^2I}{dx^2} - ZYI = 0 \tag{7}$$

The square root of ZY is the *propagation constant* γ which may have real and imaginary parts. Thus,

$$\gamma = \text{propagation constant} = \sqrt{ZY} = \alpha + j\beta,\ \text{rad m}^{-1}$$

$$\alpha = \text{attenuation constant} = \text{Re}\sqrt{ZY},\ \text{Np m}^{-1}\ (\text{see footnote}\dagger)$$

$$\beta = 2\pi/\lambda = \text{phase constant} = \text{Im}\sqrt{ZY},\ \text{rad m}^{-1}$$

$$\lambda = \text{wavelength, m}$$

A general solution for the line voltage is (see footnote ‡)

$$V = V_1 e^{\alpha x} e^{j(\omega t + \beta x)} + V_2 e^{-\alpha x} e^{j(\omega t - \beta x)} \qquad \text{(V)} \tag{8}$$

and for the line current is

$$I = \frac{V_1}{\sqrt{Z/Y}} e^{\alpha x} e^{j(\omega t + \beta x)} - \frac{V_2}{\sqrt{Z/Y}} e^{-\alpha x} e^{j(\omega t - \beta x)} \qquad \text{(A)} \tag{9}$$

The first terms of (8) and (9) represent propagation in the negative *x* direction and the second terms energy propagation in the positive *x* direction.

For energy propagation in one direction

$$Z_{\text{line}} = \left|\frac{V}{I}\right| = \sqrt{\frac{Z}{Y}} = \sqrt{\frac{R + j\omega\mathcal{L}}{G + j\omega C}} \qquad (\Omega) \qquad \textbf{\textit{Characteristic impedance of line}} \tag{10}$$

†1 Np (1 neper) attenuation indicates a reduction to $1/e$ ($= 0.368$) of the original value.

‡The quantity $e^{j(\omega t + \beta x)}$ is a *phasor* of magnitude unity and angle $\phi = \omega t + \beta x$ which may be expressed variously as $e^{j\phi} = \angle\phi = \cos\phi + j\sin\phi = \sqrt{\cos^2\phi + \sin^2\phi}\,\angle\phi$.

For a lossless line $R = G = 0$, and

$$Z_{\text{line}} = \sqrt{\frac{\mathcal{L}}{C}} \quad (\Omega) \qquad \textbf{\textit{Characteristic resistance of line}} \tag{11}$$

The *energy velocity,* or velocity with which energy is propagated, is

$$v = \frac{\omega}{\beta} = \frac{\omega}{\text{Im}\sqrt{ZY}} \quad (\text{m s}^{-1}) \tag{12}$$

where $\sqrt{ZY} = \sqrt{(R + j\omega\mathcal{L})(G + j\omega C)}$. If $R = G = 0$ (line lossless) or if R and G are small ($R \ll \omega\mathcal{L}$ and $G \ll \omega C$),

$$\sqrt{ZY} = j\omega\sqrt{\mathcal{L}C}$$

and

$$\text{Wave velocity on line} = v = \frac{\omega}{\omega\sqrt{\mathcal{L}C}} = \frac{1}{\sqrt{\mathcal{L}C}} \quad (\text{m s}^{-1}) \tag{13}$$

as in (4).

This is the circuit view of a transmission line.

Example 3-1. Input impedance and attenuation. A uniform transmission line has constants $R = 12\,\text{m}\Omega\,\text{m}^{-1}, G = 1.4\,\mu\mho\,\text{m}^{-1}, \mathcal{L} = 1.5\,\mu\text{H}\,\text{m}^{-1}$, and $C = 1.4\,\text{nF}\,\text{m}^{-1}$. At 7 kHz find (*a*) characteristic impedance and (*b*) attenuation in decibels per kilometer.

Solution. From (10)

$$Z_0 = \left[\frac{R + j\omega\mathcal{L}}{G + j\omega C}\right]^{1/2} = \left[\frac{12 \times 10^{-3} + j2\pi 7 \times 10^3 \times 1.5 \times 10^{-6}}{1.4 \times 10^{-6} + j2\pi 7 \times 10^3 \times 1.4 \times 10^{-9}}\right]^{1/2}$$

$$= \left[\frac{10^{-3} \times 67 \angle 80^\circ}{10^{-6} \times 62 \angle 88.7^\circ}\right]^{1/2} = 33 - j2.5\,\Omega \quad \textit{Ans.} \quad (a)$$

$$\alpha = Re\sqrt{(R + j\omega\mathcal{L})(G + j\omega C)} = 1.82\,\text{dB/km} \quad \textit{Ans.} \quad (b)$$

Problem 3-2-1. Impedance, velocity, and attenuation. A uniform transmission line has constants $R = 12\ \text{m}\Omega/\text{m}$, $G = 0.8\ \mu\mho/\text{m}$, $\mathcal{L} = 1.3\ \mu\text{H/m}$, and $C = 0.7$ nF/m. At 5 kHz find (*a*) impedance, (*b*) v/c velocity ratio, and (*c*) dB attenuation in 2 km. *Ans.* (*a*) $43.6 - j5.5\Omega$; (*b*) 0.11; (*c*) 2.7 dB.

3-3 FIELD THEORY

For the full electromagnetic field view, we start with Maxwell's equations (Sec. 2-16). Thus, Maxwell's equation from Ampere's law (with conduction current $\mathbf{J} = 0$) is

$$\nabla \times \mathbf{H} = \varepsilon\frac{\partial \mathbf{E}}{\partial t} \tag{1}$$

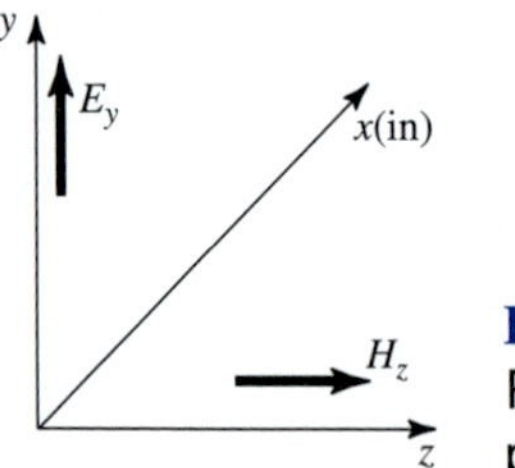

FIGURE 3-4
Rectangular coordinates for wave in x direction (into page).

In rectangular coordinates (Fig. 3-4), (1) expands (via 2-15-5) to

$$\hat{\mathbf{x}}\left[\frac{\partial H_z}{\partial y} - \frac{\partial H_y}{\partial z}\right] + \hat{\mathbf{y}}\left[\frac{\partial H_x}{\partial z} - \frac{\partial H_z}{\partial x}\right] + \hat{\mathbf{z}}\left[\frac{\partial H_y}{\partial x} - \frac{\partial H_x}{\partial y}\right]$$

$$= \varepsilon\frac{\partial}{\partial t}\left[\hat{\mathbf{x}}E_x + \hat{\mathbf{y}}E_y + \hat{\mathbf{z}}E_z\right] \tag{2}$$

For a wave traveling in the x direction along a transmission line, this reduces to

$$\frac{\partial H_z}{\partial x} = -\varepsilon\frac{\partial E_y}{\partial t} \tag{3}$$

Maxwell's equation from Faraday's law is

$$\nabla \times \mathbf{E} = -\mu\frac{\partial \mathbf{H}}{\partial t} \tag{4}$$

In rectangular coordinates this is the same as (2) with **E** and **H** interchanged. For a wave in the x direction, this reduces to

$$\frac{\partial E_y}{\partial x} = -\mu\frac{\partial H_z}{\partial t} \tag{5}$$

Differentiating (3) with respect to time and (5) with respect to x, we obtain the ***wave equation*** for the electric field E_y in terms of time t and distance x. Thus,

$$\frac{\partial^2 E_y}{\partial t^2} = \frac{1}{\mu\varepsilon}\frac{\partial^2 E_y}{\partial x^2} \qquad \textit{Wave equation} \tag{6}$$

Dimensionally (6) is

$$\frac{\text{Electric field}}{\text{time}^2} = \frac{1}{\mu\varepsilon}\frac{\text{electric field}}{\text{distance}^2} \quad \text{or} \quad \frac{1}{\mu\varepsilon} = \left(\frac{\text{distance}}{\text{time}}\right)^2 = \text{velocity}^2$$

Thus, $1/\sqrt{\mu\varepsilon}$ has the dimensions of *velocity*. Introducing the values of μ and ε for air or vacuum gives

$$v = \frac{1}{\sqrt{4\pi \times 10^{-7} \times 8.85 \times 10^{-12}}} = 3 \times 10^8 \text{ m s}^{-1} = c \tag{7}$$

where c = *velocity of light, radio, and other electromagnetic waves.*

From (3-2-11) the characteristic resistance of a losslesss transmission line is

$$Z_{\text{line}} = \sqrt{\frac{\mathcal{L}}{C}} \quad (\Omega) \qquad \textit{Characteristic resistance of transmission line} \tag{8}$$

For any field cell $\mathcal{L}/l = \mu$ and $C/l = \varepsilon$. Thus,

$$Z_{\text{cell}} = \sqrt{\frac{\mathcal{L}/l}{C/l}} = \sqrt{\frac{\mu}{\varepsilon}} = Z_0 \quad (\Omega) \qquad \textit{Intrinsic impedance of medium} \tag{9}$$

For air or vacuum,

$$Z_0 = \sqrt{\frac{\mu_0}{\varepsilon_0}} = \sqrt{\frac{4\pi \times 10^{-7}\ \text{H m}^{-1}}{8.85 \times 10^{-12}\ \text{F m}^{-1}}} = 376.731 \approx 120\pi \approx 377\ \Omega \tag{10}$$

which is the *intrinsic resistance of empty space*. This is the impedance of a square field cell, so Z_0 is more explicitly the impedance *per square*, meaning any square area. Its units are simply *ohms.*

The integral of **E** between the two conductors of the line in Fig. 3-5 (on next page) equals the voltage V between the lines. The integral of **H** around either line equals the line current I. Thus,

$$V = \int_{\text{Line 1}}^{\text{Line 2}} \mathbf{E} \cdot d\mathbf{L} \quad (\text{V}) \qquad \text{and} \qquad I = \oint_{\substack{\text{Around}\\ \text{one line}}} \mathbf{H} \cdot d\mathbf{L} \quad (\text{A}) \tag{11}$$

From Ohm's law the impedance of the transmission line is equal to the ratio of the voltage V to current I. Thus, from (11)

$$Z_{\text{line}} = \frac{V}{I} = \frac{\int \mathbf{E} \cdot d\mathbf{L}}{\oint \mathbf{H} \cdot d\mathbf{L}} \quad (\Omega) \tag{12}$$

For any field cell, such as the one with dimensions h and w in Fig. 3-5, the integrals of **E** and **H** reduce to scalar products Eh and Hw. These are the same for any curvilinear square in the map since where E is larger, h is smaller in inverse proportion. The same is true for H and w. Thus,

$$V_{\text{cell}} = Eh \qquad \text{and} \qquad I_{\text{cell}} = Hw$$

and

$$Z_{\text{cell}} = \frac{V_{\text{cell}}}{I_{\text{cell}}} = \frac{Eh}{Hw} \quad (\Omega) \tag{13}$$

Since $w = h$, and from (10)

$$Z_{\text{cell}} = \frac{E}{H} = \sqrt{\frac{\mu_0}{\varepsilon_0}} = 377\ \Omega \qquad \text{(for air or vacuum)} \tag{14}$$

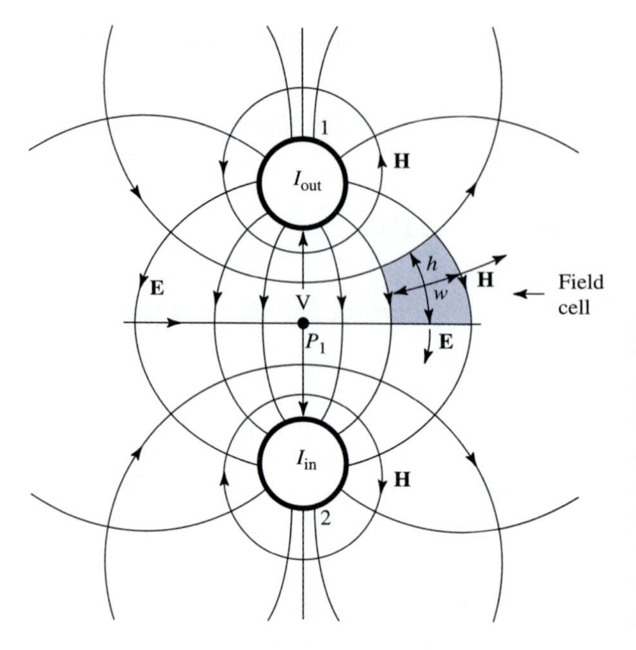

FIGURE 3-5
Two-conductor transmission line with map of **E** and **H** field lines. The map is divided into curvilinear squares with dimensions h and w shown on one of them. Each "square" or field cell carries equal power.

From the field map the voltage between the two lines is equal to the product of the cell voltage (V_{cell}) and the *number of cells in series* (N_s) between the lines, or

$$V_{line} = N_s V_{cell} = \sum N_s E h \tag{15}$$

The line current is equal to the product of the cell current (I_{cell}) and the *number of cells in parallel* (N_p) enclosing one conductor, or

$$I_{line} = N_p I_{cell} = \sum N_p H w \tag{16}$$

The line impedance is then the ratio of (15) to (16) and, for $h = w$, we have

$$Z_{line} = \frac{V_{line}}{I_{line}} = \frac{\sum N_s E h}{\sum N_p H w} = \frac{N_s E}{N_p H} = \frac{N_s}{N_p} Z_0 \quad (\Omega) \tag{17}$$

Example 3-2. Two-conductor transmission line impedance. Using the field map of Fig. 3-5, find the impedance of the line.

Solution. There are 6 curvilinear squares or cells in series ($N_s = 6$) and 10.4 cells in parallel ($N_p = 10.4$). Thus, from (17),

$$Z_{line} = \frac{6}{10.4} 377 = 218\ \Omega \qquad Ans.$$

Problem 3-3-1. Two-conductor line impedance. Using (3-3-26) calculate the impedance of the two-conductor line of Fig. 3-5 and compare with the value obtained in Example 3-2. The ratio of the center-to-center spacing to conductor diameter is 3 to 1. $\varepsilon_1 = 1$. *Ans.* 215 Ω vs. 218 Ω.

Problem 3-3-2. Cell power. If $V = 202$ V on the line of Fig. 3-5, what is the power per cell? *Ans.* 3.0 W.

The Microstrip Transmission Line

The microstrip transmission line is the most widely used line, with applications in integrated circuits and circuit boards. Although much more convenient to use than coaxial cable, it is not shielded and has a fringing field.

The dynamics of a wave traveling on a microstrip transmission line are shown in Fig. 3-6 on the next page. Figure 3-6*a* is a perspective view of the line. It shows the **E** and **H** lines and the direction of the current **I** on both strip and ground plane at three positions over a distance of one wavelength (λ) at the same instant of time (T) (time freeze). Figure 3-6*b* shows the line in cross section at three instants of time over one period at the same position on the line (position freeze).

Both **E** and **H** are everywhere perpendicular to each other and to the direction of propagation. Thus, the wave is called a *transverse electromagnetic* or *TEM wave*. The medium is air and the strip and ground plane are perfectly conducting ($\sigma = \infty$).

A complete field map of the microstrip line divided into curvilinear squares is shown in Fig. 3-7 with **H**-field lines encircling the strip and **E**-field lines connecting the strip and ground plane. The medium is air everywhere ($\varepsilon_r = 1$). The strip and ground plane are perfectly conducting ($\sigma = \infty$). The line is assumed to be either perfectly matched or infinitely long.

From (17) the line impedance

$$Z_{\text{line}} = \frac{N_s}{N_p}377 = \frac{5}{38}377 = 49.6\ \Omega$$

where N_s = cells in series and N_p = cells in parallel. Since the medium is air everywhere, the effective relative permittivity $\varepsilon_{\text{eff}} = \varepsilon_r = 1$ and the wave velocity

$$v = c = 3 \times 10^8 \text{ m s}^{-1}$$

Since each cell carries equal power, the fraction of the power in the fringing field (not under the strip) is given by the *fringing fraction:*

$$\frac{N_0}{N_t} = \frac{\text{number of cells not under strip}}{\text{total number of cells}} \tag{17.1}$$

From Fig. 3-7, we have

$$\text{Fringing fraction} = \frac{N_0}{N_t} = \frac{62}{190} = 0.33 \text{ or } 33\%$$

In a typical line, the strip rests on a dielectric substrate as in Fig. 3-8. The medium above is air ($\varepsilon_r = 1$) with the dielectric substrate $\varepsilon_r = 4$. In this map the cells are, in general, curvilinear rectangles, with height h related to the width w by

$$h = \sqrt{\varepsilon_r}w \tag{17.2}$$

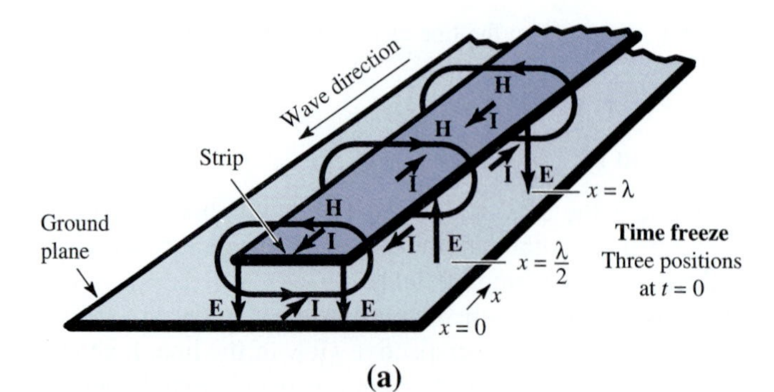

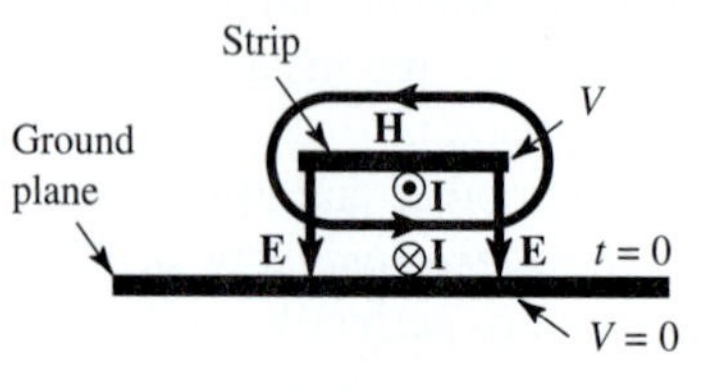

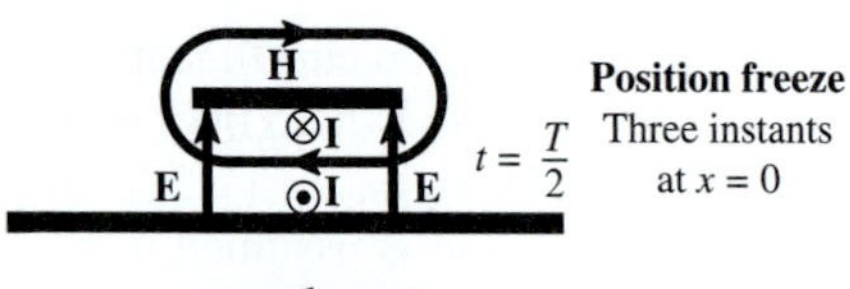

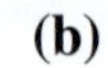

FIGURE 3-6
(*a*) Perspective view of transverse electromagnetic wave approaching on microstrip line at $t = 0$ (time freeze) with directions of the fields **E** and **H** and the current **I** shown. (*b*) Cross section of line at $x = 0$ (position freeze) at three instants of time.

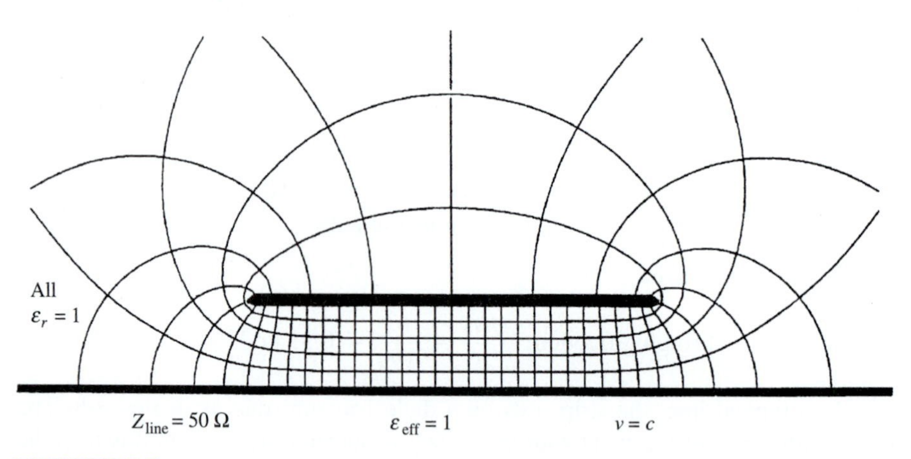

FIGURE 3-7
Field map of cross-section microstrip transmission line with **H**-field lines encircling the strip and **E**-field lines connecting the strip and ground plane. $\varepsilon_r = 1$ everywhere. (*Map by Nicolas B. Piller, ETH, Zürich.*)

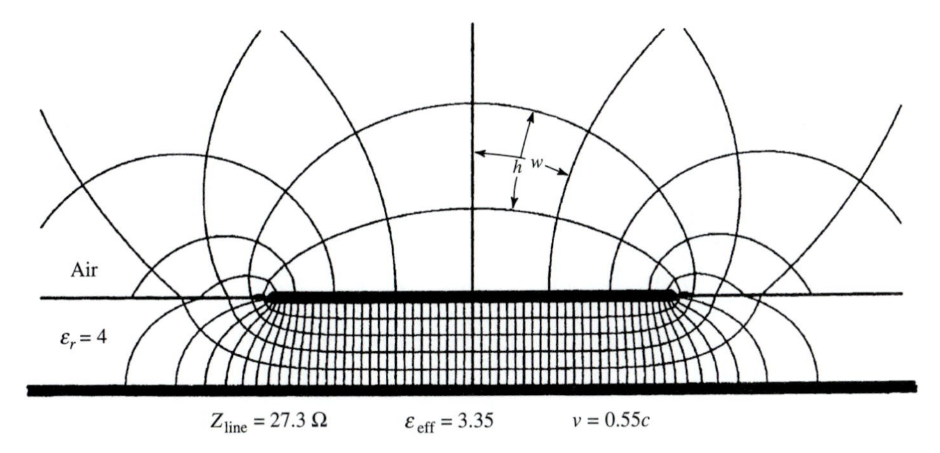

FIGURE 3-8
Microstrip transmission line with substrate dielectric $\varepsilon_r = 4$. Note the refraction of the field lines at the air-dielectric interface and the change of cell shape from curvilinear squares ($h = w$ in air) to curvilinear rectangles ($h = \sqrt{\varepsilon_r}w$ in substrate). (*Map by Nicolas B. Piller, ETH, Zürich.*)

The cells are curvilinear squares only when the cell $\varepsilon_r = 1$. With the condition of (17.2) all cells have the same impedance and the strip-line impedance is given by

$$Z_{\text{line}} = Z_{\text{cell}}\frac{N_s}{N_p} = 377\frac{N_s}{N_p} \quad (\Omega) \tag{17.3}$$

where N_s = cells in series and N_p = cells in parallel.

Example 3-3. Microstrip transmission line with substrate dielectric $\varepsilon_r = 4$. For the microstrip line of Fig. 3-8, find (*a*) the line impedance, (*b*) the effective relative permittivity, (*c*) the wave velocity and (*d*) the fringing fraction.

Solution. From (17.3)

$$Z_{\text{line}} = \frac{N_s}{N_p}377 = \frac{5}{69}377 = 27.3\ \Omega \quad \textit{Ans.} \quad (a)$$

$$Z_{\text{line}} = \frac{50}{\sqrt{\varepsilon_{\text{eff}}}}.\ \therefore\ \varepsilon_{\text{eff}} = \left(\frac{50}{27.3}\right)^2 = 3.35 \quad \textit{Ans.} \quad (b)$$

$$v = \frac{c}{\sqrt{\varepsilon_{\text{eff}}}} = \frac{c}{\sqrt{3.35}} = 0.546c = 164\ \text{Mm s}^{-1} \quad \textit{Ans.} \quad (c)$$

From (17.1) the fringing fraction is

$$\frac{N_0}{N_t} = \frac{80}{330} = 0.24 \text{ or } 24 \text{ percent} \quad \textit{Ans.} \quad (d)$$

which is a reduction from the all-air case (24 versus 33 percent).

Problem 3-3-3. Line parameters. Confirm the effective permittivity and velocity values of Fig. 3-9.

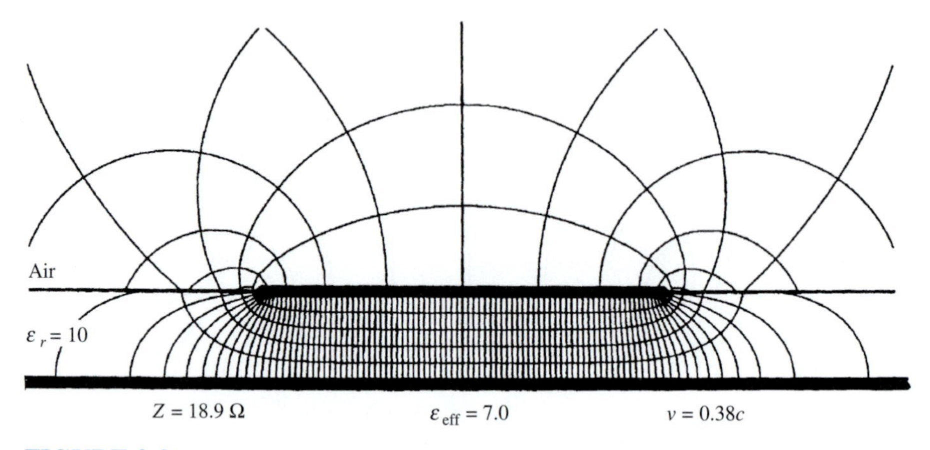

FIGURE 3-9
Microstrip transmission line with substrate dielectric $\varepsilon_r = 10$ and $h = \sqrt{10}w$ in substrate. (*Map by Nicolas B. Piller, ETH, Zürich.*)

A very simple approximate value of the line impedance is given by (28). Table 3-1 compares this and the above line impedances for a microstrip line with $W/H = 4.8$ as in Figs. 3-7, 3-8, and 3-9 for substrate $\varepsilon_r = 1$, 4, and 10. See Fig. 3-10d for meaning of W and H.

Transmission Line Impedance Formulas

The impedance Z of a transmission line is given by

$$Z_{\text{line}} = \sqrt{\frac{\mathcal{L}}{C}} = \sqrt{\frac{\mathcal{L}/l}{C/l}} \tag{17.4}$$

For the coaxial line (Fig. 3-10a),

$$\frac{C}{l} = \frac{Q/l}{\Delta V} = \frac{\rho_L}{\Delta V} \tag{18}$$

TABLE 3-1
Strip-line impedances

Substrate ε_r	Z_{line} (Ω) Eq. (17.3) $h = \sqrt{\varepsilon_r}w$	Eq. (28) (Fig. 3-10d)	Wheeler formula*
1	49.6	55.4	50.2
4	27.3	27.7	27.2
10	18.9	17.5	17.9

*H. A. Wheeler, "Transmission-line Properties of Parallel Strips Separated by a Dielectric Sheet," IEEE Transactions of Microwave Theory and Techniques, **MTT-3**, March 1965, p. 179.

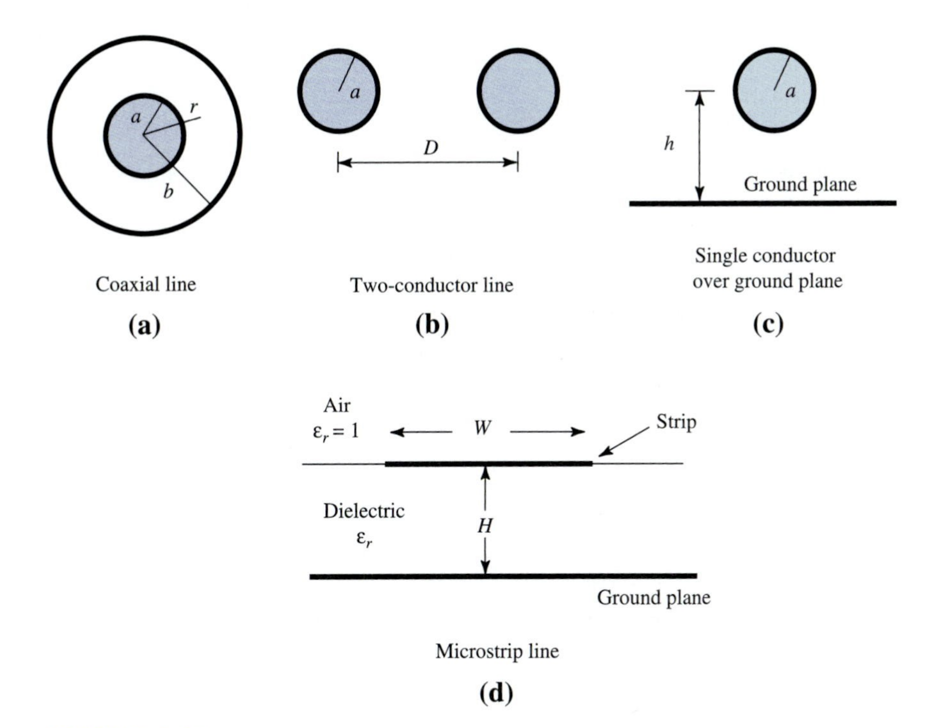

FIGURE 3-10
(*a*) Coaxial, (*b*) two-conductor, (*c*) single conductor (over ground plane), and (*d*) microstrip lines. Note in (*d*) W and H (capital letters) for the strip line as distinguished from w and h (lowercase) for a cell.

where

$$\Delta V = \int_a^b E_r\, dr$$

and, from (2-2-6),

$$E_r = \frac{\rho_L}{2\pi\varepsilon r} \tag{19}$$

Integrating gives

$$\Delta V = \frac{\rho_L}{2\pi\varepsilon} \ln \frac{b}{a} \tag{20}$$

Therefore

$$\frac{C}{l} = \frac{\rho_L}{\Delta V} = \frac{2\pi\varepsilon}{\ln \dfrac{b}{a}} \tag{21}$$

$$\mathcal{L} = \frac{\Lambda}{I} = \frac{l \displaystyle\int_a^b B_r\, dr}{I}$$

and, from (2-12-2),

$$B_r = \frac{\mu I}{2\pi r} \tag{22}$$

Integrating gives

$$\frac{\mathscr{L}}{l} = \frac{\mu}{2\pi} \ln \frac{b}{a} \tag{23}$$

Therefore

$$Z = \sqrt{\frac{\mathscr{L}/l}{C/l}} = \frac{1}{2\pi}\sqrt{\frac{\mu_0}{\varepsilon_0}}\sqrt{\frac{\mu_r}{\varepsilon_r}} \ln \frac{b}{a} \tag{24}$$

and for $\mu_r = 1$

$$Z_0 = \frac{138}{\sqrt{\varepsilon_r}} \log \frac{b}{a} \quad (\Omega) \qquad \textbf{\textit{Coaxial line}} \tag{25}$$

In a similar manner (Fig. 3-10*b* and *c*),

$$Z_0 = \frac{276}{\sqrt{\varepsilon_r}} \log \frac{D}{a} \quad (\Omega) \qquad \textbf{\textit{Two-conductor line}} \tag{26}$$

and

$$Z_0 = \frac{138}{\sqrt{\varepsilon_r}} \log \frac{2h}{a} \quad (\Omega) \qquad \textbf{\textit{One conductor over ground plane}} \tag{27}$$

An approximate formula for the impedance of a microstrip line (Fig. 3-10*d*) is

$$Z_0 \approx \frac{377}{\sqrt{\varepsilon_r}[(W/H) + 2]} \quad (\Omega) \qquad \textbf{\textit{Microstrip line}} \tag{28}$$

Example 3-4. Coaxial line impedance. A coaxial line has an impedance of 30 Ω. If the inside diameter of the outer conductor is 20 mm, how many field cells in parallel (N_p) are there? $\varepsilon_r = 1$.

Solution. Referring to a cell with midlines *h* and *w* (Fig. 3-11), the magnetic flux in the sector $b - r$ equals that in the sector $r - a$. Therefore,

$$\ln \frac{b}{r} = \ln \frac{r}{a}$$

or

$$\ln r = \frac{1}{2}(\ln b + \ln a)$$

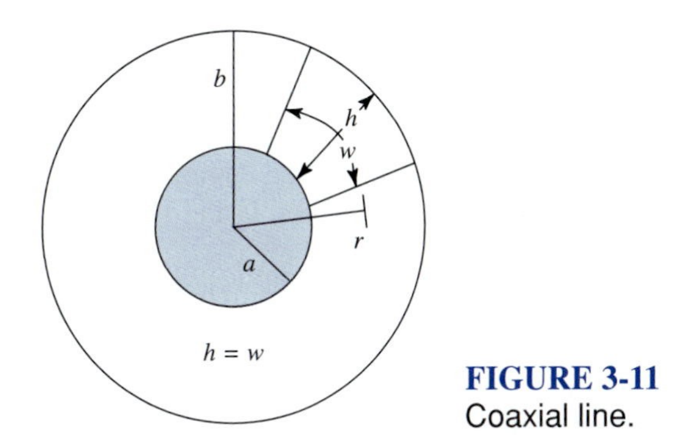

FIGURE 3-11
Coaxial line.

From (25), $a = 12.1$ mm (or larger than drawn in the figure), so $r = 15.6$ mm and since $2\pi r = N_p w = N_p h$,

$$N_p = \frac{2\pi r}{b - a} = \frac{2\pi 15.6}{20 - 12.1} = 12.4 \quad \textit{Ans.}$$

Check: $Z_0/N_p = 377/12.4 = 30.4\ \Omega$

Problem 3-3-4. Single-conductor line. A line with single conductor of 10-cm radius over a ground plane (Fig. 3-10*c*) has an impedance of 75 Ω. Find the height h. $\varepsilon_r = 3$. *Ans.* 437 mm.

Problem 3-3-5. Why 50 Ω? Why are the most common coaxial cable impedances 50 and 75 Ω? *Ans.* The theoretical impedance for the minimum attenuation of a coaxial line is 77 Ω, while the best impedance for maximum power handling capacity is 30 Ω. The average is 53.5 Ω, which rounded off is 50 Ω. A 75-Ω coaxial cable impedance is also used extensively because it is close to the impedance for minimum attenuation (77 Ω). The above is according to Mario A. Maury, Jr., "Microwave Coaxial Connector Technology: A Continuing Evolution," *Microwave Journal, 1990 State of the Art Reference*. This article provides a detailed description of coaxial connectors of all types.

Problem 3-3-6. Microstrip line impedance. Using the simple approximate impedance formula of (28) find the impedance of the microstrip lines of (*a*) Fig. 3-7, (*b*) Fig. 3-8, and (*c*) Fig. 3-9 and compare with the cell count values. $w/h = 5$. *Ans.* (*a*) 54 versus 50 Ω; (*b*) 27 versus 30 Ω; (*c*) 17 versus 20 Ω.

Energy, Power, and Poynting Vector

The power P conveyed by a transmission line is, in circuit notation,

$$P = VI \text{ (W)} \tag{29}$$

where V = line voltage, V, and I = line current, I. More generally, if V and I vary sinusoidally with time and are not in time phase, the average power

$$P_{\text{av}} = \tfrac{1}{2} V_0 I_0 \cos\theta \qquad \text{(W)} \tag{30}$$

where V_0 = peak voltage, V
I_0 = peak current, A
θ = phase angle between V and I, degrees

In field notation, the power density **S** is given by the Poynting vector, which is a cross-product of the electric and magnetic field vectors, **E** and **H**. Thus,

$$\mathbf{S} = \mathbf{E} \times \mathbf{H} \qquad (\text{W m}^{-2}) \qquad \textit{Poynting vector} \tag{31}$$

where **S** = Poynting vector = power density, W m^{-2}
E = electric field vector, V m^{-1}
H = magnetic field vector, A m^{-1}

Turning **E** into **H** and proceeding as a right-handed screw gives the direction of **S** perpendicular to both **E** and **H**. **S** is a *power density* (W m^{-2}) called the *Poynting vector*. Its value in (31) is the *instantaneous Poynting vector*. The *average Poynting vector* is obtained by integrating the instantaneous Poynting vector over one period and dividing by one period. It is also readily obtained in complex notation from

$$\mathbf{S}_{av} = \frac{1}{2}\text{Re}\,\mathbf{E} \times \mathbf{H}^* = \frac{1}{2}\hat{\mathbf{x}}|E_y||H_z|\cos\xi \qquad (\text{W m}^{-2}) \tag{32}$$

where $\mathbf{S}_{av} = \hat{\mathbf{x}}S$ = average Poynting vector, W m^{-2}
$\mathbf{E} = \hat{\mathbf{y}}E_y = \hat{\mathbf{y}}|E_y|e^{j\omega t}$, V m^{-1}
$\mathbf{H}^* = \hat{\mathbf{z}}H_z^* = \hat{\mathbf{z}}|H_z|e^{-j(\omega t-\xi)}$, A m^{-1}
ξ = time phase angle between E_y and H_z, rad or deg

$\mathbf{H}^*$ is called the *complex conjugate* of **H**, where

$$\mathbf{H} = \hat{\mathbf{z}}H_z = \hat{\mathbf{z}}|H_z|e^{j(\omega t-\xi)} \qquad (\text{A m}^{-1})$$

The quantities **H** and its complex conjugate $\mathbf{H}^*$ have the same space direction, but they differ in sign in their phase factors. Note that if E_y and H_z in (32) are root mean square (rms) values instead of (peak) amplitudes, the factor ½ in (32) is omitted.

The magnitude of the *average Poynting vector* is given by

$$S_{av} = \frac{1}{2}\text{Re}\,E_yH_z^* = \frac{1}{2}|E_y||H_z|\cos\xi \qquad (\text{W m}^{-2}) \tag{33}$$

Since the intrinsic impedance of the medium

$$Z_0 = \frac{E}{H} = \frac{|E|}{|H|}\angle\xi = |Z_0|\angle\xi$$

the magnitude of the *average Poynting vector* can also be written

$$S_{av} = \frac{1}{2}\,\text{Re}\,H_zH_z^*Z_0 = \frac{1}{2}|H_z|^2\,\text{Re}\,Z_0 \qquad (\text{W m}^{-2}) \tag{34}$$

or

$$S_{av} = \frac{1}{2}\text{Re}\frac{E_yE_y^*}{Z_0} = \frac{1}{2}|E_y|^2\,\text{Re}\,\frac{1}{Z_0} \qquad (\text{W m}^{-2}) \tag{35}$$

Further,

$$S_{av} = \frac{1}{2}|H_z|^2|Z_0|\cos\xi = \frac{1}{2}\frac{|E_y|^2}{|Z_0|}\cos\xi \tag{36}$$

When $\xi = 90°$, $S_{av} = 0$.

Similarly, for V and I, the average power is given by

$$P_{av} = \frac{1}{2}I_0^2|Z_0|\cos\theta = \frac{1}{2}\frac{V_0^2}{|Z_0|}\cos\theta \tag{37}$$

The bars | | on E_y and H_z indicate peak magnitude, which can also be indicated by a zero subscript. Thus,

$$|E_y| = E_{y0} \text{ and } |H_z| = H_{z0}$$

Another option is to use rms values. Thus,

$$V_{rms} = 0.707V_0; \quad I_{rms} = 0.707I_0; \quad E_y\text{ (rms)} = 0.707|E_y|; \quad H_z\text{ (rms)} = 0.707|H_z|$$

Equation (36) thus becomes

$$S_{av} = H_z^2(\text{rms})|Z_0|\cos\xi = \frac{E_y^2(\text{rms})}{|Z_0|}\cos\xi \tag{38}$$

and (37) becomes

$$P_{av} = I_{rms}^2|Z_0|\cos\theta = \frac{V_{rms}^2}{|Z_0|}\cos\theta \tag{39}$$

Root-mean-square means the square root of the mean (or average) of the square. One cycle or period (T) of a sinusoidal wave is shown in Fig. 3-12 as a solid curve and its square is shown as a dashed curve. Integrating V^2 over one period ($t = T$) and dividing by the radians in one period ($= 2\pi$) yields an average value of $\frac{1}{2}|V|^2$,

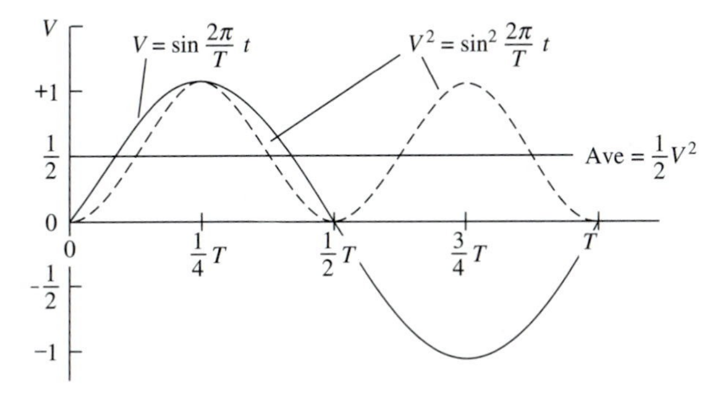

FIGURE 3-12
One cycle or period (2π radians) of a sinusoidal wave (solid curve) and its square (dashed curve).

as is obvious from the symmetry of the V^2 curve. Finally, taking the square root, we obtain the root-mean-square value,

$$V_{\text{rms}} = 0.707|V|$$

Example 3-5. Line power. (*a*) Find the average power on a transmission line for which $V = 180 \sin 2\pi 60t$ V, and $I = 600 \sin(2\pi 60t + 24°)$ mA. (*b*) If this power is confined inside a long hollow perfectly conducting tube 1 m in diameter, find the average power density, or Poynting vector.

Solution. From (28),

$$P_{\text{av}} = \frac{1}{2}(180 \times 0.6)\cos 24° = 49.3 \text{ W} \quad \textit{Ans.} \quad (a)$$

$$S_{\text{av}} = 49.3/\text{area} = 49.3/(\pi \times 0.5^2) = 62.8 \text{ W m}^{-2} \quad \textit{Ans.} \quad (b)$$

Problem 3-3-7. Line E and H values. Referring to (*b*) of Example 3-5, find the "peak-average" values of E and H, i.e., peak with respect to time and average with respect to the tube cross section. *Ans.* $E = 203$ V/m, H = 677 mA/m.

An important property of a field map divided into curvilinear cells of the same impedance is that *the same power P* (W) *is transmitted through each cell.* The Poynting vector (W m^{-2}) at each cell is then P divided by the cell area.

Example 3-6. Poynting vector on two-conductor transmission line. Referring to Fig. 3-5, the two conductors have a center-to-center spacing of 1 m and operate at a potential difference of 4160 V (rms) at 60 Hz. The ratio of the center-to-center spacing to conductor diameter is 3 to 1. Find (*a*) power transmitted and (*b*) power per cell. Neglect any line loss. $\varepsilon_r = 1$.

Solution. (*a*) From the field map, the line impedance

$$Z = \frac{N_s}{N_p}377 = \frac{6}{10.4}377 = 218 \ \Omega$$

From (26),

$$Z = \frac{276}{\sqrt{\varepsilon_r}} \log \frac{D}{a} = 276 \log 6 = 215 \ \Omega$$

From (39),

$$P_{\text{av}} = \frac{V^2(\text{rms})}{|Z_0|} = \frac{4160^2}{215} = 80.5 \text{ kW} \quad \textit{Ans.} \quad (a)$$

$$(b)\ P_{\text{cell}} = \frac{80.5}{N_s N_p} = \frac{80.5}{6 \times 10.4} = 1.3 \text{ kW} \quad \textit{Ans.} \quad (b)$$

Problem 3-3-8. Poynting vector at midpoint between two-conductor line. Referring to Example 3-6, find the Poynting vector at P_1 in Fig. 3-5. *Ans.* 70 kW/m^2.

Problem 3-3-9. Traveling wave. Poynting vector. A plane traveling wave has a peak electric field $E_0 = 4$ V/m. If the medium is lossless with $\mu_r = 1$ and $\varepsilon_r = 3.5$, find (*a*) velocity of wave, (*b*) peak Poynting vector, (*c*) average Poynting vector, (*d*) impedance of medium, and (*e*) peak value of the magnetic field H. *Ans.* (*a*) 160 Mm/s; (*b*) 79 mW/m^2; (*c*) 40 mW/m^2; (*d*) 202 Ω/square; (*e*) 20 mA/m.

Example 3-7. Field and power relations for microstrip line. For the microstrip line of Fig. 3-7, find (*a*) capacitance C per meter, (*b*) inductance $\mathcal{L}$ per meter, and (*c*) impedance Z_{line}. For an applied voltage at 1 GHz of $V = 10\sin 2\pi 10^9 t$ V, find (*d*) line current I, (*e*) line power, and (*f*) power per cell. The line is assumed uniform and infinitely long. $\varepsilon_r = \mu_r = 1$.

Solution. (*a*) $C = \varepsilon_0 \dfrac{N_p}{N_s} = \varepsilon_0 \dfrac{38}{5} = 67.3 \text{ pF m}^{-1}$ *Ans.* (*a*)

(*b*) $\mathcal{L} = \mu_0 \dfrac{N_s}{N_p} = 165 \text{ nH m}^{-1}$ *Ans.* (*b*)

(*c*) $Z_{\text{line}} = \sqrt{\dfrac{\mathcal{L}}{C}} = 49.5\ \Omega$ *Ans.* (*c*)

(*d*) $I = V/Z = 202\sin 2\pi 10^{-9} t$ mA *Ans.* (*d*)
(*e*) Peak power $= V^2/Z = 10^2/49.5 = 2.02$ W *Ans.* (*e*)
(*f*) Peak power/cell $= 2.02/(5 \times 38) = 10.6$ mW *Ans.* (*f*)

Problem 3-3-10. Microstrip transmission line power. Referring to Example 3-7, if the average power per cell is 25 mW, what is (*a*) the total power and (*b*) the power in the fringing field? *Ans.* (*a*) 4.75 W; (*b*) 1.57 W.

3-4 THE TERMINATED UNIFORM TRANSMISSION LINE AND VSWR

Thus far we have considered only lines of infinite length. Let us now analyze the situation where a line of characteristic impedance Z_0 is terminated in a load impedance Z_L, as in Fig. 3-13 on the next page. The load is at $x = 0$ and positive distance x is measured to the left along the line. The total voltage and total current are expressed as the resultant of two traveling waves moving in opposite directions as on an infinite transmission line. However, on the terminated line (Fig. 3-13) the wave to the right may be regarded as the incident wave and the wave to the left as the reflected wave, with the reflected wave related to the incident wave by the load impedance Z_L.

At a point on the line at a distance x from the load, let the voltage between the wires and the current through one wire due to the incident wave traveling to the right be designated V_0 and I_0, respectively. Let V_1 and I_1 be the voltage and current due to the wave traveling to the left that is reflected from the load. The resultant voltage V at a point on the line is equal to the sum of the voltages V_0 and V_1 at the point. That is, in phasor notation

$$V = V_0 + V_1 \qquad \text{(V)} \qquad (1)$$

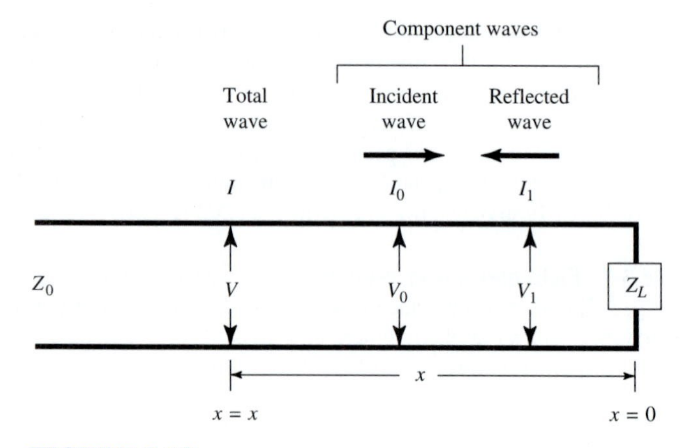

FIGURE 3-13
Terminated transmission line with incident and reflected waves.

where $V_0 = |V_0|e^{\gamma x}$
$V_1 = |V_1|e^{-\gamma x + j\xi}$
$\gamma =$ propagation constant $= \alpha + j\beta$
$\xi =$ phase shift at load

At the load ($x = 0$) we have $V_0 = |V_0|$ and $V_1 = |V_1|e^{j\xi} = |V_1|\angle\xi$, so that *at the load the ratio of the reflected and incident voltages is given by*

$$\frac{V_1}{V_0} = \frac{|V_1|}{|V_0|}\angle\xi = \rho_v \tag{2}$$

where ρ_v is the *reflection coefficient for voltage* (dimensionless). It follows that

$$V = |V_0|(e^{\gamma x} + \rho_v e^{-\gamma x}) \qquad \text{(V)} \tag{3}$$

The resultant current I at a point on the line is equal to the sum of the currents I_0 and I_1 at the point. That is,

$$I = I_0 + I_1 \qquad \text{(A)} \tag{4}$$

where $I_0 = |I_0|e^{\gamma x - j\delta}$
$I_1 = |I_1|e^{-\gamma x + j(\xi - \delta)}$
$\delta =$ phase difference between current and voltage

At the load the ratio of the reflected and incident currents is given by

$$\frac{I_1}{I_0} = \frac{|I_1|}{|I_0|}\angle\xi = \rho_i \tag{5}$$

where ρ_i is the *reflection coefficient for current* (dimensionless). It follows that

$$I = |I_0|e^{-j\delta}(e^{\gamma x} + \rho_i e^{-\gamma x}) \tag{6}$$

Now ρ_v and ρ_i may be expressed in terms of the characteristic impedance Z_0 and the load impedance Z_L. Thus, we note that at any point on the line

$$Z_0 = \frac{V_0}{I_0} = \frac{|V_0|}{|I_0|} \angle\delta = -\frac{V_1}{I_1} = -\frac{|V_1|}{|I_1|} \angle\delta \tag{7}$$

while at the load ($x = 0$)

$$Z_L = \frac{V}{I} \tag{8}$$

It follows from (4) and (7) that at the load

$$\frac{V}{Z_L} = \frac{V_0}{Z_0} - \frac{V_1}{Z_0} = \frac{V_0 - V_1}{Z_0} \tag{9}$$

But $V = V_0 + V_1$; so we have

$$\frac{V_0 + V_1}{Z_L} = \frac{V_0 - V_1}{Z_0} \tag{10}$$

Solving for V_1/V_0 yields

$$\frac{V_1}{V_0} = \frac{Z_L - Z_0}{Z_L + Z_0} = \rho_v \qquad \textit{Reflection coefficient for voltage} \tag{11}$$

For real load impedances Z_L ranging from 0 to ∞, ρ_v ranges from -1 to $+1$ in value.

In a similar way it can be shown that

$$\rho_i = -\frac{Z_L - Z_0}{Z_L + Z_0} = -\rho_v \tag{12}$$

The ratio V/I at any point x on the line gives the impedance Z_x at the point looking toward the load. Taking this ratio and introducing the relation (12) in (6) for I, we obtain

$$Z_x = \frac{V}{I} = \frac{|V_0|}{|I_0|}\angle\delta\left(\frac{e^{\gamma x} + \rho_v e^{-\gamma x}}{e^{\gamma x} - \rho_v e^{-\gamma x}}\right) \tag{13}$$

Noting (7) and (11), we can reexpress this as

$$Z_x = Z_0 \frac{Z_L + Z_0 \tanh \gamma x}{Z_0 + Z_L \tanh \gamma x} \quad (\Omega) \qquad \textit{Impedance at distance x from load} \tag{14}$$

where Z_x = impedance at distance x looking toward load, Ω
Z_0 = characteristic impedance of line, Ω
Z_L = load impedance, Ω
γ = propagation constant $= \alpha + j\beta$, m^{-1}
α = attenuation constant, Np m^{-1}
β = phase constant, rad m^{-1} or deg m^{-1}
x = distance from load, m

This is the *general expression* for the line impedance Z_x as a function of the distance x from the load.

If the line is *open-circuited*, $Z_L = \infty$ and (14) reduces to

$$Z_x = \frac{Z_0}{\tanh \gamma x} = Z_0 \coth \gamma x \tag{15}$$

If the line is *short-circuited*, $Z_L = 0$ and (14) reduces to

$$Z_x = Z_0 \tanh \gamma x \tag{16}$$

It is to be noted that, in general, γ is complex ($= \alpha + j\beta$). Thus

$$\tanh \gamma x = \frac{\sinh \alpha x \cos \beta x + j \cosh \alpha x \sin \beta x}{\cosh \alpha x \cos \beta x + j \sinh \alpha x \sin \beta x} \tag{17}$$

$$\tanh \gamma x = \frac{\tanh \alpha x + j \tan \beta x}{1 + j \tanh \alpha x \tan \beta x} \tag{18}$$

Note that the product of the impedance of the line when it is open-circuited and when it is short-circuited equals the square of the characteristic impedance Z_0. Thus,

$$Z_0^2 = Z_{oc} Z_{sc} \qquad \text{or} \qquad Z_0 = \sqrt{Z_{oc} Z_{sc}} \tag{19}$$

where $Z_{oc} = Z_x$ for open-circuited line ($Z_L = \infty$) and $Z_{sc} = Z_x$ for short-circuited line ($Z_L = 0$).

If the line is *lossless* ($\alpha = 0$), the above relations reduce to the following: *In general*,

$$Z_x = Z_0 \frac{Z_L + jZ_0 \tan \beta x}{Z_0 + jZ_L \tan \beta x} \quad (\Omega) \qquad \textbf{\textit{Impedance at distance x from load on lossless line}} \tag{20}$$

When the line is *open-circuited* ($Z_L = \infty$),

$$Z_x = \frac{Z_0}{j \tan \beta x} = -jZ_0 \cot \beta x \quad (\Omega) \qquad \textbf{\textit{Open circuit}} \tag{21}$$

When the line is *short-circuited* ($Z_L = 0$),

$$Z_x = jZ_0 \tan \beta x \quad (\Omega) \qquad \textbf{\textit{Short circuit}} \tag{22}$$

We note that (19) is also fulfilled on the lossless line. Furthermore, the impedance for an open- or short-circuited lossless line is a pure reactance.

The impedance relations developed above apply to all uniform two-conductor lines, such as coaxial lines, strip lines, microstrip lines, and two-wire lines. They give the input impedance, Z_x, of a uniform transmission line of length x and characteristic impedance Z_0 terminated in a load Z_L (see Fig. 3-14). These relations are summarized in Table 3-2. In this chapter all lines (in examples and problems) are lossless. Lossy lines are discussed in Chap. 4.

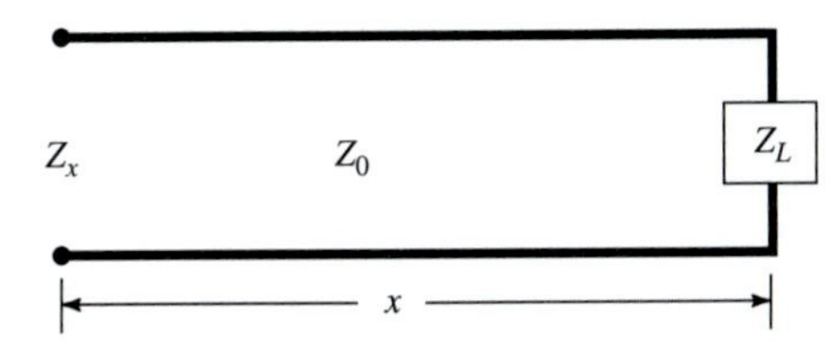

FIGURE 3-14
Terminated transmission line of length x has input impedance Z_x.

TABLE 3-2
Input impedance of terminated transmission line†

Load condition	General case ($\alpha \neq 0$)	Lossless case ($\alpha = 0$)
Any value of load Z_L	$Z_x = Z_0 \dfrac{Z_L + Z_0 \tanh \gamma x}{Z_0 + Z_L \tanh \gamma x}$	$Z_x = Z_0 \dfrac{Z_L + jZ_0 \tan \beta x}{Z_0 + jZ_L \tan \beta x}$
Open-circuited line ($Z_L = \infty$)	$Z_x = Z_0 \coth \gamma x$	$Z_x = -jZ_0 \cot \beta x$
Short-circuited line ($Z_L = 0$)	$Z_x = Z_0 \tanh \gamma x$	$Z_x = jZ_0 \tan \beta x$

†$\gamma = \alpha + j\beta$, where α = attenuation constant, Np m^{-1}, $\beta = 2\pi/\lambda$ = phase constant, rad m^{-1}, and λ = wavelength, m.

Example 3-8. $\lambda/4$ matching section. Find the line impedance Z_0 to match a load Z_L to a desired value Z_x.

Solution. Use a $\lambda/4$ section ($x = \lambda/4$). Then (20) reduces to

$$Z_x = Z_0^2/Z_L \quad \text{and}$$

$$Z_0 = \sqrt{Z_x Z_L} \quad \textit{Ans.}$$

Problem 3-4-1. Matching to 50-Ω line. Find the transmission line impedance Z_0 required to match a load $Z_L = 100\ \Omega$ to a 50-Ω line ($Z_x = 50\ \Omega$). *Ans.* 70.7 Ω.

Problem 3-4-2. Power splitter. Referring to Fig. 3-15, two loads $R_1 = 300\ \Omega$ and $R_2 = 200\ \Omega$ are to be fed in phase from a 100-Ω line with R_1 receiving twice the power of R_2. Find (*a*) l_1 and Z_1 and (*b*) l_2 and Z_2. *Ans.* (*a*) $l_1 = \lambda/4$, $Z_1 = 212\ \Omega$; (*b*) $l_2 = \lambda/4$, $Z_2 = 245\ \Omega$.

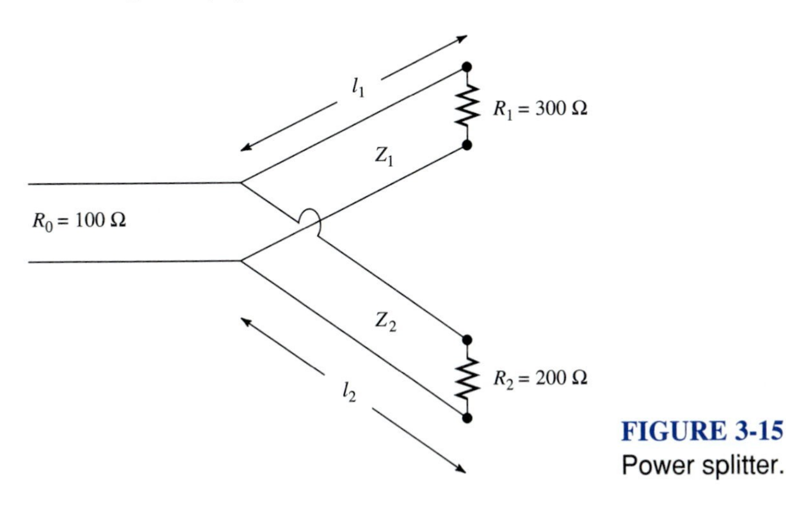

FIGURE 3-15
Power splitter.

On a lossless line the ***voltage standing-wave ratio*** (VSWR) is given by

$$\text{VSWR} = \frac{V_{\max}}{V_{\min}} = \frac{I_{\max}}{I_{\min}} \tag{23}$$

It follows that

$$\text{VSWR} = \frac{|V_0| + |V_1|}{|V_0| - |V_1|} = \frac{1 + (|V_1|/|V_0|)}{1 - (|V_1|/|V_0|)} \tag{24}$$

$$\frac{|V_1|}{|V_0|} = |\rho_v| \tag{25}$$

and so

$$\text{VSWR} = \frac{1 + |\rho_v|}{1 - |\rho_v|} \tag{26}$$

where ρ_v is the reflection coefficient for voltage.

Solving (26) for the magnitude of the *reflection coefficient* gives

$$|\rho_v| = \frac{\text{VSWR} - 1}{\text{VSWR} + 1} \qquad \textbf{\textit{Reflection coefficient}} \tag{27}$$

The relations for reflection and transmission coefficients developed in this section are summarized in Table 3-3.

TABLE 3-3
Relations for reflection and transmission coefficients

Reflection coefficient for voltage	$\rho_v = \frac{Z_L - Z_0}{Z_L + Z_0}$
Reflection coefficient for current	$\rho_i = \frac{Z_0 - Z_L}{Z_0 + Z_L} = -\rho_v$
Transmission coefficient for current	$\tau_i = \frac{2Z_0}{Z_0 + Z_L} = 1 + \rho_i$
Transmission coefficient for voltage	$\tau_v = \frac{2Z_L}{Z_0 + Z_L} = 1 + \rho_v$
Voltage standing-wave ratio	$\text{VSWR} = \frac{1 + \|\rho_v\|}{1 - \|\rho_v\|} = \frac{1 + \|\rho_i\|}{1 - \|\rho_i\|}$
Magnitude of reflection coefficient	$\|\rho_v\| = \|\rho_i\| = \frac{\text{VSWR} - 1}{\text{VSWR} + 1}$

Example 3-9. Reflections on a lossless line. A signal of 10 V is applied to a 50-Ω coaxial transmission line terminated in a 200-Ω load. Find (*a*) the voltage reflection coefficient, (*b*) the magnitude of the reflected voltage, and (*c*) the magnitude of the reflected current.

Solution. From (11),

$$\rho_v = \frac{Z_L - Z_0}{Z_L + Z_0} = \frac{200 - 50}{200 + 50} = \frac{150}{250} = 0.6 \quad \textit{Ans. (a)}$$

From (2), the reflection coefficient is equal to the ratio of the reflected voltage to the input voltage at the load. Thus

$$\frac{V_r}{V_i} = \rho_v$$

$$V_r = \rho_v V_i = 0.6 \times 10 = 6 \text{ V} \quad \textit{Ans.} \quad (b)$$

From (7), the magnitude of the reflected current is given by

$$Z_0 = -\frac{|V_r|}{|I_r|} \angle 0^\circ$$

Solving for $|I_r|$ we obtain its magnitude as

$$|I_r| = \frac{|V_r|}{Z_0} = \frac{6}{50} = 0.12 \text{ A} \quad \textit{Ans.} \quad (c)$$

Problem 3-4-3. VSWR on terminated line. A lossless 100-Ω transmission line is terminated in $50 + j75\ \Omega$. Find (*a*) voltage reflection coefficient and (*b*) VSWR. *Ans.* (*a*) $0.537 \angle -83^\circ$, (*b*) 3.32.

Example 3-10. Determination of transmission line impedance. RG-59U, a common type of transmission line for microwave applications, has an open-circuit impedance of $150 \angle 25^\circ\ \Omega$ and a short-circuit impedance of $37.5 \angle -35^\circ\ \Omega$. What is the characteristice impedance of this line?

Solution. From (19), the characteristic impedance of a line is a geometric mean of the open- and short-circuit impedances. Thus

$$Z_0 = \sqrt{Z_{oc}Z_{sc}} = \sqrt{(150 \angle 25^\circ)(37.5 \angle -35^\circ)}$$
$$= \sqrt{(150)(37.5)} \angle 0.5[25^\circ + (-35^\circ)] = 75 \angle -5^\circ\ \Omega \quad \textit{Ans.}$$

Problem 3-4-4. Coaxial line. A coaxial line has an open circuit $Z_{oc} = 105 \angle 17^\circ\ \Omega$ and a short circuit $Z_{sc} = 24 \angle -3^\circ\ \Omega$. Find the characteristic impedance of the line. *Ans.* $50.2 \angle 7^\circ\ \Omega$.

Example 3-11. Lossless line impedance at a distance from the load. Find the impedance on a 50-Ω transmission line at a distance of $\lambda/8$ from a 400-Ω load.

Solution. From (20), the transmission-line impedance as a function of distance from a load is

$$Z_x = Z_0 \frac{Z_L + jZ_0 \tan \beta x}{Z_0 + jZ_L \tan \beta x}$$

$$= 50 \frac{400 + j50 \tan\left(\frac{\pi}{4}\right)}{50 + j400 \tan\left(\frac{\pi}{4}\right)} = \frac{400 + j50}{1 + j8}$$

This may be evaluated in two ways. In the phasor approach, both the numerator and the denominator are converted to a magnitude and phase. We then divide the magnitude of the numerator by the magnitude of the denominator, and subtract the phase of the denominator from the phase of the numerator. For this case,

$$\frac{400 + j50}{1 + j8} = \frac{403.1 \angle 7.1^\circ}{8.06 \angle 82.8^\circ} = 50.0 \angle -75.9^\circ \quad \textit{Ans.}$$

An alternative approach is to convert the denominator to a real number by multiplying by its complex conjugate:

$$\frac{400 + j50}{1 + j8} \times \frac{1 - j8}{1 - j8} = \frac{400 - j3200 + j50 + 400}{1 + 64} = 12.3 - j48.5$$

$$= 50.0 \angle -75.9^\circ \quad \textit{Ans.}$$

Problem 3-4-5. Z_x on terminated line. A lossless 50-Ω transmission line is terminated in $35 + j65\ \Omega$. Find (*a*) voltage reflection coefficient, (*b*) VSWR, (*c*) impedance (Z_x) 0.35λ from load, (*d*) shortest length of line for which impedance is purely resistive; and (*e*) the value of this resistance. *Ans.* (*a*) $0.623 \angle 65.6^\circ$; (*b*) 4.31; (*c*) $11.63 + j2.66\ \Omega$; (*d*) 0.091λ; (*e*) 216 Ω.

Example 3-12. Voltage along a lossless transmission line. A lossless 75-Ω transmission line is terminated in a load of 350 Ω. (*a*) Plot the magnitude of the voltage along the line, at increments of $\lambda/4$ for a distance of 1λ from the load. Assume an input voltage of V_0. (*b*) What is the VSWR?

Solution. From (1) the voltage along the line is the sum of the incoming and reflected voltages. The reflection coefficient for voltage, from (12), is

$$\rho_v = \frac{Z_L - Z_0}{Z_L + Z_0} = \frac{350 - 75}{350 + 75} = 0.647$$

Along a lossless line, the magnitudes of the incoming and reflected voltages are constant, and the phase varies as βx. Thus, the magnitude of the total voltage is given by

$$|V| = V_0 \left| e^{j\beta x} + \rho_v e^{-j\beta x} \right|$$

as tabulated here and shown in Fig. 3-15.1. *Ans.* (*a*)

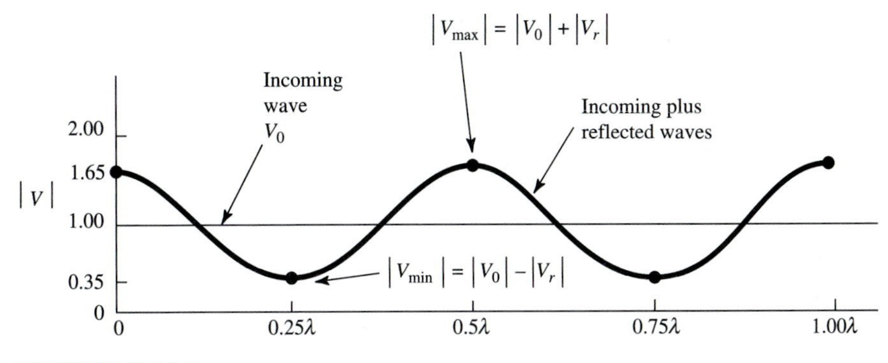

FIGURE 3-15.1
Voltage on terminated line [Ans. (*a*)].

x	$e^{j\beta x}$	V_i	V_r	$\|V\|$
0	$1 + j0$	V_0	$0.65V_0$	$1.65V_0$
$\lambda/4$	$0 + j1$	jV_0	$-j0.65V_0$	$0.35V_0$
$\lambda/2$	$-1 + j0$	$-V_0$	$-0.65V_0$	$1.65V_0$
$3\lambda/4$	$0 - j1$	$-jV_0$	$j0.65V_0$	$0.35V_0$
λ	$1 + j0$	V_0	$0.65V_0$	$1.65V_0$

$$\text{VSWR} = \frac{1.65}{0.35} = 4.71 \quad \textit{Ans.} \quad (b)$$

Problem 3-4-6. Voltage on terminated line. A lossless 100-Ω transmission line is terminated in $200 + j200$ Ω. Find: (*a*) voltage reflection coefficient, (*b*) VSWR, (*c*) impedance 0.375λ from load, (*d*) shortest length of line for which impedance is purely resistive, and (*e*) the value of this resistance. For 15 V applied to the line find: (*f*) maximum and (*g*) minimum line voltage. *Ans.* (*a*) 0.62 ∠29.5°; (*b*) 4.26; (*c*) $31 + j54$ Ω; (*d*) 0.042λ; (*e*) 42.5 Ω; (*f*) 24.3 V; (*g*) 5.7 V.

Impedance Matching; the Smith Chart

The Smith Chart is a valuable graphical tool. It helps us understand transmission line problems. We work with a chart, a pencil, a compass, and our brain—that magnificent computer that sits on seven cervical vertebra. Paper, pencil, and brain can provide innovative insights and solutions. Yes, there are computer-assisted solutions which are very useful but with them we are in a prepackaged environment. We may get a solution but be ignorant of better ones. So Smith Charts serve a dual purpose: (1) solve transmission line problems and (2) teach us how to think on our own.

The objective of most transmission line problems is to match a line to a load. Under matched conditions the reflection coefficient $\rho_v = 0$ and the VSWR $= 1$. We enter the chart at the load impedance. It is then like a game where the objective is to reach the center of the chart (Home) where $\rho_v = 0$ and the VSWR $= 1$.

Example 3-13. Smith Chart and rectangular chart compared. Find: (*a*) the length x of the 100-ohm transmission line of Fig. 3-16 which converts a load impedance $Z_L = 100 + j100\ \Omega$ to a pure resistance, (*b*) the value of the resistance R_x, and (*c*) the VSWR.

Solution. It greatly simplifies things to work in normalized impedances. Thus, to normalize the load impedance Z_L we divide it by the line impedance R_0, or

$$\frac{Z_L}{R_0} = \frac{100 + j100}{100} = 1 + j1$$

Entering this dimensionless number at $R = 1, X = 1$ on the Smith Chart and moving away from the load (clockwise along blue arc) on a constant VSWR $= 2.6$ circle [*Ans.* to (*c*)] to the R axis gives the required distance $x = 0.09\lambda$ [*Ans.* to (*a*)] and the pure resistance $R_x = 2.6 \times R_0 = 260\ \Omega$ [*Ans.* to (*b*)].

Entering $1 + j1$ on the rectangular chart gives the same answers but much less conveniently.

Using a Smith Chart

The Smith Chart transforms a rectangular chart that extends to infinite R and X to an area inside a circle. Zero R and X are at the left end of the horizontal axis and infinite R and X at the right with VSWR circles centered on $R = 1$ (Home) at the center of the chart. And a most convenient feature of the Smith Chart is that distance from the load is measured as a function of angle around the outside of the chart.

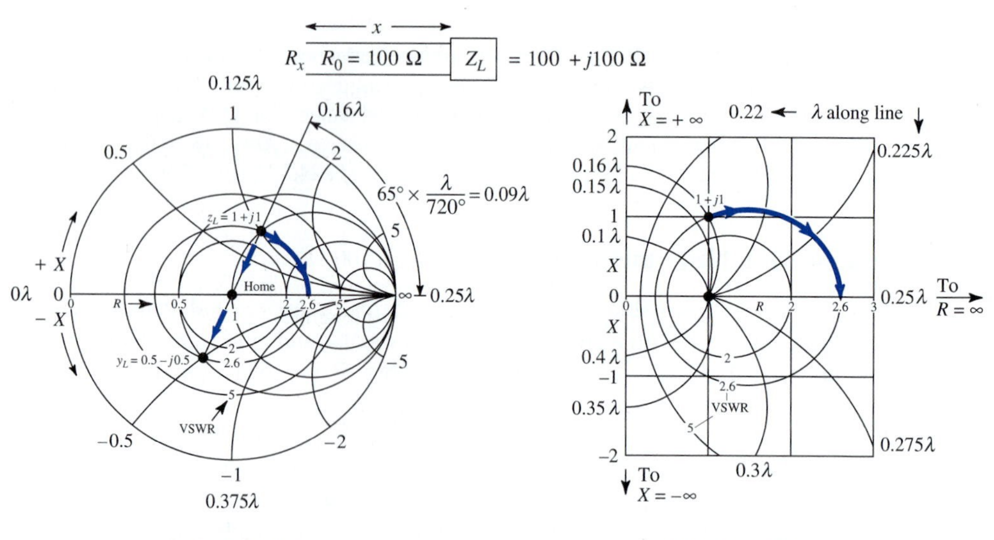

FIGURE 3-16
Smith Chart and rectangular chart compared for transmission line.

Thus, the angle from point $1 + j1$ to the R axis is 65° (Fig. 3-16). Since once around the chart (360°) is ½ wavelength ($\lambda/2$), we have

$$x = \frac{65^\circ}{720^\circ}\lambda = 0.09\lambda$$

When stubs are connected across or in parallel to a transmission line, it is convenient to work in admittances. On a Smith Chart any impedance $Z = R + jX$ is converted to an admittance $Y = G - jB$ by moving 180° on a constant VSWR circle or on a line through the center of the chart. Thus, the normalized admittance $Y_L = 1/Z = 1/(1 + j1) = 0.5 - j0.5$ *at the load*. We moved on the chart but *not* on the line. See blue arrows on Fig. 3-16.

In this way, Smith Charts can be used for either impedances or admittances. The numbers on the chart stay the same. Thus, an open circuit $R_L = \infty$ is at the right end of the horizontal axis corresponding to $G = 0$ at the left end.

Another way to think of it is that a Smith Chart performs a division of complex numbers graphically. Thus, the mathematical steps for the above example are

$$Y_L = \frac{1}{Z_L} = \frac{1}{1 + j1} = \frac{1}{\sqrt{2}\angle 45^\circ} = \frac{1\angle -45^\circ}{\sqrt{2}} = \frac{0.707 - j0.707}{\sqrt{2}} = \frac{1}{2} - j\frac{1}{2}$$

To transform Y_L back to an impedance we reverse the procedure. Thus,

$$Z_L = \frac{1}{Y_L} = \frac{1}{\frac{1}{2} - j\frac{1}{2}} = 1 + j1$$

Problem 3-4-7. Equation versus chart. Find (*b*) of Example 3-12 using (20) and Example 3-13 using (11) and (26). *Ans.* (*b*) 260 Ω, (*c*) 2.6.

Problem 3-4-8. Stub lengths. Determine the physical length x of the line of Example 3-13 at (*a*) 1 GHz and (*b*) 5 GHz. *Ans.* (*a*) 27 mm; (*b*) 5.4 mm.

Example 3-14. Single-stub matching. A uniform 100-Ω transmission line is terminated in a resistive load of 500 Ω. Use a shorted stub to match this load to the line. Find (*a*) distance d_1 from load to stub, (*b*) distance or length d_2 of the stub, (*c*) VSWR on d_1 line, and (*d*) VSWR on stub.

Solution. With a Smith Chart, shown in its complete form in Fig. 3-17, it is convenient to use normalized impedances Z_n (dimensionless). Thus, the normalized impedance of the load is

$$Z_n = \frac{Z_L}{R_0} = \frac{500 + j0}{100} = 5 + j0$$

Step 1: Enter the chart at $5 + j0$ (point A in Fig. 3-18 on page 149).

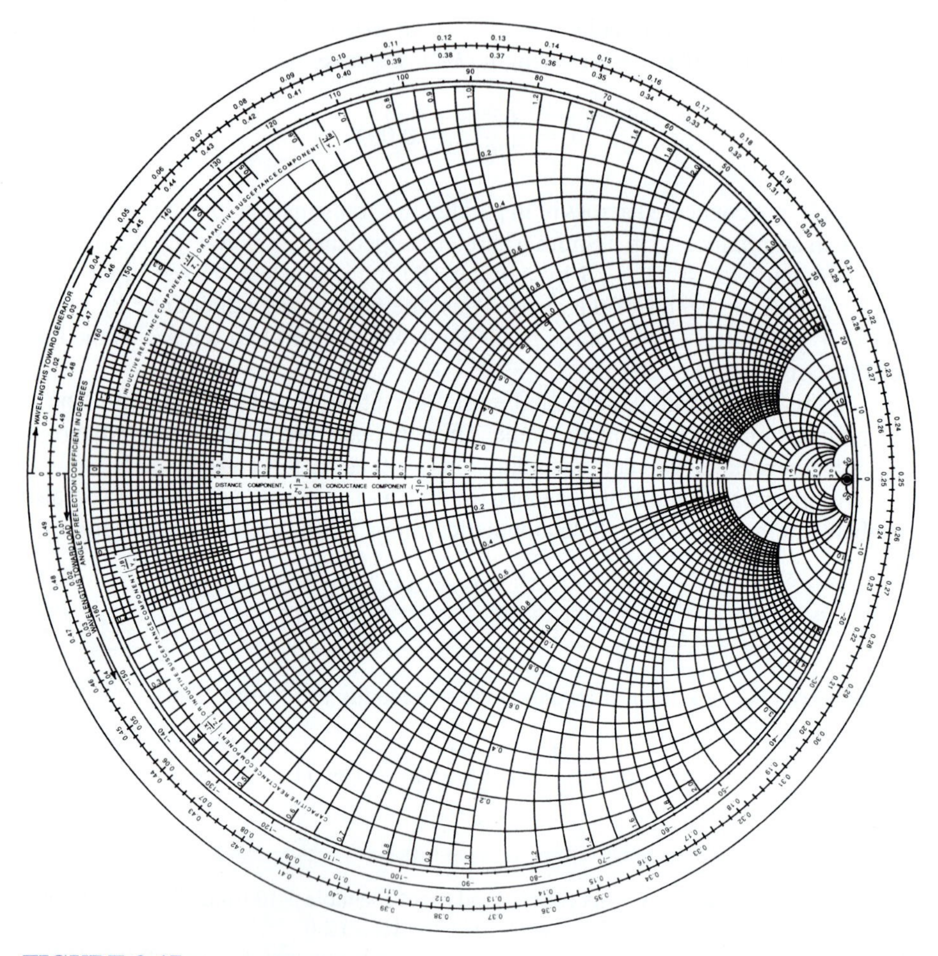

FIGURE 3-17
Smith Chart in its complete form. VSWR circles are drawn as needed. (*Smith Chart ©. Reproduced by permission of Philip H. Smith. Regular charts are 50 percent larger.*)

Step 2: Draw a VSWR circle centered on Home and passing through A. Using a Smith Chart is like playing a game. The object is to reach the center of the chart (Home) where $R_n + jX_n = 1 + j0$; the match condition.

Step 3: Since the stub and line will be connected in parallel, it is advantageous to work in admittances. To do this move $\lambda/4$ on the VSWR $= 5$ circle to point B. This converts $Z_n = R_n = 5$ to $Y_n = G_n = 0.2$. Note that we are still *at the load.*

Step 4: Now move clockwise away from the load on the VSWR $= 5$ circle [*Ans.* to (c)] until we reach point C at the R circle ($R_n = G_n = 1$) that passes through Home. This is the point where we attach the stub.

Step 5: The distance d_1 from B to C is measured along the edge of the chart as 0.182λ [*Ans.* to (a)].

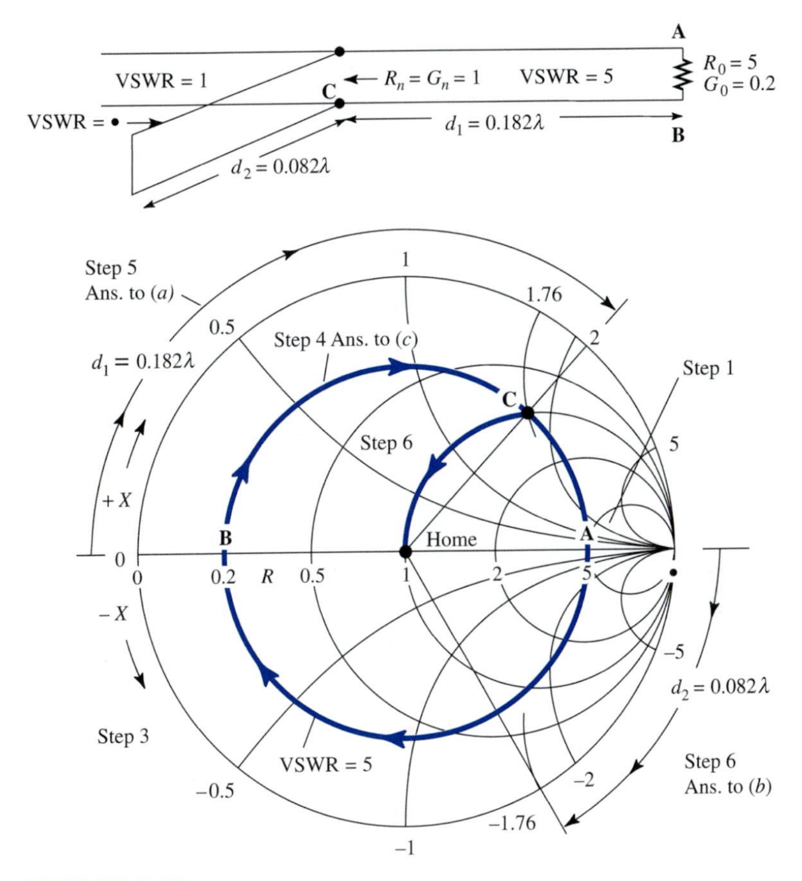

FIGURE 3-18
Line with stub and Smith Chart.

Step 6: For the distance d_2 (stub length), we note that point C is at $Y_n = 1 + j1.76$. If the stub presents a pure susceptance $-j1.76$, then the total admittance at point C on the line is $1 + j1.76 - j1.76 = 1 + j0$. On the chart we move from C to Home and the line is matched (VSWR = 1). Moving clockwise from $G_n = \infty$ (short circuit) along the edge of the chart to the $-j1.76$ curve, we obtain $d_2 = 0.082\lambda$ [*Ans.* to (*b*)]. Since the stub "load" $= G_n = \infty$ ($R_n = 0$), VSWR (stub) $= \infty$ [*Ans.* to (*d*)].

Problem 3-4-9. Single-stub matching. A uniform 100-Ω transmission line is terminated in a load impedance $Z_L = 150 + j50\ \Omega$. Find (*a*) distance d_1 from load to stub and (*b*) stub length d_2 for a match. *Ans.* $d_1 = 0.194\lambda$; $d_2 = 0.167\lambda$.

Example 3-15. λ/4 transformer matching. A 100-Ω line is terminated in a load impedance $Z_L = 200 - j100\ \Omega$. Referring to Fig. 3-19, find: (*a*) d_1, (*b*) $Z(\lambda/4)$, (*c*) VSWR on d_1 line, and (*d*) VSWR on λ/4 line.

Solution. The function of the d_1 line is to convert Z_L to a pure resistance at point B. The function of the $\lambda/4$ line is then to convert this resistance to 100 Ω for a match.

Step 1: Convert Z_L to a normalized impedance. Thus,

$$Z_n = \frac{200 - j100}{100} = 2 - j1$$

and enter the Smith Chart at $2 - j1$ (point A on the chart).

Step 2: Draw a VSWR circle through point A and move away from the load (clockwise) on the VSWR = 2.6 circle [*Ans.* to (*c*)] to point B on the R axis at $R_n = 0.38 + j0$. Note that $1/0.38 = 2.6$.

Step 3: Multiplying R_n by the line resistance $R_0 = 100\ \Omega$ and we obtain the impedance at point B as $Z_B = 38 + j0\ \Omega$. To convert this to 100 Ω, the $\lambda/4$ line impedance should be equal to the geometric mean, or square root of the product of Z_B and R_0 [from (20) when $x = \lambda/4$]. Thus,

$$Z(\lambda/4) = \sqrt{Z_B R_0} = \sqrt{38 \times 100} = 61.6\ \Omega \quad \textit{Ans.} \quad (b)$$

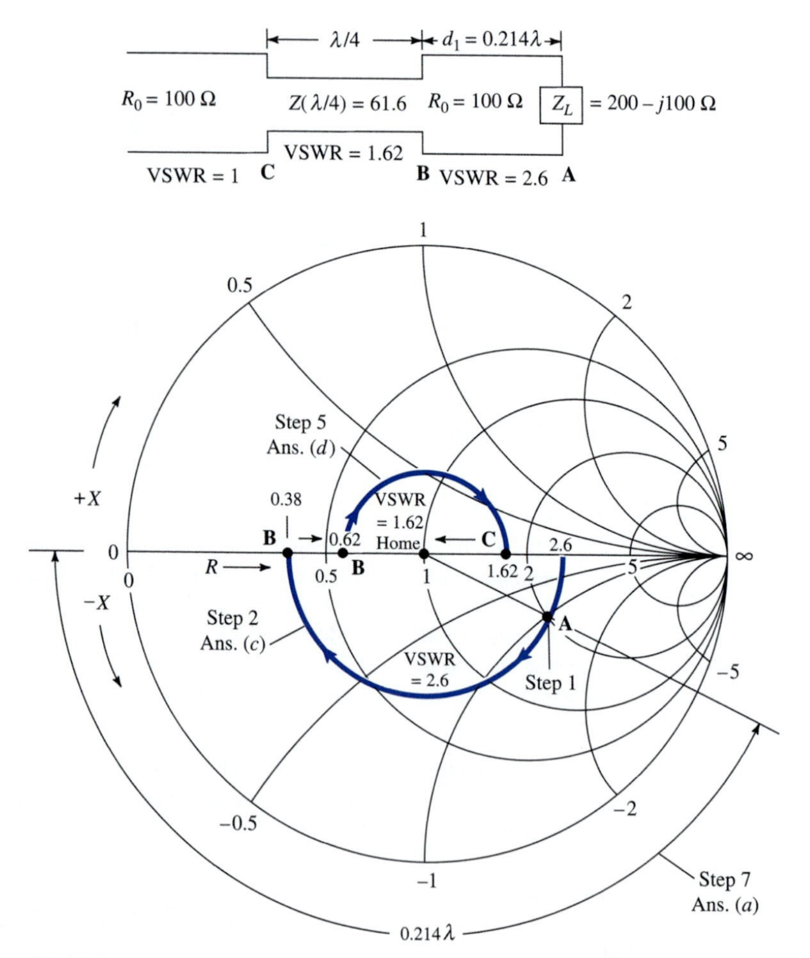

FIGURE 3-19

Step 4: To confirm this on the chart, we normalize Z_B to the $\lambda/4$ line resistance. Thus,

$$Z_n = \frac{Z_B}{Z(\lambda/4)} = \frac{38}{61.6} = 0.62$$

Step 5: We enter the chart at $R_n = 0.62$, which is still at the point B on the line, and move $\lambda/4$ (180°) clockwise away from the load on the VSWR = 1.62 circle [*Ans.* to (*d*)] to the $R = 0$ axis at $R_n = 1.62$ (point C).

Step 6: To convert this normalized resistance (1.62) to its ohmic value, we multiply it by the $\lambda/4$ line resistance of 61.6. Thus, $1.62 \times 61.6 = 100 + j0$ for a match as anticipated in Step 3.

Step 7: The line length d_1 as read along the edge of the chart is 0.214λ [*Ans.* to (*a*)].

Note that in this example we did not convert to admittances because no parallel connections were involved.

Problem 3-4-10. Quarter-wave transformer matching. A 100-Ω line is connected to a load impedance of $Z_L = 300 + j200\ \Omega$. Find: (*a*) line length d_1 required to transform this impedance to a pure resistance, (*b*) impedance of the $\lambda/4$ line required for a match, (*c*) VSWR on the d_1 line, and (*d*) VSWR on the $\lambda/4$ line. *Ans.* (*a*) 0.026λ; (*b*) 214 Ω; (*c*) 4.6; (*d*) 2.15.

Stub Tuning

If a transmission line has two wires as in Fig. 3-18, Example 3-14, it is easy to move the stub along the line to the proper position. But if the line is a coaxial cable, it requires a special telescoping section called a "line stretcher" to do this. It is often more convenient to have two stub lines attached at fixed positions and achieve a match by adjusting the lengths of the two stubs. Such double-stub tuners are very common. See Figure 3-20 on page 152.

Example 3-16. Double-stub matching. A 50-Ω coaxial line is terminated in a load impedance $Z_L = 100 + j50\ \Omega$ with two stub tuners at $\lambda/8$ spacings as shown in Fig. 3-20. Each stub is "tuned" with a sliding short-circuit plug. All lines and stubs are 50 Ω. Find: (*a*) length d_1 of stub 1, (*b*) length d_2 of stub 2, (*c*) VSWR on $\lambda/8$ line between the load and stub 1, (*d*) VSWR on $\lambda/8$ line between the stubs, (*e*) VSWR on stub 1, and (*f*) VSWR on stub 2.

Solution

Step 1: Normalize the load impedance by dividing by the line impedance. Thus,

$$Z_n = \frac{100 + j50}{50} = 2 + j1$$

and enter the chart at this value (point A).

Step 2: Draw a VSWR = 2.4 circle [*Ans.* to (*c*)] through point A and go halfway around the chart to point B, converting Z_n to the admittance $Y_n = 0.4 - j0.2$. We are still at the load.

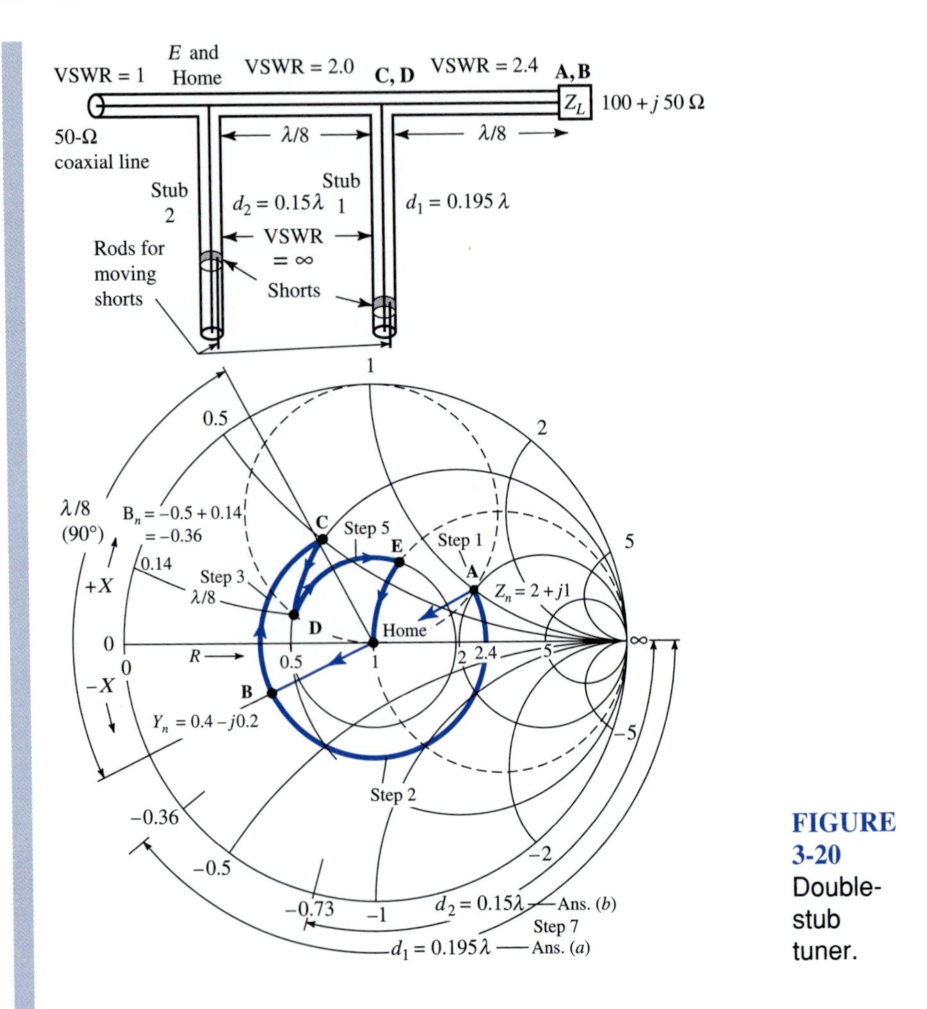

FIGURE 3-20 Double-stub tuner.

Step 3: We now move away from the load λ/8 to the first stub which is at point *C* on the chart. To reach the center of the chart from *C*, we need to think ahead. Thus, let us rotate the $G_n = 1$ circle (shown dashed) λ/8 (90°) toward the load or to the "up" position as shown in Fig. 3-20 because we want to be on this circle when we arrive at stub 2. To do this, we move on a constant G_n circle to point *D* on the chart.

Step 4: To go from $C(Y_n = 0.5 + j0.5)$ to point $D(Y_n = 0.5 + j0.14)$ requires that stub 1 present a susceptance $B_n = -0.5 + 0.14 = -0.36$ to the main line. The total admittance at point *D* is then

$$Y_n = 0.5 + j0.5 - j0.36 = 0.5 + j0.14$$

Points *C* and *D* are both at stub 1.

Step 5: Now we move λ/8 from stub 1 to stub 2, or from point *D* on the constant VSWR = 2.0 circle [*Ans.* to (*d*)], to point *E* on the $G_n = 1$ circle where the line admittance is

$$Y_n = 1 + j0.73$$

Step 6: To achieve a match, the length of stub 2 is adjusted so as to present a susceptance of $-j0.73$ to the main line taking us from E to the center of the chart (Home) where $Y_n = 1 + j0$. When $G_n = 1$, $R_n = 1$ and the impedance presented to the main line at stub 2 is equal to $R_n \times R_0 = 1 \times 50 = 50\ \Omega$ for a match.

Step 7: Reading off the stub lengths along the edge of the chart by moving away, or clockwise, from the short-circuit plug ($G_n = \infty$) to $-j0.36$ for stub 1 and to $-j0.73$ for stub 2, we obtain $d_1 = 0.195\lambda$ [*Ans.* to (*a*)] and $d_2 = 0.15\lambda$ [*Ans.* to (*b*)]. Also VSWR $= \infty$ [*Ans.* to (*e*) and (*f*)].

Problem 3-4-11. Double-stub matching. For the same coaxial line as in Example 3-16 (Fig. 3-20), find two other stub lengths which will achieve a match. *Ans.* $d_1 = 0.361\lambda$, $d_2 = 0.108\lambda$.

Problem 3-4-12. Double-stub matching. A 50-Ω coaxial line has two tuning stubs: stub 1, $\lambda/4$ from the load and stub 2, $\lambda/8$ farther. To match a load impedance $Z_L = 25 + j50\ \Omega$, find: (*a*) stub 1 length d_1, (*b*) stub 2 length d_2, (*c*) VSWR on $\lambda/4$ line, (*d*) VSWR on $\lambda/8$ line, (*e*) VSWR on stub 1 and (*f*) VSWR on stub 2. *Ans.* (*a*) 0.138 λ; (*b*) 0.15λ; (*c*) 4.2; (*d*) 2.05; (*e*) ∞; (*f*) ∞.

Problem 3-4-13. Double-stub matching. For the same coaxial line problem as above, find other stub lengths which will achieve a match. *Ans.* $d_1 = 0.364\lambda$, $d_2 = 0.443\lambda$. Note that these lengths are greater than in the preceding problem and require longer stubs.

Example 3-17. Why stubs? Since stubs add a pure reactance or susceptance to a line, might not a coil or capacitor be used instead? Compare coil or capacitor sizes with sizes of Example 3-15 at 2 GHz.

Solution. For stub 1, $B_n = -0.36$ so $X_n = 1/B_n = +2.78$ and $X = 50 \times 2.78 = 139\ \Omega = \omega\mathcal{L}$, so the required coil inductance

$$\mathcal{L} = \frac{139}{\omega} = \frac{139}{2\pi f} = \frac{139}{2\pi \times 2 \times 10^9} = 11.1 \times 10^{-9}\ \text{H}$$

From (2-12-22), the inductance for a long coil

$$\mathcal{L} = \frac{\mu_0 N^2 A}{l} \qquad \text{(H)}$$

For a 2-mm-diameter coil, 5 mm long, the required number of turns is given by

$$N^2 = \frac{\mathcal{L}l}{\mu_0 A} = \frac{11.1 \times 10^{-9} \times 5 \times 10^{-3}}{4\pi \times 10^{-7} \times \pi \times 10^{-6}} = 14.0$$

or

$N = 3.74$ turns for a coil versus a stub 1 length of 2.93 cm. *Ans.*

and

$N = 2.63$ turns for a coil versus a stub 2 length of 2.25 cm. *Ans.*

The choice here is between a linear length adjustment for a stub versus a fractional turn adjustment for a coil. The stub is usually more convenient. For an LC match see Example 3-21.

Example 3-18. 4-patch antenna array. An array of four patch antennas is mounted on a $\varepsilon_r = 4$ substrate. The patches are in a linear array with 1λ spacing between them as in Fig. 3-21. Each patch has a 50-Ω terminal resistance. The four patches are to be fed in-phase from a 10-GHz transmitter via a single 50-Ω microstrip line. Design the feed system.

Solution. To feed all four patches in-phase requires that the line lengths be identical from each patch to the feed line connection. A popular method is with a "corporate feed." The term alludes to a business organization with a president, two vice presidents, and four assistant vice presidents. Here the patches are the assistant vice presidents and the president is at the feed point *C*.

First, we connect the left-hand pair of patches in parallel at point *A* and the right-hand pair at point *B*. Two 50-Ω lines in parallel combine as 25 Ω at *A* and *B*. With $\lambda/4$ sections this is transformed to 50 Ω on the lines to *C*. At *C* the same procedure could be repeated but instead a $\lambda/4$ section transforms each 50-Ω line to 100 Ω. These then combine to 50 Ω on the line to the transmitter.

We still need to calculate the resistance of the $\lambda/4$ sections and specify the physical dimensions of the feed network. From (20), when $\beta x = \lambda/4$,

$$Z(\lambda/4) = \sqrt{R(\text{in})R(\text{out})} = \sqrt{25 \times 50} = 35.4\ \Omega \text{ at } A \text{ and } B \quad \textit{Ans.}$$

$$Z(\lambda/4) = \sqrt{50 \times 100} = 70.7\ \Omega \text{ at } C \quad \textit{Ans.}$$

At 10 GHz,

$$\lambda_0 = \frac{c}{f} = \frac{3 \times 10^8\ \text{ms}^{-1}}{10 \times 10^9\ \text{Hz}} = 30\ \text{mm} \qquad \text{(free space)}$$

and

$$\frac{\lambda_0}{4} = 7.5\ \text{mm} \qquad \text{(free space)}$$

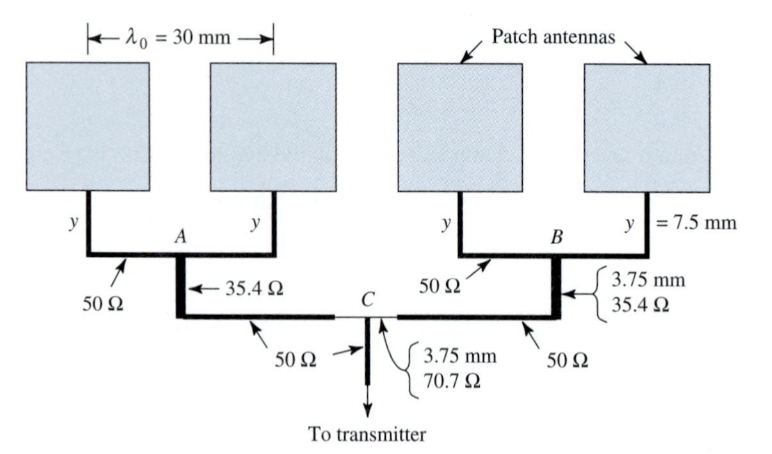

FIGURE 3-21
Four-patch antenna array.

Assuming that all of the energy is transmitted in the substrate (which it is not), the length of the $\lambda/4$ sections should be about

$$\frac{\lambda_0}{4} = \frac{7.5 \text{ mm}}{\sqrt{4}} = 3.75 \text{ mm} \quad \textit{Ans.}$$

The length of the y sections is arbitrarily made $\lambda_0/4 = 7.5$ mm. For simplicity, we have neglected mutual impedances or the effect of one patch on the impedences of the others.

Problem 3-4-14. Design a feed system for the four-patch array of Example 3-18 with feed lines from all four patches connected directly to the main feed point C. This is a "Deming feed" alluding to a Deming-type organization with direct lines of communication from bottom to top. To keep all lines equal in length this will require zigzagging of the lines from the center two patches. Match in two ways: (*a*) with $\lambda/4$ sections in the four patch lines and (*b*) with a $\lambda/4$ section in the line to the transmitter. *Ans.* (*a*) $Z(\lambda/4) = 100\ \Omega$; (*b*) $Z(\lambda/4) = 25\ \Omega$.

3-5 BANDWIDTH

As noted in Sec. 3-4, a $\lambda/4$ section is a useful matching device at the frequency for which the section is $\lambda/4$ long. At some other frequency, the length is not $\lambda/4$ and the match is imperfect. Thus, the $\lambda/4$ section is a frequency-sensitive device.

In this section we investigate how useful this and other matching devices are over a band of frequencies. We define bandwidth (BW) as

$$BW = \frac{f_{\text{high}} - f_{\text{low}}}{f_0} \times 100\%$$

where f_{high} = highest frequency of band, Hz
f_{low} = lowest frequency of band, Hz
f_0 = design or center frequency, Hz

Example 3-19. "$\lambda/4$ section" bandwidths. A load of $R_L = 400\ \Omega$ is matched to a $R_0 = 100$-Ω line with a $\lambda/4$ section at the design frequency f_0 as in Fig. 3-22, where the $\lambda/4$ section impedance is given by

$$Z(\lambda/4) = \sqrt{R_L R_0} = \sqrt{400 \times 100} = 200\ \Omega$$

If $f_0 = 300$ MHz find (*a*) the VSWR at $f_{\text{low}} = 150$ MHz, (*b*) the VSWR at $f_{\text{high}} = 450$ MHz, (*c*) the bandwidth (BW) for VSWR ≤ 1.5, and (*d*) the bandwidth for VSWR ≤ 1.2.

Solution. Using a Smith Chart (Fig. 3-17), we enter the normalized load impedance

$$Z_{Ln} = \frac{400 + j0}{200} = 2 + j0$$

at point A (the load) and move on the VSWR $= 2$ circle away from the load (clockwise) $\lambda/4$ (halfway) around the chart to point B where

$$Z_B = 0.5 + j0$$

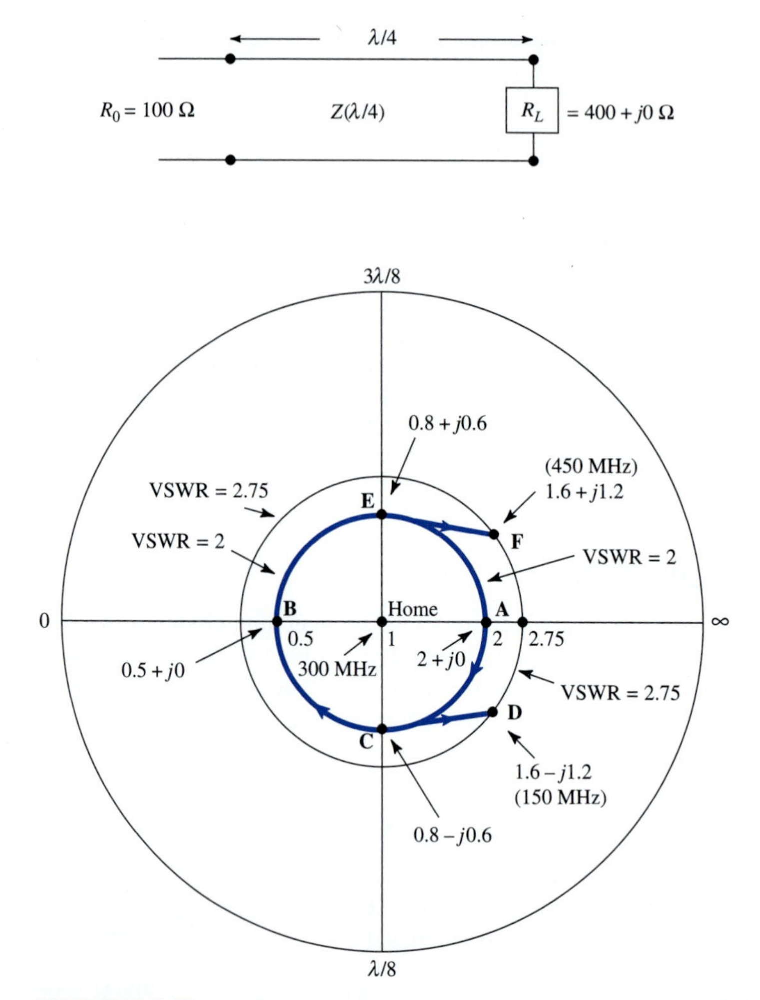

FIGURE 3-22
Line and chart for bandwidths.

We are now at the left end of the λ/4 section. Normalizing to the 100-Ω line, we have

$$R_{Bn} = \frac{Z_L(\lambda/4)}{R_0} Z_B = \frac{200}{100} 0.5 + j0 = 1 + j0$$

which is at the center of the chart (Home) for a match at the design frequency of 300 MHz.

At $f_{\text{low}} = 150$ MHz, the length of the section is

$$\frac{150}{300}\frac{\lambda}{4} = \frac{\lambda}{8} = 0.125\lambda$$

We again enter the chart at $2 + j0$ (point A) which is at the load and move on the VSWR = 2 circle clockwise (away from the load) $\lambda/8$ to point C on the chart corresponding to the left end of the "$\lambda/4$ section." Here

$$Z_C = 0.8 - j0.6$$

Normalizing this to the 100-Ω line, we have

$$Z_{Dn} = \frac{200}{100}(0.8 - j0.6) = 1.6 - j1.2$$

at point D on the chart which is on a VSWR = 2.75 circle. *Ans.* (*a*)

Proceeding in the same way for $f_{\text{high}} = 450$ MHz, the "$\lambda/4$ section" is 0.375λ long at this frequency. Moving this distance on the chart, we arrive at point E, where

$$Z_{En} = 0.8 + j0.6$$

Normalizing to the 100-Ω line puts us at point F, where

$$Z_{Fn} = 1.6 + j1.2$$

which is on the VSWR = 2.75 circle. *Ans.* (*b*)

Evaluating the VSWR at other frequencies between 200 and 400 MHz and plotting the results in Fig. 3-23 (shown on next page), we have for a VSWR ≤ 1.5

$$\text{BW} = \frac{353 - 247}{300} \times 100 = 35.3\% \quad \textit{Ans.} \quad (c)$$

For a VSWR ≤ 1.2, we have

$$\text{BW} = \frac{325 - 275}{300} \times 100 = 16.7\% \quad \textit{Ans.} \quad (d)$$

Is it possible to match the 400-Ω load to a 100-Ω line with shorter matching sections? Yes; the next two examples describe two ways.

Example 3-20. Matching using two ~λ/16 sections. As shown on the next page in Fig. 3-24 the 400-Ω load is connected to a 100-Ω section of length l. This, in turn, is connected to a 400-Ω section of the same length. This provides a match at the design frequency f_0 if

$$l = \frac{\lambda}{360^\circ} \sin^{-1}\left[\frac{\sqrt{R_L/R_0}}{(R_L/R_0) + 1}\right] \tag{1}$$

where R_L = load resistance = 400 Ω and R_0 = line resistance = 100 Ω.

Evaluating (1), $l = 0.0655\lambda$ or approximately $\lambda/16$. The steps on a Smith Chart are:

Step 1: Enter at $Z_{Ln} = 4 + j0$ and move clockwise 0.0655λ on the VSWR = 4 circle to $1.2 - j1.6$.

Step 2: Normalize to the 400-Ω section, obtaining $Z_n = 0.3 - j0.4$.

Step 3: Move 0.0655λ to $0.25 + j0$.

Step 4: Normalize to the 100-Ω section, obtaining $Z_n = 1 + j0$ for a match.

Note that the total length l of the two sections is about half a $\lambda/4$ section.

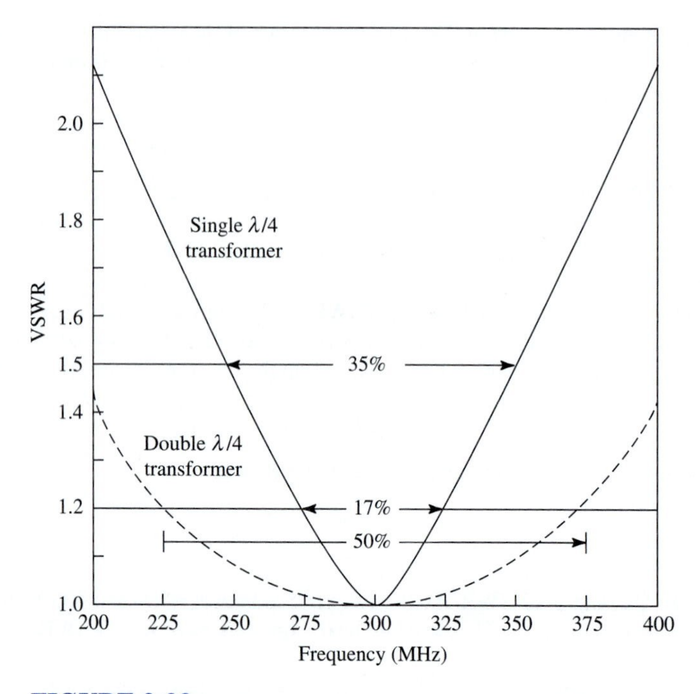

FIGURE 3-23
Graph with data from Example 3-19 and Problem 3-5-3 comparing bandwidths of single and double $\lambda/4$ transformers. The double $\lambda/4$ transformer has a much wider bandwidth, 50% versus 17% for VSWR $=$ 1.2.

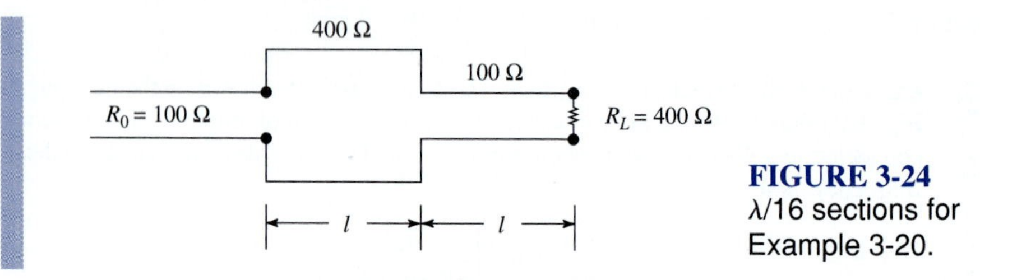

FIGURE 3-24
$\lambda/16$ sections for Example 3-20.

Problem 3-5-1. Bandwidth. Find the bandwidth of the two $\sim\lambda/16$ matching sections of Example 3-20 for a VSWR $\leq$ 1.2. *Ans.* 13 percent.

Example 3-21. Lumped $\mathscr{L}C$ match. Replacing the distributed parameters of the line sections of Example 3-20 with lumped values, as in Fig. 3-25, a capacitor C replaces the 100-Ω section and an inductance $\mathscr{L}$ replaces the 400-Ω section. A lossless line impedance $Z = \sqrt{\mathscr{L}/C}$, so a lower-impedance line is more "capacitative" and a higher-impedance line is more "inductive," as suggested in Fig. 3-25.

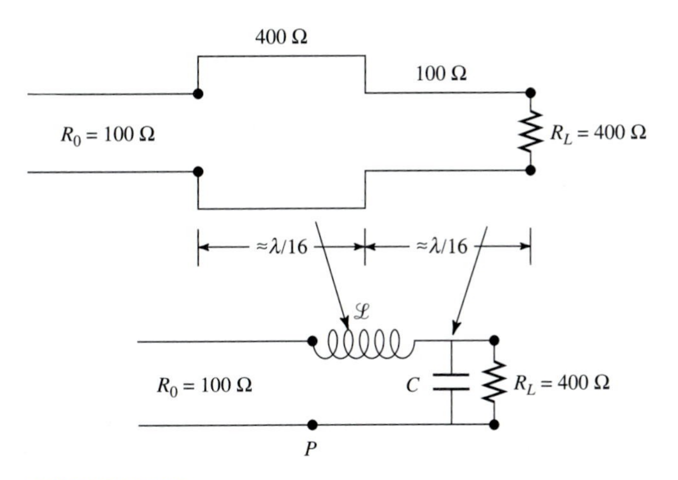

FIGURE 3-25
Lumped sections.

A match at the design frequency is obtained when

$$\omega\mathscr{L} = \sqrt{R_0(R_L - R_0)} \tag{2}$$

and

$$\omega C = \frac{\sqrt{R_0(R_L - R_0)}}{R_L R_0} \tag{3}$$

For our case, $R_L = 400\ \Omega$ and $R_0 = 100\ \Omega$ and

$$\omega\mathscr{L} = \sqrt{100(400 - 100)} = 173$$

$$\omega C = \frac{\sqrt{100(400 - 100)}}{400 \times 100} = 0.00433$$

$$G = \frac{1}{R} = \frac{1}{400} = 0.0025$$

Thus, the impedance at point P is

$$Z_P = j\omega\mathscr{L} + \frac{1}{G + j\omega C} = j\omega\mathscr{L} + 100 - j173 = 100\ \Omega$$

for a match.

Problem 3-5-2. Lumped match. Find the bandwidth of the lumped $\mathscr{L}C$ matching network of Example 3-20 for a VSWR $\leq$ 1.2. *Ans.* 11 percent.

Problem 3-5-3. Two-λ/4-section match. Find the bandwidth for a VSWR $\leq$ 1.2 of two series-connected λ/4 sections (total length λ/2) to match a 400-Ω load to a 100-Ω line. The two sections have impedances of 141.4 and 383 Ω. *Ans.* 50 percent.

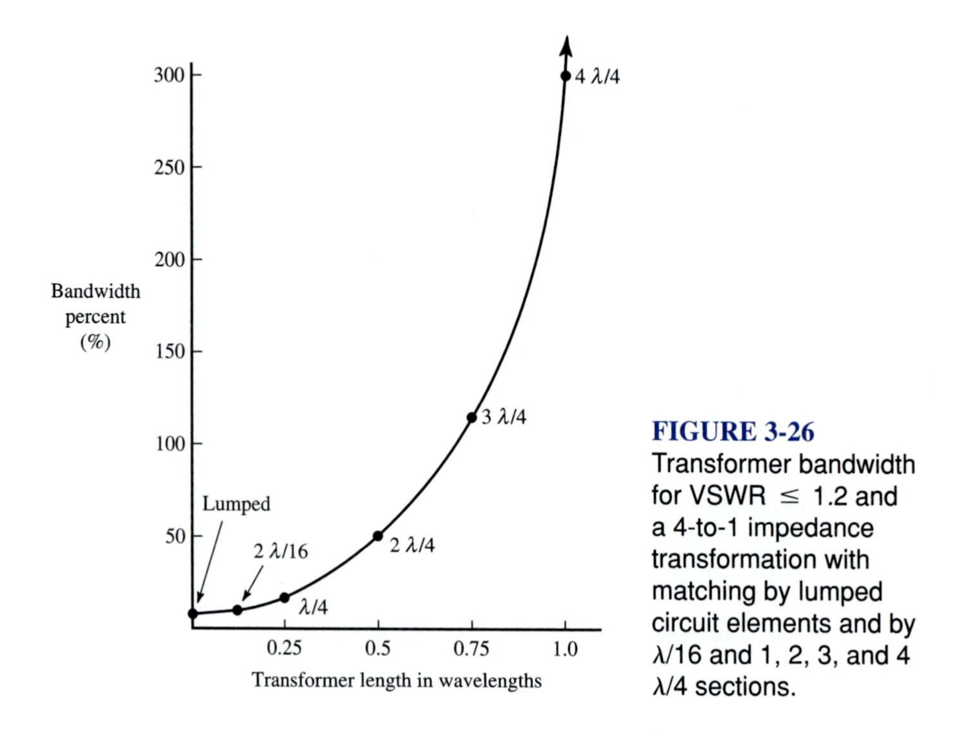

FIGURE 3-26
Transformer bandwidth for VSWR $\leq$ 1.2 and a 4-to-1 impedance transformation with matching by lumped circuit elements and by $\lambda/16$ and 1, 2, 3, and 4 $\lambda/4$ sections.

Problem 3-5-4. Three-λ/4-section match. Find the bandwidth for a VSWR $\leq$ 1.2 of three series-connected $\lambda/4$ sections (total length $3\lambda/4$) to match a 400-Ω load to a 100-Ω line. The three sections have impedances of 119, 200, and 336-Ω. See footnote.† *Ans.* 137 percent.

Problem 3-5-5. Four-λ/4-section match. Find the bandwidth for a VSWR $\leq$ 1.2 of four series-connected $\lambda/4$ sections (total length λ) to match a 400-Ω load to a 100-Ω line. *Ans.* >300 percent.

The results of Examples 3-19, 3-20, and 3-21 and the above problems are presented graphically in Fig. 3-26. Two ways of obtaining different impedances of series-connected $\lambda/4$ sections are shown in Fig. 3-27.

†To satisfy the geometric-mean requirement, the impedances Z_1, Z_2, and Z_3 of the three sections are 119, 200, and 336 Ω; i.e., the logarithms of the impedance ratios are related as the coefficients of the binomial series (1, 2, 1 for two sections; 1, 3, 3, 1 for three sections; 1, 4, 6, 4, 1 for four sections, etc.). Thus, we have the requirement that

$$3 \log \frac{Z_1}{100} = \log \frac{Z_2}{Z_1} = \log \frac{Z_3}{Z_2} = 3 \log \frac{400}{Z_3}$$

from which $Z_1 = 119$, $Z_2 = 200$, and $Z_3 = 336\ \Omega$.

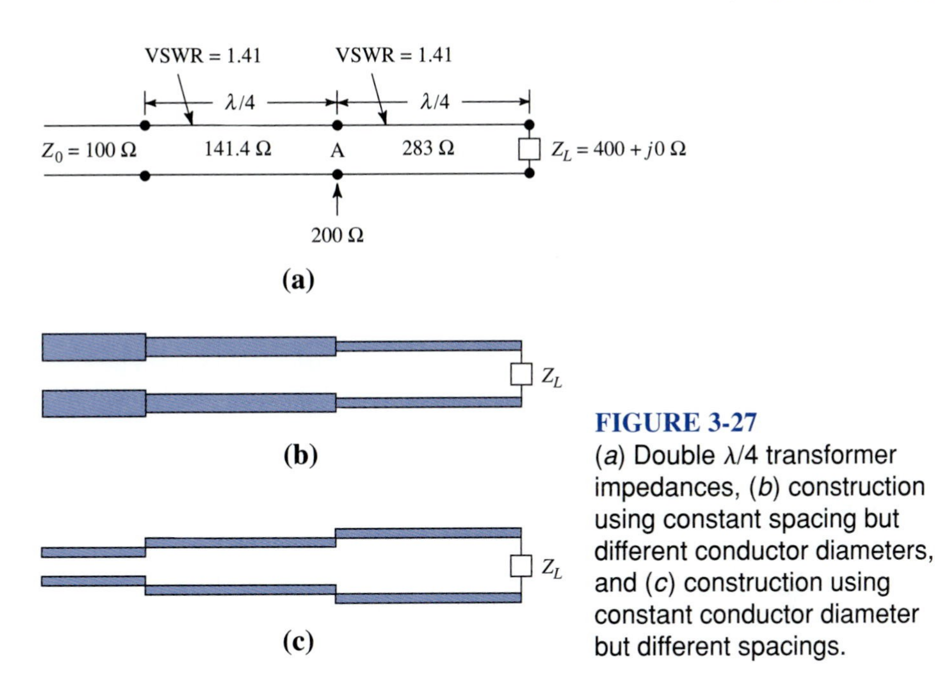

FIGURE 3-27
(*a*) Double $\lambda/4$ transformer impedances, (*b*) construction using constant spacing but different conductor diameters, and (*c*) construction using constant conductor diameter but different spacings.

It is obvious that by transforming in smaller impedance steps the bandwidth can be greatly increased. But a smooth, gradual transition is a further improvement. Two ways of tapering a line are shown in Fig. 3-28.

A gradual, tapered transition provides a broadband, nearly reflectionless transformer which has the very important additional advantage that it works for pulses. By contrast, the stepped transformers of the series-connected $\lambda/4$ sections are crude approximations of a gradual taper that work only for continuous waves or pulses that are long compared to the transformer length. Pulses on transmission lines are discussed in the next section, and Chap. 10 has more about pulses.

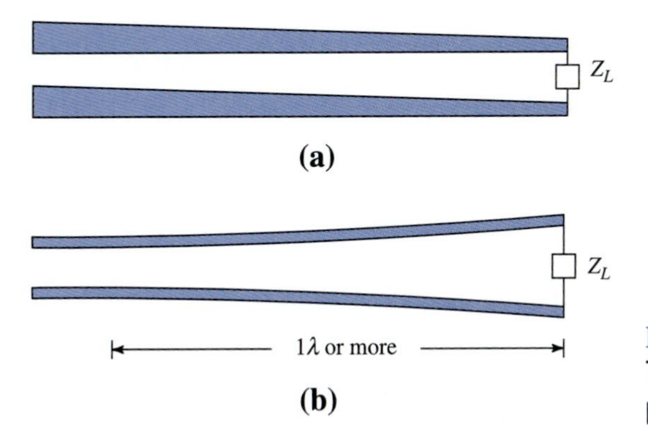

FIGURE 3-28
Tapered transformers for pulses and wide bandwidths.

3-6 PULSES AND TRANSIENTS

The λ/4 Transformer

The reflectionless property of a λ/4 transformer for continuous waves is achieved by adjusting the reflections at the two ends to balance out at the design frequency as illustrated in Example 3-22. The behavior of the same λ/4 for a short pulse is illustrated by Example 3-23. Pulses on a tapered transition are discussed in Example 3-24.

Example 3-22. Continuous waves on a λ/4 transformer. Referring to Fig. 3-29, a 400-Ω line is matched to a 100-Ω line with a λ/4 section of impedance $Z(\lambda/4) = \sqrt{400 \times 100} = 200\ \Omega$. Trace the buildup of the steady-state voltage on the λ/4 section as a function of time.

Solution. The story begins with the arrival of a 1-V wavefront from the left at junction A, which has a 2-to-1 mismatch. The reflection coefficient

$$\rho_v = \frac{Z(\lambda/4) - R_0}{Z(\lambda/4) + R_0} = \frac{200 - 100}{200 + 100} = \frac{1}{3}$$

and the transmission coefficient

$$\tau_v = \frac{2Z(\lambda/4)}{Z(\lambda/4) + R_0} = \frac{2 \times 200}{200 + 100} = \frac{4}{3}$$

Therefore, a $\frac{1}{3}$-V wave is reflected back to the left from the junction with a $1\frac{1}{3}$-V wave continuing to the right on the λ/4 section. When this wave hits the junction at B, $\frac{4}{3}$ V is transmitted and $\frac{1}{3}$ V reflected. When this wave gets back to junction A, the reflection coefficient

$$\rho_v = \frac{R_0 - Z(\lambda/4)}{R_0 + Z(\lambda/4)} = \frac{100 - 200}{100 + 200} = -\frac{1}{3}$$

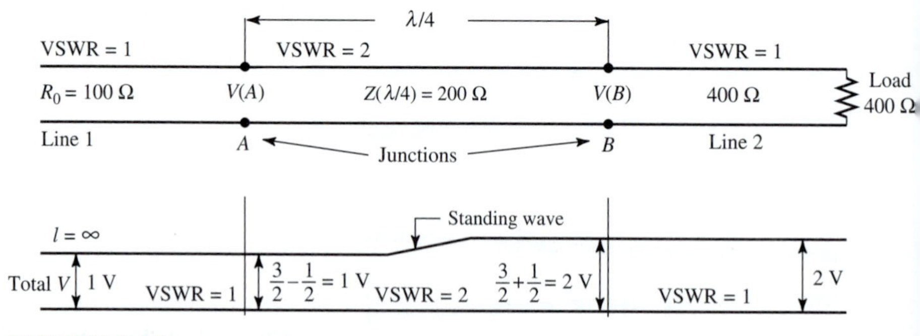

FIGURE 3-29

λ/4 section for matching 100 Ω to 400 Ω and transforming 1 V to 2 V after reaching steady state.

The wave to the left at junction A now adds up to

$$V(A) = 1 + \rho_v(A) - [\tau_v(A)\,\rho_v(B)\,\tau_v(A)]$$
$$= 1 + \frac{1}{3} - \left[\frac{4}{3} \times \frac{1}{3} \times \frac{2}{3}\right] = 1.037 \text{ V}$$

and the wave to the right of junction B is

$$V(B) = \tau_v(A)\,\tau_v(B) = \frac{4}{3} \times \frac{4}{3} = \frac{16}{9} = 1.777 \text{ V}$$

After two more back-and-forth reflections

$$V(A) = 1.004 \text{ V}$$

and

$$V(B) = 1.997 \text{ V}$$

After an infinite time, a steady state is reached with

$$V(A) = 1 \text{ V and } V(B) = 2 \text{ V} \qquad \text{End of story}$$

Problem 3-6-1. Referring to Example 3-22, find (*a*) power delivered to line 2 and (*b*) incoming power on line 1. *Ans.* (*a*) 10 mW; (*b*) 10 mW.

Example 3-23. Pulse on λ/4 transformer. Referring to Fig. 3-30 (on next page), a square pulse of duration t_p and magnitude of 1 V is applied to the line at junction A. Trace the progress of this pulse across the $\lambda/4$ transition section of length l given by

$$l = 6\,v\,t_p \qquad \text{(m)}$$

where v = velocity of wave on the line, m s^{-1}. Thus, the transition section is 6 times longer than the pulse.

Solution. The reflection and transmission coefficients are the same as in Example 3-22. Thus, as shown in Fig. 3-30, the 1-V pulse incident on junction A splits into a $\frac{1}{3}$-V reflected pulse and a $\frac{4}{3}$-V transmitted pulse. On the diagram, distance is to the right and time is down.

After a time $6t_p$ (down), the pulse has traveled a distance l to junction B. Here it splits into a $4/3 \times 4/3 = 16/9$ V transmitted pulse and a $(-4/3) \times (1/3) = -4/9$ V reflected pulse. After another time $6t_p$ (total time $12t_p$) the reflected pulse returns to junction A where $2/3 \times (-4/9) = -8/27$ V is transmitted to the left and $(-1/3) \times (-4/9) = 4/27$ V is reflected back to the right.

As time progresses (down on the diagram), the residual pulse bounces back and forth, ejecting smaller and smaller pulses to the right of B and to the left of A (back on the line).

Figure 3-31 (on next page) shows that the reflected and transmitted pulses are spaced by the length of time it takes a pulse to make one round trip on the section, or a time

$$t = 2\,l/v \qquad \text{(s)}$$

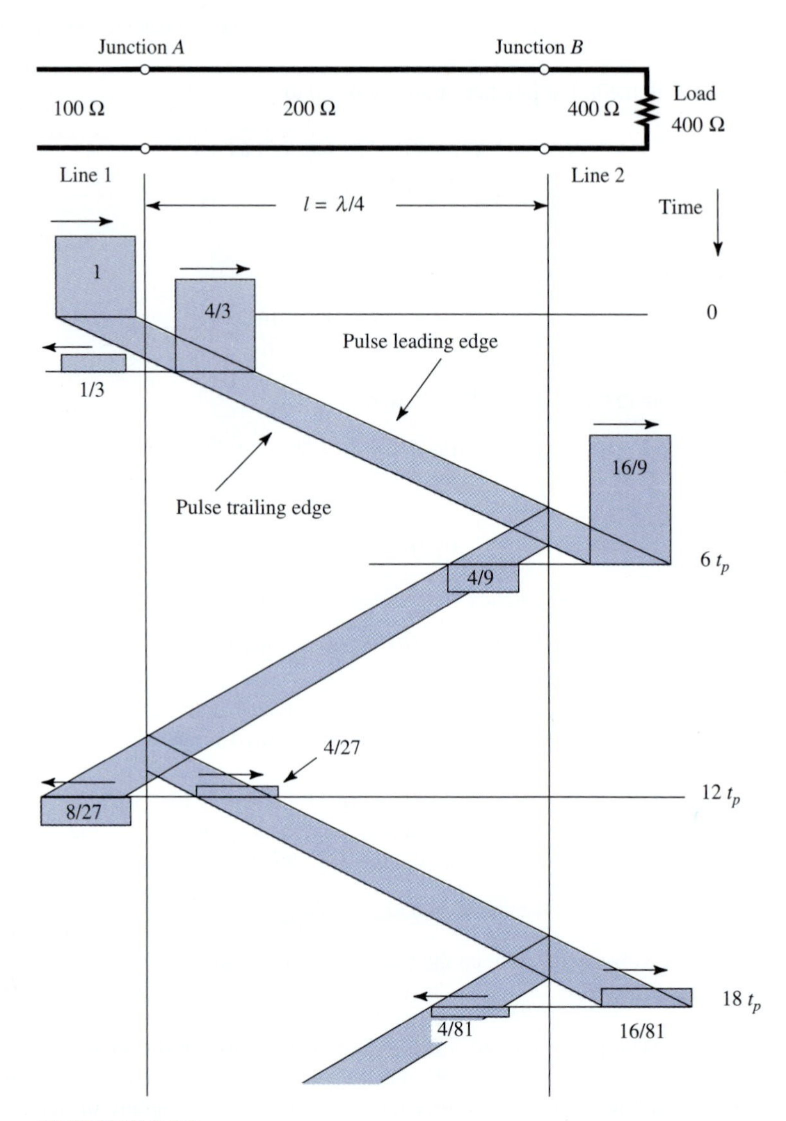

FIGURE 3-30
Progress of a short pulse bouncing on a $\lambda/4$ transformer. Time increases downward on the chart. (See oscillogram of bouncing pulse in Fig. 10-4.)

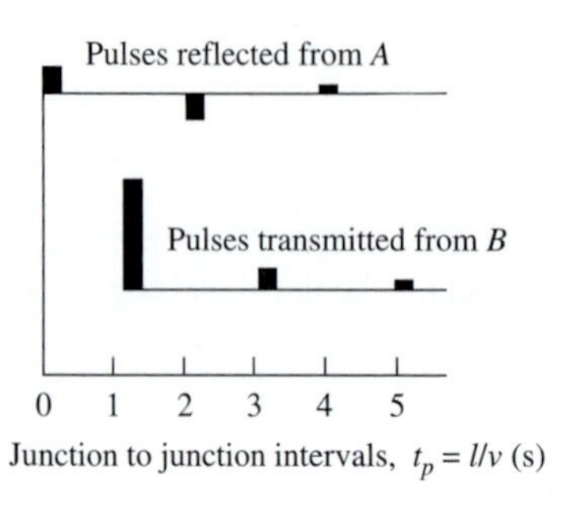

FIGURE 3-31
Pulse trains reflected and transmitted from transformer of Fig. 3-30.

Problem 3-6-2. Decay of bouncing pulse. Referring to Example 3-23, how many times must a pulse bounce between the two junctions A and B of Fig. 3-30 before the voltage back onto line 1 is reduced to less than 0.001 of the initial pulse voltage? *Ans.* 8 bounces (4 round trips); time $= 48\,t_p$.

Important Conclusions

It is clear that the transformer fails to function as a reflectionless device for a short pulse. It is not possible for the pulse reflected from B to catch up with the one reflected from A and reduce its magnitude. Only for the steady-state condition or for a very long pulse can the transformer produce a match. To avoid reflections with short pulses, a gradually tapered line is needed, as discussed in Example 3-24. Conversely, one can use short pulses to locate and measure the magnitude of discontinuities on transmission lines.

If the load is the antenna of a pulse radar, the reflected and transmitted pulses produce false targets. If the load is a digital device, the pulse trains can scramble information. As shown by Fourier analysis, a short pulse produces a wideband signal. A narrow-bandwidth transformer may not pass all the frequency components, altering the pulse shape. The wide bandwidth of a gradually tapered transition helps preserve the pulse shape. This type of transition also effectively eliminates the reflected and transmitted pulse trains that are so detrimental to the operation of pulse radars and digital devices.

Example 3-24. Pulse on tapered transition. Trace the progress of a short pulse on the tapered transition of Fig. 3-32 for matching a 100-Ω line to a 400-Ω load.

Solution. The tapered transition may be regarded as a series of small incremental steps in the line as suggested in the figure. The reflection coefficient for an incremental length Δl is proportional to

$$\rho_v \approx \frac{\Delta Z}{Z}$$

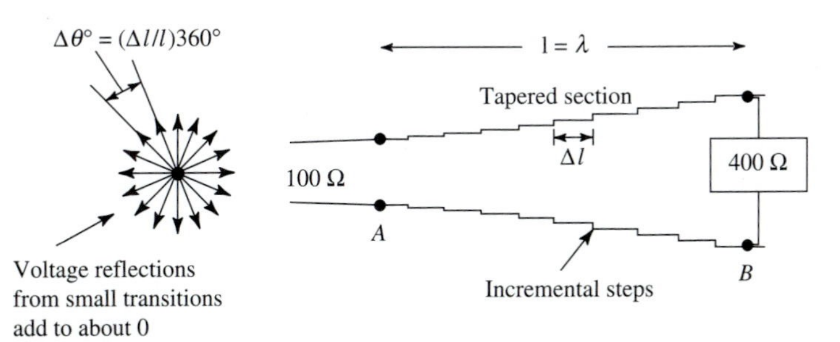

FIGURE 3-32
Principal of gradual taper illustrated by many small reflections from a series of small incremental steps in the line. The small reflections (over phase angles of 360° or more) result in a small-magnitude reflection and a wide bandwidth.

where ΔZ = impedance change over length Δl
Z = impedance of line at Δl

For a smooth gradual transition, ρ_v will be small but not zero. Thus, any reflection from the transition will be a long ($t = l/v$) pulse of small magnitude.

The transmission coefficients for each incremental length combine to double the voltage of the pulse at the load. The shape of the pulse delivered to the 400-Ω load is preserved and there are no later smaller pulses. It is assumed that the 100-Ω line is matched at the transmitter.

PROBLEMS

Answers are given in Appendix E to problems followed by the @ symbol.

3-2-2. Small *R* and *G*. Show that for *R* and *G* small but nonzero,

$$(a) \qquad \alpha = \frac{R}{2}\sqrt{\frac{C}{\mathcal{L}}} + \frac{G}{2}\sqrt{\frac{\mathcal{L}}{C}}$$

and

$$(b) \qquad \beta = \omega\sqrt{\mathcal{L}C}$$

3-2-3. Validity of approximation. For a uniform transmission line with $R = 0.25$ Ω/m, $G = 5$ m℧/m, $\mathcal{L} = 35$ μH/m, and $C = 12$ μF/m, compare the values of α and β given by $\sqrt{ZY} = \sqrt{(R + j\omega\mathcal{L})(G + j\omega C)}$ with the approximations of Problem 3-2-2 at frequencies of (*a*) 1 kHz, (*b*) 1 MHz, and (*c*) 1 GHz. @

3-2-4. Distortionless line. *Heaviside condition* for a distortionless transmission line requires

$$\frac{G}{C} = \frac{R}{\mathcal{L}}$$

Show that under this condition Z_0 is real for small *R* and *G*.

3-2-5. Distortionless line. (*a*) Is the transmission line of Example 3-1 distortionless? (*b*) What series resistance would result in a distortionless line? @

3-2-6. Line impedance. A uniform transmission line has constants $R = 800$ μΩ/m, $G = 2.1$ m℧/m, $L = 0.65$ μH/m, and $C = 14$ nF/m. Find the impedance of the line at frequencies of (*a*) 1 kHz, (*b*) 1 MHz, and (*c*) 1 GHz.

3-2-7. Velocity ratio (lossless line). A lossless transmission line has inductance $L = 150$ nH/m. Find the velocity ratio v/c for this line if the shunt capacitance is (*a*) 10 nF/m, (*b*) 100 nF/m, and (*c*) 1 μF/m. @

3-2-8. Velocity ratio (general case). If the transmission line of Problem 3-2-7 is not lossless, but instead has $R = 50$ mΩ/m and $G = 5.2$ μ℧/m find the velocity ratio v/c for each value of shunt capacitance.

3-2-9. Attenuation of line. A uniform transmission line has constant $R = 1$mΩ/m, $G = 2.5$ μ℧/m, $L = 15$ μH/m, and $C = 25$ nF/m. Find the dB attenuation of this line at frequencies of (*a*) 10 kHz, (*b*) 10 MHz, and (*c*) 10 GHz.@

3-3-11. Transmission line comparison. A coaxial line, two-conductor line, and single-conductor line above a ground plane all have the same impedance and $\varepsilon_r = 1$. Find a relationship between *a*, *b*, *D*, and *h*.

3-3-12. Poynting vector. A TEM wave on a lossless transmission line has an average Poynting vector of 125 μW/m^2. If $\mu_r = 1$ and $\varepsilon_r = 1.7$, find (*a*) the peak Poynting vector, (*b*) the peak value of the electric field E, and (*c*) the peak value of the magnetic field H. @

3-3-13. Impedance of media. The relative permittivities of many common materials are presented in Table 7-1. Using this information, find the intrinsic impedance of (*a*) styrofoam, (*b*) plywood, (*c*) quartz, (*d*) marble, (*e*) silicon, (*f*) ethyl alcohol, (*g*) glycerin, (*h*) ice, and (*i*) distilled water. Assume $\mu_r = 1$ in every case.

3-3-14. Coaxial line impedance. A coaxial transmission line has $a = 5$ mm and $b = 15$ mm. Find the impedance of the line if the dielectric is (*a*) air, (*b*) sulfur, and (*c*) silicon. @

3-3-15. Two-conductor line impedance. If the transmission line of Problem 3-3-14 is two-conductor instead of coaxial, find the line impedance for $D = 15$ mm for each of the listed dielectrics.

3-3-16. One-conductor line impedance. Repeat Problem 3-3-14 for a one-conductor line with $a = 5$ mm located 15 mm above a conducting ground plane. @

3-3-17. Microstrip line impedance. Repeat Problem 3-3-14 for a microstrip line with $H = 5$ mm and $W = 15$ mm.

3-3-18. Poynting vector. For each transmission line of Problems 3-3-14 through 3-3-17, find the power transmitted by the line if it is operated at a potential difference of 950 V (rms). @

3-4-15. Impedance along a line. Plot the magnitude of the complex impedance Z_x, as well as the resistive and reactive components of Z_x, from $x = 0$ to $x = \lambda$ for a line with $Z_0 = 100\ \Omega$ for (*a*) $Z_{load} = Z_0$, (*b*) shorted line ($Z_{load} = 0$), (*c*) open line ($Z_{load} = \infty$), (*d*) resistive load with $Z_{load} = 200\ \Omega$, and (*e*) capacitive load with $Z_{load} = -j100\ \Omega$.

3-4-16. Impedance at a distance. For the transmission line of Example 3-1, find the impedance at a distance of 1 km from a load with $Z_{load} = 140 + j35\ \Omega$. @

3-4-17. Impedance at a distance. Show that for a lossless line $Z_x = Z_0^2/Z_L$ at $x = \lambda/4$.

3-4-18. Terminated line. A 75-Ω lossless transmission line is terminated in $50 + j10\ \Omega$. Find (*a*) voltage reflection coefficient, (*b*) VSWR, (*c*) impedance 0.4λ from load, (*d*) shortest length of line for which impedance is purely resistive, and (*e*) the value of this resistance. (*Hint:* At the zero-reactance point, $Z_x = \text{VSWR} \times Z_0$.) @

3-4-19. Voltage excursion. For 30 V applied to the line of Problem 3-4-18, find (*a*) the maximum and (*b*) the minimum line voltage.

3-4-20. VSWR program. Use the VSWR program (Appendix C) to find the VSWR for the following cases: (*a*) $Z_{load} = 75 + j150\ \Omega$ on a 75-Ω line, (*b*) $Z_{load} = 150 + j0\ \Omega$ on a 75-Ω line, (*c*) $Z_{load} = 100 - j25$ on a 50-Ω line, and (*d*) $Z_{load} = 95 + j5$ on a 100-Ω line. @

3-4-21. VSWR by hand. Confirm the results of the VSWR program for the cases of Problem 3-4-20 using equations (3-4-11) and (3-4-26).

3-4-22. ZX Program. Use the ZX program (Appendix C) to find the first and second zero-reactance distances and the maximum and minimum resistance values for the following loads on a 50-Ω line: (*a*) $75 - j25\ \Omega$, (*b*) $40 + j25\ \Omega$, (*c*) $50 - j5\ \Omega$, (*d*) $25 + j0\ \Omega$, (*e*) $0 + j100\ \Omega$, and (*f*) $100 + j20\ \Omega$.

3-4-23. Periodic discontinuities. A lossless flexible 50-Ω dielectric-filled coaxial cable has small irregularities in the braid of the outer conductor. The reflection coefficient due to each irregularity is $0.001\angle 0°$. The irregularities are regularly spaced 11.25 cm apart. The wave velocity in the cable is $0.75c$. If the far end of a 50-m length of cable is connected to a matched load, find the resultant reflection coefficient measured at the

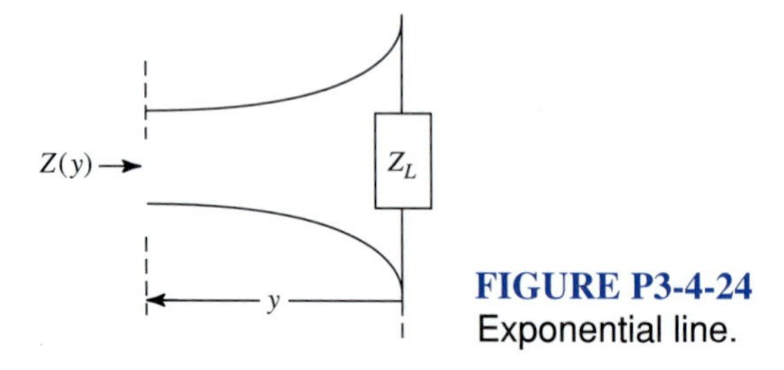

FIGURE P3-4-24
Exponential line.

input end of the cable at frequencies of (*a*) 500 MHz, (*b*) 1 GHz, and (*c*) 2 GHz. (*d*) What is the reflection coefficient for a 100-ps pulse? (*e*) What is ε_r for the cable dielectric? *Note:* Your results for (*a*) through (*d*) should not all be the same. Explain the large differences. @

3-4-24. Exponential line. The input impedance to a lossless exponential line of length y connected as shown in Fig. P3-4-24 to a load Z_L is given by

$$Z(y) = Z_0(0)e^{-\beta y}\frac{Z_L + jZ_0(0)\tan\beta y}{Z_0(0) + jZ_L\tan\beta y}$$

where $Z_0(0)$ is the characteristic impedance of the line at the load ($y = 0$). If $Z_L = 100 - j25\ \Omega$, $Z_0(0) = 75 + j0\ \Omega$ and $y = \lambda/2$, find $Z(y)$.

3-4-25. Impedance at a distance. For a transmission line with characteristic impedance of 75 Ω and attenuation of 2 Np/m, find the impedance looking into a 300-Ω load at distances of $\lambda/3$, $\lambda/2$, and 1.3λ if the frequency is 250 MHz.

3-4-26. Impedance of lossless line. Repeat Problem 3-4-25 for a lossless line.

3-6-3. Pulses. A 200-Ω line with load $Z_L = 200 + j0\ \Omega$ (at the right) is matched to a 50-Ω line (at the left) by a $\lambda/4$ section of 100-Ω line ($\sqrt{200 \times 50} = 100$). Draw a time-domain snapshot of the voltage on the line for 1 λ to the right and left of the $\lambda/4$ section at the instant a short 1-V pulse (length on line $\ll \lambda/4$) has propagated from the 50-Ω line through the $\lambda/4$ section and traveled $3\lambda/4$ beyond the 200-Ω line. (*Hint:* There should be five pulses of varying amplitudes.)

3-6-4. Bouncing pulses. For a 100-Ω line 100 m long, examine a 100-ns pulse at the mid-point of the line as a function of time and determine the decibel reduction of the second pulse with respect to the first, the third pulse with respect to the second, etc., if (*a*) $R_L = 65\ \Omega$ and $R_G = 10\ \Omega$ and (*b*) $R_L = 50\ \Omega$ and $R_G = 50\ \Omega$. Use the BOUNCING PULSES program (Appendix C).

CHAPTER 4

WAVE PROPAGATION, ATTENUATION, POLARIZATION, REFLECTION, REFRACTION, AND DIFFRACTION

4-1 INTRODUCTION

In an electromagnetic wave, a changing electric field produces a changing magnetic field, which in turn generates an electric field, and so on, with a resulting propagation of energy.

In this chapter, the propagation of waves through space and media, their polarization, reflection, refraction, and diffraction are discussed. Wherever appropriate, transmission line analogies are given.

4-2 WAVES IN SPACE

For a plane wave in space, the electric and magnetic field lines, **E** and **H**, are everywhere perpendicular to each other and perpendicular to the wave direction, as in Fig. 4-1. **E** and **H** are also in phase. A wave of this type is called a ***Transverse ElectroMagnetic (TEM) wave*** with its *wave equation* from (3-3-6) given by

$$\frac{\partial^2 E_y}{\partial t^2} = \frac{1}{\mu\varepsilon}\frac{\partial^2 E_y}{\partial x^2} \qquad \textbf{\textit{Wave equation}} \qquad (1)$$

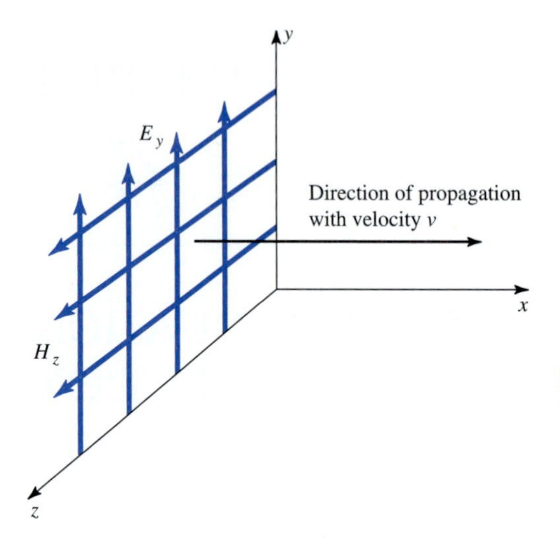

FIGURE 4-1
Transverse electromagnetic (TEM) wave propagating in the x direction with **E** in y direction (E_y) and **H** in z direction (H_z).

The ratio of E_y to H_z is an *impedance* Z_0 as given by

$$Z_0 = \frac{E_y}{H_z} = \sqrt{\frac{\mu}{\varepsilon}} \qquad (\Omega) \tag{2}$$

For air or vacuum,

$$Z_0 = \sqrt{\frac{\mu_0}{\varepsilon_0}} = 377\ \Omega \qquad \textit{Intrinsic impedance of space} \tag{3}$$

The product of E_y and H_z has the dimensions of *power per unit area* and is called the *Poynting vector* (see Sec. 4-10). Thus,

$$|\mathrm{PV}| = E_y H_z = \frac{\text{volts}}{\text{meter}}\frac{\text{amperes}}{\text{meter}} = \frac{\text{watts}}{\text{meter}^2} \qquad (\mathrm{W\ m^{-2}})$$

Example 4-1. Velocity and power of wave. The electric field E_y of a TEM wave equals 100 V $\mathrm{m^{-1}}$ rms. Find (*a*) velocity and Poynting vector magnitude |PV| in air and (*b*) velocity and |PV| in a lossless dielectric medium with $\varepsilon_r = 9$.

Solution

(*a*) $v = 1/\sqrt{\mu_0\varepsilon_0} = 3 \times 10^8 \quad \mathrm{m\ s^{-1}}$

$$|\mathrm{PV}| = E_y H_z = \frac{E_y^2}{Z_0} = \frac{100^2}{377} = 26.5\ \mathrm{W\ m^{-2}} \quad \textit{Ans.} \quad (a)$$

(*b*) $v = c/\sqrt{\varepsilon_r} = c/\sqrt{9} = 10^8\ \mathrm{m\ s^{-1}}$

$$|\mathrm{PV}| = \frac{E_y^2}{Z_0/\sqrt{\varepsilon_r}} = \frac{100^2}{\left(\frac{1}{3}\right)(377)} = 79.6 \quad \mathrm{W\ m^{-2}} \quad \textit{Ans.} \quad (b)$$

Problem 4-2-1. Wave power and fields. If the Poynting vector or power density of a TEM wave = 100 W/m^2 in a lossless medium ($\sigma = 0$) with $\varepsilon_r = 4$, find the magnitude of (*a*) the electric field and (*b*) the magnetic field. *Ans.* (*a*) 137 V/m, (*b*) 0.727 A/m.

The wave equation (1) is a linear partial differential equation of the second order. A solution is

$$E_y = E_0 \sin(\omega t - \beta x) \tag{4}$$

where ω = radian frequency = $2\pi f$, rad s^{-1}
β = phase constant = $2\pi/\lambda$, rad m^{-1}
f = frequency, Hz
λ = wavelength, m

For a constant-phase point, such as P in Fig. 4-2,

$$\omega t - \beta x = \text{constant} \tag{5}$$

Differentiating (5) with respect to time, we obtain the phase velocity of the wave as

$$\frac{dx}{dt} = \frac{\omega}{\beta} = \frac{2\pi f}{2\pi/\lambda} = \lambda f = v = \text{phase velocity} \qquad (\text{m s}^{-1}) \tag{6}$$

This velocity is a function of the constants μ and ε of the medium. Thus,

$$v = \lambda f = \frac{1}{\sqrt{\mu\varepsilon}} \qquad (\text{m s}^{-1}) \tag{7}$$

Figure 4-2 shows the magnitude of the electric field of a traveling wave as a function of distance βx at three instants of time. A constant phase point, such as P at the crest of the wave, moves to the right with a velocity v and the surfboarder rides with it.

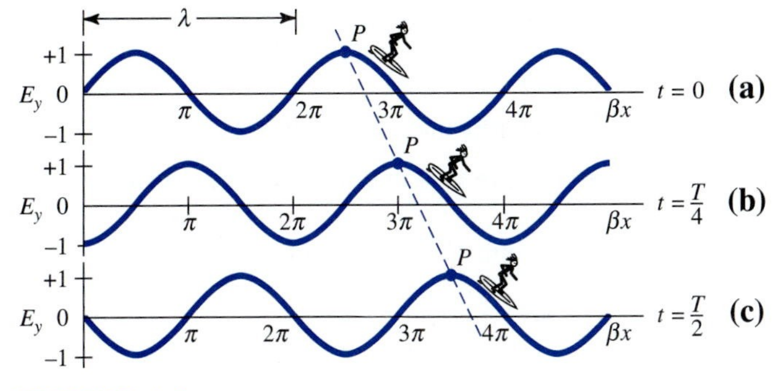

FIGURE 4-2
Curves for $E_y = \sin(\omega t - \beta x)$ at three instants of time: $t = 0$, $t = T/4$, and $t = T/2$. A constant-phase point P moves to the right as time progresses and the surfboarder rides with it.

The wave equation (1) is for a *lossless medium* ($\sigma = 0$). For a medium that is not lossless (σ finite), Maxwell's curl equations are

$$-\frac{\partial H_z}{\partial x} = \sigma E_y + \varepsilon \frac{\partial E_y}{\partial t} \tag{8}$$

and

$$\frac{\partial E_y}{\partial x} = -\mu \frac{\partial H_z}{\partial t} \tag{9}$$

or in phasor form such as $E_y = E_0 e^{jwt}$,

$$\frac{\partial H_z}{\partial x} = -(\sigma + j\omega\varepsilon)E_y \tag{10}$$

and

$$\frac{\partial E_y}{\partial x} = -j\omega\mu H_z \tag{11}$$

Differentiating (11) with respect to x and substituting (10) yields

$$\frac{\partial^2 E_y}{\partial x^2} = (j\omega\mu\sigma - \omega^2\mu\varepsilon)E_y \qquad \textbf{\textit{Wave equation in conducting medium}} \tag{12}$$

This is the wave equation in E_y for a plane wave in a *conducting medium.* As before, these equations are for a linearly polarized wave (**E** in the y direction) traveling in the x direction.

Space may be regarded as an array of field-cell transmission lines, as in Fig. 4-3. If we direct our attention to a single transmission line cell, the upper and lower surfaces of the cell can be regarded as consisting of conducting strips of width w and of infinite length in the direction of wave propagation (x direction, out of page). Recalling from earlier chapters that for field cells the inductance $\mathscr{L}$ per unit length (in x direction) is equal to the permeability μ of the medium, the capacitance C per unit length is equal to the permittivity ε of the medium, and the conductance G per unit length is equal to the conductivity σ of the medium, we can write

$$\mathscr{L} = \mu = \text{inductance per unit length, H m}^{-1}$$

$$C = \varepsilon = \text{capacitance per unit length, F m}^{-1}$$

$$G = \sigma = \text{conductance per unit length, ℧ m}^{-1}$$

where the symbols $\mathscr{L}$, C, and G are now understood to be *distributed quantities,* that is, per unit length.

Introducing $\mathscr{L}$, C, and G into (12), we obtain

$$\frac{\partial^2 E_y}{\partial x^2} - j\omega\mathscr{L}(G + j\omega C)E_y = 0 \tag{13}$$

Integrating $E_y(= |\mathrm{E}|)$ between the upper and lower strips yields the potential difference $V(= E_y h$, or $E_y = V/h)$, so that (13) can be expressed as

$$\frac{\partial^2 V}{\partial x^2} - j\omega\mathscr{L}(G + j\omega C)V = 0 \qquad \textbf{\textit{Wave equation for field-cell transmission line}} \tag{14}$$

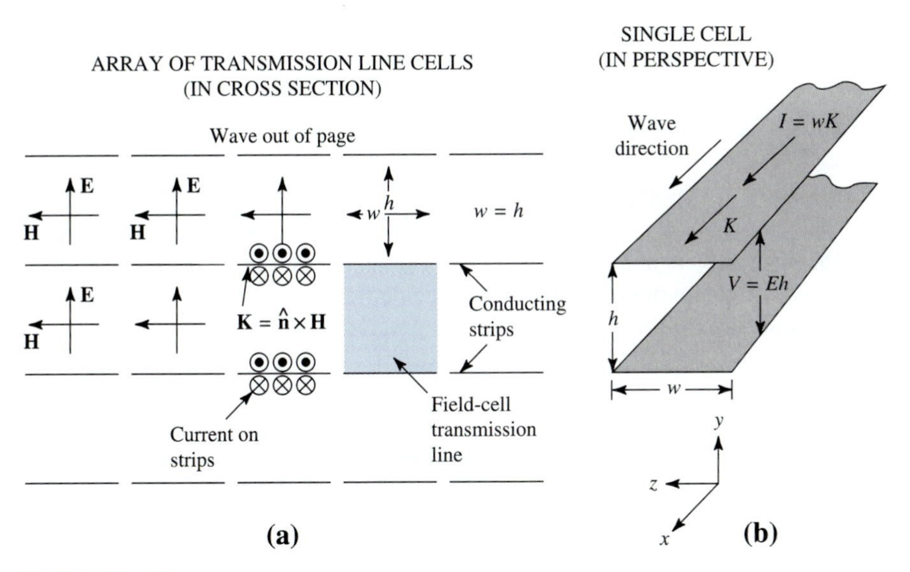

FIGURE 4-3
Field cells are a link between space and transmission lines. In (*a*) a space wave traveling out of the page is divided into an array of field-cell transmission lines. One field cell is shown in perspective in (*b*).

This is the wave equation for a field-cell transmission line in terms of the voltage V between the conducting strips.

Note that (12) and (13) have no quantity analogous to the series resistance of a transmission line because there is no conduction term in (11) like the σ term in (10). There are electric (conduction) currents in nature but no magnetic currents.

Thus, from (3-2-10),

$$\text{Line impedance } Z_0 = \sqrt{\frac{R + j\omega\mathscr{L}}{G + j\omega C}} \quad (\Omega) \tag{15}$$

and from (10) and (11)

$$\text{Medium impedance } Z_0 = \sqrt{\frac{j\omega\mu}{\sigma + j\omega\varepsilon}} \quad (\Omega) \tag{16}$$

However, if the medium is not only conducting (with loss) but also has dielectric and magnetic hysteresis losses (Chap. 7), ε and μ become complex and we have†

$$\varepsilon = \varepsilon' - j\varepsilon'' \qquad \text{and} \qquad \mu = \mu' - j\mu''$$

so

$$j\omega\mu = j\omega(\mu' - j\mu'') = \omega\mu'' + j\omega\mu'$$

†Note that the imaginary parts of these quantities, when multiplied by j, are negative real quantities and become attenuation terms.

and

$$\sigma + j\omega\varepsilon = \sigma + j\omega(\varepsilon' - j\varepsilon'') = \sigma + \omega\varepsilon'' + j\omega\varepsilon'$$

Equation (16) can now be written

$$\text{Medium impedance } Z_0 = \sqrt{\frac{\omega\mu'' + j\omega\mu'}{\sigma + \omega\varepsilon'' + j\omega\varepsilon'}} \tag{17}$$

Here the term $\omega\mu''$ is analogous to the R term in (15) and $\sigma + \omega\varepsilon''$ is analogous to G. Methods of evaluating conduction and dielectric loss are discussed in Sec. 4-5.

Impedances of lines and media are summarized in Table 4-1.

TABLE 4-1
Impedance of media and lines

Type of medium or line	Intrinsic impedance (for media) or characteristic impedance (for lines), Ω	Equation number
Free space	$Z_0 = \frac{E}{H} = \sqrt{\frac{\mu_0}{\varepsilon_0}} = 376.731 \approx 120\pi = 377$	
Dielectric medium (lossless)	$Z_0 = \frac{377}{\sqrt{\varepsilon_r}}$	
Conducting medium	$Z_0 = \sqrt{\frac{j\omega\mu}{\sigma + j\omega\varepsilon}}$	(4-2-16)
Conducting medium with conductive, magnetic, and dielectric loss	$Z_0 = \sqrt{\frac{\omega\mu'' + j\omega\mu'}{\sigma + \omega\varepsilon'' + j\omega\varepsilon'}}$	(4-2-17)
Transmission line (with loss) general case	$Z_0 = \frac{V}{I} = \sqrt{\frac{Z}{Y}} = \sqrt{\frac{R + j\omega\mathscr{L}}{G + j\omega C}}$	(4-2-15)
Transmission line (lossless)	$Z_0 = \sqrt{\frac{\mathscr{L}}{C}} = R_0$	
Transmission line (small loss, $G/C = R/\mathscr{L}$, Heaviside distortionless line)	$Z_0 = \sqrt{\frac{\mathscr{L}}{C}}\left[1 + j\left(\frac{G}{2\omega C} - \frac{R}{2\omega\mathscr{L}}\right)\right]$	
Transmission line cell map ($h = \sqrt{\varepsilon_r}w$)	$Z_0 = 377\frac{N_s}{N_p}$	
Strip line (lossless)	$Z_0 \approx \frac{377}{\sqrt{\varepsilon_r}[(W/H) + 2]}$	(3-3-28)
Coaxial line	$Z_0 = \frac{138}{\sqrt{\varepsilon_r}} \log \frac{b}{a}$	(3-3-25)
Coaxial line (air filled)	$Z_0 = 138 \log \frac{b}{a}$	(3-3-25.1)
Two-wire line ($D \gg a$)	$Z_0 = \frac{276}{\sqrt{\varepsilon_r}} \log \frac{D}{a}$	(3-3-26)
Two-wire line (in air) ($D \gg a$)	$Z_0 = 276 \log \frac{D}{a}$	(3-3-26.1)

Problem 4-2-2. Phase velocity. Find the phase velocity of an electromagnetic wave propagating in a nonmagnetic ($\mu_r = 1$) medium with (*a*) $\varepsilon_r = 1$ (air), (*b*) $\varepsilon_r = 12$ (silicon), and (*c*) $\varepsilon_r = 81$ (water). *Ans.* (*a*) 300×10^6 m/s; (*b*) 87×10^6 m/s; (*c*) 33×10^6 m/s.

Problem 4-2-3. Medium impedance. A medium has constants $\sigma = 4.2 \times 10^{-4}$ mhos/m, $\varepsilon_r = 5 - j3$, and $\mu_r = 2 + j1$. Find the impedance of the medium at a frequency 1 MHz. *Ans.* $165\angle 46°\Omega$.

4-3 TRAVELING WAVES AND STANDING WAVES

We now consider both traveling waves and standing waves in free space. A standing wave consists of two waves traveling in opposite directions. We keep in mind that both kinds of waves also occur on transmission lines. Thus, the equations developed here involving the electric and magnetic fields of a space wave are identical in form with those for the voltage and current on a lossless transmission line provided only that we replace the electric field **E** by the line voltage V(= line integral of **E** between line conductors) and the magnetic field **H** by the line current I (= line integral of **H** around one conductor). So in this sense we review wave behavior on lines but do it in the general context of waves in space.

The instantaneous values of E_y and H_z for a plane traveling wave are illustrated in Fig. 4-4 with the wave progressing in the positive x direction. Figure 4-4*a* shows the condition at the time $t = 0$, while Fig. 4-4*b* shows the condition one-quarter period later ($t = T/4$). **E** is in the y-direction and **H** in the z-direction. Both **E** and **H** are in phase.

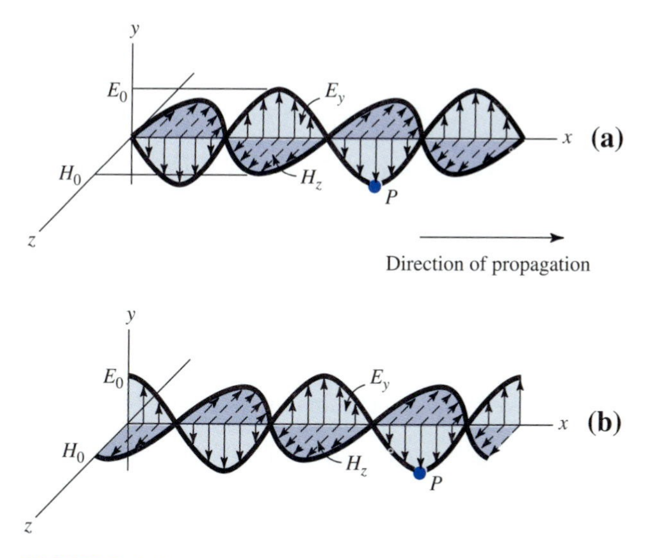

FIGURE 4-4
Instantaneous values of E_y and H_z for a traveling wave: (*a*) at time $t = 0$ and (*b*) $\frac{1}{4}$ period later. In this interval the point P has advanced $\lambda/4$ to the right.

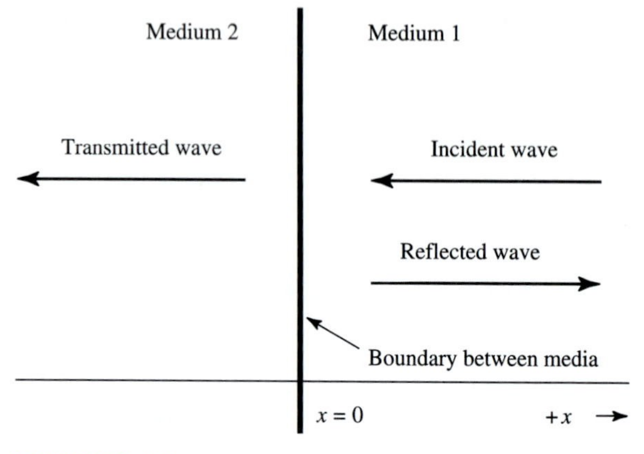

FIGURE 4-5
Relation of incident, reflected, and transmitted waves.

Referring to Fig. 4-5, assume that space is divided into two media, 1 and 2, with a plane boundary between as shown. A wave originating in medium 1 and incident on the boundary is said to be the *incident wave.* The wave reflected from the boundary back into medium 1 is called the *reflected wave.* If the reflection of the incident wave at the boundary is not complete, some of the wave continues on into medium 2 and this wave is referred to as the *transmitted wave.*

Let the incident wave (traveling to the left) be given by

$$E_{y0} = E_0 e^{j(\omega t + \beta x)} \tag{1}$$

and the reflected wave (traveling to the right) by

$$E_{y1} = E_1 e^{j(\omega t - \beta x + \delta)} \tag{2}$$

where δ = time-phase lead of E_{y1} with respect to E_{y0} at $x = 0$, that is, δ = phase shift at point of reflection
E_0 = amplitude of incident wave
E_1 = amplitude of reflected wave

Adding (1) and (2) we have

$$E_y = E_{y0} + E_{y1} \tag{3}$$

If $\delta = 0$ or 180°, we have on taking the imaginary parts of (1) and (2),

$$E_y = (E_0 + E_1) \sin \omega t \cos \beta x + (E_0 - E_1) \cos \omega t \sin \beta x \tag{4}$$

If medium 2 is a perfect conductor, the reflected wave is equal in magnitude to the incident wave. If $x = 0$ is taken to be at the boundary between media 1 and 2, the boundary relation for the tangential component of **E** requires that $E_y = 0$, so that $E_1 = -E_0$ at the boundary ($\delta = 180°$). Thus (4) becomes

$$E_y = 2E_0 \cos \omega t \sin \beta x \tag{5}$$

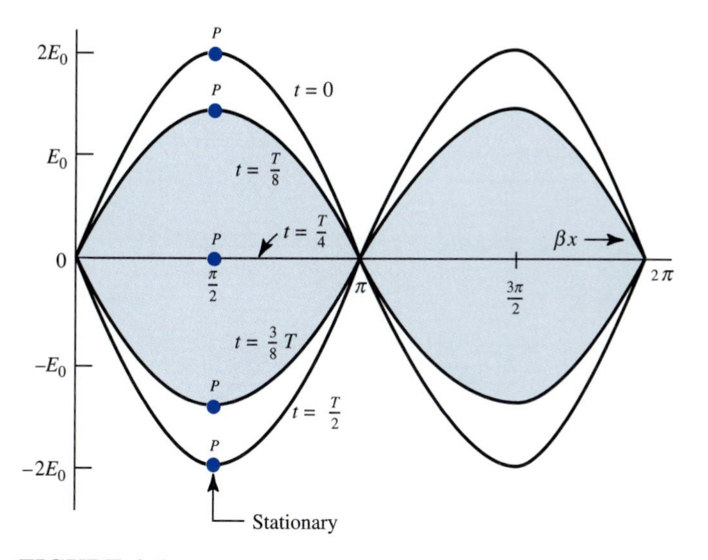

FIGURE 4-6
Two equal and opposite traveling waves result in a pure standing wave E_y shown at various instants of time. A point P is stationary in position.

This represents a wave which is stationary in space. The values of E_y at a particular instant are a sine function of x. The instantaneous values at a particular point are a cosine function of t. The peak value of the wave is the sum of the incident and reflected peak values or $2E_0$. A stationary wave of this type for which $|E_1| = |E_0|$ is a *pure standing wave.* This type of wave is associated with resonators.

The space and time variations of E_y for a pure standing wave are shown by the curves of Fig. 4-6. It is to be noted that a constant-phase point, such as P, does not move in the x direction but remains at a fixed position as time passes.

Now let us examine the conditions when the reflected wave is smaller than the incident wave, say one-half as large. Then, $E_1 = -0.5E_0$. (In the analogous transmission line case, $Z_L = 3Z_0$ and $\rho_v = \frac{1}{2}$.) Evaluating (4) for this case at four instants of time gives the curves of Fig. 4-7. The curves show the values of E_y as a function of βx at times equal to 0, $\frac{1}{8}$, $\frac{1}{4}$, and $\frac{3}{8}$ period. The peak values of E_y range from $1.5E_0$ at $t = 0$ to $0.5E_0$ at $t = \frac{1}{4}$ period. The peak values as a function of x as observed over an interval of time greater than one cycle correspond to the envelope as indicated. The *envelope remains stationary, but* focusing our attention on a constant-phase point P of the wave, we note that the *total instantaneous wave travels to the left.* It will also be noted that the velocity with which P moves is not constant. Between time 0 and $\frac{1}{8}$ period P moves about $0.05\lambda(0.1\pi)$, while in the next $\frac{1}{8}$ period P moves about 4 times as far, or about $0.2\lambda(0.4\pi)$. Although the average velocity of the constant-phase point is the same as for a pure traveling wave, its instantaneous magnitude varies between values which are greater and less.

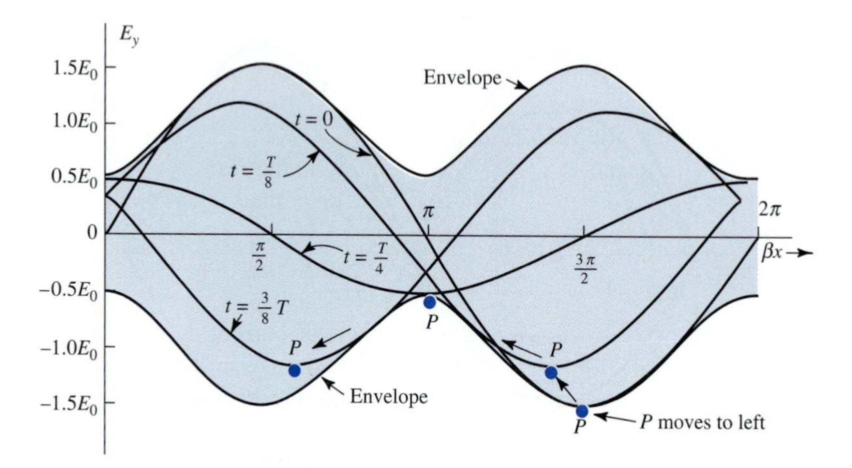

FIGURE 4-7
Standing-wave envelope for reflected wave equal to half the incident wave amplitude ($E_1 = -0.5\,E_0$) with resultant (traveling) wave at four instants of time: $t = 0$, $t = T/8$, $t = T/4$, and $t = 3T/8$. Note that the position of the constant phase point P moves to the left.

Example 4-2. VSWR and reflection coefficient ρ for a standing wave. Find: (*a*) Voltage standing wave ratio (VSWR) and (*b*) ρ for the wave of Fig. 4-7.

Solution

$$\text{VSWR} = \frac{E_{\max}}{E_{\min}} = \frac{E_0 + E_1}{E_0 - E_1} = \frac{1 + (1/2)}{1 - (1/2)} = 3 \quad \textit{Ans.} \quad (a) \tag{6}$$

Rewriting (6) gives

$$\text{VSWR} = \frac{1 + (E_1/E_0)}{1 - (E_1/E_0)} = \frac{1 + |\rho|}{1 - |\rho|}$$

Thus, the magnitude of the reflection coefficient is

$$|\rho| = \frac{\text{VSWR} - 1}{\text{VSWR} + 1} = \frac{3 - 1}{3 + 1} = \frac{1}{2} \quad \textit{Ans.} \quad (b)$$

Problem 4-3-1. Standing wave parameters. A standing wave with a VSWR = 2.2 in a lossless medium has a maximum field $E = 76$ mV/m. Find: (*a*) minimum field and (*b*) reflection coefficient. *Ans.* (*a*) 34.5 mV/m, (*b*) 0.375.

4-4 CONDUCTORS AND DIELECTRICS

According to Maxwell's curl equation from Ampère's law,

$$\nabla \times \mathbf{H} = \mathbf{J} + \frac{\partial \mathbf{D}}{\partial t} \tag{1}$$

Since $\mathbf{J} = \sigma\mathbf{E}$, (1) becomes

$$\nabla \times \mathbf{H} = \sigma\mathbf{E} + \frac{\partial \mathbf{D}}{\partial t} \tag{2}$$

For a linearly polarized plane wave traveling in the x direction with $\mathbf{E}$ in the y direction, the vector equation (2) reduces to the scalar phasor equation

$$-\frac{\partial H_z}{\partial x} = \sigma E_y + j\omega\varepsilon E_y \tag{3}$$

The terms in (3) each have the dimensions of current density, which is expressed in amperes per square meter. The term σE_y represents the *conduction-current density,* while the term $j\omega\varepsilon E_y$ represents the *displacement-current density.* Thus, according to (3) the space rate of change of H_z equals the sum of the conduction- and displacement-current densities. If the conductivity is zero, the conduction-current term vanishes and we have the condition considered in previous sections. If σ is not equal to zero, one may arbitrarily define three conditions as follows:†

1. $\omega\varepsilon \gg \sigma$
2. $\omega\varepsilon \approx \sigma$
3. $\omega\varepsilon \ll \sigma$

When the displacement current is much greater than the conduction current, as in condition 1, the medium behaves like a dielectric. If $\sigma = 0$, the medium is a perfect, or lossless, dielectric. For σ not equal to zero the medium is a lossy, or imperfect, dielectric. However, if $\omega\varepsilon \gg \sigma$, it behaves more like a dielectric than anything else and may, for practical purposes, be classified as a *dielectric.* On the other hand, when the conduction current is much greater than the displacement current, as in condition 3, the medium may be classified as a *conductor*. Under conditions midway between these two, when the conduction current is of the same order of magnitude as the displacement current, the medium may be classified as a *quasi-conductor* (not to be confused with "semiconductor").

We can be even more specific and arbitrarily classify media as belonging to one of three types according to the value of the ratio $\sigma/\omega\varepsilon$ as follows:

Conductors: $$\frac{\sigma}{\omega\varepsilon} > 100$$

Quasi-conductors: $$\frac{1}{100} < \frac{\sigma}{\omega\varepsilon} < 100$$

Dielectrics: $$\frac{\sigma}{\omega\varepsilon} < \frac{1}{100}$$

†In the case of a lossy dielectric, condition 1 would be modified to $\omega\varepsilon' \gg \sigma'$ and condition 2 to $\omega\varepsilon' \sim \sigma'$, but for condition 3 we have $\omega\varepsilon' \approx \omega\varepsilon$ and $\sigma' = \sigma$, and hence we can write $\omega\varepsilon \ll \sigma$, as indicated.

where σ = conductivity of medium, ℧ m^{-1}
ε = permittivity of medium, F m^{-1}
ω = radian frequency = $2\pi f$, where f is the frequency, Hz

The ratio $\sigma/\omega\varepsilon$ is dimensionless.

Frequency is an important factor in determining whether a medium acts like a dielectric or a conductor. For example, take the case of average rural ground (mid-west United States) for which $\varepsilon_r = 14$ (at low frequencies) and $\sigma = 10^{-2}$ ℧ m^{-1}. Assuming no change in these values as a function of frequency, the ratio $\sigma/\omega\varepsilon$ at three different frequencies for average rural ground is as tabulated:

Ratio $\sigma/\omega\varepsilon$ of rural ground at three frequencies

Frequency, Hz	Ratio $\sigma/\omega\varepsilon$	Behavior
10^3(= 1 kHz)	1.3×10^4	Conductor
10^7(= 10 MHz)	1.3	Quasi-conductor
3×10^{10} (= 30 GHz)	4.3×10^{-4}	Dielectric

At 1 kHz rural ground behaves like a conductor, while at the microwave frequency of 30 GHz it acts like a dielectric.

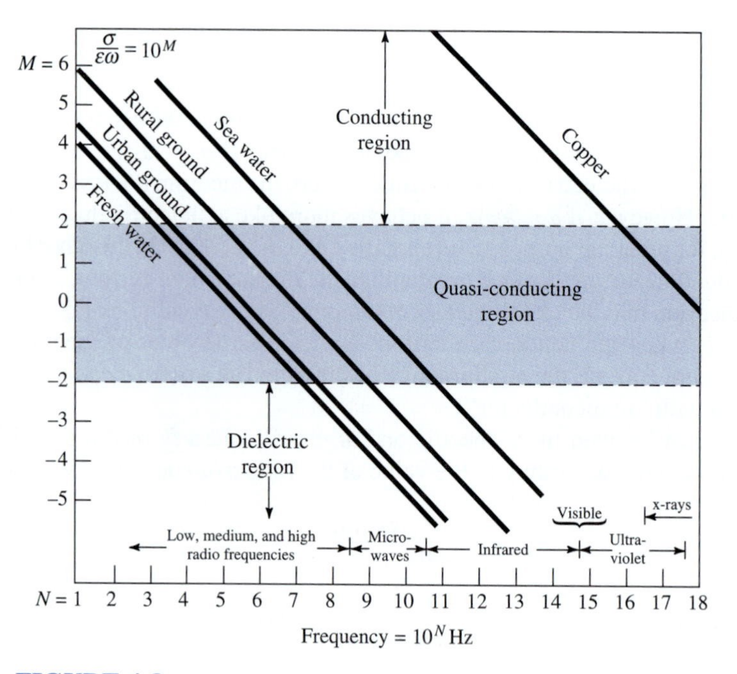

FIGURE 4-8
Ratio $\sigma/\omega\varepsilon$ as a function of frequency for some common media (log-log plot). Low-frequency constants for these media are listed in Table 4-2. Note that fresh water and urban ground act like conductors at low frequencies but like dielectrics at high radio frequencies.

TABLE 4-2
Table of constants for some common media

Medium	**Relative permittivity ε_r, dimensionless**	**Conductivity σ, ℧ m^{-1}**
Copper	1	5.8×10^7
Sea water	80	4
Rural ground	14	10^{-2}
Urban ground	3	10^{-4}
Fresh water	80	10^{-3}

Example 4-3. Impedance of ground. To match the impedance of a ground penetrating radar (GPR) antenna to the ground, it is important to know the impedance of the ground. For ground with $\varepsilon_r = 14$, $\mu_r = 1$, and $\sigma = 10^{-2}$ ℧ m^{-1}, find its impedance at 200 MHz.

Solution. From (4-2-16)

$$Z_0 = \sqrt{\frac{j\omega\mu}{\sigma + j\omega\varepsilon}} = \sqrt{\frac{\mu_0}{\varepsilon_0}}\sqrt{\frac{j}{\dfrac{\sigma}{\omega\varepsilon_0} + j\varepsilon_r}} = 377\sqrt{\frac{j}{0.9 + j14}} = 101\angle 1.8^\circ\ \Omega \quad \textit{Ans.}$$

Problem 4-4-1. Impedance of moist soil. For a sample of moist soil $\varepsilon_r = 8$, $\sigma = 1$ ℧/m, $\mu_r = 1$, find its impedance at 10 MHz. *Ans.* $8.9\angle 45^\circ\ \Omega$.

In Fig. 4-8 (on facing page) the ratio $\sigma/\omega\varepsilon$ is plotted as a function of frequency for a number of common media. In preparing Fig. 4-8 the constants were assumed to maintain their low-frequency values at all frequencies. The curves in Fig. 4-8 should therefore not be regarded as accurate above the microwave region since the constants of media may vary with frequency, particularly at frequencies of the order of 1 GHz and higher. A list of the low-frequency constants for the media of Fig. 4-8 is presented in Table 4-2.

4-5 CONDUCTING MEDIA AND LOSSY LINES

Referring to Fig. 4-9, consider a wave that is transmitted *into* the conducting medium. Let $x = 0$ at the boundary of the conducting medium, with x increasing positively into the conducting medium.

From (4-2-12), the wave equation for a conducting medium is

$$\frac{\partial^2 E_y}{\partial x^2} - \gamma^2 E_y = 0 \tag{1}$$

where

$$\gamma^2 = j\omega\mu\sigma - \omega^2\mu\varepsilon \tag{2}$$

A solution of (1) for a wave traveling in the positive x direction is

$$E_y = E_0 e^{-\gamma x} \tag{3}$$

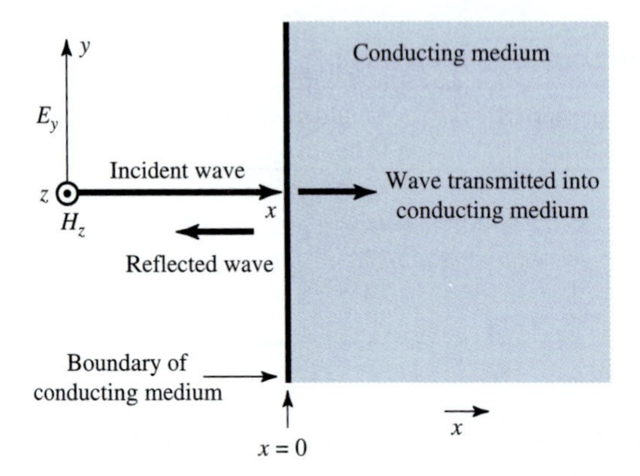

FIGURE 4-9
Plane wave entering conducting medium at normal incidence.

For conductors, $\sigma \gg \omega\varepsilon$, so that (2) reduces to

$$\gamma^2 \approx j\omega\mu\sigma \tag{4}$$

and†

$$\gamma = \sqrt{j\omega\mu\sigma} = (1+j)\sqrt{\frac{\omega\mu\sigma}{2}} \tag{5}$$

Thus, γ has a real and imaginary part. Putting $\gamma = \alpha + j\beta$, we have α, the real part, associated with attenuation and β, the imaginary part, associated with phase. Hence,

$$E_y = E_0 e^{-\alpha x} e^{-j\beta x} \tag{6}$$

where $\alpha = \operatorname{Re}\gamma = \sqrt{\dfrac{\omega\mu\sigma}{2}}$ = attenuation constant, Np m^{-1}

$\beta = \operatorname{Im}\gamma = \sqrt{\dfrac{\omega\mu\sigma}{2}}$ = phase constant, rad m^{-1}

ω = radian frequency = $2\pi f$, rad s^{-1}

μ = permeability of medium, H m^{-1}

σ = conductivity of medium, ℧ m^{-1}

x = distance, m

j = complex operator = $\sqrt{-1}$, dimensionless

Equation (6) is a solution of the wave equation for a plane wave traveling in the positive x direction in the conducting medium. It gives the variation of E_y in both magnitude and phase as a function of x. The field attenuates exponentially and is retarded linearly in phase with increasing x.

Let us now obtain a quantitative measure of the penetration of a wave into the conducting medium. Referring to Fig. 4-9, let (6) be written in the form

$$E_y = E_0 e^{-x/\delta} e^{-j(x/\delta)} \tag{7}$$

where $\delta = \sqrt{2/\omega\mu\sigma}$. At $x = 0$, $E_y = E_0$. This is the amplitude of the field at the surface of the conducting medium. Now δ in (7) has the dimension of distance. At

†*Note:* $\sqrt{j} = \sqrt{\dfrac{2j}{2}} = \sqrt{\dfrac{1+2j-1}{2}} = \sqrt{\dfrac{(1+j)^2}{2}} = \dfrac{1+j}{\sqrt{2}} = 1\angle 45°$

a distance $x = \delta$, the amplitude of the field is

$$|E_y| = E_0 e^{-1} = E_0 \frac{1}{e} \tag{8}$$

Thus, E_y decreases to l/e (36.8 percent) of its initial value, while the wave penetrates to a distance δ. Hence δ is called the ***1/e depth of penetration.*** See Fig. 4-10.

As an example, consider the depth of penetration of a plane electromagnetic wave incident normally on a good conductor, such as copper. Since $\omega = 2\pi f$, the $1/e$ depth becomes

$$\delta = \frac{1}{\sqrt{f\pi\mu\sigma}} = \frac{1}{e} \qquad \textbf{\textit{Depth of penetration}} \tag{9}$$

For copper $\mu_r = 1$, so that $\mu = 1.26\ \mu\text{H m}^{-1}$, and the conductivity $\sigma = 58\ \text{M℧ m}^{-1}$. Putting these values in (9), we obtain for copper

$$\delta = \frac{6.6 \times 10^{-2}}{\sqrt{f}} \tag{10}$$

where $\delta = 1/e$ depth of penetration, m, and f = frequency, Hz.

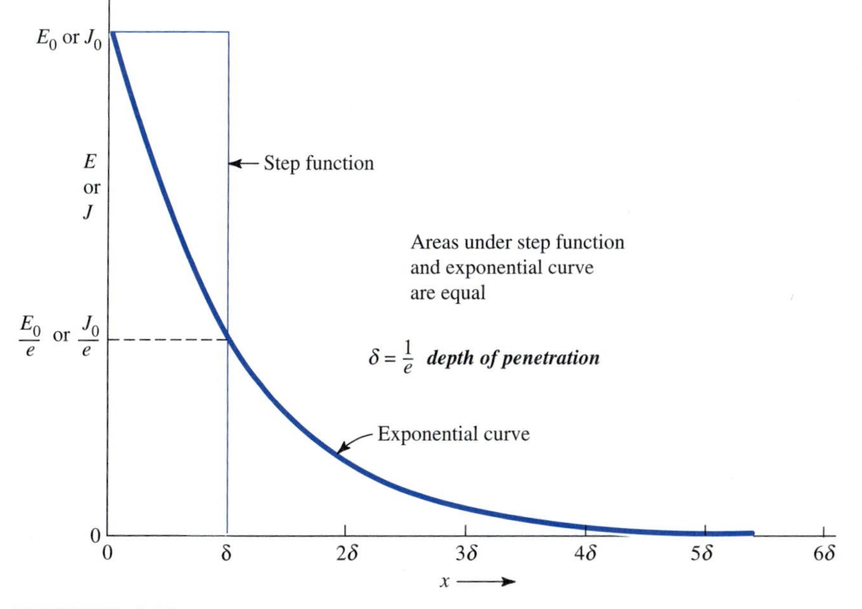

FIGURE 4-10

Relative magnitude of electric field **E** or current density **J** ($= \sigma\mathbf{E}$) as a function of depth of penetration δ for a plane wave traveling in x direction into conducting medium. The abscissa gives the penetration distance x and is expressed in $1/e$ depths (δ). The wavelength in the conductor equals $2\pi\delta$.

Evaluating (10) at specific frequencies, we find that

At 60 Hz: $\delta = 8.5 \times 10^{-3}$ m

At 1 MHz: $\delta = 6.6 \times 10^{-5}$ m

At 30 GHz: $\delta = 3.8 \times 10^{-7}$ m

Thus, at 60 Hz the $1/e$ depth of penetration is 8.5 mm. At 10-mm wavelength (30 GHz) the penetration is only 0.00038 mm, or less than $\frac{1}{2}$ μm. This small penetration is often called the *skin effect.*

Thus, a high-frequency field attenuates as it penetrates a conductor in a shorter distance than a low-frequency field.†

In addition to the $1/e$ depth of penetration, we can speak of other depths for which the electric field decreases to an arbitrary fraction of its original value. For example, consider the depth at which the field is 0.01 (1 percent) of its original value. This depth is obtained by multiplying the $1/e$ depth by 4.6 and may be called the *1 percent depth of penetration.*

Phase velocity is given by the ratio ω/β. In the present case, $\beta = 1/\delta$ so that the *phase velocity* in the conductor is

$$v_c = \omega\delta = \sqrt{\frac{2\omega}{\sigma\mu}} \tag{11}$$

Since the $1/e$ depth is small, the phase velocity in conductors is small. It is apparent from (11) that the velocity is a function of the frequency and hence of the wavelength. In this case, $dv/d\lambda$ is negative; where λ is the free-space wavelength. Conductors are anomalously dispersive media (see Sec. 4-9).

The ratio of the velocity of a wave in free space to that in the conducting medium is the index of refraction for the conducting medium. At low frequencies the index is very large for conductors.

To find the wavelength λ_c in the conductor, we have from (11) that $f\lambda_c = \omega\delta$, or

$$\lambda_c = 2\pi\delta \tag{12}$$

In (12), both λ_c and δ are in the same units of length. Hence the wavelength in the conductor is 2π times the $1/e$ depth. Since the $1/e$ depth is small for conductors, the wavelength in conductors is small.

Values of the $1/e$ depth, 1 percent depth, wavelength, velocity, and refractive index for a medium of copper are given in Table 4-3 for three frequencies. It is interesting to note that the electric field is damped to 1 percent of its initial amplitude in about $3\lambda/4$ in the metal. Compare rows 3 and 4 in the table.

Since the penetration depth is inversely proportional to the square root of the frequency, a thin sheet of conducting material can act as a lowpass filter for electromagnetic waves. (The ionosphere leaks 60-Hz power grid radiation into space.)

---

†This is analogous to the way in which a rapid-temperature variation at the surface of a thermal conductor penetrates a shorter distance than a slow-temperature variation.

TABLE 4-3
Penetration depths, wavelength, velocity, and refractive index for copper

	Frequency		
	60 Hz	**1 MHz**	**30 GHz**
Wavelength in free space, λ	5 Mm	300 m	10 mm
$1/e$ depth, m	8.5×10^{-3}	6.6×10^{-5}	3.8×10^{-7}
1 percent depth, m	3.9×10^{-2}	3×10^{-4}	1.7×10^{-6}
Wavelength in conductor λ_c, m	5.3×10^{-2}	4.1×10^{-4}	2.4×10^{-6}
Velocity in conductor v_c, m s^{-1}	3.2	4.1×10^{2}	7.1×10^{4}
Index of refraction, dimensionless	9.5×10^{7}	7.3×10^{5}	4.2×10^{3}

For the case of a conducting medium where $\sigma \gg \omega\varepsilon$, we have from (4-2-16) that the characteristic impedance

$$Z_0 = Z_c = \sqrt{\frac{j\omega\mu}{\sigma}} = \sqrt{\frac{\omega\mu}{\sigma}}\angle 45^\circ \qquad (\Omega) \tag{13}$$

Example 4-4. Ocean penetration. Calculate the ocean depths at which a 1-μV m^{-1} field will be obtained with E at the surface equal to 1 V m^{-1} at frequencies of 1, 10, 100, and 1000 kHz. What is the most suitable frequency for communication by wireless with undersea craft?

Solution. From Table 4-1, $\sigma = 4$ ℧ m^{-1} and $\varepsilon_r = 80$ for sea water. At the highest frequency (1000 kHz), $\sigma \gg \omega\varepsilon$, so that $\alpha = \sqrt{\omega\mu\sigma/2}$ can be used at all four frequencies. At 1 kHz

$$\alpha = \sqrt{\frac{2\pi 10^3 4\pi \times 10^{-7} 4}{2}} = 0.13 \text{ Np m}^{-1}$$

Since

$$\frac{E}{E_0} 10^{-6} = e^{-\alpha x}$$

we see that

$$x = \frac{6}{\alpha}\ln 10 = \frac{13.8}{\alpha}$$

and $x = 106$ m at 1 kHz, 35 m at 10 kHz, 11 m at 100 kHz, and 3.5 m at 1 MHz.

Thus, 1 kHz would appear to be the best of the above four frequencies, although an even lower frequency might be desirable depending on other factors including the efficiency of antennas used for transmitting and receiving. *Ans.*

Example 4-5. Lossy line and lossy medium. A uniform lossy line is connected to a load as in Fig. 4-11.1. The attenuation constant $\alpha = 0.3$ Np λ^{-1} and there is a 2-to-1 mismatch at the load (load resistance twice line impedance). (*a*) Draw graphs of the incident, reflected and total voltages on the line. (*b*) Do the same for a wave in a lossy medium.

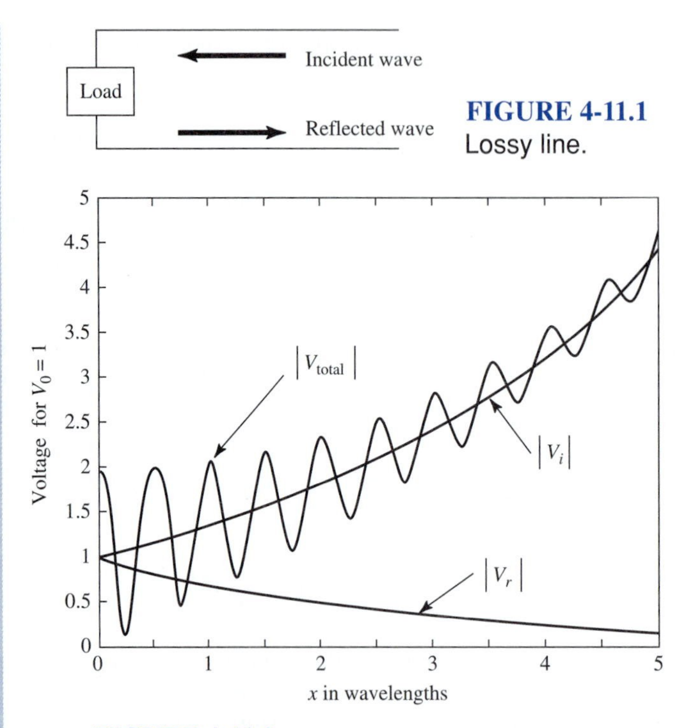

FIGURE 4-11.1
Lossy line.

FIGURE 4-11.2
Magnitudes of incident voltage $|V_i|$, reflected voltage $|V_r|$ and total voltage $|V_{total}|$ from (16) on 5-wavelength-long lossy line. Note almost pure traveling wave at 5λ and almost complete standing wave at 0 to 1λ.

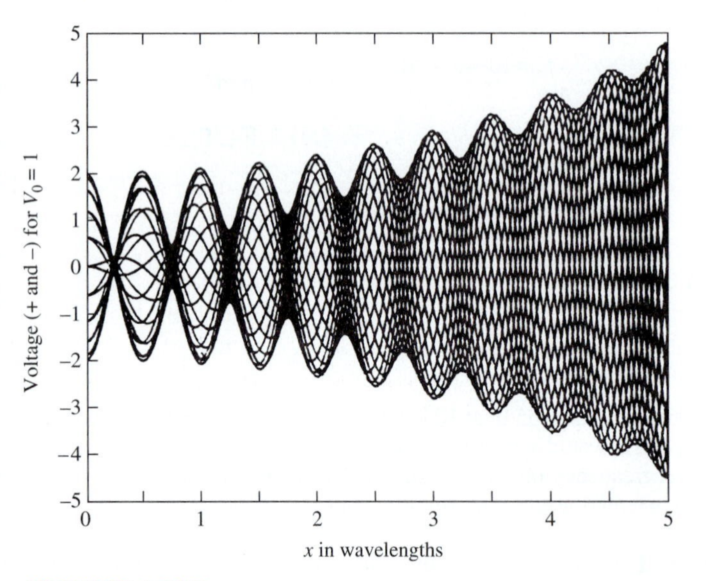

FIGURE 4-11.3
Total voltage (V_{total}) evaluated at 60 instants of time from (16) giving 12 increments per wavelength. Note almost pure traveling wave at 5λ and almost complete standing wave at 0 to 1λ *Ans.* (*a*). Produced by LossyLine program (Appendix C).

Solution. From (3-2-8)

$$V_i = V_0 e^{\alpha x} e^{j(\omega t+\beta x)} \qquad \text{(V)} \tag{14}$$

and the voltage wave reflected from the load (traveling left to right) by

$$V_r = \rho_0 V_0 e^{-\alpha x} e^{j(\omega t-\beta x)} \qquad \text{(V)} \tag{15}$$

where V_0 = voltage of incident wave at the load, V
α = attenuation constant, Np λ^{-1}
$\omega = 2\pi f$ = radian frequency, where f = frequency, Hz
$\beta = 2\pi/\lambda$ = propagation constant, λ^{-1}
ρ_0 = reflection coefficient, dimensionless
x = distance from load, λ

The total voltage on the line is the sum of (14) and (15), or

$$V_{\text{total}} = V_i + V_r = V_0 e^{\alpha x} e^{j(\omega t+\beta x)} + \rho_0 V_0 e^{-\alpha x} e^{j(\omega t-\beta x)} \qquad \text{(V)} \tag{16}$$

Graphs of (16) are shown in Figures 4-11.2 and 4-11.3. *Ans.* (*a*).

The graphs of Figs. 4-11.2 and 4-11.3 also apply to the field E in V m^{-1} for a wave traveling in a lossy medium with partial transmission and partial reflection at a plane boundary with another medium (replacing the load on the line) with an attenuation constant $\alpha = 0.3$ Np λ^{-1} and a reflection coefficient $\rho = 1/3$ at the interface. *Ans.* (*b*).

For waves in motion, see Lossy Line program (Appendix C).

Problem 4-5-1. 1/*e* depths. A medium has constants $\sigma = 10^2$ ℧/m, $\mu_r = 2$ and $\varepsilon_r = 3$. If the constants do not change with frequency, find the $1/e$ and 1 percent depths of penetration at (*a*) 60 Hz, (*b*) 2 MHz, and (*c*) 3 GHz. *Ans.* (*a*) 4.6 and 21 m, (*b*) 25 and 116 mm, (*c*) 650 μm and 3 mm.

4-6 DIELECTRIC HYSTERESIS AND DIELECTRIC LOSS

In dielectrics that are good insulators, the dc conduction current may be negligible. However, an appreciable ac current in phase with the applied field may be present because of *dielectric hysteresis.* This phenomenon is analogous to magnetic hysteresis in ferromagnetic materials. Materials, such as glass or plastics, which are good insulators under static conditions, may consume considerable energy in alternating fields. The heat generated in this way finds application in radio-frequency heating, as in the molding of plastics and in the heating of foods with microwave ovens.

Thus, the permittivity becomes complex, and Maxwell's equation

$$\nabla \times \mathbf{H} = \mathbf{J} + \frac{\partial \mathbf{D}}{\partial t} = \sigma \mathbf{E} + j\omega\varepsilon\mathbf{E} \tag{1}$$

becomes

$$\nabla \times \mathbf{H} = \sigma \mathbf{E} + j\omega(\varepsilon' - j\varepsilon'')\mathbf{E} \tag{2}$$

where $\varepsilon = \varepsilon' - j\varepsilon''$
ε' = real or lossless part of ε
ε'' = imaginary or lossy part of ε

Rearranging (2) gives

$$\nabla \times \mathbf{H} = [(\sigma + \omega\varepsilon'') + j\omega\varepsilon']\mathbf{E}$$
$$= (\sigma' + j\omega\varepsilon')\mathbf{E} = J_{\text{total}}\mathbf{E} \tag{3}$$

where $\sigma' = \sigma + \omega\varepsilon''$ = equivalent conductance.

It is apparent that ε'' (imaginary part of ε) is involved in a frequency-dependent term ($\omega\varepsilon''$) with the dimensions of conductance. At dc ($\omega = 0$ and therefore $\omega\varepsilon'' = 0$), power loss is small in a good dielectric, for which σ is small. However, at high frequencies (ω large), losses can become larger as $\omega\varepsilon''$ becomes significant. The sum of σ and $\omega\varepsilon''$ constitutes what may be called the *equivalent conductivity* σ'.

The total current density J_{total} is the sum of a conduction current density ($\sigma'E$) and a displacement current density $\omega\varepsilon'E$ in time-phase quadrature. Thus,

$$\tan\delta = \frac{\sigma'}{\omega\varepsilon'} \qquad \textbf{\textit{Loss tangent}} \tag{4}$$

The quantity $\tan\delta$ is called the *loss tangent.* Also the cosine of the angle $\theta(= 90^\circ - \delta)$ is the *power factor* (PF). Thus (for small δ)

$$\text{PF} = \cos\theta \approx \tan\delta = \frac{\sigma'}{\omega\varepsilon'} \qquad \textbf{\textit{Power factor}} \tag{5}$$

The power dissipated per unit volume is

$$p = \frac{\text{power}}{\text{volume}} = \frac{\text{current}}{\text{area}}\frac{\text{voltage}}{\text{length}} = JE = \sigma'E^2 \qquad (\text{W m}^{-3})$$

Example 4-6. Lossy dielectric medium. Find the average power dissipated per cubic meter in a nonconducting dielectric medium with relative permittivity of 4 and a loss tangent of 0.001 if $E = 1$ kV m^{-1} rms and the frequency is 10 MHz.

Solution. Since $\sigma = 0$, $\sigma' = \omega\varepsilon''$ and $\tan\delta = \sigma'/\omega\varepsilon' = \varepsilon''/\varepsilon'$, or $\varepsilon'' = \varepsilon'\tan\delta$, and since $\tan\delta$ is small, $\varepsilon'' \ll \varepsilon'$ and $\varepsilon' \approx \varepsilon$, so $\sigma' = \omega\varepsilon'\tan\delta \approx \omega\varepsilon\tan\delta$, or

$$\sigma' \approx 2\pi \times 10^7 \times 4 \times 8.85 \times 10^{-12} \times 10^{-3} = 2.22\ \mu\mho\ \text{m}^{-1}$$

The power p dissipated per unit volume is then

$$p = E^2\sigma' = 10^6 \times 2.22 \times 10^{-6} = 2.22\ \text{W m}^{-3} \qquad \textit{Ans.}$$

From (3) it is apparent that the real part (ε') of the complex permittivity is associated with the displacement-current density and, hence, with the stored-energy density ($= \varepsilon'E^2$). The imaginary part (ε'') is associated with the conduction-current density and, hence, with the power dissipated per unit volume as heat [$= (\sigma + \omega\varepsilon'')E^2 = \sigma'E^2$].

Problem 4-6-1. Capacitor. Power factor. A parallel-plate capacitor is filled with a dielectric with PF $= 0.004$ and $\varepsilon_r = 14$. The plates are 350 mm square, and the distance between them is 12 mm. If 300 V (rms) at 3 MHz is applied to the capacitor, find the power dissipated as heat. *Ans.* 8.6 W.

Problem 4-6-2. Capacitor. Dielectric heating. A parallel-plate capacitor has a plate area of 0.2 m^2 and a plate separation of 6 mm. If 50 V (rms) is applied at 100 MHz, find the power lost as heat in the dielectric between the plates if its constants are $\sigma = 5 \times 10^{-3}$ ℧/m, $\varepsilon_r' = 20$, $\varepsilon_r'' = 2$, and $\mu_r = 1$. Neglect fringing. *Ans.* 1.34 kW.

Problem 4-6-3. Capacitor. Power factor. A parallel-plate capacitor is filled with a dielectric of 0.003 power factor and $\varepsilon_r = 10$. The plates are 250 mm square, and the distance between is 10 mm. If 500 V (rms) at 2 MHz is applied to the capacitor, find the power dissipated as heat. *Ans.* 5.22 W.

4-7 PLANE WAVES AT INTERFACES AND ANALOGOUS TRANSMISSION LINES

Consider a linearly polarized wave traveling in the positive x direction with $\mathbf{E}$ in the y direction and $\mathbf{H}$ in the z direction. The wave is incident normally on the boundary between two media with intrinsic impedances Z_1 and Z_2, as in Fig. 4-12*a*. Assume that the incident traveling wave has field components E_i and H_i, at the boundary. Part of the incident wave is, in general, reflected while another part is transmitted into the second medium. The reflected traveling wave has field components E_r and H_r at the boundary. The transmitted wave has field components E_t and H_t at the boundary.

The situation here is analogous to that of a terminated transmission line in circuit theory. By analyzing the wave interaction at the interface between two media with field theory, it will be shown that we arrive at identical expressions for the reflection and transmission coefficients of waves and lines.

From the continuity of the tangential field components at a boundary

$$E_i + E_r = E_t \tag{1}$$

and

$$H_i + H_r = H_t \tag{2}$$

The electric and magnetic fields of a plane wave are related by the intrinsic impedance of the medium. Thus

$$\frac{E_i}{H_i} = Z_1$$

$$\frac{E_r}{H_r} = -Z_1$$

$$\frac{E_t}{H_t} = Z_2 \tag{3}$$

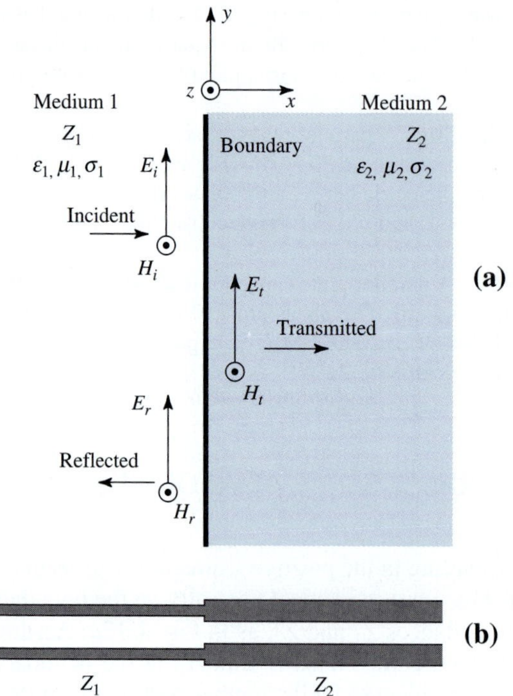

FIGURE 4-12
(*a*) Plane wave incident normally on boundary between two media and (*b*) analogous transmission line.

The impedance of the reflected wave (traveling in the negative x direction) is taken to be negative Z_1 and of the incident wave, positive Z_1. From (2) and (3)

$$H_t = \frac{E_t}{Z_2} = \frac{E_i}{Z_1} - \frac{E_r}{Z_1} \tag{4}$$

or

$$E_t = \frac{Z_2}{Z_1}E_i - \frac{Z_2}{Z_1}E_r \tag{5}$$

Multiplying (1) by Z_2/Z_1 gives

$$\frac{Z_2}{Z_1}E_t = \frac{Z_2}{Z_1}E_i + \frac{Z_2}{Z_1}E_r \tag{6}$$

Adding (5) and (6), we get

$$E_t\left(1 + \frac{Z_2}{Z_1}\right) = \frac{2Z_2}{Z_1}E_i \tag{7}$$

or

$$E_t = \frac{2Z_2}{Z_2 + Z_1}E_i = \tau E_i \tag{8}$$

where τ = *transmission coefficient.*

It follows that

$$\boxed{\tau = \frac{E_t}{E_i} = \frac{2Z_2}{Z_2 + Z_1} \qquad \textbf{\textit{Transmission coefficient}}} \tag{9}$$

Subtracting (5) from (6) gives

$$E_t\left(\frac{Z_2}{Z_1} - 1\right) = \frac{2Z_2}{Z_1}E_r \tag{10}$$

Substituting E_t from (8) into (10) and solving for E_r, we have

$$E_r = \frac{Z_2 - Z_1}{Z_2 + Z_1}E_i = \rho E_i \tag{11}$$

where ρ = *reflection coefficient.*

It follows that

$$\rho = \frac{E_r}{E_i} = \frac{Z_2 - Z_1}{Z_2 + Z_1} \qquad \textbf{\textit{Reflection coefficient}} \tag{12}$$

From (9) and (12)

$$\tau = \frac{2Z_2}{Z_2 + Z_1} = 1 + \frac{Z_2 - Z_1}{Z_2 + Z_1} = 1 + \rho \qquad \boldsymbol{\tau, \rho}\ \textbf{\textit{relation}} \tag{12a}$$

The situation (Fig. 4-12) of a plane wave incident normally on a boundary between two different media of infinite extent, with intrinsic impedances Z_1 and Z_2, is analogous to the situation of a wave on an infinite transmission line having an abrupt change in impedance from Z_1 to Z_2 (Fig. 4-12*b*). The transmission and reflection coefficients for voltage across the transmission line are identical to those given in (9) and (12) if the intrinsic impedance Z_1 of medium 1 is taken to be the characteristic impedance of the line to the left of the junction (Fig 4-12*b*) and the intrinsic impedance Z_2 of medium 2 is taken to be the characteristic impedance of the line to the right of the junction.

Returning now to the case of a plane wave incident normally on the boundary between two media of infinite extent as in Fig. 4-12*a*, let us consider several examples.

Example 4-7. Medium 1 air (Z_0), medium 2 conductor (Z_c). $Z_0 \gg Z_c$. Wave entering conductor. See Fig. E4-7. Find E_t, E_r, H_t, and H_r.

Solution. From (8),

$$E_t \approx \frac{2Z_c}{Z_0}E_i \qquad \textit{Ans.}$$

From (11),

$$E_r = \frac{Z_c - Z_0}{Z_c + Z_0}E_i \text{ and } E_r \approx -E_i \quad \textit{Ans.}$$

From (3),

$$H_t \approx 2H_i \text{ and } H_r \approx H_i \quad \textit{Ans.}$$

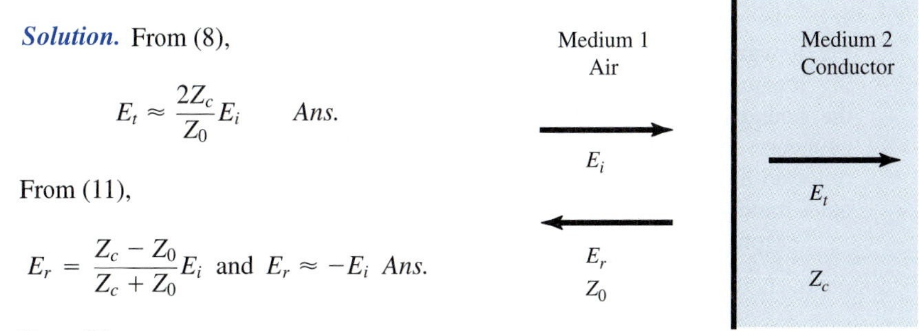

FIGURE E4-7

Example 4-8. Medium 1 air (Z_0), medium 2 perfect conductor ($Z_c = 0$). Wave entering perfect conductor. Same as Example 4-7 with $\sigma = \infty$ ($Z_c = 0$). Find E_t and E_r.

Solution. From (8),

$$E_t = 0 \quad \text{exactly} \quad \textit{Ans.}$$

From (11),

$$E_r = \frac{Z_c - Z_0}{Z_c + Z_0} E_i \quad \text{and} \quad E_r = -E_i \quad \text{exactly} \quad \textit{Ans.}$$

The wave is completely reflected and zero field is transmitted into the conductor. This situation is analogous to a short-circuited transmission line.

Example 4-9. Medium 1 conductor (Z_c), medium 2 air (Z_0). $Z_c \ll Z_0$. See Fig. E4-9. Find E_t, E_r, H_t, and H_r.

Solution. From (8),

$$E_t \approx 2E_i \quad \textit{Ans.}$$

and from (11),

$$E_r \approx E_i \quad \textit{Ans.}$$

For a wave leaving the conducting medium, E nearly doubles at the boundary. To the left of the boundary, the VSWR $= \infty$ but the reflected E_r in the conductor attenuates rapidly.

From (3),

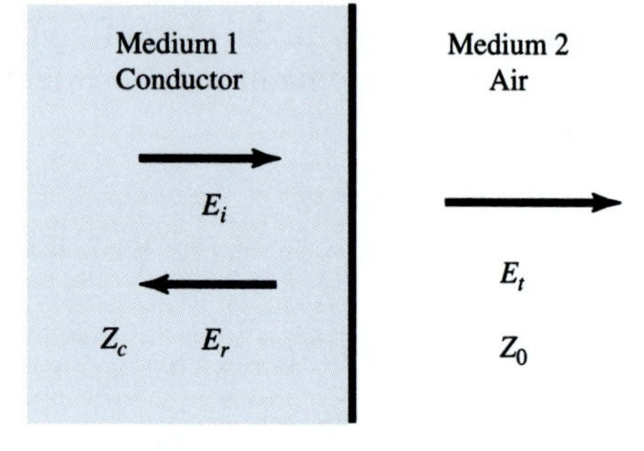

FIGURE E4-9

$$H_t \approx \frac{2Z_c}{Z_0} H_i \quad \text{and} \quad H_r \approx -H_i \quad \textit{Ans.}$$

Example 4-10. Medium 1 air (Z_0), medium 2 infinite impedance (Z_∞). Find E_t and E_r.

Solution. In this hypothetical case, we have from (11) that

$$E_r = E_i \quad Ans.$$

so that the wave is completely reflected with the field doubled at the boundary. This case is analogous to an open-circuited transmission line but for a space wave it requires a higher impedance than 377 Ω for medium 2, and this requires that $\mu_r > 1$ as in a ferromagnetic medium.

Example 4-11. Medium 1 and medium 2 lossless nonferromagnetic dielectric. ($\mu_1 = \mu_2 = \mu_0$). Medium 1 permittivity $= \varepsilon_1$, medium 2 permittivity $= \varepsilon_2$. Find E_t and E_r.

Solution. From (12),

$$E_r = \rho E_i = \frac{\sqrt{\varepsilon_1/\varepsilon_2} - 1}{\sqrt{\varepsilon_1/\varepsilon_2} + 1} E_i \quad Ans.$$

From (9),

$$E_t = \tau E_i = \frac{2}{1 + \sqrt{\varepsilon_2/\varepsilon_1}} E_i \quad Ans.$$

Thus, the reflected and transmitted fields depend on the ratio of the permittivities ε_1 and ε_2 of the two media.

Problem 4-7-1. Wave in air incident on dielectric medium. If the dielectric $\varepsilon_r = 10$, find E_r and E_t. *Ans.* $E_r = 0.520E_i$; $E_t = 0.480E_i$.

In the above problem, the wave is partly reflected and partly transmitted.

Analogously to the $\lambda/4$ matching section of a transmission line, a $\lambda/4$ thickness of one dielectric can be used to match the incident space wave into a half-space of another dielectric as in Example 4-12.

Example 4-12. $\lambda/4$ matching plate. Referring to Fig. 4-13*a*, find the permittivity and thickness of a $\lambda/4$ plate of dielectric to match a space wave into a half-space of dielectric with $\varepsilon_r = 4$.

Solution. From (3-4-20),

$$Z(\lambda/4) = \sqrt{Z_0 Z_d}$$

where Z_0 = intrinsic impedance of air and Z_d = intrinsic impedance of half-space of dielectric. Here $Z_0 = 377\ \Omega$ and $Z_d = 377/\sqrt{\varepsilon_r} = 377/\sqrt{4} = 188\ \Omega$. Therefore,

$$Z(\lambda/4) = \sqrt{377 \times 188} = 266\ \Omega$$

and

$$\varepsilon_r = \left[\frac{377}{Z(\lambda/4)}\right]^2 = \left(\frac{377}{266}\right)^2 = 2 \quad \textit{Ans.}$$

The $\lambda/4$ in the dielectric is given by

$$\lambda_d/4 = \frac{v_d}{v_0}\frac{\lambda_0}{4} = \frac{c/\sqrt{\varepsilon_r}}{c}\frac{\lambda_0}{4} = \frac{1}{\sqrt{4}}\frac{\lambda_0}{4} = 0.125\lambda_0 \quad \textit{Ans.}$$

where λ_d = wavelength in dielectric, m
λ_0 = wavelength in air, m
v_d = velocity in dielectric, m s^{-1}
v_0 = velocity in air = c, m s^{-1}

Problem 4-7-2. Phase shift. A plane 3-GHz wave is incident normally on a large sheet of polystyrene ($\varepsilon_r = 2.7$) with a large hole. How thick should the sheet be so that the wave through the sheet and the wave through the hole will be in phase? *Ans.* 156 mm.

Problem 4-7-3. Matching plate. A plane 1-GHz wave is incident normally on the plane surface of a half-space of material having constants $\mu_r = 1$ and $\varepsilon_r = 3.5$. Find (*a*) the thickness in millimeters and (*b*) the relative permittivity required for a matching plate which will eliminate reflection of the incident wave. *Ans.* (*a*) 54.8 mm, (*b*) 1.87.

An important application of the $\lambda/4$ matching plate is for eliminating reflections in many optical devices in which $\lambda/4$ coatings are deposited on the surfaces of lenses and prisms in cameras, binoculars, and telescopes to improve their efficiency, also on reading glasses to eliminate annoying reflections.

In Fig. 4-13*b* a wave is matched *through* a thick slab of $\varepsilon_r = 4$ dielectric by means of two $\lambda/4$ plates of $\varepsilon_r = 2$.

An alternative arrangement is shown in Fig. 4-14 on page 196 where pyramids match the wave, analogous to a tapered transmission line. This arrangement provides a gradual transition from air to the main body of dielectric and from the body back to air providing greater bandwidth than abrupt $\lambda/4$ transitions.

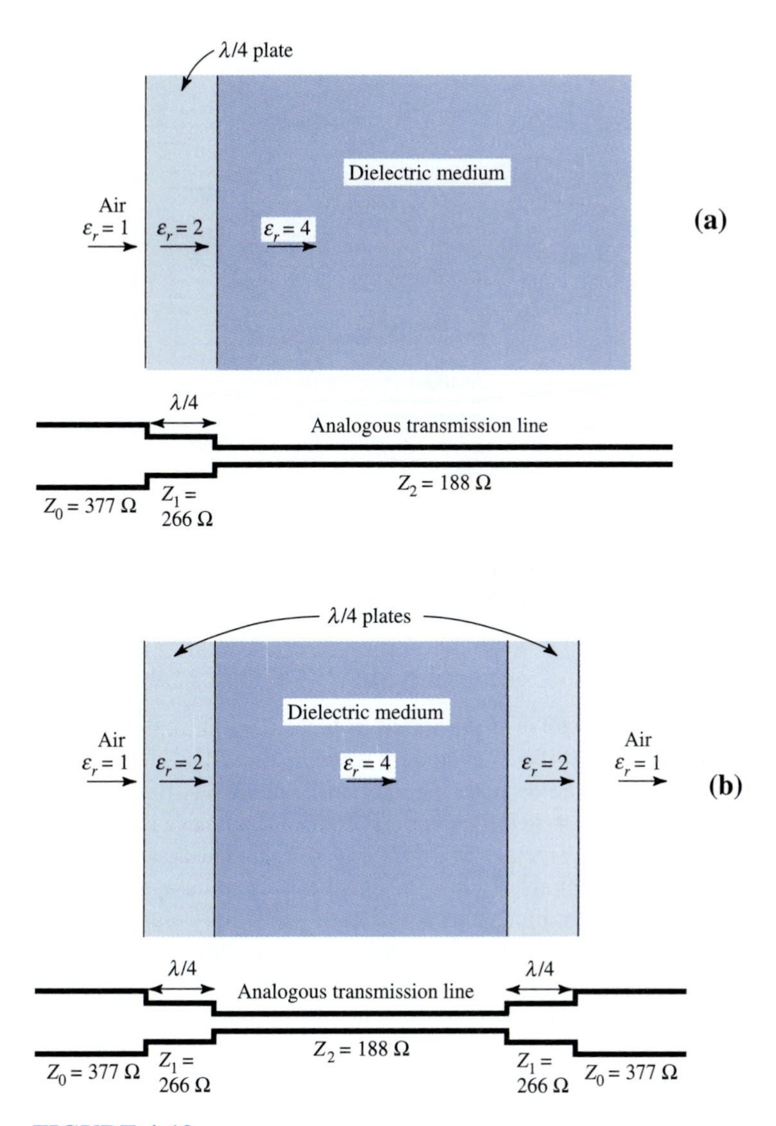

FIGURE 4-13

(*a*) Matching of wave *into* dielectric medium with $\lambda/4$ plate (Example 4-12). (*b*) Matching of wave *through* a dielectric medium with two $\lambda/4$ plates. Note that the plates of (*a*) and (*b*) are 0.25 wavelength thick in terms of the wavelength λ_d, *in the plate.* Thus, for $\varepsilon_r = 2$, $(\lambda_d/4) = 0.177\ \lambda_0$ for air. The analogous transmission lines are also shown.

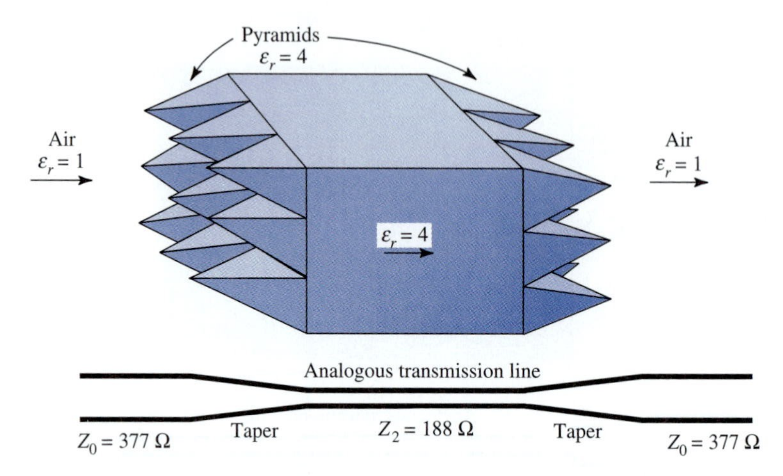

FIGURE 4-14
Matching of wave through a dielectric medium with pyramids. Note that the main body and the pyramids have the same ε_r. The analogous transmission line with tapered sections is also shown.

Wave Absorption with Conducting Sheet (Salisbury Sheet)

If a sheet of space paper of 377 Ω per square is placed $\lambda/4$ in front of a perfectly conducting sheet, as in Fig. 4-15*a*, the incident wave in air will be completely absorbed in the sheet and dissipated as heat. The action is analogous to the short-circuited transmission line in Fig. 4-15*b*. A square sheet of this paper clamped on opposite sides by high-conductivity metal strips, as in Fig. 4-15*c*, measures 377 Ω. The square may be of any convenient size. Thus, the sheet is said to have a resistance of 377 *ohms per square* although dimensionally its units are simply ohms.

The impedance presented to the incident wave at the sheet of space paper is 377 Ω, being the impedance of the sheet backed by an infinite impedance. As a consequence, this arrangement results in the total absorption of the wave by the space paper, with no reflection to the left of the space paper. There is, however, a standing wave and energy circulation between the paper and the conducting sheet. The analogous transmission-line arrangement is illustrated in Fig. 4-15*b*.

In the case of the plane wave, the perfectly conducting sheet effectively isolates the region of space behind it from the effects of the wave. In a roughly analogous manner, the shorting bar on the transmission line reduces the wave beyond it to a small value.

A transmission line may also be terminated by placing an impedance Z_0 across the line at A in Fig. 4-15*b*, and disconnecting the line beyond it. Although this provides a practical method of terminating a transmission line, there is no analogous counterpart for a space wave because it is not possible to "disconnect" the space to the right of the conducting sheet.

Problem 4-7-4. Conductivity of space paper. What conductivity σ is required for a sheet of space paper 0.1 mm thick? *Ans.* 2.65 ℧/m.

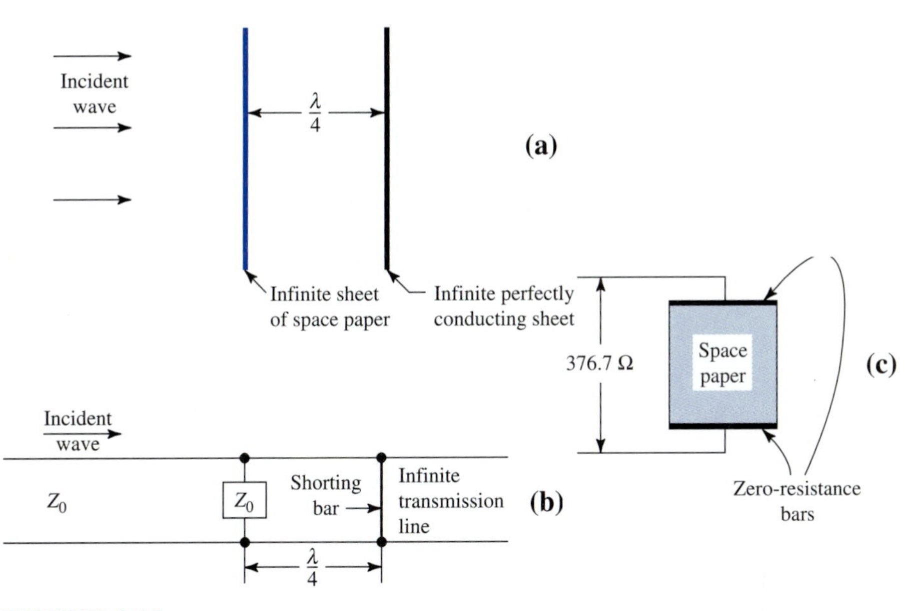

FIGURE 4-15
(*a*) Plane wave traveling to right incident normally on sheet of space paper backed by conducting sheet is absorbed without reflection with no transmission beyond conducting sheet. (*b*) Wave traveling to right on transmission line is absorbed without reflection by analogous arrangement, the short-circuited $\lambda/4$ line acting as an open (infinite-impedance) line at its right end. (*c*) Measurement of a square sheet of space paper.

Problem 4-7-5. Space paper alone. If a sheet of space paper is used alone with no conducting sheet at $\lambda/4$, find E_r, E_t, and the power absorbed per unit area of the sheet. *Ans.* $E_r = -\frac{1}{3}E_i$; $E_t = \frac{2}{3}E_i$; $1.18E_i^2 \times 10^{-3}$ W/m^2.

Wave Absorption with Ferrite-Titanate Medium

The space paper arrangement of Fig. 4-15 completely absorbs waves at only the frequency for which its spacing to the conducting sheet is $\lambda/4$. For wider bandwidth operation a lossy mixture of a high-μ (ferrite) material and a high-ε (barium titanate) material can be used effectively for wave absorption with both μ and ε being complex, and with the ratio μ/ε equal to that for free space ($\mu_r/\varepsilon_r = 1$). Although the mixture constitutes a physical discontinuity, an incident wave enters it without reflection. The velocity of the wave is reduced, and large attenuation can occur in a short distance.

Example 4-13. Lossy medium with same impedance as space. Let a plane 100-MHz (3-m) wave be incident normally on a solid ferrite-titanate slab of thickness $d = 10$ mm for which $\mu_r = \varepsilon_r = 60(2 - j1)$, as in Fig. 4-16. The medium is backed by a flat conducting sheet. How much is the reflected wave attenuated with respect to the incident wave? The medium is nonconducting ($\sigma = 0$).

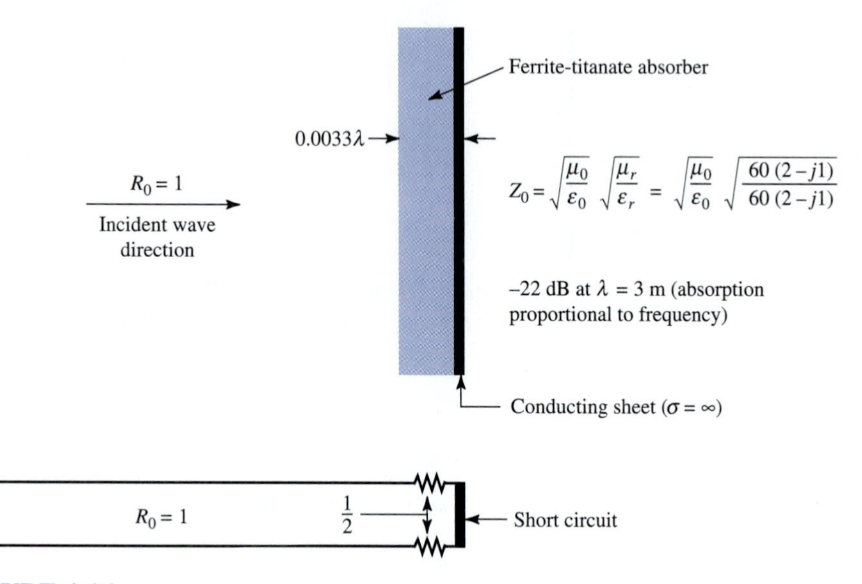

FIGURE 4-16
Ferrite-titanate slab with same impedance as space results in high attenuation and absorption of incident wave over a wide bandwidth (upper diagram). Equivalent transmission line (lower diagram).

Solution. The wave enters the ferrite-titanate slab without reflection since its intrinsic impedance Z_0 is the same as that of space. Thus,

$$Z_0 = \sqrt{\frac{\mu_0}{\varepsilon_0}}\sqrt{\frac{\mu_r}{\varepsilon_r}} = \sqrt{\frac{\mu_0}{\varepsilon_0}}\sqrt{\frac{60(2-j1)}{60(2-j1)}} = \sqrt{\frac{\mu_0}{\varepsilon_0}} = \text{intrinsic impedance of space}$$

From (4-5-2) the attenuation constant in general is given by

$$\alpha = \operatorname{Re}\gamma = \operatorname{Re}\sqrt{j\omega\mu\sigma - \omega^2\mu\varepsilon}$$

If both μ and ε are complex, so that $\mu = \mu' - j\mu''$ and $\varepsilon = \varepsilon' - j\varepsilon''$, then, in general,

$$\alpha = \operatorname{Re}\sqrt{(\omega\mu'' + j\omega\mu')[\sigma + j\omega(\varepsilon' - j\varepsilon'')]}$$

For $\sigma = 0$, we have

$$\alpha = \operatorname{Re} j\frac{2\pi}{\lambda_0}\sqrt{(\mu_r' - j\mu_r'')(\varepsilon_r' - j\varepsilon_r'')]} = \operatorname{Re} j\frac{2\pi}{\lambda_0}\sqrt{\mu_r\varepsilon_r}$$

$$= \operatorname{Re} j\frac{2\pi}{\lambda_0}60(2 - j1) = \operatorname{Re}\frac{120\pi}{\lambda_0}(1 + j2) = \frac{120\pi}{\lambda_0}$$

Therefore,

$$\text{Attenuation} = \alpha d = 120\pi\frac{d}{\lambda_0} = 120\pi \times 0.0033 = 1.24 \text{ Np} = 11 \text{ dB} \quad \textit{Ans.}$$

In traveling 0.0033λ (10 mm at $\lambda = 3$ m), the attenuation is 11 dB. After reflection from the flat conducting sheet, an equal attenuation occurs. This reflected wave passes on out into space again without reflection (intrinsic impedances matched) and is 22 dB down from the incident wave. Thus, the reflected wave power is less than 1/100 of the incident wave power. The reflected wave is not completely eliminated, but it is reduced enough to be satisfactory for many practical purposes, such as applications on moving vehicles, ships, and aircraft to reduce the likelihood of radar detection.

Whereas the space paper termination is matched only at frequencies for which the spacing is $\lambda/4$ (between it and the conducting sheet), the ferrite-titanate medium matching is independent of frequency, although the amount of attenuation is proportional to the frequency. A transmission-line equivalent of the slab is also shown in Fig. 4-16.

Problem 4-7-6. Absorbing sheet. A large flat sheet of a ferrous nonconducting dielectric medium is backed by aluminum foil. At 500 MHz the constants of the medium are $\mu_r = \varepsilon_r = 6 - j6$. How thick must the sheet be for a 500-MHz wave (in air) incident on the sheet to be reduced upon reflection by 30 dB if the wave is incident normally? *Ans.* 27.5 mm.

Problem 4-7-7. Three-sheet Jaumann sandwich. Another wide-bandwidth absorber using only resistance paper is provided by the three paper sheets called a *Jaumann sandwich* shown in Fig. 4-17. The three sheets have resistances per unit area of $4.15Z_0$, $1.66Z_0$, and $0.86Z_0$ where $Z_0 = 377\ \Omega$ per square. Confirm these values. *Hint:* Use a Smith Chart and the transmission line analogy.

Problem 4-7-8. Six-sheet Jaumann sandwich. An even wider bandwidth is provided by the six-sheet Jaumann sandwich of Fig. 4-18 on next page. Confirm the resistance values indicated.

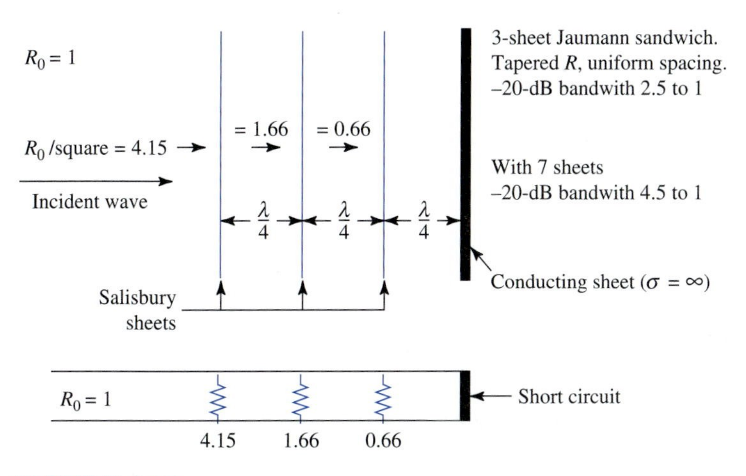

FIGURE 4-17
Three-sheet Jaumann sandwich wave absorber provides −20 dB reflection over 2.5-to-1 bandwidth centered on frequency for which spacing is $\lambda/4$. It is called a "*sandwich*" because the space between the resistance sheets is filled with polyfoam having $\varepsilon_r \approx 1$, making it a "paper-poly sandwich."

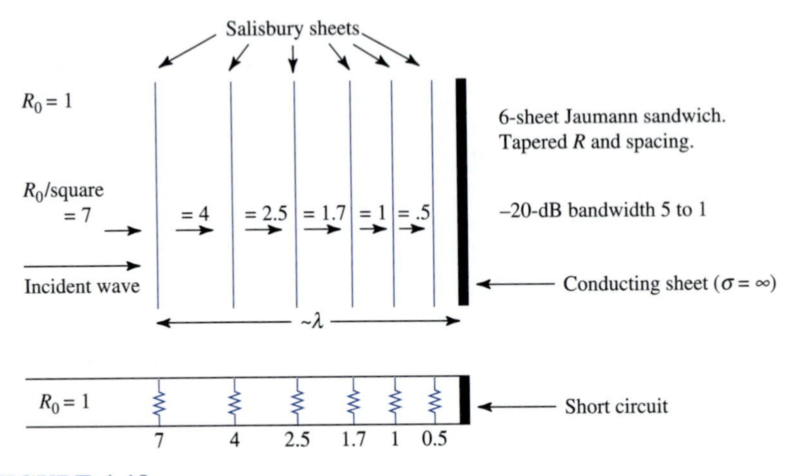

FIGURE 4-18
Six-sheet Jaumann sandwich wave absorber. Provides −20-dB reflection over 5-to-1 bandwidth.

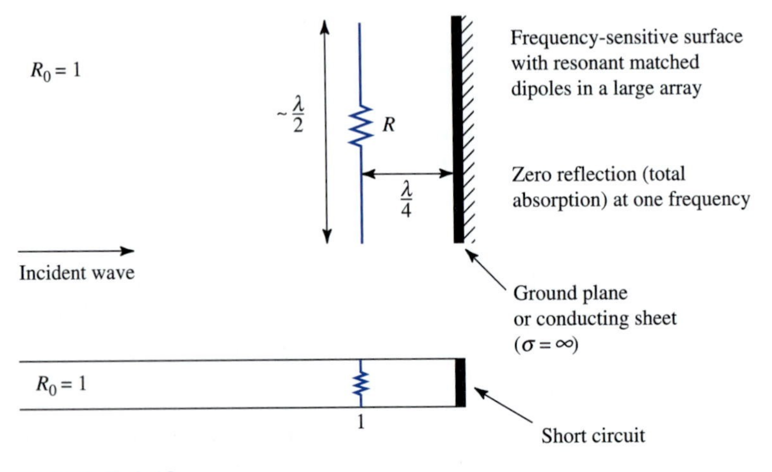

FIGURE 4-19
Resonant dipoles as wave absorber or receiving antenna.

Another method for absorbing an incident wave is with an array of resonant dipoles as suggested in Fig. 4-19. Here the incident power is dissipated as heat in the resistances R. If, instead of terminating the dipoles in the resistances R, they are connected by transmission lines as an antenna array, all of the incident power can be delivered to a receiver.

Yet another method for absorbing an incident plane wave is with an array of lossy dielectric pyramids as in Fig. 4-20*a*. With constants $\mu_r = 1$ and $\varepsilon_r = 2 - j1$, the 377 Ω impedance of space is only approximated but the taper of the pyramids results in low reflection at nose-on incidence over a wide bandwidth. The analogous or equivalent transmission line configuration is shown in Fig. 4-20*b*. Arrays of such

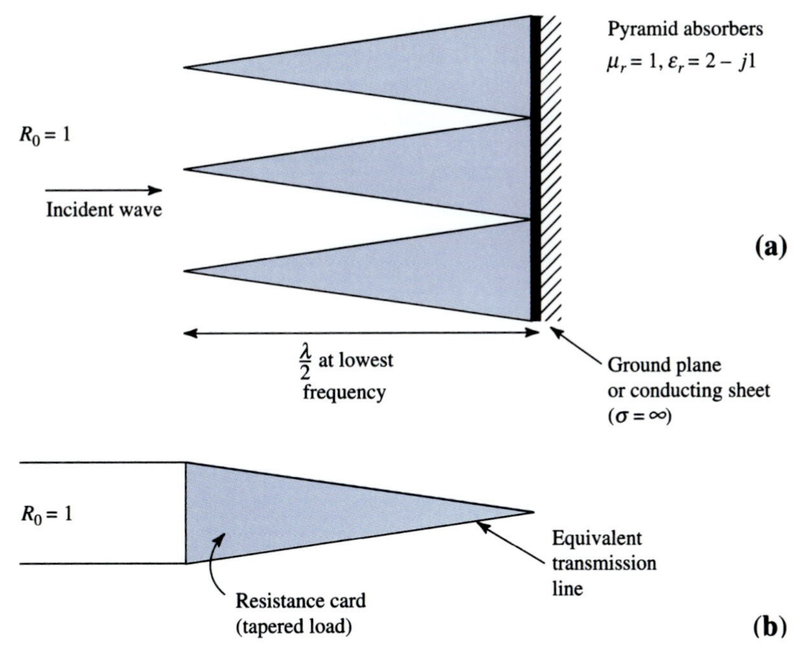

FIGURE 4-20
(*a*) Array of lossy pyramids for wave absorption. (*b*) Analogous transmission line.

pyramids are commonly used on the walls and ceilings of electromagnetic anechoic chambers. See Figs. 10-14 and 10-15.

Problem 4-7-9. Absorbing slab. (*a*) Find the reflection coefficient for a 3-mm-thick absorbing slab backed by a perfectly conducting metal plate at 3 GHz if the constants of the sheet are $\sigma = 0$ and $\mu_r = \varepsilon_r = 10 - j10$. (*b*) What is the attenuation in decibels? *Ans.* (*a*) 0.023; (*b*) 33 dB.

Problem 4-7-10. Absorbing pyramid. (*a*) Find the normal (nose-on) reflection coefficient at 3 GHz for an array of dielectric pyramids 30 cm from tip to base backed by a perfectly conducting metal plate (Fig. 4-20) if the constants are $\sigma = 0$ and $\mu_r = 1$, and $\varepsilon_r = 2 - j1$. (*b*) What is the attenuation in decibels? *Hint:* Consider pyramid height 3 times base width. *Ans.* (*a*) 0.0123; (*b*) 38.

4-8 RELATIVE PHASE VELOCITY AND INDEX OF REFRACTION

The phase velocity relative to the velocity of light, or the *relative phase velocity,* is

$$p = \frac{v}{c} = \frac{\sqrt{\mu_0 \varepsilon_0}}{\sqrt{\mu \varepsilon}} = \frac{1}{\sqrt{\mu_r \varepsilon_r}} \quad \text{(dimensionless)} \tag{1}$$

where μ_r = relative permeability of medium and ε_r = relative permittivity of medium.

The phase velocity of a plane wave in an unbounded lossless medium is equal to or less than the velocity of light ($p \leq 1$). In general, however, the phase velocity may have values both greater and less than the velocity of light. For example, in a hollow metal waveguide v is always equal to or greater than c (see Chap. 8).

In optics the *index of refraction* η is defined as the reciprocal of the relative phase velocity p. That is,

$$\eta = \frac{1}{p} = \frac{1}{v/c} = \frac{c}{v} = \sqrt{\mu_r \varepsilon_r} \qquad \textbf{\textit{Index of refraction}} \tag{2}$$

For nonferrous media μ_r is very nearly unity so that

$$\eta = \sqrt{\varepsilon_r} \tag{3}$$

Example 4-14. Dielectric medium. A dielectric medium has a relative permittivity $\varepsilon_r = 6$. Find (*a*) the index of refraction and (*b*) the phase velocity for a wave in an unbounded medium of this dielectric.

Solution. (*a*) Index of refraction

$$\eta = \sqrt{6} = 2.45 \quad \textit{Ans.} \quad (a)$$

(*b*) Phase velocity

$$v = \frac{c}{\sqrt{6}} = 1.22 \times 10^8 \text{ m s}^{-1} = 122 \text{ Mm s}^{-1} \quad \textit{Ans.} \quad (b)$$

Example 4-15. Water medium. Distilled water has the constants $\sigma \approx 0$, $\varepsilon_r = 81$, $\mu_r = 1$. Find η and v.

Solution

$$\eta = \sqrt{81} = 9 \quad \textit{Ans.}$$

$$v = \frac{c}{\sqrt{81}} = 0.111c = 33.3 \text{ Mm s}^{-1} \quad \textit{Ans.}$$

The index of refraction for water in the above example is the value at low frequencies ($f \to 0$). At light frequencies, say, for sodium light ($\lambda = 589$ nm), the index of refraction is observed to be about 1.33 instead of 9 as calculated on the basis of the relative permittivity. This difference was at one time cited as invalidating Maxwell's theory. The explanation for the difference is that the permittivity ε is not a constant but is a function of frequency. At zero frequency $\varepsilon_r = 81$, but at light frequencies $\varepsilon_r = 1.33^2 = 1.77$. The index of refraction and permittivity of many other substances also vary as a function of the frequency.

Thus, the term *permittivity* is more appropriate than the older term *dielectric constant* because it is *not* always a constant.

Over very wide frequency ranges (radio to ultraviolet) materials may exhibit several resonances and associated permittivity changes. Ice is a good example of this variation of permittivity with frequency, as illustrated in Fig. 4-21.

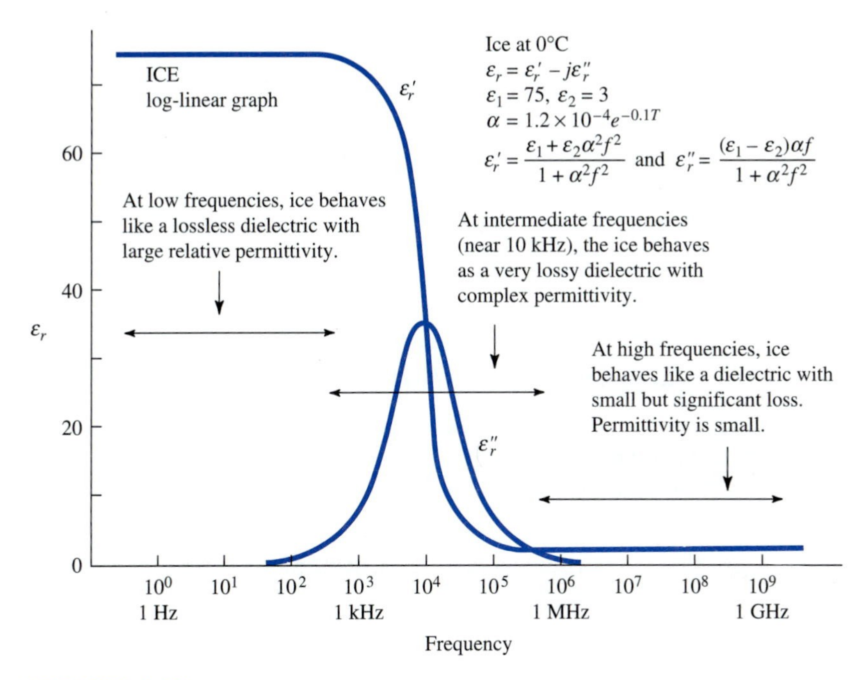

FIGURE 4-21
Complex relative permittivity of ice as a function of frequency.

Problem 4-8-1. Impedance of ice. For radar measurements of ice masses in Antarctica, Greenland, the world's glaciers, and frozen lakes, it is important to know the impedance of the ice for proper matching of the antenna of ice-penetrating radar (IPR) to the ice. Using Fig. 4-21, find the impedance of ice at (*a*) 100 Hz, (*b*) 1.2 kHz, (*c*) 300 kHz, and (*d*) 1 GHz. Take $\sigma = 10^{-8}$ ℧/m. *Ans.* (*a*) $43.5 \angle 1.4° \ \Omega$; (*b*) $44.2 \angle 8° \ \Omega$; (*c*) $182 \angle 45° \ \Omega$; (*d*) $218 \ \Omega$.

Problem 4-8-2. Index of refraction. The measured phase velocity of a dielectric medium is 186 Mm/s at frequency f_1 and 223 Mm/s at f_2. Find the index of refraction at the two frequencies. *Ans.* 1.6 at f_1 and 1.34 at f_2.

4-9 GROUP VELOCITY

Consider a plane wave traveling in the positive x direction, as in Fig. 4-1. Let the total electric field be given by

$$E_y = E_0 \cos(\omega t - \beta x) \tag{1}$$

Suppose now that the wave has not one but two frequencies of equal amplitude expressed by

$$\omega_0 + \Delta\omega$$

and

$$\omega_0 - \Delta\omega$$

It follows that the β values corresponding to these two frequencies are

$$\beta_0 + \Delta\beta \qquad \text{corresponding to } \omega_0 + \Delta\omega$$

and

$$\beta_0 - \Delta\beta \qquad \text{corresponding to } \omega_0 - \Delta\omega$$

For frequency 1

$$E'_y = E_0 \cos [(\omega_0 + \Delta\omega)t - (\beta_0 + \Delta\beta)x] \tag{2}$$

and for frequency 2

$$E''_y = E_0 \cos [(\omega_0 - \Delta\omega)t - (\beta_0 - \Delta\beta)x] \tag{3}$$

Adding (2) and (3) gives the total field

$$E_y = E'_y + E''_y \tag{4}$$

and rearranging, we get

$$E_y = 2E_0 \cos (\omega_0 t - \beta_0 x) \cos (\Delta\omega t - \Delta\beta x) \tag{5}$$

The two cosine factors indicate the presence of beats, i.e., a slow variation superimposed on a more rapid one.

For a *constant-phase* point

$$\omega_0 t - \beta_0 x = \text{constant}$$

and

$$\frac{dx}{dt} = \frac{\omega_0}{\beta_0} = v = f_0\lambda_0 \qquad \textbf{\textit{Phase velocity}} \tag{6}$$

where v is the *phase velocity.*

Setting the argument of the second cosine factor equal to a constant, we have

$$\Delta\omega t - \Delta\beta x = \text{constant}$$

and

$$\frac{dx}{dt} = \frac{\Delta\omega}{\Delta\beta} = u = \Delta f\, \Delta\lambda \qquad \textbf{\textit{Group velocity}} \tag{7}$$

where u is the phase velocity of the wave envelope, which is usually called the *group velocity.*

In the above development we can consider $\omega_0 + \Delta\omega$ and $\omega_0 - \Delta\omega$ as the two sideband frequencies caused by the modulation of a carrier frequency ω_0 by a frequency $\Delta\omega$, the carrier frequency being suppressed.

In nondispersive media the group velocity is the same as the phase velocity. Free space is an example of a lossless, nondispersive medium, and in free space $u = v = c$. However, in dispersive media the phase and group velocities differ.

A *dispersive medium* is one in which the phase velocity is a function of the frequency (and hence of the free-space wavelength). Dispersive media are of two types:

1. *Normally dispersive.* In these media, the change in phase velocity with wavelength is positive; that is, $dv/d\lambda > 0$. For these media $u < v$.
2. *Anomalously dispersive.* In these media, the change in phase velocity with wavelength is negative; that is, $dv/d\lambda < 0$. For these media $u > v$.

The terms *normal* and *anomalous* are arbitrary, the significance being simply that anomalous dispersion is different from the type of dispersion described as normal.

For a particular frequency (bandwidth vanishingly small)

$$u = v + \beta \frac{dv}{d\beta} \qquad \textbf{\textit{Group velocity}} \tag{8}$$

or

$$u = v - \lambda \frac{dv}{d\lambda} \qquad \textbf{\textit{Group velocity}} \tag{9}$$

Equations (8) and (9) are useful in finding the group velocity for a given phase-velocity function.

Example 4-16. Dispersive medium. A 1-MHz (300-m-wavelength) plane wave traveling in a normally dispersive, lossless medium has a phase velocity at this frequency of 300 Mm s^{-1}. The phase velocity as a function of wavelength is given by $v = k\sqrt{\lambda}$ where k is a constant. Find the group velocity.

Solution. From (9) the group velocity is

$$u = v - \lambda \frac{dv}{d\lambda} = v - \frac{k}{2}\sqrt{\lambda}$$

or

$$u = v(1 - 1/2)$$

Hence, the group velocity is

$$u = \frac{v}{2} = 150 \text{ Mms}^{-1} \qquad \text{or} \qquad c/2 \quad \textit{Ans.}$$

The group velocity u is one-half the phase velocity v. Any intelligence conveyed by the modulation moves with the velocity of the envelope, i.e., at the group velocity.†

The difference between phase and group velocities is also illustrated by a crawling caterpillar. The humps on its back move forward with phase velocity, while the caterpillar as a whole progresses with group velocity.

Example 4-17. Distance to pulsar. A pulsar is a rapidly rotating, very dense neutron star that transmits broadband pulses. Typically the pulses are strongest in the frequency range of 100 to 500 MHz. Electrons make the interstellar medium dispersive and this provides a means of measuring a pulsar's distance.

The group or energy velocity of the pulsar radiation is given by

$$v_g = c\sqrt{1 - (f_0/f)^2} \qquad (\text{m s}^{-1}) \tag{10}$$

†In a lossless medium the energy is also conveyed at the group velocity.

where f = pulsar frequency, Hz
$f_0 = 9\sqrt{N}$, critical or plasma frequency, Hz
N = electron density, number m^{-3}
c = velocity of light $= 3 \times 10^8\ \text{m s}^{-1}$

For $N = 3 \times 10^4\ \text{m}^{-3}$, $f_0 = 1.56$ kHz so at pulsar frequencies (100 to 500 MHz), $f \gg f_0$ and (10) can be rewritten as

$$v_g = c\left[1 - \frac{1}{2}\left(\frac{f_0}{f}\right)^2\right] \quad (\text{m s}^{-1}) \qquad (11)$$

As a result of the dispersion of the interstellar medium, a pulse arrives earlier at a higher frequency and later at a lower frequency. The difference in arrival time Δt of a pulse transmitted simultaneously at two frequencies f_1 and f_2 is

$$\Delta t = \frac{L}{2c}\left[\left(\frac{f_0}{f_2}\right)^2 - \left(\frac{f_0}{f_1}\right)^2\right] \quad (\text{s}) \qquad (12)$$

where L = path length, m.

Expressing the distance as the light travel time gives

$$\frac{L}{c} = \frac{2\Delta t}{\left(\frac{f_0}{f_2}\right)^2 - \left(\frac{f_0}{f_1}\right)^2} = \frac{2}{81N}\frac{\Delta t}{f_2^{-2} - f_1^{-2}} \quad (\text{s}) \qquad (13)$$

Referring to the pulsar recording of Fig. 4-22, find the distance.

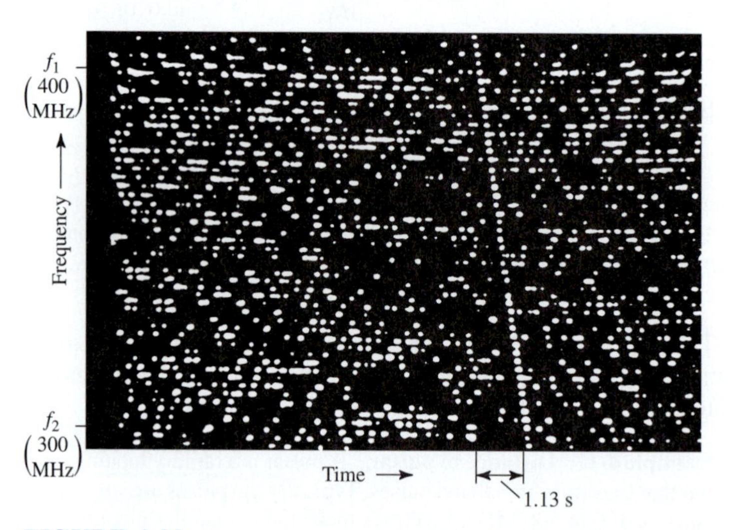

FIGURE 4-22
Raster display of 48-channel recording of a pulsar, illustrating the delayed arrival of pulses due to dispersion of the interstellar medium. On a single channel, the pulse is lost in the noise, but in the simultaneous 48-channel display the pulses connect and form a very distinctive diagonal line. (*Recording courtesy of Martin Ewing.*)

Solution. The difference in arrival time of the pulse ($\Delta t = 1.13$ s) for frequencies $f_1 = 400$ MHz and $f_2 = 300$ MHz, with $N = 3 \times 10^4$ m^{-3} (30 per liter), we have from (13)

$$\frac{L}{c} = 8.23 \times 10^{-7} \times 1.13 \times 10^{16}/0.0486$$

$$= 191 \times 10^9 \text{ light-seconds} = 6060 \text{ light-years (LY)} \quad \textit{Ans.}$$

Thus, the pulses received today were transmitted by the pulsar over 6 millennia ago!

If the distance to a pulsar is known independently by another means, the above example can be turned around and the observations used to determine the average electron density of the interstellar medium.

Problem 4-9-1. Velocity difference of pulsar pulses. In the pulsar example, what is the difference in velocity of the pulsar pulses at 300 and 400 MHz? *Ans:* About 1 part in 100 billion, measurable only because of the great distance.

Problem 4-9-2. Velocities, phase and group. Find the group velocity for a 100-MHz wave in a normally dispersive, lossless medium for which the phase velocity $v = 2 \times 10^7 \lambda^{2/3}$ m/s. *Ans.* 13.9 Mm/s.

4-10 POWER AND ENERGY RELATIONS

Consider a region of space represented by an array of field-cell transmission lines of total width W and total height H, as in Fig. 4-23 (on next page) with a plane wave traveling from left to right. The electric field E_y is vertical, and the magnetic field H_z is horizontal. The voltage $V = E_y H$ and the current $I = H_z W$. By analogy to circuits, the power conveyed is

$$P = VI = E_y H_z H W = E_y H_z A \qquad \text{(W)} \tag{1}$$

where $A = HW$ = area of field-cell array. The power (surface) density is then

$$S = \frac{P}{A} = E_y H_z \qquad (\text{W m}^{-2}) \tag{2}$$

Equation (2) relates the scalar magnitudes. The power flow is perpendicular to **E** and **H**, and it can be shown that in vector notation the *power density* is given by

$$\boxed{S = \mathbf{E} \times \mathbf{H} \qquad (\text{W m}^{-2}) \qquad \textbf{\textit{Poynting vector}}} \tag{3}$$

Turning **E** into **H** and proceeding as with a right-handed screw gives the direction of **S** perpendicular to both **E** and **H**. **S** is a *power surface density* called the *Poynting vector.* Its value in (3) is the *instantaneous Poynting vector.* The *average Poynting vector* is obtained by integrating the instantaneous Poynting vector over one period and dividing by one period. It is also readily obtained in complex notation from

$$\mathbf{S}_{\text{av}} = \tfrac{1}{2}\text{Re}\, \mathbf{E} \times \mathbf{H}^* = \tfrac{1}{2}\hat{\mathbf{x}}\, |E_y|\, |H_z| \cos\xi \qquad (\text{W m}^{-2}) \tag{4}$$

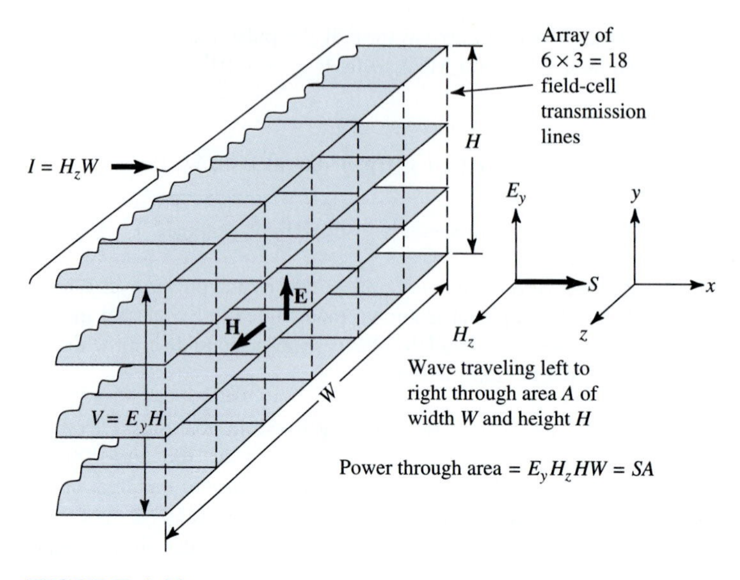

FIGURE 4-23
Power flow of wave traveling left to right through area of width W and height H is equal to E_yH_zHW.

where $\mathbf{S}_{av} = \hat{\mathbf{x}}S$ = average Poynting vector, W m^{-2}
$\mathbf{E} = \hat{\mathbf{y}}E_y = \hat{\mathbf{y}}|E_y|e^{j\omega t}$, V m^{-1}
$\mathbf{H}^* = \hat{\mathbf{z}}H_Z^* = \hat{\mathbf{z}}|H_z|e^{-j(\omega t-\xi)}$, A m^{-1}
ξ = time phase angle between E_y and H_z, rad or deg

$\mathbf{H}^*$ is called the *complex conjugate* of $\mathbf{H}$, where

$$\mathbf{H} = \hat{\mathbf{z}}H_z = \hat{\mathbf{z}}|H_z|e^{j(\omega t-\xi)} \qquad (\text{A m}^{-1})$$

The quantities $\mathbf{H}$ and its complex conjugate $\mathbf{H}^*$ have the same space direction, but they differ in sign in their phase factors. Note that if E_y and H_z in (4) are rms values instead of (peak) amplitudes, the factor $\frac{1}{2}$ in (4) is omitted.

The magnitude of the *average Poynting vector*

$$S_{av} = \tfrac{1}{2}\text{Re}E_yH_z^* = \tfrac{1}{2}|E_y||H_z|\cos\xi \qquad (\text{W m}^{-2}) \tag{5}$$

The relation corresponding to (5) for the average power of a traveling wave on a transmission line is

$$P_{av} = \tfrac{1}{2}\text{Re}VI^* = \tfrac{1}{2}|V||I|\cos\theta \qquad (\text{W}) \tag{6}$$

where V = voltage between conductors of transmission line, V
I = current through one conductor, A
I^* = complex conjugate of I
θ = time phase angle between V and I, rad or deg

Since the intrinsic impedance of the medium

$$Z_0 = \frac{E}{H} = \frac{|E|}{|H|}\angle\xi = |Z_0|\angle\xi \tag{6.1}$$

the magnitude of the *average Poynting vector* can also be written

$$S_{av} = \frac{1}{2}\text{Re}\, H_z H_z^* Z_0 = \frac{1}{2}|H_z|^2\,\text{Re}\, Z_0 \qquad (\text{W m}^{-2}) \tag{7}$$

or†

$$S_{av} = \frac{1}{2}\text{Re}\,\frac{E_y E_y^*}{Z_0} = \frac{1}{2}|E_y|^2\,\text{Re}\,\frac{1}{Z_0} \qquad (\text{W m}^{-2}) \tag{8}$$

Equation (7) is very useful, since if the intrinsic impedance Z_0 of a conducting medium and also the magnetic field H_z at the surface are known, it gives the average Poynting vector (or average power per unit area) into the conducting medium.

Example 4-18. Power into copper sheet. A plane 1-GHz traveling wave in air with peak electric field intensity of 1 V m^{-1} is incident normally on a large copper sheet. Find the average power absorbed by the sheet per square meter of area.

Solution From (4-5-13), the intrinsic impedance of copper at 1 GHz is

$$Z_0 = \sqrt{\frac{\omega\mu}{\sigma}}\angle 45^\circ$$

For copper $\mu_r = \varepsilon_r = 1$ and $\sigma = 58$ M℧ m^{-1}. Hence the real part of Z_0 is

$$\text{Re}\, Z_0 = \cos 45^\circ\sqrt{\frac{2\pi\times 10^9\times 4\pi\times 10^{-7}}{5.8\times 10^7}} = 8.2\ \text{m}\Omega$$

Next we find the value of H at the sheet (tangent to the surface). This is very nearly double H for the incident wave. Thus,

$$H = 2\frac{E}{Z} = \frac{2\times 1}{377}\ \text{A m}^{-1}$$

From (7), we find that the average power per square meter into the sheet is tiny:

$$S_{av} = \frac{1}{2}\left(\frac{2}{377}\right)^2 8.2\times 10^{-3} = 115\ \text{nW m}^{-2} \qquad \textit{Ans.}$$

Problem 4-10-1. Poynting vector into aluminum sheet. A 3-GHz wave is incident on a large sheet of aluminum ($\sigma = 3.5\times 10^{-7}$℧/m). If the field $E = 15$ V/m, find the average power absorbed by the sheet (W/m^2). *Ans.* 366 nW/m^2.

The relations corresponding to (7) and (8) for the average power of a traveling wave on a transmission line are

$$P_{av} = \frac{1}{2}\,\text{Re}\, II^* Z_0 = \frac{1}{2}|I|^2\,\text{Re}\, Z_0 \qquad (\text{W}) \tag{9a}$$

and

$$P_{av} = \frac{1}{2}\,\text{Re}\,\frac{VV^*}{Z_0} = \frac{1}{2}|V|^2\,\text{Re}\,\frac{1}{Z_0} \qquad (\text{W}) \tag{9b}$$

†Note that in general $\text{Re}(1/Z_0) \neq 1/(\text{Re}Z_0)$.

When Z_0 is real ($\xi = 0$) and E and H are rms values, we have for the *traveling space wave*

$$S_{av} = EH = H^2 Z_0 = \frac{E^2}{Z_0} \qquad (\text{W m}^{-2}) \tag{10}$$

and for the *traveling wave on a transmission line* ($\theta = 0$ and V and I rms)

$$P_{av} = VI = I^2 Z_0 = \frac{V^2}{Z_0} \qquad (\text{W}) \tag{11}$$

From (2-9-9) the energy density w_e at a point in an electric field is

$$w_e = \tfrac{1}{2}\varepsilon E^2 \qquad (\text{J m}^{-3}) \tag{12}$$

where ε = permittivity of medium, Fm^{-1}, and E = electric field intensity, Vm^{-1}

From (2-12-25.1) the energy density w_m at a point in a magnetic field is

$$w_m = \tfrac{1}{2}\mu H^2 \qquad (\text{J m}^{-3}) \tag{13}$$

where μ = permeability of medium, H m^{-1}, and H = magnetic field, A m^{-1}.

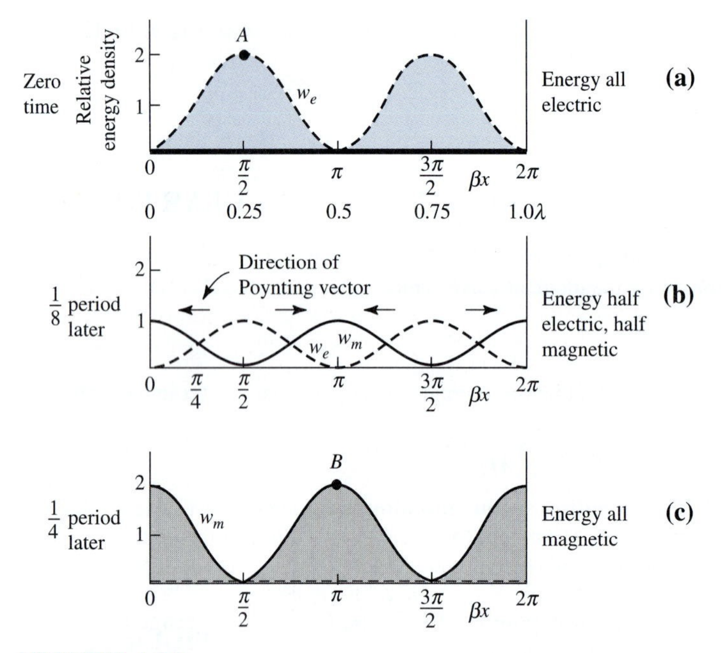

FIGURE 4-24
Total electric and magnetic energy densities at three instants of time for a pure standing wave. Conditions are shown over a distance of 1λ ($\beta x = 2\pi$). There is no net transmission of energy in a pure standing wave, but locations of energy oscillate back and forth. The situation here (pure standing wave) is identical with that in a short-circuited transmission line or in a resonator.

In a traveling wave in an unbounded, lossless medium

$$\frac{E}{H} = \sqrt{\frac{\mu}{\varepsilon}} \tag{14}$$

Substituting for H from (14) in (13), we have

$$w_m = \tfrac{1}{2}\mu H^2 = \tfrac{1}{2}\varepsilon E^2 = w_e \tag{15}$$

Thus the electric and magnetic energy densities in a plane traveling wave are equal, and the total energy density w is the sum of the electric and magnetic energies. Thus,

$$w = w_e + w_m = \tfrac{1}{2}\varepsilon E^2 + \tfrac{1}{2}\mu H^2 \qquad (\text{J m}^{-3}) \qquad \textbf{\textit{Energy density}} \tag{16}$$

or

$$w = \varepsilon E^2 = \mu H^2 \qquad (\text{J m}^{-3}) \tag{17}$$

Two waves of equal magnitude traveling in opposite directions produce a standing wave. There is no net transfer of energy in a pure standing wave but energy does oscillate back and forth like water slops in a pail. At one instant the energy is all electric as in Fig. 4-24*a* with maximum at point A. One-quarter of a period later, the energy is all magnetic, as in Fig. 4-24*c*, with maximum at B, which is at a distance of $\lambda/8$ from A. One-quarter period later the energy is all back at A. At intermediate times the energy is moving and is half electric and half magnetic as in Fig. 4-24*b*.

4-11 LINEAR, ELLIPTICAL, AND CIRCULAR POLARIZATION†

Consider a plane wave traveling out of the page (positive z direction), as in Fig. 4-25*a*, with the electric field at all times in the y direction. This wave is said to be *linearly polarized* (in the y direction). As a function of time and position, the

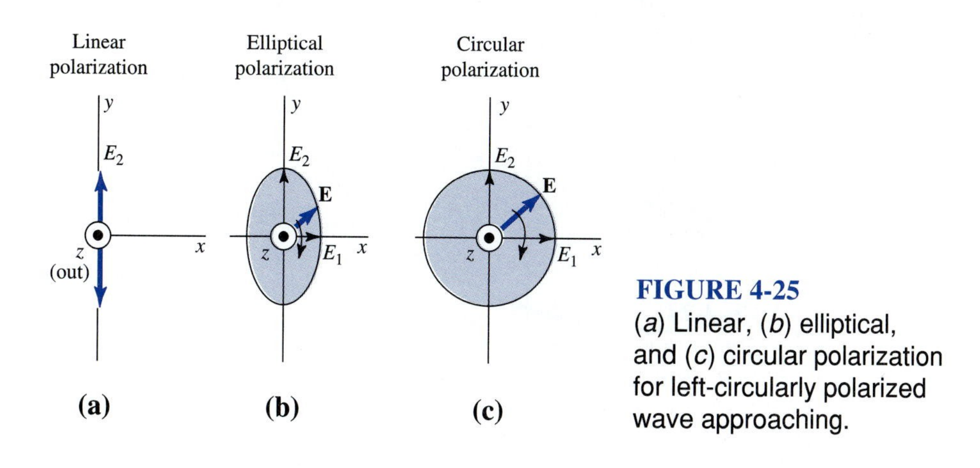

FIGURE 4-25
(*a*) Linear, (*b*) elliptical, and (*c*) circular polarization for left-circularly polarized wave approaching.

†A much more detailed and complete discussion of wave polarization is given by J. D. Kraus, "Radio Astronomy," 2d ed., Cygnus-Quasar, Powell, Ohio, 1986.

electric field is given by

$$E_y = E_2 \sin(\omega t - \beta z) \tag{1}$$

In general, the electric field of a wave traveling in the z direction may have both a y component and an x component, as suggested in Fig. 4-25b. In this more general situation, with a phase difference δ between the components, the wave is said to be *elliptically polarized.* At a fixed value of z the electric vector **E** rotates as a function of time, the tip of the vector describing an ellipse called the ***polarization ellipse.*** The ratio of the major to minor axes of the polarization ellipse is called the ***axial ratio*** (AR). Thus, for the wave Fig. 4-25b, AR $= E_2/E_1$. Two extreme cases of elliptical polarization correspond to *circular polarization,* as in Fig. 4-25c, and *linear polarization,* as in Fig. 4-25a. For circular polarization $E_1 = E_2$ and AR $= 1$, while for linear polarization $E_1 = 0$ and AR $= \infty$.

In the most general case of elliptical polarization, the polarization ellipse may have any orientation, as suggested in Fig. 4-26. The elliptically polarized wave may be expressed in terms of two linearly polarized components, one in the x direction and one in the y direction. Thus, if the wave is traveling in the positive z direction (out of the page), the electric field components in the x and y directions are

$$E_x = E_1 \sin(\omega t - \beta z) \tag{2}$$

$$E_y = E_2 \sin(\omega t - \beta z + \delta) \tag{3}$$

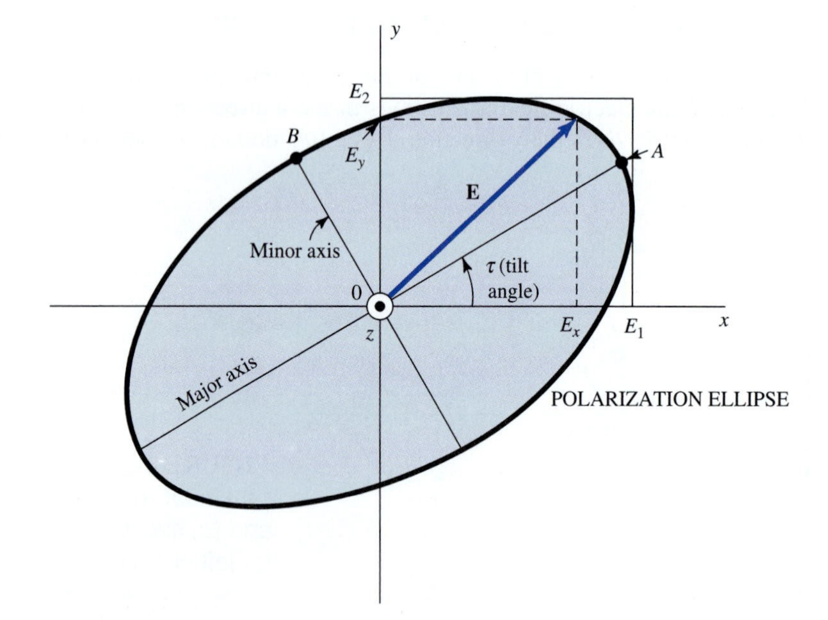

FIGURE 4-26
Polarization ellipse at tilt angle τ showing instantaneous components E_x and E_y and amplitudes (or peak values) E_1 and E_2.

where E_1 = amplitude of wave linearly polarized in x direction
E_2 = amplitude of wave linearly polarized in y direction
δ = time-phase angle by which E_y leads E_x

Combining (2) and (3) gives the instantaneous total vector field $\mathbf{E}$:

$$\mathbf{E} = \hat{\mathbf{x}}E_1 \sin(\omega t - \beta z) + \hat{\mathbf{y}}E_2 \sin(\omega t - \beta z + \delta) \quad (4)$$

At $z = 0$, $E_x = E_1 \sin \omega t$ and $E_y = E_2 \sin(\omega t + \delta)$. Expanding E_y yields

$$E_y = E_2(\sin \omega t \cos \delta + \cos \omega t \sin \delta) \quad (5)$$

From the relation for E_x we have $\sin \omega t = E_x/E_1$ and $\cos \omega t = \sqrt{1 - (E_x/E_1)^2}$. Introducing these in (5) eliminates ωt, and, on rearranging, we obtain

$$\frac{E_x^2}{E_1^2} - \frac{2E_xE_y \cos \delta}{E_1E_2} + \frac{E_y^2}{E_2^2} = \sin^2 \delta \quad (6)$$

or

$$aE_x^2 - bE_xE_y + cE_y^2 = 1 \quad (7)$$

where $\quad a = \dfrac{1}{E_1^2 \sin^2 \delta} \qquad b = \dfrac{2 \cos \delta}{E_1E_2 \sin^2 \delta} \qquad c = \dfrac{1}{E_2^2 \sin^2 \delta}$

Equation (7) describes a (polarization) ellipse, as in Fig. 4-26. The line segment OA is the semimajor axis, and the line segment OB is the semiminor axis. The tilt angle of the ellipse is τ. The axial ratio is

$$\text{AR} = \frac{OA}{OB} \qquad (1 \leq \text{AR} \leq \infty) \qquad \textit{Axial ratio} \quad (8)$$

If $E_1 = 0$, the wave is linearly polarized in the y direction. If $E_2 = 0$, the wave is linearly polarized in the x direction. If $\delta = 0$ and $E_1 = E_2$, the wave is also linearly polarized but in a plane at an angle of 45° with respect to the x axis ($\tau = 45°$).

If $E_1 = E_2$ and $\delta = \pm 90°$, the wave is circularly polarized. When $\delta = +90°$, the wave is *left circularly polarized,* and when $\delta = -90°$, the wave is *right circularly polarized.* For the case $\delta = +90°$ and for $z = 0$ and $t = 0$, we have from (2) and (3) that $\mathbf{E} = \hat{\mathbf{y}}E_2$, as in Fig. 4-27*a*. One-quarter cycle later ($\omega t = 90°$), $\mathbf{E} = \hat{\mathbf{x}}E_1$, as in Fig. 4-27*b*. Thus, at a fixed position ($z = 0$) the electric field vector rotates clockwise (viewing the wave approaching). According to the IEEE definition, this

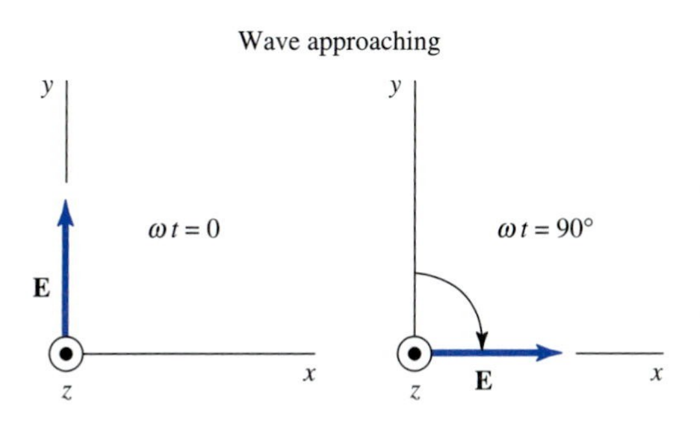

FIGURE 4-27
Instantaneous orientation of electric field vector **E** at two instants of time for a left-circularly polarized wave which is approaching (out of page).

corresponds to left circular polarization. The opposite rotation direction ($\delta = -90°$) corresponds to right circular polarization.

If the wave is viewed receding (from negative z axis in Fig. 4-27), the electric vector appears to rotate in the opposite direction. Hence, clockwise rotation of **E** with the wave approaching is the same as counterclockwise rotation with the wave receding. Thus, unless the wave direction is specified, there is a possibility of ambiguity as to whether the wave is left- or right-handed. This can be avoided by defining the polarization with the aid of an axial-mode helical antenna. Thus, a right-handed helical antenna radiates (or receives) right circular (IEEE) polarization.† A right-handed helix, like a right-handed screw, is right-handed regardless of the position from which it is viewed. There is no possibility here of ambiguity.

The Institute of Electrical and Electronics Engineers (IEEE) definition is opposite to the classical optics definition which had been in use for centuries. The intent of the IEEE Standards Committee was to make the IEEE definition agree with the classical optics definition, but it got turned around so now we have two definitions. In this book we use the IEEE definition, which has the advantage of agreement with helical antennas as noted above.

4-12 POYNTING VECTOR FOR ELLIPTICALLY AND CIRCULARLY POLARIZED WAVES

In complex notation the Poynting vector is

$$\mathbf{S} = \frac{1}{2}\mathbf{E} \times \mathbf{H}^* \tag{1}$$

The average Poynting vector is the real part of (1), or

$$\mathbf{S}_{\text{av}} = \text{Re}\,\mathbf{S} = \frac{1}{2}\text{Re}\,\mathbf{E} \times \mathbf{H}^* \tag{2}$$

We can also write

$$\mathbf{S}_{\text{av}} = \frac{1}{2}\hat{\mathbf{z}}\frac{E_1^2 + E_2^2}{Z_0} = \frac{1}{2}\hat{\mathbf{z}}\frac{E^2}{Z_0} \qquad \textit{Average Poynting vector} \tag{3}$$

where $E = \sqrt{E_1^2 + E_2^2}$ is the amplitude of the total **E** field.

Example 4-19. Elliptically polarized wave power. An elliptically polarized wave traveling in the positive z direction in air has x and y components:

$$E_x = 3\sin(\omega t - \beta x) \qquad (\text{V m}^{-1})$$
$$E_y = 6\sin(\omega t - \beta x + 75°) \qquad (\text{V m}^{-1})$$

Find the average power per unit area conveyed by the wave.

Solution. The average power per unit area is equal to the average Poynting vector, which from (3) has a magnitude

$$S_{\text{av}} = \frac{1}{2}\frac{E^2}{Z} = \frac{1}{2}\frac{E_1^2 + E_2^2}{Z}$$

†A left-handed helical antenna radiates (or receives) left circular (IEEE) polarization.

From the stated conditions, the amplitude $E_1 = 3\ \text{V m}^{-1}$ and the amplitude $E_2 = 6\ \text{V m}^{-1}$. Also for air $Z = 377\ \Omega$. Hence

$$S_{\text{av}} = \frac{1}{2}\frac{3^2+6^2}{377} = \frac{1}{2}\frac{45}{377} \approx 60\ \text{mW m}^{-2} \quad \textit{Ans.}$$

Problem 4-12-1. EP wave power. An elliptically polarized (EP) wave in a medium with constants $\sigma = 0$, $\mu_r = 2$, $\varepsilon_r = 5$ has H-field components (normal to the direction of propagation and normal to each other) of amplitudes 3 and 4 A/m. Find the average power conveyed through an area of 5 m^2 normal to the direction of propagation. *Ans.* 14.9 kW.

4-13 THE POLARIZATION ELLIPSE AND THE POINCARÉ SPHERE

In the Poincaré sphere† representation of wave polarization, the ***polarization state*** is described by a point on a sphere where the longitude and latitude of the point are related to parameters of the polarization ellipse (see Fig. 4-28) as follows:

$$\begin{aligned}\text{Longitude} &= 2\tau \\ \text{Latitude} &= 2\varepsilon\end{aligned} \tag{1}$$

where τ = tilt angle, $0° \le \tau \le 180°$ (see footnote‡) and $\varepsilon = \tan^{-1}(1/\mp \text{AR})$, $-45° \le \varepsilon \le +45°$. The axial ratio (AR) and angle ε are negative for right-handed and positive for left-handed (IEEE) polarization.

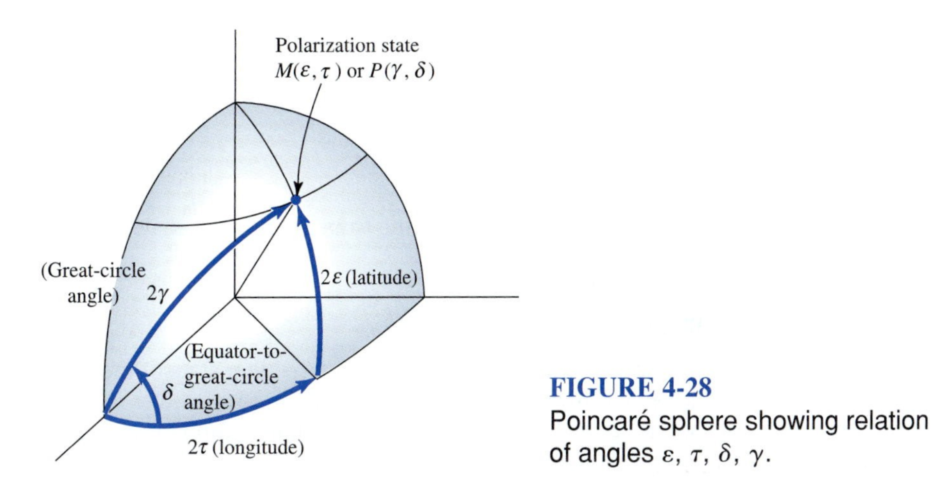

FIGURE 4-28
Poincaré sphere showing relation of angles ε, τ, δ, γ.

†H. Poincaré, "Théorie mathématique de la lumière," G. Carré, Paris, 1892; G. A. Deschamps, Geometrical Representation of the Polarization of a Plane Electromagnetic Wave, *Proc. IRE,* **39**: 540 (May 1951).

‡Note that when the Greek letter τ is used in this section to denote the *tilt angle,* it is associated with the incident, reflected, or transmitted wave (that is, τ_i, τ_r, or τ_t), but when it is used to denote the *transmission coefficient,* it is associated with either a parallel or perpendicular case (that is, $\tau_\parallel$ or $\tau_\perp$).

The polarization state described by a point on a sphere here can also be expressed in terms of the angle subtended by the great circle drawn from a reference point on the equator and the angle between the great circle and the equator (see Fig. 4-28) as follows:

$$\begin{aligned} \text{Great-circle angle} &= 2\gamma \\ \text{Equator-to-great-circle angle} &= \delta \end{aligned} \tag{2}$$

where $\gamma = \tan^{-1}(E_2/E_1)$, $0° \leq \gamma \leq 90°$, and δ = phase difference between E_y and E_x, $-180° \leq \delta \leq +180°$.

The geometric relation of τ, ε, and γ to the polarization ellipse is illustrated in Fig. 4-29. The trigonometric interrelations of τ, ε, γ, and δ are as follows: †

$$\begin{aligned} \cos 2\gamma &= \cos 2\varepsilon \cos 2\tau \\ \tan \delta &= \frac{\tan 2\varepsilon}{\sin 2\tau} \\ \tan 2\tau &= \tan 2\gamma \cos \delta \\ \sin 2\varepsilon &= \sin 2\gamma \sin \delta \end{aligned} \qquad \textbf{\textit{Polarization parameters}} \tag{3}$$

Knowing ε and τ, one can determine γ and δ or vice versa. It is convenient to describe the ***polarization state*** by either of the two sets of angles (ε, τ) or (γ, δ) which describe a point on the Poincaré sphere (Fig. 4-28). Let the polarization state as a function of ε and τ be designated by $M(\varepsilon, \tau)$, or simply M, and the polarization state as a function of γ and δ be designated by $P(\gamma, \delta)$, or simply P, as in Fig. 4-29.

As an application of the Poincaré sphere representation (see Fig. 4-30) it may be shown that the voltage response V of an antenna to a wave of arbitrary polarization is given by‡

$$V = k \cos \frac{MM_a}{2} \qquad \textbf{\textit{Antenna voltage response}} \tag{4}$$

where MM_a = angle subtended by great-circle line from polarization state M to M_a
M = polarization state of wave
M_a = polarization state of antenna
k = constant

The polarization state of the antenna is defined as the polarization state of the wave radiated by the antenna when it is transmitting. The factor k in (4) involves the field strength of the wave and the size of the antenna. An important result to note is that, if $MM_a = 0°$, the antenna is matched to the wave (polarization state of wave same as for antenna) and the response is maximized. However, if $MM_a = 180°$, the response is zero. This can occur, for example, if the wave is linearly polarized in the y

†These relations involve spherical trigonometry. See M. Born and E. Wolf, "Principles of Optics," pp. 24–27, Macmillan, New York, 1964.

‡G. Sinclair, The Transmission and Reception of Elliptically Polarized Waves, *Proc. IRE,* **38:** 151 (1950).

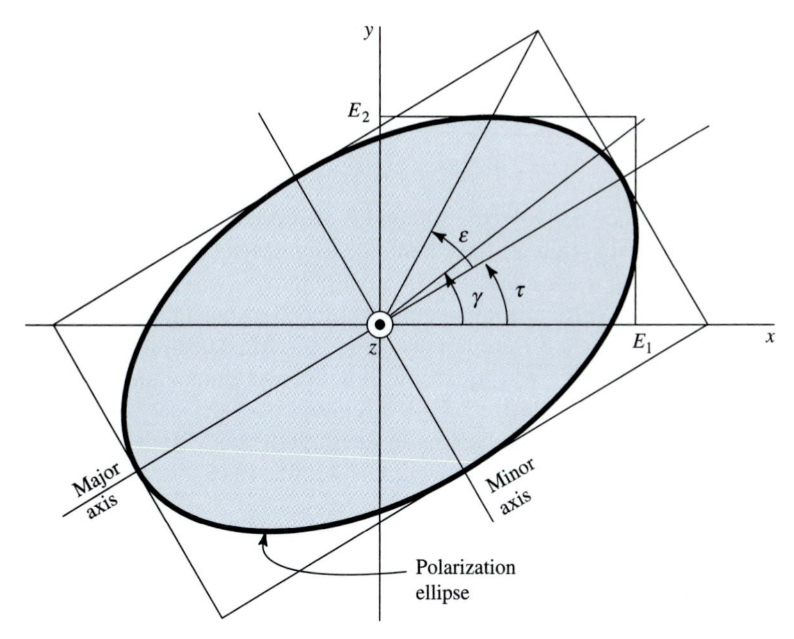

FIGURE 4-29

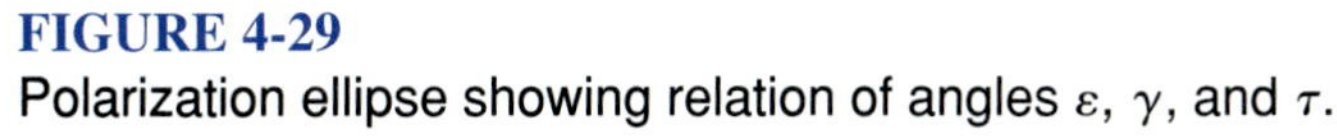

Polarization ellipse showing relation of angles ε, γ, and τ.

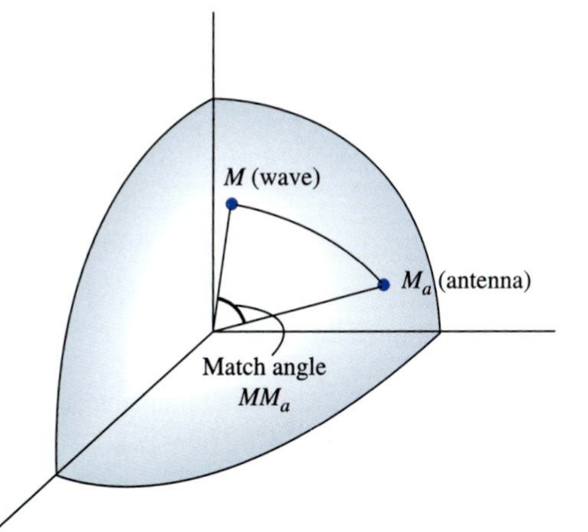

FIGURE 4-30

The match angle MM_a between the polarization state of wave (M) and antenna (M_a). For $MM_a = 0°$, the match is perfect. For $MM_a = 180°$, the match is zero.

direction while the antenna is linearly polarized in the x direction; or if the wave is left circularly polarized while the antenna is right circularly polarized. More generally we may say that *an antenna is blind to a wave of opposite (or antipodal) polarization state.*

Referring to (4), a *polarization matching factor F* (for power) is given by

$$F = \cos^2 \frac{MM_a}{2} \tag{5}$$

Thus, for a perfect match the match angle $MM_a = 0°$ and $F = 1$ (states of wave and antenna the same). For a complete mismatch the match angle $MM_a = 180°$ and $F = 0$ (Fig. 4-30).

For *linear polarization*, $MM_a/2 = \Delta\tau$ and (5) reduces to

$$F = \cos^2 \Delta\tau \tag{6}$$

where $\Delta\tau$ = difference between the tilt angles of wave and antenna.

In the above discussion we have assumed a completely polarized wave, that is, one where E_x, E_y, and δ are constants. In an unpolarized wave they are not. Such a wave results when the vertical component is produced by one noise generator and the horizontal component by a different noise generator. Most cosmic radio sources are unpolarized and can be received equally well with an antenna of any polarization. If the wave is completely unpolarized, $F = \frac{1}{2}$ regardless of the state of polarization of the antenna. For a more general discussion, see J. D. Kraus, "Radio Astronomy," 2d ed., Cygnus-Quasar, P.O. Box 85, Powell, OH 43065, 1986, Sec. 4-4.

Example 4-20. Polarization matching. Find the polarization matching factor F for a left elliptically polarized wave (w) with AR = 4 and $\tau = 15°$ incident on a right elliptically polarized antenna (a) with AR = −2 and $\tau = 45°$.

Solution. From (1), $2\varepsilon(w) = 28.1°$ and $2\varepsilon(a) = -53.1°$. Thus, the wave polarization state M is at latitude $+28.1°$ and longitude 30° while the antenna polarization state M_a is at latitude $-53.1°$ and longitude 90°. Locate these positions on a globe and measure MM_a with a string. The globe and string not only give the total great-circle angle MM_a but also illustrate the geometry. Then compare this result with an analytical one as follows: From proportional triangles obtain $2\tau(w) = 20.7°$ along the equator and $2\tau(a) = 39.3°$ further along the equator. Next from (3), obtain $2\gamma(w) = 34.3°$ and $2\gamma(a) = 62.4°$. Thus, the total great-circle angle $MM_a = 2\gamma(w) + 2\gamma(a) = 96.7°$ and the polarization matching factor

$$F = \cos^2\left(\frac{96.7}{2}\right) = 0.44$$

or the received power is 44 percent of the maximum possible value. *Ans.*

Problem 4-13-1. Antenna response. Find the relative voltage response for an antenna oriented to receive a wave traveling in the $+x$ direction if the wave is given by $\mathbf{E} = \hat{\mathbf{z}} \sin(\omega t - \beta x)$ mV(rms)/m and the parameters for the antenna are: (*a*) AR = −1; (*b*) AR = ∞, $\tau = 0°$ (with respect to y direction); (*c*) AR = ∞, $\tau = 45°$; (*d*) AR = ∞, $\tau = 90°$; and (*e*) AR = 1.5, $\tau = 67.5°$. *Ans.* (*a*) 0.707; (*b*) 0; (*c*) 0.707; (*d*) 1; (*e*) 0.79.

Problem 4-13-2. Polarization matching factor. Find the polarization matching factor F for the following cases: (*a*) wave VLP, antenna HLP; (*b*) wave SLP ($\tau = 60°$), antenna HLP; (*c*) wave RCP, antenna LCP; (*d*) wave RCP, antenna VLP; (*e*) wave RCP, antenna HLP; (*f*) wave RCP, antenna REP (AR = −3, $\tau = 0°$); and (*g*) wave LEP (AR = 4, $\tau = 0°$), antenna REP (AR = −4, $\tau = 45°$). VLP = vertical linear polarization, HLP = horizontal linear polarization, SLP = slant linear polarization, RCP = right circular polarization, LCP = left circular polarization, REP = right elliptical polarization, and LEP = left elliptical polarization. *Ans.* (*a*) 0; (*b*) 0.25; (*c*) 0; (*d*) 0.5; (*e*) 0.5; (*f*) 0.8; (*g*) 0.39.

4-14 OBLIQUE INCIDENCE; REFLECTION AND REFRACTION

Consider a linearly polarized plane wave obliquely incident on a boundary between two media, as shown in Fig. 4-31. The incident wave (from medium 1) makes an angle of θ_i with the y axis, the *reflected wave* (medium 1) makes an angle of θ_r with the y axis, and the *transmitted wave*† makes an angle of θ_t with the negative y axis.

Consider two cases: (1) the electric field perpendicular to the *plane of incidence* (the xy plane) and (2) the electric field parallel to the plane of incidence. These waves are said to be *perpendicularly polarized* and *parallel polarized,* respectively. The field vectors shown in Fig. 4-31 are for the case of *perpendicular polarization.* It is clear that any arbitrary plane wave can be resolved into perpendicular and parallel components.

Perpendicular Case ($\mathbf{E}_\perp$)

From the boundary conditions,

$$\eta_1 \sin\theta_i = \eta_1 \sin\theta_r = \eta_2 \sin\theta_t \tag{1}$$

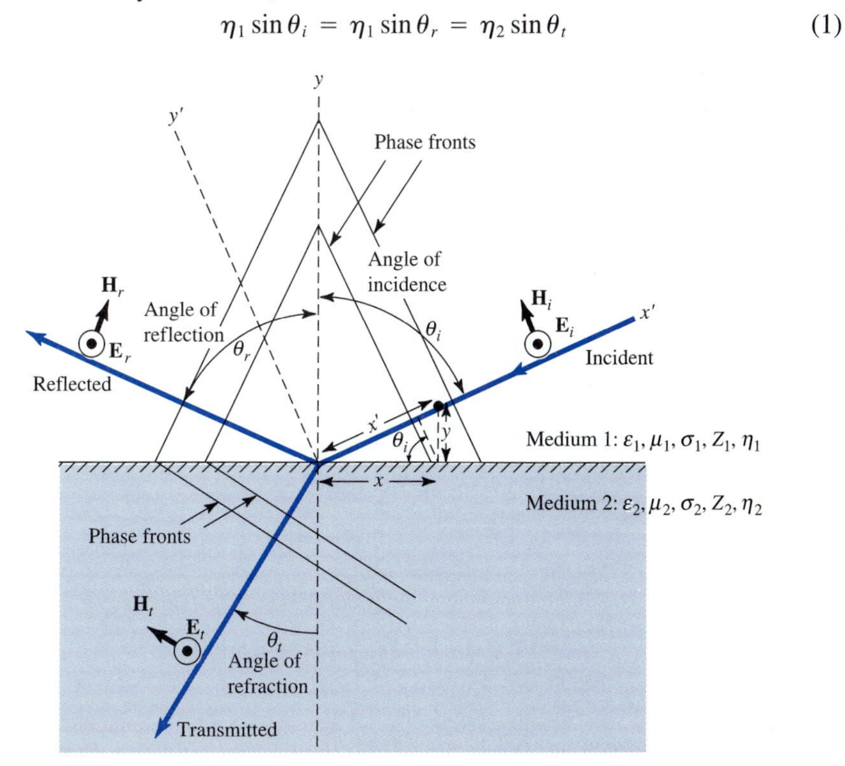

FIGURE 4-31
Geometry in the plane of incidence (x-y plane or plane of the page) for linearly polarized wave at oblique incidence and for perpendicular polarization. The z direction is outward from the page.

†The transmitted wave is also called the *refracted wave.* Hence, θ_t is called the *angle of refraction.*

From the first equality

$$\theta_r = \theta_i \tag{1.1}$$

i.e., ***the angle of reflection is equal to the angle of incidence***. From the second equality,

$$\sin\theta_t = \frac{\eta_1}{\eta_2}\sin\theta_i \qquad \textbf{\textit{Snell's law}} \tag{2}$$

where η_1 and η_2 are the indices of refraction of medium 1 and medium 2, respectively. Equation (2) is known as ***Snell's law***†and is a relation of fundamental importance in geometrical optics. For a lossless medium the index of refraction η can be written as equal to $\sqrt{\mu_r\varepsilon_r}$, and Snell's law becomes

$$\sin\theta_t = \sqrt{\frac{\mu_1\varepsilon_1}{\mu_2\varepsilon_2}}\sin\theta_i \qquad \textbf{\textit{Snell's law}} \tag{3}$$

Example 4-21. Polystyrene-air interface. Polystyrene has a relative permittivity of 2.7. If a wave is incident at an angle of $\theta_i = 30°$ from air onto polystyrene, (*a*) calculate the angle of transmission θ_t and (*b*) interchange polystyrene and air and repeat the calculation.

Solution. From air onto polystyrene, $\varepsilon_1 = \varepsilon_0$, $\mu_1 = \mu_0$, $\varepsilon_2 = 2.7\varepsilon_0$, and $\mu_2 = \mu_0$. From (3)

$$\sin\theta_t = \sqrt{\frac{1}{2.7}}(0.5) = 0.304$$

$$\theta_t = 17.7° \quad \textit{Ans.} \quad (a)$$

From polystyrene onto air, $\varepsilon_1 = 2.7\varepsilon_0$, $\mu_1 = \mu_0$, $\varepsilon_2 = \varepsilon_0$, and $\mu_2 = \mu_0$.

$$\sin\theta_t = \sqrt{2.7}(0.5) = 0.822$$

$$\theta_t = 52.2° \quad \textit{Ans.} \quad (b)$$

We have also

$$-\cos\theta_i + \rho_\perp\cos\theta_i = -\tau\frac{Z_1}{Z_2}\cos\theta_t \tag{4}$$

and

$$1 + \rho_\perp = \tau_\perp \tag{5}$$

and on substituting (5) into (4) and solving for the Fresnel reflection coefficient $\rho_\perp$, we have

$$\rho_\perp = \frac{Z_2\cos\theta_i - Z_1\cos\theta_t}{Z_2\cos\theta_i + Z_1\cos\theta_t} \tag{6}$$

†For complete development of these equations see the fourth edition of this book (Chap. 13).

where Z_1 and Z_2 are the impedances of medium 1 and medium 2, respectively. It is seen that the previously derived reflection coefficient for normal incidence, (4-7-12), is obtained as a special case of (6) when $\theta_i = 0$.

If medium 2 is a perfect conductor, $Z_2 = 0$ and $\rho_\perp = -1$. If both media are *lossless nonmagnetic dielectrics,* (6) becomes

$$\rho_\perp = \frac{\cos\theta_i - \sqrt{(\varepsilon_2/\varepsilon_1) - \sin^2\theta_i}}{\cos\theta_i + \sqrt{(\varepsilon_2/\varepsilon_1) - \sin^2\theta_i}} \qquad \textbf{\textit{Reflection coefficient}} \perp \tag{7}$$

Provided medium 2 is a more dense dielectric than medium 1 ($\varepsilon_2 > \varepsilon_1$), the quantity under the square root will be positive and $\rho_\perp$ will be real. If, however, the wave is incident from the more dense medium onto the less dense medium ($\varepsilon_1 > \varepsilon_2$), and if $\sin^2\theta_i \geq \varepsilon_1/\varepsilon_2$, then $\rho_\perp$ becomes complex and $|\rho_\perp| = 1$. Under these conditions, the incident wave is *totally internally reflected* back into the more dense medium.† The incident angle for which $\rho_\perp = 1\angle 0°$ is called the *critical angle* θ_{ic}. From (7) it is seen that this happens when the radical is zero, so that

$$\theta_{ic} = \sin^{-1}\sqrt{\frac{\varepsilon_2}{\varepsilon_1}} \qquad \textbf{\textit{Critical angle}} \tag{8}$$

defines the critical angle. For all angles greater than the critical angle, $|\rho_\perp| = 1$. Using Snell's law, we see that when $\theta_i > \theta_{ic}$, then $\sin\theta_t > 1$,‡ and $\cos\theta_t$ must be imaginary; i.e.,

$$\cos\theta_t = \sqrt{1 - \sin^2\theta_t} = jA \tag{9}$$

where $A = \sqrt{(\varepsilon_1/\varepsilon_2)\sin^2\theta_i - 1}$ is a real number.

The electric field in the less dense medium can now be written

$$\mathbf{E}_t = \hat{\mathbf{z}}\tau_\perp E_0 \exp(-\alpha y)\exp(j\beta_2 x \sin\theta_t) \tag{10}$$

where

$$\alpha = \beta_2 A = \omega\sqrt{\mu_2\varepsilon_2}\sqrt{\frac{\varepsilon_1}{\varepsilon_2}\sin^2\theta_i - 1} \tag{11}$$

Thus, $E_\perp$ in the less dense medium has a magnitude $\tau_\perp E_0$, decaying exponentially away from the surface (y direction) and propagating without loss in the $-x$ direction. Waves whose fields are of the form of (10) are called *surface waves.* These results can be summarized by the ***principle of total internal reflection*** as follows. *When a wave is incident from the more dense onto the less dense medium at an angle equal to or exceeding the critical angle, the wave will be totally internally reflected and will also be accompanied by a surface wave in the less dense medium.*

Example 4-22. Total internal reflection with surface wave. Referring to Fig. 4-32, a linearly polarized plane wave is incident from water onto the water-air interface at

†It can be shown that this is true for either perpendicular or parallel polarization.

‡Although the sine is greater than unity and the cosine is imaginary, there is a simple interpretation of the resulting field in such a case. See Example 4-22.

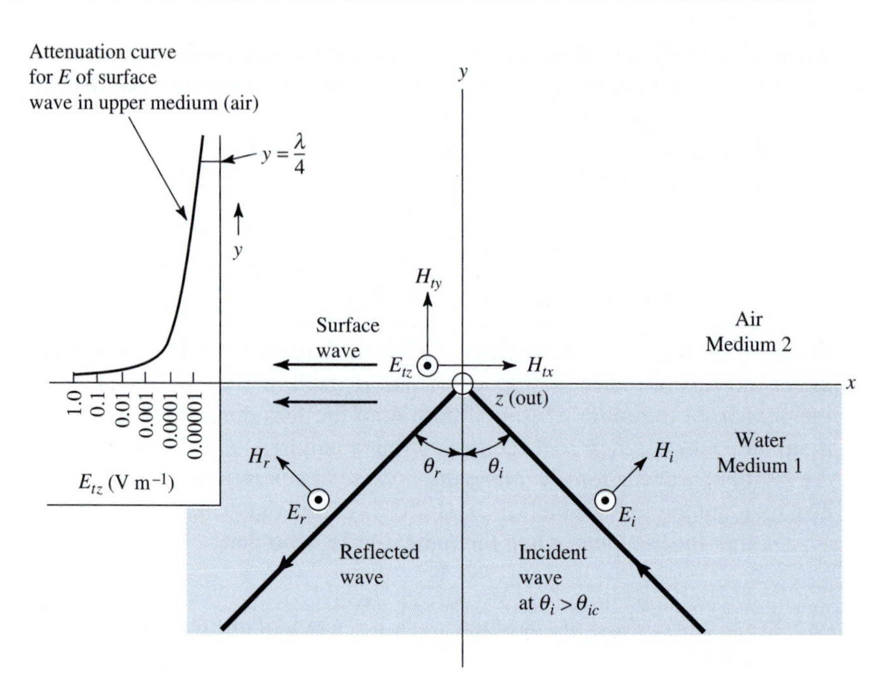

FIGURE 4-32
Total internal reflection of incident wave with accompanying surface wave which attenuates exponentially above surface (y direction) as shown by graph at left. No power is transmitted in y direction (up).

45°. Calculate the magnitude of the electric field in air (*a*) at the interface and (*b*) $\lambda/4$ above the surface if the incident electric field $E_i = 1\ \text{Vm}^{-1}$. Take the water constants to be those of distilled water: $\varepsilon_r = 81$, $\mu_r = 1$, $\sigma \approx 0$.

Solution. From (8), the critical angle

$$\theta_{ic} = \sin^{-1}\sqrt{\frac{1}{81}} = 6.38°$$

Thus, the angle of incidence θ_i ($=45°$) exceeds the critical angle and the wave will be totally internally reflected (see Fig. 4-32). From (3),

$$\sin\theta_t = \sqrt{81}(0.707) = 6.36$$

From (9),

$$\cos\theta_t = jA = \sqrt{1 - 6.36^2} = j6.28$$

From (11),

$$\alpha = \beta_2 A = \frac{2\pi}{\lambda_0}6.28 = \frac{39.49}{\lambda_0}\ \text{Np m}^{-1}$$

From (5) and (7),

$$\tau_{\perp} = 1 + \rho_{\perp} = 1 + \frac{0.707 - \sqrt{\frac{1}{81} - 0.5}}{0.707 + \sqrt{\frac{1}{81} - 0.5}} = 1.42 \angle -44.64^{\circ}$$

Therefore, the magnitude of the field strength is
(*a*) At the interface:

$$|E_t| = 1.42 \text{ V m}^{-1}$$

(*b*) $\lambda/4$ away from the interface:

$$|E_{\tau}| = 1.42 \exp\left(-\frac{39.49}{\lambda_0}\frac{\lambda_0}{4}\right) = 73.2\,\mu\text{V m}^{-1} \quad \textit{Ans.} \quad (a)$$

Thus, the field $\lambda/4$ above the surface is

$$20 \log \frac{73.2 \times 10^{-6}}{1.42} = -85.8 \text{ dB} \quad \textit{Ans.} \quad (b)$$

less than the field at the surface. Recalling that a power ratio of 1 billion equals 90 dB, it is evident that the field attenuates very rapidly above the surface (in the y direction) (see Fig. 4-32), meaning that the surface wave is very tightly bound to the water surface. Note that $\sin\theta_t$ is greater than 1 but real, while $\cos\theta_t$ is imaginary. From (10),

$$\mathbf{E}_t = \hat{\mathbf{z}}\tau_{\perp} E_0 e^{-(\beta_2 A)y} e^{j\beta_2 x \sin\theta_t} \tag{12}$$

and

$$\mathbf{H}_t = (-\hat{\mathbf{x}} jA + \hat{\mathbf{y}} \sin\theta_t)\tau_{\perp} \frac{E_0}{Z_2} e^{-(\beta_2 A)y} e^{j\beta_2 x \sin\theta_t} \tag{13}$$

where $A = -\sqrt{\sin^2\theta_t - 1}$.

From (12) and (13), the average Poynting vector of the wave in the y direction in air (above the water surface) is

$$S_{y(\text{av})} = \tfrac{1}{2}\text{Re}\mathbf{E} \times \mathbf{H}^* = \hat{\mathbf{y}}\tfrac{1}{2}E_{tz}H_{tx} \sin\phi \cos\phi \tag{14}$$

where ϕ = space angle between **E** and **H** (= 90°) and θ = time-phase angle between **E** and **H**.

The exponentials in (12) and (13) are identical. However, H_{tx} has a j factor whereas E_{tz} does not, indicating a 90° time-phase difference between **E** and **H**. Hence, $\theta = 90^{\circ}$, and since $\sin\theta = \sin 90^{\circ} = 1$,

$$S_{y(\text{av})} = \tfrac{1}{2}E_{tz}H_{tx} \cos 90^{\circ} = 0 \tag{15}$$

Thus, no power is transmitted in the y direction (wave reactive). Both E_t and H_t decay exponentially with y. Similar waves, called *evanescent waves,* exist in hollow conducting waveguides at wavelengths too long to propagate through the guide (see Chap. 8).

The waves in medium 2 (air) involving E_{tz} and H_{ty} propagate without attenuation as a surface wave in the $-x$ direction with a velocity v_x equal to the wave velocity in the water (medium 1) as observed parallel to the x axis ($v_x = v_{\text{water}}/\sin\theta_i$). The traveling wave is simply the matching field at the boundary. Total internal reflection with a surface wave can also occur for $\mathbf{E}_{\parallel}$, but the details differ.

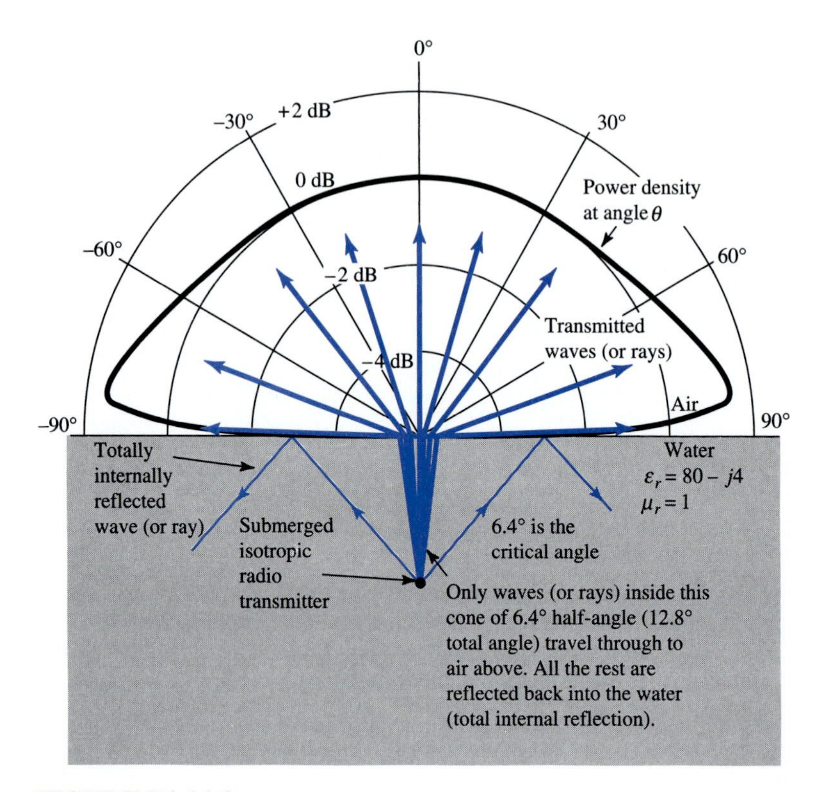

FIGURE P4-14-2

Problem 4-14-1. Submerged transmitter. (*a*) A 60-MHz radio transmitter which radiates isotropically is situated 20 m below the surface of a deep freshwater lake with constants $\mu_r = 1$ and $\varepsilon_r = 80 - j4$. If the polarization is linear (**E** parallel to surface), what is the relative field strength at a radius of 1 km at angles of 0° to 90° from the zenith? Take 0 dB at zenith. The radial distance is measured from a point on the water surface directly above the transmitter. *Ans.* See Fig. P4-14-2.

Parallel Case ($\mathbf{E}_{\|}$)

Consider now the case of *parallel* (||) *polarization.* The geometry is the same as in Fig. 4-31 but with E_i, E_r, and E_t parallel to the plane of incidence as would be obtained by replacing $\mathbf{H}_i$ by $\mathbf{E}_i$, $\mathbf{H}_r$ by $\mathbf{E}_r$, and $\mathbf{H}_t$ by $\mathbf{E}_t$. By matching boundary conditions, as before, it is found that the angle of incidence equals the angle of reflection and that Snell's law (2) holds. It can also be shown that

$$1 + \rho_{\|} = \frac{\cos\theta_t}{\cos\theta_i}\tau_{\|} \tag{16}$$

The Fresnel reflection coefficient is found to be

$$\rho_{\|} = \frac{Z_2\cos\theta_t - Z_1\cos\theta_i}{Z_1\cos\theta_i - Z_2\cos\theta_t} \tag{17}$$

which *for lossless nonmagnetic dielectrics* becomes

$$\rho_{\parallel} = \frac{-(\varepsilon_2/\varepsilon_1)\cos\theta_i + \sqrt{(\varepsilon_2/\varepsilon_1) - \sin^2\theta_i}}{(\varepsilon_2/\varepsilon_1)\cos\theta_i + \sqrt{(\varepsilon_2/\varepsilon_1) - \sin^2\theta_i}} \qquad \textbf{\textit{Reflection coefficient}} \parallel \tag{18}$$

and reduces to $\rho_{\parallel} = -1$ if medium 2 is a perfect conductor.

It is of especial interest that, for parallel polarization, it is possible to find an incident angle so that $\rho_{\parallel} = 0$ and the wave is *totally transmitted* into medium 2. This angle, called the *Brewster angle* θ_{iB}, can be found by setting the numerator of (18) to zero, giving

$$\theta_{iB} = \sin^{-1}\sqrt{\frac{\varepsilon_2/\varepsilon_1}{1 + (\varepsilon_2/\varepsilon_1)}} = \tan^{-1}\sqrt{\frac{\varepsilon_2}{\varepsilon_1}} \qquad \textbf{\textit{Brewster angle}} \tag{19}$$

The Brewster angle is also sometimes called the *polarizing angle* since a wave composed of both perpendicular and parallel components and incident at the Brewster angle produces a reflected wave with only a perpendicular component. Thus, a circularly polarized wave incident at the Brewster angle becomes linearly polarized on reflection.

Example 4-23. Brewster angle. A parallel-polarized wave is incident from air onto (*a*) distilled water ($\varepsilon_r = 81$), (*b*) flint glass ($\varepsilon_r = 10$), and (*c*) paraffin ($\varepsilon_r = 2$). Find the Brewster angle for each of these cases.

Solution.

$$\theta_{iB} = \tan^{-1}\sqrt{81} = 83.7^\circ \quad \textit{Ans.} \quad (a)$$

$$\theta_{iB} = \tan^{-1}\sqrt{10} = 72.4^\circ \quad \textit{Ans.} \quad (b)$$

$$\theta_{iB} = \tan^{-1}\sqrt{2} = 54.7^\circ \quad \textit{Ans.} \quad (c)$$

Example 4-23.1. Effect of ground reflection on antenna pattern. A linear in-phase antenna in free space radiates equally in all directions perpendicular to its length. Above a perfectly conducting ground the field may double or go to zero depending on the relative phase of the direct and ground-bounce waves. If the antenna height $h = \lambda$, what is the field pattern?

Solution. Referring to Fig. E4-23.1*a* the antenna is horizontal and perpendicular to the page. The pattern is the same as that of the antenna and its image as given by

$$E(\theta) = 2E_0 \sin\left(2\pi\frac{h}{\lambda}\sin\theta\right)$$

where E_0 = free space field, V m^{-1}, and h = height above ground, m. No mutual coupling of antenna and image is assumed. In practice this is small when h is large ($> \lambda/2$).

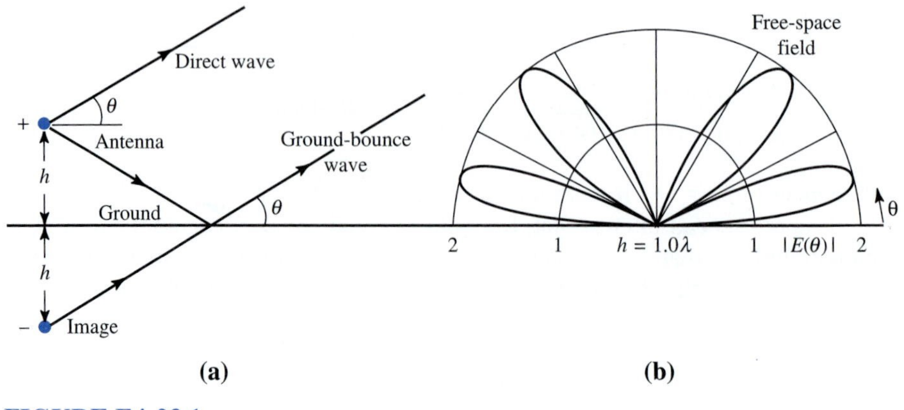

FIGURE E4-23.1

The pattern is shown in Fig. E4-23.1*b*. *Ans.*

Note that this pattern applies to any in-phase linear antenna. Thus, it applies to a $\lambda/2$ antenna, a 1λ antenna, or longer linear in-phase antenna. Note also that the field is a function of elevation angle θ in the *far field,* whereas in Project P4-2 it is a function of height in the *near field.*

Problem 4-14-2. Angle of maximum field. Find the angles for which the field in Fig. E4-23.1*b* is maximum. *Ans.* 14.5° and 48.6°.

Problem 4-14-3. Angle of maximum field. Find the angles for which the field is maximum if $h = 2\lambda$. *Ans.* 7.2°, 22.0°, 38.7°, and 61.0°.

4-15 ELLIPTICALLY POLARIZED PLANE WAVE, OBLIQUE INCIDENCE

We now consider an elliptically polarized plane wave obliquely incident on a boundary. The problem is to find the magnitude and polarization of the reflected and transmitted waves.

The incident electric field is composed of both parallel ($E_{i\|}$) and perpendicular ($E_{i\perp}$) components, as shown in inset *A* of Fig. 4-33. As viewed from the origin, the orthogonal field components are

$$E_z = E_{i\perp} \tag{1}$$

$$E_{y'} = E_{i\|} e^{j\delta_i} \tag{2}$$

where δ_i is the angle by which $E_{y'}$ leads E_z and $E_{i\perp}$ and $E_{i\|}$ are the component amplitudes. Thus, as in Sec. 4-13, the incident wave has a polarization ellipse specified by (γ_i, δ_i), where

$$\gamma_i = \tan^{-1} \frac{E_{i\|}}{E_{i\perp}} \tag{3}$$

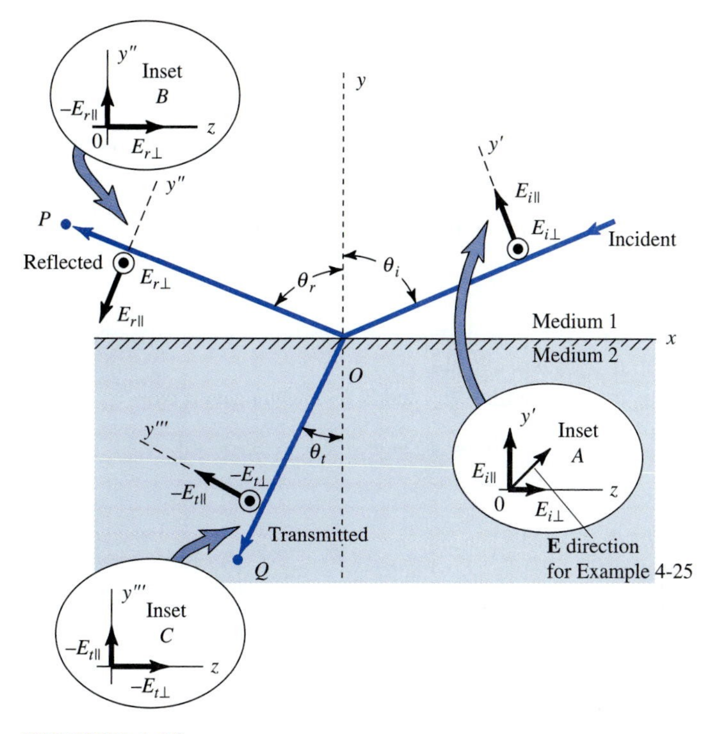

FIGURE 4-33
Geometry in the plane of incidence (x-y plane) for elliptically polarized wave incident obliquely on a plane surface. The incident electric field has parallel ($E_{i\parallel}$) and perpendicular ($E_{i\perp}$) components which appear as in inset *A*, when viewed from the origin 0. The reflected and transmitted fields shown in insets *B* and *C* are viewed looking toward the origin from points *P* and *Q*, respectively.

The polarization ellipse can also be specified in terms of its axial ratio AR_i and tilt angle τ_i.†From (4-13-3) we can relate (AR_i, τ_i) to (γ_i, δ_i):

$$\tan 2\tau_i = \tan 2\gamma_i \cos \delta_i \tag{4}$$

$$\sin 2\varepsilon_i = \sin 2\gamma_i \sin \delta_i \tag{5}$$

$$\cos 2\gamma_i = \cos 2\varepsilon_i \cos 2\tau_i \tag{6}$$

$$\tan \delta_i = \frac{\tan 2\varepsilon_i}{\sin 2\tau_i} \tag{7}$$

†Note that when the Greek letter τ is used in this section to denote the *tilt angle,* it is associated with the incident, reflected, or transmitted wave (that is, τ_i, τ_r, or τ_t), but when it is used to denote the *transmission coefficient*, it is associated with either a parallel or perpendicular case (that is, $\tau_\parallel$ or $\tau_\perp$).

where
$$\varepsilon_i = \cot^{-1}(\mp \text{AR}_i) \tag{8}$$

If the wave is right elliptically polarized, the minus sign is used; if it is left elliptically polarized, the plus sign is used.

Considering the reflected wave as viewed from point P (see inset B), the orthogonal field components are

$$E_z = E_{r\perp} = |\rho_\perp| E_{i\perp} \exp(j\phi_\perp) \tag{9}$$

$$E_{y''} = -E_{r\parallel} = |\rho_\parallel| E_{i\parallel} \exp[j(\phi_\parallel + \delta_i + \pi)] \tag{10}$$

where $\phi_\parallel$ and $\phi_\perp$ are the phase angles of the parallel and perpendicular reflection coefficients $\rho_\parallel$ and $\rho_\perp$, respectively.

The phase angle by which $E_{y''}$ leads E_z is thus given by

$$\delta_r = \delta_i + \pi + (\phi_\perp - \phi_\parallel) \tag{11}$$

and similarly [see (3)]

$$\gamma_r = \tan^{-1}\left(\frac{\rho_\parallel}{\rho_\perp} \tan\gamma_i\right) \tag{12}$$

The relations (11) and (12) specify the polarization state of the reflected wave, as seen by an observer at point P. To find the axial ratio and tilt angle of the polarization ellipse of the reflected wave, the relations (4), (5), and (8) are used with the subscript i replaced by the subscript r.

In similar fashion, it can be shown that an observer at point Q (see inset C) will see a transmitted wave with polarization ellipse given by

$$\delta_t = \delta_i + (\xi_\perp - \xi_\parallel) \tag{13}$$

$$\gamma_t = \tan^{-1}\left(\frac{\tau_\parallel}{\tau_\perp} \tan\gamma_i\right) \tag{14}$$

where $\xi_\parallel$ and $\xi_\perp$ are the phase angles of the parallel and perpendicular transmission coefficients $\tau_\parallel$ and $\tau_\perp$, respectively.

Example 4-24. Circular polarized wave reflection. A right circularly polarized wave (RCP) is incident at an angle of 45° from air onto (*a*) a perfect conductor and (*b*) polystyrene ($\varepsilon_r = 2.7$). What is the polarization state of the reflected wave for these two cases?

Solution.

(*a*) When medium 1 is air and medium 2 is a perfect conductor,

$$\rho_\parallel = 1\angle 180^\circ \qquad \rho_\perp = 1\angle 180^\circ$$

For a right circularly polarized wave, $\gamma_i = 45^\circ$, $\delta_i = -90^\circ$, so that from (11), $\delta_r = -90^\circ + 180^\circ + (180^\circ - 180^\circ) = 90^\circ$ and from (12),

$$\gamma_r = \tan^{-1}(\tfrac{1}{1} \times 1) = 45^\circ$$

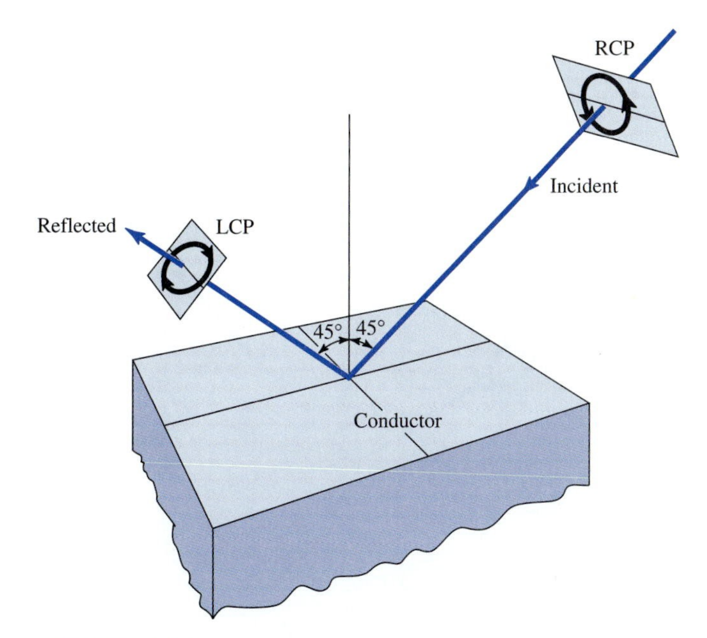

FIGURE 4-34
[Example 4-24 (*a*)] Right circularly polarized (RCP) wave incident on a perfect conductor. The reflected wave is left circularly polarized (LCP). No wave is refracted (transmitted).

Therefore the reflected wave is *left circularly polarized* (LCP), as shown in Fig. 4-34. Conversely, if the incident wave had been LCP, the reflected wave would be RCP. *Ans.* (*a*).

(*b*) When medium 1 is air and medium 2 is polystyrene, *then* from (4-14-18)

$$\rho_{\parallel} = \frac{-2.7(0.707) + \sqrt{2.7 - 0.5}}{2.7(0.707) + \sqrt{2.7 - 0.5}} = 0.126\angle 180^\circ$$

and from (4-14-7)

$$\rho_{\perp} = \frac{0.707 - \sqrt{2.7 - 0.5}}{0.707 + \sqrt{2.7 - 0.5}} = 0.354\angle 180^\circ$$

Therefore,
$$\delta_r = -90^\circ + 180^\circ + (180^\circ - 180^\circ) = 90^\circ$$

$$\gamma_r = \tan^{-1}\frac{0.126}{0.354} = 19.6^\circ$$

Substituting these values into (4), (5), and (8) (with subscripts i replaced by r), we find $\tau_r = 0^\circ$ and $\mathrm{AR}_r = 2.81$. Thus the reflected wave is left elliptically polarized (LEP), as shown in Fig. 4-35. *Ans.* (*b*).

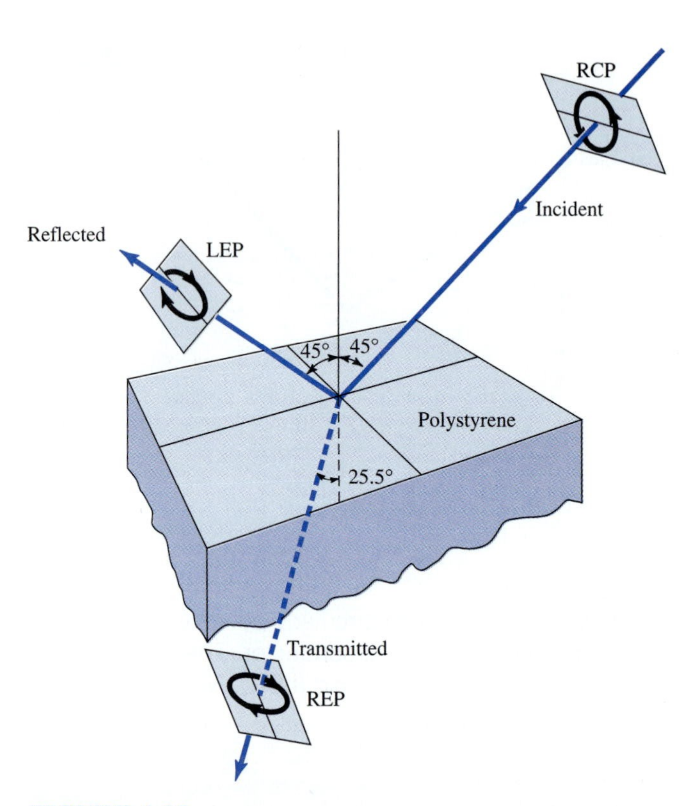

FIGURE 4-35
[Example 4-24 (*b*)] Right circularly polarized (RCP) wave incident on dielectric slab of polystyrene. The reflected wave is left elliptically polarized (LEP) with major axis of the polarization ellipse horizontal and the transmitted wave right elliptically polarized (REP). If the angle of incidence is increased to 58.7° (= Brewster angle), the angle of reflection increases to 58.7° and the angle of refraction increases from 25.5 to 31.3°. Also the polarization ellipse of the reflected wave collapses to a straight horizontal line (linear ⊥ polarization).

Example 4-25. Linearly polarized wave with equal $E_\perp$ and $E_\parallel$ components. A linearly polarized wave whose electric field vector bisects the zy' axis (see Fig. 4-33, inset *A*) is incident at an angle of 45° from air onto (*a*) a perfect conductor and (*b*) polystyrene ($\varepsilon_r = 2.7$). What is the polarization state of the reflected wave for these two cases? (*c*) If the angle of incidence onto polystyrene is 58.7° (Brewster angle), what is the polarization state of the reflected wave?

Solution.

(*a*) Since $\rho_{\parallel} = 1\angle 180°$ and $\rho_{\perp} = 1\angle 180°$,

$$\delta_r = 0° + 180° + (180° - 180°) = 180°$$

$$\gamma_r = \tan^{-1}(\tfrac{1}{1} \times 1) = 45°$$

From (4), $\tan 2\tau_r = -\infty$, so that $\tau_r = 135°$; and from (5) and (8), $\text{AR}_r = \infty$. Therefore the reflected wave is *linearly polarized* with the electric field vector bisecting the $-z$ and y'' axes. *Ans.* (*a*).

(*b*) From Example 4-24, part (*b*),

$$\rho_{\parallel} = 0.126\angle 180° \qquad \text{and} \qquad \rho_{\perp} = 0.354\angle 180°$$

so that $\gamma_r = \tan^{-1}(0.126/0.354) = 19.6°$ and $\delta_r = 180°$. Then from (4), $\tan 2\tau_r = \tan 39.2° \cos\pi = -0.815$, so that $\tau_r = 70.4°$. From (5) and (8), $\text{AR}_r = \infty$. Therefore the reflected wave is *linearly polarized* with the electric field vector making an angle in the z-y'' plane of 70.4° with the z axis. *Ans. (b).*

(*c*) When $\theta_i = \theta_{ib} = 58.7°$,

$$\rho_{\parallel} = 0$$

$$\rho_{\perp} = \frac{0.520 - 1.40}{0.520 + 1.40} = 0.46\angle 180°$$

Therefore the reflected wave is perpendicularly polarized. *Ans.* (*c*).

Problem 4-15-1. CP wave at Brewster angle. A circularly polarized 1-GHz wave in air is incident on a half-space of lossless dielectric medium ($\varepsilon_r = 5$) at the Brewster angle. Find the axial ratio AR for (*a*) transmitted wave and (*b*) reflected wave. (*c*) Find angle of transmitted wave. *Ans.* (*a*) 1.34; (*b*) ∞; (*c*) 24.1°.

Problem 4-15-2. CP wave, Brewster angle. A circularly polarized wave in air is incident on a flat medium with constants $\sigma = 0$, $\mu_r = 1$, and $\varepsilon_r = 5$ at the Brewster angle. Find AR and τ (with respect to plane of incidence) for (*a*) transmitted wave and (*b*) reflected wave. *Ans.* (*a*) 1.34, $\tau_{\perp} = 0.334$; (*b*) ∞, $\tau_{\parallel} = 0.447$.

4-16 HUYGENS' PRINCIPLE AND PHYSICAL OPTICS; DIFFRACTION

Huygens' (Hoy-gens) principle† states that *each point on a primary wave front can be considered to be a new source of a secondary spherical wave and that a secondary wave front can be constructed as the envelope of these secondary spherical waves,* as suggested in Fig. 4-36. This fundamental principle of physical optics can be used to explain the apparent bending of radio waves around obstacles, i.e., the diffraction of

†C. Huygens, "Traité de la lumière," Leyden, 1690. English translation by S. P. Thompson, London, 1912, reprinted by the University of Chicago Press.

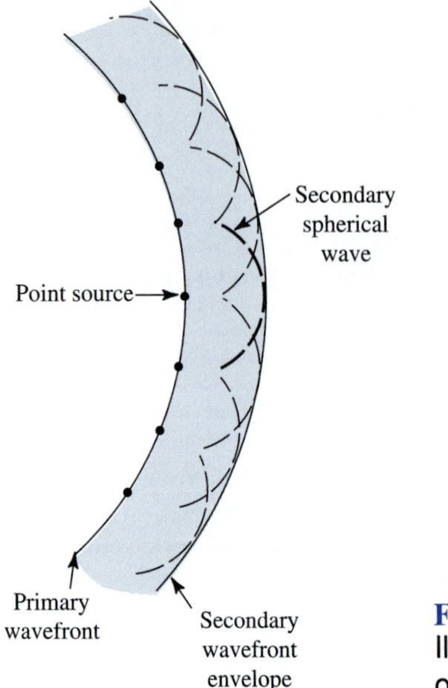

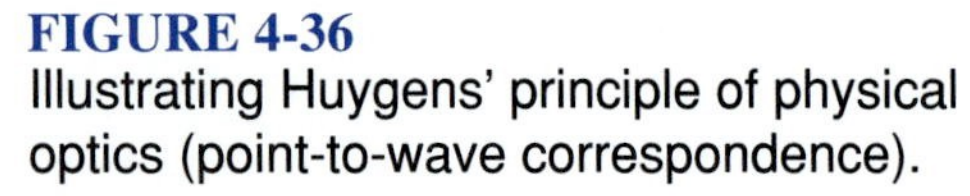
FIGURE 4-36
Illustrating Huygens' principle of physical optics (point-to-wave correspondence).

waves. A *diffracted* ray is one that follows a path that cannot be interpreted as either reflection or refraction. The difference between *diffraction* and *refraction* may also be stated in connection with a transparent prism. Waves bent in traveling through the prism are *refracted* while waves scattered off the edges are *diffracted*.

As an example, consider a uniform plane wave incident on a conducting half-plane, as in Fig. 4-37. We want to calculate the electric field at point P by using Huygens' principle; i.e.,

$$E = \int_{\substack{\text{over} \\ x\text{ axis}}} dE \tag{1}$$

where dE is the electric field at P due to a point source at a distance x from 0, as in Fig. 4-37*b*, i.e.,

$$dE = \frac{E_0}{r} e^{-j\beta(r+\delta)}\, dx \tag{2}$$

so that

$$E = \frac{E_0}{r} e^{-j\beta r} \int_a^{\infty} e^{-j\beta\delta}\, dx \tag{3}$$

If $\delta \ll r$, it follows that

$$\delta = \frac{x^2}{2r} \tag{4}$$

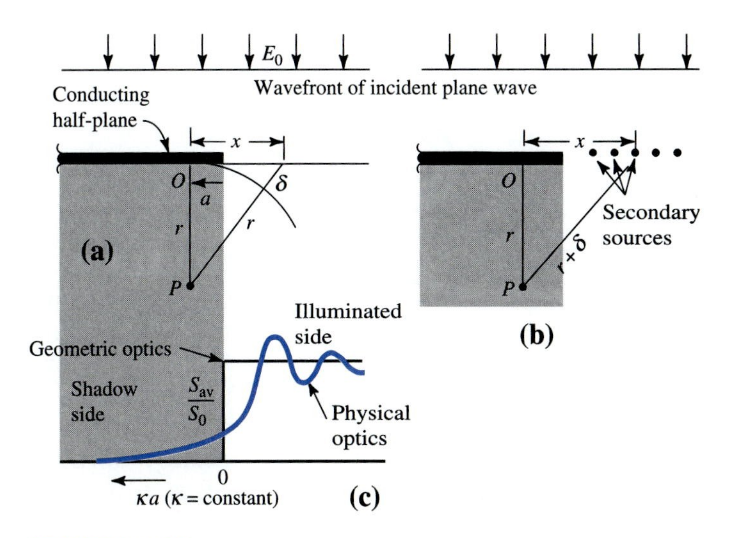

FIGURE 4-37
Plane wave incident from above onto a conducting half-plane with resultant power-density variation below the plane as obtained by physical optics.

When we let $\kappa^2 = 2/r\lambda$ and $u = \kappa x$, (3) becomes

$$E = \frac{E_0}{\kappa r} e^{-j\beta r} \int_{\kappa a}^{\infty} e^{-j\pi u^2/2}\, du \tag{5}$$

which can be rewritten as

$$E = \frac{E_0}{\kappa r} e^{-j\beta r} \left(\int_0^{\infty} e^{-j\pi u^2/2}\, du - \int_0^{\kappa a} e^{-j\pi u^2/2}\, du \right) \tag{6}$$

The integrals in (6) have the form of Fresnel integrals, so that (6) can be written

$$E = \frac{E_0}{\kappa r} e^{-j\beta r} \left\{ \frac{1}{2} + j\frac{1}{2} - [C(\kappa a) + jS(\kappa a)] \right\} \tag{7}$$

where
$$C(\kappa a) = \int_0^{\kappa a} \cos \frac{\pi u^2}{2}\, du = \text{Fresnel cosine integral} \tag{8}$$

$$S(\kappa a) = \int_0^{\kappa a} \sin \frac{\pi u^2}{2}\, du = \text{Fresnel sine integral} \tag{9}$$

A graph of $C(\kappa a)$ and $S(\kappa a)$ yields Cornu's spiral (Fig. 4-38). Since $C(-\kappa a) = -C(\kappa a)$, and $S(-\kappa a) = -S(\kappa a)$, the spiral for negative values of κa is in the third quadrant and is symmetrical with respect to the origin for the spiral in the first quadrant.

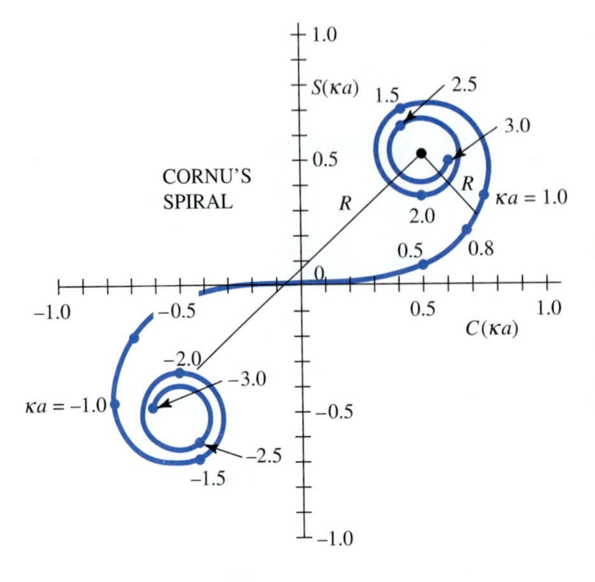

FIGURE 4-38 Cornu's spiral showing $C(\kappa a)$ and $S(\kappa a)$ as a function of κa values along the spiral. For example, when $\kappa a = 1.0$, $C(\kappa a) = 0.780$ and $S(\kappa a) = 0.338$. When $\kappa a = \infty$, $C(\kappa a) = S(\kappa a) = 1/2$.

The power density as a function of κa is then

$$S_{av} = \frac{EE^*}{2Z} = S_0 \frac{1}{2}\left\{\left[\frac{1}{2} - C(\kappa a)\right]^2 + \left[\frac{1}{2} - S(\kappa a)\right]^2\right\} \qquad (\text{W m}^{-2}) \qquad (10)$$

where

$$S_0 = \frac{E_0^2 \lambda}{2Zr} \qquad (\text{W m}^{-2}) \qquad (11)$$

The power density variation (10) as a function of κa (with r, λ, and κ constant) is shown in Fig. 4-37*c*. Note that when $\kappa a = -\infty$, which corresponds to no edge present, $S_{av} = S_0$; when $\kappa a = 0$, $S_{av} = S_0/4$; and when $\kappa a = +\infty$, which corresponds to complete obscuration of the plane wave by the half-plane, $S_{av} = 0$. Furthermore, the power density does not go to zero abruptly as the point of observation goes from the illuminated side ($\kappa a < 0$) to the shadow side ($\kappa a > 0$); rather, there are fluctuations followed by a gradual decrease in power density (see Fig. 4-37*c*).

From (10) and (11), the relative power density as a function of κa is

$$S_{av}(\text{rel}) = \frac{S_{av}}{S_0} = \frac{1}{2}\left\{\left[\frac{1}{2} - C(\kappa a)\right]^2 + \left[\frac{1}{2} - S(\kappa a)\right]^2\right\} \qquad (12)$$

The relative power density (12) is equal to $\frac{1}{2}R^2$, where R is the distance from a κa value on the Cornu spiral to the point $(\frac{1}{2}, \frac{1}{2})$. (See Fig. 4-38.) For large positive values of κa, R is small and approaches $1/\pi\kappa a$, so that (12) reduces approximately to

$$S_{av}(\text{rel}) = \frac{1}{2}\left(\frac{1}{\pi \kappa a}\right)^2 = \frac{r\lambda}{4\pi^2 a^2} \qquad (13)$$

where r = distance from obstacle (conducting half-plane), m
λ = wavelength, m
a = distance into shadow region, m

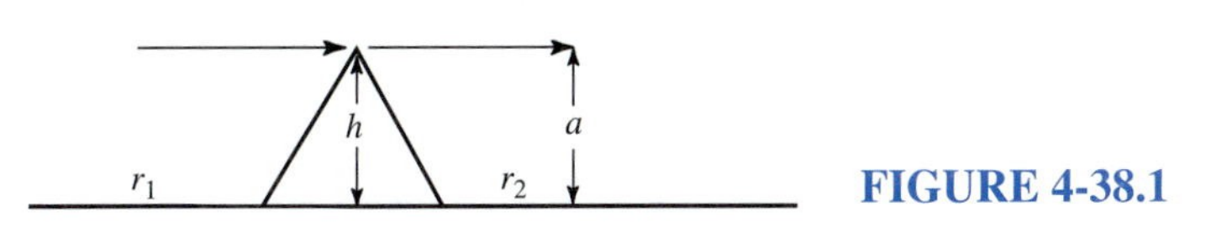

FIGURE 4-38.1

Equation (13) gives the relative power density for large $\kappa a(>3)$ (well into the shadow region). For this condition it is apparent that the power flux density (Poynting vector) due to diffraction increases with wavelength and with distance r (below edge) but decreases as the square of the distance a into the shadow region.

Referring to Fig. 4-38.1, a more general approximation ($r \gg h$) is

$$S = \frac{\lambda}{4\pi^2 h^2}\left(\frac{r_1 r_2}{r_1 + r_2}\right) \tag{14}$$

When $r_1 \gg r_2$, (14) reduces to (13). For the case $r_1 = r_2$, see Problem 4-16-2.

Another illustration of Huygens' principle occurs in *holography*. In an ordinary photograph only amplitude information is recorded. In a *hologram* both amplitude information and phase information are recorded. Thus, when the hologram is illuminated with coherent light, it generates waves (in both amplitude and phase) which, in accord with Huygens' principle, produce a three-dimensional picture.

Example 4-26. Radio wave diffraction over a mountain. Referring to Fig. 4-39, the direct path between a transmitting antenna and a receiving antenna is blocked by a mountain. Assuming that the ridge of the mountain acts as a knife edge, calculate the level of the signal diffracted over the mountain with respect to the direct path level at a frequency of 30 MHz. Transmitting and receiving antennas are separated by 10 km. The mountain ridge extends 1 km above the reference baseline (see Fig. 4-39).

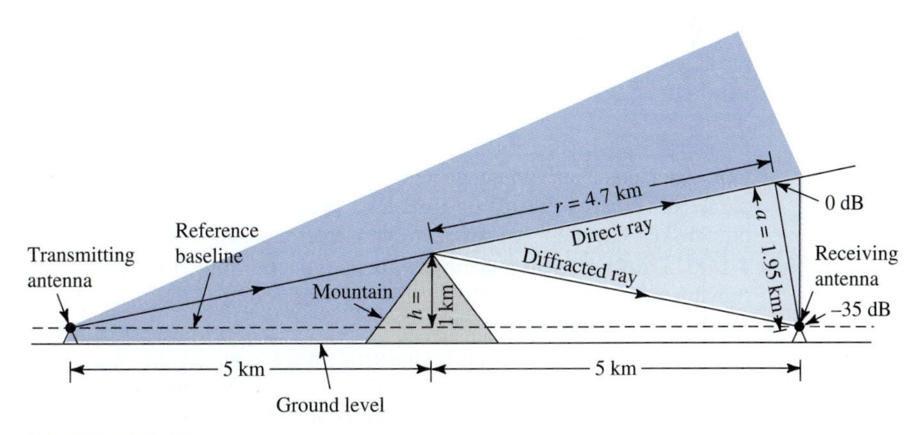

FIGURE 4-39
Radio-wave diffraction over knife-edge obstacle (mountain). The diffracted ray is 35 dB below the direct ray level.

Solution. From geometry the receiving antenna distance r behind the mountain ridge is 4.7 km and the distance a into the shadow region is 1.95 km. From $\lambda = c/f$, the wavelength is 10 m. From (13)

$$S_{av}\,(\text{rel}) = \frac{r\lambda}{4\pi^2 a^2} = \frac{4700 \times 10}{4\pi^2 1950^2} = \frac{1}{3194} \text{ or } -35 \text{ dB} \quad \textit{Ans.}$$

Thus, the mountain causes 35-dB attenuation as compared to a direct path signal. Since an actual mountain ridge may not be knife-edge sharp with respect to dimensions of the order of a wavelength (= 10 m), the actual attenuation may be 10 or 20 dB more (round edges diffract less power into shadow region than sharp edges).

Problem 4-16-1. Flying saucer on demand. When you hold your forefinger close to but not quite touching your thumb and look through the gap at a lamp or bright area, you see what appears to be a saucer suspended in midair (Fig. P4-16-1). Explain the saucer.

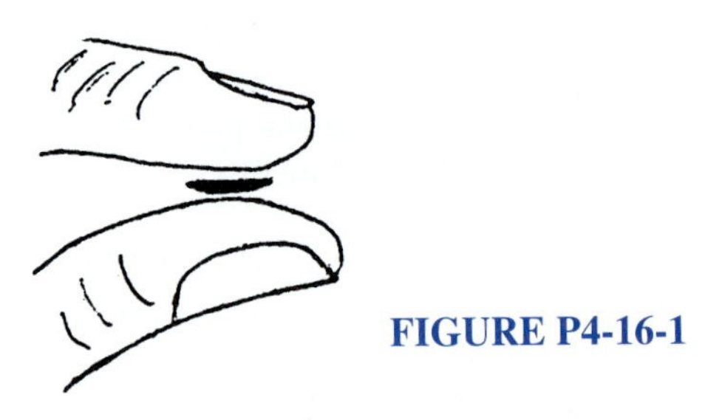

FIGURE P4-16-1

Problem 4-16-2. Alternative mountain formula. For a mountain or object of height h at a horizontal distance r (= $r_1 = r_2$) from both transmitting and receiving antennas (that is, half-way between) (14) reduces to

$$S = \frac{r\lambda}{8\pi^2 h^2}$$

Since $r_1 = r_2$ in Fig. 4-39, this is the better approximation. Using this equation calculate the additional loss with respect to line-of-sight for Example 4-26. *Ans.* −32 dB or 3 dB less. [Equation (14) and the one above were provided by Christopher Haslett, Aircom International, U.K.]

Problem 4-16-3. Transmitter at a large distance. If the transmitter in Fig. 4-39 is 140 km from the mountain, what is the added attenuation compared to a direct line-of-sight path? *Ans.* −29 dB.

Problem 4-16-4. Diffraction. A 3-km-high mountain ridge blocks the direct line-of-sight transmission path from an 85-MHz (Channel 6) TV transmitting antenna on a 150-m tower and a TV receiving antenna on a 15-m mast. The transmitting and receiving sites are 30 km apart and the mountain ridge is half-way between. Find the added attenuation due to the ridge as compared to a direct line-of-sight path. *Ans.* −41 dB.

PROJECTS

Project P4-1. Polarization rotation. Using 10-GHz (3-cm) transmitting and receiving antennas as shown, perform the following tests:

Test 1:

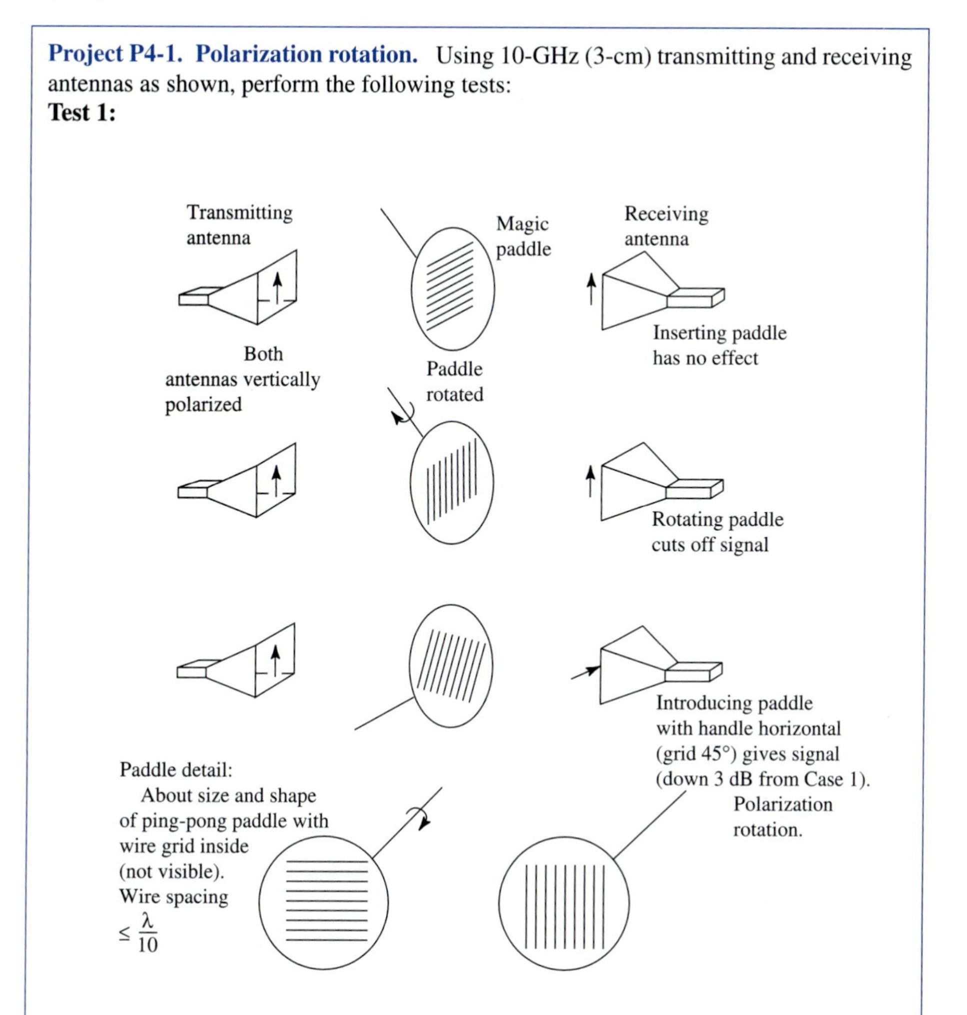

With both antennas vertically polarized (VP), insert the "magic" paddle with handle at 45° so grid wires are horizontal. No effect. Rotate handle so grid wires are vertical. Signal is cut off.

With transmitting antenna VP and receiving antenna horizontally polarized (HP) so signal is zero, insert the paddle with handle horizontal so grid of wires is at 45° and note that the signal is restored, although weaker. The paddle has "rotated" the polarization.

Test 2: With right circularly polarized (RCP) helical transmitting and receiving antennas, note signal strength. Then insert paddle and note that the signal continues regardless of whether the wire grid is vertical, horizontal, or other angle although weaker. The grid "filters" one of the RCP signal's linear components.

(*a*) What is the dB reduction in Test 1 when the vertically polarized wave is rotated completely horizontal? (*b*) What is the reduction in Test 2?

Solution. **Test 1.** The field at 45° is $0.707E$ (vertical) and the relative power $= 0.707^2 = 0.5$, or 3 dB less. With the grid of a second paddle horizontal the field can be rotated another 45°, producing pure horizontal polarization. This adds another 3-dB reduction so that the total reduction = 6 dB. *Ans.* (*a*)

Test 2. In a circularly polarized wave the vertical and horizontal components each carry half the total power. Therefore, the reduction = 3 dB. *Ans.* (*b*)

This project can be expanded by using other objects instead of the grid. Thus, with receiving and transmitting antennas cross polarized, what polarization occurs with a disk, sphere, cube, helix, single wire, etc.? Also note that the fingers of a hand form a satisfactory grid. See Appendix D for equipment used in the project.

Problem P4-1-1. Polarization rotation in 30° step. If Test 1 of Project 4-1 is repeated with a grid at 30° from vertical and a second one horizontal, what is the reduction? *Ans.* 7.3 dB.

Project P4-2. Multipath by reflection and diffraction; absorption. Refer to Appendix D for the equipment components required. Then, referring to Fig. P4-2, cover a table with a flat metal sheet as a ground plane and place the 10-GHz (3-cm wavelength) transmitting antenna at point 1. The antenna at 1 could represent the transmitting antenna on a cell or a TV or FM tower.

Perform the following tests:

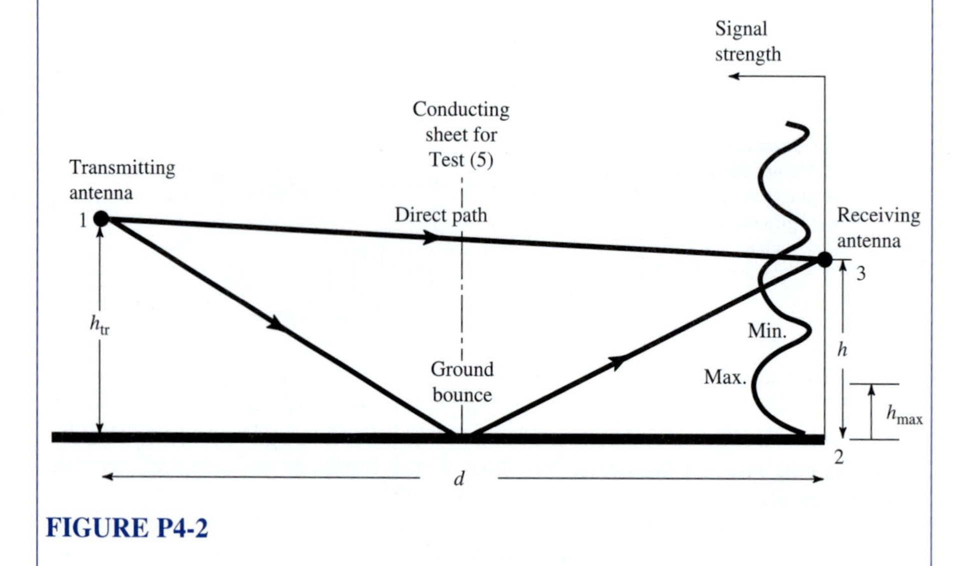

FIGURE P4-2

1. With transmitting antenna horizontally polarized, start at point 2 at table level and move upward with the receiving antenna also horizontally polarized noting the locations of nulls and maxima. Note that the nulls occur when the direct and ground-bounce paths are both an integral number of wavelengths (signals in opposite phase because of phase reversal at the ground bounce) and that the maxima occur when the paths differ by $\frac{1}{2}$ wavelength (signals in phase) (6-dB gain). Compare the

observed height of maxima h_{max} with the formula

$$h_{max} = \frac{d\lambda}{4h_{tr}} \qquad \text{(m)}$$

where d = distance as in Fig. P4-2, m
h_{tr} = height of transmitting antenna, m
λ = wavelength, m

2. Turn both transmitting antennas to vertical polarization and note that nulls and maxima are reversed from those in Test 1.
3. Replace both transmitting and receiving antennas with right-handed (right circularly polarized) helical antennas and note the absence of nulls. This is the big advantage of circular polarization in a complex multipath situation. It may avoid the catastrophic loss of a signal in a null ($-\infty$ dB).
4. Replace the receiving helix with a left-handed helix and note again the absence of nulls but that the signal strength decreases as the height h of the receiving antenna increases.
5. Place a flat conducting sheet vertically at the position of the dashed line in Fig. P4-2 and repeat tests 1 through 4, noting diffraction effects. If the vertical sheet does not make tight metal-to-metal contact with the ground plane, radiation can leak through so that there will be radiation coming to point 3 from both top and bottom edges of the sheet. To eliminate leakage see Problem P4-2-2.
6. Remove the vertical sheet and place sheets of paper, cardboard, cloth, rubber, soil, or other materials on the ground plane and note effects of absorption . If the coefficients of the absorbing materials are known, compare the results with those of computer program "Ground Bounce" in Appendix C.

With the transmitting-receiving equipment of this project there are an almost endless number of tests or experiments which can be performed, limited only by your imagination.

Problem P4-2-1. Derive the formula in Project P4-2. *Hint:* Consider the image of point 1 h_{tr} below the ground plane with opposite phase.

Problem P4-2-2. λ/4 choke. If the vertical sheet in Test 5 of Project P4-2 does not make tight metal-to-metal contact with the ground plane, radiation can leak through the slot-like opening. One way to eliminate this is to bend a $\lambda/4$ lip at the bottom of the sheet forming a $\lambda/4$ choke. See sketch. The doors of microwave ovens may use this feature as an rf seal. Why does it work? Consider the impedance $\lambda/4$ from the open end of a transmission line.

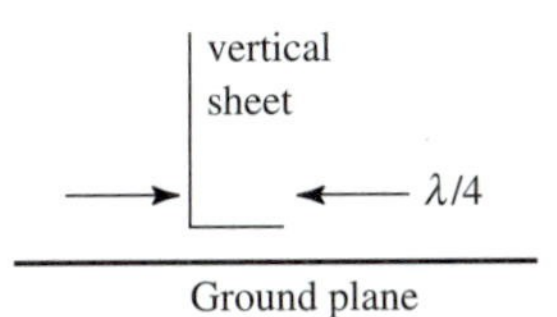

PROBLEMS

Answers are given in Appendix E to problems followed by the @ symbol.

4-2-4. Attenuation in lossy medium. A medium has constants $\sigma = 1.112 \times 10^{-2}$ ℧/m, $\mu_r = 5 - j4$, and $\varepsilon_r = 5 - j2$. At 100 MHz find (*a*) impedance of the medium and (*b*) distance required to attenuate a wave by 20 dB after entering the medium. @

4-2-5. Index of refraction. The measured phase velocity of a dielectric medium is 186 Mm/s at f_1 and 223 Mm/s at f_2. Find: Index of refraction at the two frequencies. @

4-2-6. Field intensity. Find the magnetic field intensity for a TEM wave with electric field intensity of 1 μV/m in (*a*) air, (*b*) a lossless dielectric with $\varepsilon_r = 5$, and (*c*) a lossless dielectric with $\varepsilon_r = 14$. @

4-2-7. Medium impedance. What is the impedance of a medium with $\sigma = 10^{-2}$ ℧/m, $\varepsilon_r = 3$, and $\mu_r = 1$ (*a*) at 1 MHz, (*b*) at 50 MHz, and (*c*) at 1 GHz?

4-2-8. Medium impedance. Find the impedance of a conducting medium with $\sigma = 10^{-4}$ ℧/m, $\mu_r = 1 + j0.5$, and $\varepsilon_r = 12 - j4$ at a frequency of 800 MHz. @

4-3-2. Traveling waves. The normalized voltage on a transmission line resulting from an incident traveling wave to left and a reflected wave to right is given by

$$V = \sin(\omega t + \beta x) + \rho_v \sin(\omega t - \beta x)$$

where $\omega = 2\pi f$ (f = frequency), $\beta = 2\pi/\lambda$. Set $2\pi = \mathrm{B} = 6.2832$, $ft = C$, $x/\lambda = D$, and the reflection coefficient $\rho_v = R$ in TRAVELING WAVES program (see Appendix C). The program then plots the wave at tenth-period intervals over 2λ of line. Thus, *we observe time-lapse pictures* at tenth-period intervals. When $R = 0$, we see clearly the incident wave moving one-tenth wavelength to the left for each tenth of a cycle. Plots for 2λ at 0.1-period intervals are shown in Fig. P4-3-2 (on facing page) for reflection coefficients $\rho_v = R = 0$, 0.5, and 1.0. Obtain plots for (*a*) $R = 1/4$ and (*b*) $R = 3/4$.

4-3-3. Standing wave. A standing wave has a maximum field of 150 μV/m and a minimum field of 30 μV/m. Find (*a*) the VSWR and (*b*) the reflection coefficient for this wave.

4-4-2. Impedance of common media. Using the conductivity and permittivity values of Tables 4-2 and 7-1, find the impedance of (*a*) copper, (*b*) sea water, (*c*) rural ground, (*d*) urban ground, and (*e*) freshwater at 500 MHz.

4-5-2. Lossy medium. Complex constants. A medium has constants $\sigma = 3.34$ ℧/m and $\mu_r = \varepsilon_r = 5 + j2$. Find the $1/e$ depth of penetration at 30 GHz. @

4-5-3. Lossy medium. At 200 MHz a solid ferrite-titanate medium has constants $\sigma = 0$, $\mu_r = 15(1 - j3)$ and $\varepsilon_r = 50(1 - j1)$. Find (*a*) Z/Z_0, (*b*) λ/λ_0, (*c*) v/v_0, (*d*) the $1/e$ depth, (*e*) dB attenuation for a 5-mm thickness, and (*f*) reflection coefficient ρ for a wave in air incident normally on the flat surface of the medium. The zero subscripts refer to parameters for air (or vacuum). @

4-5-4. Wavelengths. Find the ratio of free-space wavelength to the wavelength in a conductor with $\sigma = 10^5$ ℧/m and $\mu_r = 5$ at (*a*) 10 kHz, (*b*) 10 MHz, and (*c*) 10 GHz. Assume constant σ and μ_r with frequency.

4-5-5. $1/e$ and 1 percent depths. Find the $1/e$ and 1 percent depths for a wave with frequency of 1 GHz in a medium with $\sigma = 3 \times 10^4$ ℧/m and $\mu_r = 1$.

4-5-6. Line reflections. A 50-Ω coaxial transmission line has a small discontinuity at $\lambda/2$ intervals. If the reflection coefficient $\rho_0 = 0.01\angle 0°$ at each discontinuity, find the net reflection for a 3λ section of line (*a*) for a pulse which is short compared to the travel

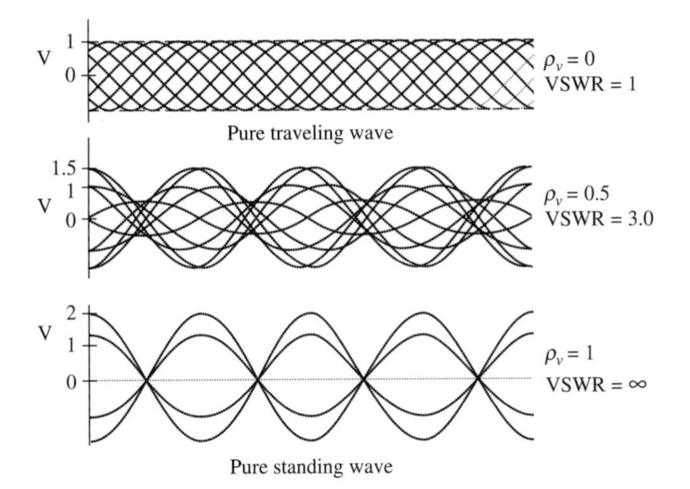

FIGURE P4-3-2
Voltage over 2λ on a transmission line with traveling waves at tenth-period intervals. During the plotting process, the waves move as in a time-lapse film. When $R = 0$, we see the incident wave moving one-tenth wavelength to the left for each tenth of a cycle.

time for $\lambda/2$ and (*b*) for a continuous steady-state wave. The far end of the line is matched. Note that the pulse reflection will consist of a succession of pulses. Assume that the line is lossless. @

4-6-4. Capacitor. Dielectric heating. A parallel-plate capacitor with 10-mm plate separation is filled with a medium for which $\sigma = 10^{-3}$ ℧/m, $\mu_r = 1$, and $\varepsilon_r = 20 - j0.2$. Find the heat lost in the capacitor if 100 V (rms) is applied at 300 MHz. The volume of the capacitor is 200 ml. Neglect fringing.

4-6-5. Capacitor. Dielectric heating. A capacitor consists of two parallel square plates 500 mm on a side and separated by 2 mm. If the medium filling the capacitor has constants $\sigma = 0$, $\mu_r = 1$, and $\varepsilon_r = 15 - j2$, find the power lost as heat in the capacitor at 1 GHz if 100 V (rms) is applied to the plates. Neglect fringing.

4-6-6. Capacitor current and heating. A parallel-plate capacitor 1 m square with 100-mm plate separation is filled with a medium for which $\sigma = 0.005$ ℧/m, $\varepsilon_r = 20 - j10$, and $\mu_r = 1$. If 100 V (rms) at 10 MHz is applied, find (a) total current and (*b*) power lost as heat. Assume a uniform field in the capacitor. Neglect fringing.

4-6-7. Capacitor. Power factor. A parallel-plate capacitor is filled with a dielectric of 0.001 power factor and $\varepsilon_r = 2.5$. The plates are 150 mm square, with separation of 25 mm. If 15 V (rms) at 1 MHz is applied to the capacitor, find the power dissipated as heat. @

4-6-8. Capacitor. Dielectric heating. A parallel-plate capacitor with 1-mm plate separation is filled with a medium for which $\sigma = 10^{-4}$ ℧/m, $\mu_r = 1$, and $\varepsilon_r = 5 - j0.1$. Find the heat loss in the capacitor if 10 V (peak) is applied at 1 GHz. The volume of the capacitor is 10 ml. Neglect fringing.

4-6-9. Capacitor. Dielectric heating. A capacitor consists of two parallel square plates 2 cm on a side and separated by 12 mm. If the medium filling the capacitor has constants $\sigma = 10^{-6}$ ℧/m, $\mu_r = 1$, and $\varepsilon_r = 4 + j1.5$, find the power lost as heat in the capacitor at 10 MHz if 25 V (rms) is applied to the plates. Neglect fringing. @

4-7-11. Matching plate. A plane 9-GHz wave is incident normally on a plane surface of a half-space of material with constants $\mu_r = 1$ and $\varepsilon_r = 12$. Find (*a*) the relative permittivity and (*b*) the thickness of a matching plate which will eliminate reflection of the incident wave. @

4-7-12. Lossy medium. At 350 MHz a solid ferrite-titanate medium has constants $\sigma = 0$, $\mu_r = 8\ (1 - j5)$, and $\varepsilon_r = 37\ (1 - j2)$. Find (*a*) Z/Z_0, (*b*) λ/λ_0, (*c*) v/v_0, (*d*) $1/e$ depth, (*e*) dB attenuation for 3-mm thickness, and (*f*) reflection coefficient for a wave in air normally incident on the flat surface of the medium. The zero subscripts refer to parameters for air (or vacuum).

4-7-13. Attenuation by lossy slab. A nonconducting slab 150 mm thick has constants $\mu_r = \varepsilon_r = 5 - j3$. Find the dB attenuation of the slab to a 1-GHz wave. @

4-7-14. Attenuation by lossy sheet. A nonconducting sheet 2 mm thick has constants $\mu_r = \varepsilon_r = 7 - j2$. Find the dB attenuation of the sheet to a 180-MHz wave.

4-7-15. Attenuation by lossy sheet. A nonmagnetic conducting sheet 5 mm thick has a conductivity $\sigma = 2 \times 10^2$ ℧/m. Find the dB attenuation of the sheet to a 450-MHz wave. @

4-7-16. Attenuation in lossy medium. A medium has constants $\sigma = 4.5 \times 10^{-3}$℧/m, $\mu_r = 2 - j3$, and $\varepsilon_r = 3 - j1$. At 350 MHz find (*a*) the impedance of the medium and (*b*) the distance required to attenuate a wave by 20 dB after entering the medium.

4-7-17. Reflection from dielectric medium. A plane 800-MHz wave is incident normally from air onto a half-space of dielectric with constants $\sigma = 0$, $\mu_r = 1$, and $\varepsilon_r = 5 - j7$. Find the dB reduction of the reflected wave. @

4-7-18. Absorbing sheet. A large flat sheet of a ferrous nonconducting dielectric medium is backed by aluminum foil. At 500 MHz the constants of the medium are $\mu_r = \varepsilon_r = 6 - j6$. How thick must the sheet be for a 500-MHz wave (in air) incident on the sheet to be reduced upon reflection by 30 dB if the wave is incident normally? @

4-7-19. Reflection from dielectric medium. A plane 3-GHz wave is incident normally from air onto a half-space of dielectric with constants $\sigma = 0$, $\mu_r = 1$, and $\varepsilon_r = 2 - j2$. Find the dB value of the reflected power. @

4-7-20. Conducting slab. A slab 3 mm thick with constants $\sigma = 1.5 \times 10^{-3}$ ℧/m and $\mu_r = \varepsilon_r = 1$ is backed by a perfectly conducting metal plate. (*a*) Find the reflection coefficient for a 900-MHz plane wave incident normally on the slab. (*b*) What is the decibel attenuation?

4-7-21. Lossy medium. A slab with constants $\sigma = 10^{-2}$ ℧/m, $\mu_r = 5 - j3$, and $\varepsilon_r = 5 + j4$ is backed by a perfectly conducting metal plate. For a 125-MHz wave at normal incidence, how thick must the slab be to reduce the reflected wave by 35 dB? @

4-7-22. Poynting vector. A plane wave is traveling in a medium for which $\sigma = 0$, $\mu_r = 1$, and $\varepsilon_r = 3$. If E(peak) = 5 V/m, find (*a*) peak Poynting vector, (*b*) average Poynting vector, (*c*) peak value of H, (*d*) the phase velocity, and (*e*) the impedance Z of the medium.

4-7-23. Poynting vector. A plane 200-MHz wave is traveling in a medium for which $\sigma = 0$, $\mu_r = 2$, and $\varepsilon_r = 4$. If the average Poynting vector is 5 W/m^2 find (*a*) rms E; (*b*) rms H; (*c*) phase velocity; and (*d*) impedance of the medium. @

4-8-3. Phase velocity. What is the relative permittivity of a nonferrous medium for which the phase velocity is 150 Mm/sec? @

4-8-4. Ice. Permittivity as a function of temperature and frequency. The expressions for ε_r' and ε_r'' provided in Figure 4-21 are the Drude-Debye relations. These relations are applicable from DC to 10 GHz and from 0 to $-70°$C (E. R. Pounder, "Physics of Ice," Pergamon Press, New York, 1965). Calculate the complex relative permittivity of water ice at 0°C, $-35°$C, and $-70°$C for several frequencies between 10 Hz and 10 GHz, and compare with the curves of Fig. 4-21.

4-9-3. Group velocity. A 10-GHz plane wave traveling in a normally dispersive, lossless medium has a phase velocity at this frequency of 300 Mm/s. The phase velocity varies with wavelength as $v = k\lambda^{3/4}$. Find the group velocity. @

4-10-2. Poynting vector. A plane traveling wave has a peak electric field $E_0 = 15$ V m^{-1}. If the medium is lossless with $\mu_r = 1$ and $\varepsilon_r = 12$, find (*a*) velocity of the wave, (*b*) peak Poynting vector, and (*c*) impedance of medium. @

4-10-3. Poynting vector. A plane traveling 800-MHz wave has an average Poynting vector of 8 mW/m^2. If the medium is lossless with $\mu_r = 1.5$ and $\varepsilon_r = 6$, find (*a*) velocity of wave, (*b*) wavelength, (*c*) impedance of medium, (*d*) rms electric field E, and (*e*) rms magnetic field H.

4-10-4. Poynting vector. A plane wave propagating in free space has a peak electric field of 750 mV/m. Find the average power through a square area 120 cm on a side perpendicular to the direction of propagation. @

4-10-5. Energy density. Find the energy density in a plane traveling wave with electric field intensity $E = 5$ V/m in a nonmagnetic medium with impedance $Z = 100\ \Omega$.

4-10-6. Solar power on Mercury. The planet Mercury receives approximately 12.5 gcal/min/cm^2 of sunlight. (*a*) What is the Poynting vector in watts per square meter? (*b*) What is the power output of the sun in sunlight assuming that the sun radiates isotropically? (*c*) What is the rms electric field at Mercury assuming that the sunlight is all at a single frequency? (*d*) How long does it take the sunlight to reach Mercury? Take the Mercury-sun distance as 60 Gm (1 W = 14.3 gcal/min). @

4-11-1. Polarization ellipse. Draw the polarization ellipse for a plane wave with electric field components $E_1 = 5$ V/m and $E_2 = 3$ V/m if $\delta = 20°$. What is the axial ratio in this case? @

4-12-2. CP wave power. A circularly polarized wave in a medium with constants $\sigma = 0$, $\mu_r = 1$, and $\varepsilon_r = 12$ has an E-field amplitude of 100 mV/m (peak). Find the average power conveyed through a circular area of l-m radius normal to the propagation direction. @

4-13-3. HP wave. A wave traveling in the $+y$ direction is given by $\mathbf{H} = \hat{\mathbf{z}}60\sin(\omega t - \beta y)$ μA(rms)/m. Find (*a*) axial ratio AR (*b*) tilt angle τ, (*c*) $\mathbf{E}$, and (*d*) average Poynting vector $\mathbf{S}_{av}$. @

4-13-4. CP waves. A wave traveling in the $+x$ direction has two components given by $\mathbf{E}_1 = \hat{\mathbf{z}}16\cos(\omega t - \beta x)$ and $\mathbf{E}_2 = \hat{\mathbf{y}}16\sin(\omega t - \beta x)$ mV(rms)/m. For the resultant wave, find (*a*) axial ratio AR, (*b*) $\mathbf{E}$, (*c*) $\mathbf{H}$, and (*d*) average Poynting vector $\mathbf{S}_{av}$. (*e*) Is wave left or right-handed? @

4-13-5. Matching factor. Find the matching factor F for a linear vertically polarized antenna to a wave of (*a*) vertical polarization, (*b*) circular polarization, (*c*) horizontal polarization, (*d*) 45° linear polarization and (*e*) elliptical polarization with vertical ellipse axis half the horizontal axis. @

4-13-6. Poincaré sphere. Figure P4-13-6 is a flat projection of the Poincaré sphere. Identify the polarization states at points 1 through 9.

4-13-7. Polarization matching factors. Find the polarization matching factors to HLP at each of the 9 points on the Poincaré sphere in Fig. P4-13-6. @

4-14-4. Angles of reflection and transmission. A plane wave is incident from air onto a medium with $\varepsilon_r = 5$ at an angle of 30°. Find (*a*) the angle of reflection and (*b*) the angle of transmission. (*c*) Repeat with the materials interchanged. @

4-14-5. Reflection coefficient, perpendicular polarization. Find the reflection coefficient for a plane wave with polarization perpendicular to the plane of incidence from air onto a medium with permittivity $\varepsilon_r = 5$ at an angle of 30°. @

4-14-6. Critical angle. If the media Problem 4-14-5 were interchanged, find the critical angle at which total internal reflection occurs.

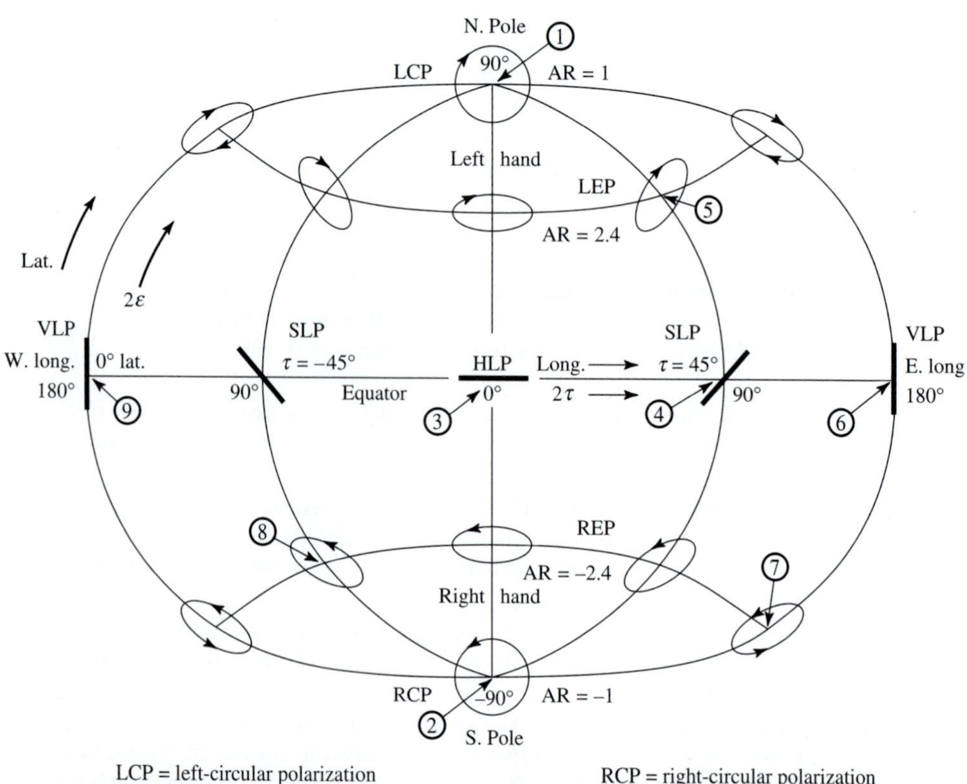

FIGURE P4-13-6

4-14-7. Reflection coefficient, parallel polarization. Repeat Problem 4-14-5 for parallel polarization.

4-14-8. Brewster angle. Find the Brewster angle for the conditions of Problem 4-14-7.

4-14-9. Ground-bounce range reflection coefficient. Many outdoor radar cross-section (RCS) measurement ranges use reflections from the ground to increase the signal at the object under test. A typical configuration is shown in Fig. P4-14-9. The following values of ε' and ε'' for the soil at 1.5 GHz as a function of moisture content were measured at the University of New Mexico RCS range. (F. Bush, "Evaluation of Ground-bounce Range," Report No. PE00911, Physical Sciences Laboratory, New Mexico State University, August 1978.)

Soil Moisture %	ε'	ε''
0	3.2	0.0
5	3.5	0.8
10	4.6	1.7
15	6.4	2.5
25	11.5	4.4

Find the reflection coefficient at each moisture level for parallel (horizontal) and perpendicular (vertical) polarization. Assume a range distance $d = 2.3$ km, antenna height $h_a = 15$ m, object height $h = 25$ m, and frequency of 1.5 GHz.

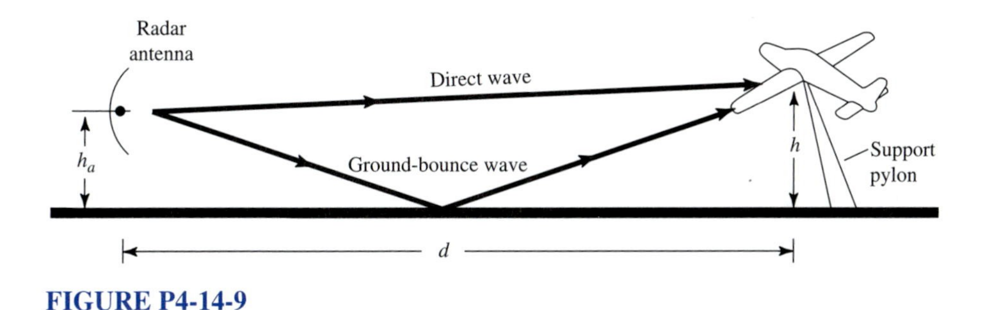

FIGURE P4-14-9

4-15-3. Circular polarization at radar range. Find the axial ratio of the ground-reflected signal for the conditions of Problem 4-14-9 if right circularly polarized waves are transmitted.

4-16-5. Over-mountain TV. A TV receiving antenna is 25 km from a 60-MHz transmitter. A 2-km-high mountain ridge is situated halfway between. The receiving antenna is at a height of 10 m and the transmitting antenna at a height of 200 m. Find the added attenuation caused by the mountain as compared to the signal level without the mountain. @

4-16-6. Diffraction ridge. To reduce the illumination of the base of the support pylon used in RCS measurements, a "diffraction ridge" such as that shown in Fig. P4-16-6 is used at some facilities. For a ridge height $h_r = 10$ m, range-to-ridge distance $r_1 = 2$ km, and range-to-object distance $r_2 = 2.5$ km, find the relative power density at the base of the pylon at (*a*) 18 GHz, (*b*) 5 GHz, and (*c*) 500 MHz. $h_a = 0.4$ m at 18 GHz, 1.4 m at 5 GHz, and 14.4 m at 500 MHz. @

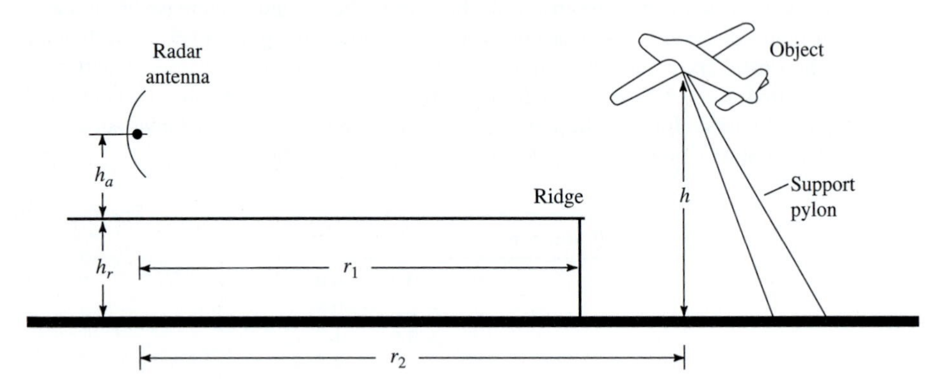

FIGURE P4-14-6

CHAPTER

5

ANTENNAS, RADIATION, AND WIRELESS SYSTEMS

5-1 INTRODUCTION

In Chapters 3 and 4 we discussed transmission lines for guiding electromagnetic energy with a minimum of radiation. Now, we consider antennas which are designed to radiate (or receive) energy as effectively as possible. Antennas act as the transition between space and circuitry. They convert photons to electrons or vice versa.†

We begin this chapter with a discussion of the basic antenna parameters. Next comes antenna array theory. Then the short dipole and the $\lambda/2$ dipole are considered. This is followed by a description of the principal antenna types: dipoles, slots, patches, loops, Yagis, helixes, horns, spirals, log-periodics, dishes, and arrays. Finally, there are discussions of frequency-selective surfaces, atmospheric absorption, wireless links, radar (pulse and doppler), antenna temperature, remote sensing, communication satellites, and global-positioning satellites (GPS).‡

5-2 BASIC ANTENNA PARAMETERS

Regardless of antenna type, all involve the same basic principle that radiation is produced by accelerated (or decelerated) charge. The *basic equation of radiation*

†A photon is a quantum unit of electromagnetic energy equal to hf, where h = Planck's constant ($= 6.63 \times 10^{-34}$ J s) and f = frequency (Hz).

‡For a much more complete coverage of all types of antennas, see J. D. Kraus, "Antennas," 2d ed., McGraw-Hill, New York, 1988.

may be expressed simply as

$$\dot{I}L = Q\dot{v} \qquad (\mathrm{A\ m\ s^{-1}}) \qquad \textbf{\textit{Basic radiation equation}} \tag{1}$$

where $\dot{I}$ = time-changing current, A s^{-1}
L = length of current element, m
Q = charge, C
$\dot{v}$ = time change of velocity which equals the acceleration of the charge, m s^{-2}

Thus, time-changing current radiates and accelerated charge radiates. For steady-state harmonic variation, we usually focus on current. For transients or pulses, we focus on charge.† The radiation is perpendicular to the acceleration, and the radiated power is proportional to the square of $\dot{I}L$ or $Q\dot{v}$.

A two-wire transmission line is shown in Fig. 5-1, connected to a radio-frequency generator (or transmitter). Along the uniform part of the line, energy is guided as a plane Transverse ElectroMagnetic Mode (TEM) wave with little loss. The spacing between wires is assumed to be a small fraction of a wavelength. Further on, the transmission line opens out in a tapered transition. As the separation approaches the order of a wavelength or more, the wave tends to be radiated so that the opened-out line acts like an antenna which launches a free-space wave. The currents on the transmission line flow out on the antenna and end there, but the fields associated with them keep on going.

The transmitting antenna in Fig. 5-1*a* is a region of transition from a guided wave on a transmission line to a free-space wave. The receiving antenna (Fig. 5-1*b*) is a region of transition from a space wave to a guided wave on a transmission line. Thus, *an antenna is a transition device, or transducer, between a guided wave and a free-space wave, or vice versa.* The antenna is a device which interfaces a circuit and space.

From the circuit point-of-view, the antennas appear to the transmission lines as a resistance R_r, called the *radiation resistance.* It is not related to any resistance in the antenna itself but is a resistance coupled from space to the antenna terminals.

In the transmitting case, the radiated power is absorbed by objects at a distance: trees, buildings, the ground, the sky, and other antennas. In the receiving case, passive radiation from distant objects or active radiation from other antennas raises the apparent temperature of R_r. For lossless antennas this temperature has nothing to do with the physical temperature of the antenna itself but is related to the temperature of distant objects that the antenna is "looking at," as suggested in Fig. 5-1.1. In this sense, a receiving antenna (and its associated receiver) may be regarded as a remote-sensing, temperature-measuring device (Sec. 5-11).

As pictured schematically in Fig. 5-1.1, the radiation resistance R_r may be thought of as a "virtual" resistance that does not exist physically but is a quantity

†A pulse radiates with a broad bandwidth (the shorter the pulse, the broader the bandwidth). A sinusoidal variation results in a narrow bandwidth (theoretically zero at the frequency of the sinusoid if it continues indefinitely).

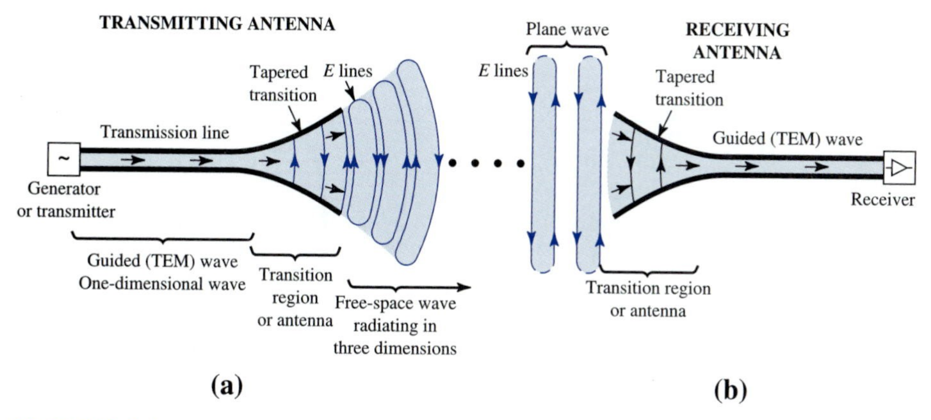

FIGURE 5-1
(a) Radio (or wireless) communication link with transmitting antenna and (b) receiving antenna. The receiving antenna is remote from the transmitting antenna so that the spherical wave radiated by the transmitting antenna arrives as an essentially plane wave at the receiving antenna.

coupling the transmission line terminals to distant regions of space via a "virtual" transmission line.

Both the radiation resistance R_r and its temperature T_A are simple scalar quantities. The radiation patterns, on the other hand, are three-dimensional quantities involving the variation of field or power (proportion to the field squared) as a function of the spherical coordinates θ and ϕ.† Figure 5-2a on page 250 shows a

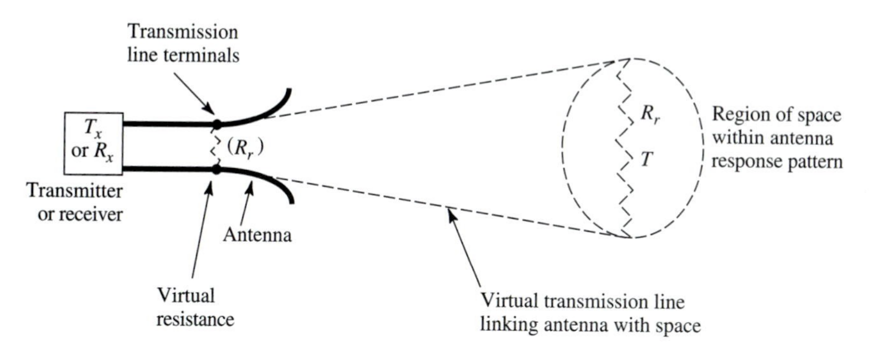

FIGURE 5-1.1
Schematic representation of region of space at temperature T linked via a "virtual" transmission line and antenna to the terminals of a transmission line.

†It is to be noted that the radiation resistance, the antenna temperature, and the radiation patterns are functions of the frequency. In general, the patterns are also functions of the distance at which they are measured, but at distances which are large compared to the size of the antenna and large compared to the wavelength, the pattern is independent of distance. Usually the patterns of interest are for this far-field condition (see Sec. 5-14).

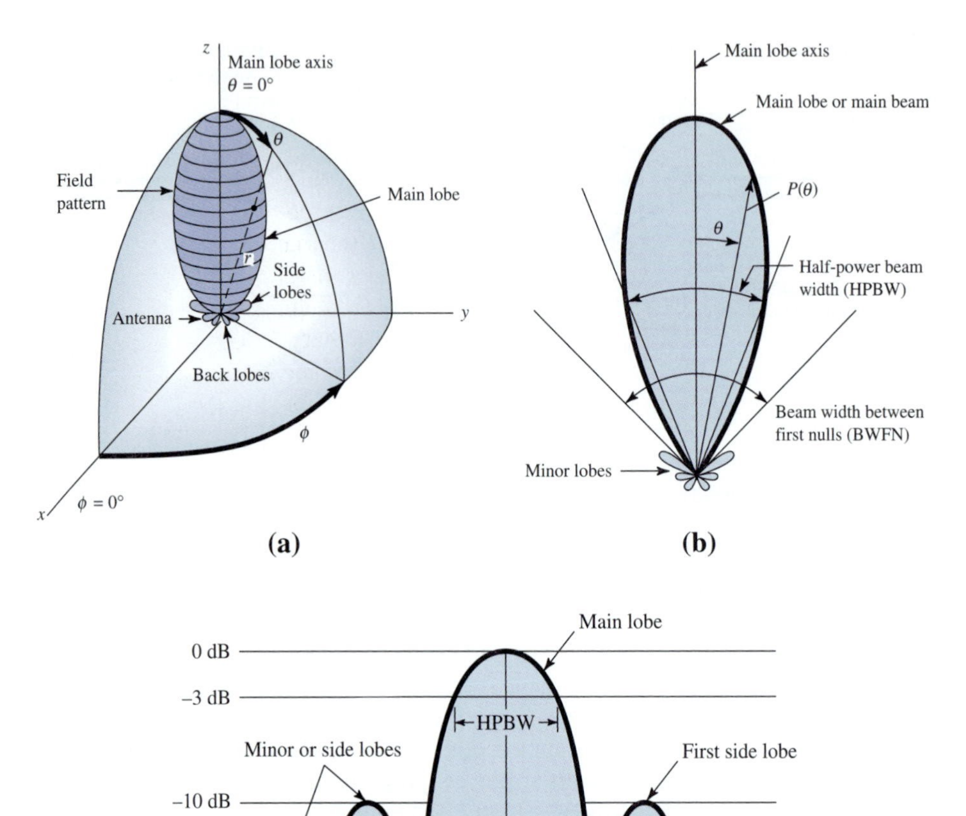

FIGURE 5-2
(*a*) Antenna field pattern with coordinate system. (*b*) Antenna power pattern in polar coordinates (linear scale). (*c*) Antenna pattern in rectangular coordinates and decibel (logarithmic) scale. Patterns (*b*) and (*c*) are the same.

three-dimensional field pattern with pattern radius r (from origin to pattern boundary at the dot) proportional to the field intensity in the direction θ, ϕ. The pattern has its *main lobe* (maximum radiation) in the z direction ($\theta = 0$) with *minor lobes* (side and back) in other directions.

To completely specify the radiation pattern with respect to field intensity and polarization requires three patterns:

1. The θ component of the electric field as a function of the angles θ and ϕ or $E_\theta(\theta, \phi)$ (V m^{-1}).

2. The ϕ component of the electric field as a function of the angles θ and ϕ or $E_\phi(\theta, \phi)$ (V m^{-1}).
3. The phases of these fields as a function of the angles θ and ϕ or $\delta_\theta(\theta, \phi)$ and $\delta_\phi(\theta, \phi)$ (rad or deg).

Any of these field patterns can be presented in three-dimensional spherical coordinates, as in Fig. 5-2*a*, or by plane cuts through the main-lobe axis. Two such cuts at right angles, called the *principal plane patterns* (as in the xz and yz planes in Fig. 5-2*a*) may suffice and if the pattern is symmetrical around the z axis, one cut is sufficient. For a linearly polarized antenna, it is customary for one principal plane pattern to be in the **E** plane and the other pattern in the plane perpendicular to **E**.

Figure 5-2*b* is a principal plane pattern in polar coordinates, and, to show the minor lobes in more detail, the same pattern is presented in Fig. 5-2*c* in rectangular coordinates on a logarithmic, or decibel, scale. If the pattern is symmetrical, the three-dimensional pattern is a figure of revolution of Fig. 5-2*b* around the main-lobe axis similar to the pattern in Fig. 5-2*a*.

Although the radiation pattern characteristics of an antenna involve three-dimensional vector fields for a full representation, there are several simple single-valued scalar quantities which can provide the pattern information required for most engineering applications. Three of these are the *beam area* Ω_A, the *directivity* D (or gain G), and the *effective aperture* A_e.†

Another way of describing the pattern of an antenna is in terms of the angular width of the main beam at a particular level. Referring to Fig. 5-2*b*, the angular beam width at the half-power level or *half-power beam width* (HPBW) (also called the 3-dB beam width) is the one usually given. The *beam width between first nulls* (BWFN) is also sometimes used. ‡

Dividing a field component by its maximum value, we obtain a *normalized field pattern* which is a dimensionless number with maximum value of unity. Thus, the normalized field pattern for the θ component of the dielectric field is given by

$$E_\theta(\theta, \phi)_n = \frac{E_\theta(\theta, \phi)}{E_\theta(\theta, \phi)_{\max}} \qquad \text{(dimensionless)} \qquad (2)$$

The half-power level occurs at those angles θ and ϕ for which $E_\theta(\theta, \phi)_n = 1/\sqrt{2}$.

If an antenna is linearly polarized with only an E_θ component, then the variation of E_θ as a function of the angle θ and ϕ, or $E_\theta(\theta, \phi)$ suffices to describe the pattern. In general, however, both E_θ and E_ϕ components may be present (antenna elliptically or circularly polarized) so that we need to know $E_\theta(\theta, \phi)$, $E_\phi(\theta, \phi)$, $\delta_\theta(\theta, \phi)$, and $\delta_\phi(\theta, \phi)$. A practical and useful simplification of this pattern information is to deal with the *power pattern* $P(\theta, \phi)$ expressed as the radial component of the average Poynting vector $S_r(\theta, \phi)$ multiplied by the square of the distance r at which it is

---

†Distinguish carefully between Ω_A for beam solid angle or beam area and Ω for ohms.

‡In patterns for which no clearly defined minimum exists, the extent of the main lobe may be somewhat indefinite and an arbitrary low level used to delineate it.

measured. Thus, regardless of polarization,

$$P(\theta, \phi) = \frac{E_\theta^2(\theta, \phi) + E_\phi^2(\theta, \phi)}{Z_0} r^2 = S_r(\theta, \phi) r^2 \qquad (\text{W sr}^{-1}) \tag{3}$$

where $P(\theta, \phi)$ = power per steradian, W sr^{-1}
$E_\theta(\theta, \phi)$ = θ component of E, V(rms) m^{-1}
$E_\phi(\theta, \phi)$ = ϕ component of E, V(rms) m^{-1}
Z_0 = intrinsic impedance of space, 377 Ω
$S_r(\theta, \phi)$ = radial component of Poynting vector at distance r, W m^{-2}
r = distance from antenna to point of measurement, m

The quantity $P(\theta, \phi)$ has the dimensions of power per solid angle (W sr^{-1}) and is often called the *radiation intensity.*

Dividing $P(\theta, \phi)$ by its maximum value, we obtain the *normalized power pattern* with maximum value unity. Thus,

$$P_n(\theta, \phi) = \frac{P(\theta, \phi)}{P(\theta, \phi)_{max}} \qquad (\text{dimensionless}) \tag{4}$$

The arc of a circle as seen from its center subtends an angle. Thus, in Fig. 5-3*a* the arc length θr subtends the angle θ. The total angle in the circle is 2π (= 360°), and the total arc length is $2\pi r$ (= circumference).

An area A of the surface of a sphere as seen from the center of the sphere subtends a solid angle Ω (Fig. 5-3*b*). The total solid angle or area subtended by the sphere is 4π steradians (or square radians),† abbreviated sr, and the sphere's total area = $4\pi r^2$ (m^2).

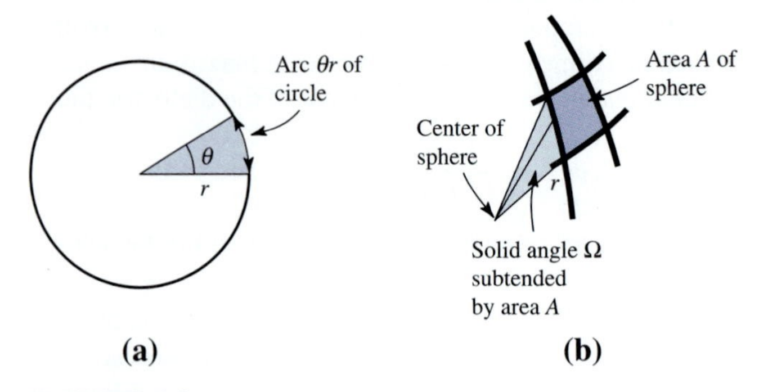

FIGURE 5-3
(*a*) Arc length $r\theta$ of circle of radius r subtends an angle θ. (*b*) The area A of a sphere of radius r subtends a solid angle Ω.

$$\dagger 4\pi \text{ sr} = 4\pi \times 1 \text{ rad}^2 = 4\pi \left(\frac{180}{\pi}\right)^2 (\text{deg}^2) = 4\pi \times 3282.8064 (\text{deg}^2)$$
$$= 41252.96 \approx 41253 \text{ square degrees}$$
$$= \text{solid angle in sphere}$$

Now the *beam area* (or *beam solid angle*) Ω_A of an antenna (Fig. 5-4 on page 254) is given by the integral of the normalized power pattern over a sphere (4π sr), or

$$\Omega_A = \int_{\phi=0}^{\phi=2\pi} \int_{\theta=0}^{\theta=\pi} P_n(\theta, \phi) \sin\theta d\theta d\phi \tag{5a}$$

and

$$\Omega_A = \iint_{4\pi} P_n(\theta, \phi) d\Omega \qquad \text{(sr)} \qquad \textbf{Beam area} \tag{5b}$$

where $d\Omega = \sin\theta d\theta d\phi$, sr.

The beam area Ω_A is the solid angle through which all of the power radiated by the antenna would stream if $P(\theta, \phi)$ maintained its maximum value over Ω_A and was zero elsewhere. Thus, the *power radiated* $= P(\theta, \phi)_{\max}\Omega_A$ watts.

The beam area can sometimes be usefully *approximated* by the product of the half-power beamwidths in the two principal planes. Thus

$$\Omega_A \approx \theta_{\text{HP}}\phi_{\text{HP}} \qquad \text{(sr)} \tag{6}$$

where θ_{HP} and ϕ_{HP} are the half-power beam widths (HPBW) in the two principal planes, minor lobes being neglected.

It is sometimes convenient to separate the Ω_A into the component area Ω_M (due to the main lobe) and the component area Ω_m (due to the minor lobes). Thus,

$$\Omega_A = \Omega_M + \Omega_m$$

This leads to the definition of *main beam efficiency* as

$$\varepsilon_M = \frac{\Omega_M}{\Omega_A} \qquad \textbf{Main beam efficiency} \tag{6.1}$$

The *directivity* D of an antenna is given by the ratio of the maximum power density to its average value over a sphere. The power density may be expressed in terms of the Poynting vector $S(\theta, \phi)$ (W m^{-2}) or of the power per steradian (radiation intensity) $P(\theta, \phi)$ (W sr^{-1}). Thus,

$$D = \frac{S(\theta, \phi)_{\max}}{S(\theta, \phi)_{\text{av}}} = \frac{P(\theta, \phi)_{\max}}{P(\theta, \phi)_{\text{av}}} \tag{7}$$

with the power density measured in the far field. The directivity is a dimensionless ratio ≥ 1.

The average power density over a sphere is given by

$$P(\theta, \phi)_{\text{av}} = \frac{1}{4\pi}\int_{\phi=0}^{\phi=2\pi} \int_{\theta=0}^{\theta=\pi} P(\theta, \phi) \sin\theta d\theta d\phi$$

$$= \frac{1}{4\pi}\iint_{4\pi} P(\theta, \phi)\, d\Omega \qquad (\text{W sr}^{-1}) \tag{8}$$

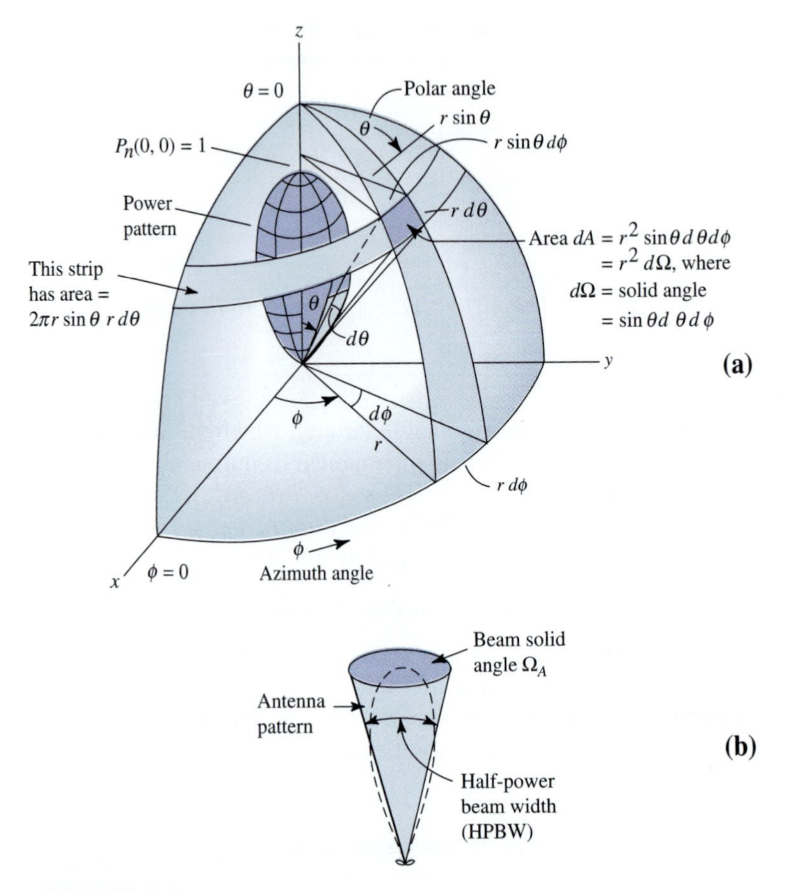

FIGURE 5-4

(a) Normalized antenna power pattern aligned with the $\theta = 0$ direction (zenith) with spherical coordinates and solid angles. (b) Antenna power pattern and its equivalent solid angle or beam area Ω_A.

Therefore,

$$D = \frac{P(\theta,\phi)_{\max}}{(1/4\pi)\iint_{4\pi} P(\theta,\phi)\,d\Omega} = \frac{1}{(1/4\pi)\iint_{4\pi} [P(\theta,\phi)/P(\theta,\phi)_{\max}]\,d\Omega} \tag{9a}$$

and

$$D = \frac{4\pi}{\iint_{4\pi} P_n(\theta,\phi)\,d\Omega} = \frac{4\pi}{\Omega_A} \quad \text{(dimensionless)} \qquad \textbf{\textit{Directivity}} \tag{9b}$$

The smaller the beam area Ω_A, the greater the directivity.

If an antenna were isotropic (radiating the same in all directions), $P_n(\theta, \phi) = 1$ for all θ and ϕ, this would result in a beam area $\Omega_A = 4\pi$ and $D = 1$. Thus, Ω_A must always be equal to or less than 4π and the directivity equal to or greater than 1. Neglecting the effect of minor lobes we have from (6) the simple *approximation*†

$$D \approx \frac{4\pi}{\theta_{HP}\phi_{HP}} \approx \frac{41\,000}{\theta^\circ_{HP}\phi^\circ_{HP}} \qquad \textbf{\textit{Approximate directivity}} \tag{10}$$

where θ_{HP} = half-power beam width in one principal plane, rad
ϕ_{HP} = half-power beam width in other principal plane, rad
θ°_{HP} = half-power beam width in one principal plane, deg
ϕ°_{HP} = half-power beam width in other principal plane, deg

Equation (10) is an *approximation* and should be used in this context.

If an antenna has a main beam with both half-power beam widths (HPBWs) = 20°, its directivity, from (10), is *approximately*

$$D \approx \frac{41\,000}{20^\circ \times 20^\circ} = 103 \text{ or } 20 \text{ dBi} \qquad \text{(dB above isotropic)}$$

which means that the antenna radiates a power density in the direction of the main beam maximum which is about 100 times as much as would be radiated by a nondirectional (isotropic) antenna with the same power input at the same distance.

The directivity is an important quantity, giving us in a simple single-valued scalar number a valuable insight into the antenna's performance. If the antenna is lossless its *gain* G equals the directivity. However, if the antenna is not 100 percent efficient, the gain is less than the directivity. Thus,

$$G = kD \qquad \text{(dimensionless)} \tag{11}$$

where k = efficiency factor of antenna ($0 \le k \le 1$), dimensionless. The efficiency has to do only with ohmic losses in the antenna. In transmitting, these losses involve power fed to the antenna which is not radiated but heats the antenna structure.

†4π sr = 41 253 square degrees. Since (10) is an approximation, 41 253 is rounded off to 41 000.

Example 5-1. Directivity. The normalized field pattern of an antenna is given by $E_n = \sin\theta \sin\phi$, where θ = zenith angle (measured from z axis) and ϕ = azimuth angle (measured from x axis) (Fig. E5-1). E_n has a value only for $0 \le \theta \le \pi$ and $0 \le \phi \le \pi$ and is zero elsewhere (pattern is unidirectional with maximum in $+y$ direction). Find (*a*) the exact directivity, (*b*) the approximate directivity from (10), and (*c*) the decibel difference.

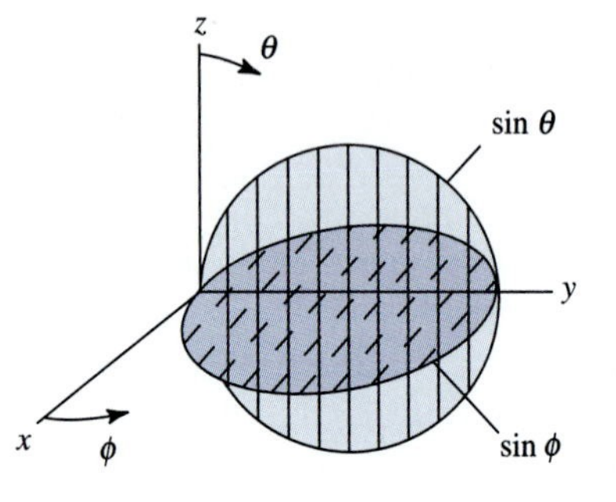

FIGURE E5-1
Unidirectional sin θ and sin ϕ field patterns.

Solution.

$$D = \frac{4\pi}{\int_0^\pi \int_0^\pi \sin^3\theta \sin^2\phi \, d\theta d\phi} = \frac{4\pi}{2\pi/3} = 6 \quad Ans. \quad (a)$$

$$D \approx \frac{41\,000}{90^2} = 5.1 \quad Ans. \quad (b)$$

$$10 \log \frac{6.0}{5.1} = 0.7 \text{ dB} \quad Ans. \quad (c)$$

Problem 5-2-1. Directivities. Calculate the exact directivity of a unidirectional antenna if the normalized power pattern is given by: (*a*) $P_n = \cos\theta$, (*b*) $P_n = \cos^2\theta$, (*c*) $P_n = \cos^3\theta$, and (*d*) $P_n = \cos^n\theta$. In all cases these patterns are unidirectional (in $+z$ direction) with P_n having a value only for zenith angles $0° \le \theta \le 90°$ and with $P_n = 0$ for $90° \le \theta \le 180°$. The patterns are independent of the azimuth angle ϕ. *Ans.* (*a*) 4; (*b*) 6; (*c*) 8; (*d*) $2(n+1)$.

A receiving antenna extracts power from an incident wave and delivers it to a load or terminating impedance Z_L (Fig. 5-5*a*). If the power delivered is P and the Poynting vector of the incident wave is S (W m^{-2}) their ratio is an area A. Thus,†

$$\frac{P \text{ (W)}}{S \text{ (W m}^{-2})} = A \qquad (\text{m}^2) \tag{12}$$

The significance of A is that it is an aperture over which the antenna extracts power from the passing wave. In the circuit of Fig. 5-5*b*, the antenna is replaced by an equivalent or Thévenin generator having a voltage V and internal (antenna) impedance Z_A. The voltage V (induced by the passing wave) produces a current I

†Distinguish between P for power (W) and $P(\theta, \phi)$ for power per steradian (W sr^{-1}).

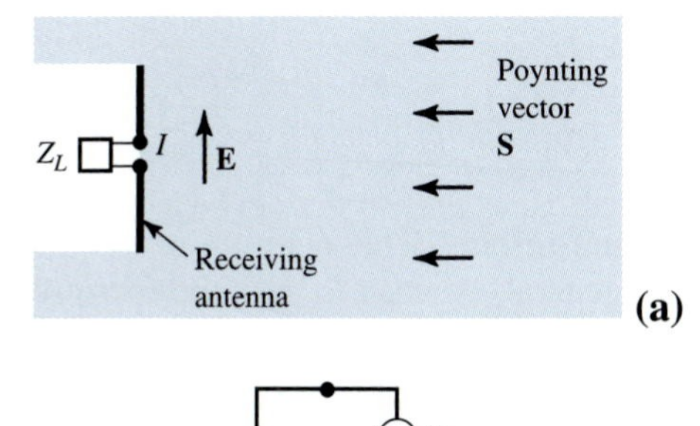

FIGURE 5-5
(*a*) Terminated receiving antenna immersed in field of plane traveling wave. (b) Equivalent circuit.

through the load Z_L given by

$$I = \frac{V}{Z_L + Z_A} \tag{13}$$

where I and V are rms or effective values.

In general,

$$Z_A = R_r + R_l + jX_A \tag{14}$$

where R_r = antenna radiation resistance, Ω
R_l = antenna loss resistance, Ω
X_A = antenna reactance, Ω

When the antenna is transmitting, *the power I^2R_r equals the power radiated* while *the power I^2R_l equals the power lost as heat in the antenna structure*. For the load impedance

$$Z_L = R_L + jX_L \tag{15}$$

Thus, the magnitude of current I is given by

$$I = \frac{V}{\sqrt{(R_r + R_l + R_L)^2 + (X_A + X_L)^2}} \tag{16}$$

and the power delivered to Z_L by

$$P = I^2R_L = \frac{V^2R_L}{(R_r + R_l + R_L)^2 + (X_A + X_L)^2} \tag{17}$$

For a lossless antenna ($R_l = 0$) and also a conjugate match between Z_L and the antenna, $R_L = R_r$ and $X_L = -X_A$, (17) becomes

$$P = \frac{V^2}{4R_r} \tag{18}$$

and the *aperture,* from (12), becomes

$$A = \frac{P}{S} = \frac{V^2}{4SR_r} \qquad (\text{m}^2 \text{ or } \lambda^2) \tag{19}$$

Assuming the antenna is oriented for maximum response, the aperture of (19) is called the *maximum effective aperture* A_{em}. In general, as when $R_l \neq 0$, the aperture is less and is designated simply as the *effective aperture* A_e.

For horn antennas or parabolic dish antennas which have a well-defined physical aperture A_p, we can define the ratio of A_e to A_p as the *aperture efficiency* ε_{ap}. Thus,

$$\varepsilon_{ap} = \frac{A_e}{A_p} \qquad (\text{dimensionless}) \qquad \textit{Aperture efficiency} \tag{20}$$

Consider now that an antenna with effective aperture A_e has the beam solid angle Ω_A, as suggested in Fig. 5-6. If the field E_a is constant over the aperture, the power radiated is

$$P = \frac{E_a^2}{Z_0} A_e \qquad (\text{W}) \tag{21}$$

where Z_0 is the intrinsic impedance of the medium.

Let the field at a radius r be E_r. Then the power radiated is given by

$$P = \frac{E_r^2}{Z_0} r^2 \Omega_A \qquad (\text{W}) \tag{22}$$

Equating (21) and (22) and substituting $E_r = E_a A_e / r\lambda$ yields the important relation

$$\lambda^2 = A_e \Omega_A \qquad (\text{m}^2) \qquad \textit{Aperture-beam area relation} \tag{23}$$

where λ = wavelength, m
A_e = effective aperture, m^2
Ω_A = beam area, sr

According to (23), the product of the effective aperture and the beam area is equal to the wavelength squared. If A_e is known, we can determine Ω_A (or vice versa) at a given wavelength.

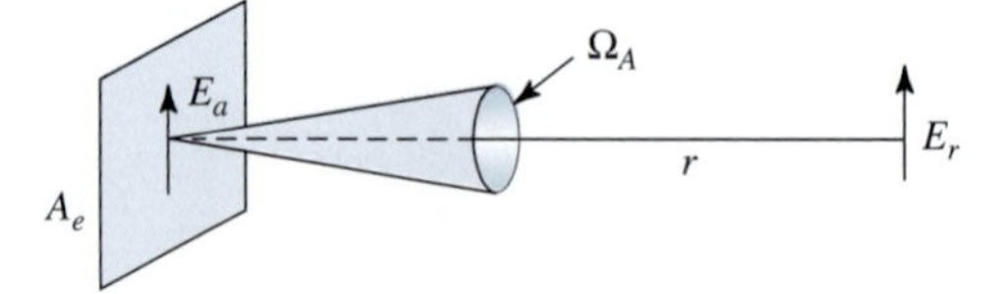

FIGURE 5-6
Radiation from aperture A_e with uniform field E_a across the aperture.

From (23) and (9*b*) it follows that

$$D = 4\pi \frac{A_e}{\lambda^2} \qquad \textbf{\textit{Directivity}} \tag{24}$$

All antennas have an effective aperture, which can be calculated or measured. From (24) we note that even the hypothetical isotropic antenna (for which $D = 1$) has an effective aperture

$$A_e = \frac{D\lambda^2}{4\pi} = \frac{\lambda^2}{4\pi} = 0.0796\lambda^2 \tag{25}$$

All lossless antennas must have an effective aperture equal to or greater than this.

Three expressions have now been given for the *directivity D*. They are

$$D = \frac{P(\theta, \phi)_{\text{max}}}{P(\theta, \phi)_{\text{av}}} \qquad \textbf{\textit{Directivity from pattern}} \tag{26}$$

$$D = \frac{4\pi}{\Omega_A} \qquad \textbf{\textit{Directivity from pattern}} \tag{27}$$

$$D = 4\pi \frac{A_e}{\lambda^2} \qquad \textbf{\textit{Directivity from aperture}} \tag{28}$$

To summarize this section, we have discussed (see Fig. 5-7) the space parameters of an antenna, namely, field and power patterns, beam area, directivity, gain, and effective aperture. The radar cross section will be discussed in Sec. 5-12. We have also discussed the circuit quantity of radiation resistance and alluded to antenna temperature, which is discussed further in Sec. 5-11.

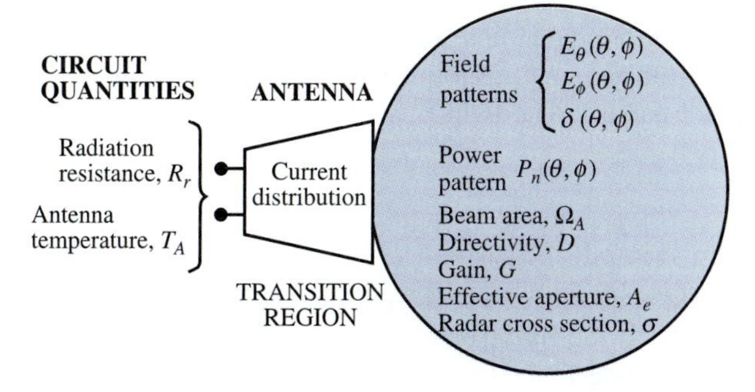

FIGURE 5-7
Schematic diagram of basic antenna parameters, illustrating the duality of an antenna, being a circuit device (with a resistance and temperature) on the one hand and a space device (with radiation patterns, beam angles, directivity, gain, aperture, and radar cross-section) on the other.

Example 5-2. Power. An antenna radiates isotropically over a half-space above a perfectly conducting flat ground plane. If $E = 50$ mV m^{-1} (rms) at a distance of 1 km, find (*a*) power radiated and (*b*) the radiation resistance if the antenna terminal current $I = 3.5$ A.

Solution.

$$P = 2\pi r^2 S = 2\pi(10^3)^2\, 0.05^2/377 = 41.7 \text{ W} \quad \textit{Ans.} \quad (a)$$

$$P = I^2 R \qquad R = P/I^2 = 41.7/3.5^2 = 3.4\ \Omega \quad \textit{Ans.} \quad (b)$$

Example 5-3. Radiation resistance. Find the radiation resistance of an antenna with unidirectional power pattern $P = 8\sin^2\theta\sin^3\phi$ W sr^{-1}, where $0° \le \theta \le 180°$ and $0° \le \phi \le 180°$, if the antenna terminal current is 3 A.

Solution.

$$R = \frac{P\Omega_A}{I^2} = \frac{8\int_0^\pi \sin^3\theta\, d\theta \int_0^\pi \sin^3\phi\, d\phi}{3^2} = \frac{8(4/3)^2}{3^2} = 1.6\ \Omega \quad \textit{Ans.}$$

Problem 5-2-2. Directional pattern in θ and ϕ. An antenna has a uniform field pattern for zenith angles (θ) between 45 and 90° and for azimuth (ϕ) angles between 0 and 120°. If $E = 3$ V/m at distance of 500 m from the antenna and the terminal current is 5 A, find the radiation resistance of the antenna. $E = 0$ except within the angles given above. *Ans.* 354 Ω.

Problem 5-2-3. Directional pattern in θ and ϕ. An antenna has a uniform field $E = 2$ V/m (rms) at distance of 100 m for zenith angles between 30 and 60° and azimuth angle ϕ between 0 and 90° with $E = 0$ elsewhere. The antenna terminal current is 3 A (rms). Find (*a*) directivity, (*b*) effective aperture, and (*c*) radiation resistance. *Ans.* (*a*) 21.9; (*b*) 1.74 λ^2; (*c*) 6.8 Ω.

Problem 5-2-4. Isotropic antenna. Resistance. An omnidirectional (isotropic) antenna has a field pattern given by $E = 10\, I/r$ (V/m), where I = terminal current (A) and r = distance (m). Find the radiation resistance. *Ans.* 3.33 Ω.

5-3 ARRAYS

Much of antenna theory (and almost all of array theory) involves little more than the proper addition of the *field* contributions from all parts of an antenna. We must deal with the field (not power). The field is a vector quantity with both magnitude *and* phase.

Two Isotropic Point Sources

Consider two point sources separated by a distance d as in Fig. 5-8. The sources radiate isotropically in the plane of the page.† (They could represent $\lambda/2$ dipoles perpendicular to the page.)

Let the two sources have the same amplitude and phase. If the reference point for phase is taken halfway between the sources, the far field in the direction θ is

†By reciprocity the pattern of the array for transmitting is the same as the pattern for receiving.

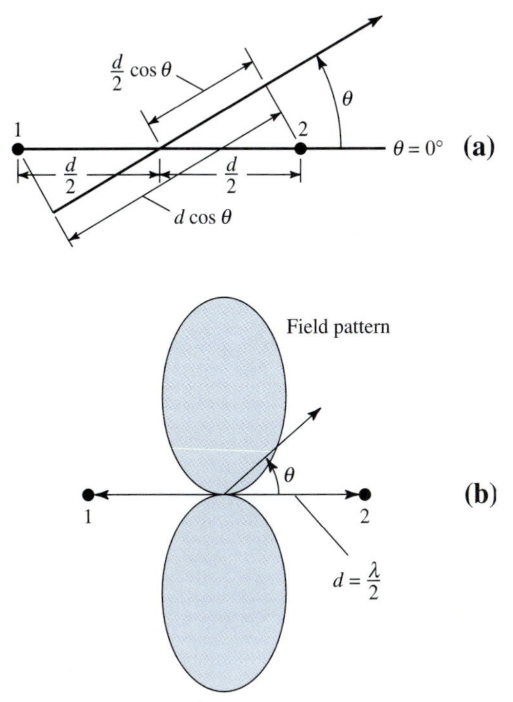

FIGURE 5-8
(a) Two isotropic point sources separated by a distance d. (b) Field pattern when sources are of equal amplitude and same phase with $\lambda/2$ separation.

given by

$$E = E_2 e^{j\psi/2} + E_1 e^{-j\psi/2} \tag{1}$$

where E_1 = electric far field at distance r due to source 1
E_2 = electric far field at distance r due to source 2
$\psi = \beta d \cos\theta = (2\pi d/\lambda)\cos\theta$

The quantity ψ is the phase-angle difference between the fields from the two sources as measured along the radial line at the angle θ (see Fig. 5-8a). When $E_1 = E_2$, we have

$$E = 2E_1 \frac{e^{j\psi/2} + e^{-j\psi/2}}{2} = 2E_1 \cos\frac{\psi}{2} \tag{2}$$

For a spacing of $\lambda/2$, the field pattern is as shown in Fig. 5-8b.

If the reference for phase for the two sources in Fig. 5-8b had been taken at source 1 (instead of midway between the sources), the resultant far-field pattern would be

$$E = E_1 + E_2 e^{j\psi} \tag{3}$$

and for $E_1 = E_2$,

$$E = 2E_1 \cos\frac{\psi}{2} e^{j\psi/2} = 2E_1 \cos\frac{\psi}{2} \angle \psi/2 \tag{4}$$

The field (amplitude) pattern is the same as before, but the phase pattern is not. This is because the reference was taken at the *phase center* (midpoint of array) in developing (2) but at one end of the array in developing (4).

Pattern Multiplication

We assumed above that each point source was isotropic (nondirectional). If the individual point sources have directional patterns which are identical, the resultant pattern is given by (2), where E_1 is now also a function of angle $[E_1 = E(\theta)]$. The pattern $E(\theta)$ may be called the *primary pattern*, and $\cos \psi/2$ the *secondary pattern*, *array pattern*, or *array factor*. This is an example of the *principle of pattern multiplication*, which may be stated more generally as follows: *The total field pattern of an array of nonisotropic but similar sources is the product of the individual source pattern and the pattern of an array of isotropic point sources each located at the phase center of the individual source with the relative amplitude and phase of the source, while the total phase pattern is the sum of the phase patterns of the individual sources and the array of isotropic point sources.*†

Thus:
$$E\ (\text{total}) = \underset{\substack{\text{Primary}\\ \text{pattern}}}{E\ (\text{source})} \times \underset{\substack{\text{Array}\\ \text{pattern}}}{E\ (\text{isotropic})} \tag{4.1}$$

Binomial Array

From (2), the normalized far-field pattern of two identical in-phase isotropic point sources spaced $\lambda/2$ apart is given by

$$E = \cos\left(\frac{\pi}{2}\cos\theta\right) \tag{5}$$

This pattern, shown in Fig. 5-8*b*, has no minor lobes. If a second identical array of two sources is placed $\lambda/2$ from the first, we obtain four sources arranged as in Fig. 5-9*a*. The two sources at the center should be superimposed but are shown separated for clarity. By the principle of pattern multiplication, the resultant pattern is given by

$$E = \cos^2\left(\frac{\pi}{2}\cos\theta\right) \tag{6}$$

as shown in Fig. 5-9*a*. If this three-source array with amplitudes 1:2:1 is arrayed with an identical one at a spacing $\lambda/2$, the source arrangement of Fig. 5-9*b* is obtained with the pattern

$$E = \cos^3\left(\frac{\pi}{2}\cos\theta\right) \tag{7}$$

†It is assumed that the pattern of the individual source is the same when it is in the array as when it is isolated.

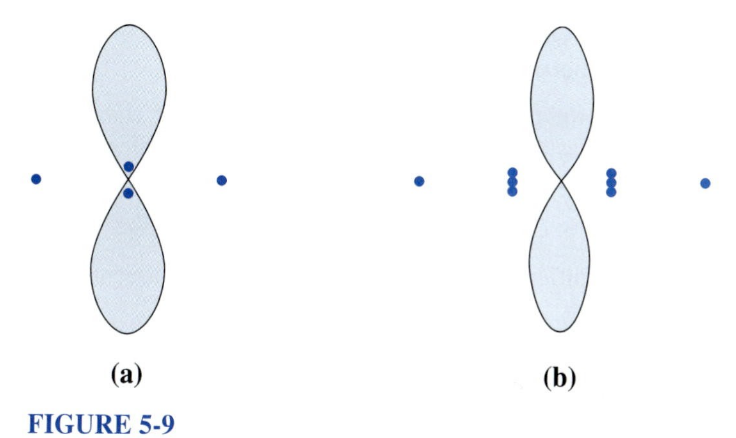

FIGURE 5-9
(*a*) Binomial array with source amplitudes 1:2:1 and (*b*) binomial array with source amplitude 1:3:3:1. The spacing between sources is $\lambda/2$. These arrays have no minor lobes.

as shown in Fig. 5-9*b*. This four-source array has amplitudes 1:3:3:1, and it also has no minor lobes. Continuing this process, it is possible to obtain a pattern with arbitrarily high directivity and no minor lobes if the amplitudes of the sources in the array correspond to the coefficients of a binomial series.† These coefficients are conveniently displayed by Pascal's triangle (Table 5-1). Each internal integer is the sum of the adjacent ones above. The pattern of the array is then

$$E = \cos^{n-1}\left(\frac{\pi}{2}\cos\theta\right) \tag{8}$$

where n = the total number of sources.

Although the above array has no minor lobes, its directivity is less than that of an array of the same size with equal amplitude sources. In practice most arrays are designed as a compromise between these extreme cases (binomial and uniform).

TABLE 5-1
Pascal's triangle

						1						
					1		1					
				1		2		1				
			1		3		3		1			
		1		4		6		4		1		
	1		5		10		10		5		1	
1		6		15		20		15		6		1

†John Stone Stone, U.S. Patents 1,643,323 and 1,715,433.

Linear Arrays of *n* Isotropic Point Sources of Equal Amplitude and Spacings†

Let us consider now the case of n isotropic point sources of equal amplitude and spacing in a linear array (Fig. 5-10) where n is any positive integer. The total field E at a large distance in the direction θ is given by

$$E = 1 + e^{j\psi} + e^{j2\psi} + e^{j3\psi} + \cdots + e^{j(n-1)\psi} \tag{9}$$

where ψ is the phase difference of the field radiated in the θ direction from adjacent sources as given by

$$\psi = \frac{2\pi d}{\lambda}\cos\theta + \delta \tag{10}$$

where d is spacing between sources and δ is the phase difference between adjacent sources, i.e., source 2 with respect to 1, 3 with respect to 2, etc.

The amplitudes of the fields from the sources are all equal and taken as unity. Source 1 (Fig. 5-10) is the phase reference. Thus, at a distant point in the direction θ, the field from source 2 is advanced in phase with respect to source 1 by ψ, the field from source 3 is advanced in phase with respect to source 1 by 2ψ, etc.

Equation (9) is a geometric series (a polynomial of degree $n-1$). Each term represents a phasor, and the amplitude of the total field E and its phase angle ξ can be obtained by phasor (vector) addition as in Fig. 5-11. Analytically, E can be expressed in a simple trigonometric form as follows:

Multiplying (9) by $e^{j\psi}$ gives

$$Ee^{j\psi} = e^{j\psi} + e^{j2\psi} + e^{j3\psi} + \cdots + e^{jn\psi} \tag{11}$$

Subtracting (11) from (9) and dividing by $1 - e^{j\psi}$ yields

$$E = \frac{1 - e^{jn\psi}}{1 - e^{j\psi}} \tag{12}$$

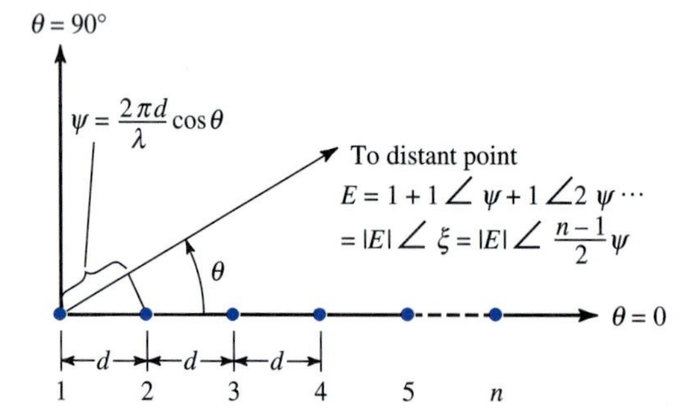

FIGURE 5-10
Linear array of n isotropic in-phase point sources.

†S. A. Schelkunoff, "Electromagnetic Waves," Van Nostrand, New York, 1943, p. 342.; J. A. Stratton, "Electromagnetic Theory," McGraw-Hill, New York, 1941, p. 451.

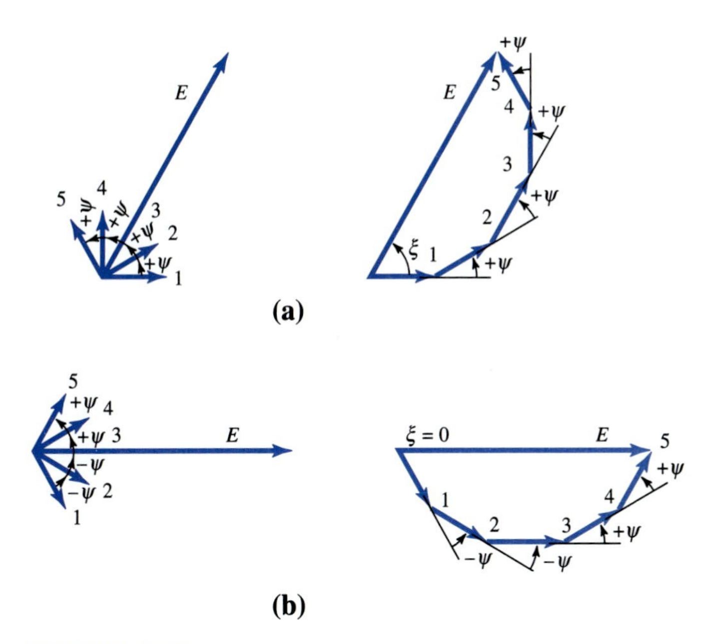

FIGURE 5-11
(*a*) Addition of fields of different phase ψ arriving at a distant point from a linear array of five isotropic sources of equal amplitude with phase of source 1 as reference. (*b*) Same, but with phase of source 3, at midpoint of array, as reference.

Equation (12) may be rewritten as

$$E = \frac{e^{jn\psi/2}}{e^{j\psi/2}}\left(\frac{e^{jn\psi/2} - e^{-jn\psi/2}}{e^{j\psi/2} - e^{-j\psi/2}}\right) \tag{13}$$

from which the total field

$$E = e^{j\xi}\frac{\sin(n\psi/2)}{\sin(\psi/2)} = \frac{\sin(n\psi/2)}{\sin(\psi/2)}\angle\xi \tag{14}$$

where ξ is the phase angle of E referred to the field from source 1.

The value of ξ is given by

$$\xi = \frac{n-1}{2}\psi \tag{15}$$

If the phase is referred to the center point of the array, $\xi = 0$ and (14) becomes

$$E = \frac{\sin(n\psi/2)}{\sin(\psi/2)} \qquad \textbf{\textit{Array pattern}} \tag{16}$$

In this case the phase pattern is a step function as given by the sign of (16). The phase of the field is constant wherever E has a value but changes sign when E goes through zero.

When $\psi = 0$, (14) or (16) is indeterminate so that for this case E must be obtained as the limit of (16) as ψ approaches zero. Thus, for $\psi = 0$, we have the relation that

$$E = n \tag{17}$$

This is the maximum value that E can attain. Hence, the normalized value of the total field for $E_{\max} = n$ is

$$E_n = \frac{1}{n}\frac{\sin(n\psi/2)}{\sin(\psi/2)} \qquad \textit{Normalized array pattern} \tag{18}$$

For an array of nonisotropic but similar radiators, E_n of (18) is the *array pattern* of (4.1).

Values of E_n from (18) for various numbers of sources (1 to 20) are presented in Fig. 5-12*a*. If the phase angle ψ is known as function of the space angle θ, then E_n can be read from Fig. 5-12*a*.

Example 5-4. Five isotropic-source end-fire array. Five sources in Fig. 5-12*d* (on page 268) have equal amplitudes and are spaced $\lambda/4$. The maximum field is to be in line with the sources (at $\theta = 0°$). Plot the field pattern of the array in polar coordinates and indicate the phase referred to the center of the array.

Solution. For $\psi = 0$, the fields from all sources arrive at a distant point in phase. Thus, for a maximum field in the end-fire direction ($\theta = 0°$) equation (10) becomes

$$0 = \frac{2\pi d}{\lambda}\cos 0° + \delta = \frac{2\pi}{4} + \delta = 90° + \delta \tag{18.1}$$

or

$$\delta = -90° \qquad \text{and} \qquad \psi = 90°(\cos\theta - 1) \tag{18.2}$$

The array pattern as a function of ψ is shown in Fig. 5-12*b*. Introducing (18.2) into (18) yields the E field pattern as a function of θ as shown in Fig. 5-12*d* by $E(\theta)$. Phase is indicated by + and − signs, the phase being constant over each lobe.

Problem 5-3-1. Gain of five-isotropic-source end-fire array. (*a*) Calculate the HPBW for the five-source array of Example 5-4 and, using (5-2-10), its approximate directivity. (*b*) Compare this with the directivity obtained using ARRAYPATGAIN (Appendix C). Note that since the array is assumed lossless the directivity and gain are equal. *Ans.* (*a*) 4.1 or 6.1 dBi (*b*) 4.32 dBi.

Example 5-5. Five-dipole end-fire array. If the five isotropic sources of Example 5-4 are replaced by five short dipoles (see insert in Fig. 5-12*d*), plot the amplitude pattern and indicate the phase referred to the center of the array. (*Continued on page 268.*)

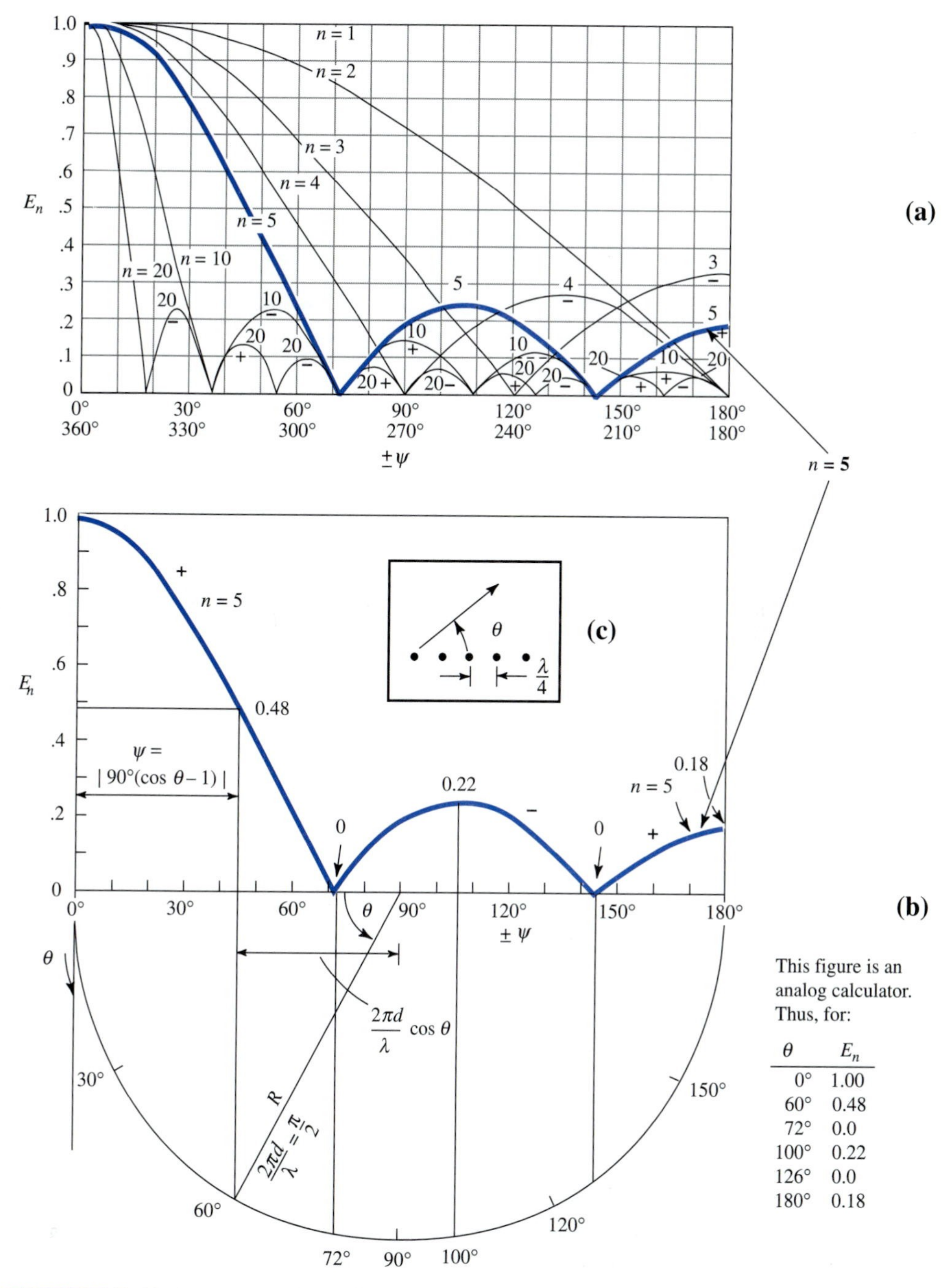

θ	E_n
0°	1.00
60°	0.48
72°	0.0
100°	0.22
126°	0.0
180°	0.18

FIGURE 5-12
(a) Universal field-pattern chart giving normalized field E_n for linear arrays of various numbers n of isotropic point sources of equal amplitude and spacing as a function of ψ. (b) Diagram relating space angle θ to phase angle ψ and in turn to E_n for five-source ordinary end-fire array shown in (c) with $d = \lambda/4$ and progressive phasing $\delta = -90°$.

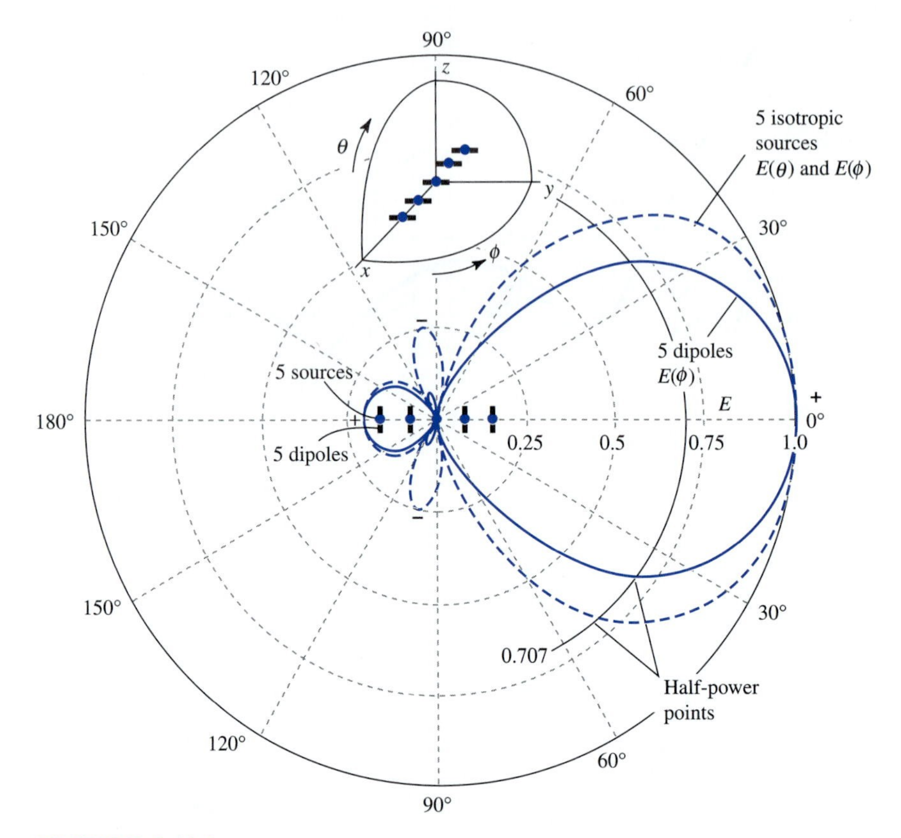

FIGURE 5-12*d*
Five-isotropic-source and five-dipole end-fire array field patterns. Note that for convenience both θ and ϕ are measured from the x axis (see insert).

Solution. The pattern is the same as that for the five isotropic sources of Example 5-4 multiplied by the pattern of the short dipole ($\cos\phi$). This is an example of pattern multiplication. Thus, from (18),

$$E_n = \frac{1}{5}\frac{\sin(5\psi/2)}{\sin(5\psi)}\cos\phi$$

The pattern $E(\phi)$ is shown in Fig. 5-12*d* superimposed on the pattern of the five isotropic sources and is significantly narrower. Phase is indicated by + and − signs, the phase being constant over individual lobes.

Problem 5-3-2. Gain of five-dipole end-fire array. (*a*) Calculate the HPBWs for the five-dipole array of Example 5-5 and, using (5-2-10), its approximate directivity. (*b*) Compare this with the directivity obtained using ARRAYPATGAIN. *Ans.* (*a*) $\theta_{HP} = 100°$, $\phi_{HP} = 76°$, $D = 5.4$ or 7.3 dBi; (*b*) 4.69 dBi.

Example 5-6. Four-isotropic-source broadside array. Four sources in Fig. 5-13 have equal amplitudes, are spaced $\lambda/2$ apart, and are in phase. (*a*) Plot the amplitude in polar coordinates and (*b*) plot both amplitude and phase in rectangular coordinates with phase referred to the midpoint of the array and also to source 1. (*Continued on page 270.*)

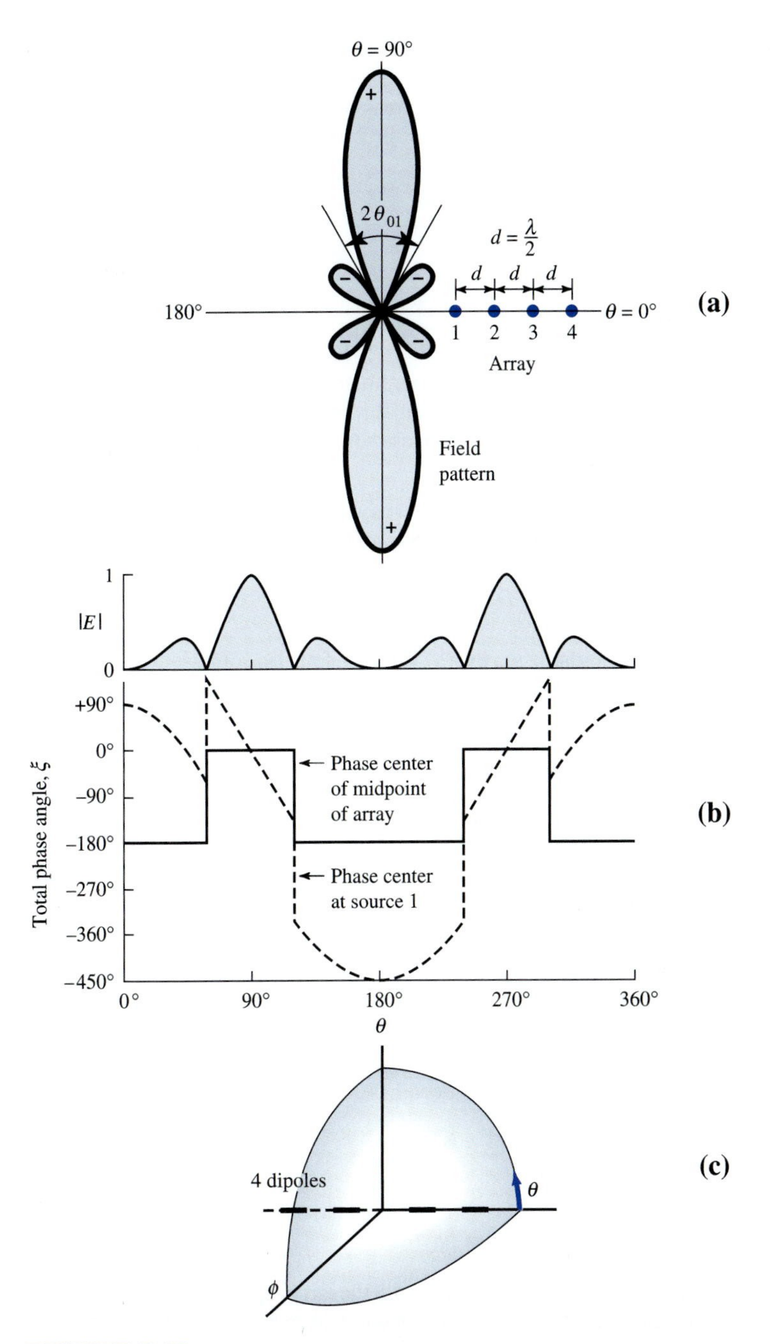

FIGURE 5-13
(a) Field pattern in polar plot of broadside array of four isotropic point sources of same amplitude and phase. Spacing is $\lambda/2$. (b) Field pattern in rectangular plot with phase pattern referred to midpoint and to source 1. The reference direction for phase is at $\phi = 90°$. (c) Orientation of four-dipole array which replaces the four isotropic sources for Problem 5-3-4.

Solution. For $\psi = 0$, the fields from all sources arrive at a distant point in phase. Thus, for a maximum field broadside ($\theta = 90°$ or $270°$), equation (10) becomes

$$0 = \frac{2\pi d}{\lambda}\cos 90° + \delta \tag{19}$$

or

$$\delta = 0$$

The pattern in polar coordinates is shown in Fig. 5-13*a* [*Ans.* (*a*)]. The amplitude and phase pattern are shown in rectangular coordinates in Fig. 5-13*b* [*Ans.* (*b*)].

The maximum field is in a direction normal (perpendicular) to the array. Hence, this condition, which is characterized by in-phase sources ($\delta = 0$), results in a "broadside" array.

Problem 5-3-3. Four-isotropic-source broadside array. (*a*) Calculate the HPBWs for the four-source broadside array of Example 5-6 and, using (5-2-10), its approximate directivity. (*b*) Compare this with the directivity using ARRAYPATGAIN. The array is assumed lossless so $G = D$. *Ans.* (*a*) $\theta_{HP} = 25°$, $\phi_{HP} = 360°$, 4.6 or 6.6 dBi, (*b*) 6.02 dBi.

Problem 5-3-4. Four-dipole broadside array. (*a*) With the four-isotropic-sources of Example 5-6 replaced by four in-line short dipoles as in Fig. 5-13*c*, calculate the HPBWs and approximate directivity. (*b*) Compare with the directivity using ARRAYPATGAIN. (*c*) What is the gain increase over the four-isotropic-source array of Example 5-6, calculated in the previous problem? *Ans.* (*a*) $\theta_{HP} = 23°$, $\phi_{HP} = 360°$, 5.0 or 6.9 dBi, (*b*) 6.33 dBi, (*c*) 0.3 dB.

Problem 5-3-5. Array with flat reflector. How much more gain would be obtained by placing the four-dipole broadside array $\lambda/4$ in front of a flat-sheet reflector? *Ans.* 3 dB.

Example 5-7. Four-isotropic-source end-fire array with $\lambda/2$ spacing. Four sources shown in Fig. 5-14 have equal amplitudes and are in-phase. (*a*) Plot the amplitude in polar coordinates and (*b*) plot both amplitude and phase in rectangular coordinates with phase referred to the midpoint of the array and also to source 1.

Solution. For $\psi = 0$ at $\theta = 0°$, (10) becomes

$$0 = 180° \cos 0° + \delta$$

and

$$\delta = -180°$$

The patterns are shown in Fig. 5-14. Note that with the larger spacing of $\lambda/2$, the pattern is bidirectional and end-fire whereas with $\lambda/4$ spacing, as in Examples 5-4 and 5-5, the patterns are unidirectional end-fire.†

For this end-fire array, the phase between sources is retarded progressively by the same amount as the spacing between sources in radians. Thus, with $\lambda/2$ spacing, the phase of source 2 lags source 1 by 180°, source 3 lags source 2 by 180°, etc.

---

†If the spacing between elements exceeds λ, sidelobes appear which are equal in amplitude to the main (center) lobe. These are called *grating lobes*.

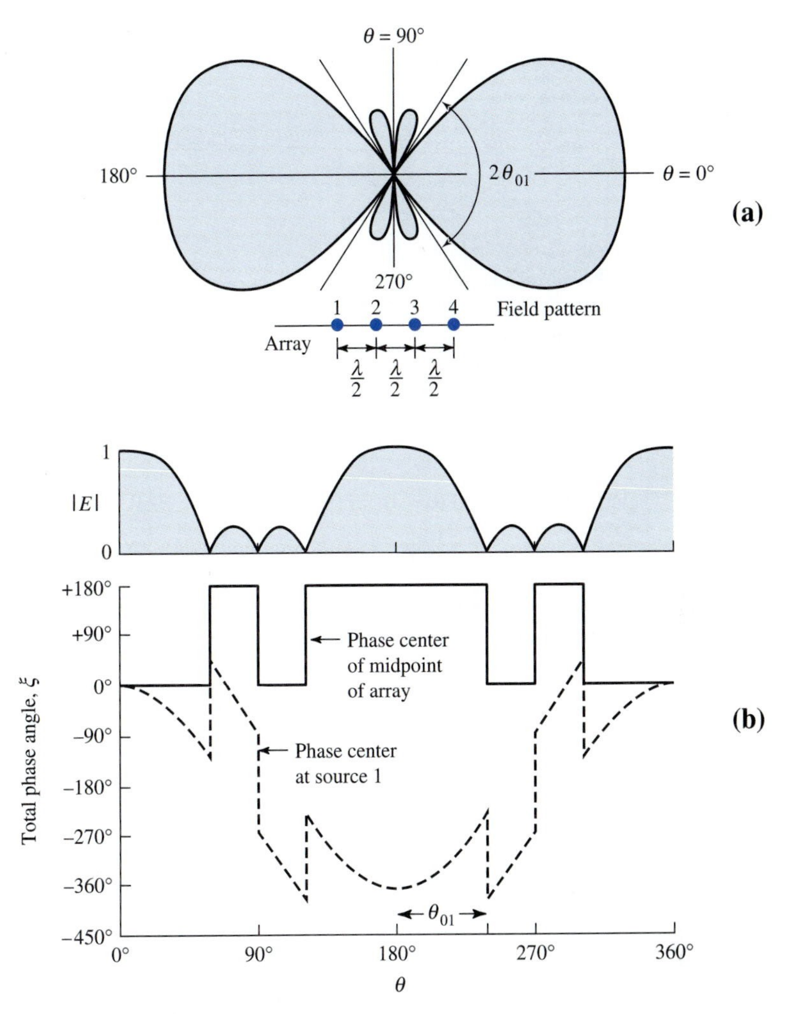

FIGURE 5-14
(a) Field pattern in polar plot of ordinary end-fire array of four isotropic point sources of same amplitude. Spacing is $\lambda/2$, and progressive phase angle $\delta = -\pi$. (b) Field pattern in rectangular plot with phase pattern referred to midpoint and to source 1. The reference direction for phase is at $\theta = 0°$. The first nulls are at $\pm\theta_{01}$.

Problem 5-3-6. Ordinary end-fire directivity. Obtain the directivity of the four-source ordinary end-fire array of Example 5-7 using (a) (5-2-10) and (b) ARRAYPATGAIN. *Ans.* (a) 3.2 or 5.1 dBi, (b) 6.02 dBi.

Example 5-8. Ten-isotropic-source ordinary end-fire array with $\lambda/4$ spacing. The 10 sources have equal amplitudes. The maximum field is to be end-fire (in line with the array) at $\theta = 0°$. Plot the field pattern in polar coordinates and indicate the phasing referred to the midpoint of the array.

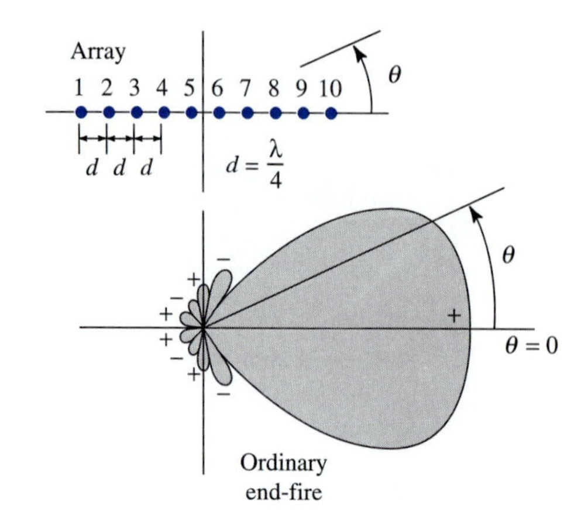

FIGURE 5-15
Field pattern of 10-source array with $\lambda/4$ spacing and $\delta = -90°$ phasing.

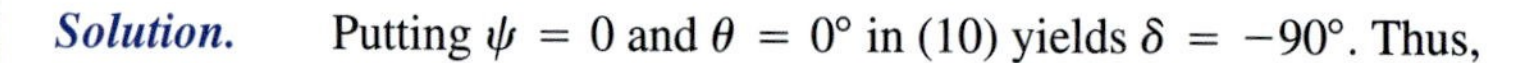

Solution. Putting $\psi = 0$ and $\theta = 0°$ in (10) yields $\delta = -90°$. Thus,

$$\psi = 90°(\cos\theta - 1)$$

Introducing this value of ψ in (18), we obtain the field pattern of Fig. 5-15. Phase is indicated by + and − signs, the phase being constant over each lobe.

Although the phasing $\delta = -90°$ in Example 5-8, called *ordinary end-fire*, produces a maximum field in the end-fire direction, it does not give the maximum directivity. It has been shown by Hansen and Woodyard† that a larger directivity is obtained by increasing the phase change between sources so that

$$\delta = -\left(\frac{2\pi d}{\lambda} + \frac{\pi}{n}\right) \tag{20}$$

where d = spacing between sources, m
λ = wavelength, m
n = number of sources

This is referred to as the *increased directivity end-fire* condition. Thus, for the phase difference of the fields at a large distance, we have

$$\psi = \frac{2\pi d}{\lambda}(\cos\theta - 1) - \frac{\pi}{n} \tag{21}$$

The axial-mode helical antenna has inherent phasing between successive turns that obeys the increased-directivity condition making it a "supergain" antenna (Sec. 5-9).

Example 5-9. Ten-isotropic-source end-fire array with $\lambda/4$ spacing and increased directivity. The 10 sources have equal amplitudes. Plot the field pattern in polar coordinates and indicate the phasing referred to the midpoint of the array.

†W. W. Hansen and J. R. Woodyard, A New Principle in Directional Antenna Design, *Proc. IRE*, **26**: 333–345 (March 1938).

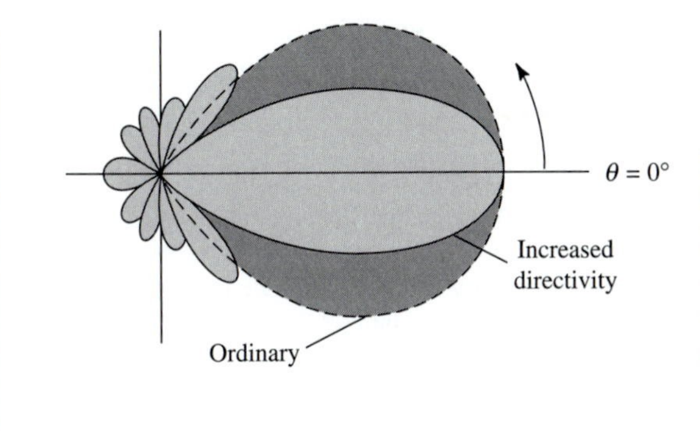

FIGURE 5-16
Field pattern of 10-source increased directivity array with $\lambda/4$ spacing. The "ordinary" main lobe is shown for comparison.

 *Solution.* For this case, (21) becomes

$$\psi = 90°(\cos\theta - 1) - 18°$$

Introducing this into (18), we obtain the polar field pattern shown in Fig. 5-16. The main-lobe beam width is reduced at the expense of increased minor lobes.

The beam widths and directivities for the two patterns of Example 5-9 are compared in Table 5-2.

The maximum of the field pattern of Fig. 5-16 occurs at $\theta = 0$ and $\psi = -\pi/n$. In general, any *increased directivity end-fire array,* with maximum at $\psi = -\pi/n$, has a normalized field pattern given by

$$E = \sin\left(\frac{\pi}{2n}\right)\frac{\sin(n\psi/2)}{\sin(\psi/2)} \qquad \textit{Array pattern (increased directivity)} \qquad (22)$$

TABLE 5-2

	Ordinary end-fire array	End-fire array with increased directivity
Beam width between half-power points	69°	38°
Beam width between first nulls	106°	74°
Directivity	11	19

Problem 5-3-7. Beam width comparison. Compare the HPBWs (half-power beam widths) and the BWFNs (beam width between first nulls) for the 10-source end-fire arrays of Examples 5-8 and 5-9. *Ans*: HPBW (ord.) = 69°, HPBW (inc. dir.) = 40°; BWFN (ord) = 106°, BWFN (inc. dir.) = 74°.

Problem 5-3-8. Directivity comparison. Using (*a*) (5-2-10) and (*b*) ARRAYPATGAIN, compare the directivities of the 10-source arrays of Examples 5-8 and 5-9. *Ans.* (*a*) D (ord.) = 8.6 or 9.4 dBi, D (inc. dir.) = 26 or 14 dBi, (*b*) D (ord.) = 11 or 10.4 dBi, D (inc. dir.) = 19 or 12.8 dBi, or about 1 dB difference between the approximate (5-2-10) formula and ARRAYPATGAIN.

By adjusting the phasing of an array, it is possible to not only maximize the field in broadside and end-fire directions but to maximize it in other directions.

Furthermore, it is possible to steer a null to a specific direction. This is useful for reducing or eliminating interference to a receiving antenna. Example 5-10 is a simple introduction to beam steering.

Example 5-10. Four-isotropic-source array with beam steering. Four sources (Fig. 5-17) have equal amplitude with $\lambda/2$ spacing. (*a*) Find the phase angle δ required to maximize the field in the direction $\theta = 60°$ and using ARRAYPATGAIN plot the field pattern and determine the directivity of the array. (*b*) Find the phase angle δ required to place a pattern null at $\theta = 60°$ and, using ARRAYPATGAIN, plot the field pattern and determine the directivity of the array.

Solution.

(*a*) Start with (10):

$$\psi = \frac{2\pi d}{\lambda}\cos\theta + \delta \tag{10}$$

Put $d = \lambda/2$. Then for a maximum field at $\theta = 60°$, set $\psi = 0$ (sources in phase) and $\theta = 60°$ in (10), obtaining $\delta = -90°$. Enter d and this value of δ in (10) so it reads

$$\psi = 180°\cos\theta - 90°$$

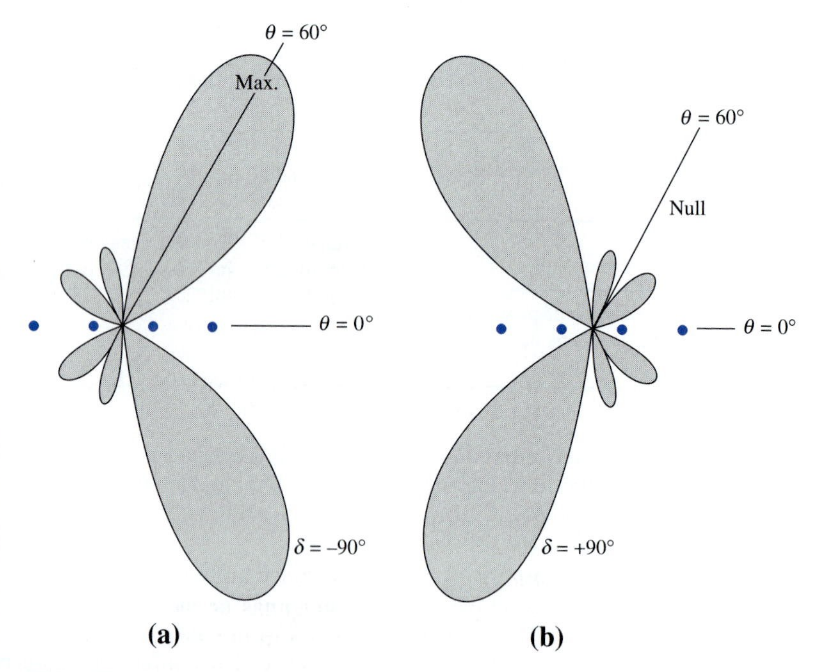

FIGURE 5-17
Field patterns of four-source beam-steering array with $d = \lambda/2$. There is a maximum at $\theta = 60°$ when the phase angle $\delta = -90°$ (*a*) and a null when $\delta = +90°$ (*b*).

Enter this value of ψ and $n = 4$ in (18) and run ARRAYPATGAIN, obtaining the pattern shown in Fig. 5-17*a* with maximum at 60°. $D = 6.02$ dBi. *Ans.* (*a*)

(*b*) For a null (zero field) at $\theta = 60°$ set $\psi = 180°$ (fields out-of-phase) and $\theta = 60°$ in (10), obtaining $\delta = +90°$. Enter d and this value of δ in (10) for ψ. Then enter ψ and $n = 4$ in (18) and run ARRAYPATGAIN, producing the pattern of Fig. 5-17*b* with a null at $\psi = 60°$. $D = 6.02$ dBi. *Ans.* (*b*)

Note: From (5-2-9*b*), $D = 4\pi/\Omega_A$ where Ω_A = pattern solid angle. (The solid angle of an isotropic source = 4π sr.) For a lossless array, D equals the gain in the direction of pattern maximum over an isotropic source. The directivities for (*a*) and (*b*) are equal, as one might expect from the symmetries of the two patterns, but from the field pattern of (*b*) we note that the radiation or response of the array at $\theta = 60°$ is zero.

This is a very simple, crude example but it illustrates the principle of beam steering. In its more sophisticated form, each antenna element of a large array is equipped with its own mixer, voltage-controlled oscillator (VCO), intermediate frequency amplifier and phase detector. Then by controlling the phase and amplitude of each element, the main beam can be steered toward the desired signal while one or more nulls are steered toward interfering signals, making for a "smart antenna."

Problem 5-3-9. Beam steering array. For the four-source array of Example 5-10, (*a*) find δ for a maximum at $\theta = 30°$, plot the pattern, and determine the gain G; (*b*) repeat for a null at $\theta = 30°$. *Ans.* (*a*) $\delta = -156°$; $G = 6.02$ dB; (*b*) $\delta = +24°$; $G = 0$.

Problem 5-3-10. Scanning array. A linear array of six isotropic sources of equal amplitude spaced a distance d apart is fed from one end by a lossless transmission line ($v = c$). If $d = \lambda = 2$ m for a broadside beam (at $\theta = 0°$), find the frequencies for which θ equals (*a*) 0°, (*b*) ±5°, (*c*) ±10°, (*d*) ±15°, (*e*) ±20°, and (*f*) ±25°. This type of array has no moving parts, no phase shifters, and no switches, making it one of the simplest types of scanning arrays. The beam can be shifted or swept rapidly by simply changing the frequency. This technique finds application in radio astronomy where most celestial sources radiate over very wide bandwidths. *Ans*: (*a*) 150 MHz, (*b*) +5°, 137 MHz; −5°, 163 MHz; (*c*) +10°, 124 MHz; −10°, 176 MHz; (*d*) +15°, 111 MHz; −15°, 188 MHz; (*e*) +20°, 99 MHz; −20°, 200 MHz; (*f*) +25°, 87 MHz; −25°, 213 MHz. Note that $\Delta f = f(-\theta) - f(+\theta)$ increases from 0 at $\theta = 0°$ to 63 MHz at $\theta = \pm 25°$.

5-4 RETARDED POTENTIALS

Propagation time is important for plane waves in space and waves on transmission lines. Propagation time is also important with antennas or radiating systems, the difference here being that we deal with propagation in two or three dimensions as compared to the one-dimensional propagation of plane waves along a transmission line.

Accordingly, instead of writing the time-varying current I in a radiating current element as

$$I = I_0 \cos \omega t \tag{1}$$

which implies instantaneous propagation of the effect of the current, we take into account the time of propagation (or retardation time) by introducing the term r/c, and write

$$[I] = I_0 \cos \omega \left(t - \frac{r}{c}\right) \tag{2}$$

where $[I]$ is called the *retarded current*. The brackets [ ] indicate explicitly that the current is retarded.

Equation (2) is a statement of the fact that the disturbance at a time t and at a distance r from the element is caused by a current $[I]$ that occurred at an earlier time $t - (r/c)$. The time difference r/c is the interval required for the disturbance to travel the distance r in free space at the velocity of light $c = 300 \text{ Mm s}^{-1}$.

When we discussed propagation time for lines and waves we were dealing with retarded quantities although we did not use the term. For example, a solution of the wave equation was given that involves $\cos(\omega t - \beta x)$, which is similar in form to the trigonometric function in (2) since†

$$\cos \omega \left(t - \frac{r}{c}\right) = \cos(\omega t - \beta r) \tag{3}$$

where $\beta(= \omega/c = 2\pi/\lambda)$ is the phase constant.

5-5 THE SHORT DIPOLE ANTENNA AND ITS RADIATION RESISTANCE

A short linear conductor is often called a short *dipole*. In the following discussion, a short dipole is always of finite length even though it may be very short. If the dipole is vanishingly short, it is an infinitesimal dipole.

Any linear antenna may be regarded as being composed of a large number of short dipoles connected in series. Thus, a knowledge of the properties of the short dipole is useful in determining the properties of longer dipoles or conductors of more complex shape such as are commonly used in practice.

Let us consider a short dipole like that shown in Fig. 5-18*a*. The length l is very short compared with the wavelength ($l \ll \lambda$). Plates at the ends of the dipole provide capacitative loading. The short length and the presence of these plates result in a uniform current I along the entire length l of the dipole. The dipole may be energized by a balanced transmission line, as shown. It is assumed that the transmission line does not radiate, and its presence will therefore be disregarded. Radiation from the end plates is also considered to be negligible. The diameter d of the dipole is small compared with its length ($d \ll l$). Thus, for the purpose of analysis we may consider that the short dipole appears as in Fig. 5-18*b*. Here it consists simply of a thin conductor of length l with a uniform current I and point charges q at the ends.

†The expression $\cos(\omega t - \beta x)$ refers to a plane wave traveling in the x direction. The relation $\cos \omega[t - (r/c)]$ or $\cos(\omega t - \beta r)$ refers to a spherical wave traveling in the radial or r direction. An important point of difference between a plane and a spherical wave is that a plane wave suffers no attenuation (in a lossless medium) but a spherical wave does because it expands over a larger and larger region as it propagates.

The continuity of current and charge requires that

$$\frac{dq}{dt} = I \tag{1}$$

Let us now obtain the fields everywhere around the dipole. Let the dipole of length l be placed coincident with the z axis and with its center at the origin, as in Fig. 5-19. At any point P the electric field has, in general, three components, E_θ, E_ϕ, and E_r. It is assumed that the medium surrounding the dipole is air or vacuum.

The dipole has no **E** component in the ϕ direction so $E_\phi = 0$. It may be shown that the other components E_r and E_θ are given by†

$$E_r = \frac{I_0 l e^{j(\omega t - \beta r)} \cos\theta}{2\pi\varepsilon_0}\left(\frac{1}{cr^2} + \frac{1}{j\omega r^3}\right) \tag{2}$$

$$E_\theta = \frac{I_0 l e^{j(\omega t - \beta r)} \sin\theta}{4\pi\varepsilon_0}\left(\frac{j\omega}{c^2 r} + \frac{1}{cr^2} + \frac{1}{j\omega r^3}\right) \tag{3}$$

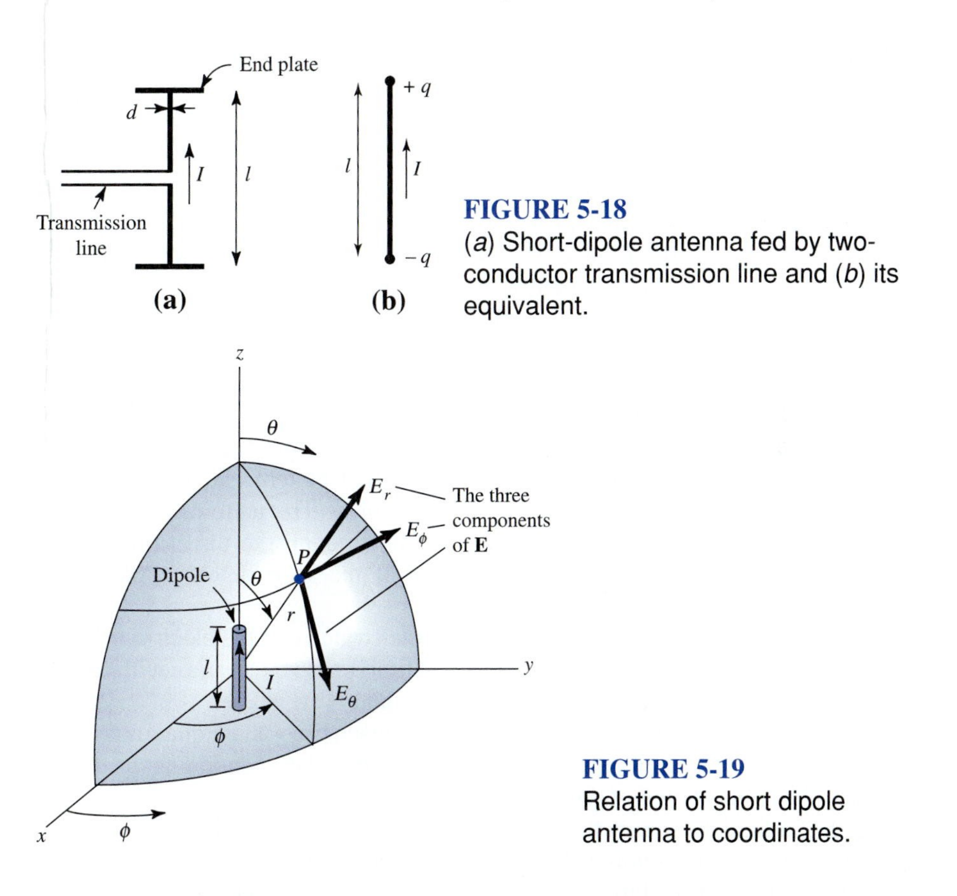

FIGURE 5-18
(*a*) Short-dipole antenna fed by two-conductor transmission line and (*b*) its equivalent.

FIGURE 5-19
Relation of short dipole antenna to coordinates.

†For a detailed development see J. D. Kraus, "Antennas," 2d ed., McGraw-Hill, New York, 1988, p. 200.

The magnetic field has only a ϕ component, as given by

$$H_\phi = |\mathbf{H}| = \frac{I_0 l e^{j(\omega t - \beta r)} \sin\theta}{4\pi}\left(\frac{j\omega}{cr} + \frac{1}{r^2}\right) \tag{4}$$

Thus, the dipole has three field components: E_r, E_θ, and H_ϕ. E_r has terms involving $1/r^2$ and $1/r^3$. E_θ has terms involving $1/r$, $1/r^2$, and $1/r^3$. H_ϕ has terms involving $1/r$ and $1/r^2$.

At a large distance from the dipole only the $1/r$ terms involving E_θ and H_ϕ are significant. Thus, in the *far field*

$$E_\theta = \frac{j\omega I_0 l e^{j(\omega t - \beta r)} \sin\theta}{4\pi\varepsilon_0 c^2 r} = j\frac{30 I_0 \beta l}{r} e^{j(\omega t - \beta r)} \sin\theta \tag{5}$$

$$H_\phi = \frac{j\omega I_0 l e^{j(\omega t - \beta r)} \sin\theta}{4\pi c r} = j\frac{I_0 \beta l}{4\pi r} e^{j(\omega t - \beta r)} \sin\theta \tag{6}$$

Far fields of dipole

Taking the ratio of E_θ to H_ϕ, we obtain

$$\frac{E_\theta}{H_\phi} = 30 \times 4\pi = 376.99 \approx 377\ \Omega \tag{7}$$

Intrinsic impedance of space

which is the *intrinsic impedance of free space.*

It is to be noted that E_θ and H_ϕ are in time phase in the far field. Thus, **E** and **H** in the far field of the spherical wave from the dipole are related in the same manner as in a *plane traveling wave*. Both are also proportional to $\sin\theta$. That is, both are a maximum when $\theta = 90°$ and a minimum when $\theta = 0$ (in the direction of the dipole axis). This variation of E_θ (or H_ϕ) with angle can be portrayed by a *field pattern* as in Fig. 5-20, the length ρ of the radius vector being proportional to the value of the far field (E_θ or H_ϕ) in that direction from the dipole. The pattern in Fig. 5-20*a* is one-half of a three-dimensional pattern and illustrates that the fields are a function of θ but are independent of ϕ. The pattern in Fig. 5-20*b* is two-dimensional and represents a cross section through the three-dimensional pattern. The three-dimensional far-field pattern of the short dipole is doughnut-shaped.

From (2), (3), and (4) we note that for a small value of r the electric field has two components, E_r and E_θ, both of which are in time-phase quadrature with respect to the magnetic field H_ϕ. Thus, in the *near field*, **E** and **H** are related as in a *standing wave*. At intermediate distances, E_θ and E_r can approach time-phase quadrature with each other so that the total electric field vector rotates in a plane containing the dipole, exhibiting the phenomenon of *cross-field*.

In the far field the energy flow is real. That is, the energy flow is always radially outward. This *energy is radiated*. As a function of angle, it is maximum at the equator ($\theta = 90°$). In the near field the *energy flow is largely reactive*. That is, energy flows out and back twice per cycle without being radiated. There is also *angular energy*

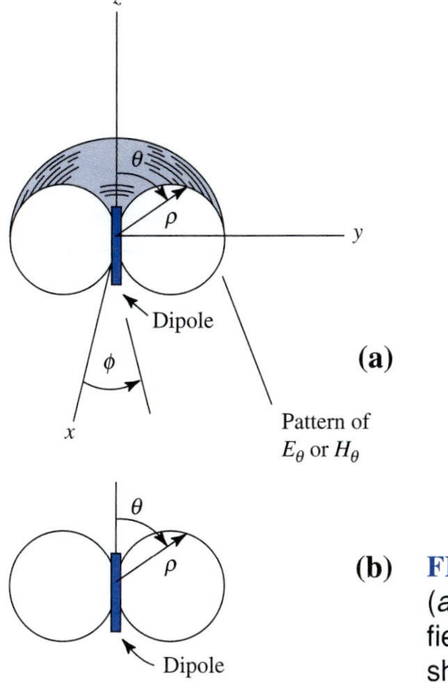

FIGURE 5-20
(*a*) Three-dimensional and (*b*) two-dimensional field pattern of far field (E_θ and H_θ) from a short dipole.

flow (in the θ direction). This energy picture is suggested by Fig. 5-21, where the arrows represent the direction of energy flow at successive instants.†

At very low frequencies, we have the *quasi-stationary* (or almost dc) case where the three field components given by (2), (3), and (4) reduce to

$$E_r = \frac{q_0 l \cos\theta}{2\pi\varepsilon_0 r^3} \tag{8}$$

$$E_\theta = \frac{q_0 l \sin\theta}{4\pi\varepsilon_0 r^3} \tag{9}$$

$$H_\phi = \frac{I_0 l \sin\theta}{4\pi r^2} \tag{10}$$

The electric field components, (8) and (9), are the same as for a static electric dipole, while the magnetic field component H_ϕ in (10) is equivalent to that for a current element. Since these fields vary as $1/r^2$ or $1/r^3$, they are effectively confined to the vicinity of the dipole and radiation is negligible. Significant radiation requires higher frequencies with fields varying as $1/r$ as given by (5) and (6).

The expressions for the fields from a short dipole, developed above, are summarized in Table 5-3. In the table the restriction applies that $r >> l$ and $\lambda >> l$. The three field components not listed are everywhere zero; that is, $E_\phi = H_r = H_\theta = 0$.

†The instantaneous direction and time rate of energy flow per unit area is given by the instantaneous Poynting vector ($= \mathbf{E} \times \mathbf{H}$).

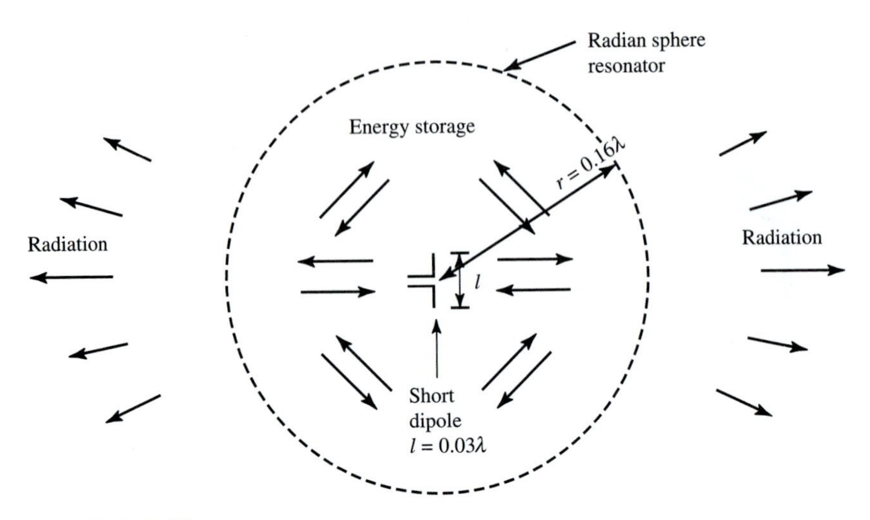

FIGURE 5-21
Sketch suggesting that within the radian sphere ($r = \lambda/2\pi$) the situation is like that inside a resonator with high-density pulsating energy accompanied by leakage which is radiated. The arrows show the direction of energy flow.

Referring to (5) for the far E_θ field of a short dipole, it is instructive to separate the expression into its six basic factors. Thus

$$E_\theta = \underset{\text{Magnitude}}{60\pi} \quad \underset{\text{Current}}{I_0} \quad \underset{\text{Length}}{\frac{l}{\lambda}} \quad \underset{\text{Distance}}{\frac{1}{r}} \quad \underset{\text{Phase}}{je^{j(\omega t - \beta r)}} \quad \underset{\text{Pattern}}{\sin\theta} \tag{11}$$

where 60π is a constant (*magnitude*) factor, I_0 is the dipole *current*, l/λ is the dipole *length* in terms of wavelengths, $1/r$ is the *distance* factor, $je^{j(\omega t-\beta r)}$ is the *phase* factor, and $\sin\theta$ is the *pattern* factor giving the variation of the field with angle. In general, the expression for the *field of any antenna involves these six factors*.

The field relations in Table 5-3 are those for a short dipole. Longer linear antennas or large antennas of other shape may be regarded as being made up of many

TABLE 5-3
Fields of a short dipole

Component	General expression	Radiation or far field	Quasi-stationary or near field
E_r	$\dfrac{I_0 l e^{j(\omega t-\beta r)}\cos\theta}{2\pi\varepsilon_0}\left(\dfrac{1}{cr^2}+\dfrac{1}{j\omega r^3}\right)$	0	$\dfrac{q_0 l\cos\theta}{2\pi\varepsilon_0 r^3}$
E_θ	$\dfrac{I_0 l e^{j(\omega t-\beta r)}\sin\theta}{4\pi\varepsilon_0}\left(\dfrac{j\omega}{c^2 r}+\dfrac{1}{cr^2}+\dfrac{1}{j\omega r^3}\right)$	$\dfrac{j60\pi I_0 e^{j(\omega t-\beta r)}\sin\theta}{r}\dfrac{l}{\lambda}$	$\dfrac{q_0 l\sin\theta}{4\pi\varepsilon_0 r^3}$
H_ϕ	$\dfrac{I_0 l e^{j(\omega t-\beta r)}\sin\theta}{4\pi}\left(\dfrac{j\omega}{cr}+\dfrac{1}{r^2}\right)$	$\dfrac{jI_0 e^{j(\omega t-\beta r)}\sin\theta}{2r}\dfrac{l}{\lambda}$	$\dfrac{I_0 l\sin\theta}{4\pi r^2}$

such short dipoles. Hence, the fields of these larger antennas can be obtained by integrating the field contributions from all the small dipoles making up the antenna.

By taking the surface integral of the average Poynting vector over any surface enclosing an antenna, the total power radiated by the antenna is obtained. Thus

$$P = \int_s \mathbf{S}_{av} \cdot d\mathbf{s} \qquad \text{(W)} \tag{12}$$

where P = power radiated, W; and $\mathbf{S}_{av}$ = average Poynting vector, W m^{-2}.

The simplest surface is a sphere with the antenna at the center. Since the far-field equations for an antenna are simpler than the near-field relations, it is to our advantage to make the radius of the sphere large compared with the dimensions of the antenna. In this way the surface of the sphere lies in the far field, and only the far-field components need to be considered.

Assuming no losses, the power radiated by the antenna is equal to the average power delivered to the antenna terminals. This equals $\frac{1}{2}I_0^2 R_r$ where I_0 is amplitude (peak value in time) of the current at the terminals and R_r is the *radiation resistance* appearing at the terminals (where the dipole is opened at its center point). Thus, $P = \frac{1}{2}I_0^2 R_r$ and the radiation resistance is

$$R_r = \frac{2P}{I_0^2} \qquad (\Omega) \qquad \textbf{\textit{Radiation resistance}} \tag{13}$$

where P is the radiated power in watts.

Let us now carry through the calculation, as outlined above, in order to find the radiation resistance of a short dipole. The power radiated is

$$P = \int_s \mathbf{S}_{av} \cdot d\mathbf{s} = \frac{1}{2}\int_s \text{Re}(\mathbf{E} \times \mathbf{H}^*) \cdot d\mathbf{s} \qquad \textbf{\textit{Radiated power}} \tag{14}$$

In the far field only E_θ and H_ϕ are not zero, so that (14) reduces to

$$P = \frac{1}{2}\int_s \text{Re} E_\theta H_\phi^* \hat{\mathbf{r}} \cdot d\mathbf{s} \tag{15}$$

where $\hat{\mathbf{r}}$ is the unit vector in the radial direction. Thus, the power flow in the far field is entirely radial (normal to surface of sphere of integration). But $\hat{\mathbf{r}} \cdot d\mathbf{s} = ds$, so

$$P = \frac{1}{2}\int_s \text{Re} E_\theta H_\phi^* ds \tag{16}$$

where E_θ and H_ϕ^* are complex, H_ϕ^* being the complex conjugate of H_ϕ. Now $E_\theta = H_\phi Z$, so (16) becomes

$$P = \frac{1}{2}\int_s \text{Re} H_\phi H_\phi^* Z\, ds = \frac{1}{2}\int_s |H_\phi|^2 \text{Re} Z\, ds \tag{17}$$

Since $\text{Re}Z = \sqrt{\mu_0/\varepsilon_0}$ and $ds = r^2 \sin\theta d\theta d\phi$,†

$$P = \frac{1}{2}\sqrt{\frac{\mu_0}{\varepsilon_0}} \int_0^{2\pi}\int_0^{\pi} |H_\phi|^2 r^2 \sin\theta \, d\theta \, d\phi \tag{18}$$

where the angles θ and ϕ are as shown in Fig. 5-19 and $|H_\phi|$ is the absolute value (or amplitude) of the H field. From (4) this is

$$|H_\phi| = \frac{\omega I_{\text{av}} l \sin\theta}{4\pi c r} \tag{19}$$

Substituting this into (18), we have

$$P = \frac{1}{32}\sqrt{\frac{\mu_0}{\varepsilon_0}} \left(\frac{\beta I_{\text{av}} l}{\pi}\right)^2 \int_0^{2\pi}\int_0^{\pi} \sin^3\theta \, d\theta \, d\phi \tag{20}$$

Upon integration (20) becomes

$$P = \sqrt{\frac{\mu_0}{\varepsilon_0}} \frac{(\beta I_{\text{av}} l)^2}{12\pi} \quad \text{(W)} \qquad \textbf{\textit{Short dipole power}} \tag{21}$$

This is the power radiated by the short dipole where I_{av} is the average current on the dipole.

Substituting the power P from (21) into (13) yields for the *radiation resistance of the short dipole*

$$R_r = \sqrt{\frac{\mu_0}{\varepsilon_0}} \frac{(\beta l)^2}{6\pi} \left(\frac{I_{\text{av}}}{I_0}\right)^2 \quad (\Omega) \tag{22}$$

Since $\sqrt{\mu_0/\varepsilon_0} = 377\ \Omega$, (22) reduces to

$$R_r = 20(\beta l)^2 \left(\frac{I_{\text{av}}}{I_0}\right)^2 = 80\pi^2 \left(\frac{l}{\lambda}\right)^2 \left(\frac{I_{\text{av}}}{I_0}\right)^2 \quad (\Omega) \qquad \textbf{\textit{Dipole radiation resistance}} \tag{23}$$

Example 5-11. Radiation resistance of short dipole antenna and short vertical antenna. Calculate the radiation resistance of (*a*) a center-fed $\lambda/10$ dipole antenna and (*b*) half of the same dipole erected vertically over a flat conducting ground plane.

Solution.

(*a*) The word *short* as used here refers, of course, to the dipole's *electrical length* or length in wavelengths. Physically, the dipole could be long if the wavelength is long [see part (*b*)].

Referring to Fig. 5-22*a*, the current is zero at the ends of the dipole and increases almost linearly to its maximum value at the terminals. Thus, the average current is

---

†Since $\sqrt{\mu_0/\varepsilon_0} = E_\theta/H_\phi \approx 120\pi$, we may also write

$$P = \frac{1}{240\pi}\int_0^{2\pi}\int_0^{\pi} |E_\theta|^2 r^2 \sin\theta d\theta d\phi$$

one-half the terminal current I_0 and from (23) the radiation resistance

$$R_r = 80\pi^2 \left(\frac{l}{\lambda}\right)^2 \left(\frac{1}{2}\right)^2 \approx 2\ \Omega \quad \textit{Ans.} \quad (a)$$

This is the resistive or real part of the terminal impedance, which will also include a large negative (capacitive) reactance of about 1900 Ω (the dipole is *not* resonant), so the terminal impedance of the dipole is†

$$Z \approx 2 - j1900\ \Omega$$

As the dipole length is increased, this reactance decreases until for a dipole length slightly less than $\lambda/2$, the reactance becomes zero (dipole resonant). At the same time the radiation resistance increases to a value near 70 Ω. When the dipole (assumed to be thin) is exactly $\lambda/2$ long, it terminal impedance is given by

$$Z = 73 + j42.5\ \Omega$$

See Sec. 5-6.

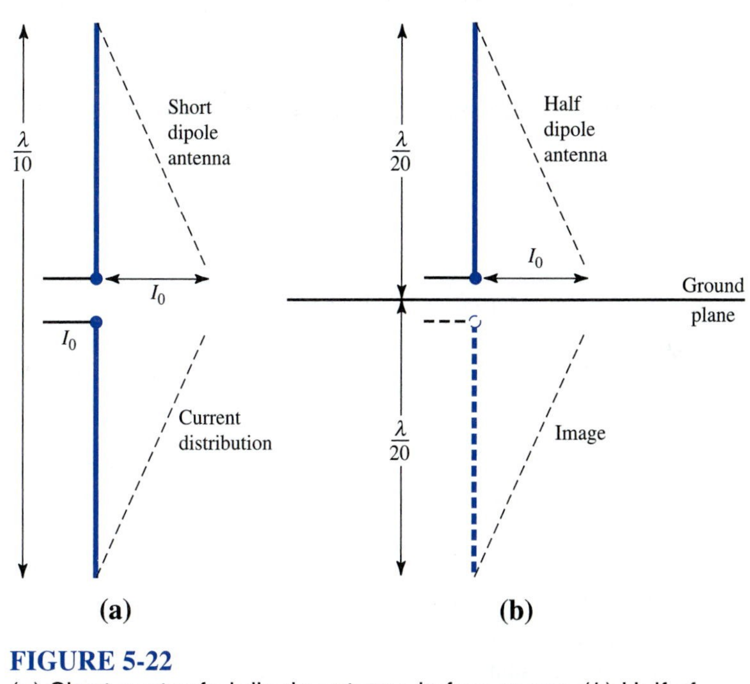

FIGURE 5-22
(*a*) Short center-fed dipole antenna in free space. (*b*) Half of short dipole erected above a flat conducting ground plane with its image.

†See J. D. Kraus, "Antennas," 2d ed., McGraw-Hill, New York, 1988, p. 407.

(*b*) Half of the dipole of part (*a*), erected vertically over a ground plane, has an image as suggested in Fig. 5-22*b*. Thus, the problem is the same as for (*a*) except that the terminal impedance (measured from upper half of dipole to ground) is one-half the value for (*a*). Thus,

$$R_r = 1\ \Omega \quad \textit{Ans.} \quad (b)$$

and

$$Z \approx 1 - j950\ \Omega$$

If this vertical antenna operates at $\lambda = 600$ m (used in the AM broadcast band) its height is $600/20 = 30$ m.

Problem 5-5-1. Equal fields. At what distance from a short dipole are the magnitudes of all three terms for E_θ in (3) equal? *Ans.* At the radian distance: $\lambda/2\pi$.

Although antennas are usually very efficient (radiated power >> power lost as heat in the antenna structure), this situation requires a radiation resistance $R_r >> R_l$, where R_l is the equivalent loss resistance. Thus, the total terminal resistance R_t of the antenna is given by

$$R_t = R_r + R_l \qquad (\Omega) \tag{24}$$

Suppose, for instance, that $R_l = 1\ \Omega$, a not unlikely value for the vertical half-dipole of part (*b*) of the above example, making the terminal resistance $R_t = 1 + 1 = 2\ \Omega$. The antenna efficiency is then given by

$$\frac{\text{Power radiated}}{\text{Power input}} = \frac{I_0^2 R_r}{I_0^2 (R_r + R_l)} = \frac{1}{1+1} = 50\% \tag{25}$$

A taller antenna (height > $\lambda/20$) with larger radiation resistance would be more efficient, provided R_l remained small.

The radiation resistance of antennas other than the short dipole can be calculated as above provided the far field is known as a function of angle. Thus, from (13) and (17), the *radiation resistance at the terminals of an antenna* is given by

$$R_r = \frac{120\pi}{I_0^2}\int_s |H|^2 ds = \frac{120\pi}{I_0^2}\int_s \frac{|E|^2}{Z_0^2} ds \qquad (\Omega) \qquad \textbf{\textit{Radiation resistance}} \tag{26}$$

where $|H|$ = amplitude of far H field, A m^{-1}
$|E|$ = amplitude of far E field, V m^{-1}
I_0 = amplitude of terminal current, A
Z_0 = intrinsic impedance of space = 377 Ω

If we integrate the complex Poynting vector (= ½**E** × **H***) over a surface enclosing an antenna, we obtain, in general, both a real part which is equal to the power radiated and an imaginary part equal to the reactive power. Whereas the real part, or radiated power, is the same for *any* surface enclosing the antenna, the imaginary, or reactive, power obtained depends on the location and shape of the surface enclosing the antenna. For a large surface lying only in the far field, the reactive power is zero, but for a surface lying in the near field, it may be of considerable magnitude. In the case of a very thin linear antenna, it turns out that if the surface of integration is collapsed so as to coincide with the surface of the antenna, the complex power so obtained divided by the square of the terminal current yields the terminal impedance $R + jX$, where R is the radiation resistance.

5-6 PATTERN AND RADIATION RESISTANCE OF $\lambda/2$ AND 3 $\lambda/2$ DIPOLES

In Sec. 5-5 the dipole antenna was analyzed for the case where its length was small compared to the wavelength (say $\lambda/10$ or less). Let us now consider linear dipole antennas of any length L. It is assumed that the current distribution is sinusoidal. This is a satisfactory approximation provided the conductor diameter is small, say $\lambda/200$ or less.

Referring to Fig. 5-23, we calculate the far-field pattern for a symmetrical thin linear center-fed dipole antenna of length L. It is done by regarding the antenna as being made up of a series of elemental short dipoles of length dy and current I and integrating their contribution at a large distance. The form of the current distribution on the antenna is found from experimental measurements to be given approximately by

$$I = I_0 \sin\left[\frac{2\pi}{\lambda}\left(\frac{L}{2} \pm y\right)\right] \qquad \text{(A)} \qquad (1)$$

where I_0 is the value of the current at the current maximum point, and where $(L/2)+y$ is used when $y < 0$ and $(L/2) - y$ is used when $y > 0$.

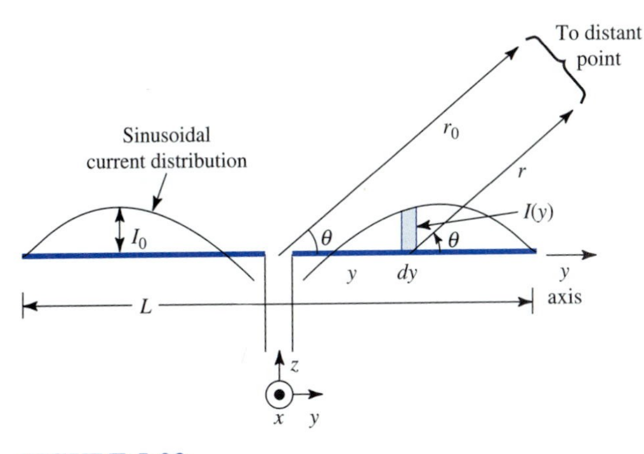

FIGURE 5-23
Geometry for linear center-fed dipole antenna of length L with sinusoidal current distribution for calculating pattern and radiation resistance.

Integrating the E_θ far field of a short dipole, we find the far (electric) field of an antenna of length L is

$$E_\theta = k \sin\theta \int_{-L/2}^{L/2} I \exp(j\beta y \cos\theta) dy \tag{2}$$

where $k = \{j60\pi \exp[j(\omega t - \beta r_0)]\}/r_0\lambda$. Substituting (1) for I in (2) and integrating yields

$$E_\theta = \frac{j60[I_0]}{r_0} \left\{ \frac{\cos[(\beta L \cos\theta)/2] - \cos(\beta L/2)}{\sin\theta} \right\} \tag{3}$$

where $[I_0] = I_0 \exp\{j\omega[t - (r_0/c)]\}$ is the retarded current.

The shape of the far-field pattern is given by the factor in braces { } in (3). For a $\lambda/2$ dipole antenna this factor reduces to

$$E_\theta(\theta) = \frac{\cos[(\pi/2)\cos\theta]}{\sin\theta} \qquad \textbf{\lambda/2 dipole pattern} \tag{4}$$

with the shape shown in Fig. 5-24*a*. This pattern is only slightly more directional than the $\sin\theta$ pattern of an infinitesimal or short dipole (shown dashed). The HPBW of the $\lambda/2$ antenna is 78°, compared with 90° for a short dipole.

For a 1.5λ dipole the pattern factor is

$$E_\theta(\theta) = \frac{\cos\left(\frac{3}{2}\pi\cos\theta\right)}{\sin\theta} \qquad \textbf{1.5\lambda dipole pattern} \tag{5}$$

as shown in Fig. 5-24*b*. With the midpoint of the antenna as phase center, the phase of the far field (at a constant r_0) shifts 180° at each null, the relative phase of the

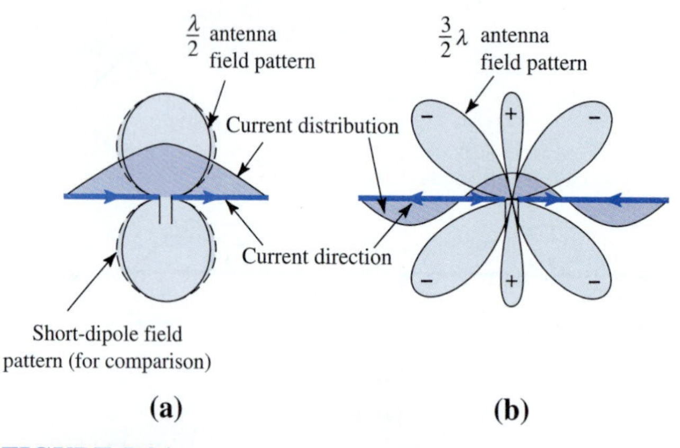

FIGURE 5-24
(*a*) Far-field pattern of $\lambda/2$ center-fed dipole antenna.
(*b*) Far-field pattern of 1.5λ center-fed dipole antenna.
Current distributions are also shown.

lobes being indicated by the plus and minus signs. The three-dimensional pattern is a figure of revolution of the one shown in Fig. 5-24*b* with the dipole as axis.

Example 5-12. Directivity of λ/2 dipole. Find the directivity of a λ/2 linear dipole.

Solution. For the λ/2 dipole, we have from (4) and (5-2-9*b*) that

$$D = \frac{4\pi}{\iint P_n(\theta,\phi)\,d\Omega} = \frac{4\pi}{2\pi \int_0^{\pi} \frac{\cos^2[(\pi/2)\cos\theta]}{\sin^2\theta} \sin\theta d\theta} = 1.64\ (= 2.15 \text{ dBi}) \quad \textit{Ans.} \tag{6}$$

This is 1.64/1.5 = 1.09 (= 0.4 dB) greater than the directivity of an infinitesimal or short dipole. *Note:* The integration of (6) involves the sine integral (Si) and cosine integral (Ci).

Problem 5-6-1. HPBW of λ/2 dipole. Find (*a*) the HPBW of a λ/2 dipole from (4) and (*b*) calculate the approximate directivity using the approximate formula of (5-2-10). Compare with (6) of Example 5-12. *Ans.* (*a*) 78°; (*b*) 1.46 or 1.46/1.64 = 0.89 of exact value.

To find the radiation resistance R_r of the λ/2 dipole, the power density is integrated over a large sphere, yielding the power radiated. This power is then equated to $I^2 R_r$, where I is the rms terminal current. Performing this calculation for the λ/2 dipole yields

$$R_r = 30 \text{ Cin } 2\pi = 30 \times 2.44 = 73\ \Omega \qquad \textbf{\textit{λ/2 dipole resistance}} \tag{7}$$

where Cin is the modified cosine integral. This is the resistance presented by a λ/2 dipole to the transmission line terminals (Fig. 5-24*a*). The antenna impedance Z also contains a positive reactance of 42.5 Ω ($Z = 73 + j42.5\ \Omega$). To make the antenna resonant ($X = 0$), the length can be shortened a few percent. However, this also results in a slight reduction of the radiation resistance (to approximately 70 Ω). The radiation resistance for the 1.5λ linear dipole is about 100 Ω.

5-7 BROADSIDE ARRAY

In Sec. 5-3 we discussed patterns of radiating arrays of isotropic sources and of antenna elements where source amplitudes or antenna currents were specified. However, we did not consider how the array elements might be fed in order to produce these currents. Let us now look at the very practical problem of how to do this and at the same time calculate the directivity (or gain) of the array using circuit analysis.

Consider the two-element array of Fig. 5-25 consisting of two side-by-side λ/2 dipole antennas which we wish to feed with equal in-phase currents from a common point F by a two-wire 300-Ω transmission line. Dipole 1 has a self-impedance

$$Z_{11} = R_{11} + jX_{11} \qquad (\Omega) \tag{1}$$

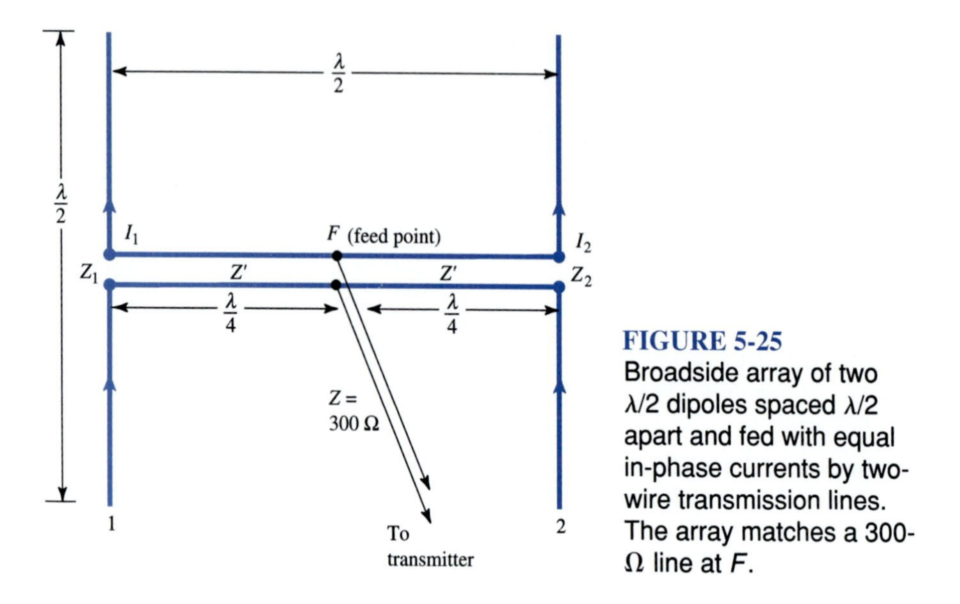

FIGURE 5-25
Broadside array of two λ/2 dipoles spaced λ/2 apart and fed with equal in-phase currents by two-wire transmission lines. The array matches a 300-Ω line at *F*.

Dipole 2 has a self-impedance

$$Z_{22} = R_{22} + jX_{22} \qquad (\Omega) \tag{2}$$

Each dipole couples to the other via a mutual impedance

$$Z_{12} \text{ (dipole 1 to 2)} = R_{12} + jX_{12} \qquad (\Omega) \tag{3}$$

and

$$Z_{21} \text{ (dipole 2 to 1)} = R_{21} + jX_{21} \qquad (\Omega) \tag{4}$$

The mutual impedances can be calculated by an extension of the analysis for self-impedance.

Since we are involved with the input side of the antennas, our problem now becomes one of simple circuit analysis. Thus, the voltages V_1 and V_2 at the terminals of antennas 1 and 2 are given by

$$V_1 = I_1Z_{11} + I_2Z_{12} \qquad \text{(V)} \tag{5}$$

and

$$V_2 = I_2Z_{22} + I_1Z_{21} \qquad \text{(V)} \tag{6}$$

But since $I_1 = I_2$, $Z_{11} = Z_{22}$, and $Z_{12} = Z_{21}$, we have

$$V_1 = I_1(Z_{11} + Z_{12}) = V_2 \tag{7}$$

and the input or driving impedances Z_1 and Z_2 of the antennas are

$$Z_1 = \frac{V_1}{I_1} = Z_{11} + Z_{12} = Z_2 \tag{8}$$

or

$$Z_1 = Z_2 = R_{11} + R_{12} + j(X_{11} + X_{12}) \tag{9}$$

If the reactance is tuned out with a series reactance of opposite sign at the dipole terminals,†

$$Z_1 = Z_2 = R_{11} + R_{12} \tag{10}$$

and the input power P to the array is given by

$$P = 2I_1^2(R_{11} + R_{12}) \tag{11}$$

from which

$$I_1 = \sqrt{\frac{P}{2(R_{11} + R_{12})}} \tag{12}$$

and the field broadside to the array is given by

$$E(\text{array}) = 2kI_1 = k\sqrt{\frac{2P}{R_{11} + R_{12}}} \tag{13}$$

where k = a dimensional constant involving distance.

Suppose we feed a single $\lambda/2$ dipole with the same power. Then its field at the same distance is given by

$$E(\lambda/2) = kI_0 = k\sqrt{\frac{P}{R_{11}}} \tag{14}$$

Now the ratio of (13) to (14) gives us the gain in field, G(field), of the array over the single $\lambda/2$ dipole as

$$G(\text{field}) = \sqrt{\frac{2R_{11}}{R_{11} + R_{12}}} \qquad \text{(dimensionless)} \tag{15}$$

Introducing R_{11} and R_{12} (see Tables 5-4 and 5-5) gives

$$G(\text{field}) = \sqrt{\frac{2 \times 73}{73 - 13}} = 1.56 \tag{16}$$

From (5-6-6) the directivity of a $\lambda/2$ dipole is 1.64. Therefore, the directivity of the broadside array of two $\lambda/2$ dipoles is

$$D = 1.56^2 \times 1.64 = 4.0\ (= 6.0 \text{ dBi or } 6.0 \text{ dB over isotropic}) \tag{17}$$

This is the same directivity that would be obtained by integrating the power pattern of the array over 4π sr, and yet we have carried out no such integration, at least not *explicitly*. However, such an integration or its equivalent is *implicit* in the self- and mutual resistance values.

Table 5-4 lists some directivity, gain, beam width, effective aperture, and self-impedance values for reference.

---

†It is often simpler to resonate the dipole elements by shortening them a bit. This modifies the resistance and pattern slightly.

TABLE 5-4
Dipole directivities, beam widths, etc.

	Directivity	Gain,† dBi	HPBW	A_e, λ^2	R_r, Ω
Isotropic antenna	1.0	0		0.08	
Short dipole	1.5	1.76	90°	0.12	$80\pi^2\left(\frac{l}{\lambda}\right)^2\left(\frac{I_{av}}{I_0}\right)^2$
λ/2 dipole	1.64	2.15	78°	0.13	73

† Antenna assumed to be lossless. dBi = dB over isotropic antenna.

Example 5-13. Feed system for array of two λ/2 dipoles. Referring to Fig. 5-25, what is the required impedance Z' of the λ/4 sections in order to match the array to a 300-Ω two-conductor transmission line?

Solution. To feed the array with a two-wire 300-Ω transmission line requires that each λ/4 line between F and the dipole terminals transforms the 73 − 13 = 60 Ω resistance at the terminals to 600 Ω at F. Thus, the required λ/4 line impedance

$$Z' = \sqrt{60 \times 600} = 190\ \Omega \quad \textit{Ans.} \tag{18}$$

The two 600-Ω resistances in parallel provide a match to the 300-Ω line to the transmitter. Thus, with lines connected as in Fig. 5-25, the two λ/2 dipoles are fed with equal in-phase currents and the 300-Ω feed line is matched.

Problem 5-7-1. Broadside array of four λ/2 dipoles. Four λ/2 dipoles are arranged in a 2 × 2 broadside array as in Fig. P5-7-1. All are fed in-phase with equal currents I. If the mutual impedances $Z_{12} = -13\ \Omega$, $Z_{13} = 26\ \Omega$ and $Z_{14} = -12\ \Omega$, find the directivity. *Ans.* 6.5 or 8.1 dBi.

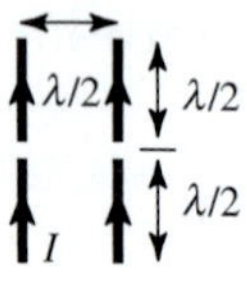

FIGURE P5-7-1

We have assumed that the array is lossless. To account for losses, a loss resistance R_l is added to (10) so that

$$Z_1 = R_{11} + R_l + R_{12} \tag{19}$$

If the comparison λ/2 dipole is lossless, then (15) becomes

$$G(\text{field}) = \sqrt{\frac{2R_{11}}{R_{11} + R_l + R_{12}}} \tag{20}$$

If, for example, $R_l = 1\ \Omega$, the array directivity is 3.93 or 0.08 dB less than for a lossless array, a very small difference.

We have assumed in our analysis that the diameter of the dipole conductor is much less than its length.

TABLE 5-5
Mutual impedances for λ/2 dipoles

Parallel side-by-side		Colinear	
Spacing, λ	Z(mutual), Ω	Spacing between centers, λ	Z(mutual), Ω
0.5	$-13 - j29$	0.5	$26 + j18$
1.0	$+4 + j18$	1.0	$-4 - j2$
1.5	$-2 - j12$	1.5	$+2 + j0$

Table 5-5 gives the mutual impedances for thin λ/2 dipoles in a variety of configurations.

5-8 FIELDS OF λ/2 DIPOLE ANTENNA

To illustrate the configuration of the fields radiated by an antenna, the electric field lines of a λ/2 antenna are shown in Fig. 5-26. Although the fields were computer-generated (and plotted) for a prolate spheroid with λ/2 interfocal distance, they are substantially the same as for a thin λ/2 center-fed dipole except in the immediate proximity of the antenna.

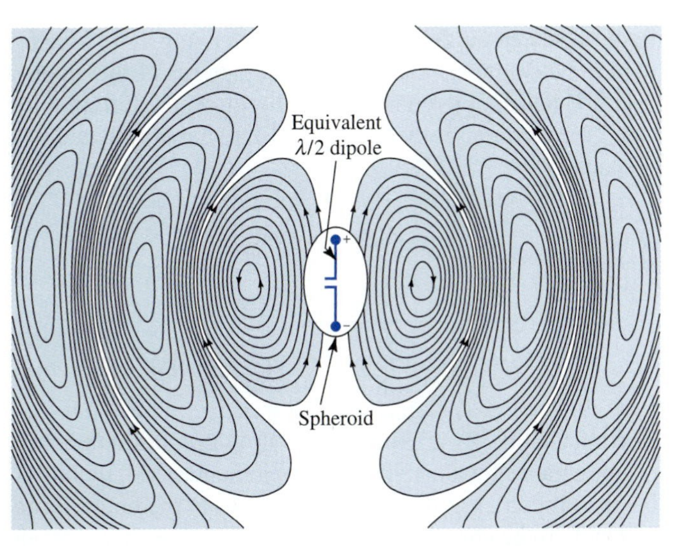

FIGURE 5-26
Electric field configuration for a λ/2 antenna. For the fields in motion see the book's Web site. (*From "A Resonant Spheroid Radiator," produced at the Ohio State University for the National Committee for Electrical Engineering Films; Project Initiator, Prof. Edward M. Kennaugh; diagrams courtesy of Prof. John D. Cowan, Jr.*)

5-9 ANTENNA TYPES

The variety of antennas is enormous. It is beyond the scope of this edition to do more than take a quick look at some of them.† Although we avoid detailed analyses, we can nevertheless obtain many useful insights.

Twenty-three different antennas are presented in Figs. 5-27 to 5-38 grouped into basic types, opened-out coaxial, two-wire and waveguide types, reflector and aperture types, end-fire, and broadband types. The coaxial and two-wire types (Figs. 5-29 and 5-30) are arranged as an evolutionary sequence from broad to narrow bandwidth. Directivities, bandwidths, and field patterns are indicated. These, with the dimensions given, are sufficient in many cases to construct the antenna and determine its gain and beam widths.

The figures give an overview of many antenna types and relate them in a kind of genus and species classification that helps in understanding how one type carries over or evolves into another. Many of the examples and problems relate to the antennas in these figures.

Loops, Dipoles, and Slots

Referring to Fig. 5-27 the small loop antenna at (*a*) may be regarded as the magnetic counterpart of the short electric dipole at (*b*). With loop axis parallel to the dipole, as in the figure, the loop and dipole have identical field patterns but with **E** and **H** interchanged. Both have the same directivity, $D = 1.5$.

If the dipole has been cut from a metal sheet, leaving a slot as at (*c*), the dipole and slot are said to be *complementary*. The field patterns of the dipole and slot are the same but with **E** and **H** interchanged. Furthermore, the terminal impedance Z_d of the dipole and the terminal impedance Z_s of the slot are related to the intrinsic impedance of space Z_0 ($= 377\ \Omega$) as given by

$$Z_d Z_s = \frac{Z_0^2}{4} \tag{1}$$

from which

$$Z_s = \frac{Z_0^2}{4Z_d} = \frac{Z_0^2}{4} Y_d \qquad \textbf{\textit{Slot impedance}} \tag{2}$$

so that the *slot impedance* Z_s is proportional to the *dipole admittance* Y_d. If the dipole requires inductance for a match, the complementary slot requires capacitance. Thus, knowing the properties of the dipole enables us to predict the properties of the complementary slot. To be completely complementary, the sheet containing the slot should be large (ideally infinite) and perfectly conducting.

Slot antennas are typically $\lambda/2$ long, and these are then complementary to a $\lambda/2$ dipole antenna (Fig. 5-27). Note that the field patterns in Fig. 5-27 are *not* field lines. However, the directions of **E** and **H** at a point are shown.

---

†For a detailed treatment of all types of antennas, see J. D. Kraus, "Antennas," 2d ed., McGraw-Hill, New York, 1988.

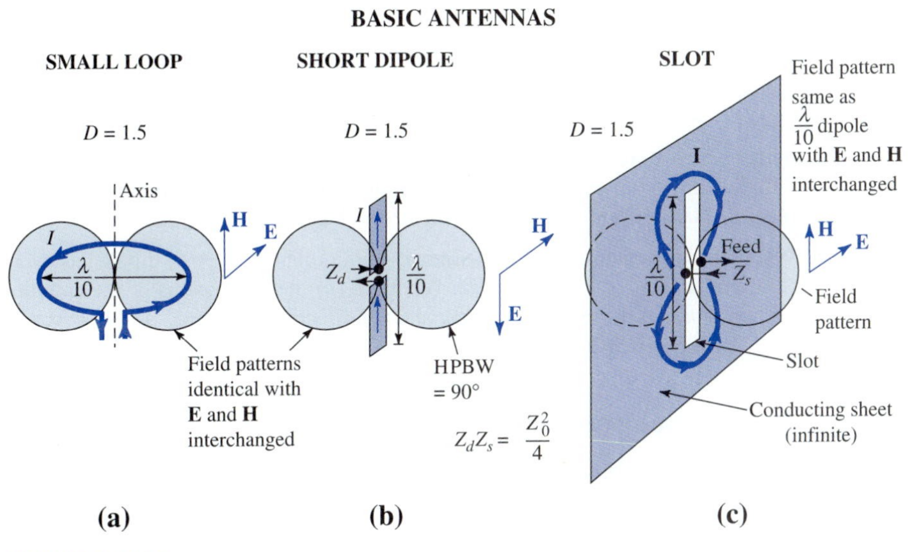

FIGURE 5-27
Three basic types of antennas: the small loop (a), short dipole (b), and slot antenna (c). The small loop and short dipole (loop axis parallel to dipole) have identical field patterns with **E** and **H** interchanged. The slot and dipole have the same field patterns with **E** and **H** interchanged. Directivities *D* are identical, as indicated.

Example 5-14. One-wavelength dipole and slot antenna. In Fig. 5-28, a λ-long cylindrical dipole and its complementary slot are compared. The actual length $L = 0.925\ \lambda$. The dipole cylinder has a diamter $D = L/28 = 0.033\ \lambda$ and a terminal impedance $Z_d = 710 + j0\ \Omega$. The complementary slot has a width $w = 2D$. Find the terminal impedance Z_s of the slot.

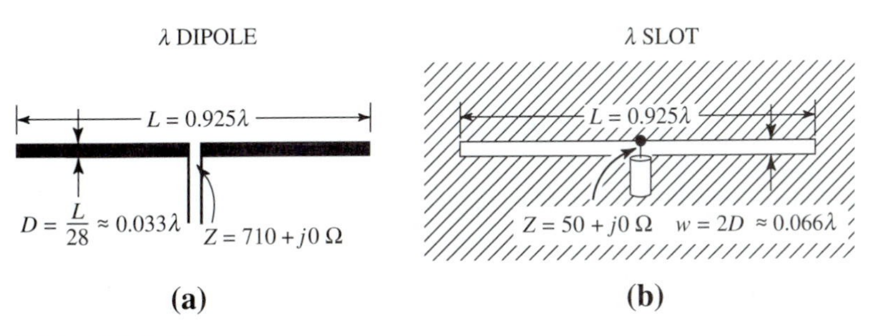

FIGURE 5-28
Comparison of impedances of cylindrical dipole antenna with complementary slot antenna. The slot in (*b*) matches directly the 50-Ω coaxial line.

Solution. From (2),

$$Z_s = \frac{377^2}{4 \times 710} = 50\ \Omega \quad \textit{Ans.}$$

for an impedance match to a 50-Ω coaxial line as indicated in Fig. 5-28*b*.

Radiation from a patch antenna. (Fig. 5-40) occurs as though from two slots, so this example is useful for an understanding of patch antennas.

Problem 5-9-1. Slot dimensions and impedance. A thin (diameter → 0) dipole has a terminal impedance of $73 + j42.5\ \Omega$. What are (*a*) the dimensions of the complementary slot, (*b*) the slot impedance, and (*c*) directivity? *Ans*: (*a*) $L = 0.5\ \lambda$, $w \to 0$; (*b*) $Z_s = 363 - j211\ \Omega$; (*c*) 1.64.

Opened-Out Coaxial Antennas

Figure 5-29 shows the evolution of a $\lambda/4$ monopole or stub antenna at (*c*) from the gradually tapered broad-bandwidth "volcano smoke" antenna (*a*) with the intermediate conical type (*b*).

Opened-Out Two-Conductor Antennas

Figure 5-30 on page 296 shows the evolution of two-conductor antennas from the twin Alpine horn antenna (*a*) via biconical (*b*) to the $\lambda/2$ dipole (*c*).

Opened-Out Waveguide Antennas

Opened-out waveguide antennas are shown in Fig. 5-31 on page 297. The directivity of both pyramidal and conical horns is proportional to the aperture area of the horn.

Example 5-15. Optimum pyramidal horn. Ideally the phase of the field across the horn mouth should be a constant. This requires a very long horn. However, for practical convenience the horn should be as short as possible. Referring to Fig. 5-32 on page 297, an *optimum horn* is a compromise in which the difference in the path length δ along the edge and the center of the horn is made 0.25λ or less in the E plane. However, in the H plane, δ can be larger since the field goes to zero at the horn edges (boundary condition, $E_t = 0$ satisfied). From Fig. 5-32 the horn flare angle θ is given by

$$\theta = 2\cos^{-1}\frac{L}{L+\delta} \tag{3}$$

For a horn with $L = 10\lambda$, find the largest flare angle for which $\delta = 0.25\lambda$.

Solution. From (3), $\theta = 2\cos^{-1}\dfrac{10}{10.25} = 25.4^\circ$ *Ans.*

Problem 5-9-2. Horn antenna. (*a*) Find the physical aperture of a pyramidal horn antenna with $L = 10\lambda$, δ (E plane) $= 0.25\lambda$, δ (H plane) $= 0.4\lambda$; (*b*) Assuming a hypothetical 100 percent aperture efficiency, find the directivity. This value is an upper limit. (*c*) Realistically, the aperture efficiency might be 60 percent, resulting in what directivity? *Ans.* (*a*) $24.7\lambda^2$, (*b*) 310 (24.9 dBi), (*c*) 186 (22.7 dBi).

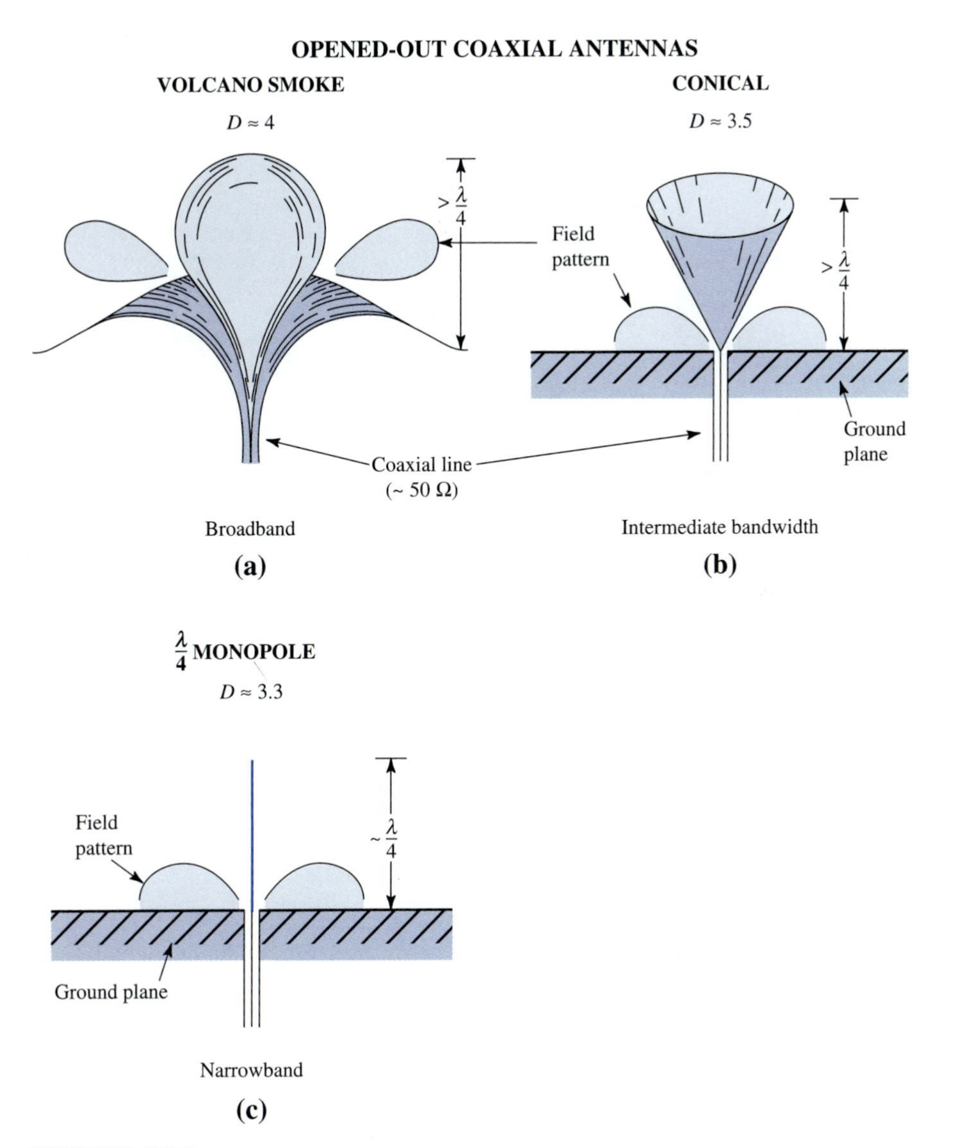

FIGURE 5-29
Opened-out coaxial antennas showing evolution of narrowband $\lambda/4$ monopole (or stub) antenna (*c*) from broad bandwidth gradually tapered constant-impedance "volcano smoke" antenna (*a*). The conical antenna (*b*) is of intermediate bandwidth. The directivity D is indicated for each antenna.

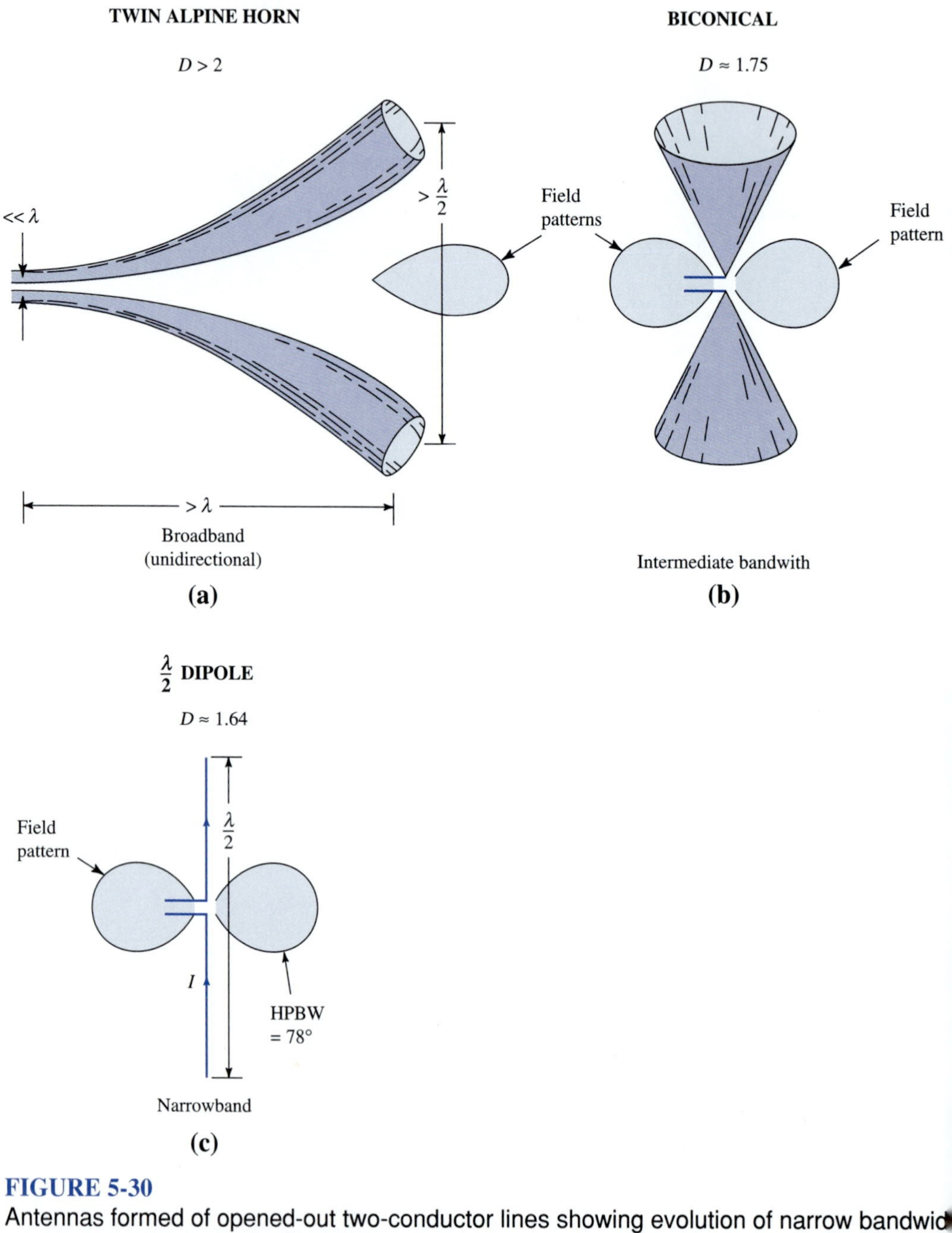

FIGURE 5-30
Antennas formed of opened-out two-conductor lines showing evolution of narrow bandwidth $\lambda/2$ dipole (*c*) from gradually tapered constant-impedance broad-bandwidth twin Alpine horn antenna (*a*). The biconical antenna (*b*) is of intermediate bandwidth. The directivity *D* is indicated for each antenna.

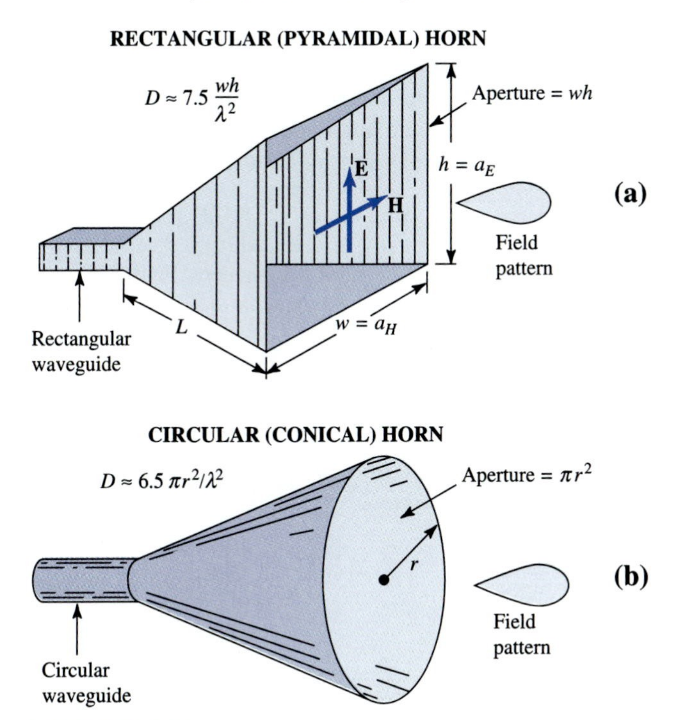

FIGURE 5-31
Opened-out waveguide antennas with pyramidal type from rectangular waveguide at (*a*) and conical type from circular waveguide at (*b*). These are both aperture-type antennas with effective aperture and directivity proportional to the area of the horn opening.

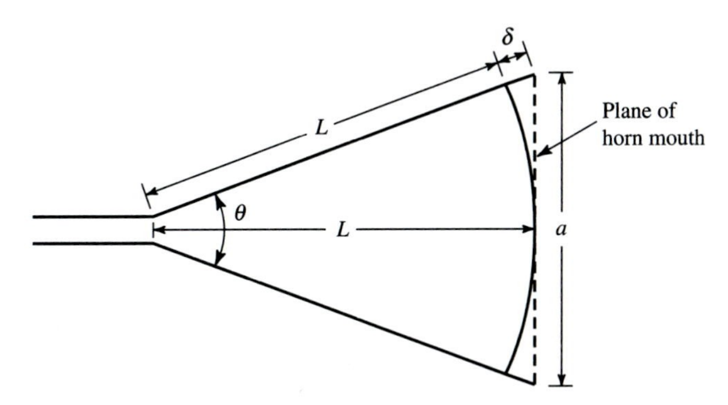

FIGURE 5-32
Cross section of pyramidal horn antenna with dimensions used in analysis. The diagram can be for either E plane or H plane cross sections. For the E plane the flare angle is θ_E and aperture dimension a_E. For the H plane the flare angle is θ_H and the aperture dimension a_H. See Fig. 5-31*a*.

Flat-Sheet Reflector Antennas

Reflector-type antennas are presented in Fig. 5-33. The directivity of a $\lambda/2$ dipole can be increased by placing it in front of a flat conducting reflector as in Fig. 5-33*a*. An array of two $\lambda/2$ dipoles in front of a flat reflector (*b*) produces higher directivity. Even more directivity is obtained by folding the flat reflector into a 90° or square corner as at (*c*). To reduce wind resistance and the amount of material required, the flat reflector can be replaced by a grid of parallel wires as shown.

Example 5-16. Power received by square-corner reflector. A U.S. channel 35 (599 MHz) TV station produces a field strength of 1 μV m^{-1} at a square-corner receiving antenna, as in Fig. 5-33*c*, with optimum dimensions for this channel. Find the power delivered to the receiver assuming it is matched to the antenna.

Solution.

$$\lambda = \frac{c}{f} = \frac{3 \times 10^8 \text{ m s}^{-1}}{599 \times 10^6 \text{ Hz}} = 0.501 \text{ m}$$

From Fig. 5-33*c*, $D = 20$ so the effective aperture

$$A_e = \frac{D\lambda^2}{4\pi} = \frac{20 \times 0.251}{4\pi} = 0.4 \text{ m}^2$$

and the received power is

$$\frac{E^2}{Z_0}A_e = \frac{(10^{-6})^2}{377}0.4 = 1.06 \times 10^{-15} \text{ W} = 1.06 \text{ fW} \quad \textit{Ans.}$$

Problem 5-9-3. Corner reflector. Assuming a directivity $D = 20$ for the corner reflector of Fig. 5-33*c*, what are the HPBWs? The *H* plane field is given by†

$$\cos(S_r \cos\phi) - \cos(S_r \sin\phi)$$

where $S_r = 2\pi S/\lambda$ and S = dipole to corner spacing. *Hint:* Use (5-2-10). *Ans.* $\theta(E \text{ plane}) = 45°$, $\theta(H \text{ plane}) = 45°$ for a symmetrical pattern.

Parabolic Dish and Dielectric Lens Antennas

A parabolic dish-shaped reflector can provide a high directivity (proportional to its aperture) but for efficient operation requires a suitable feed antenna, as shown in Fig. 5-34*a* on page 300. By contrast a simple dipole is adequate to feed a corner reflector as may be shown by image theory. Figure 5-34*c* shows a dielectric lens antenna analogous to its optical counterpart, and, like the parabolic dish, it has a directivity proportional to its aperture. Both the lens and parabolic reflector antennas (*Continued on page 301.*)

---

†Equation (12-3-6) in J. D. Kraus, "Antennas," 2d ed., McGraw-Hill, New York, 1988.

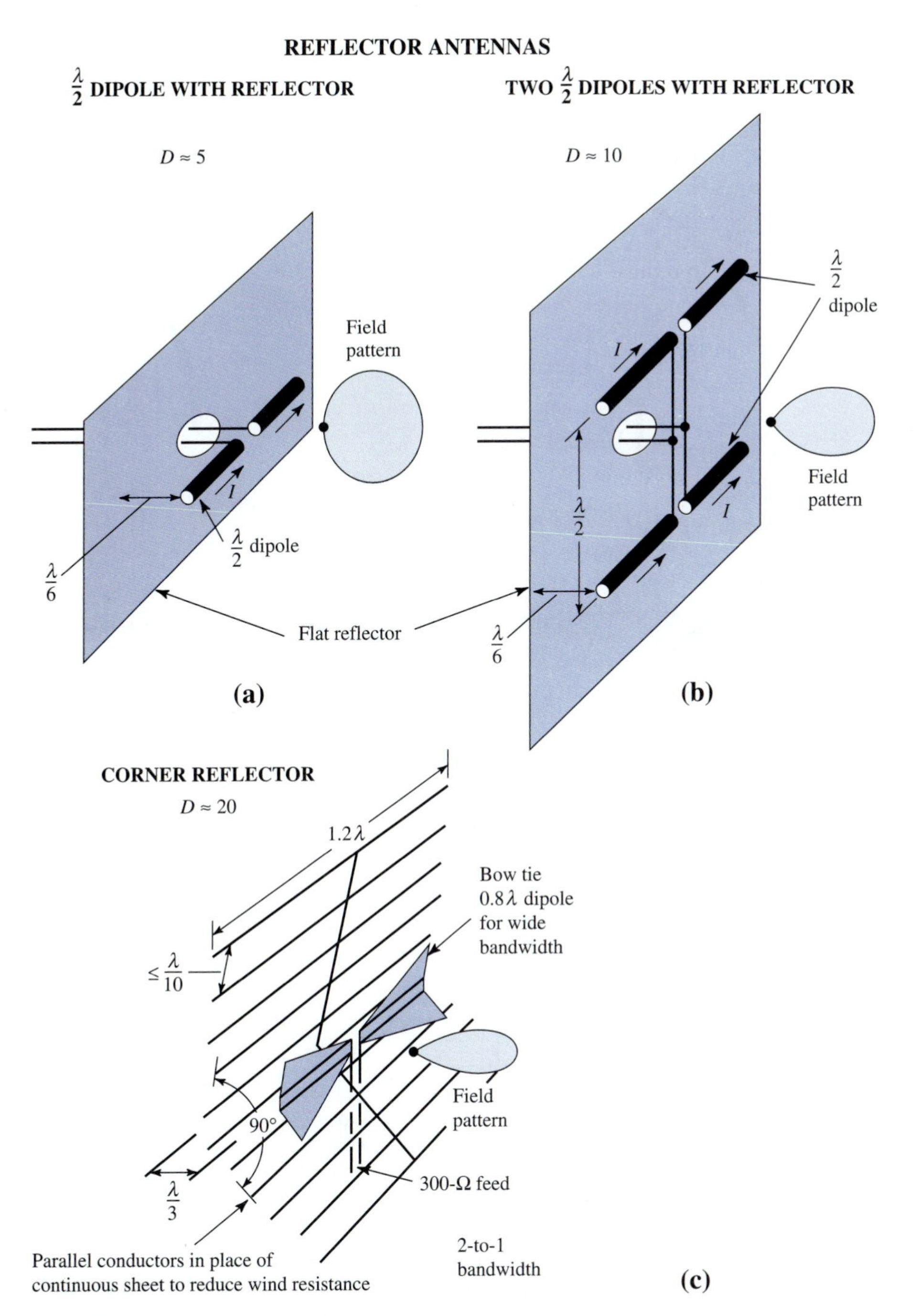

FIGURE 5-33
Reflector antennas showing (*a*) half-wave dipole with flat reflector, (*b*) array of two $\lambda/2$ dipoles with flat reflector, and (*c*) 90° (or square) corner reflector, in order of increasing directivity *D*. The corner reflector has a 2-to-1 bandwidth. The dimensions are for the center frequency. The corner reflector was invented by John Kraus in 1939.

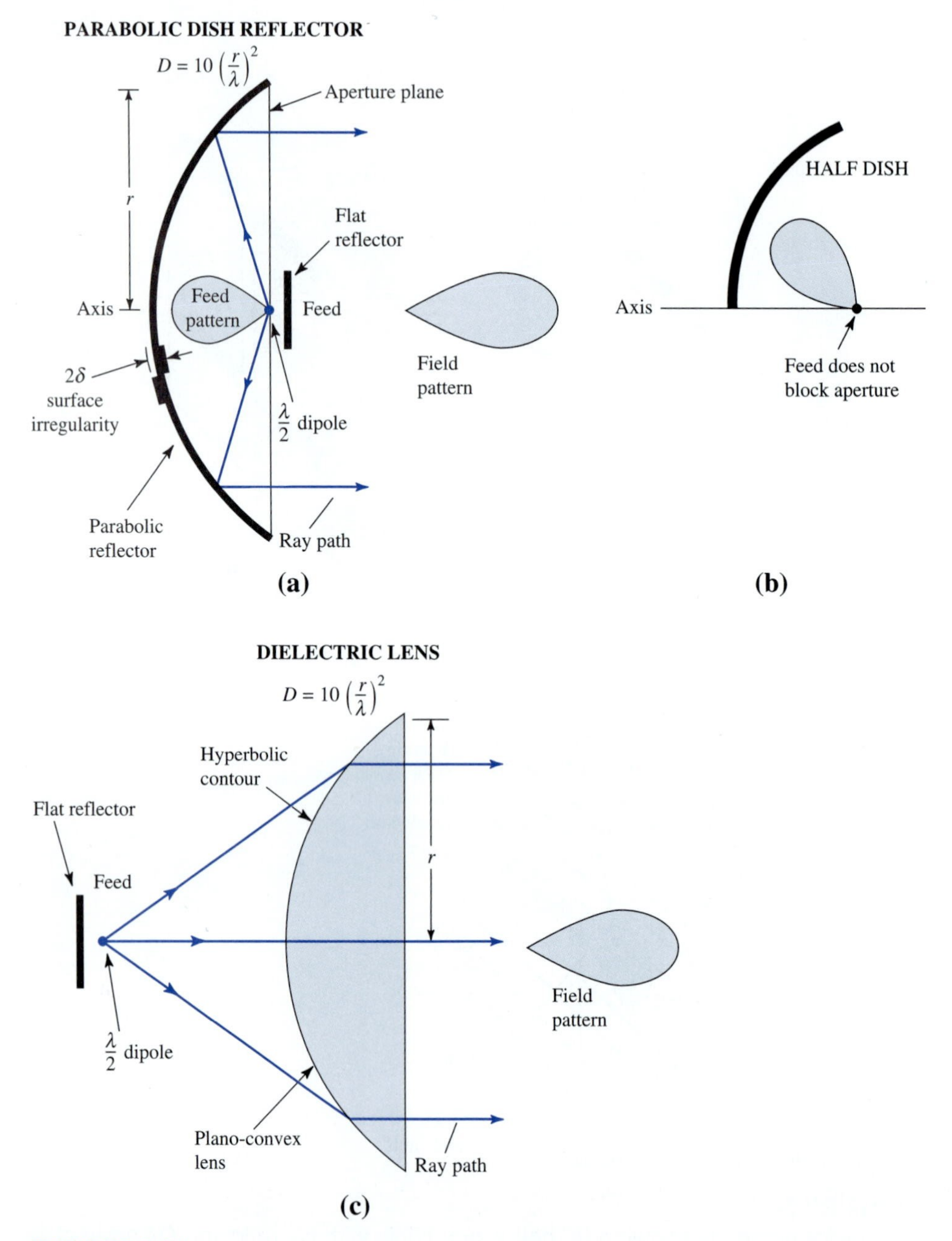

FIGURE 5-34
(*a*) Circular parabolic dish antenna. (*b*) Half dish to avoid aperture blocking. (*c*) Circular dielectric lens antenna. These are all aperture-type antennas with directivity *D* proportional to the aperture area.

require ray theory or optics in their design. Arrows in the figure trace ray paths. The feeds of both types radiate a spherical wave. The parabola converts the spherical wave into a plane wave by *reflection* while the lens does it by *refraction*. The plane wave forms the antenna beam.

Example 5-17. Parabolic dish design. The directivity (or gain) of a parabolic dish antenna depends on many factors:

1. The pattern of the feed antenna. If its pattern is too broad and spills over the edge of the dish, the gain is reduced. On the other hand, if the pattern is too narrow, the dish is not fully "illuminated" by the feed and the aperture is not fully utilized.
2. The accuracy of the dish surface relative to an ideal parabola. For example, if the surface departs a distance $\delta = \lambda/4$ (or 90° electrical degrees) from the parabolic curve, the reflected field is phase shifted 180°, which reduces the aperture efficiency. See dish surface of Fig. 5-34*a*.
3. The feed system blocks the center of the dish further reducing efficiency. By using only part of a full dish, as in Fig. 5-34*b*, blocking is avoided. The feed is still on the axis of the parabola but no longer blocks the aperture being used.
4. Many other factors are also involved. The aperture efficiency therefore varies widely depending on the specific design.†

Assuming an aperture efficiency of 70 percent, what is the directivity of a parabolic dish antenna as a function of its radius?

Solution.

$$D = \varepsilon_{ap}\frac{4\pi A_p}{\lambda^2} = 0.7\frac{4\pi\pi r^2}{\lambda^2} = 8.8\frac{\pi r^2}{\lambda^2} = 28\left(\frac{r}{\lambda}\right)^2 \quad Ans.$$

Problem 5-9-4. 3-m satellite earth-station parabolic dish. What is the directivity of the dish at 4 GHz (C band) assuming an aperture efficiency of 75 percent? *Ans.* $D = 1.18 \times 10^4$ (40.7 dBi).

End-Fire Antennas: Polyrod, Yagi-Uda, and Helical

Three end-fire antennas are shown in Fig. 5-35, the polyrod (*a*), Yagi-Uda (*b*), and axial-mode helix (*c*). The patterns of all three can be calculated to a good approximation as an end-fire array of isotropic sources spaced $\lambda/4$ with 90° phasing for the polyrod and Yagi-Uda and increased-directivity phasing for the axial-mode helical antenna. All three end-fire antennas may be regarded as rudimentary lens antennas which collect wave power over an aperture much larger than the physical cross-section perpendicular to the incoming wave. In the Yagi-Uda antenna only one element is driven (fed by transmission line), the rest being parasitic elements energized by mutual coupling, the reflector having a lagging phase and the directors leading phases.

†For a more complete discussion see J. D. Kraus, "Antennas," 2d ed., McGraw-Hill, New York, 1988.

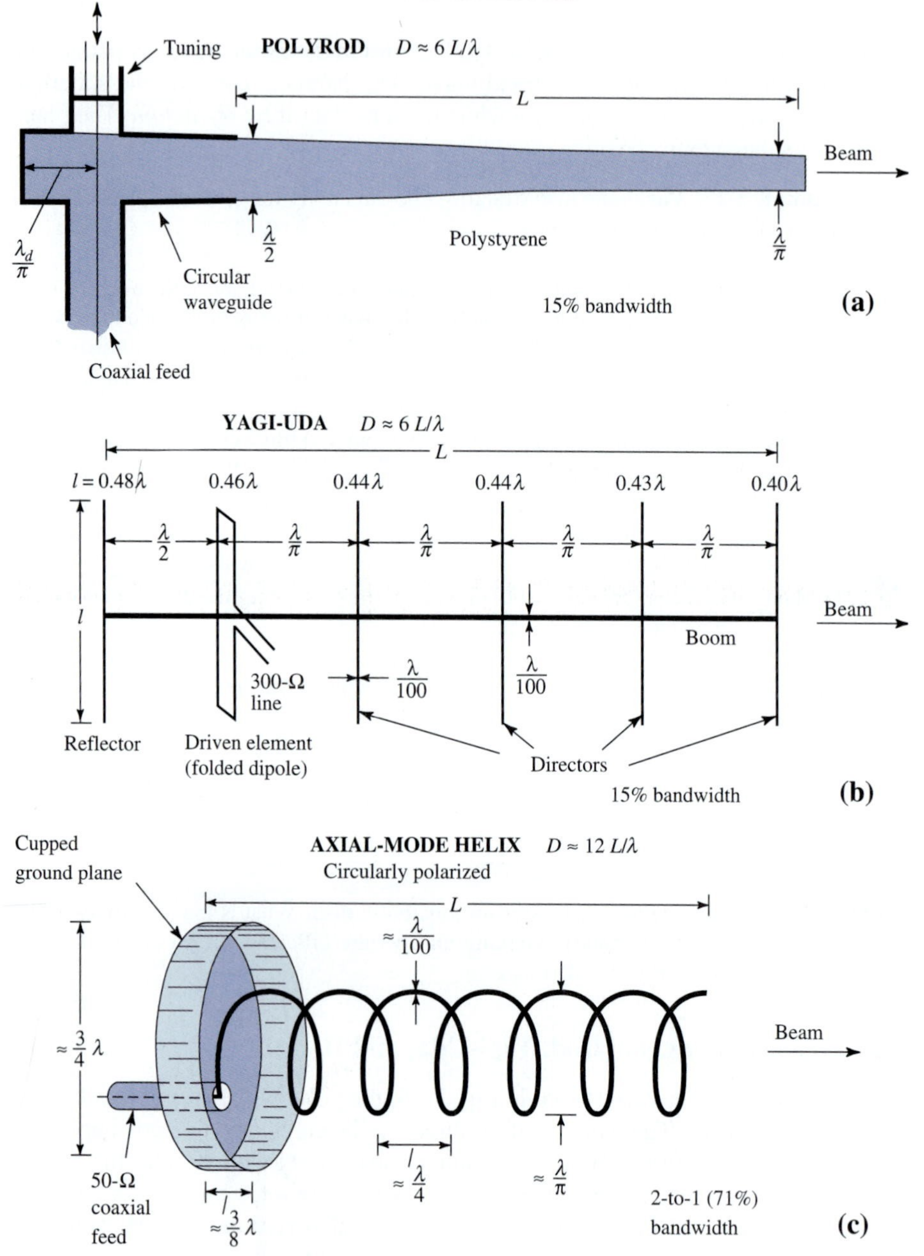

FIGURE 5-35

Three types of end-fire antennas: (*a*) polyrod, (*b*) Yagi-Uda, and (*c*) axial-mode helix. The directivity D of each is proportional to the length L with a higher directivity for the helical antenna because it operates in the increased directivity mode. The rods and cones of the retina of the eye (Fig. 9-3) are analogous to a polyrod antenna. Dimensions given are for the center frequency. Because of its high directivity, circular polarization, wide bandwidth, and noncritical dimensions, the axial-mode helical antenna is widely employed in space applications. It was invented by John Kraus in 1946.

Example 5-18. Design of quad-helix earth station antenna. Figure 5-36 shows an array of four right-handed axial-mode helical antennas for communication with satellites. Since the fields hug the helixes, there is minimal coupling or "cross talk" between adjacent helixes and the terminal impedance of each helix is approximately 50 Ω in the array, the same as when used alone.

Determine (*a*) the best spacing based on the effective apertures of the helixes, (*b*) the directivity of the array, and (*c*) connections for feeding all helixes equally and in phase.

Solution. The directivity of an axial-mode helix with circumference equal to λ at the center frequency is approximately

$$D = 12\, n S_\lambda$$

where n = number of turns and S_λ = spacing between turns in wavelengths.

For each helix in the array, $n = 10$ and $S_\lambda = 0.236$. Thus,

$$D = 12 \times 10 \times 0.236 = 28.3$$

The effective aperture of each helix is then

$$A_e = \frac{D\lambda^2}{4\pi} = \frac{28.3}{4\pi}\lambda^2 = 2.25\lambda^2$$

This is the area of a square equal to $\sqrt{2.25} = 1.5\lambda$ on a side. Therefore, a spacing between helixes of 1.5λ is appropriate. For a smaller spacing, the effective apertures of adjacent helixes overlap, decreasing the gain. A larger spacing does not increase the total aperture or the gain and may introduce grating lobes. *Ans.* (*a*)

(*Continued on next page.*)

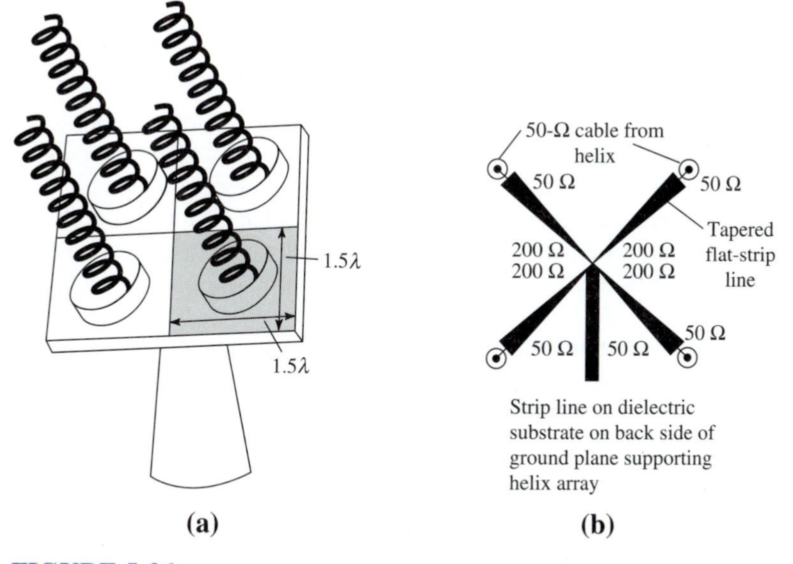

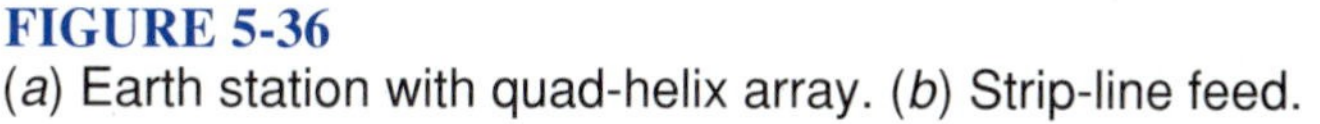

FIGURE 5-36
(*a*) Earth station with quad-helix array. (*b*) Strip-line feed.

At 1.5 λ spacing the four helixes have a total effective aperture of $2.25 \times 4 = 9\ \lambda^2$ for an array directivity of

$$D = \frac{4\pi A_e}{\lambda^2} = 113\ (20.5\ \text{dBi}) \quad \textit{Ans.} \quad (b)$$

A strip-line feed system is shown in Fig. 5-36*b*, which has 50- to 200-Ω tapered transitions joined to match a 50-Ω line. The tapered transitions have wide bandwidths which preserve the wide 2-to-1 bandwidth of the helixes. *Ans.* (c)

Problem 5-9-5. Four Yagi-Uda antennas. With four Yagi-Uda antennas instead of four helixes, as in Example 5-18, determine (*a*) best spacing and (*b*) directivity. Each antenna is 2.5λ long. *Ans.* (*a*) 1.26λ, (*b*) 60 (17.8 dBi).

Problem 5-9-6. Helix array. If the number of turns on the four-helix array of Fig. 5-36 is increased to 15, the spacing between turns to 0.25 λ, and the spacing between helixes to 1.7λ, find the directivity. *Ans.* $D = 144$ (21.6 dBi).

Problem 5-9-7. Helical antenna circular polarization. One of the reasons that helical antennas are used in space applications is that they are circularly polarized. With linearly polarized antennas, a change in orientation of the space vehicle could result in cross-polarization and loss of signal. With circular polarization this problem does not occur, provided both antennas are of the same hand. Thus, the right-handed helixes of the four-helix array match right-handed helixes on a satellite but are mismatched to left-handed helixes. Another advantage of circular polarization is in eliminating ground-bounce fading. (*a*) What is the signal loss when a right circularly polarized (RCP) antenna receives linear polarization? (*b*) What signal rejection will an RCP 10-turn helix have to a pure left circularly polarized (LCP) wave? Note that the axial ratio of an axial-mode helical antenna is

$$\text{AR} = \frac{2n+1}{n}$$

where n = number of turns. *Hint:* Use (4-11-1) and (4-11-4). *Ans.* (*a*) −3 dB; (*b*) −32 dB.

The "helical beam antenna" or axial-mode helix is conveniently fed via a ground plane as in Figs. 5-35 and 5-36. It may also be fed without a ground plane as shown in Fig. 5-37 where the ground plane is replaced by two loops.†

The first radio telescope, dubbed "Big Ear," which John Kraus built in 1952 at the Ohio Sate University, had an array of 96 helical antennas, each almost 3 m long. This telescope, operating at 250 MHz was used to produce some of the first maps of the radio sky.

The helix is the antenna of choice for most satellites. Each of the 24 GPS (Global Positioning Satellites) built by the Aerospace Corp. has an array of 12 axial-mode helixes (Fig. 5-52) and each of the Russian Ekran satellites has an array of 96 helixes. These are just two of many examples—and helical antennas are on Mars and the moon.

†J. D. Kraus, A Helical-Beam Antenna without a Ground-plane, *IEEE Antennas and Propagation Magazine,* **37**(2): 45 (April 1995).

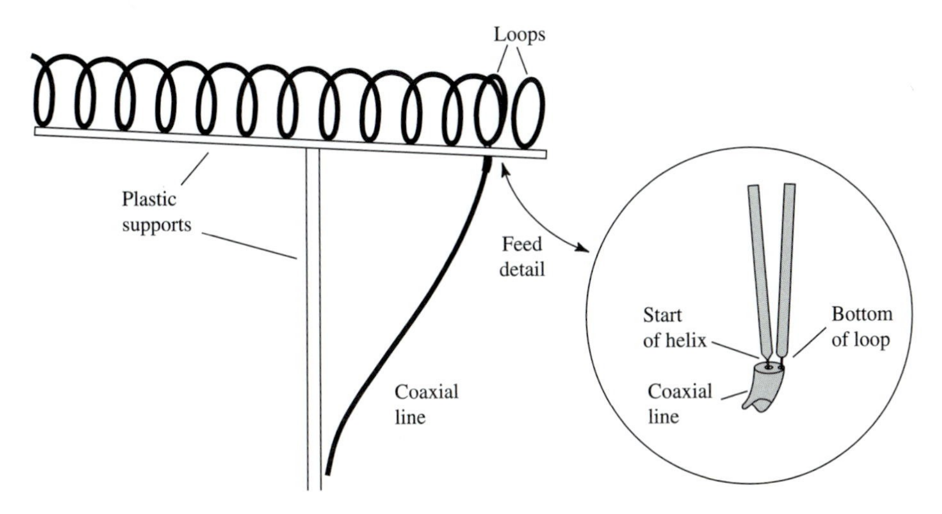

FIGURE 5-37
Ten-turn helical beam antenna with two loops instead of a ground plane. The loops are the same diameter as the helix. The right-hand loop is about 0.4λ from the loop at the feed point. The inset shows feed detail with tapered gap for broadband match.

Broad-Bandwidth Antennas: Conical Spiral, Log-Periodic, and 3-in-1

Figure 5-38 shows three very broad-bandwidth type antennas: the conical spiral, the log-periodic, and the three-in-one antenna. The conical spiral (*a*) may be regarded as a flat spiral which has been wrapped around a dielectric cone. The conical spiral is fed by a coaxial cable bonded to one conducting strip, with its inner conductor joined to the other strip at the apex, as indicated in the sketch of the flat spiral. The lower-frequency limit of the conical spiral occurs when the base diameter is $\lambda/2$. The high-frequency limit occurs when the apex diameter is $\lambda/4$. Thus, the bandwidth is in the ratio ½ base diameter to apex diameter, which, for the cone in (*a*), is about 7 to 1. The bandwidth of the log-periodic antenna depends on the ratio of the next-to-longest to next-to-shortest dipoles, which for the array in (*b*) is about 4 to 1. The 3-in-1 antenna of (*c*) combines a log-periodic, a square corner reflector, and a Yagi-Uda antenna to cover a 16-to-1 bandwidth (U.S. VHF TV, FM, and UHF TV bands). The directivities D are as indicated.

For most antennas, losses are small so that the *gain* is nearly equal to the *directivity*. Thus, as a practical matter, gain and directivity are often used interchangeably.

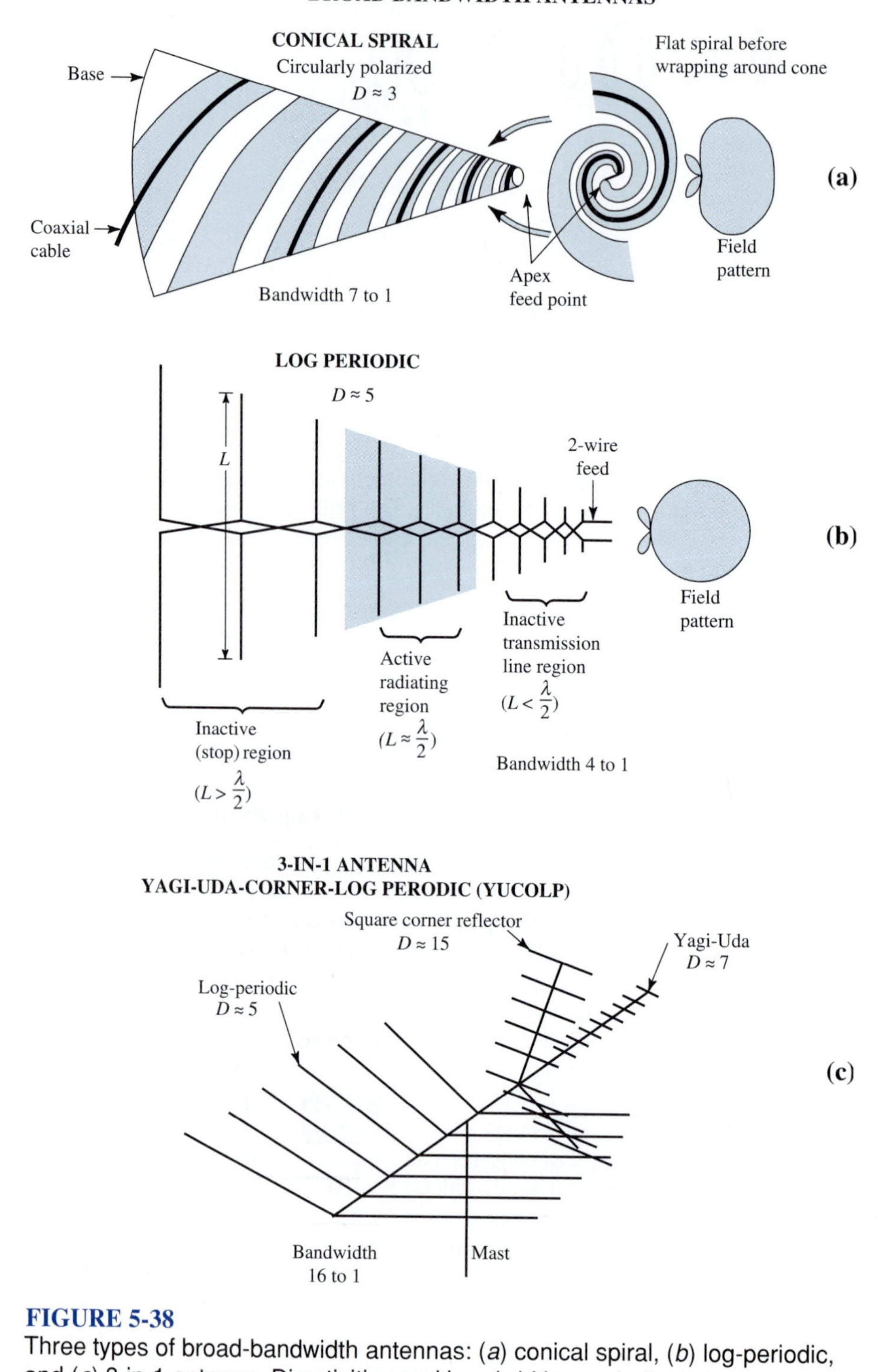

FIGURE 5-38
Three types of broad-bandwidth antennas: (*a*) conical spiral, (*b*) log-periodic, and (*c*) 3-in-1 antenna. Directivities and bandwidths are indicated.

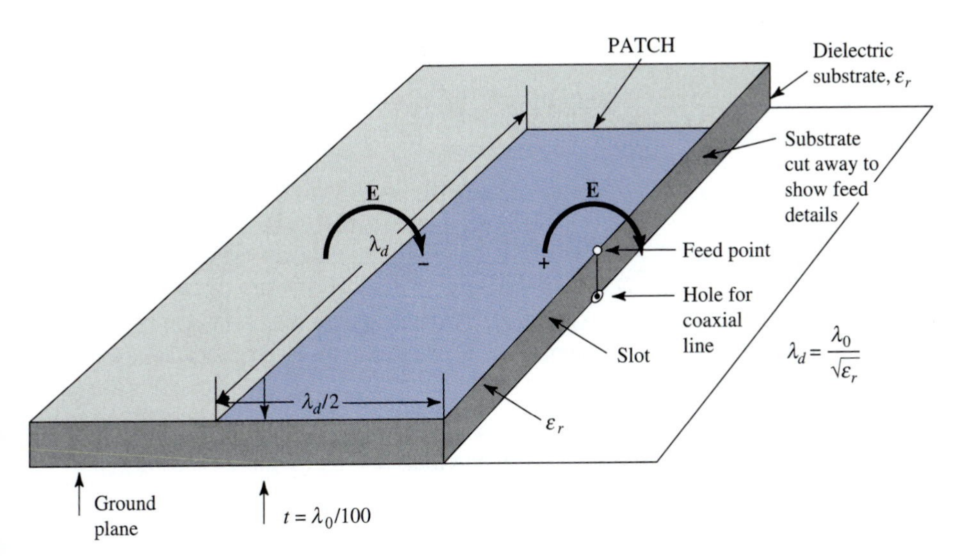

FIGURE 5-39
Single patch antenna. Arrows show the direction of **E** at the slots.

Patch Antennas

The "patch" is a low-profile, low-gain, narrow-bandwidth antenna. It is used alone and in large flat-panel arrays. Aerodynamic considerations require low-profile antennas on aircraft and many kinds of vehicles.

Figure 5-39 shows a patch antenna with its dielectric substrate partially cut away to show the feed point. Typically a patch consists of a thin conducting sheet about 1 by $\frac{1}{2}\lambda_0$ mounted on the substrate.

Radiation from the patch is like radiation from two slots, at the left and right edges of the patch. The "slot" is the narrow gap between the patch and the ground plane. The patch-to-ground-plane spacing is equal to the thickness t of the substrate and is typically about $\lambda_0/100$, as indicated in the figure.

Example 5-19. Four-patch array. Find (*a*) the directivity and (*b*) beam area of the four-patch array of Fig. 5-40. The detailed design of the strip-line feed was worked out in Example 3-18.

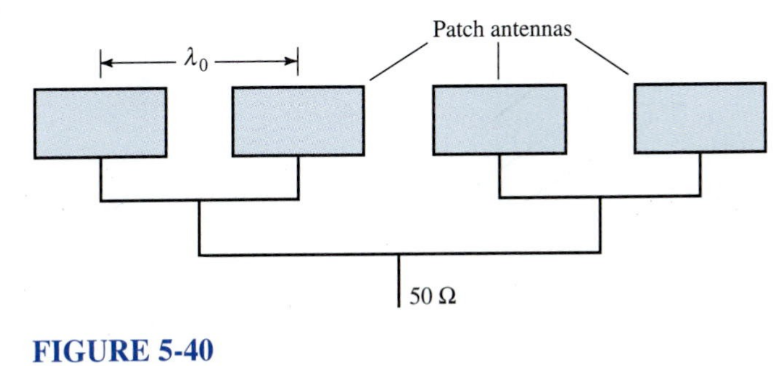

FIGURE 5-40
Four-patch array.

Solution.

$$A_e \simeq 4\lambda \times \frac{\lambda}{2} = 2\lambda^2$$

$$D = \frac{4\pi A_e}{\lambda^2} = 8\pi \simeq 25\ (14\text{dBi}) \quad Ans. \quad (a)$$

$$\Omega_A = \frac{4\pi}{D} = \frac{4\pi}{25} \simeq 0.50\ \text{sr} \quad Ans. \quad (b)$$

Problem 5-9-8. Patch array. Use ARRAYPATGAIN to plot the field pattern of the four-patch array of Fig. 5-40 in the plane perpendicular to the patches and through the long axis of the array.

The preceding pages provide a very brief introduction to a variety of antennas. A much more complete description of these antennas and many more types is given by Kraus.†

Arrays of Dipoles and Slots; Frequency Selective Surfaces (FSS).

An array of dipoles can reflect an incident wave as completely as a solid conducting sheet. Conversely, an array of slots in a conducting sheet can make the sheet completely transparent. These effects are frequency dependent and surfaces exhibiting this behavior are called *frequency selective surfaces* (FSS). They are used in hybrid radomes and other special purpose surfaces. For an introduction to the subject see B. A. Munk.‡

5-10 RADIO LINK AND FRIIS FORMULA

Consider the radio-link communication circuit of Figure 5-41 consisting of a transmitter T with antenna of effective aperture A_{et} and receiver R with antenna of effective aperture A_{er}. The distance between transmitting and receiving antennas is r. If

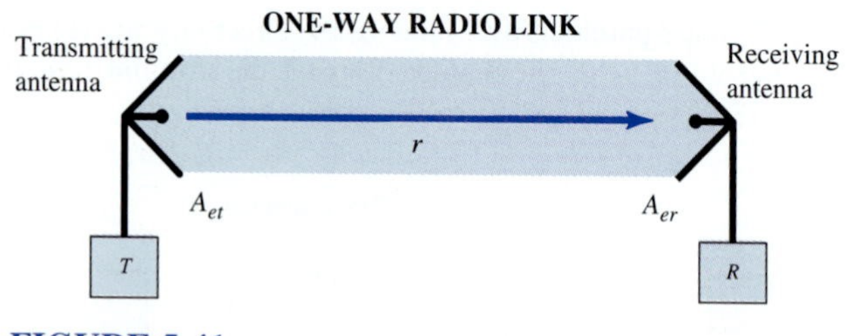

FIGURE 5-41
Single transmission path and parameters used in Friis formula.

†J. D. Kraus, "Antennas," 2d ed., McGraw-Hill, New York, 1988.

‡Benedikt A. Munk, "Frequency-sensitive (or selective) surfaces," Sec. 12–13 in J. D. Kraus, "Antennas," 2d ed., McGraw-Hill, 1988, p. 600.

the transmitter power P_t is radiated by an isotropic source, the power received per unit area at the receiving antenna is

$$S = \frac{P_t}{4\pi r^2} \qquad (\mathrm{W\ m^{-2}}) \tag{1}$$

and the power available to the receiver is

$$P_r = SA_{er} \qquad (\mathrm{W}) \tag{2}$$

The transmitting antenna has an effective aperture A_{et} and hence a directivity $D = 4\pi A_{et}/\lambda^2$, so that the power available at the receiver is D times greater, or

$$P_r = \frac{SA_{er}4\pi A_{et}}{\lambda^2} \qquad (\mathrm{W}) \tag{3}$$

Substituting (1) in (3) gives

$$P_r = \frac{P_t A_{er}}{4\pi r^2}\frac{4\pi A_{et}}{\lambda^2} \qquad (\mathrm{W})$$

or

$$\frac{P_r}{P_t} = \frac{A_{er}A_{et}}{r^2\lambda^2} \qquad \text{(dimensionless)} \qquad \textbf{\textit{Friis transmission formula}} \tag{4}$$

where P_r = received power, W
P_t = transmitted power, W
A_{er} = effective aperture of receiving antenna, $\mathrm{m^2}$
A_{et} = effective aperture of transmitting antenna, $\mathrm{m^2}$
r = distance between receiving and transmitting antennas, m
λ = wavelength, m

This is the *Friis transmission formula,* which gives the ratio of the received to transmitted power for a direct path.

5-11 ANTENNA TEMPERATURE, SIGNAL-TO-NOISE RATIO, AND REMOTE SENSING†

The *noise power per unit bandwidth* available at the terminals of a resistor of resistance R at a temperature T_r (Fig. 5-42*a*) is given by the Nyquist‡ relation as

$$p = kT_r \qquad (\mathrm{W\ Hz^{-1}}) \tag{1}$$

where p = power per unit bandwidth, $\mathrm{W\ Hz^{-1}}$
k = Boltzmann's constant = $1.38 \times 10^{-23}\ \mathrm{J\ K^{-1}}$
T_r = absolute temperature, K

†For a more detailed treatment, see J. D. Kraus, "Radio Astronomy," 2d ed., Cygnus-Quasar, Powell, Ohio, 1986, pp. 3-39 to 3-45.

‡H. Nyquist, Thermal Agitation of Electric Charge in Conductors, *Phys. Rev.,* **32**:110–113 (1928).

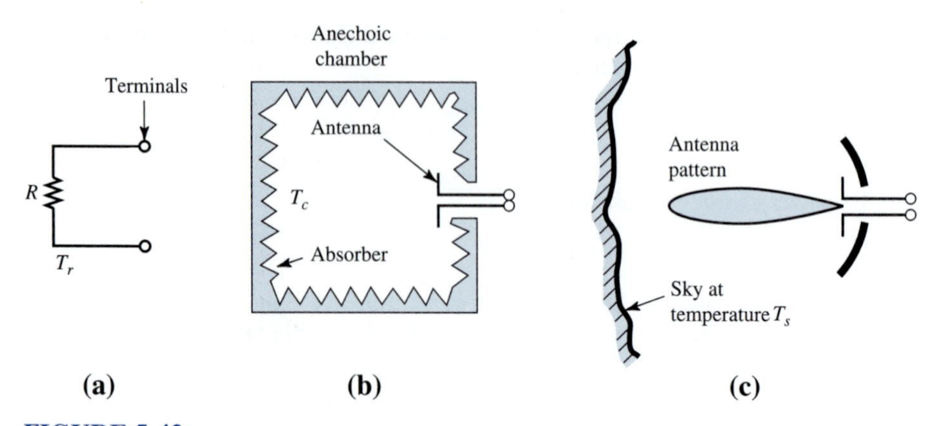

FIGURE 5-42
(*a*) Resistor at temperature T_r. (*b*) Antenna in an anechoic chamber at temperature T_C. (*c*) Antenna observing sky at temperature T_S. The same noise power per unit bandwidth is available at the terminals in all three cases if $T_r = T_C = T_S$.

If the resistor R is replaced by a *lossless* antenna of radiation resistance R in an anechoic chamber at temperature T_c (Fig. 5-42*b*), the noise power per unit bandwidth available at the antenna terminals is the same as in (1) provided $T_c = T_r$.

Now if the antenna is removed from the anechoic chamber and pointed at a sky of temperature T_s (Fig. 5-42*c*), the noise power per unit bandwidth is again the same as in (1) (provided $T_s = T_r$), and we can say that the antenna has a *noise temperature* T_A equal to the sky temperature T_s.†

Using an antenna to measure a distant temperature in this way is *passive remote sensing,* and the antenna in this application may be called a ***radio telescope.*** This passive remote sensing is in contrast to the *active remote sensing* of radar.

To measure the distant or sky temperature T_s, the antenna noise temperature may be compared with that of a resistor at an adjustable temperature T_r by alternately connecting antenna (directed at sky) and resistor to a receiver. When the receiver detects no difference, $T_A = T_s = T_r$.

The noise temperature T_A of the antenna (assumed *lossless*) is equal to the sky temperature T_s, and *not* the physical temperature of the antenna. This contrasts to the resistor of Fig. 5-42*a,* which is completely lossy and, therefore, has a noise temperature equal to its physical temperature. Thus, for a radio-telescope antenna the noise power per unit bandwidth is given by

$$p = kT_A \qquad (\text{W Hz}^{-1}) \tag{2}$$

where T_A is the *antenna* (*noise temperature*) or temperature of the antenna's radiation resistance, determined by the sky temperature at which the antenna beam is directed.

†It is assumed that the entire antenna pattern "sees" the sky of temperature T_s.

Hence, a radio telescope antenna (and receiver) may be regarded as a radiometer (or temperature-measuring device) for remote sensing the temperature of distant regions coupled to the system through the radiation resistance of the antenna. An extreme view is to imagine that with a radio telescope we can, in effect, stretch the wires from the terminals in Fig. 5-42*a*, until the resistor R comes into contact with the distant regions. Recall Fig. 5-1.1.

In the above we have assumed that the antenna has no thermal losses and also that all of its pattern is encompassed by the region being observed (negligible side and back lobes).

Multiplying (2) by the bandwidth B, we obtain the total power available as

$$\boxed{P = kT_A B \qquad \text{(W)}} \tag{3}$$

where B = receiver bandwidth, Hz.

It is often convenient to express the received power per unit bandwidth in terms of a *flux density* S. Thus, dividing (2) by the effective aperture A_e of the antenna, we have

$$S = \frac{p}{A_e} = \frac{kT_A}{A_e} \qquad (\text{W m}^{-2}\ \text{Hz}^{-1}) \tag{4}$$

In the above development we have assumed a single celestial source whose extent is greater than the antenna beam. In practice the antenna temperature may include contributions from several sources, or the source under observation may be superimposed on a background temperature region. To measure the temperature of a source under these circumstances, the radio telescope beam is moved onto and then off of the source, and an incremental or difference temperature ΔT_A is measured. Thus, (4) for the *source flux density* can be rewritten as

$$\boxed{S = \frac{k\Delta T_A}{A_e} \qquad (\text{W m}^{-2}\ \text{Hz}^{-1}) \qquad \textbf{\textit{Source flux density}}} \tag{5}$$

Note that the units for the flux density S (W m^{-2} Hz^{-1}) are the same as for the Poynting vector per unit bandwidth, so we may regard the flux density as a measure of the Poynting vector (per unit bandwidth) received from distant regions. In radio astronomy observations, flux densities are very small, and the unit of convenience is the jansky (Jy) = 10^{-26} W m^{-2} Hz^{-1}, after Karl G. Jansky who made the first radio astronomy observations in 1933.

If the remote source is small compared to the antenna beamwidth, all of ΔT_A is due to the source, and (5) gives the correct flux density, but ΔT_A is *not* equal to the source temperature. However, if the source solid angle Ω_s and the antenna beam solid angle Ω_A (Fig. 5-43) are known, the source temperature is given very simply by

$$T_s = \frac{\Omega_A}{\Omega_s}\Delta T_A \qquad \text{(K)} \tag{6}$$

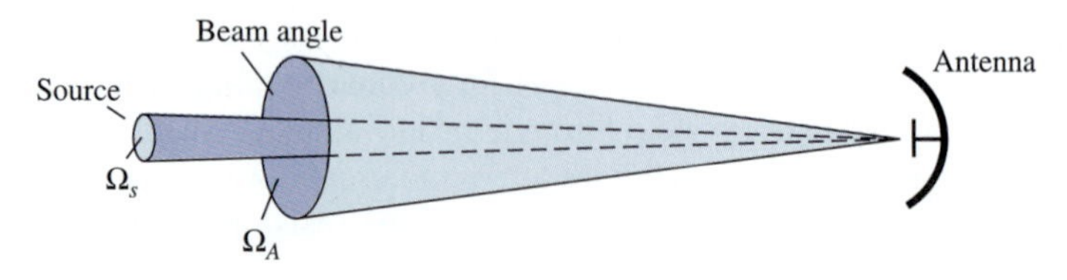

FIGURE 5-43
Situation where source Ω_s is smaller than the beam area Ω_A.

where T_s = source temperature, K
ΔT_A = incremental antenna (noise) temperature, K
Ω_s = source solid angle (see Fig. 5-43), sr
Ω_A = antenna beam solid angle (see Fig. 5-43), sr

Let us now apply (6), by way of an example, to a classic, historic remote-sensing observation.

Example 5-20. Mars temperature. The incremental antenna temperature for the planet Mars measured with the U.S. Naval Research Laboratory† 15-m radio telescope antenna at 31.5-mm wavelength was found to be 0.24 K. Mars subtended an angle of 0.005° at the time of the measurement. The antenna HPBW = 0.116°. Find the average temperature of Mars at 31.5-mm wavelength.

Solution. Assuming that Ω_A is given by the solid angle within the HPBW, we have from (6) that the Mars temperature

$$T_s = \frac{\Omega_A}{\Omega_s}\Delta T_A \approx \frac{0.116^2}{\pi(0.005^2/4)}0.24 = 164 \text{ K} \quad \textit{Ans.}$$

This temperature is less than the infrared temperature measured for the sunlit side (250 K), implying that the 31.5-mm radiation may originate farther below the Martian surface than the infrared radiation. This is an example of *passive remote sensing* of the surface of another planet from the earth.

Problem 5-11-1. Antenna temperature. An end-fire array is directed at the zenith. The array is located over flat nonreflecting ground. If 0.9 Ω_A is within 45° of the zenith and 0.08 Ω_A, between 45° and the horizon, calculate the antenna temperature. The sky brightness temperature is 5 K between the zenith and 45° from the zenith, 50 K between 45° from the zenith and the horizon, and 300 K for the ground (below the horizon). The antenna is 99 percent efficient and is at a physical temperature of 300 K. *Ans.* 14.5 K.

Problem 5-11-2. Earth-station antenna temperature. An earth-station dish of 100-m^2 effective aperture is directed at the zenith. Calculate the antenna temperature assuming that the sky temperature is uniform and equal to 6 K. Take the ground temperature equal to 300 K and assume that one-third of the minor-lobe beam area is in the back direction. The wavelength is 75 mm, and the beam efficiency is 0.8. *Ans.* 25.6 K.

†C. H. Mayer, T. P. McCullough, and R. M. Sloanaker, "Observations of Mars and Jupiter at a Wavelength of 3.15 cm," *Astrophys. J.*, **127**: 11–16 (January 1958).

The source temperature in the above discussion and example is an *equivalent temperature.* It may represent the physical temperature of a planetary surface, as in the example, but, on the other hand, a celestial plasma cloud with oscillating electrons which is at a physical temperature close to absolute zero may generate radiation with an equivalent temperature of thousands of kelvins. The temperatures we are discussing are thermal (noise) temperatures like those of a perfect emitting-absorbing object called a *blackbody.* A hot object filling the beam of a receiving antenna will ideally produce an antenna temperature equal to its thermometer-measured temperature.† However, the oscillating currents of a radio transmitting antenna can produce an equivalent temperature of millions of kelvins even though the antenna structure is at normal outdoor temperature. It may be said that the antenna (and its currents) have an equivalent blackbody (or noise) temperature of millions of kelvins.

All objects not at absolute zero produce radiation which, in principle, may be detected with a radio antenna-receiver. A few objects are shown in Fig. 5-44 with the equivalent temperatures measured when the horn antenna is pointed at them. Thus, the temperature of a distant quasar is over 10^6 K, of Mars 164 K, of a radio transmitter on the earth 10^6 K, of a human 310 K, of the ground 290 K,‡ while the empty sky at the zenith is 3 K. This temperature, called the 3 K sky background, is the residual temperature of the primordial fireball believed to have created the universe and is the *minimum possible temperature* of any antenna looking at the sky.

An assumption was made in the above discussion which requires comment. It was assumed that the antenna and source polarizations were matched (same

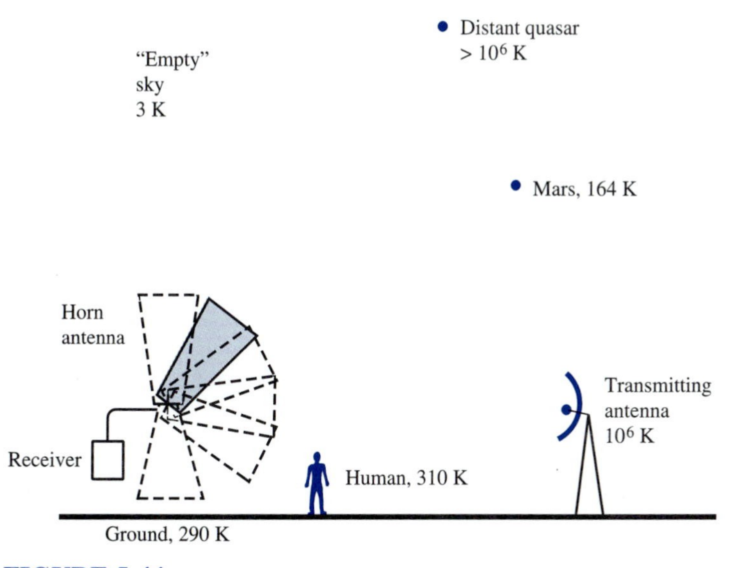

FIGURE 5-44
Horn antenna directed at various objects senses different temperatures as suggested.

†Assuming the object's intrinsic impedance = 377 Ω.

‡Due to reflection, more realistic values for a human and the ground might be less.

polarization states on the Poincaré sphere; see Sec. 4-13). Although this is possible for the transmitting antenna in Fig. 5-44, it is not possible for the other sources because their radiation is *unpolarized*† and *any* antenna, whether linearly or circularly polarized, receives only *half* of the available power. Hence, for such sources the flux density of the source at the antenna is given by twice (5), or

$$S = \frac{2k\Delta T_A}{A_e} \quad (\text{W m}^{-2}\text{ Hz}^{-1}) \qquad \textbf{Source flux density} \tag{7}$$

The Mars measurement is an example of *passive remote sensing* in which the temperature of a very distant object is determined with a radio telescope. As another example of passive remote sensing, many satellites circling the earth carry radio telescopes which are directed down at the earth for measuring the temperature of the earth's surface in great detail in the same way that the telescope in the worked example was used to determine the surface temperature of Mars. See Example 5-29. By contrast, radar detection is *active remote sensing* since a signal is transmitted and an echo received.

The sensitivity of a receiving system (antenna and receiver) depends not only on the antenna temperature T_A but also on the temperature or noise contribution of the receiver and the transmission line connecting the antenna to the receiver. The resultant of these temperatures is called the *system temperature* T_{sys}, which is a factor in the *signal-to-noise ratio* (S/N or SNR) for a *radio link* as given by

$$\frac{S}{N} = \frac{P_t A_{et} A_{er}}{r^2 \lambda^2 B k T_{\text{sys}}} \quad (\text{dimensionless}) \qquad \textbf{Signal-to-noise ratio SNR of radio link} \tag{8}$$

where P_t = transmitter power, W
A_{et} = effective aperture of transmitting antenna, m^2
A_{er} = effective aperture of receiving antenna, m^2
r = distance between transmitter and receiver, m
λ = wavelength, m
B = bandwidth, Hz
k = Boltzmann's constant $= 1.38 \times 10^{-23}\text{ J K}^{-1}$
T_{sys} = system temperature, K

In (8) matched polarizations and bandwidths are assumed.

Note: Ratios and Decibels

It is often convenient to express voltage, power, and noise power ratios in decibels (dB). Thus,

$$\text{dB} = 20\log_{10}\frac{V}{V_0}$$

where V_0 = reference voltage

†See J. D. Kraus, "Radio Astronomy," 2d ed., Cygnus-Quasar, Powell, Ohio, 1986, Secs. 4-4, 4-5, and 4-6.

and
$$\text{dB} = 10\log_{10}\frac{P}{P_0} = 10\log_{10}\frac{\text{signal power}}{\text{noise power}}$$

where P_0 = reference power. Both expressions give the same number of decibels since

$$\frac{P}{P_0} = \left(\frac{V}{V_0}\right)^2$$

Example 5-21. Horn absorber. If a perfect absorber with the impedance of space (= 377 Ω/square) is placed so as to completely cover the front of a horn antenna, it will ideally produce an antenna temperature equal to the absorber's thermometer-measured temperature (Fig. E5-21). If the absorber is completely shielded from the outside (open only to the horn) and is cooled, it can provide calibration temperatures for the radio telescope. Thus, if cooled to liquid helium temperature it will give a 4.1 K calibration. Or if the absorber temperature can be controlled so that when the absorber is in front of the horn the radio telescope response is the same as with the absorber removed, the temperature of the object or region being observed by the radio telescope is equal to the absorber temperature. This null type of measurement was used on the Cosmic Background Explorer (COBE) satellite to remeasure the 3-K, 4-GHz sky background originally discovered by Penzias and Wilson.† The temperature has now been evaluated more accurately as 2.73 K. This temperature is believed to be from the remnant of the primordial Big Bang and is the lowest possible antenna temperature for a sky-scanning antenna.

If a narrow-beam 4-GHz antenna looking at essentially empty sky at the zenith has an antenna temperature of 4.73 K, how much is due to side lobes or antenna loss?

Solution. $4.73 - 2.73 = 2°\text{C}$. *Ans.*

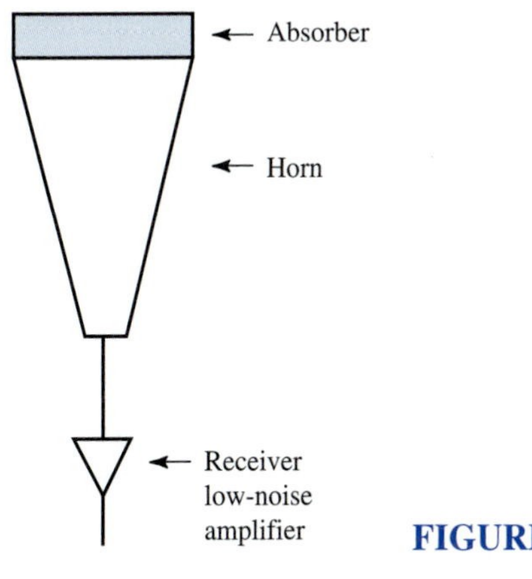

FIGURE E5-21

†R. W. Wilson, "Discovery of the Cosmic Microwave Background," pp. 185–195 in *Serendipitous Discoveries in Radio Astronomy,* NRAO, Green Bank, W. Va., 1983.

Example 5-22. Downlink signal-to-noise ratio. The criterion of detectability of a signal is the *signal-to-noise* ratio (S/N) (8). For a transmitter power of 1 W and an isotropic (nondirectional) antenna, the signal-to-noise ratio of a lossless line-of-sight radio link is given by

$$\frac{S}{N} = \frac{\lambda^2}{16\pi^2 r^2 k T_{\text{sys}} B}$$

where λ = wavelength, m
r = distance from transmitter to receiver, m
k = Boltzmann's constant = 1.38×10^{-23} J K^{-1}
T_{sys} = system temperature, K
B = bandwidth, Hz

For a Clarke-orbit geostationary satellite C-band transponder downlink to an earth station (Fig. 5-45), the transponder power = 5 W, distance = 36,000 km, λ = 7.5 cm, and antenna gain = 30 dB. If the earth station antenna has 38-dB gain and the earth station receiver a system temperature of 100 K, find the earth station S/N. The bandwidth B = 30 MHz for frequency-modulated (FM) TV signals.

Note: The S/N ratio as used in this example actually represents the carrier-to-noise (C/N) ratio, defined as the ratio of the power in the unmodulated carrier to the noise power. For FM video signals as employed by the Clarke-orbit satellites, the S/N at the output of the FM demodulator can exceed the carrier-to-noise ratio (C/N) by 35 dB or more, provided that the C/N is above the "FM threshold" of approximately 10 dB. The amount of enhancement depends on the signal's *modulation index,* the ratio of the peak frequency deviation to the modulation frequency. In practice, satellite designers strive for a C/N of 13 to 20 dB, corresponding to 3 to 10 dB of *link margin* to allow for antenna misalignment, attenuation along the propagation path, attenuation due to water or snow in the earth-station dish, transponder power variations, and demodulator inefficiency.

Solution. For 1 W and isotropic antenna,

$$\frac{S}{N} = \frac{\lambda^2}{16\pi^2 r^2 k T_{\text{sys}} B}$$

$$= \frac{0.075^2}{16\pi^2(36^2 \times 10^{12})(1.38 \times 10^{-23})(100)(30 \times 10^6)}$$

$$= 6.64 \times 10^{-7} = -61.8 \text{ dB}$$

For transponder antenna gain = 30 dB and transponder power = 5 W = 7 dB,

ERP (effective radiated power) of transponder = 30 + 7
= 37 dB (over 1 W isotropic)

For earth station gain = 38 dB,

$$S/N \text{ (downlink)} = 37 + 38 - 61.8 = 13.2 \text{ dB} \quad \textit{Ans.}$$

This is 3.2 dB more than the minimum S/N ratio considered to be acceptable.

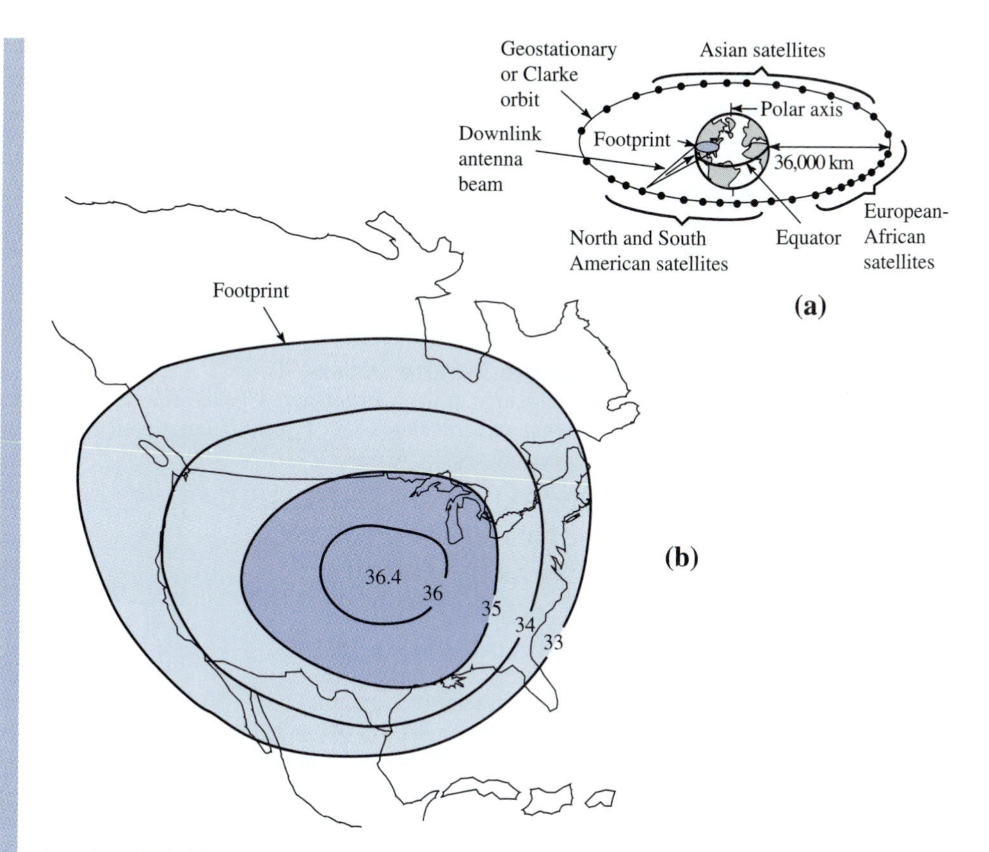

FIGURE 5-45
(*a*) Geostationary TV and communication relay satellites in Clarke orbit around the earth. (*b*) Effective radiated power contours, or footprint, in decibels above 1-W isotropic of typical transponder downlink over North America. Note that the footprint gives radiation contours over a spherical surface (the earth's) and, in most cases, at an oblique angle. Only when the satellite beams directly down on the earth's equator with a relatively narrow pattern will the footprint contours approximate the true antenna pattern contours.

Problem 5-11-3. Signal-to-noise ratio. What is the S/N ratio of a communication link operating with 50-MHz bandwidth (for 5 TV channels at 10 MHz each) over a distance of 1000 km if the parabolic dish antennas are 1-m diameter operating at 3 GHz? The transmitter power is 10 W and the receiver system temperature is 100 K. *Ans.* 33 dB.

Problem 5-11-4. Signals from Pioneer 10 beyond the solar system. In March 1992 Pioneer 10 was 8 billion kilometers (8×10^{12} m) distant, well outside the solar system into deep space. Launched by NASA 20 years earlier, on March 2, 1972, it was the first man-made object to escape from the solar system. Signals from its 8-W transmitter were still being received, informing us, among other things, that the solar wind was still observable. Pioneer 10 gave us our first close-up views of Jupiter and has become our first interstellar probe. But even at its velocity of 40,000 km/h it will be nearly a million years before it reaches the nearest stars. Although its radio may presently fail, the craft carries a gold-anodized aluminum plaque with an engraved symbolic message from us. If the Pioneer 10 antenna gain is 36 dBi and the NASA earth-station antenna gain is 66 dBi, find the following as of the year 2000 assuming that Pioneer 10 travels at a constant speed away from the earth and that the transmitter is still operating: (*a*) the signal time delay (Pioneer 10 to earth), (*b*) the received power, and (*c*) the maximum bandwidth for a signal-to-noise ratio of 5 (or 7 dB) at a wavelength of 10 cm and system temperature equal to 20 K. *Ans.* (*a*) 10 hrs, (*b*) 6.9×10^{-20} W, (*c*) 50 Hz.

Problem 5-11-5. Solar interference to earth stations. Twice a year the sun passes through the apparent declination of the geostationary Clarke-orbit satellites, causing solar-noise interference to earth stations. A typical forecast notice appearing on U.S. satellite TV screens reads:

ATTENTION CHANNEL USERS:
WE WILL BE EXPERIENCING
SOLAR OUTAGES FROM
OCTOBER 15 TO 26
FROM 12:00 TO 15:00 HOURS

(*a*) If the equivalent temperature of the sun at 4 GHz is 50,000 K, find the sun's signal-to-noise ratio (in decibels) for an earth station with a 3-m, 50 percent efficient parabolic dish antenna at 4 GHz. Take the sun's diameter as 0.5° and the earth station system temperature as 100 K. (*b*) Compare this result with that for the carrier-to-noise ratio calculated in Example 5-22 for a typical Clarke-orbit TV transponder. (*c*) How long does the interference last? Note that the relation $\Omega_A = \lambda^2/A_e$ gives the solid beam angle in steradians and not in square degrees. (*d*) Why do the outages occur between October 15 and 26 and not at the autumnal equinox around September 20 when the sun is crossing the equator? (*e*) How can satellite services work around a solar outage? *Ans.* (*a*) 16.8 dB; (*b*) +3.6 dB; (*c*) about 10 min; (*d*) because the U.S. is not on the equator, (*e*) Shift programming to satellites ahead of or behind the sun.

Problem 5-11-6. Voyager 2 at Neptune. On August 24, 1989, Voyager 2 made a close encounter with Neptune sending back close-up pictures of the planet. With its 2.5-m diameter parabolic dish antenna and 10-W, 10-GHz transmitter, what maximum earth station system temperature is permissible to provide a signal-to-noise (S/N) ratio of 3 dB for reception of a picture with 3×10^6 pixels (picture elements) in 3 minutes if the earth station antenna diameter is 70 m? Assume aperture efficiencies of 70 percent. The earth-Neptune distance = 4 light-hours. *Ans.* 24.2 K.

5-12 RADAR AND RADAR CROSS-SECTION

Consider now the situation shown in Fig. 5-46 where either a transmitter or a receiver can be connected to an antenna. Now, with transmitter connected to the antenna, a pulse is sent out that strikes a passive reflecting-scattering object as in Fig. 5-46. The power intercepted by the object is given by the Friis transmission

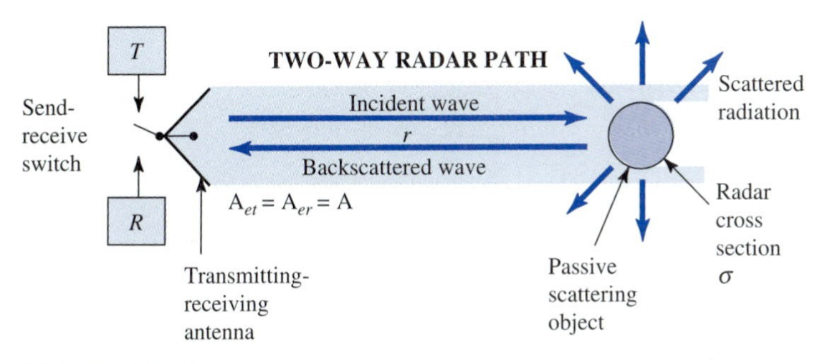

FIGURE 5-46
Double-path geometry (wave out and scattered wave back) used in obtaining radar equation. The antenna can be switched between transmitter and receiver.

formula (5-10-4) as

$$P_{\text{int}}\text{ (by object)} = \frac{P_t A_{et}}{r^2\lambda^2}\sigma \qquad \text{(W)} \tag{1}$$

where σ = *radar cross-section* of the object, m^2.

The term *radar* is an acronym for *ra*dio *d*irection *a*nd *r*ange. Assuming that the object scatters isotropically ($D = 1$), its effective aperture from (5-2-25) is $\lambda^2/4\pi$, so the scattered power from the object which is received back *at* the transmitter location is, by another application of the Friis transmission formula (5-10-4), given by

$$P_r\text{ (by antenna)} = \frac{P_{\text{int}}\text{ (by object)}A_{et}}{r^2\lambda^2}\frac{\lambda^2}{4\pi} \qquad \text{(W)} \tag{2}$$

Now, connecting the antenna to the receiver (before the backscattered pulse or echo arrives), the ratio of the backscattered power collected by the antenna to the transmitted power yields the *radar equation* as given by

$$\frac{P_r\text{ (by antenna)}}{P_t} = \frac{A^2\sigma}{4\pi r^4\lambda^2} \qquad \text{(dimensionless)} \qquad \textbf{\textit{Radar equation}} \tag{3}$$

where $A = A_{et}$ = effective aperture of antenna (same for transmitting and receiving), m^2.

It is assumed in (3) that polarizations are matched (no cross-polarized component of the backscattered wave), i.e., $F = 1$ or $MM_a = 0$ and also that the antenna and receiver are matched. If polarizations are not matched, then (3) should be multiplied by F from (4-13-5).

From (3), the radar cross-section (RCS) is

$$\sigma = \frac{P_r\text{ (by antenna)}4\pi r^4\lambda^2}{P_t A^2} \tag{4}$$

or

$$\sigma = \frac{S_r}{S_{\text{inc}}/4\pi r^2} = \frac{4\pi r^2 S_r}{S_{\text{inc}}} = \frac{\text{scattered power}}{\text{incident power density}} \quad (\text{m}^2) \qquad \textbf{\textit{Radar cross-section}} \tag{5}$$

where S_r = isotropically backscattered power density at distance r
$= [P_r \text{ (by antenna)}]/A$, W m^{-2}
S_{inc} = power density incident on object
$= (P_t A)/(r^2\lambda^2)$, W m^{-2}

In words, the *radar cross-section* σ of an object is its effective area intercepting the incident power density S_{inc} which, if scattered isotropically, would result in the backscattered power density S_r.

For a large perfectly reflecting metal sphere of radius a, the radar cross-section is equal to the physical cross-section πa^2. For imperfectly reflecting spheres, the radar cross-section is smaller. For example, at meter wavelengths, the radar cross-section of the moon is about 0.1 of its physical cross-section.

To measure accurate far-field values of the *radar cross-section* (RCS) of objects, the radar antenna should be at a distance well in excess of w^2/λ, where w = maximum width (or height) of the object being measured, in order that the radar wave front be sufficiently planar. At high frequencies and/or for big objects, this may require an impractically large distance. A solution is to use a parabola to convert the spherical wave front from the radar to a plane wave. This greatly reduces the distance required and results in what is called a "compact range." The Ohio State University 110-m radio telescope (Fig. 5-47) doubled as a compact range for RCS measurements of large objects as suggested in Fig. 5-48.

FIGURE 5-47
110-m "Big Ear" radio telescope of the Ohio State University radio observatory is larger than three football fields. Designed by John Kraus in 1954 to provide the largest aperture per unit cost, it was constructed by his students working part time. The telescope was one of the world's largest and is famous for discovering the most distant known objects in the universe. In 1998 it was demolished for a golf course.

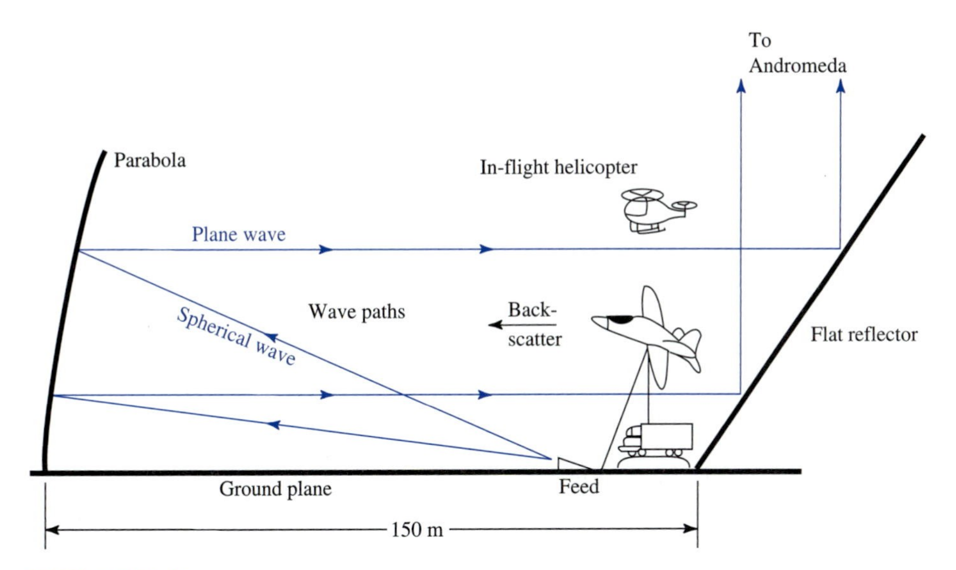

FIGURE 5-48
The 110-m "Big Ear" as it was used as a compact range. With radar feed at the focus, RCS measurements could be made of actual full-size vehicles, including an in-flight helicopter, as suggested. Between feed and parabola, wave fronts are spherical; between parabola and objects, wave fronts are plane. As a radio telescope "Big Ear" listened to signals that were billions of years old, while as a compact range it detected signals only a microsecond old.

In *pulse radar* the time Δt between the transmission of the pulse and reception of its echo gives the *distance* d of the object as

$$d = \frac{1}{2} c \Delta t \qquad \text{(m)} \tag{6}$$

where $c = 300$ Mm s^{-1} (for air) and $\Delta t =$ delay time of echo, s.

In *doppler radar* the change in frequency Δf of the echo with respect to the transmitted frequency f_0 gives the *velocity* v of the object as

$$v = \frac{1}{2} \frac{\Delta f}{f_0} c \qquad (\text{m s}^{-1}) \tag{7}$$

where $\Delta f =$ (doppler) shift of frequency, Hz
$f_0 =$ transmitting frequency, Hz
$c = 300$ Mm s^{-1} (for air)

For positive values of Δf (increase in echo frequency), the object is approaching; for negative values (decrease in echo frequency), it is receding (going away).

By moving or scanning the antenna beam, a radar can provide information on the *direction*, *distance*, and *velocity* of objects within its view. Also, since a returning pulse carries a characteristic signature, different objects may be recognized, a very short pulse response being the Fourier transform of the object's frequency response.

TABLE 5-6
Radar cross-sections†

Object	Radar cross-section σ
Sphere, radius a	πa^2
Square plate, area L^2	$4\pi L^4/\lambda^2$
Cylinder, radius a, length L	$2\pi a L^2/\lambda$

†Objects perfectly conducting and large compared to the wavelength (a and $L \gg \lambda$). Plate and cylinder at normal incidence. Cylinder length L parallel to plane of polarization of radar wave. For the general case where the object's dimensions may also be smaller than the wavelength, see R. J. Kouyoumjian in J. D. Kraus, "Antennas", 1988, pp. 791–799.

Notice in (3) that the backscattered power at the radar receiver is inversely proportional to the fourth power of the distance. This means that if you have a receiver as sensitive and an antenna as large as the radar's, you should be able to detect the radar at greater distances than it can detect you. The signal needs to cover the path to you only once ($1/r^2$ attenuation), but it must cover it twice to get back to the radar ($1/r^4$ attenuation).

In *pulse radar* the antenna is connected to the transmitter while the pulse is sent. It is then switched to the receiver which listens for the echoes. The greater the range being observed, the longer the time needed between pulses. In *doppler radar* the antenna is connected continuously to both transmitter and receiver through a circulator which isolates the receiver from the transmitter, and the transmitter is on all the time. The term CW (continuous wave) *doppler* is used to describe this mode of operation, which measures only velocity. By measuring both time delay and frequency shift of the echoes from a pulse radar, we have what is called *pulse doppler radar* which measures both distance and velocity.

Radars have wide application for ship and aircraft navigation; for harbor, airport and highway surveillance; for weather forecasting (storms, rain, hail, etc.); for terrain mapping; for measuring the distance to the moon or the rotation of Venus (whose surface is hidden by clouds); for monitoring the speed of a pitcher's fast ball; or determining the velocity of a hummingbird.

The radar cross-sections of several objects are listed in Table 5-6 where it is assumed that the objects are large compared to the wavelength. Note that, whereas the RCS of a sphere is equal to its cross-sectional area, the RCS of a plate or sheet is larger than its area. (See problem 5-12-15.)

Problem 5-12-1. Radar cross-sections. What are the radar cross-sections at $\lambda = 10$ cm for (*a*) a 1-m diameter sphere, (*b*) a 1-m^2 flat disk, and (*c*) a cylinder 1 m long and 4 mm in diameter with L parallel to the plane of polarization. All are perfectly conducting and at normal incidence. (*d*) Same as (*c*) but cross-polarized. *Ans.* (*a*) π m^2; (*b*) 40π m^2; (*c*) 0.04π m^2; (*d*) ~ 0 m^2.

Problem 5-12-2. Radar cross-section. What is the radar cross-section of an object at a distance of 1 km which gives a return of 1 nW for a transmitted power of 1 kW? The transmit-receive antenna is a 1-m-diameter dish operating at 3 GHz. Take aperture efficiency as 50 percent. *Ans.* 0.8 m^2.

Pulse Doppler Weather Radar

Coherent pulse doppler radars measure both the amplitude and the phase of radar echoes. The phase of the return is related to the distance r to the scattering object and the wavelength λ by

$$\phi = \frac{4\pi r}{\lambda} \tag{8}$$

This differs from the one-way phase shift $2\pi r/\lambda$ by a factor of 2, since radar scattering is a two-way process.

If the source of the radar echo is moving, the phase of the return changes with time as

$$\phi(t) = \frac{4\pi(r + v_r t)}{\lambda} \tag{9}$$

where v_r = velocity component along the radar line of sight.

The time derivative of this changing phase is an angular frequency $\Delta\omega$:

$$\Delta\omega = \frac{\partial\phi}{\partial t} = \frac{4\pi v_r}{\lambda} \tag{10}$$

The doppler frequency shift due to motion of the scatterer is, therefore,

$$\Delta f = \frac{\Delta\omega}{2\pi} = \frac{2v_r}{\lambda} \qquad \text{(Hz)} \tag{11}$$

In a pulse doppler radar, both the distance to the echo source and its velocity are measured by transmitting a series of pulses and measuring the phase of the return for each pulse. From the Nyquist requirement of 2 samples per cycle, the maximum doppler frequency shift which may be unambiguously measured is

$$\Delta f_{\max} = \frac{1}{2T} \qquad \text{(Hz)} \tag{12}$$

where T = pulse repetition interval (PRI), s.

If a total of N pulses is used to determine the velocity, the total measurement time is NT seconds. Two frequencies may be resolved in this time if one of them undergoes an additional cycle of phase change in the observation period. Thus, the frequency resolution of a pulse doppler radar is

$$\Delta f_{\min} = \frac{1}{NT} \tag{13}$$

where N = number of pulses observed and T = pulse repetition interval, s.

Pulse doppler weather radar uses scattering from water droplets and fluctuations in refractive index to measure the intensity of rain and wind velocity.

Example 5-23. Weather radar. For an X-band (10-GHz) weather radar, find: (*a*) the minimum pulse repetition frequency (PRF = 1/PRI) which may be used to unambiguously measure the wind velocity in a tornado with a wind speed of 350 km hr^{-1} (Fig. E5-23). (*b*) At this PRF, how many pulses must be sampled to resolve in frequency two portions of the tornado with a differential velocity of 1 km hr^{-1}?

Solution

(*a*) Using (11), the doppler shift for a velocity of 350 km hr^{-1} is

$$\Delta f = \frac{2v_r}{\lambda} = \frac{2\left(\dfrac{3.5 \times 10^5}{3600}\right)}{0.03} = 6.5 \text{ kHz}$$

From (12),

$$\Delta f_{\max} = \frac{1}{2T}$$

$$T = \frac{1}{2\Delta f_{\max}} = \frac{1}{2(6.5 \times 10^3)} = 7.7 \times 10^{-5} = 77\,\mu\text{s}$$

$$\text{PRF} = \frac{1}{T} = \frac{1}{7.7 \times 10^{-5}} = 13 \text{ kHz} \quad \textit{Ans.} \quad (a)$$

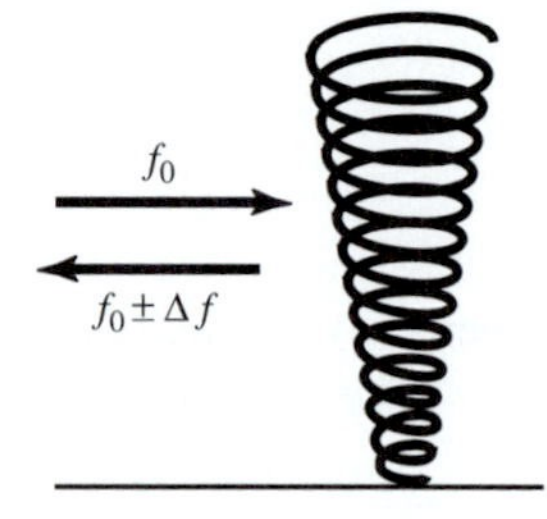

FIGURE E5-23

(*b*) Rearranging (13) to solve for N gives

$$N = \frac{1}{\Delta f_{\min} T}$$

For two scatterers with $\Delta v_r = 1$ km hr^{-1}, we have from (11)

$$\Delta f_1 - \Delta f_2 = \frac{2v_{r1}}{\lambda} - \frac{2v_{r2}}{\lambda} = \frac{2(v_{r1} - v_{r2})}{\lambda} = \frac{2\left(\dfrac{1000}{3600}\right)}{0.03} = 18.5 \text{ Hz}$$

Therefore,

$$N = \frac{1}{\Delta f_{\min} T} = \frac{1}{(18.5)(7.7 \times 10^{-5})} = 702 \text{ pulses} \quad \textit{Ans.} \quad (b)$$

Problem 5-12-3. Tornado radar. For a maximum doppler shift of 5.3 kHz with the weather radar of Example 5-23 scanning a tornado, find (*a*) the maximum wind speed in the tornado funnel. (*b*) If the lowest doppler shift observable is 920 Hz, how fast is the tornado approaching or receding from the weather radar antenna? *Ans.* (*a*) 79.5 m/s; (*b*) 46.7 m/s toward radar.

The Corner Reflector

A square-corner dihedral (or two-surface) reflector, as in Fig. 5-49*a*, with internal angle = 90°, acts as a retroreflector over a wide, almost 90° angle, as suggested in Figs. 5-49*b* and *c*.

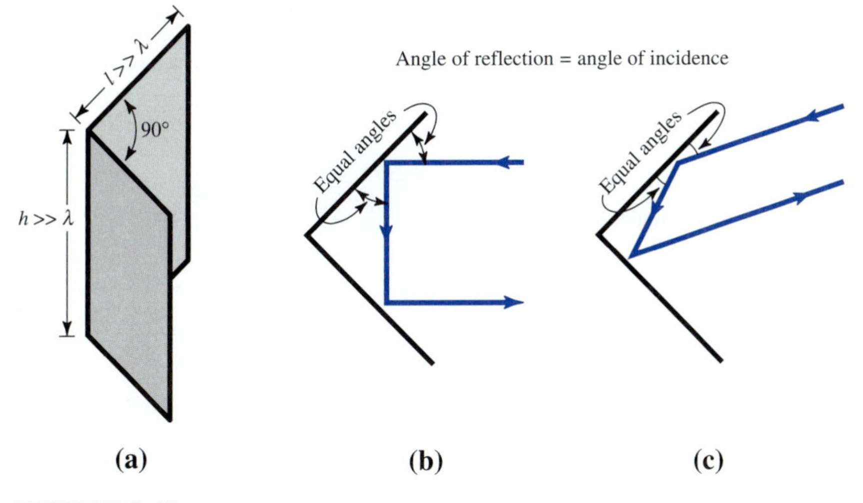

FIGURE 5-49
Square-corner reflector at (*a*) with ray diagrams at (*b*) and (*c*) showing retroreflective action at different angles of incidence.

To distinguish this type of corner reflector from the one in Fig. 5-33, this one may be called a ***passive corner*** and the one of Fig. 5-33 an ***active corner.***

The measured radar field strength return from a square-corner reflector is shown in Fig. 5-50 on next page. The corner surfaces are large compared to the wavelength. There is strong reflection from the front side of the reflector over a broad angle (almost 90°). There are also four strong but narrow spikes of reflection from the flat sides at normal incidence. But there is only a tiny spike return from the sharp edge of the back side of the corner. Thus, the front sides of corners and flat surfaces at normal incidence are good to *enhance* detection, but they should be avoided to *escape* detection.

Adding a third side to the dihedral corner results in a trihedral corner. Clustering eight of them together as in Fig. 5-51, a retroreflector is obtained that provides a strong reflection over almost 4π sr. Such reflectors are widely used to enhance radar return.

For example, small watercraft commonly carry one (usually with reflecting surfaces of wire mesh) on a tall mast to make the craft's presence more visible on radar screens and reduce chances of being rammed in a fog. To be most effective, the reflector dimensions should be many wavelengths and the periphery of the mesh hole less than $\lambda/2$. The surfaces should be flat to better than $\lambda/12$. And, to increase the probability that the radar echo will be noticed, the reflector can be rotated to avoid a persistent low return in the directions of the three planes of the reflector. Such a retroreflector may not give an echo like the Queen Elizabeth II but it can make a little watercraft appear like a sizable ship.

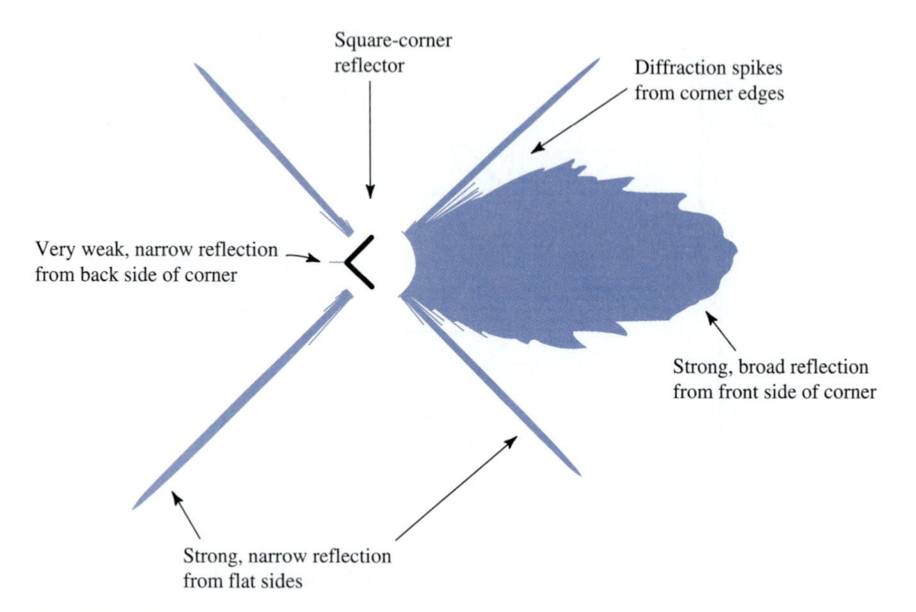

FIGURE 5-50
Measured radar field strength return from corner reflector. The corner dimensions are much larger than the wavelength. This figure is "stealth technology" in a nutshell.

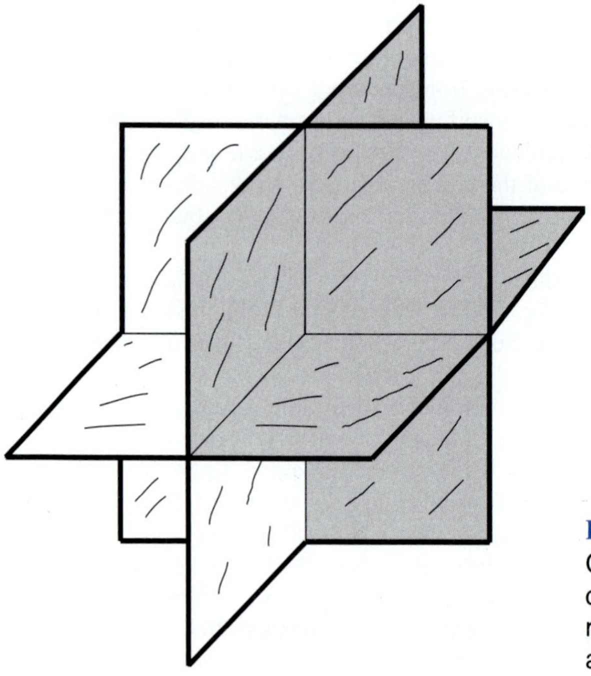

FIGURE 5-51
Cluster of eight trihedral corner reflectors providing retroreflective action over almost 4π sr.

5-13 GLOBAL POSITIONING SATELLITES AND RELATIVITY

Twenty-four Global Positioning Satellites (GPS) are in six transpolar orbits at 20,000 km above the earth as shown in Fig. 5-52(a). The satellites provide longitude (ϕ), latitude (θ), and elevation or radius (r) to receivers at any point P on (or above) the earth in any weather, both day and night.

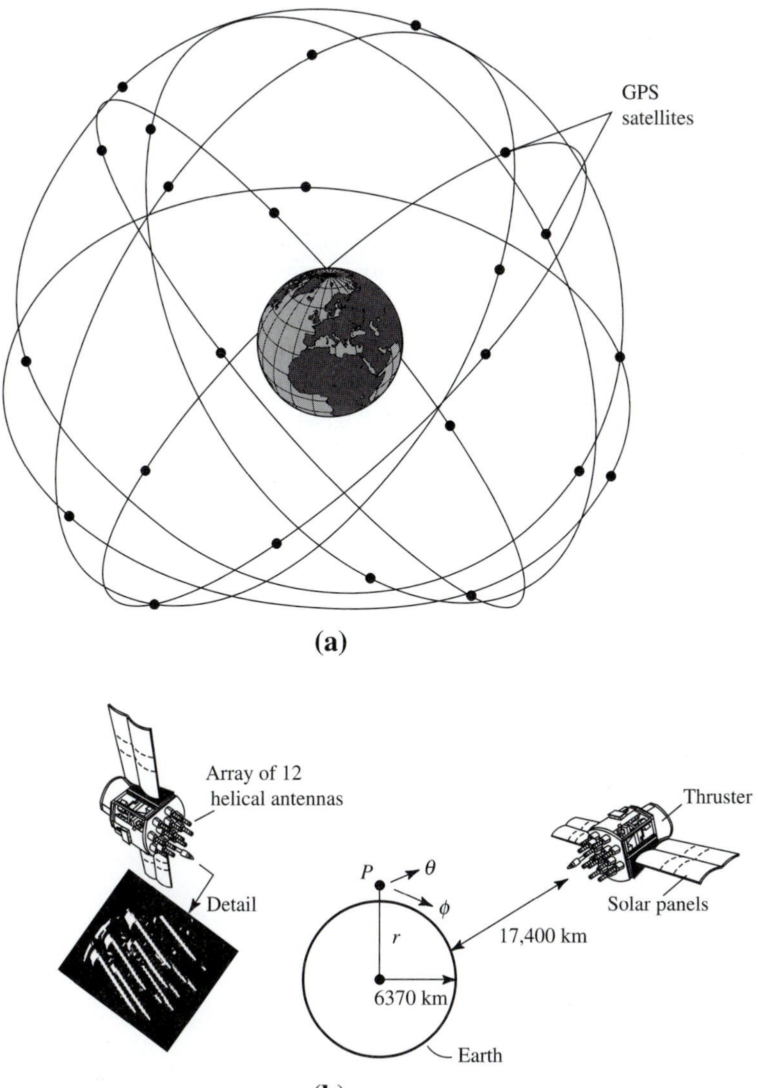

FIGURE 5-52
(a) Twenty-four GPS satellites in 12-hour medium earth orbit (MEO) at 20 000 km. They are deployed in six orbits of four satellites each. (b) Two satellites and their relation to a point P at the earth. Both drawings are to scale except for the satellite size in (b).

Each satellite continuously transmits its position and time on two frequencies: 1.23 and 1.58 GHz. Since the satellites are in motion, orbiting the earth every 12 hours, the system is relativistic or four-dimensional (three positions plus time: x, y, z, t). To achieve this, all satellites carry atomic clocks accurate to better than 1 ps while the receivers have clocks accurate to better than 1 ns.

Other global position systems with their own groups of satellites include the Russian GLONASS and the international MARISAT.

Example 5-24. Position by GPS. Consider a GPS receiver at point P above the earth's equator with two GPS satellites, 1 and 2, also above the equator at different longitudes as shown in Fig. 5-53. The receiver detects a pulse from GPS-1 with code that informs the receiver it was sent at a time Δt_1 earlier from a position at radius r_1 and longitude 20° W. Simultaneously GPS-2 sends a pulse coded to tell the receiver it was sent Δt_2 earlier from a radius r_2 and longitude 100° W.

If $\Delta t_1 = 64$ ms, and $\Delta t_2 = 60$ ms, find the longitude and elevation of the GPS receiver at point P.

Solution

$$r_3 = c\Delta t_1 = 3 \times 10^8 \times 0.064 = 19{,}200 \text{ km}$$

$$r_4 = c\Delta t_2 = 3 \times 10^8 \times 0.060 = 18{,}000 \text{ km}$$

From the geometry of the two triangles, we obtain 66° W. long., elevation $=$ 1180 km (a LEO satellite, see Sec. 5-15). *Ans.*

To illustrate the principles as simply as possible, this example was reduced to two dimensions with the earth assumed to be round. To provide longitude, latitude, and elevation, the problem becomes three-dimensional and at least three GPS satellites must be acquired by the receiver. Seven or eight are usually above the horizon at any time, with the receiver computer solving a matrix of seven or eight simultaneous equations.

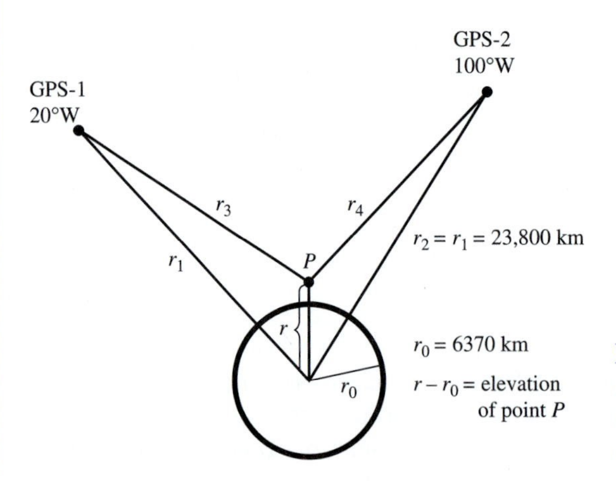

FIGURE 5-53
Geometry for determining the position of P. Drawing is to scale.

Problem 5-13-1. Position by GPS. If GPS-1 is at 30° W. long. with $\Delta t_1 = 60.3$ ms and GPS-2 is at 85° W. long. with $\Delta t_2 = 62.1$ ms, find the longitude and elevation of the GPS receiver at point P (Fig. 5-53). The GPS receiver is at 0° latitude and both GPS satellites are above the equator as in Example 5-24. *Ans.* 53° 15′ W. long.; 30 km.

Problem 5-13-2. GPS satellite power. If the satellite antenna has 30 dBi gain, how much satellite transmitter power is required to provide an SNR at the earth receiver of 20 dB. The earth receiver antenna is sensitive to right circular polarization but with a gain = 0 dBi. $T_{sys} = 300$ K, $f = 1.6$ GHz, B (eff.) = 1 kHz. *Ans.* 750 mW.

5-14 FAR FIELD, NEAR FIELD, AND FOURIER TRANSFORM†

To measure an accurate field pattern of a transmitting antenna, the measuring unit should be in the *far field*. At this distance the path lengths from all parts of the antenna are in phase or nearly so. Or if the measuring unit is replaced by an isotropic radiator, its wavefront will depart by no more than a specified distance δ at any part of the antenna, as in Fig. 5-54.

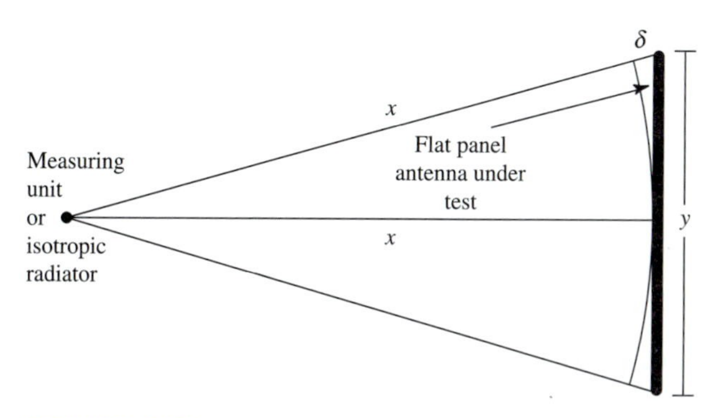

FIGURE 5-54
Geometry for far-field measurements.

Example 5-25. Far-field measurement. If the antenna dimension $y = 20$ m and $\delta = \lambda/10$, find the required distance x at a frequency of 12 GHz.

Solution. From the geometry of the triangle (Fig. 5-54),

$$x^2 + 2\delta x + \delta^2 = x^2 + \frac{y^2}{4}$$

For $x >> \delta$, the required distance $x = y^2/8\delta$. For $f = 12$ GHz and $\lambda = 0.025$ m,

$$x = \frac{20^2}{8 \times 0.0025} = 20 \times 10^3 \text{ m} = 20 \text{ km} \quad \textit{Ans.}$$

† Baron Jean Baptiste Joseph Fourier (1768–1830), French mathematician and physicist.

Problem 5-14-1. Airport radar. At an airport, a rotating radar antenna is 10 m across by 2 m tall and operates at 6 GHz. For phase differences of less than $\lambda/20$, at what distance should the field-strength measurements be made? *Ans.* 5 km.

Problem 5-14-2. Deep-space antenna. For $\delta = \lambda/20$, find the far-field distance requirement of a 100-m-diameter antenna for deep-space communication at 15 GHz. *Ans.* 1250 km.

The large distance requirement of Problem 5-14-2 makes a compact range attractive for doing the measurements if the antenna is portable. But some are not. For example, the 100-m deep-space antenna of Problem 5-14-2 is too big and it has a distance requirement of over 1000 km.

For such antennas there are a couple of options. One is to use a celestial radio source at distances of thousands, millions, or even billions of light-years. These celestial radio sources in the sky definitely satisfy the distance requirement and are easy to use. The antenna and its receiver-recorder act as a radio telescope. With the antenna pointed at the proper elevation angle, the recorder plots the antenna field pattern as the celestial source transits the beam. A requirement of the source, of course, is that its angular extent is small compared to the antenna HPBW.

A second option is to measure the near-field right at the antenna across its aperture with a small movable antenna and receiver, and then obtain the far-field pattern by a Fourier transform. Thus, referring to Fig. 5-55, the far-field pattern of a linear antenna of length a is given by

$$E(\phi) = \int_{-a_\lambda/2}^{+a_\lambda/2} E(x_\lambda) e^{j2\pi x_\lambda \sin\phi}\, dx_\lambda \tag{1}$$

where $E(x_\lambda)$ = near field as a function of position x_λ across the aperture and $x_\lambda = x/\lambda$. (An inverse transform gives the near-field distribution from the far-field pattern.)

Figure 5-56 compares the Fourier-transform far-field patterns for uniform, cosine-squared, and Gaussian near-field aperture distributions of a line source. The Gaussian near-field aperture distribution has the unique property of transforming into a Gaussian far-field pattern.

The uniform distribution has the narrowest beam width and highest gain but the first side lobe is only 13 dB down. The tapered cosine-squared and Gaussian

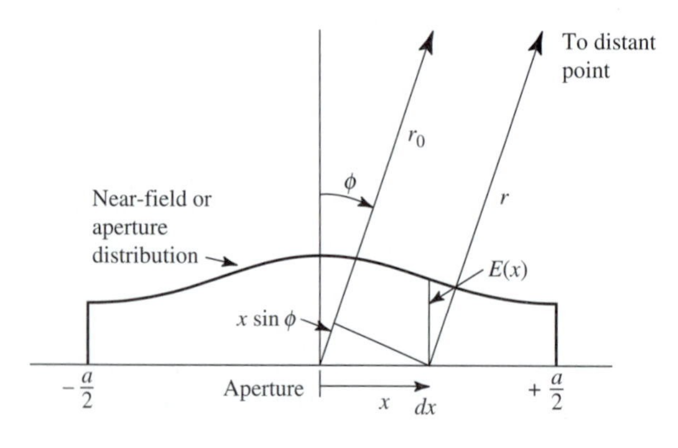

FIGURE 5-55
Aperture of width a and amplitude or near-field distribution $E(x)$.

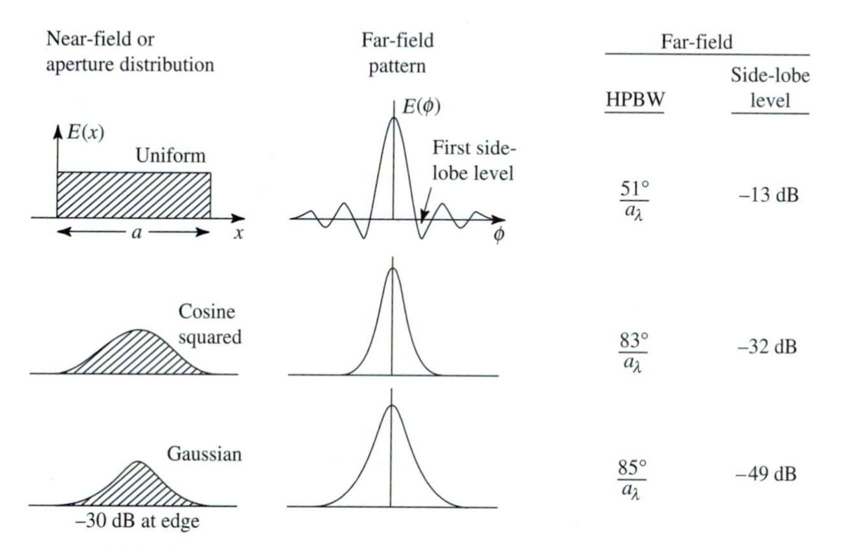

FIGURE 5-56
Field pattern, HPBW, and side-lobe level of three line source distributions. The Gaussian distribution is 30 dB down at the edges.

distributions have wider beamwidths and lower gains but very low or no side lobes. Aperture distributions used in practice are a tradeoff or compromise between gain and side-lobe level.†

For circular dish antennas as in Fig. 5-34, the relation between aperture, beamwidth, and first side-lobe level is shown in Fig. 5-57.

Example 5-26. 100-m dish for deep-space communication. The aperture distribution of a 100-m-diameter dish antenna is tapered to one-third at the edge or about 10 dB down (a typical value). See Fig. 5-57. At 10 GHz, what is its (*a*) HPBW, (*b*) gain from beam width, (*c*) gain from effective aperture, and (*d*) first side-lobe level?

Solution. HPBW $= 66°/D_\lambda = 66° \times 0.03/100 = 0.02°$. *Ans.* (*a*)

From (5-2-10), $$\text{Gain} = \frac{41,000}{\text{HPBW}^2} = 1.05 \times 10^8 \text{ or } 80 \text{ dBi} \quad \textit{Ans.} \quad (b)$$

From (5-2-20) and (5-2-24), for gain from effective aperture,

$$\text{Gain} = \frac{4\pi A_p \varepsilon_{ap}}{\lambda^2} = \frac{4\pi^2(50)^2(0.725)}{(0.03)^2}$$

$$= 8.0 \times 10^7 = 79 \text{ dBi} \quad \textit{Ans.} \quad (c)$$

First side-lobe level: -23 dB. *Ans.* (*d*)

Note that (5-2-10) gives an approximate gain from the HPBW (80 vs. 79 dBi).

†For a much more complete and detailed discussion of aperture-distribution far-field relations and celestial source measurements see J. D. Kraus, "Antennas," 2d ed., McGraw-Hill, New York, 1988, and J. D. Kraus, "Radio Astronomy," 2d ed., Cygnus-Quasar, Powell, Ohio, 1986.

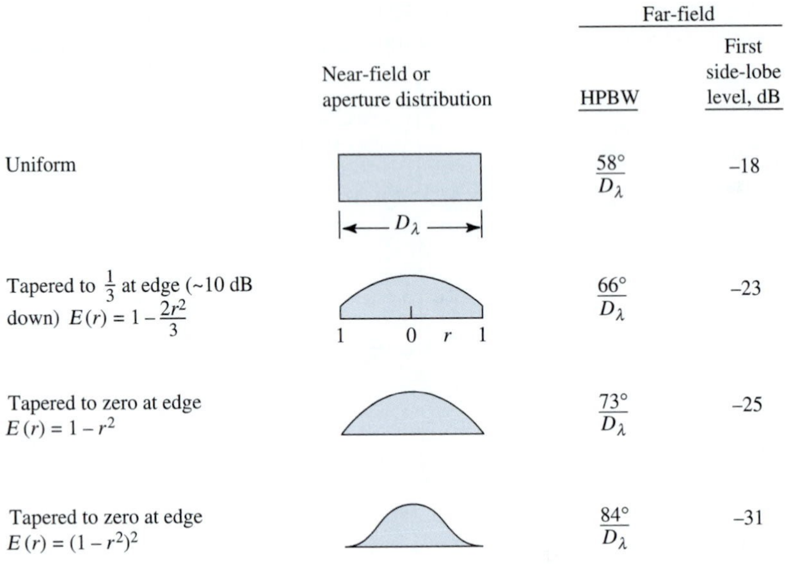

	Near-field or aperture distribution	Far-field HPBW	Far-field First side-lobe level, dB
Uniform		$\frac{58°}{D_\lambda}$	−18
Tapered to $\frac{1}{3}$ at edge (~10 dB down) $E(r) = 1 - \frac{2r^2}{3}$		$\frac{66°}{D_\lambda}$	−23
Tapered to zero at edge $E(r) = 1 - r^2$		$\frac{73°}{D_\lambda}$	−25
Tapered to zero at edge $E(r) = (1 - r^2)^2$		$\frac{84°}{D_\lambda}$	−31

FIGURE 5-57
Beamwidth (HPBW) and first side-lobe level for four circular aperture field distributions.

Problem 5-14-3. Dish with full taper. If the 100-m-diameter dish of Example 5-26 has the full-tapered aperture distribution $E(r) = (1 - r^2)^2$, find (*a*) HPBW, (*b*) gain from beam width, (*c*) gain from effective aperture, and (*d*) first side-lobe level. *Ans.* (*a*) 0.025°; (*b*) 78 dBi; (*c*) 76 dBi; and (*d*) −31 dB.

Problem 5-14-4. Dish with uniform aperture. What is the maximum hypothetical gain of the 100-m-diameter dish of Example 5-26 assuming a uniform aperture distribution (*a*) from beam width and (*b*) from effective aperture. (*c*) What is the first side-lobe level? *Ans.* (*a*) 81.3 dBi; (*b*) 80 dBi; (*c*) −18 dB.

5-15 EARTH-BASED, AIRBORNE, AND SPACEBORNE CELLULAR SYSTEMS

Originally cellular systems used antenna-studded towers, each serving a localized area or cell of varying size depending on topography and building density. Typically a cell might be 20 km in diameter with each tower having 360° coverage by means of multiple antenna arrays. A personal hand-held unit can communicate with the tower anywhere within the cell. If the personal unit leaves the cell it is automatically transferred to the tower of an adjacent cell.

This ground-based system is being expanded into space by using satellites in various orbits as shown in Fig. 5-58.

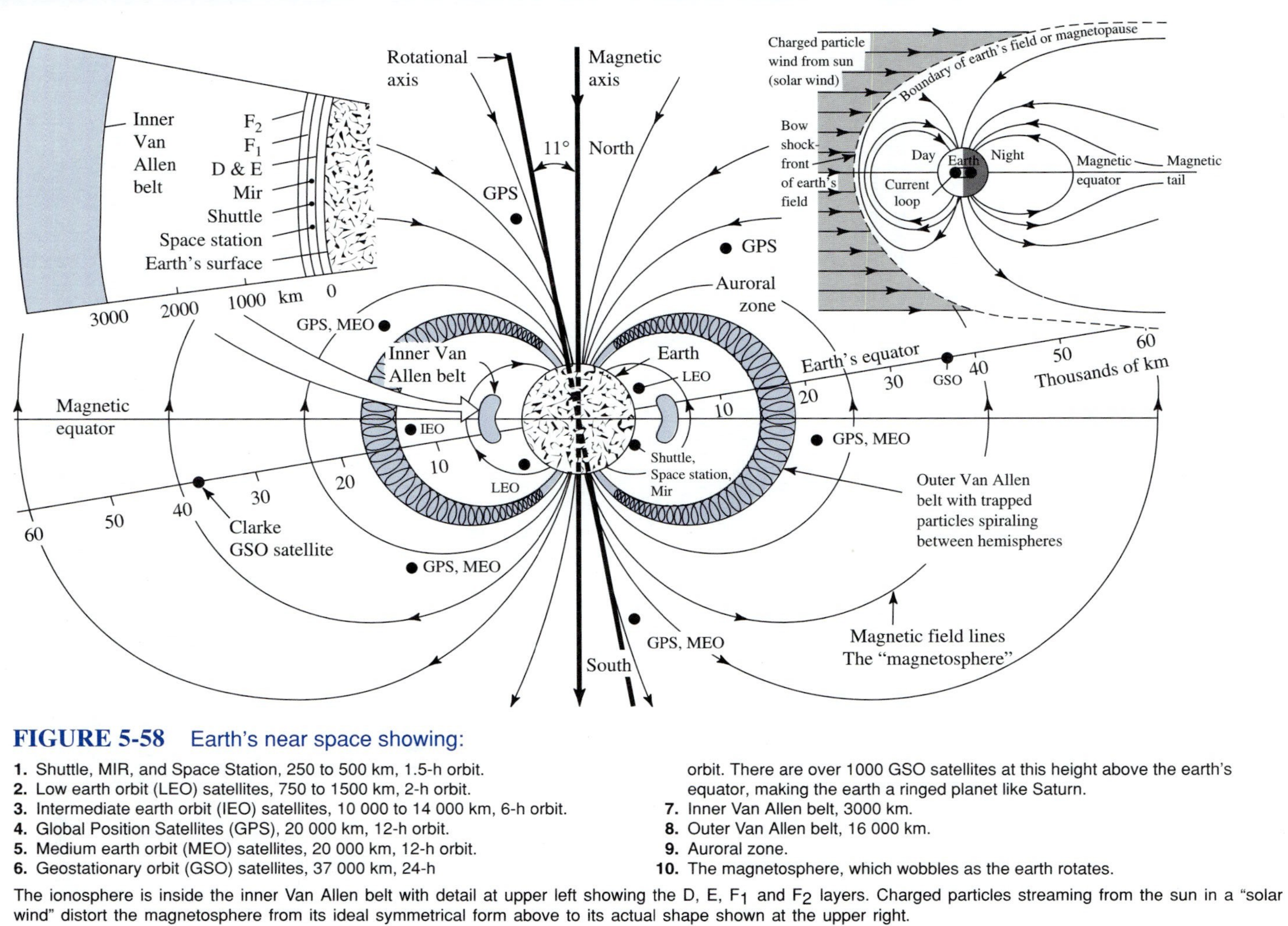

FIGURE 5-58 Earth's near space showing:

1. Shuttle, MIR, and Space Station, 250 to 500 km, 1.5-h orbit.
2. Low earth orbit (LEO) satellites, 750 to 1500 km, 2-h orbit.
3. Intermediate earth orbit (IEO) satellites, 10 000 to 14 000 km, 6-h orbit.
4. Global Position Satellites (GPS), 20 000 km, 12-h orbit.
5. Medium earth orbit (MEO) satellites, 20 000 km, 12-h orbit.
6. Geostationary orbit (GSO) satellites, 37 000 km, 24-h orbit. There are over 1000 GSO satellites at this height above the earth's equator, making the earth a ringed planet like Saturn.
7. Inner Van Allen belt, 3000 km.
8. Outer Van Allen belt, 16 000 km.
9. Auroral zone.
10. The magnetosphere, which wobbles as the earth rotates.

The ionosphere is inside the inner Van Allen belt with detail at upper left showing the D, E, F_1 and F_2 layers. Charged particles streaming from the sun in a "solar wind" distort the magnetosphere from its ideal symmetrical form above to its actual shape shown at the upper right.

Hundreds of satellites in low earth orbit (LEO) now provide personal communication to hand-held units anywhere on the earth—on oceans or deserts—dispensing entirely with cellular towers. Other systems operate in a dual mode, using both cell towers and satellites. Your call goes via a tower if you are close enough, otherwise via satellite. Thus, regardless of where you are, you can always "phone home." Still other systems use orbits at intermediate, medium, and geostationary height. At these greater heights more power and larger antennas are required, as illustrated in the following example and problems.

Example 5-27. Cell-tower system. A cell-tower system operates at 850 MHz for communication with hand-held phones. The cell-tower antenna array has 360° horizontal and 10° HPBW vertical coverage. However, the particular antenna in use has 90° horizontal HPBW coverage and, therefore, 4 times more gain. Find (*a*) tower power and (*b*) hand-held power to provide a SNR $= 40$ dB for a voice and high-speed data 1-MHz bandwidth at 12 km distance. The cell-tower receiver temperature $T = 30$ K and the hand-held $T = 100$ K.

Solution. From the Friis transmission formula,

$$P_t = \frac{S}{N}\frac{r^2\lambda^2 kTB}{A_{er}A_{et}}$$

From tower to hand-held phone,

$$A_{er} = \frac{\lambda^2}{4\pi}; \qquad D_t = \frac{41\,000}{360 \times 10} = 11.4; \qquad A_{et} = \frac{D_t\lambda^2}{4\pi} = 0.9\lambda^2; \qquad \lambda = 0.35 \text{ m}$$

$$P_t = 10^4 \frac{(12^2 \times 10^6)\lambda^2(1.38 \times 10^{-23})(100 \times 10^6)}{\frac{\lambda^2}{4\pi}0.9\lambda^2}$$

$$= 0.227 \text{ W} = 227 \text{ mW} \quad \textit{Ans.} \quad (a)$$

From hand-held phone to tower,

$$P_t = 0.3 \times 227 = 68 \text{ mW} \quad \textit{Ans.} \quad (b)$$

Problem 5-15-1. LEO to hand-held link. Find (*a*) LEO satellite power and (*b*) hand-held power for a LEO to hand-held link with SNR $= 40$ dB and bandwidth $= 1$ MHz operating at 1.6 GHz. T(hand-held) $= 100$ K, T(LEO) $= 10$ K, distance $= 1200$ km. The LEO antenna has a 30° HPBN conical pattern. The hand-held antenna has beam area $\Omega_A = 2$ sr. *Ans.* (*a*) 310 W; (*b*) 31 W. These powers are several orders of magnitude greater than for the tower to hand-held link of Example 5-27.

Problem 5-15-2. Lower power LEO to hand-held unit. Repeat the above problem with a voice and lower bandwidth of 10 kHz. *Ans.* LEO power $= 3.1$ W, hand-held power $= 310$ mW.

5-16 ABSORPTION BY ATMOSPHERE AND FOLIAGE

At frequencies of 50 GHz and above, the atmosphere acts as a shield to a wireless link reducing cross talk with other links. Between frequencies of 50 and 60 GHz the attenuation of electromagnetic waves by the atmosphere increases from 0.5 dB km^{-1} to almost 20 dB km^{-1}. This attenuation provides a natural shield for electromagnetic waves that is useful in many wireless applications. A good analogy is a laser beam penetrating fog.

The atmospheric attenuation is due to absorption by water vapor (H_2O) and molecular oxygen (O_2) as indicated in Fig. 5-59. Attenuation by scattering and absorption of rain drops is also indicated from heavy rain (100 mm h^{-1}) to very light rain (1 mm h^{-1}).

In transmitting through an absorbing atmosphere (or other medium) the signal is attenuated and a temperature observed which is a function of the atmospheric temperature and the absorption. It is given by

$$T = T_a(1 - e^{-\tau}) \tag{1}$$

where T_a = atmospheric temperature, K
τ = absorption coefficient, dimensionless = $x\alpha$
x = distance through absorbing medium, m
α = absorption coefficient, Np m^{-1}

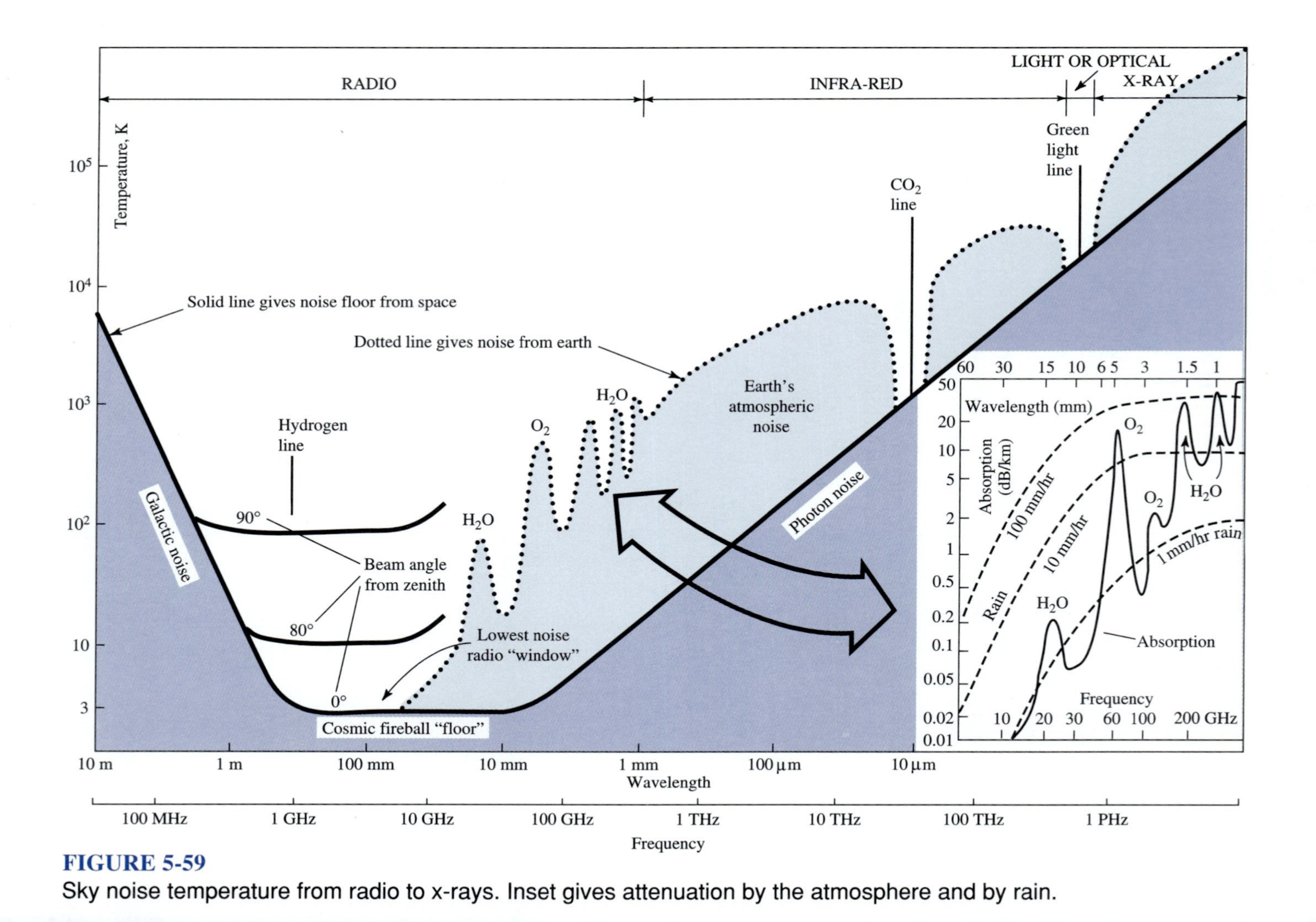

FIGURE 5-59
Sky noise temperature from radio to x-rays. Inset gives attenuation by the atmosphere and by rain.

Example 5-28. 1.5-km interbuilding high-data-rate wireless link. Two links each 1.5 km long are to be operated parallel to each other in close proximity as in Fig. 5-60. They are to have a SNR = 50 dB and a cross-talk isolation of 50 dB. For a bandwidth = 5 MHz, power = 1 W, parabolic dish diameter = 50 cm, aperture efficiency = 60 percent, frequency = 55 GHz, receiver T = 53 K, atmospheric temperature = 20° C, what is the minimum permissible separation of the two links?

Solution. From Fig. 5-59, attenuation at 55 GHz = 2 dB km^{-1}.
1 neper = 1 Np = 4.34 dB (= $10 \log e$). Therefore,

$$\alpha = 0.23 \times 2 = 0.46 \text{ Np km}^{-1} \qquad \text{and} \qquad \tau = 0.46 \times 1.5 = 0.69$$

The atmospheric temperature = 273 + 20 = 293 K, so

$$T = 293\,(1 - e^{-0.69}) = 147 \text{ K}$$

This adds to the receiver temperature giving a total

$$T = 53 + 147 = 200 \text{ K}$$

The effective apertures of the antennas

$$A_{er} = A_e = 0.6\pi r^2 = 0.12 \text{ m}^2$$

From the Friis transmission formula,

$$\frac{S}{N} = \frac{P_t A_{et} A_{er}}{r^2 \lambda^2 kTB} = \frac{1 \times 0.12^2}{1.5^2 \times 10^6 \times 5.45^2 \times 10^{-6} \times 1.38 \times 10^{-23} \times 200 \times 5 \times 10^6}$$

$$= 1.56 \times 10^{10} = 102 \text{ dB}$$

which is 52 dB more than required for fair weather and, from Fig. 5-59, is 22 dB more than needed during heavy rain (100 mm h^{-1}). (*Continued on next page.*)

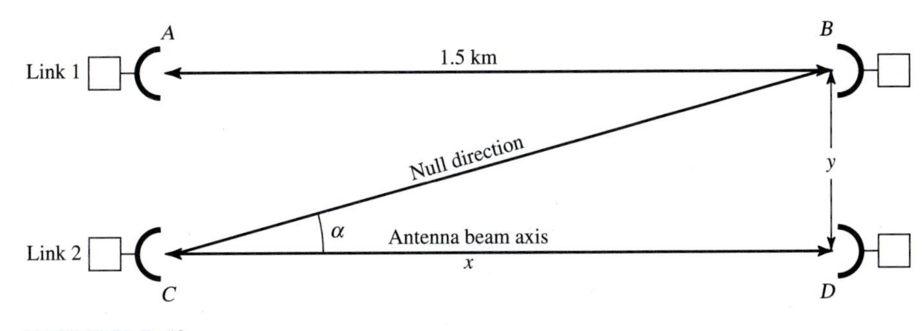

FIGURE 5-60
High data-rate wireless links.

Thus, the closest permissible spacing between the two links depends on the antenna pattern. For a uniform circular aperture the beam width between first nulls is

$$\text{BWFN} = \frac{140°}{d/\lambda} = \frac{140°}{0.5/5.45 \times 10^{-3}} = 1.53°$$

Half this angle is $= \alpha = 0.76°$, and $\tan 0.76° = 0.0133$.

From the geometry (Fig. 5-60), the distance from the beam axis to the first null at a distance of 1.5 km is

$$y = x \tan \alpha = 1500 \times 0.0133 = 20 \text{ m} \quad \textit{Ans.}$$

The null alignment is critical and in practice would need to be adjusted experimentally. Thus, with links operating, antenna C should be adjusted to place its null on antenna B and antenna B adjusted to place its null on antenna C, producing a *double null* isolation, with the same procedure for antennas A and D.

Problem 5-16-1. Rain attenuation. At what distance x in Fig. 5-60 will the system SNR fall below the specified 50 dB for (*a*) 10 mm/h and (*b*) 100 mm/h rainfall? *Ans.* (*a*) 4.4 km; (*b*) 1.1 km.

Example 5-29. Forest absorption. An earth-resource satellite's passive remote-sensing antenna directed at the Amazon River Basin measures a nighttime temperature $T_A = 21°$ C. If the earth temperature $T_e = 27°$ C and the Amazon forest temperature $T_f = 15°$ C, find the forest absorption coefficient τ_f.

Solution. The radio telescope in the satellite observes the earth (temperature T_e) through an emitting-absorbing forest (temperature T_f). See Fig. E5-29. The incremental satellite antenna temperature is then

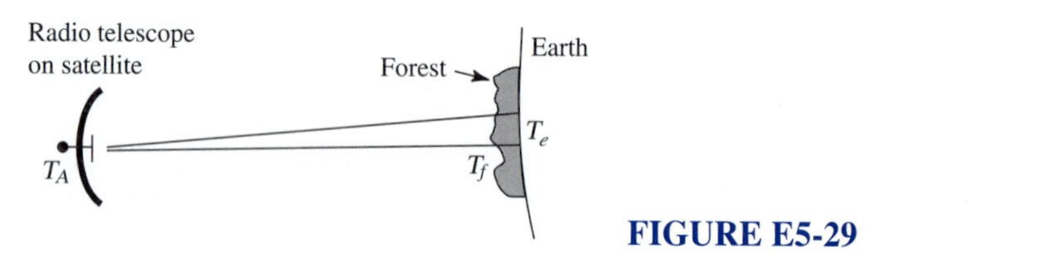

FIGURE E5-29

$$\Delta T_A = T_f(1 - e^{-\tau_f}) + T_e e^{-\tau_f} \quad \text{(K)} \tag{2}$$

where τ_f = absorption coefficient of forest.

Knowing T_e and τ_f, we can determine the temperature of the forest, or knowing T_e and T_f, we can deduce the absorption coefficient. It is by such a technique that the whole earth can be surveyed and much information obtained about the temperatures of land and water areas and, from absorption coefficients, the nature of the surface cover.

From (2),

$$e^{-\tau_f} = \frac{\Delta T_A - T_f}{T_e - T_f} = \frac{21 - 15}{27 - 15} = 0.5$$

and

$$\tau_f = 0.69 \quad \textit{Ans.}$$

Problem 5-16-2. FO-PEN (forest penetration). If the forest depth in Example 5-29 is 60 m, what is the forest (foliage) attenuation in dB/km? *Ans.* 50 dB/km.

TABLE 5-7
Antenna and antenna systems relations

Directivity, $D = \dfrac{4\pi A_e}{\lambda^2} = \dfrac{4\pi}{\Omega_A} = \dfrac{P(\theta,\phi)}{P_{\text{av}}}$ (dimensionless)

Directivity (approx.), $D = \dfrac{4\pi(\text{sr})}{\theta_{\text{HP}}\phi_{\text{HP}}(\text{sr})} = \dfrac{41\,000(\deg^2)}{\theta^\circ_{\text{HP}}\phi^\circ_{\text{HP}}}$

Directivity of short dipole, $D = 1.5$ (1.76 dBi)

Directivity of $\lambda/2$ dipole, $D = 1.64$ (2.15 dBi)

Gain, $G = kD$ (dimensionless)

Effective aperture, $A_e = \lambda^2/\Omega_A$ (m^2)

Aperture efficiency, $\varepsilon_{\text{ap}} = \dfrac{A_e}{A_p} = \dfrac{E^2_{\text{av}}}{(E^2)_{\text{av}}}$ (dimensionless)

Beam solid angle, $\Omega_A = \displaystyle\iint_{4\pi} P_n(\theta,\phi)\,d\Omega$ (sr)

Beam solid angle (approx.), $\Omega_A = \theta_{\text{HP}}\phi_{\text{HP}}\ (\text{sr}) = \theta^\circ_{\text{HP}}\phi^\circ_{\text{HP}}\ (\deg^2)$

Radiation resistance of short dipole, $R_r = 80\pi^2\left(\dfrac{l}{\lambda}\right)^2\left(\dfrac{I_{\text{av}}}{I_0}\right)^2$ (Ω)

Self-impedance of $\lambda/2$ dipole, $R_r + jX = 73 + j42.5$ (Ω)

Friis formula, $\dfrac{P_r}{P_t} = \dfrac{A_{er}A_{et}}{r^2\lambda^2}$ (dimensionless)

Radar equation, $\dfrac{P_r}{P_t} = \dfrac{A^2\sigma}{4\pi\,\lambda^2 r^4}$ (dimensionless)

Nyquist power, $p = kT$ (W Hz^{-1})

Flux density, $S = \dfrac{2kT_A}{A_e}$ ($\text{W m}^{-2}\text{ Hz}^{-1}$)

Minimum detectable flux density, $\Delta S_{\min} = \dfrac{2k\Delta T_{\min}}{A_e}$ ($\text{W m}^{-2}\text{ Hz}^{-1}$)

Signal-to-noise ratio of radio link $= \dfrac{S}{N} = \dfrac{P_t A_{et} A_{er}}{r^2\lambda^2 BkT_{\text{sys}}}$ (dimensionless)

PROJECTS

Project P5-1. Antenna pattern range. The setup is shown in P5-1. See Appendix D for equipment required.

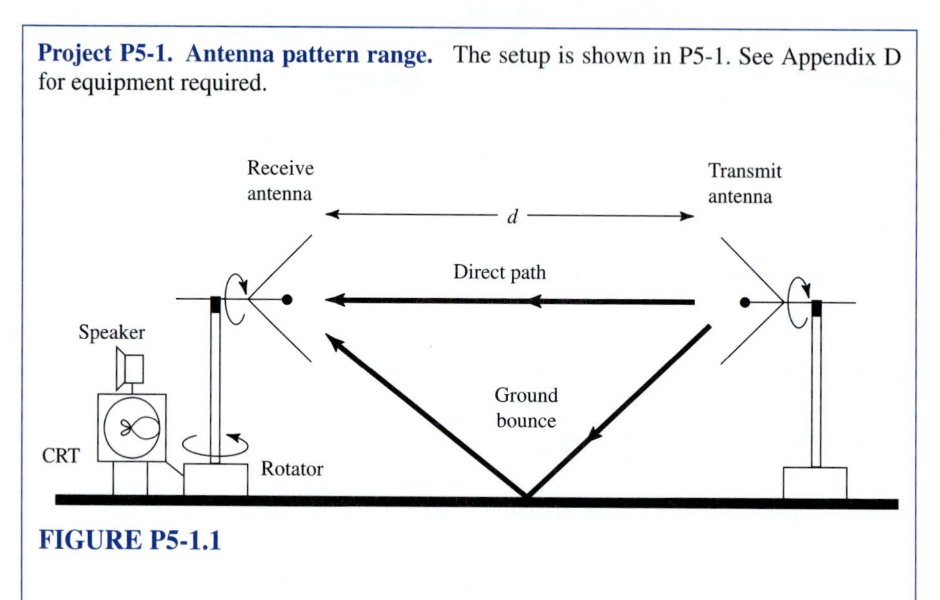

FIGURE P5-1.1

With corner reflector antennas:

Observe: HP(receive) from HP(transmit) as above
Observe: VP(receive) from VP(transmit)
Observe: VP(receive) from HP(transmit) (cross-polarization)
Observe: HP(receive) from VP(transmit) (cross-polarization)
Observe: 45°(receive) from HP(transmit) (slant polarization)
Observe: 45°(receive) from VP(transmit) (slant polarization)
Etc.

Do the same observations as with corner reflector on the other antennas. Vary d for near-field and far-field effects. Note effect of ground bounce.

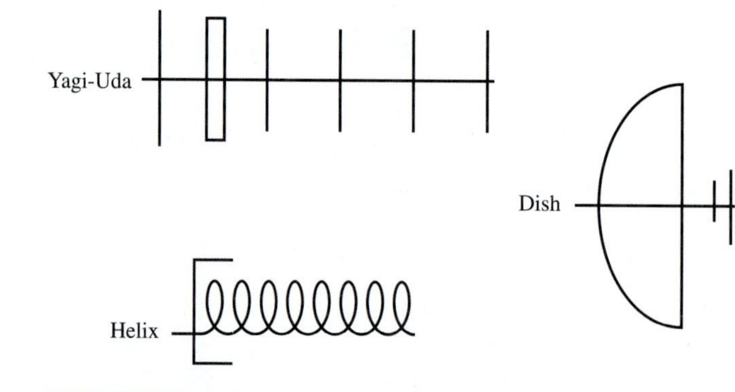

FIGURE P5-1.2

Try Yagi-Uda antenna with loops or disks in place of dipoles.

Try helix with different number of turns, with cupped ground plane as shown and with loops instead of ground plane. Try helix with 4 turns RH and 4 turns LH in series for LP.

Try dish antenna with dipole feed as shown, with corner reflector and with helix and ground plane as feed.

Try other antennas.

(HP = horizontal polarization, VP = vertical polarization, RH = right hand, LH = left hand, LP = linear polarization.)

Project P5-2. Jupiter-Io. It has been established that the earth, Jupiter, Saturn, Uranus, and Neptune emit radio noise bursts generated by the interaction of energetic charged particles with the planet's magnetosphere. Particles trapped in the magnetic field spiral at a gyrofrequency dependent on the charge e and mass m of the particle and the magnetic field B as given by equation (6-4-13). See Problem 6-4-4.

Jupiter noise bursts may sound like ocean waves crashing on a beach. At other times they have the staccato sound of hail on a tin roof. These bursts are easily observed with a shortwave receiver tuned to about 20 MHz and connected to a folded $\lambda/2$ dipole aligned east-west $\lambda/4$ above ground as in Fig. P5-2. The sporadic occurrences of the bursts are under the control of Jupiter's volcanically active moon Io. At night, to minimize man-made interference and with Jupiter in the sky, noise bursts of 1 to 10 seconds or more are observed at times that can be predicted from Io's position. These are available at the Web site given below. Recording the noise bursts on a tape unit with time superposed from a voice clock, the signals can be studied at one's convenience and something learned about the complex dynamics of planetary magnetospheres. This is useful information for anyone interested in wireless systems. See book's Web site *www.elmag5.com* for more information.

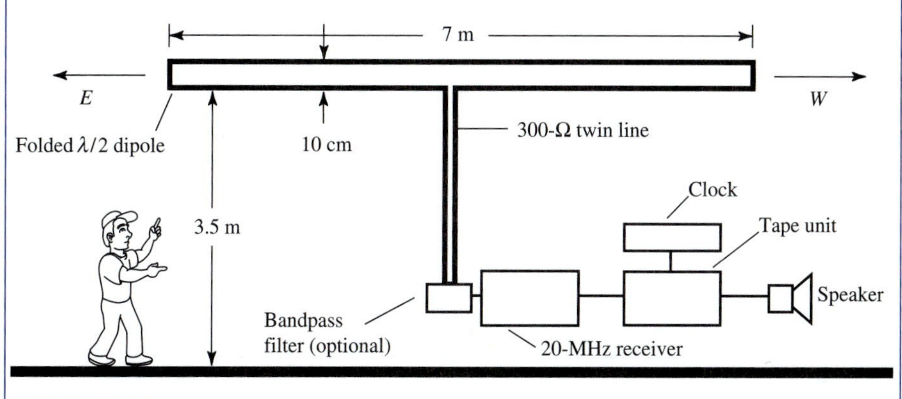

FIGURE P5-2
Earth station for recording Jupiter-Io noise bursts. No dimensions are critical.

For more information see J. D. Kraus "Radio Astronomy," 2d ed., 1986, Cygnus-Quasar, Powell, Ohio, pp. 8–61 to 8–75. Also see M. H. Wilkinson et al., "Comparison of Ground-Based Observations of Radio Source Io-B with Results from Voyager," *Proc. 12th National Australian Convention of Amateur Astronomers,* Hobart, Tasmania, 1986.

Problem P5-2-1. Jupiter power. Jupiter-Io bursts produce a peak flux density of over 6×10^6 Jy with a bandwidth of 10 MHz. Assuming that the source at Jupiter radiates isotropically, what is the power of the Jupiter-Io "transmitter"? Take Jupiter's distance as 35 light-minutes. *Ans.* 3 TW (3 million million watts).

Problem P5-2-2. Antenna direction. It is suggested in Project P5-2 that the folded $\lambda/2$ dipole be east-west. (*a*) Why? (*b*) Are there any situations where north-south might be better? *Ans.* (*a*) To ensure maximum antenna response when Jupiter crosses the meridian. (*b*) For an earth station on or near the earth's equator at times near the vernal or autumnal equinoxes, a north-south orientation might achieve a longer period of satisfactory reception before and after the meridian crossing (better low-angle response to east and west).

Project P5-3. Reflections and radar. With horizontally polarized (HP) transmitting antenna and an HP dipole probe as in Fig. P5-3, perform these tests:

1. With large flat sheet at point 2, move the probe back and forth in front of the sheet and observe maxima and nulls of standing waves.
2. Place sheet of 377 Ω per square "space paper" $\lambda/4$ in front of the metal sheet and note that the probe detects zero or nearly zero signal. See Fig. 4-15.

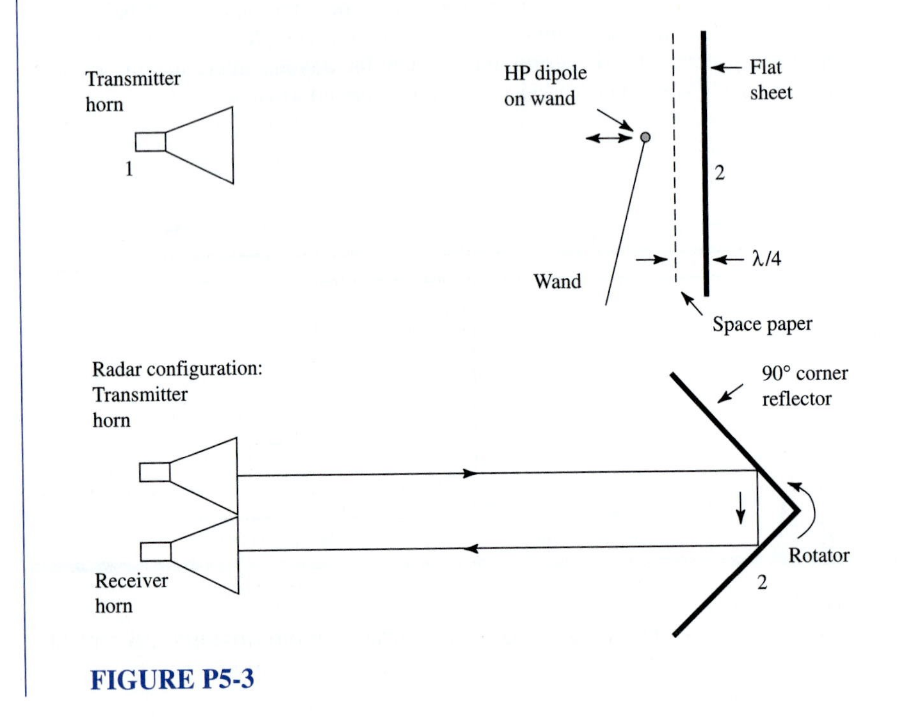

FIGURE P5-3

3. Replace above sheets with a large metal sheet with a $\lambda/2$ slot and observe the strong signal on the back side of the sheet that "pours" through the slot provided the slot is perpendicular to **E**. Now put the receiving antenna alongside the transmitting antenna in a radar configuration as shown and perform the following tests:
4. Place a large flat metal sheet at point 2 and rotate it slowly, noting the critical dependence on angle. Maximum signal occurs at normal incidence, and it decreases rapidly at angles away from normal.
5. Try the flat metal sheet with "space paper" which should make the metal sheet "invisible."
6. Try a passive 90° corner reflector at point 2 and slowly rotate it, observing the variation in reflected signal as shown in Fig. 5-50.
7. And now the fun begins. As reflecting objects at point 2, try a metal or dielectric cube or sphere, observing the variations in reflected signal with angle. For these observations it may be necessary to support the object with fine threads or dielectric strands (fish line) to ensure that the measurements involve only the object and not its support. Or make the transmit-receive antenna configuration portable and observe the radar return from trees, houses, buildings, automobiles, trucks, aircraft, or other objects.

More problems and projects are found on the book's Web site: *www.elmag5.com.*

PROBLEMS

Answers are given in Appendix E to problems followed by the @ symbol.

5-2-5. Radiation of pulse on PC trace. When a 1-pC capacitor charged to 5 V discharges onto a computer board trace in a 1-ns pulse, what power is radiated when the pulse encounters a short circuit? Assuming $v = c$ on the trace, take the radiated power $= 80\, q^2/\Delta t^2$, where q = charge, C, and Δt = pulse length, s. @

5-2-6. Number of electrons. How many electrons are contained in the 1-ns pulse of Problem 5-2-5? @

5-2-7. Exact versus approximate directivities. (*a*) Calculate the exact directivities for three unidirectional antenna having power patterns as follows:

$$P(\theta, \phi) = P_m \sin\theta \sin^2\phi$$

$$P(\theta, \phi) = P_m \sin\theta \sin^3\phi$$

$$P(\theta, \phi) = P_m \sin^2\theta \sin^3\phi$$

$P(\theta, \phi)$ has a value only for $0 \le \theta \le \pi$ and $0 \le \phi \le \pi$ and is zero elsewhere. (*b*) Calculate the approximate directivities from the half-power beam widths of the patterns. (*c*) Tabulate the results for comparison. @

5-2-8. Sun power. If the earth receives a power from the sun of 2.2 gcal/min/cm^2: (*a*) Translate this into W/m^2. (*b*) Assuming the sun is an isotropic source, what is the sun's power output? Note that the sun is a nuclear power plant that we consider to be sited at a safe distance. (*c*) How long can the sun continue to radiate this power if all its mass is converted into energy via Einstein's relation Energy $= mc^2$? (*d*) What is the sun's rms field (V/m) if all its power is radiated at a single frequency? *Note:* 14.3 gcal/min $= 1$ W. Earth-sun distance $=$ 500 light-minutes. See inside back cover for other data. @

5-2-9. Directivity approximation. Calculate the approximate values for the directivities of the antennas with the patterns provided in Problem 5-2-1 using (5-2-10). Find the dB difference from the exact values.

5-2-10. Main-beam efficiency. For an antenna with field pattern

$$E_n = \frac{\sin\theta}{\theta}\frac{\sin\phi}{\phi}$$

where $\theta =$ zenith angle (radians) and $\phi =$ azimuth angle (radians), (*a*) plot the normalized power pattern as a function of θ, (*b*) using your graph, estimate the main-beam efficiency of this antenna, and (*c*) estimate the directivity of this antenna.

5-2-11. Directivity and gain. (*a*) Estimate the directivity of an antenna with $\theta_{\text{HP}} = 2°$, $\phi_{\text{HP}} = 1°$, and (*b*) find the gain of this antenna if efficiency $k = 0.5$. @

5-2-12. Beamwidth and directivity. For most antennas, the half-power beamwidth (HPBW) may be estimated as HPBW $= \kappa\lambda/D$, where λ is the operating wavelength, D is the antenna dimension in the plane of interest, and κ is a factor which varies from 0.9 to 1.4, depending on the field amplitude taper across the antenna. Using this approximation, find the directivity and gain for the following antennas: (*a*) circular parabolic dish with 2-m radius operated at 6 GHz, (*b*) elliptical parabolic dish with dimensions of 1 m $\times$ 10 m operated at 1 GHz, and (*c*) a rectangular array with sides 0.25 m $\times$ 0.5 m operated at 35 GHz. Assume $\kappa = 1$ and 50 percent efficiency in each case.

5-3-11. Directional broadcast array. (*a*) An AM broadcasting station is to be located south of the area it is to serve. Design an antenna for this station which gives a broad coverage to the north (from NW through N to NE) with reduced field intensity in other directions. However, to obtain Federal Communications Commission approval, the pattern must have a null SE (135° from N) in order to protect another station on the same frequency in that direction. The antenna is to consist of an in-line array of $\lambda/4$ vertical elements oriented along a north-south line with equal spacing between elements. The minimum number of elements should be used. *Hint:* In plan view the problem reduces to a linear array of isotropic point sources. (*b*) Repeat part (*a*) with the additional requirement of another null to the west (90° from N). *Hint:* Apply pattern multiplication. What is the pattern equation? @

5-3-12. Array patterns, isotropic sources. Plot the field amplitude and phase patterns for two isotropic point sources with equal amplitude and phase at spacings of (*a*) $\lambda/16$, (*b*) $\lambda/8$, (*c*) $\lambda/4$, and (*d*) λ.

5-3-13. Array patterns, dipoles. Repeat Problem 5-3-12 if the sources are short dipoles with $E(\phi) = \cos\phi$.

5-3-14. Array patterns, various elements. Repeat Problem 5-3-12 if the sources have field patterns given by (*a*) $\cos^2\phi$, (*b*) $\cos^3\phi$, and (*c*) $(\sin\phi)/\phi$.

5-3-15. Binomial array pattern. (*a*) Compute and plot the field pattern of a seven-source binomial array of isotropic sources, (*b*) find the HPBW of the array, and (*c*) estimate the directivity of this array using (5-2-10).

5-5-2. Short dipole fields. A dipole antenna of length 5 cm is operated at a frequency of 100 MHz with terminal current $I_0 = 120$ mA. At time $t = 1$ s, angle $\theta = 45°$, and distance $r = 3$ m, find (*a*) E_r, (*b*) E_θ, and (*c*) H_ϕ. @

5-5-3. Short dipole far fields. For the dipole antenna of Problem 5-5-2, at a distance $r = 100$ m, use the general expressions of Table 5-3 to find (*a*) E_r, (*b*) E_θ, and (*c*) H_ϕ. Compare these results to those obtained using the far-field expressions of Table 5-3.

5-5-4. Short dipole quasi-stationary fields. For the dipole antenna of Problem 5-5-2, at a distance of 1 m, use the general expressions of Table 5-3 to find (*a*) E_r, (*b*) E_θ, and (*c*) H_ϕ. Compare these results to those obtained using the quasi-stationary expressions of Table 5-3. @

5-5-5. Short dipole fields. At what distance from a 1-m dipole antenna operated at 15 MHz are the amplitudes of the fields E_θ and H_ϕ within 1 percent of their far-field values?

5-5-6. Short dipole power. (*a*) Find the power radiated by a 10-cm dipole antenna operated at 50 MHz with an average current of 5 mA. (*b*) How much (average) current would be needed to radiate a power of 1 W? @

5-5-7. Short dipole radiation resistance. (*a*) Find the radiation resistance of a 10-m dipole antenna operated at 500 kHz (assume $I_{av} = \frac{1}{2}I_0$). (*b*) How long must this antenna be for a radiation resistance of 1 Ω?

5-6-2. Radiation resistance. An antenna measured at a distance of 500 m is found to have a far-field pattern of $|E| = E_0(\sin\theta)^{1.5}$ with no ϕ dependence. If $E_0 = 1$ V/m and $I_0 = 650$ mA, find the radiation resistance of this antenna. @

5-6-3. Directivity of 3λ/2 dipole. Use (5-6-5) and (5-2-9) to calculate the directivity of a $3\lambda/2$ dipole antenna.

5-7-2. Shortwave broadcasting array. Many shortwave broadcast stations beaming to all parts of the world use large "curtain" arrays of horizontal $\lambda/2$ dipoles supported by tall towers. A single unit of the curtain may consist of eight horizontal $\lambda/2$ dipoles arrayed as collinear pairs stacked at $\lambda/2$ intervals as in Fig. P5-7-2. The arrows show

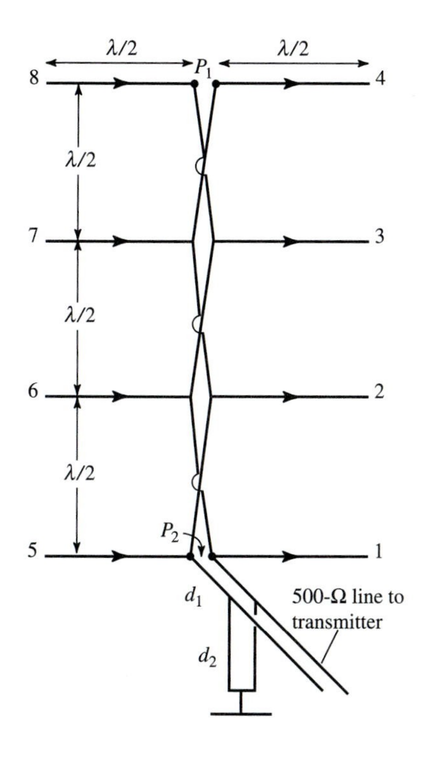

FIGURE P5-7-2

the current directions. It is assumed that all dipoles have equal in-phase currents. (*a*) If each dipole has an impedance $= 73 + j0\ \Omega$ and mutual impedances are $Z_{12} = -13\ \Omega$, $Z_{13} = 4\ \Omega$, $Z_{14} = -2\ \Omega$, $Z_{15} = 26\ \Omega$, $Z_{16} = -12\ \Omega$, $Z_{17} = 9\ \Omega$, and $Z_{18} = -6\ \Omega$, calculate the gain. (*b*) If a reflector curtain is placed $\lambda/4$ behind the array, what is the gain? @

5-7-2.1. Feed for eight-dipole array. Design a feed system for the eight-dipole array of Fig. P5-7-2 that is matched to a 500-Ω line. The dipoles are slightly less than 0.5 λ long and have a length-to-diameter ratio = 1000 resulting in an impedance at point $P_1 = 3000\ \Omega$. At $\lambda/4$ from this point on either dipole the impedance is assumed to be approximately 73 Ω if the dipole is opened at that point. @

5-7-2.2. Stub match for eight-dipole array. Another method of matching the array of Fig. P5-7-2 is to connect the 500-Ω line directly to point P_2 and use a stub. See Fig. P5-7-2. If the array operates at 10 MHz, find d_1 and d_2. *Hint:* Use a Smith Chart (see Example 3-14). @

5-7-3. Shortwave array with 256 dipoles. If a second eight-dipole curtain unit with reflector is placed alongside the one of Problem 5-7-2, a gain of approximately 3 dB may be realized. Doubling the size again for a total of 32 dipoles with reflector the gain increases approximately another 3 dB. (*a*) For such an array with 256 dipoles with reflector, what is the gain? (*b*) At a frequency of 10 MHz, what is the horizontal dimension of the array? @

5-7-4. Shortwave broadcast field strength. If the array of Problem 5-7-3 has a power input of 25 000 W, the antenna has an effective radiated power (ERP) of 20 million watts. Note that when the ionosphere is reflective, the signals are trapped in a layer between the ionosphere and the earth instead of spreading out in three dimensions. See Fig. 5-58, shown also inside back cover. Thus, the signal may attenuate more like the first power instead of the square of the distance. For the 20-MW station calculate the field strength at 12 000 km in the direction of maximum radiation assuming (*a*) first power and (*b*) square of distance attenuation. @

5-7-5. Lossy parallel array. Find the gain in field (relative to single lossless dipole) for a two-element array parallel $\lambda/2$ dipoles spaced by λ if the loss resistance is (*a*) 0 Ω, (*b*) 2 Ω, and (*c*) 10 Ω. @

5-7-6. Lossy collinear array. Repeat Problem 5-7-5 for a collinear array.

5-9-9. Dish antenna. What is the required diameter for a parabolic dish antenna operating at 2 GHz with 35 dBi gain? Aperture efficiency is 50 percent. @

5-9-10. Yagi-Uda antenna. Design a Yagi-Uda six-element antenna for operation at 500 MHz with a folded dipole feed. What are the lengths of (*a*) reflector element, (*b*) driven element, and (*c*) four director elements? What is the spacing (*d*) between reflector and driven element and (*e*) between director elements? What is (*f*) frequency bandwidth (upper and lower frequencies) and (*g*) gain? @

5-9-11. Log-periodic antenna. Design a log-periodic antenna to operate between 100 and 400 MHz with 11 elements. Give (*a*) length of longest element, (*b*) length of shortest element, and (*c*) gain.

5-9-12. Axial-mode helical antenna. Design a right circularly polarized (RCP) axial-mode helical antenna with 16 dBi gain for operation at 1600 MHz with turn spacing λ/π. Find (*a*) number of turns, (*b*) turn diameter, and (*c*) axial ratio. @

5-9-13. Rectangular horn antenna. What is the required aperture area for an optimum rectangular horn antenna operating at 2 GHz with 18 dBi gain? @

5-9-14. Conical horn antenna. What is the required diameter of a conical horn antenna operating at 3 GHz with a 12-dBi gain? @

5-9-15. Alpine-horn antenna. Referring to Fig. 5-30*a*, the low frequency limit occurs when the open end spacing $\approx \lambda/4$ and the high frequency limit when the transmission line spacing $d \approx \lambda/4$. If $d = 2$ mm and the open end spacing $= 1000\,d$, what is the bandwidth?

5-9-16. Alpine-horn antenna. If d = transmission line spacing, what open end spacing is required for a 200-to-1 bandwidth? @

5-10-1. Round-the-world signals. With a rotatable horizontally polarized bidirectional antenna operating at 15 to 20 MHz, it is often possible to observe round-the-world signals in certain directions alerting the radio operator that long-distance (DX) paths are open. If the time delay measured with an oscilloscope is 0.138 s, what is the average height that the radio waves traveled around the earth? @

Example E5-10.1 Cell phone link via "double Friis." To provide reliable 1-GHz cellular phone service in the ground-floor dining room of a 52-story (160-m) building in the center of a metropolis, a passive link via coaxial cable was installed between the top of the building and the dining room (see Fig. E5-10.1). Without the link, signals via multipath reflections were unreliable.

The top-of-the-building antenna is a $\lambda/2$ dipole (a $\lambda/4$ vertical with $\lambda/4$ conical skirt). The dining room antenna is a $\lambda/4$ stub projecting down from the ceiling. The hand-held cell phone with $\lambda/4$ stub projecting up has equivalent aperture.

If the dining room cellular phone is 10 m from the ceiling antenna and the top-of-the-building antenna is 1.2 km line-of-sight from the cell tower antenna, which has a 10 dB gain, what is the total dB path loss? The cable loss is 2 dB/100 m.

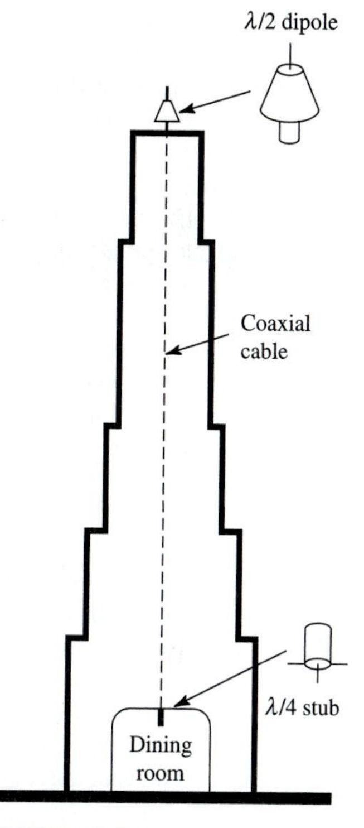

FIGURE E5-10.1

Solution. Apply the Friis transmission formula (5-10-4) twice. From (5-6-6), $D(\lambda/2) = 1.64$ and $D(\lambda/4) = 1.5$. Thus,

$$A_e(\lambda/2) = \frac{1.64\lambda^2}{4\pi}; \qquad A_e(\lambda/4) = \frac{1.5\lambda^2}{4\pi}; \qquad A_e(\text{tower}) = \frac{10\lambda^2}{4\pi}$$

Outside link:

$$\frac{P_r}{P_t} = \frac{A_e(\lambda/2)A_e(\text{tower})}{r^2\lambda^2} = \frac{16.4\lambda^2}{(4\pi)^2 r^2} = \frac{16.4 \times 10^9}{(4\pi)^2 1.44 \times 10^6}$$

$$= 6.5 \times 10^{-9} \text{ or } -82 \text{ dB}$$

Inside link:

$$\frac{P_r}{P_t} = \frac{[A_e(\lambda/4)]^2}{(4\pi)^2 r^2 \lambda^2} = \frac{2.25 \times \lambda^2}{1.58 \times 10^4} = 1.28 \times 10^{-5} \text{ or } -49 \text{ dB}$$

Coaxial cable: $\frac{160}{100} \times 2 = -3.2$ dB

Total loss $= -(82 + 49 + 3.2) = -134$ dB *Ans.*

5-10-2. Field at cell phone. If the cell tower antenna in Example E5-10.1 radiates 100 W, find (*a*) the field strength at the cellular phone in the dining room and (*b*) the SNR (signal-to-noise ratio) if the bandwidth = 10 kHz and $T_{sys} = 300$ K. @

5-10-3. Top-of-building antenna with more gain. If the top-of-the-building $\lambda/2$ dipole of Example E5-10.1 is replaced by a directional antenna pointed at the cell tower, what improvement results with (*a*) a 2.5λ Yagi-Uda antenna, (*b*) a corner reflector, and (*c*) a quad helix array as in Fig. 5-36? The cell tower antenna is vertically polarized. @

5-10-4. Intercontinental signal time. Calculate the signal time between Chicago (41.8° N lat.; 87.6° W long.) and Geneva (46.4° N lat.; 6.2° E long.): (*a*) via a GSO (geostationary orbit) satellite at 20° W long., (*b*) via LEO (low-earth orbit) satellites, and (*c*) with a fiber-optic cable via a great circle path.

5-10-5. Mars radio link. To communicate with a spacecraft on Mars, an antenna in California, Australia, or Spain is used, depending on which location has Mars above the Earth's horizon. If one of these has an effective aperture of 1200 m^2 and the spacecraft has an antenna with effective aperture of 10 m^2, (*a*) determine the transmit power needed at 6 GHz (C band) to produce a power level of 10^{-11} mW (-110 dBm) at the receiving antenna terminals when Earth is closest to Mars (distance = 77 million kilometers) and (*b*) determine the time required for a radio signal to reach Mars. @

5-10-6. Transmission line vs. radio link. The best available transmission lines have losses of approximately 0.1 dB/m at 6 GHz. (*a*) Determine the length of line which would produce losses equal to those of the earth-Mars radio link of Problem 5-10-5. (*b*) What is the loss of a coaxial cable as long as the Mars link? (*c*) The best fiber-optic cable at $\lambda = 1300$ nm has losses of approximately 0.0003 dB/m. What is the loss of a fiber-optic cable of the same length?

5-11-7. Interstellar wireless link. If an extraterrestrial civilization (ETC) transmits 10^6 W, 10-s pulses of right-hand circularly polarized 5-GHz radiation with a 100-m-diameter dish, what is the maximum distance at which the ETC can be received with an SNR = 3. Assume the receiving antenna on the earth also has a 100-m-diameter antenna responsive to right circular polarization, that both antennas (theirs and ours) have 50 percent aperture efficiency, and that the earth station has a system temperature of 10 K and bandwidth of 0.1 Hz. @

5-11-8. Quasar radio power. What is the radio power radiated by quasar OH471? It is at a distance of 14 billion light-years. It has a flux density of about 1 Jy from 1 to 100 GHz. Assume radiation is isotropic. @

5-11-9. Backpacking penguin. This penguin participated in a study of Antarctic penguin migration habits. Its backpack radio with $\lambda/4$ antenna transmitted data on its body temperature and its heart and respiration rates. It also provided information on its location as it moved with its flock across the ice cap. The backpack operated at 100 MHz with a peak power of 1 W and a bandwidth of 10 kHz of tone-modulated data signals. If $T_{sys} = 1000$ K and SNR = 30 dB, what is the maximum range? The transmitting and receiving antennas are $\lambda/4$ stubs. @

5-11-10. Satellite carrier-to-noise ratio. Since the signal-to-noise ratio as defined in (5-11-8) depends on the bandwidth of the receiver, which in turn depends on the modulation applied to the signal, communications satellite engineers use a related quantity called the *carrier-to-noise ratio* or C/N. This represents the ratio of the power in the carrier signal to the noise power per hertz of bandwidth and is given by

$$\frac{C}{N} = \frac{P_t A_{et} A_{er}}{r^2 \lambda^2 k T_{sys}}$$

where C = power density of carrier (W/Hz) and N = noise power density (W/Hz). Find the carrier-to-noise ratio for the system described in Example 5-22, page 316.

5-11-11. Full-path C/N. Show that the full-path or "circuit" (up- and downlink) carrier-to-noise ratio is given by

$$(C/N)_{\text{circuit}}^{-1} = (C/N)_{\text{up}}^{-1} + (C/N)_{\text{down}}^{-1}$$

where $(C/N)_{\text{up}}$ and $(C/N)_{\text{down}}$ are the uplink and downlink carrier-to-noise ratios.

5-11-12. Low earth orbit communications satellite. A communications satellite in low earth orbit (LEO) has $r_{\text{up}} = 1500$ km and $r_{\text{down}} = 1000$ km, with an uplink frequency of 14.25 GHz and a downlink frequency of 12 GHz. Find the full-circuit C/N if the transmitting earth station ERP is 60 dBW and the satellite ERP is 25 dBW. Assume the satellite receiver G/T is 5 dB/K and the earth-station G/T is 30 dB/K. @

5-11-13. Direct broadcast satellite (DBS). Direct broadcast satellite services provide CD-quality audio to consumers via satellites in geosynchronous orbit. The World Administrative Radio Conference (WARC) has established these requirements for such services.

Frequency band	11.7 to 12.2 GHz (K_u band)
Channel bandwidth	27 MHz
Minimum power flux density	-103 dBW/m^2
Receiver figure of merit (G/T)	6 dB/K
Minimum carrier-to-noise ratio	14 dB

(*a*) Find the effective radiated power (ERP) over 1 W isotropic needed to produce the specified flux density at the Earth's surface from a DBS satellite in a 36 000-km orbit. (*b*) If the satellite has a 100-W transmitter and is operated at 12 GHz, what size circular parabolic dish antenna must be used to achieve the required ERP? Assume 50 percent efficiency. (*c*) Does a consumer receiver with circular 1-m dish antenna with 50 percent efficiency and system noise temperature of 1000 K meet the specified G/T? (*d*) By how much does the system specified in parts (*a*) through (*c*) exceed the required carrier-to-noise ratio? @

5-11-14. Signal-to-noise ratio of Mars link. Find the signal-to-noise ratio of the Mars radio link of Problem 5-10-5 for a receiver with noise temperature of 600 K and bandwidth of 100 kHz. @

5-11-15. Simplified expression for C/N. The expression for C/N provided in Problem 5-11-10 may be simplified by making the following substitutions:

$$\text{Effective isotropic radiated power} = \text{ERP} = P_t G_t \text{ (W)}$$

$$\text{Link path loss} = L_{\text{link}} = 4\pi r^2/\lambda^2$$

The carrier-to-noise ratio may then be written as

$$\frac{C}{N} = \text{ERP}\frac{1}{L_{\text{link}}}\frac{1}{k}\frac{G_r}{T_{\text{sys}}}$$

where G_r/T_{sys} is the receive antenna gain divided by the system noise temperature. This ratio, referred to as "*G over T*," is commonly used as a figure of merit for satellite and earth-station receivers. Find the C/N ratio for the uplink to a satellite at the Clarke orbit ($r = 36\,000$ km) equipped with a 1-m parabolic dish antenna with efficiency of 50 percent and a receiver with noise temperature of 1500 K. Assume that the transmitting earth station utilizes a 1-kW transmitter and a 50 percent efficient 10-m dish antenna and operates at a frequency of 6 GHz. @

5-11-16. Carrier-to-noise and maximum data rate. The importance of carrier-to-noise density ratio (C/N_d) in communication links was established in 1949 by C. E. Shannon.† According to Shannon's theorem for the *information capacity* of a communication channel, the maximum data rate of a channel with bandwidth B is given by

$$M = B\log_2\left(1 + \frac{C}{N_d B}\right)$$

where M = channel capacity (bits/s), B = channel bandwidth (Hz), C = carrier signal power (W), and N_d = noise power density (W/Hz). Show that the maximum data rate for any channel, even if infinite bandwidth is used, is

$$M_{B\to\infty} = \frac{C}{N_d}\log_2(e) = 1.44\frac{C}{N_d}$$

Hint: Let $x = C/(N_d B)$ and use the relation

$$\lim_{x\to\infty}(1 + x)^{1/x} = e$$

5-11-17. Satellite communications relay system. A proposed global communication system uses a ring of medium earth orbit (MEO) satellites to relay signals from one side of the earth to the other and to relay data from distant spacecraft to ground stations. The ground stations use receivers with G/T of 20 dB/K and have transmit power of 10 W. The uplink frequency is 14 GHz, downlink frequency is 12 GHz, and intersatellite frequency is 100 GHz (since there is no atmospheric absorption between satellites). Each satellite has an ERP of 5000 W and receive G/T of 5 dB/K. (*a*) Find the full-link carrier-to-noise ratio and maximum data rate if three satellites are used to link two earth stations. (*b*) Find the full-link carrier-to-noise ratio and maximum data

†C. E. Shannon, "Communication in the Presence of Noise," *Proceedings of the IRE*, **37** (January 1949).

rate between a probe at Saturn (distance $= 1.5 \times 10^9$ km) with ERP of 500 W if a single satellite is used to relay the signal to the receiving ground station.

5-11-18. Galileo's uncooperative antenna. When the Galileo spacecraft arrived at Jupiter in 1995, ground controllers had been struggling for 3 years to open the spacecraft's 5-m high-gain communications dish, which was to operate at 10 GHz (X band). Unable to deploy this antenna because of prelaunch loss of lubricant, a low-directivity ($G = 10$ dB) S-band antenna operating at 2 GHz had to be used to relay all pictures and data from the spacecraft to the Earth. For a spacecraft transmit power of 20 W, distance to Earth of 7.6×10^{11} m, and 70-m dish with 50 percent efficiency at the receiving station, find (*a*) the maximum achievable data rate if the 5-m X-band antenna had deployed and (*b*) the maximum data rate using the 1-m S-band antenna. @

5-11-19. Cellular base-station antenna. A 930-MHz cellular base-station antenna for three cells has a triangular ground plane 400 mm on a side with vertical center-fed in-phase $\lambda/2$ dipoles mounted on each of the three sides as shown in Fig. P5-11-19. The dipoles carry equal in-phase currents. (*a*) What is the required dipole spacing d from the ground plane required to produce the overlapping field patterns shown? (*b*) How many $\lambda/2$ dipoles must be stacked vertically to produce a 20° vertical HPBW? (*c*) What is the gain of each of the three stacked arrays? @

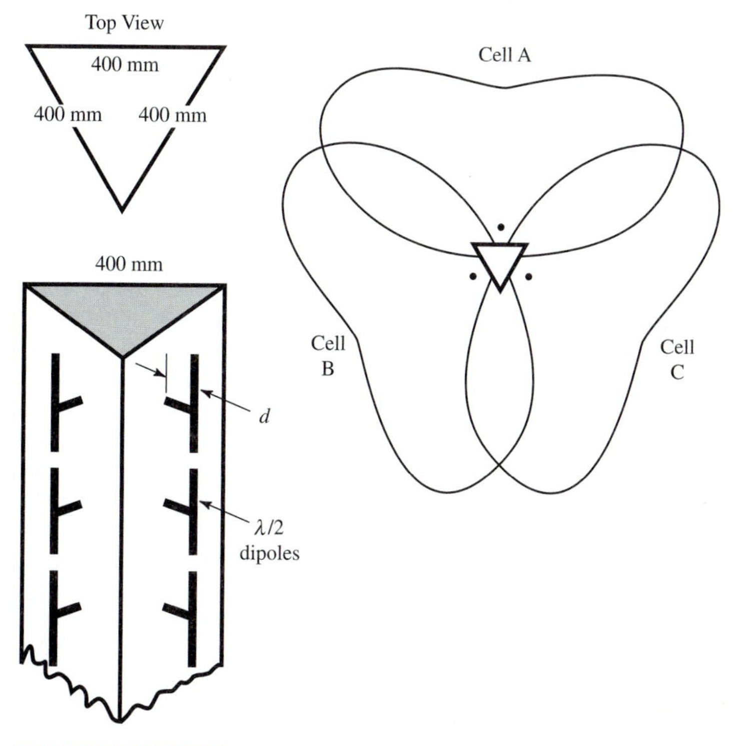

FIGURE P5-11-19
Cellular base-station antenna and field pattern.

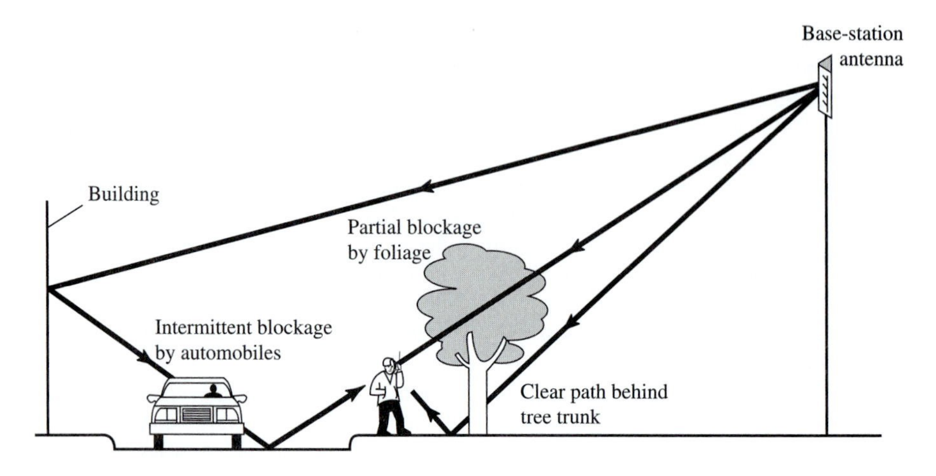

FIGURE P5-11-20
Typical multipath situation with a cellular phone. Sketch not to scale.

5-11-20. Cellular phone multipath situation. Referring to Fig. P5-11-20, the direct path from the 930-MHz vertically polarized base-station antenna 100 m above ground to the hand-held cellular phone is 5 km. Via the ground-bounce path behind the tree trunk the distance is 806 mm farther. The foliage absorption reduces the signal by 2 dB and the ground bounce near the tree also results in a 2-dB loss. Assume that the ground by the tree is perfectly conducting but rough. The path via building and pavement is 250 m farther with 8-dB loss at each reflection.
(*a*) What is the SNR at the hand-held phone with and without blockage by an automobile? The base-station power is 10 W with 10 dBi antenna gain. The cell phone bandwidth is 10 kHz and its system temperature 300 K. Assume a cell phone antenna gain of 1.5. *Hint*: See Project P4-1.
(*b*) If the base-station antenna is circularly polarized, what are the benifits? @

5-12-4. Fastball velocity. A 20-GHz radar measures a doppler shift of 6 kHz on a baseball pitcher's fastball. What is the fastball's velocity? @

5-12-5. Radar power for fastball measurement. To measure the velocity of the fastball of Problem 5-12-4 with the 20-GHz radar at a distance of 100 m, what power is required for an SNR = 30 dB? The radar uses a conical horn with diameter = 8 cm and aperture efficiency $\varepsilon_{ap} = 0.5$. The ball diameter = 7 cm and it has a radar cross-section (RCS) half that of a perfectly conducting sphere of the same diameter. @

5-12-6. Anticollision radar. To provide anticollision warnings, forward-looking radars on automobiles, trucks and other vehicles (*see Figure P5-12-6*) can alert the driver of vehicles ahead that are decelerating too fast or have stopped. The brake light on the vehicle ahead may not be working or it may be obscured by poor visibility. To warn of clear-distance decrease rates of 9 m/s or more, what doppler shift must a 20-GHz radar be able to detect? @

5-12-7. Beam width and power of anticollision radar. For the anticollision 20-GHz radar of Problem 5-12-6 to avoid false warnings from parked cars and bridge abutments along the side of the road, the beam width between first nulls should be 10 m at a range of 250 m. To avoid returns from bridges crossing the highway, the vertical

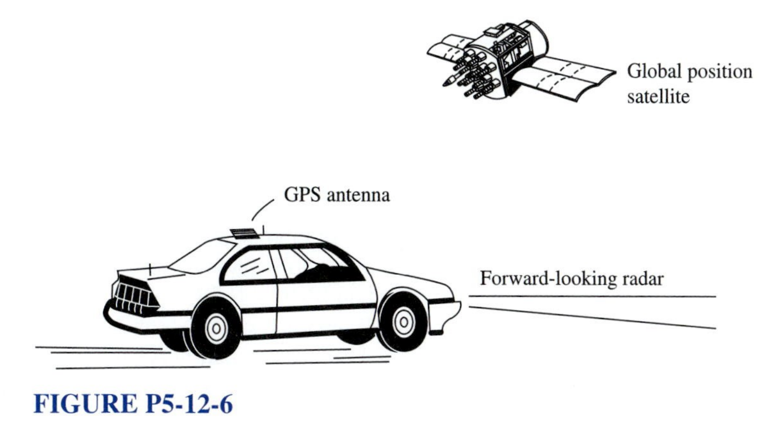

FIGURE P5-12-6

beamwidth should be one-half as much. Assume that the radar antenna is a flat-panel broadside array of patch antennas with uniform aperture distribution. (*a*) What are the antenna dimensions? (*b*) To detect a 1-m^2 object at a range of 500 m with an SNR = 30 dB, what power is required?

5-12-8. Anti-CFIT radar. For aircraft to avoid flying into mountainous terrain during normal flight in poor visibility, called *controlled flight into terrain* (CFIT), a forward-looking radar is required that is similar to that in Problem 5-12-6 for highway vehicles but with different range and closure rates. Thus, for a range of 15 km and rate of distance closure of 100 m/s or more, find (*a*) pulse rate and (*b*) power required for a 10-GHz radar with 37-dB gain antenna and SNR = 30 dB.

5-12-9. Radar altimeter. For a 10-GHz pulse radar altimeter to measure altitudes of 100 m to 10 km, (*a*) what power and (*b*) pulse rate are required? The antenna is a 3×30-cm downward-looking flat-panel patch array with uniform aperture distribution. A frequency of 10 GHz has been chosen to avoid the water-absorption band at 20 GHz (see Fig. 5-59). Although an altimeter is a valuable navigation aid, an anti-CFIT radar is also needed to warn of rapidly rising or steep terrain. The optimum system is to have both altimeter and CFIT radars working in conjunction with the aircraft's navigational radar and a GPS display that indicates the aircraft's position and the elevation of the terrain in the vicinity.

5-12-10. RCS of electron. The alternating field of a passing electromagnetic wave causes an electron (initially at rest) to oscillate (Fig. P5-12-10). The oscillation makes it equivalent to a short dipole antenna with $D = 1.5$. What is the electron's radar cross-section? *Hint*: Calculate the ratio of the power scattered to the incident Poynting vector. The reradiation is called *Thompson scatter*. @

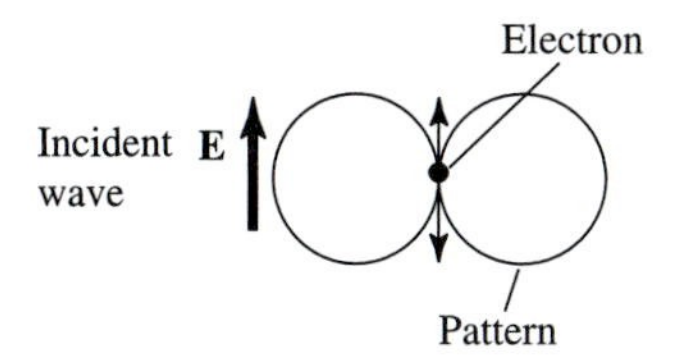

FIGURE P5-12-10

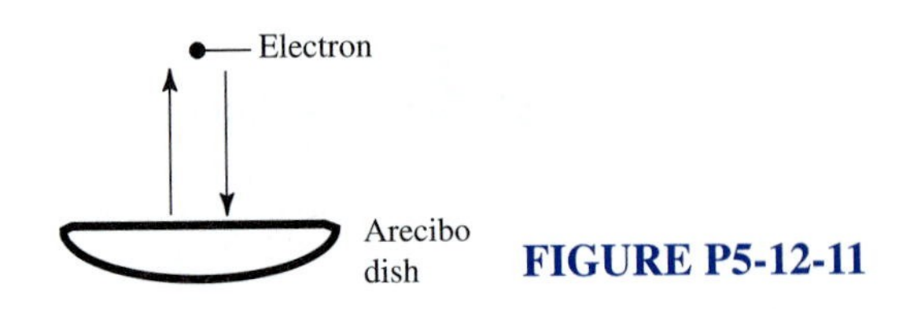

FIGURE P5-12-11

5-12-11. Detecting one electron at 10 km. If the Arecibo ionospheric 300-m-diameter antenna operates at 100 MHz, how much power is required to detect a single electron at a height (straight up) of 10 km with an SNR = 0 dB? See Fig. P5-12-11. The bandwidth is 1 Hz, T_{sys} = 100 K and the aperture efficiency = 50 percent. @

5-12-12. Pulsed radar range resolution. Pulsed radars use short bursts of electromagnetic energy rather than continuous waves to detect and track targets of interest. Use of pulse modulation has two benefits for radar systems: the transmitter is turned off while the receiver's sensitive detection circuits operate, and returns from objects at different distances arrive at the receiver at different times. The timing of returns from various distances is shown in Fig. P5-12-12, which is the radar equivalent of Fig. 3-30 for pulses on transmission lines.

As shown in the diagram, if the radar transmitter is turned on for a time period of τ seconds, returns from the object at range R_1 are present at the receiver from time $t = 2R_1/c$ until time $t = 2R_1/c + \tau$. If a second scattering object is located at range R_2, its returns are received from time $t = 2R_2/c$ until time $t = 2R_2/c + \tau$. Show that for a pulsed radar the *range resolution* (the minimum range difference between objects for which the returns do not overlap in time) is given by

$$\Delta R = R_2 - R_1 = \frac{c\tau}{2}$$

where τ is the width of the transmitted pulse in seconds. Note that this equation applies to propagation in air or vacuum; for other materials the appropriate velocity of propagation must be used.

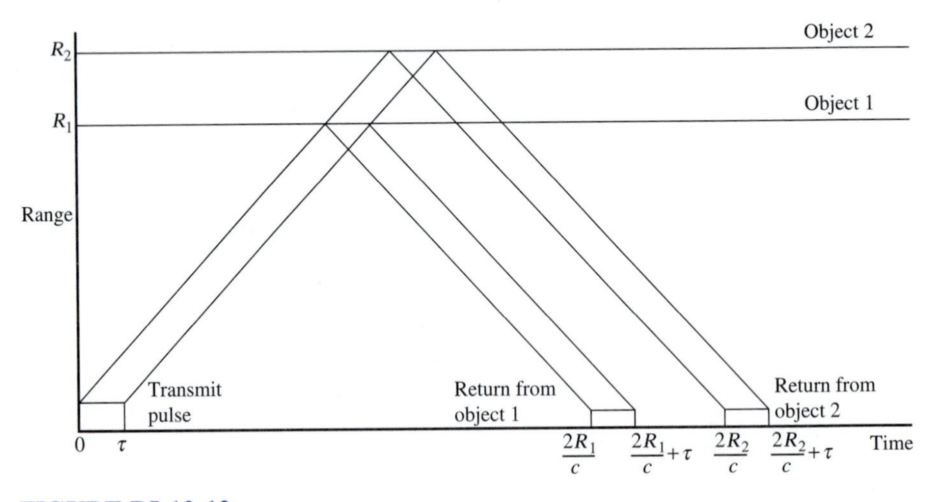

FIGURE P5-12-12

5-12-13. Police radar. A pulsed speed-measuring radar must be able to resolve the returns from two cars separated by 10 m. Find the maximum pulsewidth that can be used to prevent overlapping of the returns from the two vehicles. @

5-12-14. Ground-penetrating radar resolution. For a ground-penetrating radar operating in sandy soil with $\varepsilon_r = 8$ and $\mu_r = 0.01$, what pulsewidth is required to resolve two buried objects at depths of 10 and 15 m?

5-12-15. RCS of flat plate and sphere. Different objects with identical physical cross-section may have very different radar cross-section (RCS). For example, a conducting flat plate of physical area A_p has an RCS given by

$$\sigma_{\text{plate}} = \frac{4\pi A_p^2}{\lambda^2}$$

while the RCS of a conducting sphere of radius a is

$$\sigma_{\text{sphere}} = \pi a^2$$

In both cases, the wavelength is assumed to be small relative to the size of the illuminated object. A square flat plate with 2-m sides and a sphere with radius of 1.13 m both have a physical cross-section of 4 m^2. Calculate the RCS of the flat plate at a frequency of 10 GHz ($\lambda = 3$ cm) and compare with the RCS of the sphere. @

5-12-16. Sea clutter. Search-and-rescue aircraft using radar to locate lost vessels must contend with backscatter from the surface of the ocean. The amplitude of these returns (known as *sea clutter*) depends on the frequency and polarization of the radar waveform, the size of the illuminated patch on the surface, the angle of incidence, and the sea state. The scattering geometry is shown in Fig. P5-12-16. To characterize sea clutter independently of the radar footprint on the surface, the scattering cross-section of the ocean may be specified per unit area. This parameter, designated σ_0, has dimensions of square meters or dB above a square meter (dBsm). The total RCS of a patch of ocean surface is found by multiplying σ_0 by the area of the patch. For

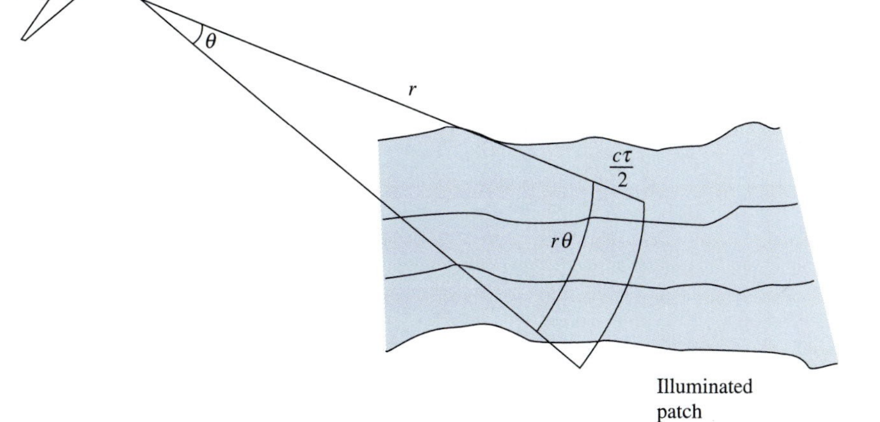

FIGURE P5-12-16
Sea search-and-rescue geometry.

area scattering, the radar equation is written

$$P_r = P_t \frac{A_e^2 \lambda^2}{4\pi r^4} \sigma_0 A_{\text{patch}}$$

For a pulsed radar with pulsewidth τ and 3-dB antenna beamwidth of θ rad, the illuminated area for low grazing angle is approximately $(c\tau/2)(r\theta)$. The radar equation may therefore be written as

$$P_r = P_t \frac{A_e^2 \lambda^2}{4\pi r^4} \sigma_0 \left(\frac{c\tau}{2}\right)(r\theta) = P_t \frac{A_e^2 \lambda^2 \sigma_0 c\tau\theta}{8\pi r^3}$$

(*a*) Determine the received power from sea clutter at a range of 10 km for a monostatic pulsed radar transmitting a 1-μs pulse with 1-kW peak power at a frequency of 6 GHz. Assume a 1.5-m circular dish antenna with 50 percent efficiency and sea state 4 ($\sigma_0 = -30$ dBsm/m^2 at C band). (*b*) For a receiver bandwidth of 10 kHz and noise figure of 3.5 dB, find the receiver noise power. (*c*) if this radar is used to search for a ship with RCS = 33 dBsm under these conditions, what signal-to-noise and signal-to-clutter ratios can be expected? @

5-12-17. Asteroid detection system. Find the transmit power needed to achieve a signal-to-noise ratio of 10 dB on a spherical asteroid 1 km in diameter at a distance of 1 million kilometers using a transmit/receive antenna with $G = 60$ dB at $\lambda = 0.3$ m. Assume a reflection coefficient of 0.1 for the asteroid, a bandwidth of 10 Hz for the receiver, and transmit/receive losses of 4 dB.

5-14-5. Airport radar. A 10-GHz airport radar dish antenna with 5° vertical HPBW must resolve a 50-m aircraft at 3 km. What are the required dish dimensions for a cosine-squared aperture distribution? @

5-16-3. Wireless through rain. What power is required for a 1-km outdoor 10-MHz-bandwidth wireless link operating at 50 GHz during a heavy rainfall of 100 mm/h with an SNR = 40 dB. The send and receive antennas each have a 30 dB gain. $T_{\text{sys}} = 1000$ K. For rain absorption see Fig. 5-59. @

5-16-4. Wireless-friendly buildings (WFB). What design features should new buildings have to facilitate installing 50-GHz wireless links within the building? Consider openings in walls covered by radio-transparent but optically-opaque panels.

CHAPTER

6

ELECTRODYNAMICS

6-1 INTRODUCTION

Previous chapters discuss static electric and magnetic fields, transmission lines, antennas, and waves in space and media. This chapter treats the interaction of charged particles with electric and magnetic fields and the operation of mechanical devices such as motors and generators. It also includes a section on the Hall generator.

6-2 CHARGED PARTICLES MOVING IN ELECTRIC FIELDS

If a charged particle is set free in an electric field, it is accelerated by a force proportional to the electric field and charge on the particle as given by

$$\boxed{\mathbf{F} = e\mathbf{E} \qquad \text{(N)}} \tag{1}$$

where $\mathbf{E}$ = electric field intensity, V m^{-1}
e = charge of particle, C.

The symbol e is used here for the charge (instead of Q or q) since e is usually employed for the charge of a particle, commonly an electron.

According to ***Newton's second law***, the rate of change of momentum ($m\mathbf{v}$) of the particle equals the applied force $\mathbf{F}$ or

$$\boxed{\mathbf{F} = \frac{d(m\mathbf{v})}{dt} = m\frac{d\mathbf{v}}{dt} + \mathbf{v}\frac{dm}{dt} \qquad \text{(N)} \qquad \textit{Newton's second law}} \tag{2}$$

where m = mass of particle, kg
$\mathbf{v}$ = velocity of particle, m s^{-1}.

If the velocity is very small compared to the velocity of light ($v \ll c$), the rate of change of mass is small (mass essentially constant), and the last term in (2) may be neglected (see last paragraph of this section). Thus, (2) reduces to

$$\mathbf{F} = m\frac{d\mathbf{v}}{dt} = m\mathbf{a} \tag{2a}$$

where $\mathbf{a}$ = acceleration of particle, m s^{-2}. Equating (1) and (2a), we have

$$m\mathbf{a} = e\mathbf{E} \tag{3}$$

which has the dimensions of force. Integrating this force over the distance the particle moves yields the energy W acquired by the particle. Thus,

$$W = m\int_1^2 \mathbf{a}\cdot d\mathbf{L} = e\int_1^2 \mathbf{E}\cdot d\mathbf{L} \qquad \text{(J)} \tag{4}$$

We recognize the line integral of $\mathbf{E}$ between points 1 and 2 as the potential difference V between the points. We also know that the acceleration is the rate of change of velocity $\mathbf{v}$ or $\mathbf{a} = d\mathbf{v}/dt$, also that the velocity $\mathbf{v}$ is the rate of change of distance $\mathbf{L}$, or $\mathbf{v} = d\mathbf{L}/dt$. Substituting these derivatives into (4), we have

$$W = m\int_1^2 \mathbf{v}\cdot d\mathbf{v} = eV \tag{5}$$

or

$$W = \tfrac{1}{2}m(v_2^2 - v_1^2) = eV \tag{6}$$

where W = energy acquired by particle, J
v_2 = velocity of particle at point 2, or final velocity, m s^{-1}
v_1 = velocity of particle at point 1, or initial velocity, m s^{-1}
e = charge on particle, C
m = mass of particle, kg
V = magnitude of potential difference between points 1 and 2, V or J C^{-1}

If the particle starts from rest, the initial velocity is zero, so

$$W = eV = \tfrac{1}{2}mv^2 \qquad \textit{Particle energy} \tag{7}$$

where v is the final velocity. Equation (7) has the dimensions of energy. The dimensional relation in SI units is

$$\text{Joules} = \text{coulombs} \times \text{volts} = \text{kilograms}\frac{\text{meters}^2}{\text{seconds}^2}$$

Thus, the energy acquired by a particle of charge e starting from rest and passing through a potential difference V is given either by the product of the charge and the potential difference or by one-half the product of the mass of the particle and the square of the final velocity.

Solving (7) for the velocity gives

$$v = \sqrt{\frac{2eV}{m}} \qquad (\text{m s}^{-1}) \qquad \textit{Particle velocity} \tag{8}$$

The energy acquired by an electron ($|e| = 1.6 \times 10^{-19}$ C) in "falling" through a potential difference of 1 V is 1.6×10^{-19} J. This amount of energy is a convenient unit for designating the energies of particles and is called one ***electron volt*** (eV).

For an electron $|e| = 1.6 \times 10^{-19}$ C and $m = 0.91 \times 10^{-30}$ kg, so that (8) becomes

$$v = 5.9 \times 10^5 \sqrt{V} \qquad (\text{m s}^{-1}) \tag{9}$$

Thus, if $V = 1$ V, the velocity of the electron is 0.59 Mm s^{-1}, or 590 km s^{-1}. It is apparent that a relatively small voltage imparts a very large velocity to an electron. If $V = 2.5$ kV, the velocity is 30 Mm s^{-1}, or about one-tenth the velocity of light. The above relations are based on the assumption that the particle velocity is small compared with that of light. The mass of a particle approaches an infinite value as the velocity approaches that of light (***relativistic effect***), whereas the above relations are based on a constant mass. Actually, however, the mass increase is of negligible consequence for most applications unless the velocity is at least 10 per cent that of light. The relation between the mass m of the particle and its mass m_0 at low velocities (*rest mass*) is given by

$$m = \frac{m_0}{\sqrt{1 - (v/c)^2}} \tag{10}$$

where v = velocity of particle, m s^{-1}
c = velocity of light = 300 Mm s^{-1}.

If the velocity is one-tenth that of light, the mass is only one-half of 1 percent greater than the rest mass.

6-3 THE CATHODE-RAY TUBE (CRT); ELECTRICAL DEFLECTION

A cathode-ray tube, sometimes called a *Braun tube* after Carl F. Braun, forms the screen of TV receivers, computer monitors, and oscilloscopes. Charged particles (electrons) are accelerated and deflected so as to strike a fluorescent screen in a desired pattern and produce a visible image. The particular usefulness of the CRT is that the small inertia of the electron beam allows it to follow very rapid changes in the applied deflecting field. What follows is a somewhat oversimplified description of a CRT, but it will suffice for a brief analysis of some of its characteristics.

A positive potential V_a is applied to the accelerating electrode, as in Fig. 6-1 on next page. This produces an accelerating field $\mathbf{E}_a$ that imparts a velocity v_x to the electrons as given by

$$v_x = \sqrt{\frac{2eV_a}{m}} \qquad (\text{m s}^{-1}) \tag{1}$$

After an electron leaves the accelerating electrode, or ***electron gun***, it maintains this velocity v_x in the vacuum of the tube.

Now with the electron on its way to the screen, let us determine the amount of deflection we can produce with a potential V_d between two deflection plates arranged

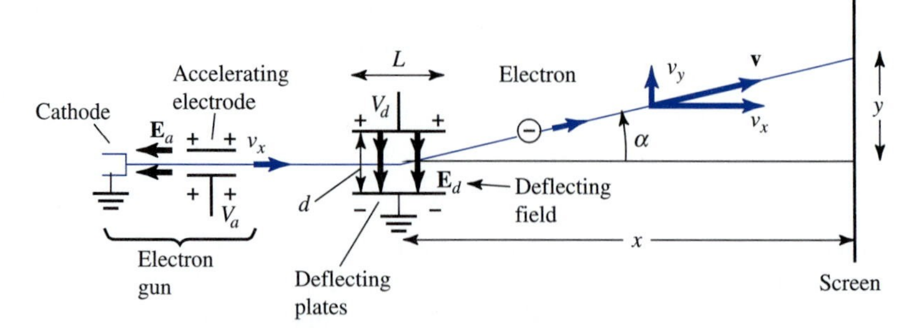

FIGURE 6-1
Cathode-ray tube showing electron path from cathode to screen with electrostatic accelerating field $\mathbf{E}_a$ and deflecting field $\mathbf{E}_d$.

like a parallel-plate capacitor, as in Fig. 6-1. The path of an electron in the transverse deflecting field is a parabola. Neglecting fringing of the field at the edges of the plates, the electron is subjected to the deflecting field E_d ($= V_d/d$) for a distance L or for a time $t = L/v_x$. The field $\mathbf{E}_d$ produces an acceleration a_y in the y direction which is

$$a_y = \frac{eV_d}{md} \tag{2}$$

Thus, the electron acquires a velocity component v_y in the y direction given by

$$v_y = a_y t = \frac{eV_d L}{mv_x d} \tag{3}$$

The deflection angle α (Fig. 6-1) is then

$$\alpha = \tan^{-1}\frac{v_y}{v_x} = \tan^{-1}\frac{eV_d L}{mv_x^2 d} \tag{4}$$

or

$$\alpha = \tan^{-1}\frac{V_d L}{2V_a d} \tag{5}$$

From the tube geometry, assuming $x >> L$, the angle α is also given by

$$\alpha = \tan^{-1}\frac{y}{x} \tag{6}$$

Equating (5) and (6), we obtain the deflection distance of the electron beam at the screen as

$$y = \frac{V_d L x}{2V_a d} \quad \text{(m)} \qquad \textbf{\textit{Deflection distance}} \tag{7}$$

where y = deflection distance at screen, m
V_d = deflection potential, V
L = length of deflection plates, m

x = distance from deflection plates to screen, m
V_a = accelerating potential, V
d = spacing of deflection plates, m

Solving for the volts per meter of deflection (ratio V_d/y), we have

$$\frac{V_d}{y} = \frac{2V_a d}{Lx} \tag{8}$$

Knowing this ratio, we can determine what voltage is required to deflect the beam a desired distance.

Example 6-1. Cathode-ray tube. A CRT with electrostatic deflection has an accelerating potential $V_a = 1500$ V, a deflecting-plate spacing $d = 10$ mm, a deflecting-plate length $L = 10$ mm, and a distance $x = 300$ mm from deflecting plates to the screen. Neglecting fringing, find the potential V_d required to deflect the spot 10 mm on the screen.

Solution. From (8),

$$\frac{V_d}{y} = \frac{2 \times 1500 \times 10^{-2}}{10^{-2} \times 300 \times 10^{-3}} = 10^4 \text{ V m}^{-1}$$

or 10 V mm^{-1} and $V_d = 100$ V. *Ans.*

To increase the sensitivity, that is, to decrease the potential required per unit of deflection, V_a or d can be decreased or L and x increased. To provide deflection in the z direction (perpendicular to the page in Fig. 6-1), another pair of plates can be placed parallel to the page. Alternatively, a magnetic field parallel to $\mathbf{E}_d$ in Fig 6-1 can be used. Magnetic field deflection of charged particles is discussed in the next section.

Problem 6-3-1. Ion or proton thruster. A proton is initially at rest in a vacuum. An electric field of 10 kV/m is applied, accelerating the proton which leaves the field after traveling 10 cm as in an electron gun. How many protons are required to produce a thrust of 500 nN? *Ans.* 3.13×10^{18} protons.

Problem 6-3-2. Electron velocity. An electron is accelerated by traveling through a potential difference of 1 MV. Find its velocity. If you obtain a velocity $v > c$, you have overlooked something. @

Problem 6-3-3. Electron velocity. An electron has velocity components $v_x = 10^6$ m/s and $v_y = 10^4$ m/s after being accelerated and deflected in a cathode-ray tube (Fig. 6-1). Find (*a*) accelerating voltage and (*b*) deflecting field if the field extends 25 mm in the electron's direction of travel. *Ans.* (*a*) 2.84 V, (*b*) 2.28 V/m.

6-4 CHARGED PARTICLES MOVING IN A STATIC MAGNETIC FIELD

If a positively charged particle is set free in an electric field **E**, it is accelerated in the direction of **E**. On the other hand, if a charged particle is released in a static or steady magnetic field, nothing happens. The particle remains at rest. But if the particle is moving across a magnetic field, it is accelerated at right angles to its direction of motion.

From (2-12-13), the force on a current element of length dL in a magnetic field is given in vector form by

$$d\mathbf{F} = (\mathbf{I} \times \mathbf{B})dL \qquad \text{(N)} \qquad \textbf{\textit{Motor equation}} \tag{1}$$

where $\mathbf{I}$ = current, A
$\mathbf{B}$ = magnetic flux density, T
dL = element length, m

This is the fundamental ***motor equation*** of electrical machinery. It also applies to moving charged particles.

The current in a conductor, or in a beam of ions or electrons, can be expressed as charge q per unit time or

$$I = \frac{q}{t} \qquad \text{(A)} \tag{2}$$

Current times length is then equivalent to the charge times its velocity, or

$$IL = \frac{q}{t}L = q\frac{L}{t} = qv \qquad \text{(A m)} \tag{3}$$

where $v = L/t$ = velocity of charge, m s^{-1}.

In vector and infinitesimal form, (3) becomes

$$\mathbf{I}\,dL = dq\,\mathbf{v} \qquad \text{(A m)} \tag{4}$$

where dq = charge in a length dL of conductor or beam.

Substituting (4) in (1) gives

$$d\mathbf{F} = dq(\mathbf{v} \times \mathbf{B}) \tag{5}$$

For a single particle of charge e, we have for the Lorentz force (after H. A. Lorentz)

$$\mathbf{F} = e(\mathbf{v} \times \mathbf{B}) \qquad \text{(N)} \qquad \textbf{\textit{Lorentz force}} \tag{6}$$

Consider now the motion of a particle of charge e in a uniform magnetic field of flux density $\mathbf{B}$. The velocity of the particle is $\mathbf{v}$. From Newton 's second law, the force on the particle is equal to the product of its mass m and its acceleration $\mathbf{a}$ ($= d\mathbf{v}/dt$).† Thus

$$m\mathbf{a} = e(\mathbf{v} \times \mathbf{B}) \tag{7}$$

or

$$\mathbf{a} = \frac{e}{m}(\mathbf{v} \times \mathbf{B}) \tag{8}$$

According to (8), the acceleration is normal to the plane containing the particle path and $\mathbf{B}$. If the direction of the particle path (indicated by $\mathbf{v}$) is normal to $\mathbf{B}$, the

†We are assuming that $v \ll c$.

acceleration is maximum. If the particle is at rest, the field has no effect. Likewise, if the particle path is in the same direction as **B**, there is no effect, the particle continuing undeflected. Only when the path has a component normal to **B** does the field have an effect.

If a magnetic field of large extent is at right angles to the direction of motion of a charged particle, the particle is deflected into a circular path. Suppose that in a field-free region a positively charged particle is moving to the right, as indicated in Fig. 6-2, and that when it reaches the point P a magnetic field is applied. The direction of **B** is normal to the page (outward). According to the cross-product of **v** into **B** in (8), the acceleration **a** is downward, so that *the particle describes a circle* in the clockwise direction in the plane of the page.

Particle Radius

Let us determine the radius R of the circle. The magnitude of the force **F** (radially inward) on the particle is, by (8),

$$F = ma = evB \tag{9}$$

This is balanced by the centrifugal force

$$F = \frac{mv^2}{R} \tag{10}$$

Equating (9) and (10) yields

$$\frac{mv^2}{R} = evB \tag{11}$$

or

$$R = \frac{mv}{eB} \quad \text{(m)} \qquad \textbf{\textit{Particle radius}} \tag{12}$$

where R = radius of particle path, m
m = mass of particle, kg
v = velocity of particle, m s^{-1}
e = charge of particle, C
B = flux density, T

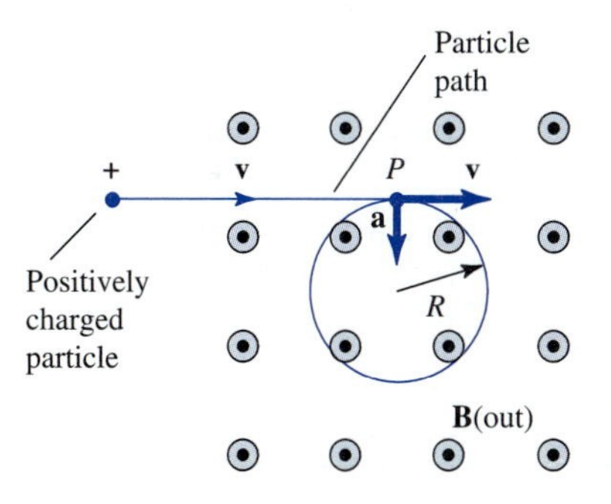

FIGURE 6-2
When a magnetic field **B** (out) is applied to a moving positively charged particle which has arrived at P, it is deflected into a circular path.

Thus, the larger the velocity of the particle or the larger its mass, the greater the radius. On the other hand, the larger the charge or the magnetic flux density, the smaller the radius.

Example 6-2. Proton curvature. A proton at rest is accelerated for 1 μs by an electric field of 2 kV m^{-1} and then moves perpendicular to a magnetic flux density $|\mathbf{B}| = 200$ μT. Find (a) the proton's velocity and (b) its radius of curvature. See Table 6-1 for particle constants.

Solution

(a) $v = (e/m)Et = (9.6 \times 10^7)(2 \times 10^3)(10^{-6}) = 19.2 \times 10^4 \quad \text{m s}^{-1}$ *Ans.*

$$\text{(b) } R = \frac{v}{(e/m)B} = \frac{(e/m)Et}{(e/m)B} = \frac{Et}{B} = \frac{(2 \times 10^3)(10^{-6})}{200 \times 10^{-6}} = 10 \text{ m} \quad \textit{Ans.}$$

Problem 6-4-1. Radius of curvature. Find the radius of curvature for the path of a proton with velocity $v = 10$ Mm/s moving perpendicular to a magnetic field with (*a*) $B = 5$ mT (solar-corona magnetic field), (*b*) $B = 100$ μT (earth's magnetic field), and (*c*) $B = 10$ nT (interstellar magnetic field). *Ans.* (*a*) 20.8 m, (*b*) 1.04 km, (*c*) 10.4 Mm.

Problem 6-4-2. Mass spectrograph. In a mass spectrograph particles are injected with a velocity $\mathbf{v}$ into a magnetic field of flux density $\mathbf{B}$ normal to $\mathbf{v}$. The difference in radius of two particles, one of which has a mass of 10^{-21} g, equals 36 mm. $B = 500$ mT. If $v = 10^6$ m/s and $e = 10^{-16}$ C for both particles, find the mass of the other particle. *Ans.* 10^{-22} g.

Gyrofrequency

The number of revolutions per second of the particle in the circular path is the **gyrofrequency** f of the particle. Introducing $v = \omega R = 2\pi fR$ in (12), gives this frequency as

$$f = \frac{v}{2\pi R} = \frac{eB}{2\pi m} \quad \text{(Hz)} \qquad \textit{Gyrofrequency} \tag{13}$$

Note that in the second equality this frequency is independent of the particle velocity and energy.

TABLE 6-1
Charge and mass of particles

Particle	Charge e, C	Rest mass m, kg	Ratio e/m, C kg^{-1}
Electron	-1.602×10^{-19}	9.11×10^{-31}	-1.76×10^{11}
Positron	$+1.602 \times 10^{-19}$	9.11×10^{-31}	$+1.76 \times 10^{11}$
Neutron	0	1.6747×10^{-27}	0
Proton (hydrogen nucleus)	$+1.602 \times 10^{-19}$	1.6725×10^{-27}	$+9.6 \times 10^7$
Deuteron (heavy-hydrogen nucleus)	$+1.6 \times 10^{-19}$	3.34×10^{-27}	$+4.8 \times 10^7$
Alpha particle (helium nucleus)	$+3.2 \times 10^{-19}$	6.644×10^{-27}	$+4.81 \times 10^7$

Example 6-3. Radius and frequency of electron. An electron has a velocity of 10 km s^{-1} normal to a magnetic field of 0.1 T flux density. Find the radius of the electron path and also its frequency. (See Table 6-1 for charge and mass of electron.)

Solution. From (12) the radius is

$$R = \frac{(9.1 \times 10^{-31})(10^4)}{(1.6 \times 10^{-19})(10^{-1})} = 569 \text{ nm}$$

This is a very small circle. The frequency is

$$f = \frac{10^4}{2\pi(5.7 \times 10^{-7})} = 2.8 \times 10^9 \text{ Hz} = 2.8 \text{ GHz} \quad \textit{Ans.}$$

If the particle in the above example had an initial velocity component parallel to **B** as well as perpendicular to **B**, the particle would move in a helical path with the axis of the helix parallel to **B**.

Problem 6-4-3. Helical path. A 2-keV electron is injected into a magnetic field $B = 1$ mT at an angle of 87° with respect to **B**, resulting in a helical path. Find (*a*) helix radius, (*b*) distance between turns of helix, (*c*) circular or orbital velocity of electron, (*d*) axial velocity of electron, and (*e*) electron's orbital frequency. *Ans.* (*a*) 0.15m; (*b*) 0.049m; (*c*) 2.65×10^7 m/s; (*d*) 1.29×10^6 m/s; (*e*) 2.8×10^7 Hz.

If, in addition to the magnetic field, there is also an electric field **E**, the total force on the particle is

$$\mathbf{F} = e\mathbf{E} + e(\mathbf{v} \times \mathbf{B}) = e[\mathbf{E} + (\mathbf{v} \times \mathbf{B})] \quad \text{(N)} \qquad \textit{Total Lorentz force} \tag{14}$$

where **F** is the total ***Lorentz force.*** This relation, together with Newton's second law (force = rate of change of momentum or mass × acceleration), is basic to calculations of charged-particle motion.

From (14) we note that **E** can exert a force on a stationary (or moving) charged particle and impart energy to it through acceleration. However, there is no force on a charged particle in a static magnetic field unless the particle is moving with a velocity **v** *across* the magnetic field **B** ($\mathbf{v} \times \mathbf{B} \neq 0$). Since the force is normal to the particle velocity, the magnetic field cannot change the particle's kinetic energy. If the particle moves in the same direction as **B**, $\mathbf{v} \times \mathbf{B} = 0$ and the magnetic field has no effect at all (exerts no force).

6-5 CATHODE-RAY TUBE; MAGNETIC DEFLECTION

Let us consider what happens when an electron with velocity $|\mathbf{v}| = v_x$ enters a magnetic field **B** (out of page), at a distance x from a screen, as in Fig. 6-3. It is acted on by a force $\mathbf{F} = e(\mathbf{v} \times \mathbf{B})$ which accelerates the charge upward in the figure (since e is negative). As an approximation, the acceleration can be taken in the y direction if

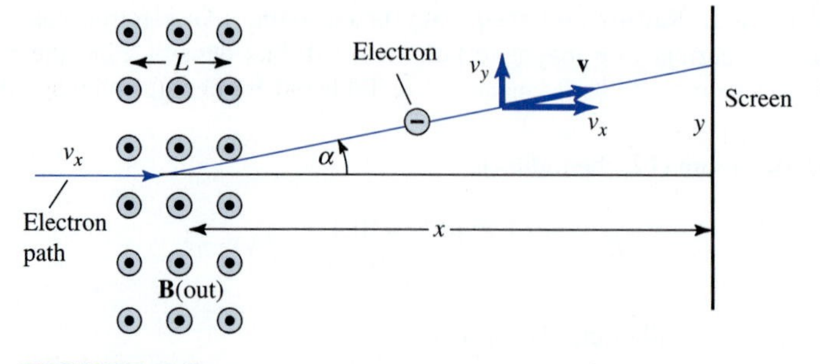

FIGURE 6-3
When a negatively charged particle (electron) enters a magnetic field **B** (out), it is deflected into a circular path (while in the magnetic field region) but on leaving continues in a straight line at an angle α with respect to the original direction.

the path L is small. Then the velocity component in the y direction is

$$v_y = a_y t = a_y \frac{L}{v_x} \tag{1}$$

From

$$\mathbf{F} = m\mathbf{a} = e(\mathbf{v} \times \mathbf{B}) \tag{2}$$

we obtain **a** as in (6-4-8), and introducing this in (1), we have

$$v_y = \frac{ev_x B}{m}\frac{L}{v_x} = \frac{eBL}{m} \tag{3}$$

The deflection angle α (Fig. 6-3) is

$$\alpha = \tan^{-1}\frac{v_y}{v_x} = \tan^{-1}\frac{eBL}{mv_x} \tag{4}$$

Assuming the electron obtained its acceleration from an electric field (in an electron gun as in Fig. 6-1), its velocity from (6-2-8) is given by

$$v_x = \sqrt{\frac{2eV_a}{m}} \tag{5}$$

where V_a = accelerating voltage, V.

Thus, the deflection angle can be expressed

$$\alpha = \tan^{-1}\left(BL\sqrt{\frac{e}{2mV_a}}\right) \tag{6}$$

But, we have also that $\alpha = \tan^{-1}(y/x)$, so

$$\pm y = xBL\sqrt{\frac{\mp e}{2mV_a}} \qquad \text{(m)} \tag{7}$$

where y = deflection distance at screen, m
x = distance to screen from magnetic field, m
$B = |\mathbf{B}|$ = magnetic flux density of deflecting field, T
e = charge of particle, C
m = mass of particle, kg
V_a = accelerating voltage, V
L = axial length of deflecting field, m

For an electron (e minus), (7) becomes

$$y = 2.96 \times 10^5 \frac{xBL}{\sqrt{V_a}} \qquad \text{(m)} \tag{8}$$

or

$$B = 3.38 \times 10^{-6} \frac{y\sqrt{V_a}}{xL} \qquad \text{(T)} \tag{9}$$

Example 6-4. *B* to deflect electron 10 mm. A cathode-ray tube with magnetic deflection has an electron gun accelerating voltage $V_a = 1500$ V, a magnetic deflecting field axial length $L = 20$ mm, and a distance $x = 300$ mm from the deflecting field to the screen. Find the magnetic flux density B required to deflect the spot of the electron beam 10 mm on the screen. See Fig. 6-3.

Solution. From (9),

$$B = 3.38 \times 10^{-6} \frac{0.01\sqrt{1500}}{0.3 \times 0.02} = 2.18 \times 10^{-4} \text{ T or } 218\ \mu\text{T} \quad \textit{Ans.}$$

Problem 6-5-1. Proton energy. Find the energy [in MeV (mega-electronvolts)] of a proton moving in a circular path of 500-mm radius in a magnetic field $B = 1$ T. *Ans.* 12 MeV.

Problem 6-5-2. Cathode-ray tube. A cathode-ray tube has electrostatic-deflection plates as in Fig. 6-1 *plus* a magnetic field between these plates. The magnetic field is oriented parallel to the electrostatic-deflection field E_d. The tube has an accelerating voltage $V_a = 7$ kV, a plate spacing $d = 20$ mm, and a plate length (and magnetic field length) $L = 20$ mm. Find (*a*) the deflection-plate voltage and (*b*) the magnetic flux density B required to deflect the electron beam 50 mm in the y direction and 100 mm in the z direction (perpendicular to page) at a screen placed at a distance $x = 500$ mm from the deflecting plates. *Ans.* (*a*) 1.4 kV; (*b*) 2.8 mT.

6-6 ROTARY MOTOR OR GENERATOR

If a rotatable rectangular loop or multiturn coil is placed in a uniform magnetic field **B** and equipped with a two-segment commutator, as in Fig. 6-4, it becomes a simple, elementary form of dc motor converting electrical energy to mechanical energy. The torque on the loop or coil is given by

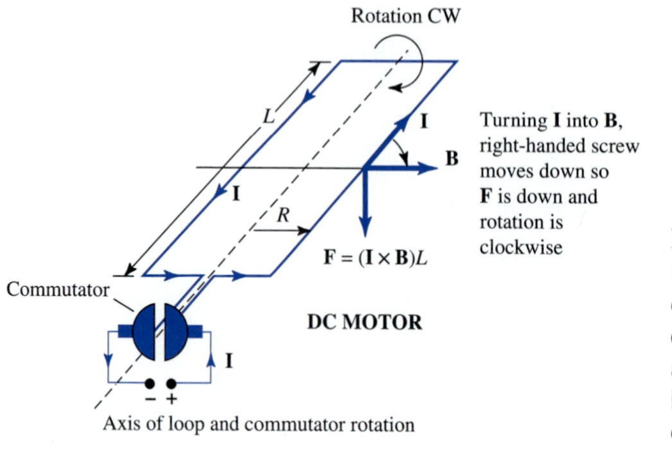

FIGURE 6-4
Rotating loop and commutator of dc motor. With applied voltage as indicated, rotation is clockwise.

$$T = IABN\sin\theta \qquad \text{(N m)} \qquad \textbf{\textit{Loop torque}} \tag{1}$$

where I = current, A
A = loop or coil area, m^2
B = magnetic field, T
N = number of turns, dimensionless
θ = angle between **B** and the normal **n** to the loop

The average torque is given by

$$T_{av} = \frac{2IABN}{\pi} \qquad \text{(N m)} \tag{2}$$

By mechanically rotating the coil, it becomes a dc generator converting mechanical energy to electrical energy. (Or with slip rings instead of a commutator, it becomes an ac generator.)

For a coil of N turns, length L, and radius R, we have from (2-14-15) that the average dc emf generated is

$$\mathcal{V} = \frac{2\omega BAN}{\pi} \qquad \text{(V)} \tag{3}$$

where $\omega = 2\pi$ revolutions/s
A = coil area = $2RL$, m^2

Example 6-5. Motor. A dc motor (Fig. 6-4) is to develop an average torque of 50 N m with 10 A in a coil measuring 80 × 120 mm in a field of 1 T. Find the required number of turns.

Solution.

$$T = \frac{2IABN}{\pi}$$

or $$N = \frac{\pi T}{2IAB} = \frac{\pi 50}{2 \times 10 \times 0.08 \times 0.12 \times 1} = 818 \text{ turns} \quad \textit{Ans.}$$

Problem 6-6-1. Deflection of indicating meter. Torque. The curved pole pieces and cylindrical iron core, as in Fig. 6-5, provide a substantially uniform radial magnetic field $B = 0.8$ T for the moving coil of an indicating meter. The coil is restrained by a spring with a torque $T = 2$ μN m/deg. If the coil has 50 turns and measures 16 mm across by 25 mm perpendicular to page, find the current required to deflect the coil through an angle of 45°. *Ans.* 5.6 mA.

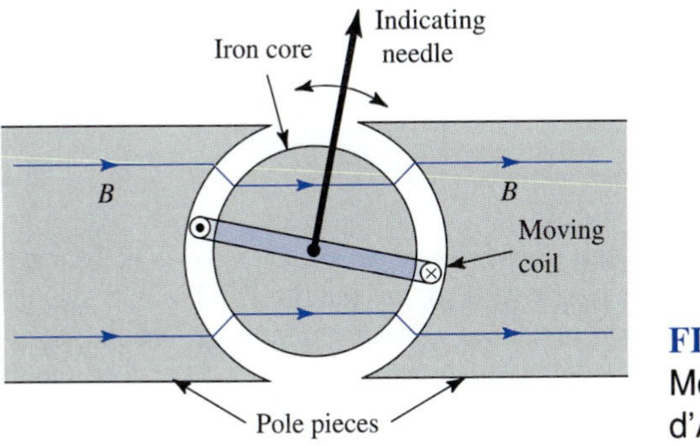

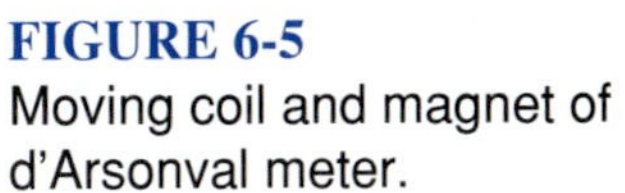

FIGURE 6-5
Moving coil and magnet of d'Arsonval meter.

6-7 LINEAR MOTOR

A wheeled vehicle rolling on tracks is propelled by the **I** × **B** force of horizontal currents and a vertical magnetic field between the tracks as suggested in Fig. 6-6.

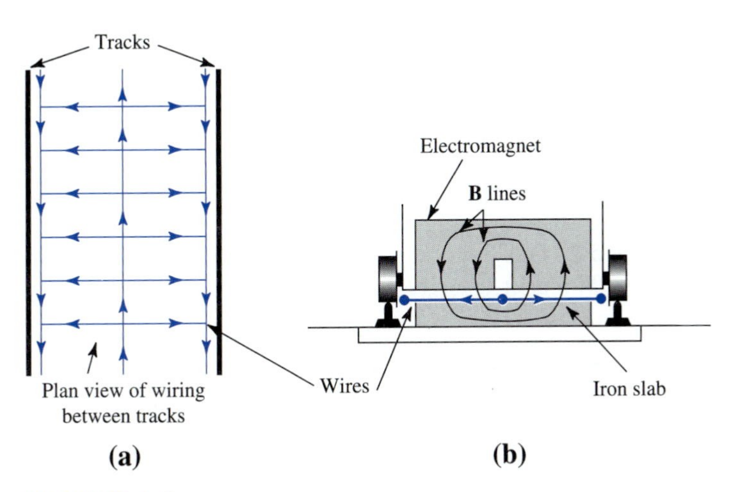

FIGURE 6-6
Linear motor of track vehicle. (*a*) Top or plan view. (*b*) End or elevation view.

Example 6-6. 50 tonnes to 100 km h^{-1} in 60 s. If the magnetic field $B = 2$ T, and the conductor width w between the tracks is 1.5 m for the arrangement shown in Fig. 6-6, find the conductor current required to accelerate a 50-tonne (50×10^3 kg) vehicle from 0 to 100 km h^{-1} in 60 s. Neglect friction.

Solution: From (6-2-2) and (6-4-1), the force required is

$$F = ma = BIL \tag{1}$$

where

$$a = \frac{v}{t_1} = \frac{10^5}{3600 \times 60} = 0.46 \text{ m s}^{-2} \tag{2}$$

Thus,

$$F = ma = 50 \times 10^3 \times 0.46 = 23 \text{ kN} \tag{3}$$

Equating (1) and (3) gives

$$I = \frac{ma}{BL} = \frac{50 \times 10^3 \times 0.46}{2 \times 1.5} = 7.7 \text{ kA} \quad \textit{Ans.} \tag{4}$$

This is the total current required in the transverse wires beneath the vehicle magnet. If the magnet has a longitudinal length of 2 m, we require 7.7/2 = 3.8 kA per meter of track length. For long track lengths the current requirement becomes prohibitive, but by segmenting the transverse wires into sections a few meters in length and only energizing those sections under the magnet as the vehicle passes, the overall current requirement might be kept below 20 kA. After 1 min and the vehicle attains its design speed, the current required goes to zero assuming a level track and no friction.

Problem 6-7-1. Track vehicle versus drag racer. For the 50-tonne track vehicle accelerated to 100 km/h in 60 s (Example 6-6) find (*a*) the power required and (*b*) the track distance in which 100 km/h is attained. (*c*) Compare the above power and distance with those for a drag racer weighing 2 tonnes accelerating to 240 km/h (150 mi/h) in 12 s. Disregard energy used in burning the tires. Note that since wheels are not used for traction, the track vehicle accelerating force is applied magnetically (no slippage) as compared to the friction-dependent action of the drag racer. *Ans.* (*a*) 322 kW; (*b*) 834 m; (*c*) 370 kW, 400 m.

6-8 HALL-EFFECT GENERATOR

When a current **I** flows in a conducting strip of width w and thickness t with a magnetic field **B** normal to the strip as in Fig. 6-7, a small voltage appears across the strip between points 1 and 2. This effect was discovered by Edwin H. Hall in 1879. An electron moving with a drift velocity v_d is acted on by a force given by (6-4-6) as

$$\mathbf{F} = e(\mathbf{v}_d \times \mathbf{B}) \qquad \text{(N)} \tag{1}$$

With **B** normal to the strip, this results in a transverse electric field

$$E_\perp = \frac{|\mathbf{F}|}{e} = v_d B \qquad (\text{N C}^{-1} \text{ or V m}^{-1}) \tag{2}$$

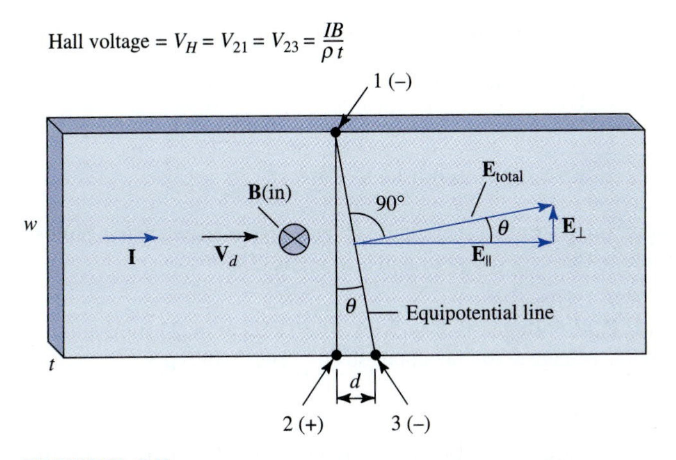

FIGURE 6-7
Current flowing transverse to magnetic field develops an emf across the strip called the *Hall voltage* with polarity as shown for an electron current but with opposite polarity for a hole current.

and the Hall voltage (actually an emf)

$$V_H = wE_\perp = wv_dB \qquad \text{(V)} \qquad \textbf{\textit{Hall voltage}} \tag{3}$$

But from (2-11-5) we have $v_d = J/\rho$, so

$$E_\perp = \frac{JB}{\rho} = \frac{IB}{\rho wt} \qquad (\text{V m}^{-1}) \tag{4}$$

and the ***Hall voltage*** between points 1 and 2 can be expressed in terms of the current I as

$$V_H = V_{21} = wE_\perp = \frac{wIB}{\rho wt} = \frac{IB}{\rho t} \qquad \text{(V)} \qquad \textbf{\textit{Hall voltage}} \tag{5}$$

where I = current in strip, A
B = magnetic field normal to strip, T
ρ = charge density, C m^{-3}
t = thickness of strip, m

Note that in the last part of (5) we end up with a relation for V_H which is independent of the width of the strip.

Example 6-7. Hall voltage for copper strip. A magnetic field **B** of 2 T is normal to a copper strip 1/2 mm thick carrying an electron current of 50 A. Take the electron density $n = 8.4 \times 10^{28}$ m^{-3}. Find the Hall voltage across the strip.

Solution: Noting that $\rho = ne$, we have from (4) that

$$V_H = V_{21} = \frac{IB}{\rho t} = \frac{IB}{net} = \frac{50 \times 2}{(8.4 \times 10^{28})(1.6 \times 10^{-19})(5 \times 10^{-4})}$$

$$= 1.5 \times 10^{-5} \text{ V} = 15 \ \mu\text{V} \quad \textit{Ans.}$$

It is important to note that this small voltage appears between two points that are exactly transverse to the strip (points 1 and 2 in Fig. 6-7).

Problem 6-8-1. Hall voltage. If $I = 50$ A, $B = 2$ T, $w = 25$ mm, and $t = 0.5$ mm, find the Hall voltage across a copper strip. *Ans.* 14.9 μV.

Referring to Fig. 6-7, an electron in the strip is subjected to two electric fields: one $E_{\parallel}$ to the strip and the other $E_{\perp}$ transverse to the strip as given by (3). Since $\mathbf{J} = \sigma\mathbf{E}$, the parallel (or longitudinal) electric field is

$$E_{\parallel} = \frac{J}{\sigma} = \frac{I}{\sigma \omega t} \tag{6}$$

The ratio of $E_{\perp}$ from (4) to $E_{\parallel}$ from (6) is

$$\frac{E_{\perp}}{E_{\parallel}} = \frac{JB\sigma}{\rho J} = \frac{\sigma}{\rho}B \tag{7}$$

For copper and the conditions of the above example,

$$\frac{E_{\perp}}{E_{\parallel}} = \frac{\sigma}{\rho}B = \frac{5.7 \times 10^7 \times 2}{1.34 \times 10^{10}} = 8.5 \times 10^{-3} \tag{8}$$

Referring to Fig. 6-7, the angle

$$\theta = \tan^{-1}\frac{E_{\perp}}{E_{\parallel}} = \tan^{-1}(8.5 \times 10^{-3}) = 0.5^{\circ} \tag{9}$$

Thus, the total electric field $\mathbf{E}_{\text{total}}$ is at an angle of 0.5° with respect to the longitudinal direction of the strip. Consequently, the equipotentials across the strip are at this angle with respect to the transverse direction, and, therefore, the voltage between points 1 and 3 is zero. Hence, the same Hall voltage appears between points 2 and 3 as between 1 and 2, and we can write

$$V_{21} = V_{23} = V_H \tag{10}$$

If the width w of the strip is 25 mm, the distance d between points 2 and 3 is given by

$$d = w\frac{E_{\perp}}{E_{\parallel}} = (25 \times 10^{-3})(8.5 \times 10^{-3}) = 2.1 \times 10^{-4} \text{ m} = 0.21 \text{ mm}$$

This small distance indicates that the electrodes to measure the Hall voltage across the strip must be placed to a precision of micrometers (μm).

Although we speak of the conduction current electrons as "free" they are constrained in their motion by the atomic lattice of the conductor and quickly transfer the momentum of their transverse motions to it. Thus, the conductor as a whole experiences the Lorentz ($\mathbf{I} \times \mathbf{B}$) force and the electrons do not flood across the conductor and pile up on one side.

Referring to Fig. 6-7, point 2 is positive with respect to points 1 and 3 for electrons but, with positive charges or holes as the conduction current, the polarity is reversed. Thus, the Hall voltage polarity can be used to distinguish between metals and semiconductors of the intrinsic and donor or *n*-type (negative) with electrons as charge carriers versus semiconductors of the acceptor or *p*-type (positive) with holes as charge carriers.

Although the Hall voltage is small, the effect can be used to measure magnetic fields. Also, if the Hall field $|\mathbf{B}|$ is made proportional to the voltage V across a piece of electrical equipment with its current I passed through the Hall-effect strip, the Hall voltage is a measure of power, and we have a Hall wattmeter [replace B by V in (5)].

6-9 MOVING CONDUCTOR IN A STATIC MAGNETIC FIELD

From (6-4-6) a conductor cutting across a magnetic field **B** with a velocity **v** as in Fig. 6-8 (on next page) generates an emf-producing field

$$\mathbf{E} = \frac{\mathbf{F}}{e} = \mathbf{v} \times \mathbf{B} \qquad (\mathrm{V\,m^{-1}}) \tag{1}$$

with an emf appearing between the endpoints 1 and 2 of the conductor given by (2-12-14) as

$$\mathcal{V}_{12} = \int_1^2 \mathbf{E} \cdot d\mathbf{L} = \int_1^2 (\mathbf{v} \times \mathbf{B}) \cdot d\mathbf{L} \tag{2}$$

where $\mathcal{V}_{12}$ = emf induced between points 1 and 2, V
$\mathbf{E}$ = electric field, V m^{-1}
$d\mathbf{L}$ = infinitesimal element of conductor length, m
$\mathbf{v}$ = velocity of conductor, m s^{-1}
$\mathbf{B}$ = magnetic flux density, T

For a straight conductor between 1 and 2 of length L at an angle θ to $\mathbf{E}$ ($= \mathbf{v} \times \mathbf{B}$)

$$\mathcal{V} = \int_0^L |\mathbf{v} \times \mathbf{B}| \cos\theta\, dL \tag{3}$$

If **v**, **B**, and the conductor are all mutually perpendicular ($\theta = 0$), (3) reduces to

$$\mathcal{V} = vBL \qquad (\mathrm{V}) \tag{4}$$

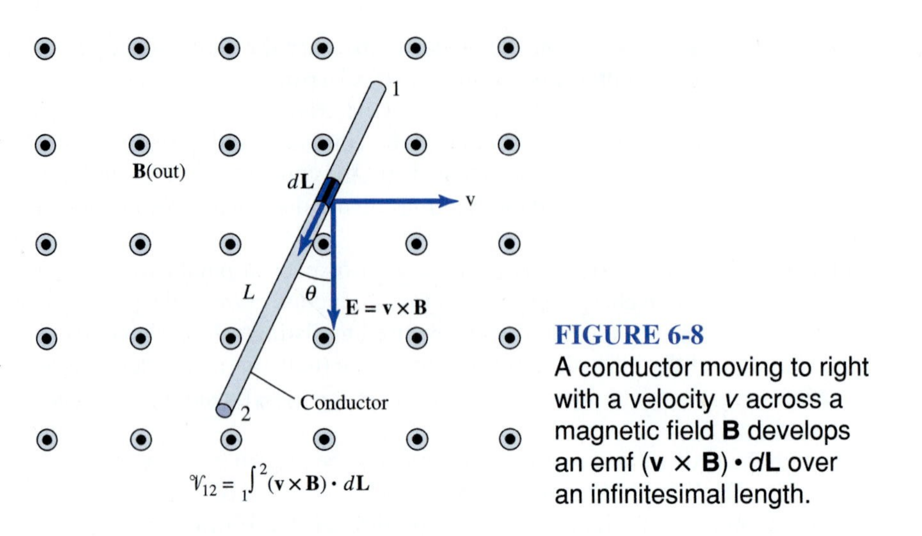

FIGURE 6-8
A conductor moving to right with a velocity v across a magnetic field **B** develops an emf $(\mathbf{v} \times \mathbf{B}) \cdot d\mathbf{L}$ over an infinitesimal length.

Equations (2), (3), and (4) are *motional-induction*, or *flux-cutting*, *laws* giving the emf induced in a conductor moving with respect to the observer. Equation (2) is the most general form, while (4) applies to a straight conductor with it, **v**, and **B** all mutually perpendicular.

The relations may be used to find the emf induced in any part of a circuit due to its motion through a magnetic field. They can also be applied to determine the total emf induced in a closed circuit that is moved or deformed in a magnetic field which does not change with time. For a closed circuit, we have for the total induced emf

$$\mathcal{V} = \oint \mathbf{E} \cdot d\mathbf{L} = \oint (\mathbf{v} \times \mathbf{B}) \cdot d\mathbf{L} \qquad \text{(V)} \qquad \textit{Total emf} \qquad (5)$$

6-10 THE MAGNETIC BRAKE

To illustrate the braking effect of a magnetic field consider a wire loop with one side in a magnetic field, as in Fig. 6-9. When we push the loop to the left with a velocity v, a field $\mathbf{E} = \mathbf{v} \times \mathbf{B}$ is generated in the wire resulting in a current **I** in the same direction, as shown. This current **I** moving in the magnetic field **B** results in a motor force $\mathbf{F} = (\mathbf{I} \times \mathbf{B})L$ which *opposes* the direction of our push. Note the vector diagrams with **v**, **B**, **E**, and **I**, **B**, **F** in the figure which relate the directions.

If instead of pushing, we try to pull the loop out, **I** is reversed, and the motor force now holds back on the loop. Thus, whether we push or pull, the magnetic field exerts a braking action which opposes our motion.

We note that if our loop is smaller and entirely inside the uniform magnetic field, the generated fields balance out, $I = 0$ and there is no braking action.

Example 6-8. Brake. A rectangular loop of copper is partway into a magnetic field $B = 2$ T, as in Fig. 6-9. The loop dimension $L = 200$ mm. If the loop conductor is 20 mm in diameter and its total length is 1 m, find the force required to pull the loop out at the slow speed of 20 mm/s. The answer indicates that the loop doesn't come out very easily. To get it out by hand, you must pull it out slowly, illustrating the braking effect of the magnetic field.

Solution

$$I = \frac{V}{R} = \frac{vBL}{R} \qquad F = BIL = \frac{v(BL)^2}{R}$$

$$R = \frac{L}{\sigma A} = \frac{1}{(5.7 \times 10^7)(\pi \times 0.01^2)} = 5.6 \times 10^{-5}\ \Omega$$

Therefore

$$F = [\underbrace{(20 \times 10^{-3})}_{v} \times \underbrace{(2^2)}_{B} \times \underbrace{(0.2^2)}_{L}]/\underbrace{(5.6 \times 10^{-5})}_{R} = 57 \text{ N} \quad \text{or} \quad 6 \text{ kg pull} \quad \textit{Ans.}$$

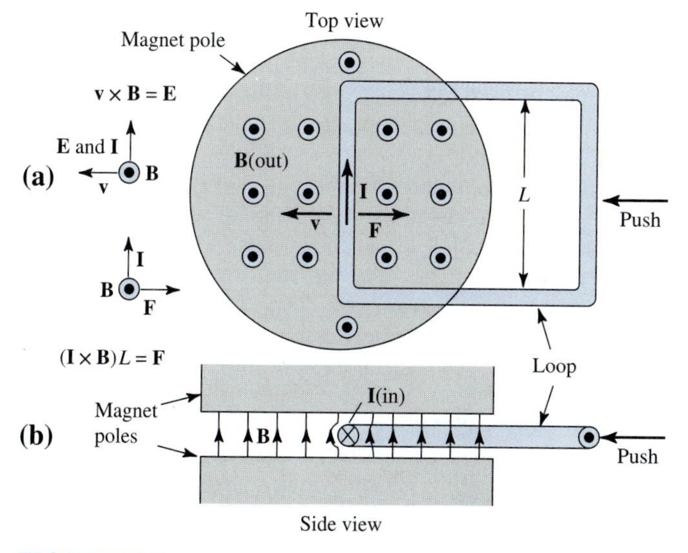

FIGURE 6-9
Pushing or pulling the loop results in a braking force which opposes the motion.

Problem 6-10-1. Loop force. If the pole diameter in Example 6-8 is 300 mm, find the force to pull the loop up (in the figure), instead of to the right, at 15 mm/s with (*a*) loop at position shown and (*b*) bottom conductor of loop at the center of the pole. *Ans.* (*a*) 0, (*b*) 24 N.

PROBLEMS

Answers are given in Appendix E to problems followed by the @ symbol.

6-3-4. Ink-jet printer. Quiet, fast printing on paper is accomplished by deflecting ink droplets (instead of electrons) in the manner of a CRT, except that a nozzle replaces the electron gun, with droplets produced in a continuous stream. Charges are then sprayed onto the ejected droplets so that they can be electrostatically deflected as in a CRT. The paper to be printed is set above the axis and a funnel placed on axis to collect undeflected droplets during pauses between characters. Another difference from the CRT is that the deflecting potential is fixed, and the amount of deflection desired is controlled by the amount of charge given the droplets. The nozzle with charging and deflecting electrodes is mounted on a carriage which moves horizontally (perpendicular to the deflection direction y) so that the printer can produce a line of characters (or type). Characters may be printed at the rate of 100 per second with 1000 droplets per character giving a speed and quietness unmatched by impact printers. The only moving parts are the nozzle (with its carriage) and the drops of ink.

If an ink drop has a mass of 50 ng (nanograms) and is given a charge of -200 fC (femtocoulombs), find its vertical (y) displacement in an ink-jet printer with 3-kV deflection potential, 2-mm deflection plate separation, and 15-mm deflection plate length. The nozzle ejects the drop at 25 m/s, and the leaving edge of the deflection plate is 15 mm from the paper. @

6-3-5. Printer droplets. The nozzle of an ink-jet printer ejects droplets at 30 m/s. For a 4-mm deflection on the sheet being printed, find the deflecting field required if the deflecting field extent in the direction of the droplet's travel is 18 mm. The sheet is 25 mm from the leaving edge of the deflecting field. Assume an inkdrop mass of 40 ng and charge of 250 nC.

6-3-6. Electron spacing. A 1-μA electron beam with electrons of 1-keV energy falls on the 300-mm-wide screen of a cathode-ray tube. If the beam scans the tube width in 100 μs, find (*a*) maximum deflecting field, (*b*) rate of change of field, and (*c*) spacing between electrons hitting the screen. The deflecting field extends 20 mm in the direction of the beam. Take $x = 400$ mm.

6-3-7. Cathode-ray tube. A CRT has an accelerating potential of 1 kV and a deflecting plate spacing of 10 mm. Find the length of the deflecting plates required to deflect the spot 5 mm at a distance of 250 mm if the deflecting voltage is 50 V. @

6-3-8. Ink-jet. If an ink drop with a charge of -250 fC and mass of 75 ng is ejected by the nozzle at 30 m/s, find the deflection potential required to displace the drop 2 mm in a distance of 18 mm. The deflecting plates have a length of 12 mm and separation of 2 mm.

6-4-4. Jovian magnetic field. What is the magnetic field strength in Jupiter's magnetosphere where the 20 MHz signals of Project P5-2 are generated? @

6-4-5. Cyclotron. Particles accelerated in a cyclotron spiral outward to some maximum radius. (*a*) Find the maximum energy (in MeV) for alpha particles in a cyclotron with maximum usable radius of 450 mm. The flux density $B = 1.2$ T in the air gap. (*b*) Repeat (*a*) if protons are used. (*c*) Repeat (*a*) if deuterons are used. *Hint:* Solve (6-4-12) for v and recall that energy $= \frac{1}{2}mv^2$.

6-4-6. Electron in loop. An electron is at the center of a 100-mm-diameter loop carrying a current of 10 A (Fig. P6-4-6). Find the force (magnitude and direction) on the electron (*a*) if the electron is at rest, (*b*) if it is moving with a velocity $v = 1$ m/s perpendicular

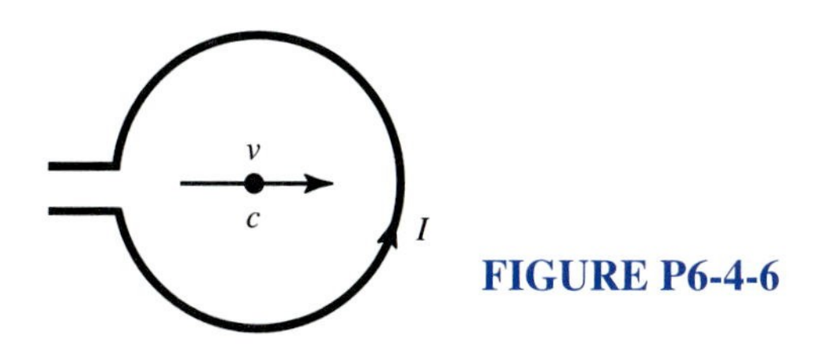

FIGURE P6-4-6

to the plane of the loop, and (*c*) if it is moving with this velocity in the plane of the loop in the direction shown. @

6-5-3. Electron with B balancing E. An electron with an energy of 1800 eV enters an electric deflecting field $E = 10^4$ V/m. A magnetic deflecting flux density **B** is perpendicular to **E**. What value should B have to balance the deflection by E so that the electron continues in a straight path?

6-5-4. Particle acceleration. A particle with a negative charge of 10^{-17} C and a mass of 10^{-26} kg is at rest in a field-free space. If a uniform electric field $E = 1$ kV/m is applied for 1 μs, find (*a*) the velocity of the particle and (*b*) the radius of curvature of the particle path if the particle enters a magnetic field $B = 2$ mT with this velocity moving normal to **B**. @

6-5-5. TV screen. The cathode-ray-tube screen of a television receiver measures 400 mm (horizontally) by 300 mm (vertically). The electron-beam spot scans 525 lines per frame and 30 frames per second. Find (*a*) the horizontal spot velocity in millimeters per microsecond and (*b*) the total distance traveled by the spot per second. (*c*) What is the rate of change of magnetic flux density B required if the tube parameters are as in Prob. 6-5-2?

6-5-6. Transmission line. Balance of forces. A transmission line consists of two parallel 10-mm-diameter conductors separated (on centers) by 500 mm. If there is a potential difference $V = 100$ kV between the conductors, is there a line current I which results in a balance between the electric and magnetic forces acting on the wires? If so, what is its value? @

6-8-2. Hall magnetometer. (*a*) Find the magnetic field B if a Hall voltage of 10 μV appears across a copper strip 10 mm wide, 1 mm thick having a current of 25 A. See Fig. 6-7. (*b*) Find the distance along one edge of the strip which has the same voltage difference.

6-10-2. Levitation. (*a*) Find the coil current required to levitate or float 3 tonnes 75 mm above a perfectly conducting sheet. The coil has 50 turns and is 800 mm wide by 1.6 m long. (*b*) Find the power loss in the coil if it is wound of copper wire of 5 mm radius. Note that this large amount of power would quickly melt the coil, making the use of superconducting wire or tape imperative. With the superconductor, power loss in the coil is ideally zero.

CHAPTER 7

DIELECTRIC AND MAGNETIC MATERIALS

7-1 INTRODUCTION

In this chapter we learn more about dielectric and magnetic materials with sections on polarization, magnetization, hysteresis, and eddy currents.

7-2 HOMOGENEITY, LINEARITY, AND ISOTROPY

A medium is ***homogeneous*** if its physical characteristics (mass density, molecular structure, etc.) do not vary from point to point. If the medium is not homogeneous, it may be described as inhomogeneous, nonhomogeneous, or heterogeneous.

A medium is ***linear*** with respect to an electrostatic field if the flux density **D** is proportional to the electric field intensity **E**. This is the case in free space, where $\mathbf{D} = \varepsilon_0 \mathbf{E}$. Here the factor ε_0, or permittivity, is a constant. In most material media the permittivity ε is constant. If it is not, the material is said to be *nonlinear.*

An ***isotropic*** material is one whose properties are independent of direction. Generally, materials whose molecular structure is randomly oriented will be isotropic. However, crystalline media and certain plasmas may have directional characteristics. Such materials are said to be nonisotropic or anisotropic.

In what follows, concepts are usually developed first for the case where the medium is homogeneous, linear, and isotropic. Later the ideas may be extended to cases where one or more of these restrictions no longer hold.

7-3 TABLE OF PERMITTIVITIES†

The relative permittivity of a few media are shown in Table 7-1. The values are for static (or low-frequency) fields and, except for vacuum or air, are approximate. For air, ε_r is so close to unity that for most problems we consider air equivalent to vacuum.

TABLE 7-1
Table of permittivities for dielectric media†

Medium	Relative permittivity ε_r	Medium	Relative permittivity ε_r
Vacuum	1‡	Formica	6
Air (atmospheric pressure)	1.0006	Lead glass	6
Styrofoam	1.03	Mica	6
Expanded PVC (polyvinyl chloride)	1.1	Sodium chloride	6
Polyfoam	1.1	Neoprene	7
Wood (dry)	2–4	Marble	8
Paraffin	2.1	Flint glass	10
Plywood	2.1	Animal muscle	10
Polystyrene	2.7	Silicon	12
PVC	2.7	Germanium	16
Amber	3	Ammonia (liquid)	22
Rubber	3	Alcohol (ethyl)	25
Paper	3	Glycerin	50
Plexiglass	3.4	Ice	75
Dry sandy soil	3.4	Water (distilled)	81
Nylon (solid)	3.8	Rutile (TiO_2)	89–173§
Silica	3.8	Barium titanate ($BaTiO_3$)	1 200¶
Sulfur	4	Barium strontium titanate	10 000¶
Quartz	5	Barium titanate stannate	20 000
Bakelite	5		

†Static or low-frequency fields at 20°C.
‡By definition.
§Crystals, in general, are nonisotropic; i.e., their properties vary with direction. Rutile is an example of such a nonisotropic crystalline substance. Its relative permittivity depends on the direction of the applied electric field with relation to the crystal axes, being 89 when the field is perpendicular to a certain crystal axis and 173 when the field is parallel to this axis. For an aggregation of randomly oriented rutile crystals $\varepsilon_r = 114$. All crystals, except those of the cubic system, are nonisotropic to electric fields; i.e., their properties vary with direction. Thus, the permittivity of many other crystalline substances may vary with direction. However, in many cases the difference is slight. For example, a quartz crystal has a relative permittivity of 4.7 in one direction and 5.1 at right angles. The average value is 4.9. The nearest integer is 5, and this is the value given in the table.
¶The permittivity of these titanates is highly temperature-sensitive. The above values are for 25°C.

†Also called *dielectric constants*. However, the permittivity is not always a constant as might be inferred from this term.

7-4 THE ELECTRIC FIELD IN A DIELECTRIC

In free space, the electric field is defined as *force per unit charge*. This implies that the electric field in free space is a *measurable quantity*. However, to measure the electric field inside a dielectric or other material medium may be very difficult or impractical. But, if we confine our attention to the external effects of the dielectric, such internal measurements become unnecessary provided that a theory can be formulated for the behavior of the dielectric which produces agreement with external measurements. Thus, a distinction should be made between an electric field as a *measurable quantity* (as in free space) and an electric field as a *theoretical quantity* (as in a dielectric). In this chapter a theory for the electrostatic field in a dielectric is developed and then related to the external field by means of boundary conditions.

7-5 THE ELECTRIC DIPOLE AND ELECTRIC-DIPOLE MOMENT

The combination of two equal point charges Q of opposite sign separated by a distance L constitutes an ***electric dipole,*** and the product QL its ***electric-dipole moment.*** The electric field lines and equipotential contours are as suggested in Fig. 7-1*a*. Dipoles are an important, basic charge configuration and are fundamental to our discussion of dielectric polarization in the next section as well as to later topics.

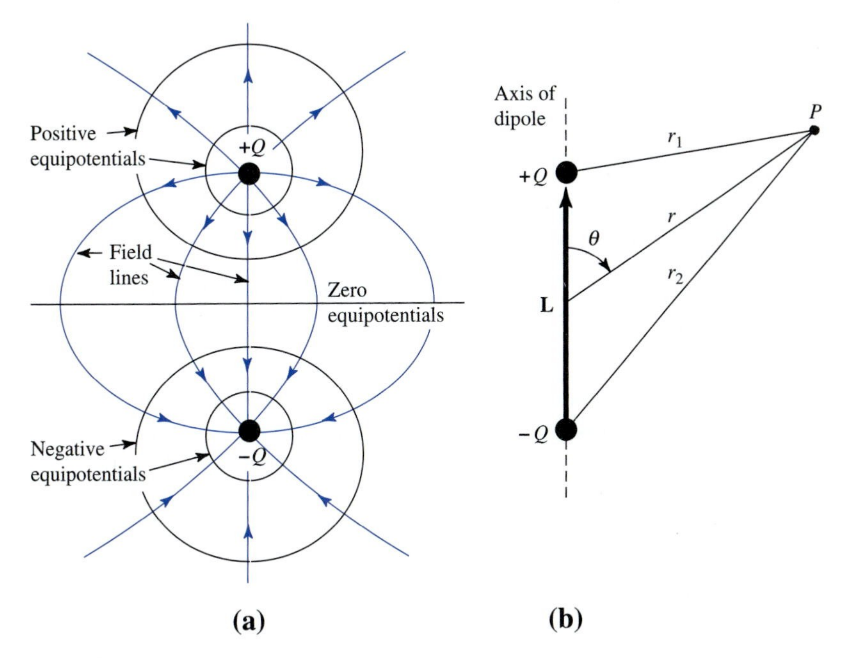

FIGURE 7-1
(*a*) Field of electric dipole. (*b*) Geometry for determining the electric potential at P.

If the two charges are superposed, the resultant field is zero, but if they are separated by even a small distance, there is a finite resultant field. Let us now determine quantitatively the value of this dipole field.

By regarding the separation between the charges as a vector **L**, pointing from the negative to the positive charge as in Fig. 7-1*b*, the dipole moment can be expressed as a vector $Q\mathbf{L}$ with the magnitude $QL\ (= |Q\mathbf{L}|)$ and the direction of **L**.

Referring to Fig. 7-1*b*, the potential of the positive charge at a point P is

$$V_1 = \frac{Q}{4\pi\varepsilon_0 r_1} \tag{1}$$

The potential of the negative charge at P is

$$V_2 = \frac{-Q}{4\pi\varepsilon_0 r_2} \tag{2}$$

The total potential V at P is then

$$V = V_1 + V_2 = \frac{Q}{4\pi\varepsilon_0}\left(\frac{1}{r_1} - \frac{1}{r_2}\right) \tag{3}$$

If the point P is at a large distance compared with the separation L, so that the radial lines r_1, r, and r_2 are essentially parallel, we have very nearly that

$$r_1 = r - \frac{L}{2}\cos\theta \qquad \text{and} \qquad r_2 = r + \frac{L}{2}\cos\theta \tag{4}$$

where r = distance from center of dipole to point P
θ = angle between axis of dipole and r.

Hence, the potential V at a distance r from an electric dipole is

$$V = \frac{QL\cos\theta}{4\pi\varepsilon_0 r^2} \quad (V) \qquad \textbf{\textit{Dipole potential}} \tag{5}$$

where it is assumed that r is much greater than $L(r \gg L)$ so that terms in L^2 can be neglected compared with those in r^2.

It is instructive to consider (5) as the product of four factors involving the dipole moment, the angle, the distance, and a constant, characteristic of the system of units employed. Thus,

$$V = \underbrace{QL}_{\text{Dipole moment}} \cdot \underbrace{\cos\theta}_{\text{Angle factor}} \cdot \underbrace{\frac{1}{r^2}}_{\text{Distance factor}} \cdot \underbrace{\frac{1}{4\pi\varepsilon_0}}_{\text{Constant}} \tag{6}$$

Expressions for the potential and also the electric field of dipoles (and quadrupoles and higher-order configurations) always contain these four kinds of factors.

To find the electric field of the dipole of Fig. 7-1*b*, it is convenient to make use of the gradient. Thus, let us take the gradient of the potential given by (5), obtaining†

$$\mathbf{E} = -\hat{\mathbf{r}}\frac{\partial V}{\partial r} - \hat{\boldsymbol{\theta}}\frac{1}{r}\frac{\partial V}{\partial \theta} = \hat{\mathbf{r}}\frac{QL\cos\theta}{2\pi\varepsilon_0 r^3} + \hat{\boldsymbol{\theta}}\frac{QL\sin\theta}{4\pi\varepsilon_0 r^3} \tag{7}$$

where $\hat{\mathbf{r}}$ = unit vector in r direction, dimensionless (see Fig. 7-2)
$\hat{\boldsymbol{\theta}}$ = unit vector in θ direction, dimensionless
L = separation of dipole charges Q, m

According to (7), the electric field has two components, as shown in Fig. 7-2, one in the r direction (E_r) and one in the θdirection (E_θ). Thus,

$$\mathbf{E} = \hat{\mathbf{r}}E_r + \hat{\boldsymbol{\theta}}E_\theta \tag{8}$$

or

$$E_r = \frac{QL\cos\theta}{2\pi\varepsilon_0 r^3} \quad \text{and} \quad E_\theta = \frac{QL\sin\theta}{4\pi\varepsilon_0 r^3} \quad (\text{V m}^{-1}) \quad \textbf{\textit{Dipole field}} \tag{9}$$

In these equations the restriction applies that $r \gg L$.

We note that whereas the potential and field of a point charge vary as $1/r$ and $1/r^2$, respectively, they vary as $1/r^2$ and $1/r^3$ for a dipole.‡

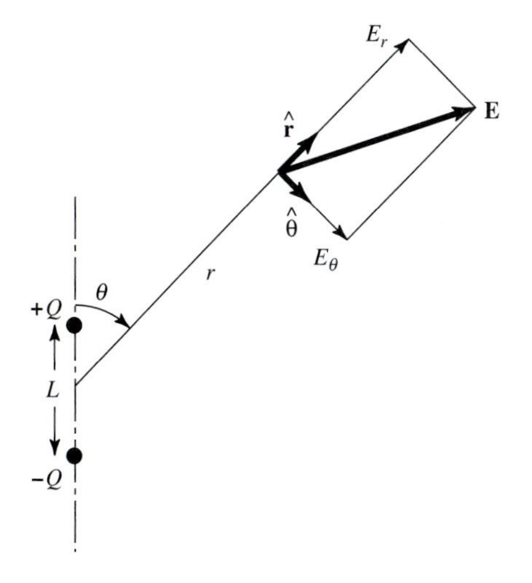

FIGURE 7-2
Component fields and total field **E** at a distance r from an electric dipole.

†See Appendix A for gradient in spherical coordinates.

‡Placing two dipoles adjacent but with one inverted, we obtain a *quadrupole*. Placing two quadrupoles side by side but with one inverted, we obtain an *octopole*. The potential and field of a quadrupole vary as $1/r^3$ and $1/r^4$, respectively, and for an octopole as $1/r^4$ and $1/r^5$.

7-6 POLARIZATION

Although there is no migration of charge when a dielectric is placed in an electric field, there does occur a slight displacement of the negative and positive charges of the dielectric's atoms or molecules so that they behave like very small *dipoles.* The dielectric is said to be *polarized* or in a state of *polarization* when the dipoles are present. For most materials,† the removal of the field results in the return of the atoms or molecules to their normal, or unpolarized, state and the disappearance of the dipoles.

As a simple example, consider that a polarized atom of a dielectric material is represented by an electric dipole, i.e., a positive point charge representing the nucleus and a negative point charge representing the electronic charge, the two charges being separated by a small distance. The electrons orbit the nucleus and act like a negatively charged cloud surrounding the nucleus. When the atom is unpolarized, the cloud surrounds the nucleus symmetrically, as in Fig. 7-3*a*, and the dipole moment is zero (the equivalent positive and negative point charges have zero displacement). Under the influence of an electric field the electron cloud becomes slightly displaced or asymmetrical, as in Fig. 7-3*b*, and the atom is polarized. According to our simple picture, the atom may then be represented by the equivalent point-charge dipole of Fig. 7-3*c* (dipole moment $p = qL$).

Consider the dielectric slab of permittivity ε in Fig. 7-4*a* situated in vacuum. Let a uniform field **E** be applied normal to the slab. This polarizes the dielectric, i.e., induces atomic dipoles throughout the slab. In the interior the positive and negative charges of adjacent dipoles annul each other's effects. The net result of the polarization is to produce a layer of negative charge on one surface of the slab and a layer

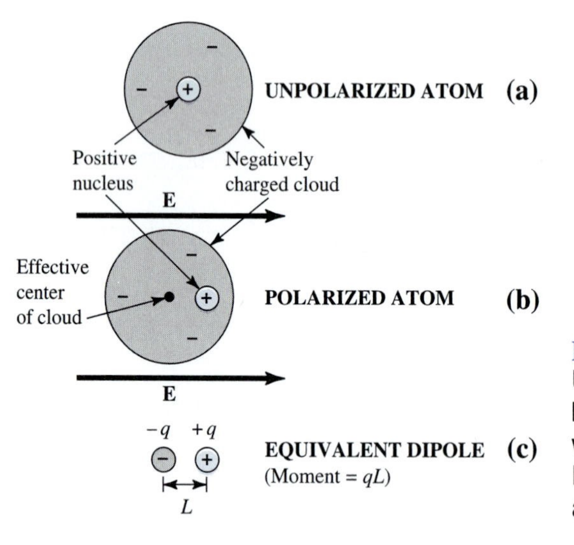

FIGURE 7-3
Unpolarized atom as in (*a*) becomes polarized as in (*b*) when electric field is applied. Equivalent dipole is shown at (*c*).

†When polarization in a dielectric persists in the absence of an applied electric field, the substance is permanently polarized and is called an *electret.* A strained piezoelectric crystal is an example of an electret.

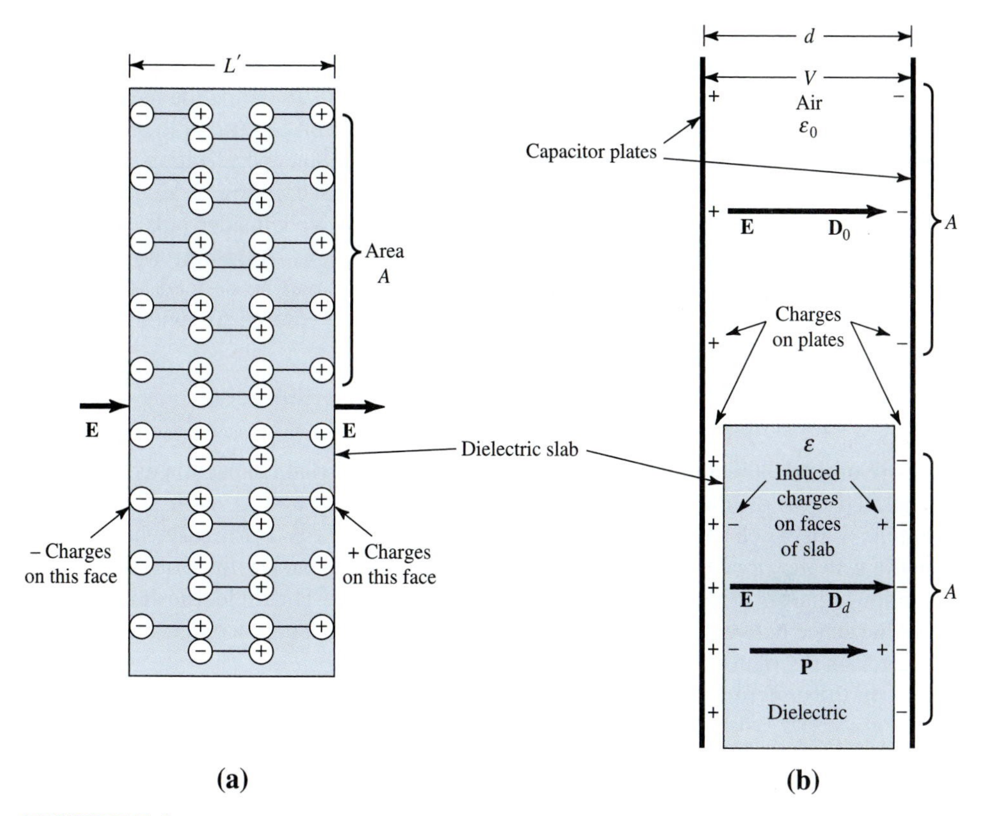

FIGURE 7-4
(*a*) Dielectric slab in uniform electric field. (*b*) Parallel-plate capacitor with dielectric slab in lower part.

of positive charge on the other, as suggested in Fig. 7-4*a,* with the charges of each resultant dipole separated by a distance L' (where L' = thickness of slab).

The effect of all this can be described by the ***polarization P,*** or ***dipole moment per unit volume.*** Thus,

$$P = \frac{n}{v} qL' = \frac{QL'}{v} \tag{1}$$

where n = number of dipoles in volume v
$Q = nq$ = charge of all dipoles
QL' = net dipole moment in volume v

For example, consider the rectangular volume of surface area A and thickness L' ($v = AL'$) in Fig. 7-4*a*. For this volume

$$P = \frac{QL'}{AL'} = \frac{Q}{A} = \rho_{sp} \qquad (\text{C m}^{-2}) \tag{2}$$

where ρ_{sp} is the surface charge density of polarization charge appearing on the slab faces. Thus, P has the dimensions of charge per area, the same as D.

The value of P in (2) is an average for the volume v. To define the meaning of P at a point, it is convenient to assume that a dielectric in an electric field has a continuous distribution of infinitesimal dipoles, i.e., a continuous polarization, whereas the dipoles actually are discrete polarized atoms. The assumption of a continuous distribution leads to no appreciable error provided that we consider only volumes containing many atoms or dipoles, i.e., macroscopic regions. Assuming now a continuously polarized dielectric, the value of P at a point can be defined as the net dipole moment QL' of a small volume Δv divided by the volume Δv, with the limit taken as Δv shrinks to zero around the point. Thus,

$$P = \lim_{\Delta v \to 0} \frac{QL'}{\Delta v} \tag{3}$$

Consider now a uniform electric field in a parallel-plate capacitor with plates separated by a distance d, as in the cross-sectional view of Fig. 7-4*b*. There is a voltage V between the plates so that the electric field $E = V/d$ everywhere. The medium in the upper part of the capacitor is vacuum (or air) with permittivity ε_0. The lower part is filled with a dielectric of permittivity ε. The dielectric completely fills the space between the plates, but in Fig. 7-4*b* there is a gap in order to show the charges on the plates.

In the upper part of the capacitor (air) (Fig. 7-4*b*) we have

$$D_0 = \varepsilon_0 E \tag{4}$$

where D_0 = electric flux density in vacuum (or air-filled) part of capacitor, C m^{-2}
ε_0 = permittivity of vacuum = 8.85 pF m^{-1}
$E = V/d$ = electric field intensity, V m^{-1}

In the lower, dielectric-filled part of the capacitor, the electric field polarizes the dielectric, causing a surface charge density ρ_{sp} to appear on both faces of the dielectric slab. These bound charges induce extra free charges of opposite sign on the capacitor plates (compare upper and lower parts of Fig. 7-4*b*). As a result the free-charge surface density on the plates is increased by ρ_{sp}. Therefore, in the dielectric we have

$$D_d = \varepsilon_0 E + \rho_{sp} \tag{5}$$

but from (2), $\rho_{sp} = P$, and so

$$D_d = \varepsilon_0 E + P \tag{6}$$

where D_d = electric flux density in dielectric, C m^{-2}
ε_0 = permittivity of vacuum = 8.85 pF m^{-1}
E = electric field intensity, V m^{-1}
P = polarization (of dielectric), C m^{-2}

Equation (6) implies the presence of dielectric (because of the P term), and so the subscript to D is redundant. Therefore, (6) can be written

$$\mathbf{D} = \varepsilon_0 \mathbf{E} + \mathbf{P} \qquad \textit{Generalized D} \tag{7}$$

Although developed for the special case of a parallel-plate capacitor, (7) is a (vector) relation which applies in general.

In the dielectric we can also write

$$\boxed{\mathbf{D} = \varepsilon \mathbf{E}} \qquad \text{or} \qquad D = D_d = \varepsilon E \tag{8}$$

where ε is the permittivity of the dielectric material in farads per meter. Equating (6) and (8) gives

$$\varepsilon E = \varepsilon_0 E + P \tag{9}$$

$$\varepsilon = \varepsilon_0 + \frac{P}{E} \qquad \text{or} \qquad \frac{P}{E} = \varepsilon - \varepsilon_0 \tag{10}$$

Example 7-1. Dielectric slab. Find the relative permittivity ε_r and polarization P for the dielectric slab of Fig. 7-4*b*. Assume each plus or minus sign represents 1 nC m^{-2} and $A = 1$ m^2.

Solution

$$\varepsilon_r = \frac{\varepsilon}{\varepsilon_0} = \frac{\varepsilon E}{\varepsilon_0 E} = \frac{D_d}{D_0} = \frac{5}{3} \qquad \text{dimensionless}$$

By inspection of Fig. 4-4*b* we can write

$$P = \rho_{sp} = 2 \text{ nC m}^{-2} \quad \textit{Ans.}$$

This result can also be obtained as follows. From (6),

$$P = D_d - \varepsilon_0 E$$

Therefore,

$$P = 5 - 3 = 2 \text{ nC m}^{-2} \quad \textit{Ans.}$$

the same as before.

Problem 7-6-1. Artificial dielectric. An artificial dielectric consists of a uniform lattice structure of conducting spheres extending in the x, y, and z directions with sphere diameters one-third their center-to-center spacing. Find ε_r. *Hint:* Consider that each sphere has a dipole moment equivalent to the dipole moment of a polarized atom of dielectric. *Ans.* $\varepsilon_r = 1.06$.

Problem 7-6-2. Polarization of dielectric. A square parallel-plate capacitor 200 mm on a side with a plate spacing of 25 mm is filled with a dielectric slab ($\varepsilon_r = 24$) of the same dimensions. If 100 V is applied to the capacitor, find (*a*) the polarization P in the dielectric and (*b*) the energy stored by the capacitor. If the voltage source is now disconnected and the dielectric slab then slipped out from between the plates, find (*c*) polarization in the dielectric, (*d*) energy stored in the dielectric, and (*e*) energy stored in the capacitor. *Ans.* (*a*) 8.1×10^{-7} C/m^2; (*b*) 1.7×10^{-10} J; (*c*) 0; (*d*) 0; (*e*) 4.1×10^{-9} J.

Example 7-2. Dielectric sandwich. A parallel-plate capacitor of 1/2 m^2 area and plate separation of 30 mm is filled with two stacked dielectric slabs (Fig. E7-2). The upper slab thickness is 20 mm with $\varepsilon_r = 6$ and the lower slab thickness is 10 mm with $\varepsilon_r = 12$.

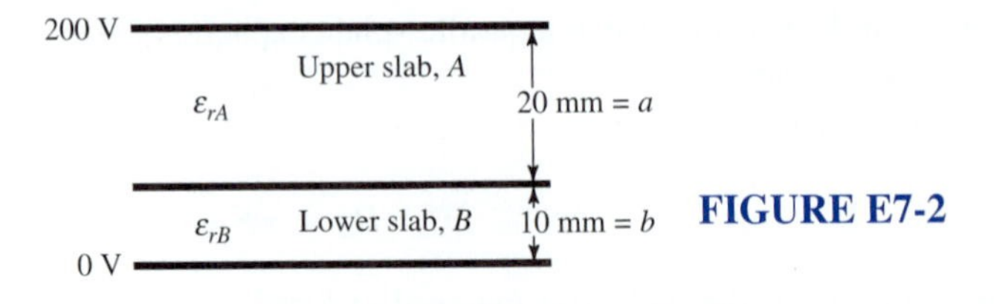

FIGURE E7-2

If a potential of 200 V is applied to the capacitor, find the values of the following quantities in each slab: (*a*) polarization *P*, (*b*) flux density D, (*c*) electric field *E*, and (*d*) capacitance C of the capacitor. Neglect fringing.

Solution

$$D_A = D_B = D = \varepsilon_0 \varepsilon_{rA} E_A = \varepsilon_0 \varepsilon_{rB} E_B$$

$$E_A = \frac{\varepsilon_{rB} E_B}{\varepsilon_{rA}} = \frac{12}{6} E_B = 2E_B$$

$$V = E_B b + E_A a = E_B b + 2E_B a = E_B(b + 2a)$$

$$E_B = \frac{V}{2a + b}, \qquad D_B = \frac{\varepsilon_0 \varepsilon_{rB} V}{2a + b}$$

(b) $$D_A = D_B = \frac{\varepsilon_0 \varepsilon_{rB} V}{2a + b} = \frac{(8.85 \times 10^{-12})(12)200}{(40 + 10)10^{-3}} = 425 \text{ nC m}^{-2} \qquad \textit{Ans.} \quad (b)$$

(c) $$E_A = 2E_B = \frac{2V}{2a + b} = \frac{2(200)}{(40 + 10)10^{-3}} = 8 \text{ kV m}^{-1}; E_B = 4 \text{ kV m}^{-1} \textit{Ans.} \quad (c)$$

(d) $$C = \frac{Q}{V} = \frac{DA}{V} = \frac{425 \times 10^{-9}(0.5)}{200} = 1.06 \text{ nF} \qquad \textit{Ans.} \quad (d)$$

(a) $$P_A = D_A - \varepsilon_0 E_A = D - \varepsilon_0 E_A = (0.425 \times 10^{-6}) - (8.85 \times 10^{-12})(8 \times 10^3)$$
$$= 354 \text{ nC m}^{-2} \qquad \textit{Ans.} \quad (a) \quad \text{Slab } A$$

$$P_B = D_B - \varepsilon_0 E_B = (0.425 \times 10^{-6}) - (8.85 \times 10^{-12})(4 \times 10^3)$$
$$= 390 \text{ nC m}^{-2} \qquad \textit{Ans.} \quad (a) \quad \text{Slab } B$$

Problem 7-6-3. Polarization in dielectric substrate slab. A plane slab of substrate ($\varepsilon_r = 6$) is situated normal to a uniform field $D = 2$ C/m². If the slab occupies a volume of 0.1 m³ and is uniformly polarized, find (*a*) *P* in the slab and (*b*) total dipole moment of the slab. *Ans.* (*a*) 1.67 C/m²; (*b*) 167 mC m.

Problem 7-6-4. Air cavities in dielectric. A uniform dielectric medium ($\varepsilon_r = 9$) of large extent has a flux density $D = 15$ pC/m² applied. (*a*) Find *D* inside a thin disk-shaped cavity cut in the dielectric with flat sides normal to **D.** (*b*) Find *D* inside a slender needle-shaped cavity with axis parallel to **D.** The cavities have air inside. *Ans.* (*a*) 15 pC/m²; (*b*) 1.67 pC/m².

In isotropic media **P** and **E** are in the same direction, so that their quotient is a scalar and hence ε is a scalar. In nonisotropic media, such as crystals, **P** and **E** are, *in general,* not in the same direction, so that ε is no longer a scalar. Thus, $\mathbf{D} = \varepsilon_0\mathbf{E}+\mathbf{P}$ is a general relation, while $\mathbf{D} = \varepsilon\mathbf{E}$ is a more concise expression, which, however, has a simple significance only for isotropic media (or certain special cases in nonisotropic media). See footnote § of Table 7-1. Thus,

$$\mathbf{D} = \underset{\substack{\uparrow \\ \text{Always}}}{\varepsilon_0 \mathbf{E} + \mathbf{P}} = \underset{\substack{\uparrow \\ \text{Isotropic,} \\ \text{linear} \\ \text{dielectric}}}{\varepsilon_r \varepsilon_0 \mathbf{E}} \qquad (\text{C m}^{-2}) \tag{11}$$

7-7 BOUNDARY RELATIONS

As discussed in Sec. 2-8, the tangential components of the electric field **E** are equal at the boundary between two dielectrics, as given by

$$E_{t1} = E_{t2} \qquad \textbf{\textit{Tangential E}} \tag{1}$$

According to (1) *the tangential components of the electric field are the same on both sides of a boundary between two dielectrics.* In other words, the tangential electric field is *continuous* across such a boundary.

If medium 2 is a conductor ($\sigma_2 \neq 0$), the field E_{t2} in medium 2 must be zero under static conditions, and hence (2) reduces to

$$E_{t1} = 0 \tag{2}$$

The difference of the normal components of the electric flux density **D** equals the surface charge, or

$$D_{n1} - D_{n2} = \rho_s \qquad \textbf{\textit{Normal D}} \tag{3}$$

According to (3), *the normal component of the flux density changes at a charged boundary between two dielectrics by an amount equal to the surface charge density* (which is usually zero at a dielectric-dielectric boundary unless charge has been placed there by mechanical means, as by rubbing).

If the boundary is free from charge, $\rho_s = 0$ and (3) reduces to

$$D_{n1} = D_{n2} \tag{4}$$

According to (4), *the normal component of the flux density is continuous across the charge-free boundary between two dielectrics.*

If medium 2 is a conductor, $D_{n2} = 0$ and (3) reduces to

$$D_{n1} = \rho_s \tag{5}$$

According to (5), *the normal component of the flux density at a dielectric-conductor boundary is equal to the surface charge density on the conductor.*

It is important to note that ρ_s in these relations refers to actual electric charge separated by finite distances from equal quantities of opposite charge and *not* to surface charge ρ_{sp} due to polarization. The polarization surface charge is produced by atomic dipoles having equal and opposite charges separated by what is assumed to be an infinitesimal distance. It is not permissible to separate the positive and negative charges of such a dipole by a surface of integration, and hence the volume must always contain an integral (whole) number of dipoles and, therefore, zero net charge. Only when the positive and negative charges are separated by a macroscopic distance (as on the opposite surfaces of a conducting sheet) can we separate them by a surface of integration. This emphasizes a fundamental difference between the polarization, or so-called *bound* charge, on a dielectric surface and the *true* charge on a conductor surface. In a similar way the boundary relation for polarization is

$$P_{n1} - P_{n2} = -\rho_{sp} \tag{6}$$

If medium 1 is free space,

$$P_{n2} = \rho_{sp} \tag{7}$$

Equations (5) and (7) are written more generally as

$$\mathbf{D} \cdot \hat{\mathbf{n}} = \rho_s \qquad \text{and} \qquad \mathbf{P} \cdot \hat{\mathbf{n}} = \rho_{sp} \tag{8}$$

Example 7-3. Boundary between a conductor and a dielectric. Suppose that medium 2 in Fig. 7-5 is a conductor. Find α_1.

Solution. Since medium 2 is a conductor, $D_2 = E_2 = 0$ under static conditions. According to the boundary relations,

$$D_{n1} = \rho_s \qquad \text{or} \qquad E_{n1} = \frac{\rho_s}{\varepsilon_1}$$

and

$$E_{t1} = 0$$

Therefore

$$\alpha_1 = \tan^{-1}\frac{E_{t1}}{E_{n1}} = \tan^{-1} 0 = 0 \quad Ans.$$

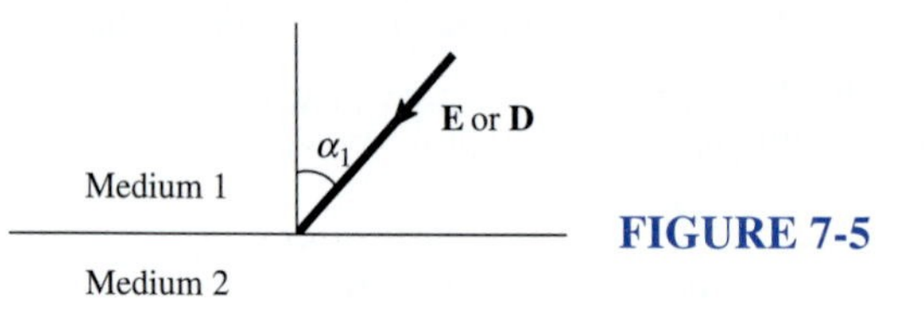

FIGURE 7-5

It follows that a static electric field line or flux tube at a dielectric-conductor boundary is always perpendicular to the conductor surface (when no currents are present). This fact is of fundamental importance in field mapping.

If a thin conducting sheet is introduced normal to an electric field as in Fig. 7-6, surface charges are induced on both sides of the sheet and the original flux density **D** is unchanged. The value of the induced surface charge density ρ_s is equal to the flux density D at the sheet. Hence, one can interpret the flux density D at a point as equal to the charge density ρ_s which would appear on a thin conducting sheet introduced

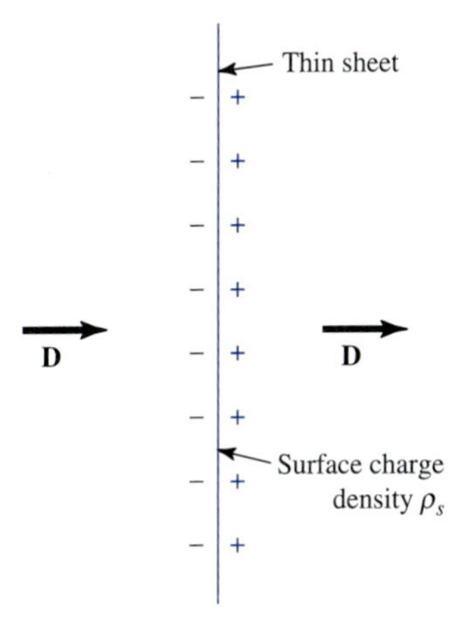

FIGURE 7-6
Thin conducting sheet placed normal to field has an induced surface charge density ρ_s equal to the flux density $|\mathbf{D}|$ of the field at the sheet. The surface charge densities on the two sides of the sheet are equal in magnitude but opposite in sign.

normal to **D** at the point. Referring, for example, to the thin conducting sheet normal to the field in Fig. 7-6, the relation of **D** and ρ_s is as follows:

$$\text{On left side:} \qquad \mathbf{D} = -\hat{\mathbf{n}}\rho_s$$
$$\text{On right side:} \qquad \mathbf{D} = +\hat{\mathbf{n}}\rho_s$$

where $\hat{\mathbf{n}}$ is the unit vector normal to the surface. Thus, **D** is normally inward on the left side and normally outward on the right. The magnitude of the flux density on each side is equal to the charge density ρ_s.

Problem 7-7-1. High-voltage bushing. A high-voltage conductor is brought through a grounded metal panel by means of the double concentric capacitor bushing shown in Fig. 7-7. The medium is air (dielectric strength = 3 MV/m). Neglect fringing and thickness of sleeves. (*a*) Find the outer sleeve length L which equalizes the voltage across each space. (*b*) Find L which results in the same value of maximum field in each space. (*c*) What is the maximum working voltage for each condition? (*d*) If the inner sleeve (200 mm long) were removed, what would be the maximum working voltage? (*e*) If the number of concentric sleeves is increased in number so that the spacing between sleeves becomes smaller, what is the ultimate working voltage? *Ans.* (*a*) 155 mm; (*b*) 150 mm; (*c*) 51.8 kV ($V_1 = V_2$), 52.7 kV (E_1 max $= E_2$ max); (*d*) 46 kV; (*e*) 60 kV.

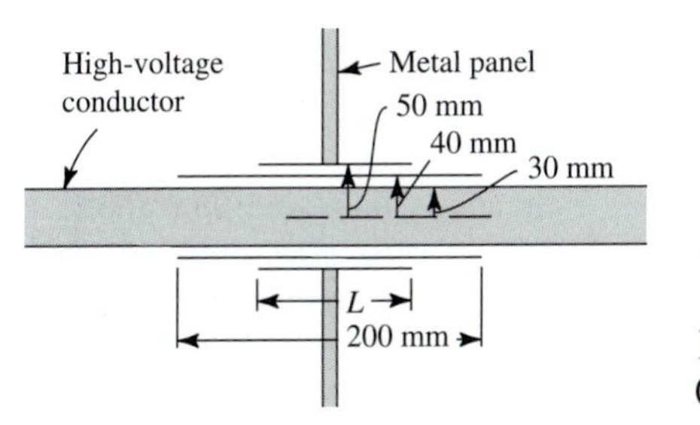

FIGURE 7-7
Concentric capacitor bushing.

7-8 TABLE OF BOUNDARY RELATIONS

Table 7-2 summarizes the boundary relations for static fields developed in the preceding section.

TABLE 7-2
Boundary relations for static electric fields†

Field component	Boundary relation		Condition
Tangential	$E_{t1} = E_{t2}$	(1)	Any two media
Tangential	$E_{t1} = 0$	(2)	Medium 1 is a dielectric; medium 2 is a conductor
Normal	$D_{n1} - D_{n2} = \rho_s$	(3)	Any two media with charge at boundary
Normal	$D_{n1} = D_{n2}$	(4)	Any two media with no charge at boundary
Normal	$D_{n1} = \rho_s$	(5)	Medium 1 is a dielectric; medium 2 is a conductor with surface charge

†Relations (1), (3), and (4) apply in the presence of currents and also for time-varying fields. The other relations, (2) and (5), also apply for time-changing situations provided $\sigma_2 = \infty$.

7-9 DIELECTRIC STRENGTH

The field intensity **E** in a dielectric cannot be increased indefinitely. If a certain value is exceeded, sparking occurs and the dielectric is said to *break down*. The maximum field intensity that a dielectric can sustain without breakdown is called its *dielectric strength.*

In the design of capacitors it is important to know the maximum potential difference that can be applied before breakdown occurs. For a given plate spacing, this breakdown is proportional to the dielectric strength of the medium between the plates. The radius of curvature of the conducting surface is another factor. The electric field E adjacent to a conductor is proportional to the electric charge density ρ_s on the conductor surface, and this charge density tends to be higher on surfaces with small radii of curvature and less on surfaces with large radii of curvature.

As E is gradually increased, sparking occurs in air almost immediately when a critical value of field is exceeded if the field is uniform (E everywhere parallel), but a corona discharge may occur first if the field is nonuniform (diverging) with sparkover following as E is increased further.

Many capacitors have air as the dielectric. These types have the advantage that if breakdown occurs, the capacitor is not permanently damaged. For applications requiring large capacitance or small physical size or both, other dielectrics are employed. The dielectric strengths of a number of common dielectric materials are listed in Table 7-3. The dielectric strengths are for a uniform field, and the materials are arranged in order of increasing strength.

TABLE 7-3
Table of dielectric strengths

Material	Dielectric strength, MV m^{-1}
Air (atmospheric pressure)	3
Oil (mineral)	15
Paper (impregnated)	15
Polystyrene	20
Rubber (hard)	21
Bakelite	25
Glass (plate)	30
Paraffin	30
Quartz (fused)	30
Mica	200

7-10 ENERGY AND ENERGY DENSITY

From (2-9-9), the energy density at a point in an electric field is given by

$$w = \lim_{\Delta v \to 0} \frac{\Delta W}{\Delta v} = \frac{1}{2}\varepsilon E^2 \qquad (\text{J m}^{-3}) \qquad \textbf{\textit{Energy density}} \qquad (1)$$

No material medium need be present for energy to be stored by a field. Thus, energy is present even in vacuum. This energy is equivalent to that required to charge the capacitor cell like the energy stored in a lifted weight. However, if a dielectric material is present, the amount of energy is increased in proportion to the permittivity ε. From (7-6-9), we can express the energy density as

$$w = \tfrac{1}{2}(\varepsilon_0 E^2 + PE) = \tfrac{1}{2}\varepsilon E^2 \qquad (\text{J m}^{-3}) \qquad (2)$$

where $\frac{1}{2}\varepsilon_0 E^2$ = energy density in vacuum, J m^{-3}
$\frac{1}{2}PE$ = energy density in dielectric medium, J m^{-3}
P = polarization, C m^{-2}

The energy in the dielectric ($= \frac{1}{2}PE$) is due to the polarization of its molecules in the electric field, like the energy stored in a stretched spring.

For vacuum, there is no polarization and the energy density is simply $\frac{1}{2}\varepsilon_0 E^2$. With a dielectric it is increased by $\frac{1}{2}PE$ to a total energy density of $\frac{1}{2}\varepsilon E^2$.

Thus, a static electric field contains energy. A magnetic field also contains energy and a moving electromagnetic field, or wave, transports energy.

The total energy W stored by a capacitor is the integral of the energy density w over the entire region in which the electric field **E** has a value

$$W = \int_v w\,dv = \frac{1}{2}\int_v \varepsilon E^2\,dv \qquad (\text{J}) \qquad (3)$$

Assuming that the field is uniform between the plates and that there is no fringing of the field at the edges of the capacitor, we have on evaluating (3)

$$W = \tfrac{1}{2}\varepsilon E^2 Ad = \tfrac{1}{2}DAEd = \tfrac{1}{2}QV \qquad \text{(J)} \tag{4}$$

where A = area of one capacitor plate, m^2
d = spacing between capacitor plates, m

This result, obtained by integrating the energy density throughout the volume between the plates of the capacitor, is identical with the relation given by (2-9-8).

Example 7-4. Earth-electrosphere capacitor. We live inside a huge spherical capacitor. The earth is the inner surface and the electrosphere† overhead is the outer surface. For a fair weather electric field of 165 V m^{-1}, find the energy density and the total energy stored in the capacitor between the earth and the electrosphere at a height of 25 km. The earth's circumference = 40 Mm.

Solution. Consider a cubical volume 1 m on a side with imaginary 1-m-sqaure conducting sheets on the top and bottom. This constitutes a capacitor of capacitance.

$$C = \frac{\varepsilon_0 A}{d} = 8.85 \times 10^{-12}\frac{1^2}{1} = 8.85 \text{ pF}$$

From (2-9-8), the energy in this cubical capacitor is

$$W = \tfrac{1}{2}CV^2 = \tfrac{1}{2}(8.85 \times 10^{-12})(165^2) = 1.2 \times 10^{-7} \text{ J}$$

Since the capacitor volume is 1 m^3, the energy density in the fair weather atmosphere is

$$w = \frac{W}{v} = \frac{1.2 \times 10^{-7}}{1} = 1.2 \times 10^{-7} \text{ J m}^{-3}$$

The total energy in the earth-electrosphere capacitor is then

$$W = wv = w4\pi r^2 h = (1.2 \times 10^{-7})(4\pi)(6.37 \times 10^6)^2(25 \times 10^3)$$
$$= 1.5 \times 10^{12} \text{ J} \quad \textit{Ans.}$$

Our calculation assumes that the fair weather field is uniform with height. Since it actually is not, the 1.5×10^{12} J result is significant only as an order of magnitude.

---

†The *electrosphere* is the dc-conducting region of the upper atmosphere with conductivity provided by *molecules* ionized by cosmic rays. Its height is variously estimated at 25 to 60 km. A radio wave–reflecting region, the *ionosphere,* is at heights of 100 km or more. It contains *free electrons* caused by the ionizing effect of solar radiation. Radio waves pass right through the electrosphere because the ions (molecules) are too heavy to interact with the high-frequency field fluctuations. The free electrons in the ionosphere, however, are light enough to interact.

Example 7-5. Energy and energy density of three devices. Let us compare the energy storage and energy density of three commercially available devices:
(*a*) A 1-kV, 1-μF power-supply capacitor measuring $3 \times 4 \times 8$ cm
(*b*) A 5-V, 0.1-F CMOS backup capacitor of 0.4 cm^3 volume
(*c*) A 100-A h, 12-V automobile storage battery measuring $15 \times 15 \times 20$ cm

Solution
(a) *Power supply capacitor energy:*

$$W = \tfrac{1}{2}CV^2 = \tfrac{1}{2}(10^{-6})(10^3)^2 = \tfrac{1}{2}\text{ J} \quad Ans. \quad (a)$$

and energy density:

$$w = \frac{W}{v} = \frac{1}{2}\frac{1}{3 \times 4 \times 8 \times 10^{-6}} = 5.2 \times 10^3 \text{ J m}^{-3} \quad Ans. \quad (a)$$

(*b*) *CMOS backup capacitor energy:*

$$W = \tfrac{1}{2}CV^2 = \tfrac{1}{2}(0.1)(5^2) = 1.25 \text{ J} \quad Ans. \quad (b)$$

and energy density:

$$w = \frac{W}{v} = \frac{1.25}{0.4 \times 10^{-6}} = 3.1 \times 10^6 \text{ J m}^{-3} \quad Ans. \quad (b)$$

(*c*) *12-V storage battery energy:*

$$W = V \times I \times t = 12 \times 100 \times 3600 = 4.3 \times 10^6 \text{ J} \quad Ans. \quad (c)$$

and energy density:

$$w = \frac{W}{v} = \frac{4.3 \times 10^6}{15 \times 15 \times 20 \times 10^{-6}} = 9.6 \times 10^8 \text{ J m}^{-3} \quad Ans. \quad (c)$$

Although the lead-acid battery achieves the highest energy density, capacitors can make their stored energy available much more rapidly.

Problem 7-10-1. Capacitor energy. The relative permittivity and dielectric strengths of two different dielectric materials are $\varepsilon_r = 6$ and $E = 50$ MV/m for material A and $\varepsilon_r = 100$ and $E = 10$ MV/m for material B. Which material is capable of storing more energy and by what ratio? Assume that the materials are to be used as fillers in a parallel-plate capacitor of fixed size. *Ans.* $A/B = 1.5$.

Problem 7-10-2. Sphere energy. A conducting sphere of radius R has a charge Q. Within what radius is 90 percent of its energy stored? *Ans.* $10R$.

Problem 7-10-3. Energy storage. If a 100-A-h, 12-V automobile storage battery can store $VIt = 12 \times 100 \times 3600 = 4.3$ MJ of energy, how many 1000-V, 2-μF capacitors are required to store the same amount of energy? *Ans.* 2×10^6.

7-11 THE ATOMIC LOOP

Magnetic fields are present around a current-carrying conductor. They also exist around a magnetized object such as an iron bar magnet. In the bar the currents flow in circuits of atomically small dimensions.

An electron circulating in an orbital cloud around the nucleus of an atom is equivalent to a tiny current loop as suggested in Fig. 7-8a. This loop has a magnetic (dipole) moment m equal to the product of the equivalent current I and the loop area A, or

$$m = IA \tag{1}$$

This moment may be expressed as a vector $\mathbf{m}$ which is perpendicular to the loop and directed in a right-hand sense (up in the figure). We note that $\mathbf{I}$ is opposite to the direction of motion of the negatively charged electron.

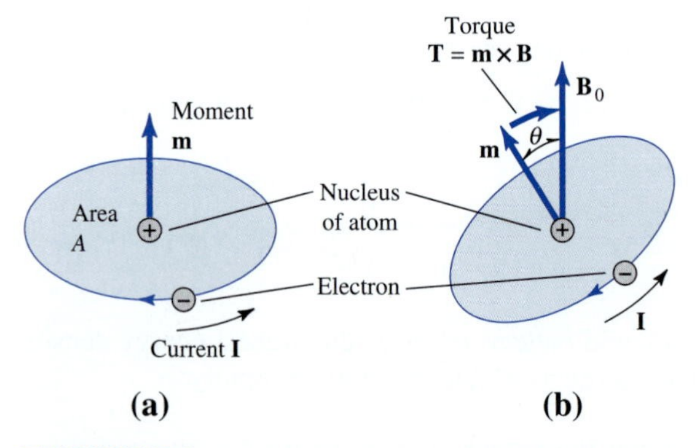

FIGURE 7-8
(*a*) Atomic current loop has a magnetic moment normal to the loop. With the fingers of the right hand in the direction of the current **I**, the thumb is in the direction of the moment (up). (*b*) When a magnetic field **B** is applied to a loop that is tilted (**m** at angle θ from **B**), a torque tends to close the angle and align **m** with **B**.

When a magnetic field **B** is applied to the atom, there is a torque **T** tending to align the moment of the atomic loop with the field as given by

$$\mathbf{T} = \mathbf{m} \times \mathbf{B} \qquad \text{(N m)} \tag{2}$$

Referring to the tilted loop in Fig. 7-8*b* , this torque tends to rotate the loop so that the moment vector **m** (perpendicular to the loop) aligns with **B**, as suggested. The magnitude of the torque

$$|\mathbf{T}| = |\mathbf{m}||\mathbf{B}| \sin \theta \tag{3}$$

When alignment is achieved ($\theta = 0$) the torque becomes zero.

Consider next the situation shown in Fig. 7-9*a* (on next page) where there are several atomic current loops oriented randomly. In Fig. 7-9*b*, a magnetic field **B** has been applied which brings the moment vectors of the loops into alignment with **B.** Viewing the configuration from the direction of **B**, we observe that a cluster of atomic loops may be considered equivalent to a single larger loop (Fig. 7-9*c*). We note that where the atomic loops touch or abut, their currents are opposed so that in the limit, if the interior of the large loop is filled with many atomic loops, the net effect is just that of a single large loop. If now we stack many of these large loops together, as in Fig. 7-9*d*, we obtain a cylindrical current sheet equivalent to that of a solenoid.

Thus, thousands of uniformly aligned atomic loops in a cylindrical iron bar are magnetically equivalent *externally* to an air-filled solenoid of the same dimensions. This equivalence is discussed further in a later section.

The foregoing presentation is a simplified introductory picture of the magnetization phenomenon. We have neglected the effect of thermal agitation which interferes with the alignment process. We have simplified an atom to a single orbiting electron, whereas there may be many electrons. Furthermore, the electron itself has a magnetic moment because of its spin. The spin of the nucleus also produces a magnetic moment, but it is usually not significant. So although incomplete, and far short of a full quantum mechanical treatment, our simplistic picture does provide some useful insights. Most of the applications we will consider are macroscopic or large scale and will not require detailed knowledge of the microscopic phenomena involved.

For the atoms of most materials, the orbiting and spin moments of the electrons tend to cancel so that the atoms show only weak magnetic effects (small net magnetic moment). In the atoms of other materials, however, the orbiting and spin moments do not cancel, and the material shows significant magnetic effects. In iron, nickel, and cobalt these effects are especially strong, and these metals are called *ferromagnetic*. These ferromagnetic materials are discussed further in Sec. 7-17.

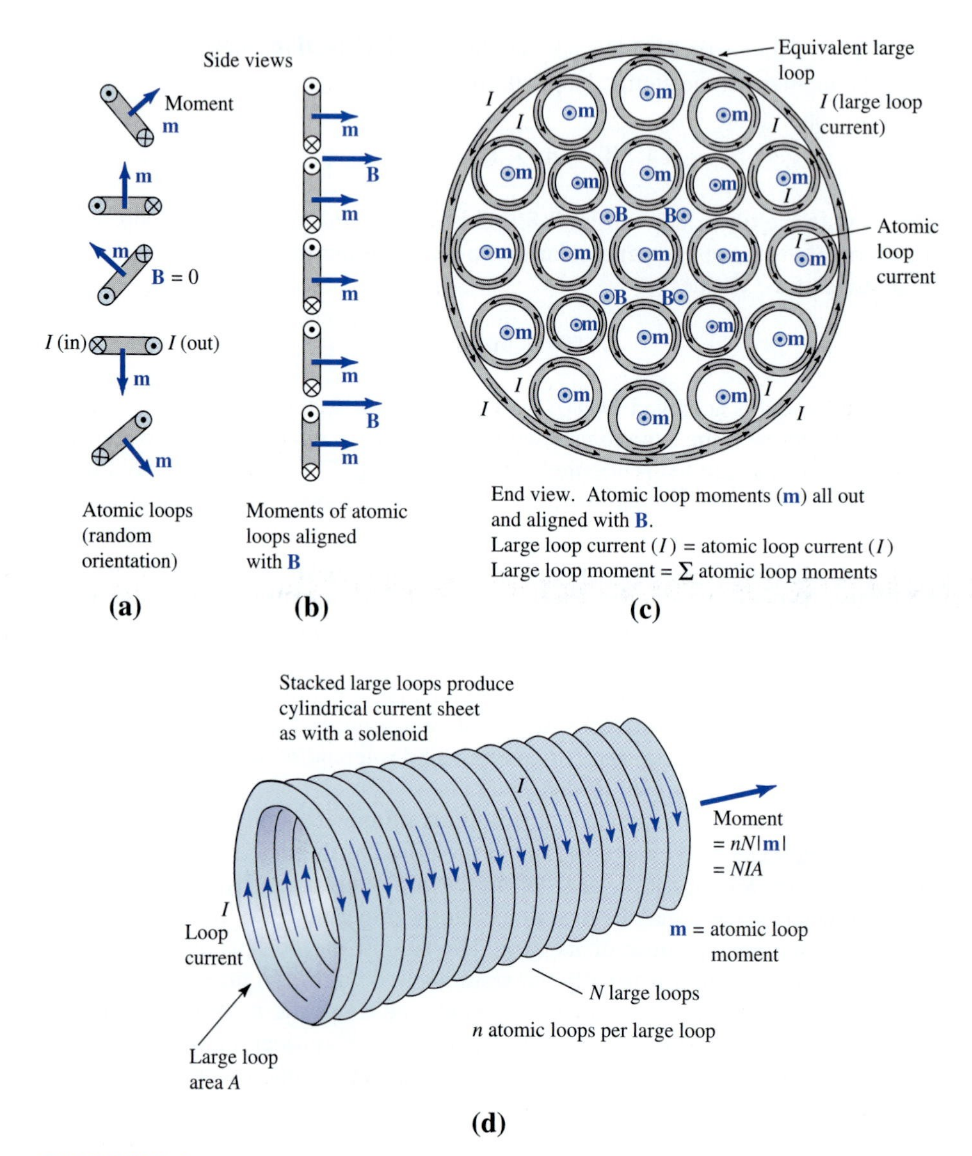

FIGURE 7-9

(*a*) Atomic loops in random orientation. (*b*) Atomic loops aligned with applied field (side view). (*c*) Same in end view showing equivalent large loop. (*d*) Stacks of large loops giving current sheet as in a solenoid. All loops are assumed circular.

7-12 MAGNETIC DIPOLES, LOOPS, AND SOLENOIDS

If a horizontal bar magnet is freely suspended, as in a compass, it turns in the earth's magnetic field so that one end points north. This end is called the *north–seeking pole* of the magnet, or simply its *north pole*. The other end is its *south pole* (of equal strength but opposite polarity).†

All magnetized bodies have both a north and a south pole. They cannot be isolated. For example, consider the long magnetized iron rod of Fig. 7-10. This rod has a north pole at one end and south pole at the other. If the rod is cut in half, new poles appear, as in Fig. 7-10*b*, so that there are two magnets. If each of these is cut in half, we obtain four magnets, as in Fig. 7-10*c*, each with a north and a south pole. As we have discussed, the ultimate source of the magnetism is an atomic current loop which is equivalent to an atomically small bar magnet. Therefore, even if the cutting process could be continued to atomic dimensions and a single iron atom isolated, it would still have a north and a south pole.

Although the existence of magnetic monopoles has been postulated, the fact that, in our experience, magnetic poles cannot be isolated is an important point of difference compared to electric charges, which can be isolated.

However, even if we can't isolate magnetic poles, an isolated pole may be approximated in some situations by confining our attention to regions close to one end of a very long needlelike magnet.

Assuming an isolated magnetic pole of strength Q_m, we can write by analogy to the electric field **E** as a force per unit electric charge Q that the magnetic field **B** is equal to the force per unit magnetic pole strength Q_m, or

$$\mathbf{B} = \frac{\mathbf{F}}{Q_m} \qquad \text{(T)} \tag{1}$$

where the force **F** is in the direction of the magnetic field **B** as suggested in Fig. 7-11 on page 400. From the motor equation $F = LIB$, we have

$$B = \frac{F}{IL} \tag{2}$$

Comparing (1) and (2), we note that the magnetic pole strength Q_m has the dimensions of a current moment (IL = current × distance) with units A m.

If a bar magnet is placed on a wooden table and covered with a sheet of paper, iron filings sprinkled on the sheet align themselves along the magnetic field lines

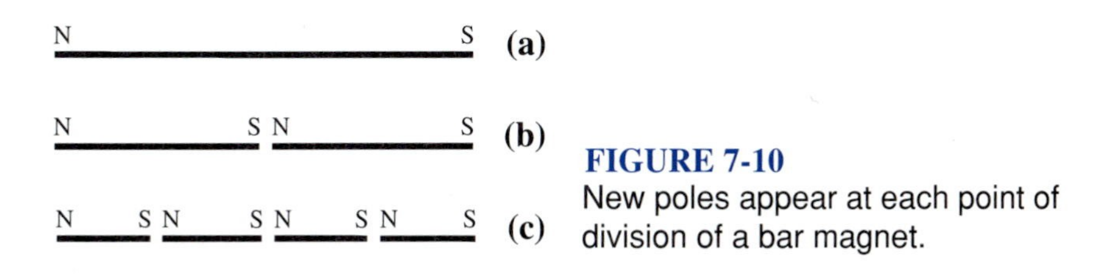

FIGURE 7-10
New poles appear at each point of division of a bar magnet.

†Note that, by this definition, the earth's north polar region contains the earth's south magnetic pole.

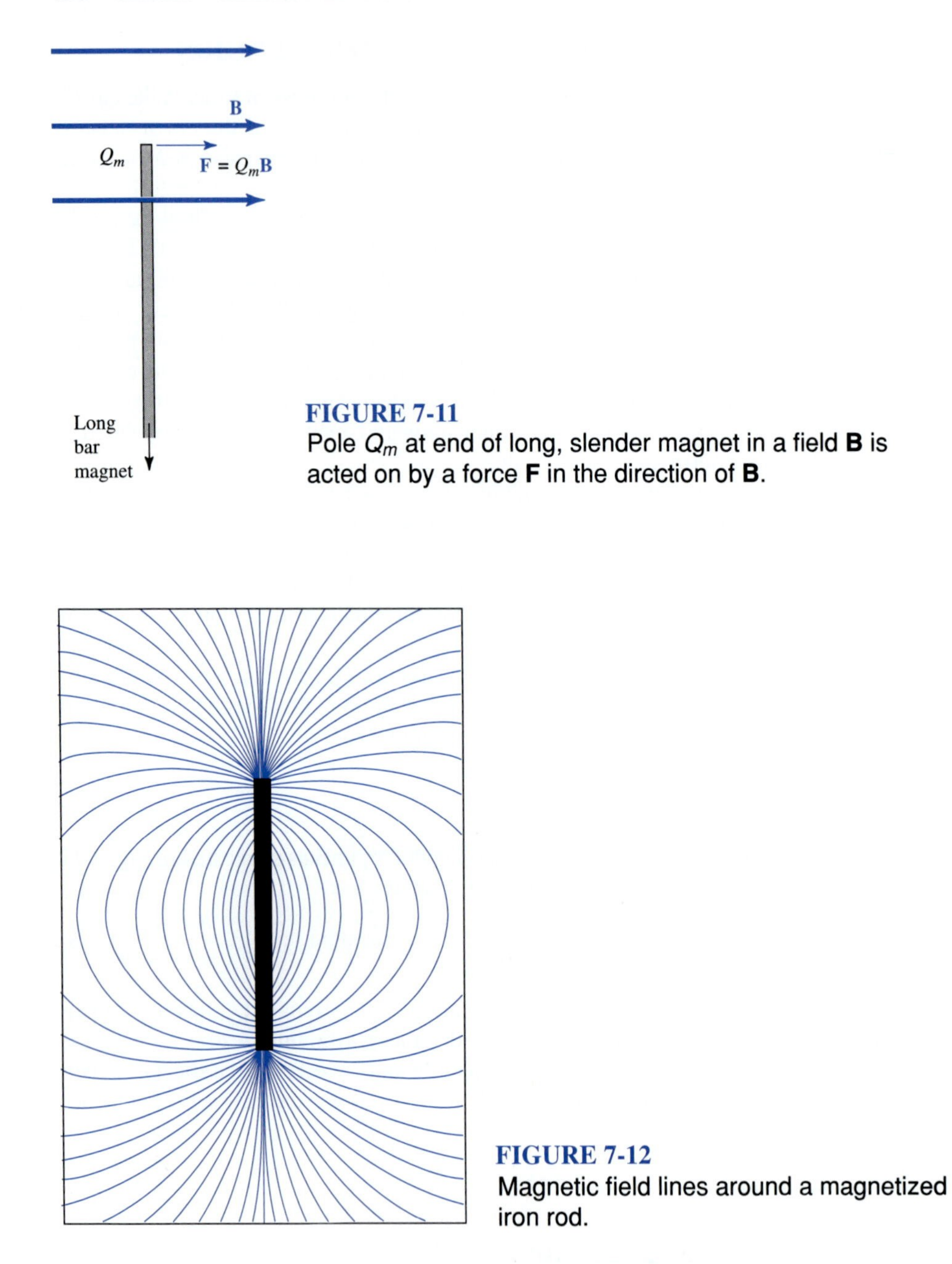

FIGURE 7-11
Pole Q_m at end of long, slender magnet in a field **B** is acted on by a force **F** in the direction of **B**.

FIGURE 7-12
Magnetic field lines around a magnetized iron rod.

of the magnet, as suggested in Fig. 7-12. The field configuration is that of a dipole with magnetic poles separated by a distance approximately equal to the length of the magnet.

A bar magnet of pole strength Q_m and length L constitutes a magnetic dipole of magnetic dipole moment Q_mL with units of A m^2, analogous to the electric dipole moment QL of two electric charges ($+Q$ and $-Q$) separated by a distance L (Fig. 7-1) with units C m. Thus,

$$Q_mL = \text{magnetic dipole moment} \qquad (\text{A m}^2)$$
$$QL = \text{electric dipole moment} \qquad (\text{C m})$$

Consider now a bar magnet, as in Fig. 7-13, of pole strength Q_m and length L (magnetic dipole moment $= m = Q_mL$) in a uniform magnetic field $\mathbf{B}$. The north (+) pole experiences a force $F = Q_mB$ to the right and the south (−) pole an equal force to the left. The torque T, or turning moment (force × distance), on the dipole is given by

$$T = 2F\frac{L}{2}\sin\theta \qquad (\text{N m}) \tag{3}$$

where $F = Q_mB$, N
L = length of dipole, m
θ = angle between dipole axis and B, rad or deg

Introducing the value of F, we have

$$T = Q_mLB\sin\theta = mB\sin\theta \tag{4}$$

In vector notation (see Fig. 7-13) the torque is given by

$$\mathbf{T} = \mathbf{m} \times \mathbf{B} = m\hat{\mathbf{n}} \times \mathbf{B} = Q_mL\hat{\mathbf{n}} \times \mathbf{B} \tag{5}$$

where $\mathbf{m} = \hat{\mathbf{n}}m = \hat{\mathbf{n}}Q_mL$ = magnetic moment, A m^2
$\hat{\mathbf{n}}$ = unit vector in the direction of $\mathbf{m}$ [in line with the magnet and directed from south (−) to north (+) pole].

Turning $\hat{\mathbf{n}}$ (or $\mathbf{m}$) into $\mathbf{B}$, the torque is into the page and tends to rotate the magnet clockwise into alignment with $\mathbf{B}$, as in Fig. 7-13 or Fig. 7-14.

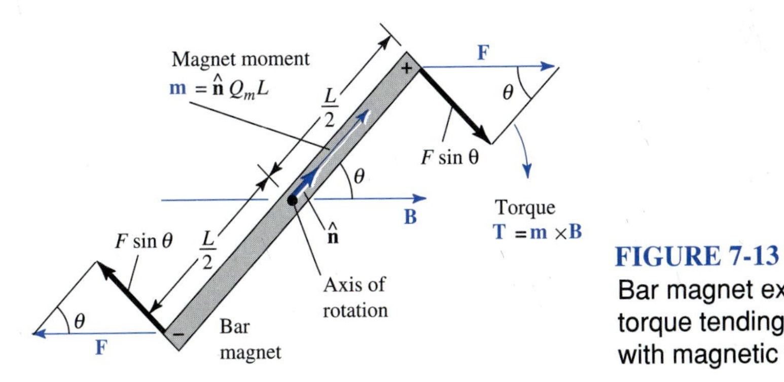

FIGURE 7-13
Bar magnet experiences torque tending to align it with magnetic field **B**.

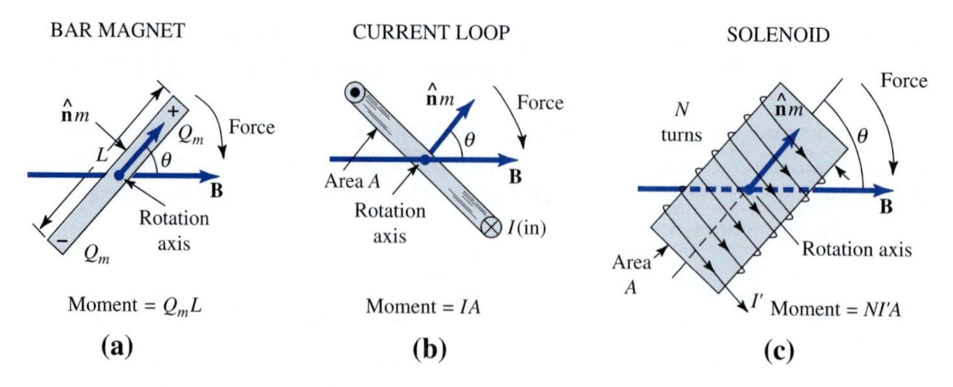

FIGURE 7-14
Bar magnet, current loop, and solenoid in a uniform field **B**. In all three cases the torque is clockwise and tends to align the magnetic moment **m** ($\hat{\mathbf{n}}m$) with **B**. If all moments are equal ($Q_mL = IA = NI'A$), the torque is the same in all three cases. The loop is shown in cross-section.

Referring to Fig. 7-14*b*, let us consider a single-turn loop in a magnetic field **B**. The torque on the loop is given by

$$\mathbf{T} = \mathbf{m} \times \mathbf{B} = m\hat{\mathbf{n}} \times \mathbf{B} = IA\hat{\mathbf{n}} \times \mathbf{B} \tag{6}$$

where $\mathbf{m} = \hat{\mathbf{n}}m = \hat{\mathbf{n}}IA =$ magnetic moment, A m^2
$I =$ loop current, A
$A =$ loop area, m^2
$\hat{\mathbf{n}} =$ unit vector normal to loop and directed in a right-handed sense

Comparing (5) and (6), we note that the magnet and loop are equivalent if their magnetic moments are equal, or

$$Q_mL = IA \tag{7}$$

Furthermore, a magnet and a solenoid are equivalent if their magnetic moments are equal, or

$$Q_mL = NI'A \tag{8}$$

where $N =$ number of turns of solenoid, dimensionless
$I' =$ solenoid current, A
$A =$ area of solenoid cross section, m^2

Thus, the magnet, loop, and solenoid are all equivalent if their magnetic moments are equal, or

$$Q_mL = IA = NI'A = m \qquad (\text{A m}^2) \qquad \textit{Magnetic moments} \tag{9}$$

with the torque on any one given by

$$\mathbf{T} = \mathbf{m} \times \mathbf{B} \qquad (\text{N m}) \qquad \textit{Torque} \tag{10}$$

A bar, loop, and solenoid are compared in Fig. 7-14. Not only will the torques be the same provided $Q_m L = IA = NI'A$, but, at a sufficient distance, the magnetic fields of all three will be the same. In fact, if the solenoid has the same dimensions as the bar magnet, the external fields of the two may be identical. Note that the positive sense of the unit vector $\hat{\mathbf{n}}$ for the bar is from the negative to positive poles while for the loop and solenoid it is determined by the right-hand rule (fingers in the direction of the current, thumb in the direction of $\hat{\mathbf{n}}$).

In all cases the torque tends to align the unit vector $\hat{\mathbf{n}}$ with $\mathbf{B}$, the torque vector $\mathbf{T}$ being coincident with the axis of rotation (perpendicular to page) and directed so that $\mathbf{m}$, $\mathbf{B}$, and $\mathbf{T}$ form a right-handed set. In other words, turning $\mathbf{m}$ or $\hat{\mathbf{n}}$ into $\mathbf{B}$, $\mathbf{T}$ is then in the direction of motion of a right-handed screw.

An important aspect of (5) or (6) is that, although the structure of a magnetic object may not be known, its strength can be described in terms of its *magnetic moment* $\mathbf{m}$, which can be determined by measuring its torque $\mathbf{T}$ in a known magnetic field $\mathbf{B}$.

Example 7-6. Magnetic moment of the earth. For many purposes the magnetic field of the earth at ionospheric heights or above may be regarded as if produced by a short magnetic dipole (or small current loop) situated at the center of the earth with $\mu = \mu_0$ everywhere (except within the dipole). See Fig. 7-14.1.

If the measured flux density $B_p = 1$ gauss ($= 1\text{G} = 10^{-4}$ T) at point P of Fig. 7-14.1, what is (*a*) the magnetic moment of the earth's "core magnet" and (*b*) the current of an equivalent loop 1 m in diameter? $\alpha = 20°$; earth circumference $= 40$ Mm.

Solution. Referring to Fig. 7-14.1, the two components of the magnetic flux density B at a distance r from a magnetic dipole m are

$$B_r = \frac{\mu_0 m \cos\theta}{2\pi r^3} \qquad \text{and} \qquad B_\theta = \frac{\mu_0 m \sin\theta}{4\pi r^3}$$

Solving for m, we have

$$m = \frac{B_r 2\pi r^3}{\mu_0 \cos\theta} = \frac{B_\theta 4\pi r^3}{\mu_0 \sin\theta} \tag{11}$$

and

$$\frac{B_\theta}{B_r} = \frac{\sin\theta}{2\cos\theta} = \frac{1}{2}\tan\theta \tag{12}$$

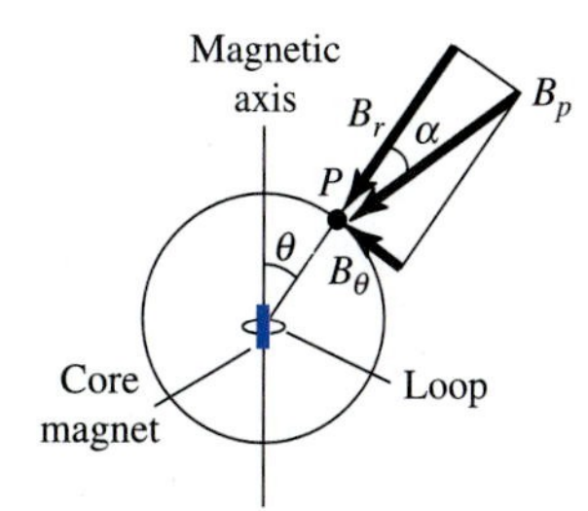

FIGURE 7-14.1
Magnetic flux density B_p at point P on the earth at angle θ from earth's magnetic axis (now 11° from the rotational axis).

From Fig. 7-14.1,

$$B_\theta = B_P \sin\alpha \text{ and } B_r = B_P \cos\alpha$$

Thus,

$$\frac{B_\theta}{B_r} = \tan\alpha \tag{13}$$

Equating (2) and (3)

$$\tan\theta = 2\tan\alpha$$

$$\theta = \tan^{-1}(2\tan 20°) = 36° \text{ from magnetic north.}$$

Thus, from (11),

$$m = \frac{B_P \cos\alpha\, 2\pi}{\mu_0 \cos\theta} r^3 = \frac{(10^{-4})(0.94)(2\pi)}{4\pi(10^{-7})(0.81)}\left(\frac{40}{2\pi}\right)^3 10^{18} = 1.5\times 10^{23}\text{ A m}^2 \quad \textit{Ans.} \quad (a)$$

Since $m = IA$

$$I = \frac{m}{A} = \frac{1.5\times 10^{23}}{\pi(0.5)^2} = 1.9\times 10^{23} A$$

or almost 200 thousand billion billion amperes! *Ans.* (*b*)

Problem 7-12-1. Magnetic moment of the earth and B above it. Using the magnetic moment of Example 7-6, calculate the magnetic flux density B at a distance of 40 Mm (*a*) above the earth's magnetic pole and (*b*) above the earth's magnetic equator. *Ans.* (*a*) 300 nT; (*b*) 150 nT.

Problem 7-12-2. Compass needle moment. If the needle of a compass placed at point P of Fig. 7-14.1 experiences a restoring torque $T = 10^{-7}$ N m when it is deflected 10° off magnetic north, what is its magnetic moment? *Ans.* 17 mA m^2.

7-13 MAGNETIC MATERIALS

All materials show some magnetic effects. In many substances the effects are so weak that the materials are often considered to be *nonmagnetic*. However, a vacuum is the only truly nonmagnetic medium.

In general, materials can be classified according to their magnetic behavior into *diamagnetic, paramagnetic, ferromagnetic, antiferromagnetic, ferrimagnetic*, and *superparamagnetic*. In *diamagnetic* materials magnetic effects are weak. Although the orbit and spin magnetic moments in such materials cancel (net magnetic moment is zero) in the absence of an external magnetic field, an applied field causes the spin moment to slightly exceed the orbital moment, resulting in a small net magnetic moment which opposes the applied field **B.** Thus, if a diamagnetic specimen is brought near either pole of a strong bar magnet, it will be *repelled,* an effect discovered by Michael Faraday in 1846. Faraday's specimen was a piece of bismuth, a substance which shows diamagnetism more strongly than most materials.

In other materials the orbit and spin magnetic moments are unequal, resulting in a net magnetic moment for the atom even with no applied field. Random orientation of the atoms may result in little net magnetic moment for a sample of the material, but when an external field is applied, the atomic dipoles experience a torque which tends to align them with the field so that the magnetic moment of the sample is increased in proportion to the number of atoms in the sample (provided perfect alignment is achieved). Internal interactions and thermal agitation tend, however, to inhibit the process so that only partial alignment may actually be achieved. Nevertheless, magnetic effects may be significant, and such substances are called *paramagnetic.* When a paramagnetic substance is brought near the pole of a strong bar magnet, it is *attracted* to it.

In a few materials, especially iron, nickel, and cobalt, a special phenomenon occurs which greatly facilitates the alignment process. In these substances, called *ferromagnetic,* there is a quantum effect known as "exchange coupling" between adjacent atoms in the crystal lattice of the material which locks their magnetic moments into a rigid parallel configuration over regions, called *domains,* which contain many atoms. However, at temperatures above a critical value, known as the *Curie temperature,* the exchange coupling disappears and the material reverts to an ordinary paramagnetic type. Ferromagnetism is discussed further in Sec. 7-17.

In *antiferromagnetic* materials the magnetic moments of adjacent atoms align in opposite directions so that the net magnetic moment of a specimen is nil even in the presence of an applied field.

In *ferrimagnetic* substances the magnetic moments of adjacent atoms are also aligned opposite, but the moments are *not* equal, so there is a net magnetic moment. However, it is less than in ferromagnetic materials. In spite of the weaker magnetic effects, some of these ferrimagnetic materials, known as *ferrites,* have a low electrical conductivity, which makes them useful in the cores of ac inductors and transformers, since induced (eddy) currents are less and ohmic (heat) losses are reduced.

TABLE 7-4
Magnetic classification of materials

Type	Characteristics and example
Nonmagnetic	Vacuum
Diamagnetic	Weakly magnetic. Moment opposes applied **B**. Repelled by bar magnet. *Example:* Bismuth
Paramagnetic	Significant magnetism. Attracted to a bar magnet. *Example:* Aluminum.
Ferromagnetic	Strongly magnetic (atomic moments aligned). Attracted to a bar magnet. Has exchange coupling and domains. Becomes paramagnetic above Curie temperature. *Examples:* Iron, nickel cobalt
Antiferromagnetic	Nonmagnetic even in presence of applied field. Moments of adjacent atoms align in opposite directions. *Example:* Manganese oxide (MnO_2)
Ferrimagnetic	Less magnetic than ferromagnetic materials. *Example:* Iron ferrite
Ferrites	Ferrimagnetic material with low electrical conductivity. Useful as inductor cores for ac applications.
Superparamagnetic	Ferromagnetic materials suspended in dielectric matrix. Used in audio and video tapes.

TABLE 7-5
Relative permeabililties

Substance	Group type	Relative permeability μ_r
Bismuth	Diamagnetic	0.99983
Silver	Diamagnetic	0.99993
Lead	Diamagnetic	0.99993
Copper	Diamagnetic	0.999991
Water	Diamagnetic	0.999991
Vacuum	Nonmagnetic	1†
Air	Paramagnetic	1.0000004
Aluminum	Paramagnetic	1.00002
Palladium	Paramagnetic	1.0008
2-81 Permalloy powder (2 Mo, 81 Ni)‡	Ferromagnetic	130
Cobalt	Ferromagnetic	250
Nickel	Ferromagnetic	600
Ferroxcube 3 (Mn-An-ferrite powder)	Ferromagnetic	1,500
Mild steel (0.2 C)	Ferromagnetic	2,000
Iron (0.2 impurity)	Ferromagnetic	5,000
Silicon Iron (4 Si)	Ferromagnetic	7,000
78 Permalloy (78.5 Ni)	Ferromagnetic	100,000
Mumetal (75 Ni, 5 Cu, 2 Cr)	Ferromagnetic	100,000
Purified iron (0.05 impurity)	Ferromagnetic	200,000
Superalloy (5 Mo, 79 Ni)§	Ferromagnetic	1,000,000

†By definition.
‡Percentage composition. Remainder is iron and impurities.
§Used in transformer applications with continuous tape-wound (gapless) cores.

A *superparamagnetic* material consists of ferromagnetic particles suspended in a dielectric (plastic) binder or matrix. Each particle may contain many magnetic domains, but exchange forces cannot penetrate to adjacent particles. With the particles suspended in a thin plastic tape it is possible to change the state of magnetization abruptly in a very small tape distance so that a tape can store large amounts of information in magnetic form in convenient lengths. Such tapes are widely used in audio, video, and data recording systems.

The eight types of magnetic materials we have mentioned are summarized in Table 7-4.

In Table 7-5, the relative permeabilities μ_r are listed for a number of substances. The substances are arranged in order of increasing permeability, and they are also classified as to group type. The value for the ferromagnetic materials is the maximum relative permeability.

7-14 MAGNETIC DIPOLES AND MAGNETIZATION

As we have discussed, a bar magnet and a loop experience equal torques in a magnetic field **B** provided their magnetic moments are equal, that is,

$$Q_m L = IA \qquad (\text{A m}^2) \qquad \textit{Magnetic moments} \tag{1}$$

where Q_m = pole strength of magnet, A m
L = pole separation of magnet, m
I = loop current, A
A = loop area, m^2

Thus, for a bar magnet and a loop, as in Fig. 7-15, not only will the torques be equal if (1) is fulfilled but the magnetic fields of the two will be identical at large distances from them.

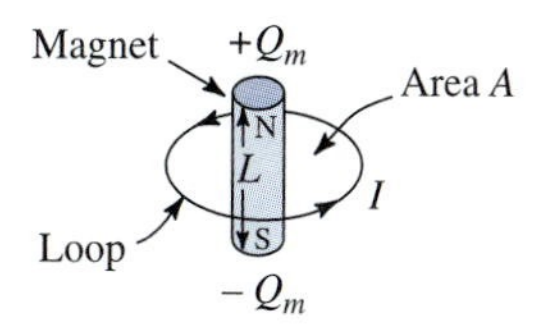

FIGURE 7-15
Bar magnet of moment $Q_m L$ and equivalent current loop of moment IA.

It was Ampère's theory that the strong magnetic effects of an iron bar occur when the magnetic moments of large numbers of the iron atoms are aligned so that their moments add. Whether we regard an iron atom as a microscopic bar magnet or as a tiny current loop is not important when we are dealing with regions containing vast numbers of them. It suffices to describe them by their magnetic moment which can be expressed either as $Q_m L$ or IA.

Consider a uniformly magnetized iron rod as shown in cross-section in Fig. 7-16. The atomic magnets are all oriented the same way and are uniformly distributed with the north pole of one atomic magnet so close to the south pole of the next magnet that the poles of adjacent magnets annul each other's effects everywhere except at the ends of the rod, leaving a surface layer of north poles on one end and south poles on the other. The situation is analogous to that for a uniformly polarized dielectric slab with the effects of its atomic electric dipoles canceling throughout the interior, but resulting in layers of opposite electric (bound) charges on the two surfaces of the slab.

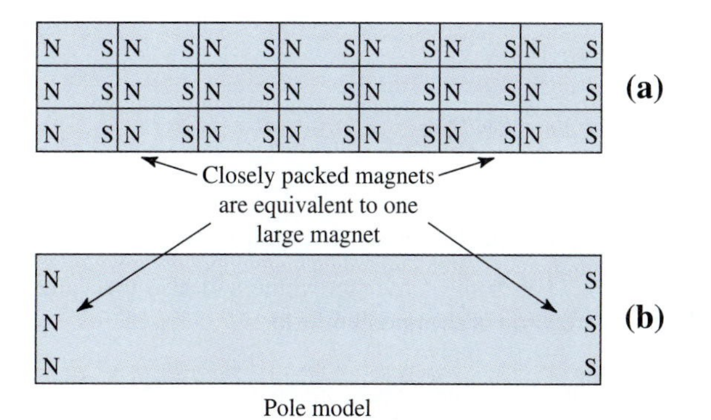

FIGURE 7-16
(*a*) The effects of the atomic dipoles of a uniformly magnetized rod annul each other except at the ends of the rod, leaving a layer of north poles at one end and south poles at the other end, as in (*b*) (pole model).

The effect of the atomic magnets can be described by a quantity called the *magnetization M,* defined as the magnetic dipole moment per unit volume, analogous to the polarization P (for the electric dipole moment per unit volume in the dielectric case). Thus, the magnetization M is

$$M = \frac{m}{v} = \frac{Q_m L}{v} \qquad (\text{A m}^{-1}) \tag{2}$$

where $m = Q_m L =$ the magnetic dipole moment in volume v, A m^2. Magnetization has the dimensions of both magnetic dipole moment per volume and of magnetic pole strength per area ($IL^2/L^3 = I/L$). It is expressed in amperes per meter (A m^{-1}).

The value of M in (2) is an average for the volume v. To define M at a point, it is convenient to assume that the iron rod has a continuous distribution of infinitesimal magnetic dipoles, i.e., a continuous magnetization, whereas the dipoles actually are of discrete, finite size. Nevertheless, the assumption of continuous magnetization leads to no appreciable error provided that we restrict our attention to volumes containing many magnetic dipoles. Then, assuming continuous magnetization, the value of M at a point can be defined as the net dipole moment m of a small volume Δv divided by the volume with the limit taken as Δv shrinks to zero around the point. Thus

$$M = \lim_{\Delta v \to 0} \frac{m}{\Delta v} \qquad (\text{A m}^{-1}) \tag{3}$$

If M is everywhere in the same direction and is known as a function of position, the total magnetic moment of the rod is given by

$$m = \int_v M\,dv \qquad (\text{A m}^2) \tag{4}$$

where the integration is carried out over the volume of the rod.

Example 7-7. Magnetized rod. If the uniformly magnetized rod of Fig. 7-16 has N' elemental magnetic dipoles of moment Δm, find the magnetization of the bar.

Solution. From (2), the magnetization is

$$M = \frac{N'}{v}\Delta m = N''\Delta m$$

where $M =$ magnetization, A m^{-1}
$N'' = N'/v =$ elemental dipoles N' per unit volume v, number m^{-3}

In this case the magnetization M is both an average value and also the value anywhere in the rod since the magnetization is assumed uniform.

Problem 7-14-1. Rod and loop. A long rod-shaped permanent magnet has a cross-section of 2.5×10^{-5} m^2 and a magnetization of $\mathbf{M} = \hat{\mathbf{z}}500$ A/m. A single-turn loop of 2×10^{-4} m^2 area carrying a current of 10 A is situated 0.15 m from one end of the magnet. Find the maximum possible torque on the loop. *Ans.* 1.11×10^{-10} N m.

7-15 UNIFORMLY MAGNETIZED ROD AND EQUIVALENT AIR-CORE SOLENOID

In Sec. 7-14 magnetization was explained by means of magnets with north and south poles (*pole model*). Now the magnetization of a bar magnet will be discussed in terms of the atomic current loops (*current model*) of Fig. 7-17 with one loop in place of each tiny magnet of Fig. 7-16 and with the moment IA of each loop equal to the moment $Q_m L$ of each magnet. Assuming that there are n loops in a single cross-section of the rod (as in the end view in Fig. 7-17), we have

$$nA' = A \tag{1}$$

where A' = area of elemental loop, m^2
A = cross-sectional area of rod, m^2

Further, let us assume that there are N such sets of loops in the length of the rod (see the side view in Fig. 7-17). Then

$$nN = N' \tag{2}$$

where n = number of loops in a cross-section of rod
N = number of such sets of loops
N' = total number of loops in rod

It follows that the magnetization M of the rod is given by

$$M = \frac{m}{v} = \frac{N'IA'}{LA} = \frac{NI}{L}\frac{nA'}{A} = \frac{NI}{L} = K' \tag{3}$$

where K' = equivalent sheet current density on outside surface of rod, A m^{-1}
L = length of rod, m.

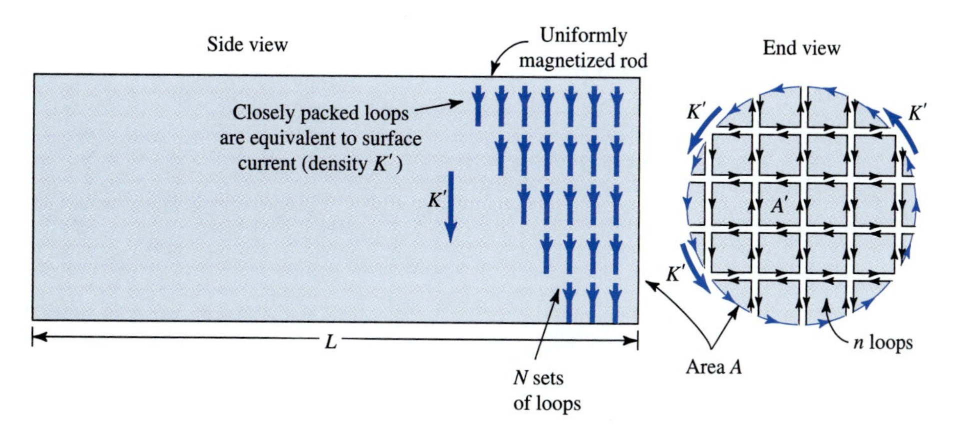

FIGURE 7-17
Effects of atomic current loops of a uniformly magnetized rod annul each other except on the cylindrical surface of the rod, resulting in a current sheet around the rod (current model).

Referring to the end view of the rod in Fig. 7-17, note that there are equal and oppositely directed currents wherever loops are adjacent, so that the currents have no net effect with the exception of the currents at the periphery of the rod. As a result there is an equivalent current sheet flowing around the rod, as suggested in Figs. 7-17 and 7-18*a*. This sheet has a linear current density K' (A m^{-1}). Although the sets of current loops are shown for clarity in Fig. 7-18 with a large spacing, the actual spacing is of atomic dimensions, so that macroscopically we can assume that the current sheet is continuous.

This type of a current sheet is effectively what we also have in the case of an air-core solenoid with many turns of fine wire, as in Fig. 7-18*b* (see also Fig. 2-45). The actual sheet-current density K for the solenoid is

$$K = \frac{NI}{L} \qquad (\text{A m}^{-1}) \tag{4}$$

where N = number of turns in solenoid, dimensionless
I = current through each turn, A
L = length of solenoid, m

The sheet-current density K is expressed in amperes per meter.

If the solenoid of Fig. 7-18b is the same length and diameter as the rod of Fig. 7-18a, and if $K = K'$, then externally the solenoid is the magnetic equivalent of the rod. In air $B = \mu_0 H$, and we have at the center of the solenoid that

$$B = \mu_0 \frac{NI}{L} = \mu_0 K = \mu_0 H \qquad (\text{T}) \qquad \textbf{\textit{Solenoid}} \tag{5}$$

At the center of the rod

$$B = \mu_0 K' = \mu_0 M \qquad (\text{T}) \qquad \textbf{\textit{Rod}} \tag{6}$$

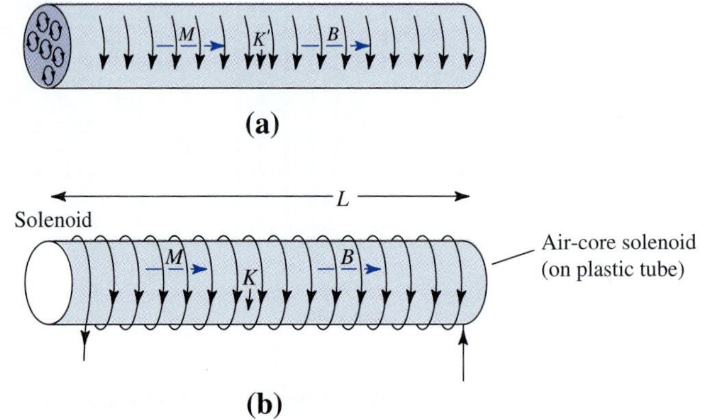

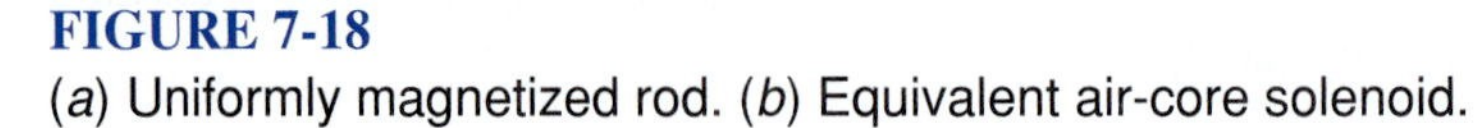

FIGURE 7-18
(*a*) Uniformly magnetized rod. (*b*) Equivalent air-core solenoid.

From (5) we have $H = K$, and from (6) we have $M = K'$. These are scalar relations. Vectorially we note that $\mathbf{H}$ is perpendicular to $\mathbf{K}$ and $\mathbf{M}$ is perpendicular to $\mathbf{K}'$.

The more general vector relations are

$$\boxed{\mathbf{K} = \hat{\mathbf{n}} \times \mathbf{H} \quad \text{and} \quad \mathbf{K}' = \mathbf{M} \times \hat{\mathbf{n}}} \tag{7}$$

where $\hat{\mathbf{n}}$ is the unit vector normal to the plane containing the field vectors.

No material medium need be present for energy to be stored by a magnetic field. Thus, energy can be present in the magnetic field of an air-core coil (or vacuum-core coil). However, if there is ferrous material (iron) inside the coil, the energy density is increased in proportion to the permeability μ,

where $$\mu = \mu_0 + \frac{\mu_0\mathbf{M}}{\mathbf{H}} \quad \text{or} \quad \mu\mathbf{H} = \mathbf{B} = \mu_0\mathbf{H} + \mu_0\mathbf{M}$$

Therefore, the energy density

$$\begin{aligned} w_m &= \tfrac{1}{2}\mu H^2 = \tfrac{1}{2}BH \\ &= \tfrac{1}{2}(\mu_0 H + \mu_o M)H \\ &= \tfrac{1}{2}(\mu_0 H^2 + \mu_0 MH) \qquad (\text{J m}^{-3}) \end{aligned}$$

where $\tfrac{1}{2}\mu_0 H^2$ = energy density in vacuum, J m^{-3}
$\tfrac{1}{2}\mu_0 MH$ = energy density in ferrous medium, J m^{-3}

The additional energy in the iron ($\tfrac{1}{2}\mu_0 MH$) is due to its magnetization. With no iron, there is no magnetization ($M = 0$), and the energy density is simply $\tfrac{1}{2}\mu_0 H^2$. With the iron, it is increased by $\tfrac{1}{2}\mu_0 MH$ to a total energy density of $\tfrac{1}{2}\mu_0\mu_r H^2 = \tfrac{1}{2}\mu H^2$. These relations are very similar in form to those for the electric energy density in a capacitor.

Example 7-8. Magnet energy. A bar magnet 0.2 m long with 10-mm diameter has a moment of 500 A m^2. Find (*a*) total energy and (*b*) energy density at the center of the bar.

Solution

$$w_m = \tfrac{1}{2}\mu_0(m/AL)^2 = \tfrac{1}{2}(4\pi \times 10^{-7})(500/\text{vol.})^2$$

$$\text{Vol.} = \pi(0.005)^2(0.2) = 15.7 \times 10^{-6}\ \text{m}^3$$

$$w_m = (2\pi \times 10^{-7})\left(\frac{500}{15.7 \times 10^{-6}}\right)^2 = 637 \times 10^6\ \text{J/m}^3 \quad \textit{Ans.} \quad (b)$$

$$W_m = w_m \times \text{vol.} = (637 \times 10^6)(15.7 \times 10^{-6}) = 10\ \text{kJ} \quad \textit{Ans.} \quad (a)$$

Problem 7-15-1. Energy density. The magnetic flux density around a long straight wire parallel to the z axis is given by $\mathbf{B} = \hat{\boldsymbol{\phi}}10/r$ T. Find the magnetic energy density at a distance of 2 m from the wire. *Ans.* 9.95 MJ/m^3.

Problem 7-15-2. Energy in solenoid. A solenoid of 1000 turns has length of 600 mm and a diameter of 40 mm. If the solenoid current is 2 A, find (*a*) the energy density of the center of the solenoid and (*b*) the total energy in the solenoid. *Ans.* (*a*) 6.98 J/m^3, (*b*) 5.26 mJ.

7-16 BOUNDARY RELATIONS

In a single medium the magnetic field is continuous. That is, the field, if not constant, changes only by an infinitesimal amount in an infinitesimal distance. However, at the boundary between two different media, the magnetic field may change abruptly both in magnitude and direction.

It is convenient to analyze this boundary situation in two parts, considering separately the relation of fields *normal* to the boundary and *tangent* to the boundary.

Taking up first the relation of fields normal to the boundary, consider two media of permeabilities μ_1 and μ_2 separated by the xy plane, as shown in Fig. 7-19. Suppose that an imaginary box is constructed, half in each medium, of area $\Delta x \Delta y$ and height Δz. Let B_{n1} be the average component of **B** normal to the top of the box in medium 1, and B_{n2} the average component of **B** normal to the bottom of the box in medium 2. B_{n1} is an outward normal (positive), while B_{n2} is an inward normal (negative). By Gauss's law for magnetic fields, the total magnetic flux over a closed surface is zero. In other words, the integral of the outward normal components of **B** over a closed surface is zero. By making the height Δz of the box approach zero, the contribution of the sides of the box to the surface integral becomes zero even though there may be finite components of **B** normal to the sides. Therefore the surface integral reduces to $B_{n1}\,\Delta x\,\Delta y - B_{n2}\,\Delta x\,\Delta y = 0$ or

$$B_{n1} = B_{n2} \qquad \textbf{\textit{Normal}}\ \mathbf{B} \tag{1}$$

According to (1), *the normal component of the flux density* **B** *is continuous across the boundary between two media.*

Turning now to the relation for magnetic fields tangent to the boundary, let two media of permeabilities μ_1 and μ_2 be separated by a plane boundary, as in Fig. 7-20. Consider a rectangular path, half in each medium, of length Δx parallel to the boundary and of length Δy normal to the boundary. Let the average value of **H** tangent to the boundary in medium 1 be H_{t1} and the average value of **H** tangent to boundary in medium 2 be H_{t2}. According to Ampère's law, the integral of **H** around a closed path

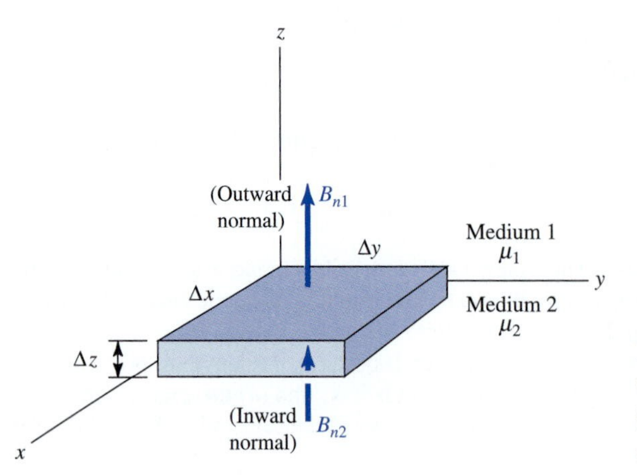

FIGURE 7-19
Construction for showing that the normal components of **B** are continuous.

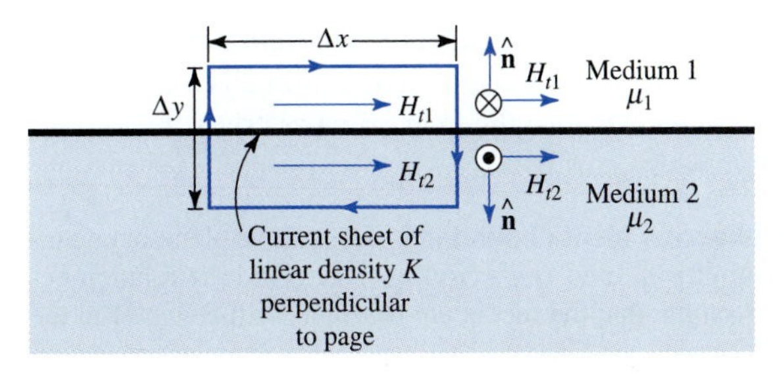

FIGURE 7-20 Construction for showing that the tangential components of **H** differ by the current density K on the surface.

equals the current I enclosed. By making the path length Δy approach zero, the contribution of these segments of the path become zero even though a finite field may exist normal to the boundary. The line integral then reduces to $H_{t1}\Delta x - H_{t2}\Delta x = I$, or

$$H_{t1} - H_{t2} = \frac{I}{\Delta x} = K \qquad (\text{A m}^{-1}) \qquad \textit{Tangential } \mathbf{H} \tag{2}$$

where K is the linear density of any current flowing in an infinitesimally thin sheet at the surface.†

According to (2), *the change in the tangential component of* **H** *across a boundary is equal in magnitude to the sheet-current density K on the boundary.* It is to be noted that **K** is normal to **H**; that is, the direction of the current sheet in Fig. 7-20 is normal to the page. This is expressed by the vector relation

$$\mathbf{K} = \hat{\mathbf{n}} \times (\mathbf{H}_{t1} - \mathbf{H}_{t2}) \tag{3}$$

where $\hat{\mathbf{n}}$ = unit vector normal to boundary, dimensionless
$\mathbf{H}_{t1}$ = magnetic field tangent to boundary on side 1, A m^{-1}
$\mathbf{H}_{t2}$ = magnetic field tangent to boundary on side 2, A m^{-1}
$\mathbf{K}$ = sheet-current density at boundary, A m^{-1}

If the field below the boundary is zero ($H_{t2} = 0$), (3) indicates that the current **K** related to H_{t1} will be into the page in Fig. 7-20 while if the field above the boundary is zero ($H_{t1} = 0$), the current **K** related to H_{t2} will be out of the page.

If $K = 0$, then

$$H_{t1} = H_{t2} \tag{4}$$

According to (4), *the tangential components of* **H** *are continuous across the boundary between two media, provided the boundary has no current sheet.*

If medium 1 is a nonconductor, and if $H_{t2} = 0$,

$$H_{t1} = K_2 \tag{5}$$

---

†If J is the current density in amperes per square meter in a thin sheet of thickness $\Delta y'$, then K is defined by

$$K = J\Delta y' \qquad (\text{A m}^{-1})$$

where $J \to \infty$ as $\Delta y' \to 0$.

where K_2 is the sheet-current density in amperes per meter in medium 2 at the boundary (into the page in Fig. 7-20). When medium 1 is air and medium 2 is a conductor, (5) is approximated at high frequencies because the skin effect restricts the current in the conductor to a very thin layer at its surface.

Example 7-9. Magnetic field across media boundary. Consider a plane boundary between two media of permeability μ_1 and μ_2, as in Fig. 7-21. Find the relation between the angles α_1 and α_2. Assume that the media are isotropic with **B** and **H** in the same direction.

Solution. From the boundary relations,

$$B_{n1} = B_{n2} \qquad \text{and} \qquad H_{t1} = H_{t2} \tag{6}$$

From Fig. 7-21,

$$B_{n1} = B_1 \cos\alpha_1 \qquad \text{and} \qquad B_{n2} = B_2 \cos\alpha_2 \tag{7}$$

$$H_{t1} = H_1 \sin\alpha_1 \qquad \text{and} \qquad H_{t2} = H_2 \sin\alpha_2 \tag{8}$$

where B_1 = magnitude of **B** in medium 1
B_2 = magnitude of **B** in medium 2
H_1 = magnitude of **H** in medium 1
H_2 = magnitude of **H** in medium 2

Substituting (7) and (8) into (6) and dividing yields

$$\frac{\tan\alpha_1}{\tan\alpha_2} = \frac{\mu_1}{\mu_2} = \frac{\mu_{r1}}{\mu_{r2}} \tag{9}$$

where μ_{r1} = relative permeability of medium 1, dimensionless
μ_{r2} = relative permeability of medium 2, dimensionless

Equation (9) gives the relation between the angles α_1, and α_2 for **B** and **H** lines at the boundary between two media† *Ans.*

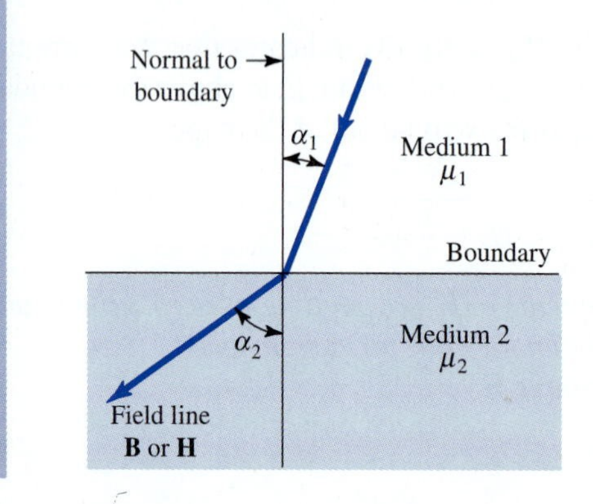

FIGURE 7-21
Boundary between two media of different permeability, showing change in direction of magnetic field line.

†This relation applies only if **B** and **H** have the same direction (μ a scalar). In the absence of magnetization, as in air, **B** and **H** have the same direction. When magnetization is present, as in a soft-iron electromagnet, **B** and **H** also tend to have the same direction. However, this is *not* the situation in a permanent magnet.

Example 7-10. B at air-iron boundary. Referring to Fig. 7-22, let medium 1 be air ($\mu_r = 1$) and medium 2 be soft iron with a relative permeability of 7000. (*a*) If **B** in the iron is incident normal on the boundary ($\alpha_2 = 0$), find α_1. (*b*) If **B** in the iron is nearly tangent to the surface at an angle $\alpha_2 = 85°$, find α_1,

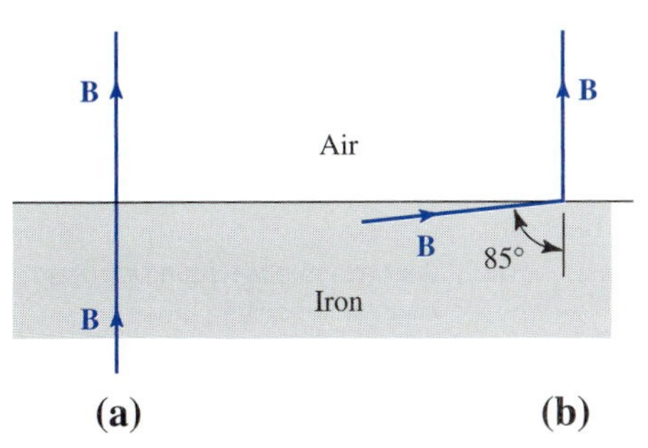

FIGURE 7-22
B lines at air-iron boundary.

Solution. (*a*) From (9),

$$\tan\alpha_1 = \frac{\mu_{r1}}{\mu_{r2}}\tan\alpha_2 = \frac{1}{7000}\tan\alpha_2 \tag{10}$$

For α_2 equal to zero, α_1 is also equal to zero. Thus the answer to (*a*) is that the **B** line in air is also normal to the boundary (see Fig. 7-22*a*).

(*b*) When $\alpha_2 = 85°$, we have, from (10), that $\tan\alpha_1 = 0.0016$, or $\alpha_1 = 0.1°$. Thus, the direction of **B** in air is almost normal to the boundary (within 0.1°) even though its direction in the iron is nearly tangent to the boundary (within 5°) (see Fig. 7-22*b*). Accordingly, for many practical purposes the *direction of* **B** *or* **H** *in air, or other medium of low relative permeability, may be taken as normal to the boundary of a medium having a high relative permeability*. This property is reminiscent of the one for **E** or **D** at the boundary of a conductor.

Problem 7-16-1. Boundary conditions. Two cavities are cut in a ferromagnetic medium ($\mu_r = 300$) of large extent. Cavity 1 is a thin disk-shaped cavity with flat faces perpendicular to the direction **B** in the ferromagnetic medium. Cavity 2 is a long needle-shaped cavity with its axis parallel to **B.** The cavities are filled with air. If $B = 1$ T, what is the magnitude of **H** at (*a*) the center of cavity 1 and (*b*) the center of cavity 2? *Ans.* (*a*) 796 kA/m, (*b*) 2.65 kA/m.

Table 7-6 (on next page) summarizes the boundary relations for magnetic fields.

7-17 FERROMAGNETISM

Ferromagnetic materials exhibit strong magnetic effects and are the most important magnetic substances. The permeability of these materials is not a constant but is a function both of the applied field and of the previous magnetic history of the specimen. In view of the variable nature of the permeability of ferromagnetic materials, special consideration of their properties is needed.

TABLE 7-6
Boundary relations for magnetic fields†

Field component	Boundary relation		Condition
Normal	$B_{n1} = B_{n2}$	(1)	Any two media
Normal	$\mu_{r1}H_{n1} = \mu_{r2}H_{n2}$	(2)	Any two media
Tangential	$H_{t1} - H_{t2} = K$	(3)	Any two media with current sheet of infinitesimal thickness at boundary.
	$\hat{\mathbf{n}} \times (\mathbf{H}_{t1} - \mathbf{H}_{t2}) = \mathbf{K}$	(3*a*)	
Tangential	$H_{t1} = H_{t2}$	(4)	Any two media with no current sheet at boundary.
Tangential	$H_{t1} = K_2$	(5)	$H_{t2} = 0$; also medium 2 has a current sheet of infinitesimal thickness at boundary; H_{t1} and K_2 are normal to each other.

†These relations apply for both static and time-varying fields.

In ferromagnetic substances, as already explained, the atomic dipoles tend to align in the same direction over regions, or *domains,* containing many atoms. The size of a domain varies but usually contains millions of atoms. In some substances the shape appears to be like a long, slender rod with a transverse dimension of microscopic size but lengths of the order of a millimeter or so. Thus, a domain acts like a small, but not atomically small, bar magnet.

In an unmagnetized iron crystal the domains are parallel to a direction of "easy magnetization," but equal numbers have north poles pointing one way as the other, so the external field of the crystal is zero. In an iron crystal there are six directions of easy magnetization. That is, there is a positive and negative direction along each of the three mutually perpendicular crystal axes (Fig. 7-23). Therefore the polarity of the domains in an unmagnetized iron crystal may be as suggested by the highly

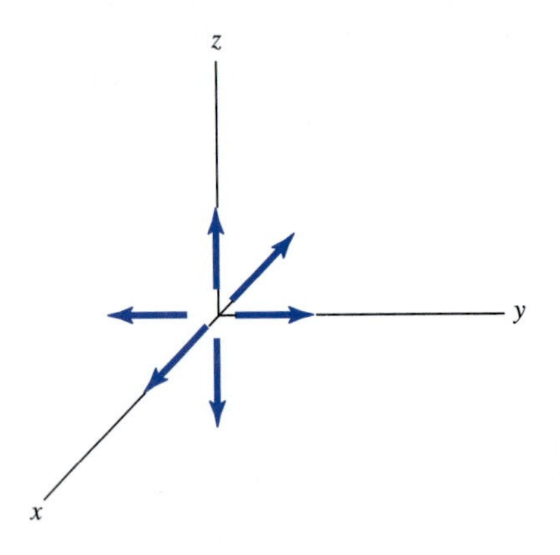

FIGURE 7-23
Six directions of easy magnetization along the three crystal axes in an iron crystal.

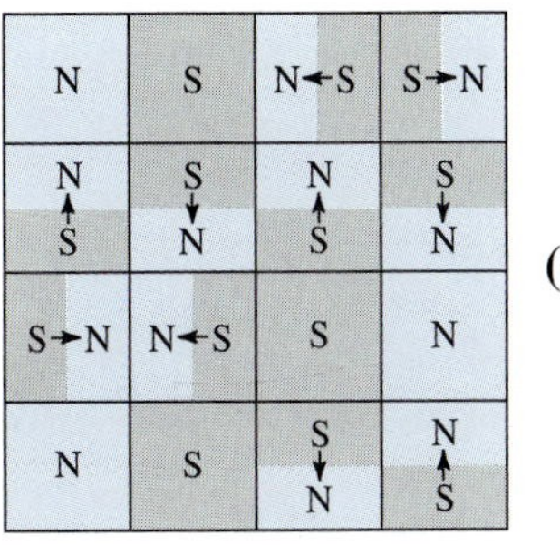

(a)

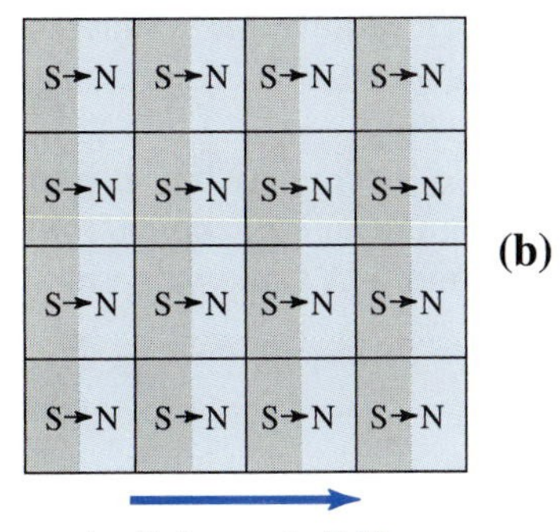

(b)

FIGURE 7-24
(*a*) Domain polarity in an unmagnetized iron crystal. Arrows indicate direction of magnetization. A single N represents a domain with a north pole pointing out of the page; a single S represents a domain with a south pole pointing out of the page. (*b*) Condition after crystal is saturated by a magnetic field directed to the right.

schematic diagram of Fig. 7-24*a*. A single N represents a domain with a north pole pointing out of the page and a single S a domain with a south pole pointing out of the page.

Now if the crystal is placed in a magnetic field parallel to one of the directions of easy magnetization, the domains with polarity opposing or perpendicular to the field become unstable and a few of them may rotate so that they have the same direction as the field. With further increase of the field more domains change over, each as an individual unit, until when all the domains are in the same direction, *magnetic saturation* is reached, as suggested by Fig. 7-24*b*. The crystal is then magnetized to a maximum extent. If the majority of the domains retain their directions after the applied field is removed, the specimen is said to be *permanently magnetized.* Heat and mechanical shock tend to return the crystal to the original unmagnetized state, and if the temperature is raised sufficiently high, the domains themselves are demagnetized (exchange coupling disappears) and the substance changes from ferromagnetic to paramagnetic. For iron this transition temperature, or Curie point, is 770°C. The residual magnetism is so weak compared to the ferromagnetic case that the material is usually considered to be unmagnetized.

Magnetization which appears only in the presence of an applied field may be spoken of as *induced magnetization*, as distinguished from *permanent magnetization*, which is present in the absence of an applied field.

7-18 MAGNETIZATION CURVES

The permeability μ of a substance is given by

$$\mu = \frac{B}{H} = \mu_0 \mu_r$$

where B = magnitude of flux density, T
H = magnitude of field $\mathbf{H}$, A m^{-1}
μ_0 = permeability of vacuum = 400π nH m^{-1}
μ_r = relative permeability of substance, dimensionless

To illustrate the relation of B to H, a graph showing B (ordinate) as a function of H (abscissa) is used. The line or curve showing B as a function of H on such a BH chart is called a *magnetization curve.*

To measure a magnetization curve for an iron sample, a ring may be cut from the sample. A uniform winding is placed over the ring, forming an iron-cored toroid, as in Fig. 7-25. If the number of ampere-turns in the toroid is NI, the value of H applied to the ring is

$$H = \frac{NI}{l} \qquad \text{A-turns m}^{-1}$$

where $l = 2\pi R$ and R = mean radius of the ring or toroid.

This value of H applied to the ring may be called the *magnetizing force.* Hence, in general, H is sometimes called by this name. The flux density B in the ring may be regarded as the result of the applied field H and is measured by placing another (secondary) coil over the ring, as in Fig. 7-25, and connecting it to a fluxmeter.† For a given change in H, produced by changing the toroid current I, one measures the change in magnetic flux ψ_m through the ring. The change in the flux density B in the

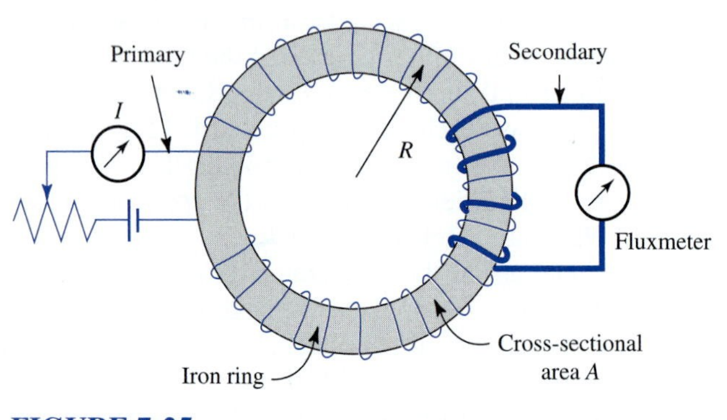

FIGURE 7-25
Rowland ring method of obtaining magnetization curves.

†The *fluxmeter* operates on the emf induced in the secondary when the magnetic flux changes.

ring is then equal to ψ_m/A, where A is the cross-sectional area of the ring. This ring method of measuring magnetization curves was used by Rowland in 1873.

A typical magnetization curve for a ferromagnetic material is shown by the solid curve in Fig. 7-26*a*. The specimen in this case was initially unmagnetized, and the change in B noted as H was increased from 0. By way of comparison, four dashed lines are also shown in Fig. 7-26*a*, corresponding to constant relative permeabilities μ_r, of 1, 10, 100, and 1000. The relative permeability at any point on the

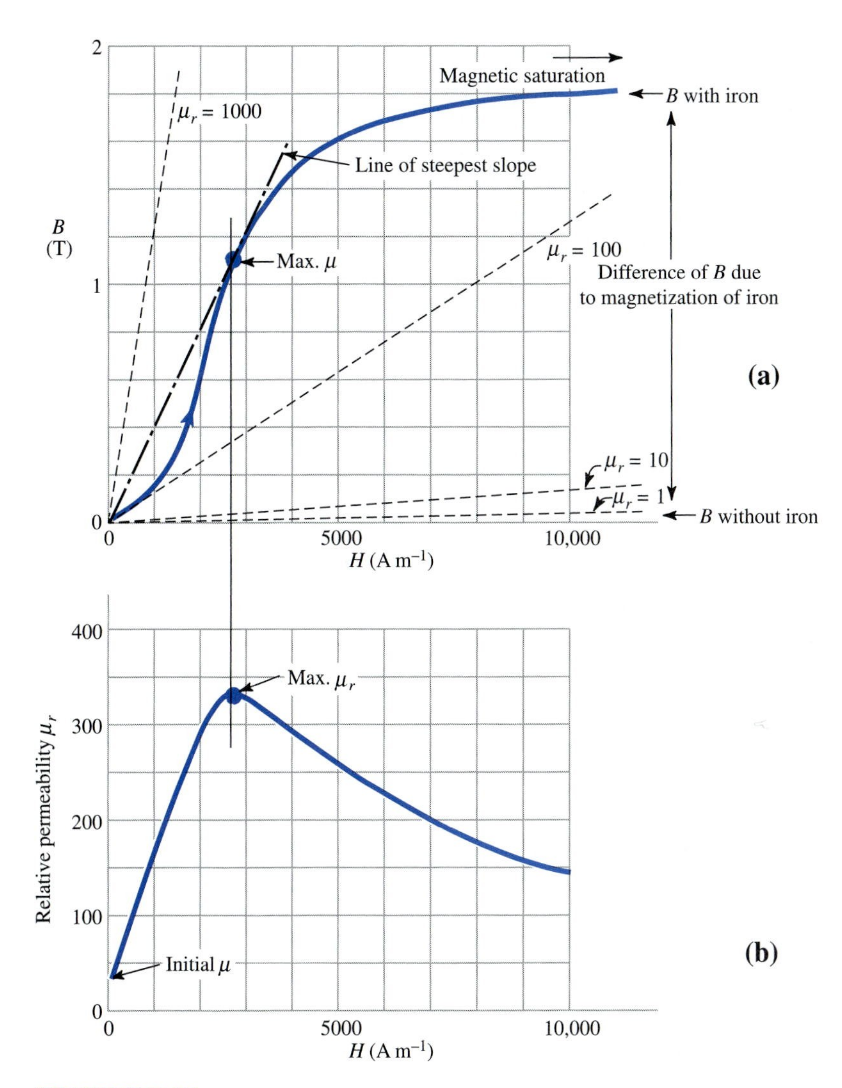

FIGURE 7-26
(*a*) Typical initial magnetization curve. (*b*) Corresponding relation of relative permeability μ_r to applied field **H**.

magnetization curve is given by

$$\mu_r = \frac{B}{\mu_0 H} = 7.96 \times 10^5 \frac{B}{H} \qquad \text{(dimensionless)}$$

where B = ordinate of the point, T
H = abscissa of the point, A m^{-1}

Note that μ_r is not proportional to the slope of the curve (dB/dH) but to the ratio B/H.

A graph of the relative permeability μ_r as a function of the applied field H, corresponding to the magnetization curve in Fig. 7-26*a*, is presented in Fig. 7-26*b*. The maximum relative permeability, and therefore the *maximum permeability,* is at the point on the magnetization curve with the largest ratio of B to H. This is designated "Max. μ." It occurs at the point of tangency with the straight line of steepest slope that passes through the origin and also intersects the magnetization curve (dash-dot line in Fig. 7 -26*a*). The magnetization curve for air or vacuum is given by the dashed line for $\mu_r = 1$ (almost coincident with the H axis) in Fig. 7-26*a.*

The magnetization curve shown in Fig. 7-26*a* is an *initial-magnetization curve.* That is, the material is completely demagnetized before the field H is applied. As H is increased, the value of B rises rapidly at first and then more slowly. At sufficiently high values of H the curve tends to become flat, as suggested by Fig. 7-26*a.* This condition is called *magnetic saturation.*

Although the B/H ratio (or permeability) has significance for the *initial magnetization curve*, and the *normal magnetization curve* discussed later, this is not the case for magnetization loops and some other magnetization curves we consider presently, where the ratio B/H may become infinite.

The magnetization curve starting at the origin has a finite slope giving an *initial permeability*. Therefore, the relative-permeability curve in Fig. 7-26*b* starts with a finite permeability for infinitesimal fields.

The initial-magnetization curve may be divided into two sections: (1) the steep section and (2) the flat section, the point P of division being on the upper bend of

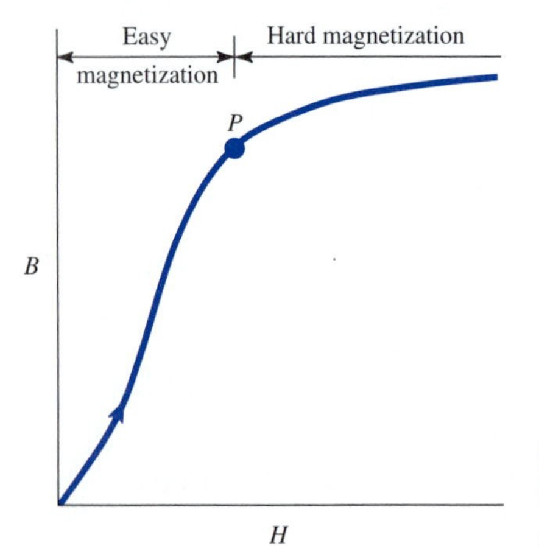

FIGURE 7-27
Regions of easy and hard magnetization of initial-magnetization curve.

the curve (Fig. 7-27). The steep section corresponds to the condition of *easy magnetization*, while the flat section corresponds to the condition of difficult, or *hard, magnetization.*

Ordinarily a piece of iron consists not of a single crystal but of an aggregate of small crystal fragments with axes oriented at random. The situation in a small piece of iron may be represented schematically as in Fig. 7-28. Here a number of crystal fragments are shown, each with a number of magnetic domains, represented in most cases by a small square. The boundaries between crystal fragments are indicated by the heavy lines, and domain boundaries by the light lines, which also indicate the

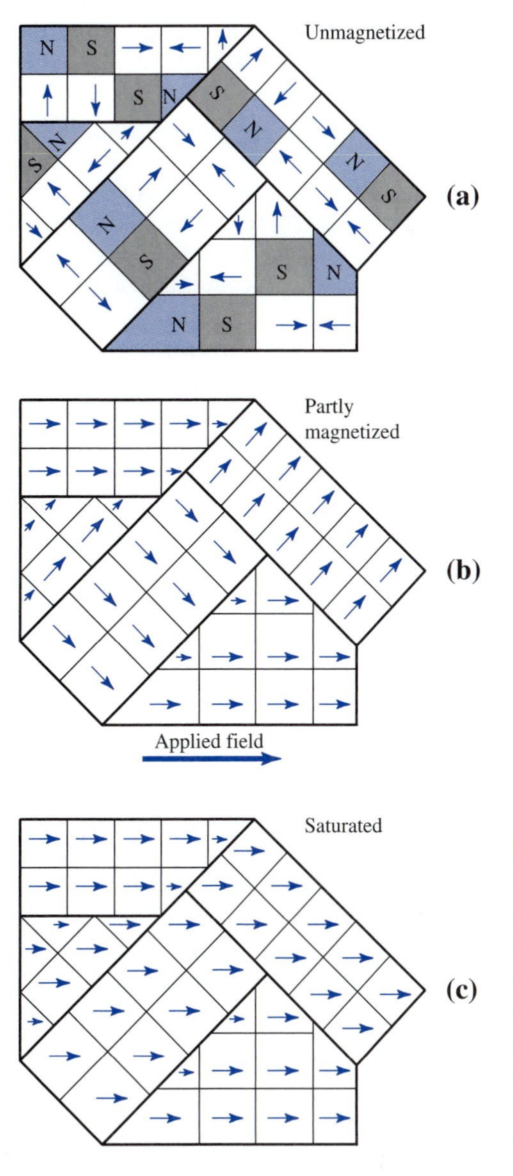

FIGURE 7-28
Successive stages of magnetization of a polycrystalline specimen with increasing field. Arrows indicate direction of magnetization of domains. An N represents a domain with a north pole pointing out of the page; an S represents a domain with a south pole pointing out of the page.

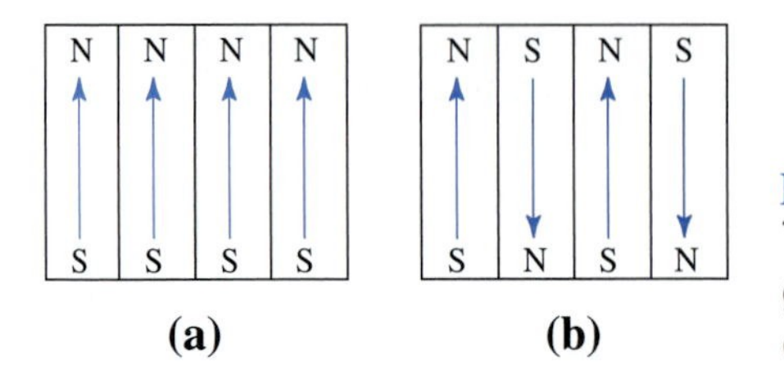

FIGURE 7-29
Total energy is reduced when domains go from aligned condition (*a*) to oppositely oriented condition (*b*).

direction of the crystal axes. In Fig. 7-28*a*, not only is the piece of iron unmagnetized, but also the individual crystal fragments are unmagnetized. The domains in each crystal are magnetized along the directions of easy magnetization, i.e., along the three crystal axes. However, the polarity of adjacent domains is opposite, so that the total magnetization of each crystal is negligible.

With the application of a magnetic field **H** in the direction indicated by the arrow (Fig. 7-28*b*), some domains with polarities opposed to or perpendicular to the applied field become unstable and rotate quickly to another direction of easy magnetization in the same direction as the field, or more nearly so. These changes take place on the steep (easy) part of the magnetization curve. The result, after all domains have changed over, is as suggested in Fig. 7-28*b*. This condition corresponds roughly to that at the point P on the magnetization curve (Fig. 7-27).

With further increase in the applied field, the direction of magnetization of the domains not already parallel to the field is rotated gradually toward the direction of **H**. This increase in magnetization is more difficult, and very high fields may be required to reach saturation, where all domains are magnetized parallel to the field, as indicated in Fig. 7-28*c*. This accounts for the flatness of the upper (hard) part of the magnetization curve.

The tendency of adjacent magnetic domains to be oppositely magnetized can be understood from energy considerations. Thus, when adjacent domains are oriented the same, as in Fig. 7-29*a*, the total energy is increased. When all domains

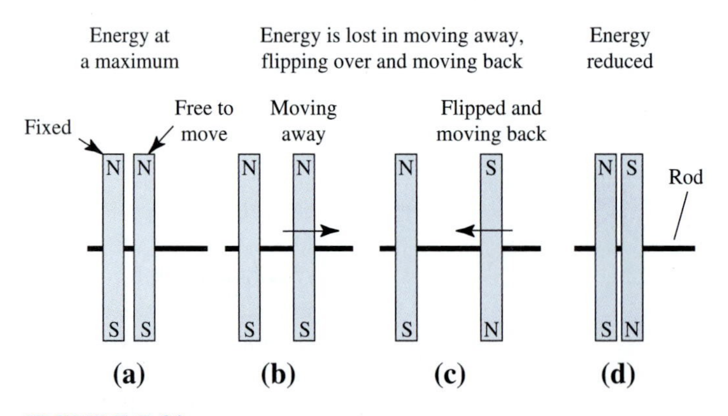

FIGURE 7-30
Experiment of bar magnet which slides away from fixed magnet, flips over, and comes back with reduced energy.

are oppositely oriented, as in Fig. 7-29*b*, the energy is decreased. The situation can be illustrated by performing an experiment with two bar magnets arranged to slide easily on a rod. Let both magnets be placed side by side and oriented the same, as in Fig. 7-30*a*. If the left magnet is held but the right magnet is released, it moves to the right, as in Fig. 7-30*b*, since the adjacent like poles repel. As the right magnet moves farther away, it rotates on the rod to the position shown in Fig. 7-30*c*. The opposite poles now attract, and the right magnet moves back to the left until it comes to rest against the left magnet, as in Fig. 7-30*d*. The pair of magnets now has less energy than at the start (Fig 7-30*a*). The decrease in total energy accounts for the work done by the right magnet in moving away, rotating, and moving back.

Example 7-11. Diametrical spoke. A circular iron ring with diametrical spoke has identical coils of N turns carrying 2 A on each half of the ring (Fig. E7-11). The current directions in the coils are such that the mmfs of both coils add. Ring and spoke have a uniform cross-section of 10 cm^2. The mean radius of the ring is 400 mm. The flux in the ring at one of the coils is 100 μWb. The iron characteristics are as in Fig. 7-26. Find (a) N and (b) the flux ψ_m in the spoke.

Solution

$$\psi_m = BA$$

$$B = \frac{\psi_m}{A} = \frac{100 \times 10^{-6}}{10 \times 10^{-4}} = 0.1 \text{ T}$$

From Fig. 7-26, $H_t = 500 \text{ A m}^{-1}$

$$2NI = H_i L = H_i 2\pi r$$

$$N = \frac{H_i 2\pi r}{2I} = \frac{500 \times 2\pi \times 0.4}{2 \times 2} = 314 \text{ turns} \qquad Ans. \qquad (a)$$

$$\psi_m(\text{spoke}) = 0 \qquad Ans. \quad (b)$$

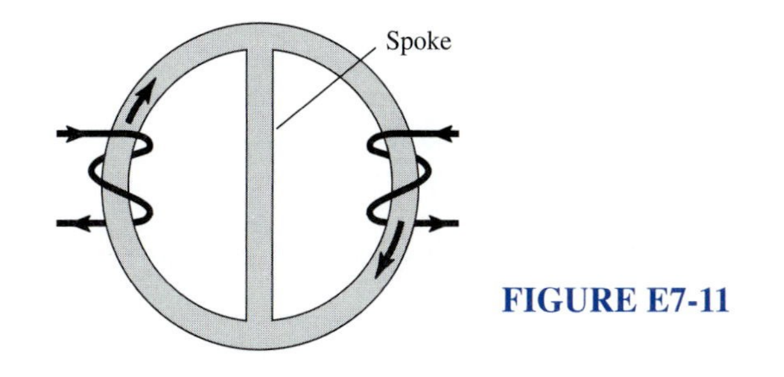

FIGURE E7-11

Problem 7-18-1. Iron ring flux. An iron ring has a mean radius of 300 mm and a uniform cross-section of 10 cm^2. A wire coil of 1000 turns carrying 5 A is on the ring. Find the magnetic flux ψ_m in the ring. The iron characteristics are as given in Fig. 7-26. *Ans.* 1 mWb.

7-19 HYSTERESIS

Starting with an unmagnetized iron specimen in a Rowland ring (Fig. 7-25), let us trace what happens to the flux density B as we change the applied field H. Starting at the origin (at 1) in Fig. 7-31, B follows the initial magnetization curve as H is increased to a value H_m where the curve flattens off and saturation is reached (at 2). Now on reducing H to zero, B does not go to zero but has a *residual flux density* or *remanence* B_r (at 3). If now we reverse H, by reversing the battery polarity in Fig. 7-25, and increase H negatively, B comes to zero at a negative field $-H_c$ called the *coercive force* (at 4). As H is increased still more in the negative direction, the specimen becomes magnetized further with negative polarity, the magnetization at first being easy and then hard as saturation is reached when the field equals $-H_m$ (at 5). Bringing the applied field H to zero again leaves a residual magnetization with flux density $-B_r$ (at 6). Reversing H and increasing it in the positive direction, B comes to zero at a positive field (or coercive force) H_c (at 7). With further increase in H the specimen reaches saturation with the original polarity. When the field equals $+H_m$ it completes (back at 2) our "tour" of what is called a *hysteresis loop.*

The phenomenon which causes B to lag behind H, so that the magnetization curve for increasing and decreasing applied fields is not the same, is called *hysteresis*, and the loop traced out by the magnetization curve, as in Fig. 7-31, is the *hysteresis loop.* If the iron specimen is carried to saturation at both ends of the magnetization curve, as implied in Fig. 7-31, the loop is called the *saturation,* or *major*, *hysteresis loop.* The residual flux density B_r on the saturation loop is called the *retentivity,* and the coercive force H_c on this loop is called the *coercivity.* Thus, the

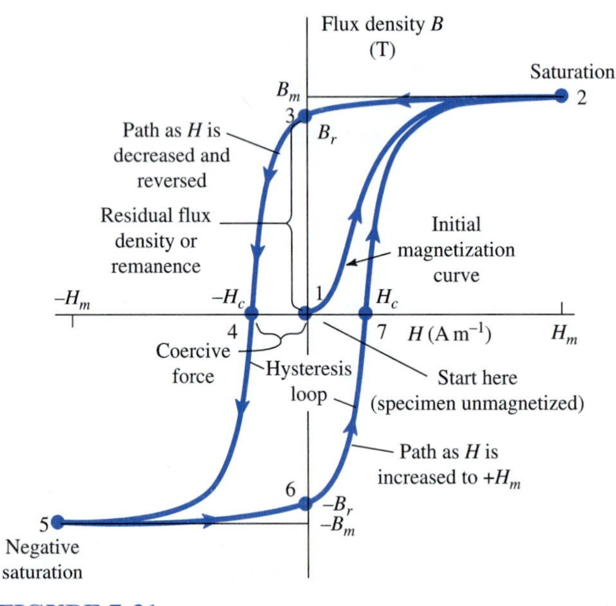

FIGURE 7-31
Hysteresis loop showing path of B as H is changed.

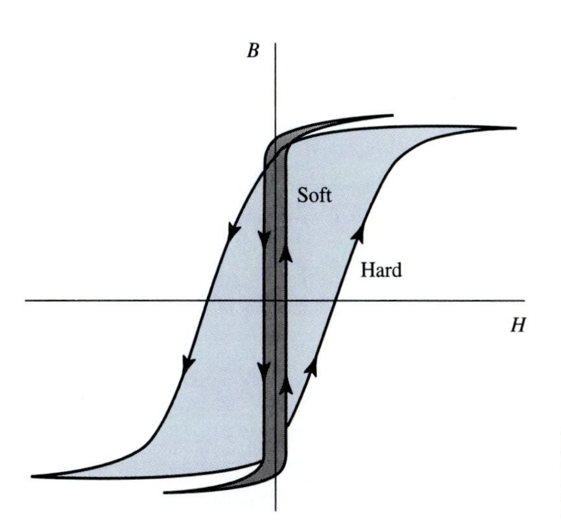

FIGURE 7-32
Hysteresis loops for soft and hard magnetic materials.

retentivity† of a substance is the maximum value which the residual flux density can attain and the coercivity the maximum value which the coercive force can attain. *For a given specimen, no points can be reached on the BH diagram outside the saturation hysteresis loop, but any point inside can.*

In soft, or easily magnetized, materials the hysteresis loop is thin, as suggested in Fig. 7-32, with a small area enclosed. By way of comparison, the hysteresis loop of a hard magnetic material is also shown, the area enclosed in this case being greater.

Suppose we cycle the applied field H over a large enough range to saturate the specimen and then cycle it over successively smaller ranges obtaining a series of hysteresis loops of decreasing size, as in Fig. 7-33. The process may be continued until

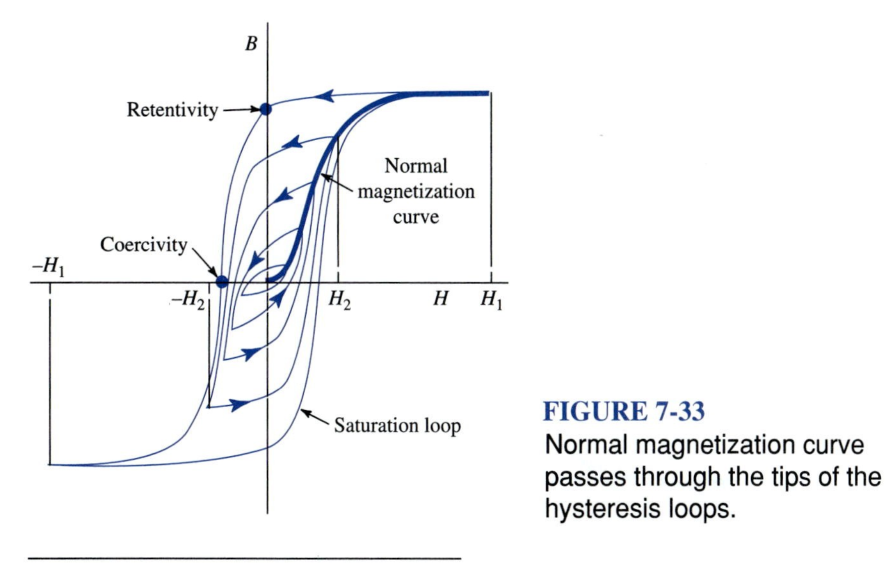

FIGURE 7-33
Normal magnetization curve passes through the tips of the hysteresis loops.

†The term *retentivity* is also sometimes used to mean the ratio of the residual flux density B_r to the maximum flux density B_m.

the excursion of H approaches zero, leaving the specimen essentially demagnetized. This process of *demagnetization by reversals* of H is widely used for demagnetizing, or *deperming,* magnetic materials.

The curve passing through the tips of the hysteresis loops in Fig. 7-33 is the *normal magnetization curve.* This curve is useful since it is reproducible and is characteristic of the particular type of magnetic material. The normal magnetization curve is very similar to the initial magnetization curve we discussed earlier. The ratio B/H (or permeability μ) has significance for both normal and initial magnetization curves.

7-20 ENERGY IN A MAGNET

A specimen of iron with residual magnetization contains energy since work has been performed in magnetizing it. The magnetic energy w_m per unit volume of a specimen brought to saturation from an originally unmagnetized condition is given by the integral of the initial-magnetization curve expressed by

$$w_m = \int_0^B H\,dB \qquad (\text{J m}^{-3}) \qquad \textbf{\textit{Initial magnetization integral}} \tag{1}$$

The dimensional relation for (1) is

$$\frac{I}{L}\frac{M}{IT^2} = \frac{M}{LT^2}$$

where M/LT^2 has the dimensions of energy density, which is expressed in joules per cubic meter.

Thus, the area between the curve and the B axis is a measure of the energy density. This is indicated in Fig. 7-34*a* for an easily magnetized (magnetically soft)

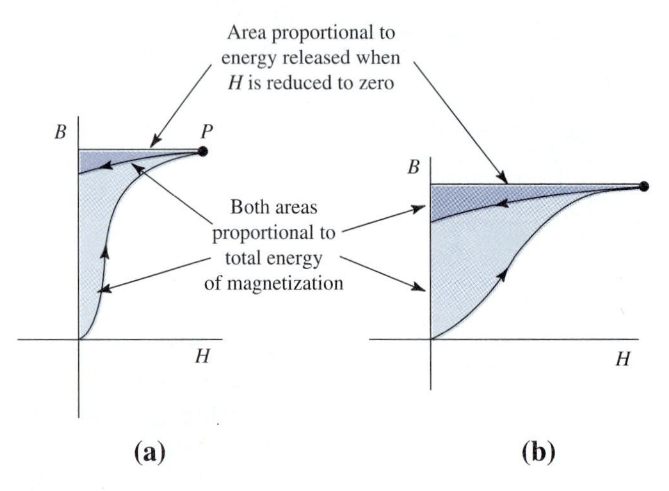

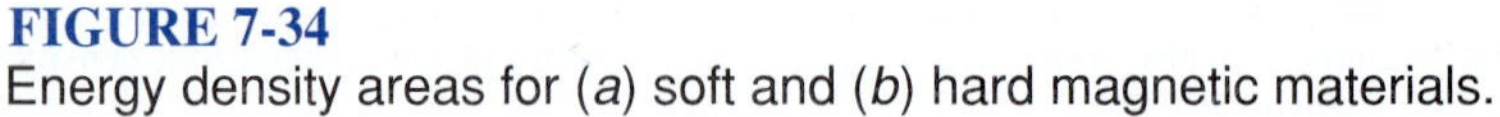
FIGURE 7-34
Energy density areas for (*a*) soft and (*b*) hard magnetic materials.

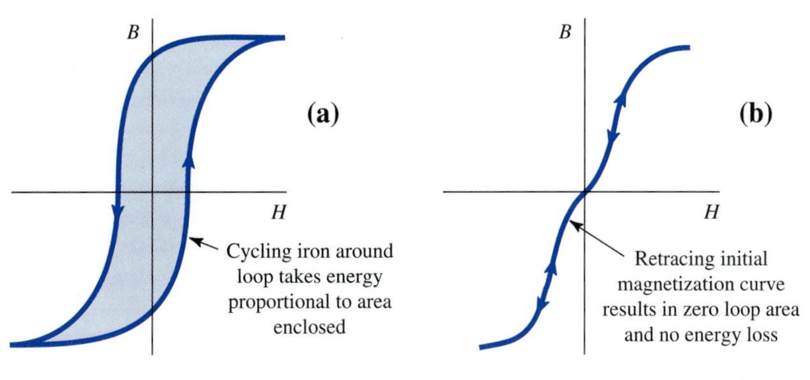

FIGURE 7-35
Energy lost in a magnetization cycle is proportional to the area enclosed by the hysteresis loop.

substance which has been carried to the point P in the magnetization process. A magnetically hard substance takes more work to magnetize, as indicated by the larger colored area in Fig. 7-34*b*. On bringing H to zero some energy is released, as indicated by the upper darker areas in Fig. 7-34.

If H is increased and decreased, so that the magnetization of a specimen repeatedly traces out a hysteresis loop, as in Fig. 7-35*a*, the area enclosed by this loop represents the energy per unit volume expended in the magnetization-demagnetization process in one complete cycle. In general, the specimen retains some energy in stored magnetic form at any point in the cycle. However, in going once around the hysteresis loop and back to this point, at which the energy will again be the same, energy proportional to the area of the loop is lost. The energy, which appears as heat, is expended reversing the magnetization zones and stressing the atomic lattice of the specimen. If no hysteresis is present and the initial magnetization curve is retraced, the area of the loop is zero (Fig. 7-35*b*). This magnetization-demagnetization process is then accomplished with no loss of energy as heat in the specimen, assuming that eddy currents are negligible.

7-21 PERMANENT MAGNETS

In many applications permanent magnets play an important part. In dealing with permanent magnets, the section of the hysteresis loop in the second quadrant of the BH diagram is of particular interest. If the loop is a saturation or major hysteresis loop, the section in the second quadrant is called the *demagnetization curve* (Fig. 7-36*a* on next page). This curve is characteristic for a given magnetic material. The intercept of the curve with the B axis is the maximum possible residual flux density B_r, or the retentivity, for the material, and the intercept with the H axis is the maximum coercive force, or the coercivity. It is usually desirable that permanent-magnet materials have a high retentivity, but it is also important that the coercivity be large so that the magnet will not be easily demagnetized.

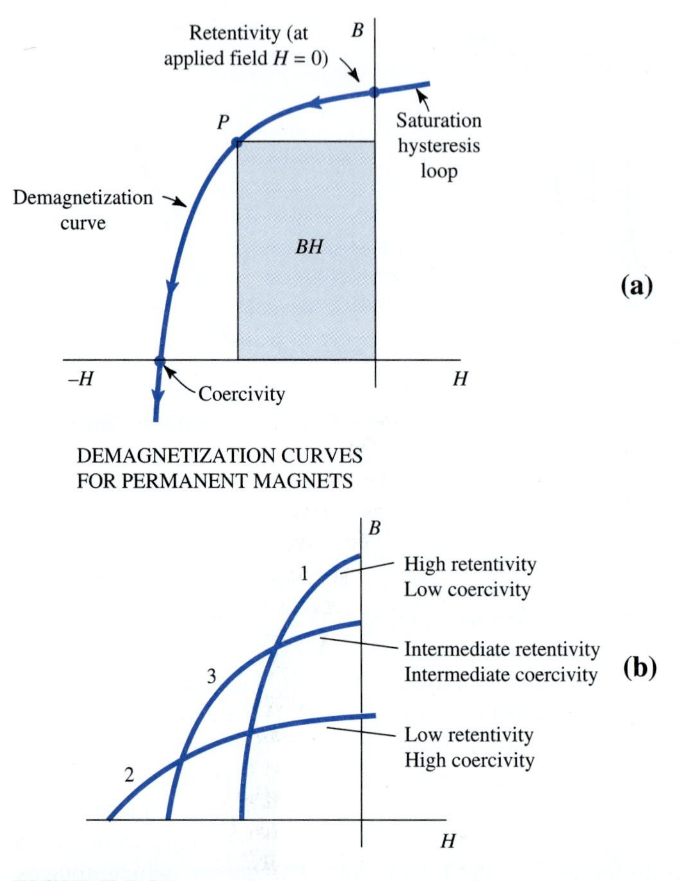

FIGURE 7-36
Demagnetization curves (B is positive, and H is negative.)

In Fig. 7-36*b*, three demagnetization curves are shown. Curve 1 represents a material having a high retentivity but low coercivity, while curve 2 represents a material which is just the reverse; i.e., it has a low retentivity and high coercivity. Curve 3 represents a material which is a compromise between the other two, having relatively high retentivity and coercivity.

The maximum BH product, abbreviated $BH_{\max}$, is also a quantity of importance for a permanent magnet. In fact, it is probably the best single figure of merit, or criterion, for judging the quality of a permanent magnet material. Referring to Fig. 7-36*b*, it is apparent that $BH_{\max}$ is greater for curve 3 than for either curves 1 or 2. The maximum BH product for a substance indicates the maximum energy density (in joules per cubic meter) stored in the magnet. A magnet at $BH_{\max}$ delivers a given flux with a minimum of magnetic material.

Since the product BH has the dimensions of energy density, it is sometimes called the *energy product* and its maximum value the *maximum energy product*.

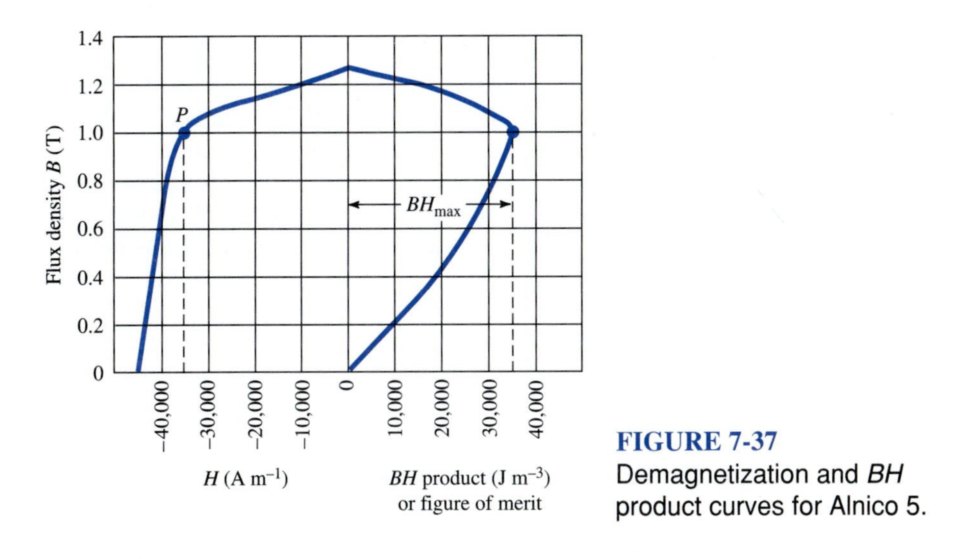

FIGURE 7-37
Demagnetization and *BH* product curves for Alnico 5.

The product *BH* for any point *P* on the demagnetization curve is proportional to the area of the shaded rectangle, as shown in Fig. 7-36*a*.

Figure 7-37 shows the demagnetization curve for Alnico 5, one of the best permanent-magnet materials, which is an alloy containing iron, cobalt, nickel, aluminum, and copper. A curve showing the *BH* product is also presented. The maximum *BH* product is about 36 000 J m^{-3} and occurs at a flux density of about 1 T (see point *P*).

7-22 TABLE OF PERMANENT MAGNETIC MATERIALS

Representative materials for permanent magnets are given in Table 7-7. The materials are listed in the order of increasing maximum *BH* product. Magnets of cobalt, copper, iron, and either cerium or samarium have been cast with coercivities of over 2 MA m^{-1}.

TABLE 7-7
Permanent magnetic materials

Material†	Retentivity, T	Coercivity, A m^{-1}	BH_{max}, J m^{-3}
Chrome steel (98 Fe, 0.9 Cr, 0.6 C, 0.4 Mn)	1.0	4 000	1 600
Oxide (57 Fe, 28 O, 15 Co)	0.2	72 000	4 800
Alnico 12 (33 Fe, 35 Co, 18 Ni, 8 Ti, 6 Al)	0.6	76 000	12 000
Alnico 2 (55 Fe, 12 Co, 17 Ni, 10 Al, 6 Cu)	0.7	44 800	13 600
Alnico 5 (Alcomax) (51 Fe, 24 Co, 14 Ni, 8 Al, 3 Cu)	1.25	44 000	36 000
Platinum cobalt (77 Pt, 23 Co)	0.6	290 000	52 000

†Compositions in percent.

7-23 DEMAGNETIZATION

A bar of ferromagnetic material with residual flux density tends to become demagnetized spontaneously. The phenomenon is illustrated by Fig. 7-38*a* which shows a bar magnetized so that a north pole is at the left and a south pole at the right. The orientation of a single domain is indicated, and it is evident that the external field of the bar magnet opposes this domain and, hence, will tend to turn it, or reverse its polarity, thereby partially demagnetizing the bar. The tendency for this demagnetization is reduced if the magnet is in the form of a U, as in Fig. 7-38*b*, since in this case there is but little demagnetizing field along the side of the magnet. The demagnetizing effect can be still further reduced by means of a soft-iron *keeper* placed across the poles, as in Fig. 7-38*c*.

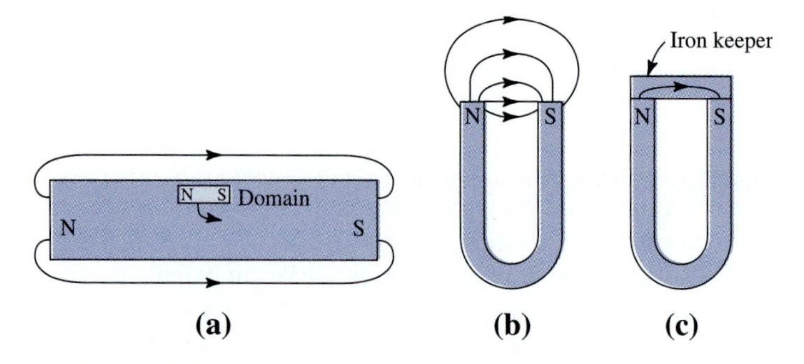

FIGURE 7-38
(*a*) Demagnetization of domain by bar-magnet field. (*b*) U-shaped magnet with less tendency for demagnetization. (*c*) U-shaped magnet with keeper (ideally no tendency for demagnetization).

The process of removing the permanent magnetization of a specimen so that the residual flux density is zero under conditions of zero **H** field is called *demagnetization* or *deperming*. It is evident that B can be reduced to zero by the application of the coercive force H_c, but on removing this field the residual flux density will rise to some value B_0 as suggested in Fig. 7-39. Although it might be possible to end up at $B = 0$ and $H = 0$ by increasing $-H$ to slightly more than the coercive force and

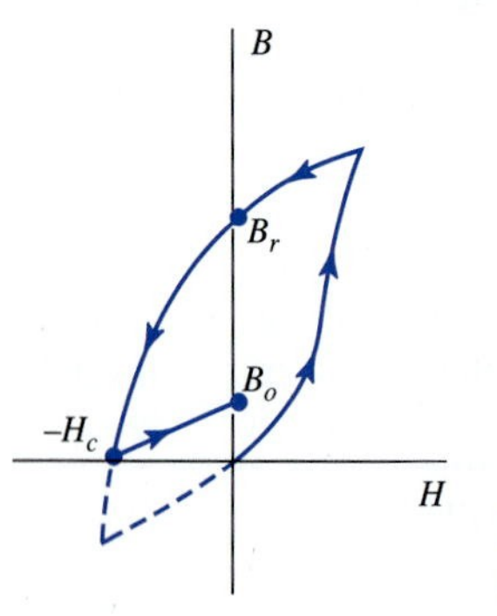

FIGURE 7-39
Partial hysteresis loop.

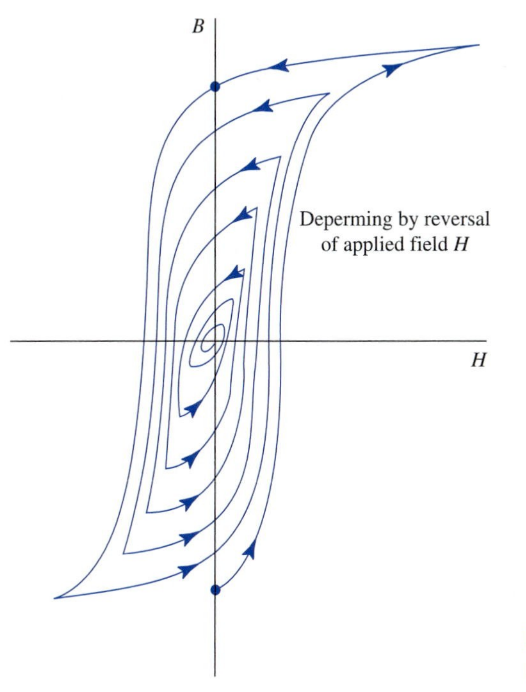

FIGURE 7-40
Demagnetization by reversals.

then decreasing it to zero, as suggested by the dashed lines, the process requires an accurate knowledge of B and H and the hysteresis loop.

A longer but more simply applied method is called *demagnetization* or *deperming by reversals.* In this method, $\pm H$ is brought to a smaller maximum amplitude on each reversal so that eventually the specimen is left in a demagnetized state at zero field, as suggested by Fig. 7-40. Although such a demagnetization procedure can be completely carried out in a matter of seconds with small magnetic specimen such as a watch (using ac fields), many seconds or even minutes may be required *for each reversal* for large magnetic objects because of the slow decay of the induced eddy currents (Sec. 7-29) and the reluctance of the domains to change polarity. With the object to be demagnetized inside a coil, the process can be accomplished with low enough ac frequency by either gradually reducing the ac coil current or by slowly removing the object from the coil with the ac current constant.

Example 7-12. Degaussing a ship. To protect ships from mines triggered by changes in the earth's magnetic field produced by a ship passing over them, the ship's field may be reduced by a field-bucking coil wound around it, a process called *degaussing.* Consider two idealized situations.

Case 1: If the ship is modeled by a short vertical bar magnet of moment 5×10^5 A m^2 with midpoint at the water surface, find (*a*) **B** at a depth of 12 m and (*b*) the value of the ampere-turn area product *NIA* required for a horizontal coil around the bar to neutralize the bar's field. (*c*) If the earth's field is 70 μT vertical, what are the allowable maximum and minimum values of *NIA* for the bar with coil to produce a field no more than 5 percent of the earth's field?

Case 2: If the ship is modeled by a short horizontal bar magnet of moment 5×10^5 A m^2 with centerline at the water surface, find (*d*) **B** at a depth of 12 m and (*e*) the *NIA* value of a vertical coil around the bar to neutralize the bar's field. (*f*) If the earth's field is 50 μT horizontal, what are the allowable maximum and minimum values of *NIA* for the bar with coil to produce a field no more than 5 percent of the earth's field? An actual ship is much more complex than is suggested by these idealized cases and may require many coils wound vertically and horizontally.

(*a*) $$B_r = \frac{\mu_0 IA}{2\pi r^3} = \frac{(4\pi \times 10^{-7})(5 \times 10^5)}{2\pi \times 12^3} = 58\ \mu\text{T} = 0.58\ \text{G} \quad \textit{Ans.}\ (a)$$

(*b*) and (*e*) $NIA = 5 \times 10^5$ A m^2 *Ans.* (*b*) and (*e*)

(*d*) $$B_\theta = \frac{\mu IA}{4\pi r^3} = \frac{(4\pi \times 10^{-7})(5 \times 10^5)}{4\pi \times 12^3} = 29\ \mu\text{T} = 0.29\ G \quad \textit{Ans.}\ (d)$$

Problem 7-23-1. Degaussing a ship. Consider two idealized situations. ***Case 1:*** Let the ship be modeled by a vertically oriented bar magnet of moment 10^5 A m^2. (*a*) At a depth of 10 m below the ship, what is *B* from the ship? Assume that the magnet is short compared to 10 m and that its middle point is at the water surface. (*b*) If the earth's field is vertical and equal to 60 μT, how accurately must the coil current be set so that the ship produces less than a 3 percent change in total field? ***Case 2:*** Let the ship be modeled by horizontally oriented bar magnet of moment 10^5 A m^2. (*c*) At a depth of 10 m below the ship what is *B* from the ship? Assume that the middle of the magnet is at the water surface. (*d*) If the earth's field is horizontal and equal to 40 μT, how accurately must the coil current be set so that the ship produces less than 3 percent change in total field? *Ans.* (*a*) 20 μT (0.2 G); (*b*) 9 percent; (*c*) 10 μT (0.1 G); (*d*) 12 percent.

Example 7-13. Demagnetization curve. (*a*) Find the maximum *BH* product if the retentivity of a permanent magnet is 0.8 T, the coercivity is 16 kA/m, and the demagnetization curve is a straight line (Fig. E7-13). (*b*) Show that this is the maximum value.

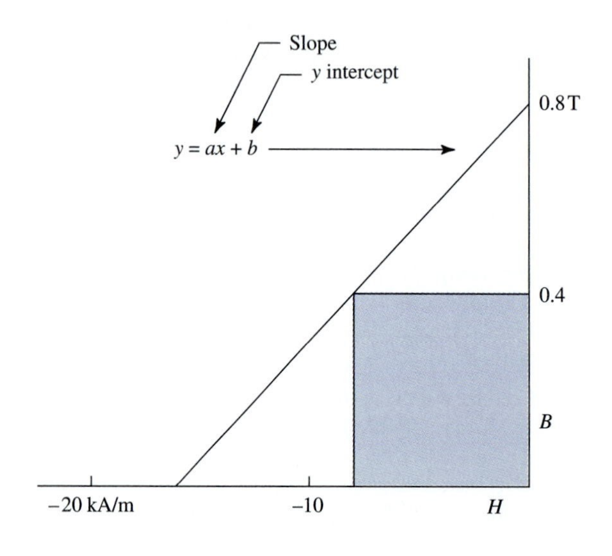

FIGURE E7-13

Solution

(*a*) $B = aH + b = (5 \times 10^{-5})H + 0.8$

$$BH = 5 \times 10^{-5}H^2 + 0.8H$$

Maximum occurs where slope = 0

or $$\frac{\partial(BH)}{\partial H} = 2(5 \times 10^{-5})H + 0.8 = 0$$

$$H = -0.8/10^{-4} = -8\ \text{kA m}^{-1}$$

Therefore

$$(BH)_{\text{max}} = (5 \times 10^{-5})(64 \times 10^6) - 6.4 \times 10^3 = -3.2\ \text{kJ m}^{-3} \quad \textit{Ans.}\ (a)$$

(*b*) $-(16000/2) \times (0.8/2) = -3.2$ kJ = maximum rectangular area under curve. *Ans.* (*b*)

Problem 7-23-2. Demagnetization curve of permanent magnet. (*a*) Assuming that the demagnetization curve of a certain ferromagnetic material is a straight line, what is the maximum BH product if the retentivity is 1 T and the coercivity is 20 kA/m? (*b*) Prove that this is the maximum value. *Ans.* (*a*) -5 kJ/m^3; (*b*) -5 kJ/m^3.

Example 7-14 Degaussing the earth. A single loop of wire laid around the earth at its magnetic equator could, in principle, neutralize the earth's field at a distance from earth, that is, degauss the earth. If the earth's magnetic dipole moment is 5×10^{21} A m^2, find (*a*) loop current for neutralization, (*b*) the distance from the earth at which the loop field balances the earth's field to $\frac{1}{2}$ percent or less, and (*c*) **B** 500 m from the loop. (*d*) Would this value of **B** be environmentally acceptable? (See Chap. 9.) Earth circumference = 40 Mm.

Solution

(*a*) $$Q_m L = 5 \times 10^{21}\ \text{A m}^2 = IA$$

$$I = \frac{Q_m L}{A} = \frac{5 \times 10^{21}}{\pi(6.4 \times 10^6)} = 39\ \text{MA} \quad \textit{Ans.}\ (a)$$

(*b*) Assume magnetic dipole is at center of earth with

$$\underset{\theta = 90^\circ}{B_\theta\ (\text{in equatorial plane})} = \frac{\mu_0 Q_m L}{4\pi r^3} = \frac{\mu_0 I d^2}{4\pi r^3} \tag{1}$$

Assume circular loop at magnetic equator is equivalent to a square loop of equal area ($A = 2\pi R^2 = d^2$, where R = earth radius) with field given by four current dipoles of length d giving

$$B_\theta = \frac{\mu_0 I d}{4\pi}\left\{\frac{1}{[r-(d/2)]^2} - \frac{1}{[r+(d/2)]^2} - \frac{2\sin(\tan^{-1}(d/2r))}{r^2}\right\} \tag{2}$$

Evaluate (1) and (2) for different r/d ratios to determine the distance for which the two cases agree with $\frac{1}{2}$ percent. The correct value is *between 22 and 28* earth radii.

Show that this distance for the differences of the B_θ components in the equatorial plane also satisfies the requirement for the difference of the B_r components of the loop around the equator and a small loop at the center of the earth as measured in the direction of the earth's magnetic field axis.

(*c*)
$$B = \frac{\mu I}{2\pi r} = \frac{(4\pi \times 10^{-7})(39 \times 10^6)}{2\pi \times 500} = 15.6 \text{ mT} = 156 \text{ G} \quad \textit{Ans.} \quad (c)$$

(*d*) The earth's magnetic field is of the order of 1 G. A value of 156 times this at 500 m from the conductor could present environmental problems. *Ans.* (*d*)

Problem 7-23-3. Magnetic field of earth. If the horizontal component of **B** at the earth's surface is 20 μT at a point where the dip angle of the field (angle from horizontal) is 72°, (*a*) find the magnetic moment required for the dipole (or loop) at the center of the earth. (*b*) Using this model, calculate B at a distance of 40 Mm above the earth's surface in the magnetic equatorial and polar directions. Assume $\mu = \mu_0$ everywhere (except within the dipole). Earth's circumference = 40 Mm. *Ans.* (*a*) 9.5×10^{22} A m^2; (*b*) 95.2 nT (equatorial); 190.4 nT (polar).

7-24 GAPLESS CIRCUIT

Consider the magnetic circuit of a closed ring of iron of uniform cross-section A and mean length L. Suppose that a coil of insulated wire is wound uniformly around the ring and that we wish to know how large the product NI (number of turns times current) must be to produce a flux density B in the ring.

The coil on the ring in Fig. 7-41 forms a toroid. In the toroid, we have,

$$B = \frac{\mu NI}{L} = \frac{\mu NI}{2\pi R} \quad \text{(T)} \tag{1}$$

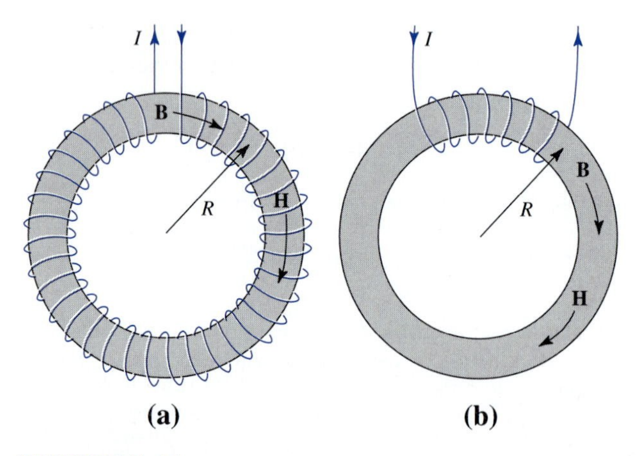

FIGURE 7-41
(*a*) Closed iron ring with uniform winding and (*b*) with concentrated winding.

where μ = permeability (assumed uniform) of medium inside of toroid, H m^{-1}
N = number of turns, dimensionless
I = current, A
L = mean length of toroid, m
R = mean radius of toroid, m

Dividing by μ, we have $NI = HL$ ampere-turns. If a certain flux density B is desired in the ring, the corresponding H value is taken from a BH curve for the ring material and the required number of ampere-turns calculated from $NI = HL$.

Example 7-15. Required NI to give $B = 1$ T in iron ring. An iron ring has a cross-sectional area $A = 1000$ mm^2 and a mean length $L = 600$ mm. Find the number of ampere-turns required to produce a flux density $B = 1$ T. Form a BH curve for the iron, $H = 1000$ A m^{-1} at $B = 1$ T.

Solution. From $NI = HL$,

$$NI = 1000 \times 0.6 = 600 \text{ A turns} \quad \textit{Ans.}$$

The coil could be 100 turns with a current of 6 A or 1000 turns with 600 mA. The coil may be uniformly distributed around the ring, as in Fig. 7-41*a*, or concentrated in a small sector, as in Fig. 7-41*b*.

Problem 7-24-1. Iron ring. An iron ring of mean radius 200 mm and 150-mm^2 cross-sectional area is wound with 100 turns of wire. If $B = 0.5$ T and $\mu_r = 250$, find the wire current. *Ans.* 20.1 A.

7-25 MAGNETIC CIRCUIT WITH AIR GAP

Let a narrow air gap of thickness g be cut in the iron ring of Fig. 7-41, as shown in Fig. 7-42*a* (on next page). The gap detail is presented in Fig. 7-42*b*. By continuity of the normal component of B, the flux density in the gap is the same as in the iron if fringing is neglected. Neglecting the fringing involves but little error where the gap is narrow, as assumed here. The field H_g in the gap is then $H_g = B/\mu_0$, while the field H_i in the iron is

$$H_i = \frac{B}{\mu} = \frac{B}{\mu_r \mu_0} = \frac{H_g}{\mu_r} \tag{1}$$

from which

$$\frac{H_g}{H_i} = \mu_r \tag{2}$$

The number of ampere-turns required to produce a certain flux density B in a magnetic circuit with gap, as in Fig. 7-42*a* is a problem for which the solution can be obtained directly. For instance, the line integral of H once around the magnetic circuit equals the total mmf F, or ampere-turns enclosed. That is,

$$\oint \mathbf{H} \cdot d\mathbf{L} = F = NI \tag{3}$$

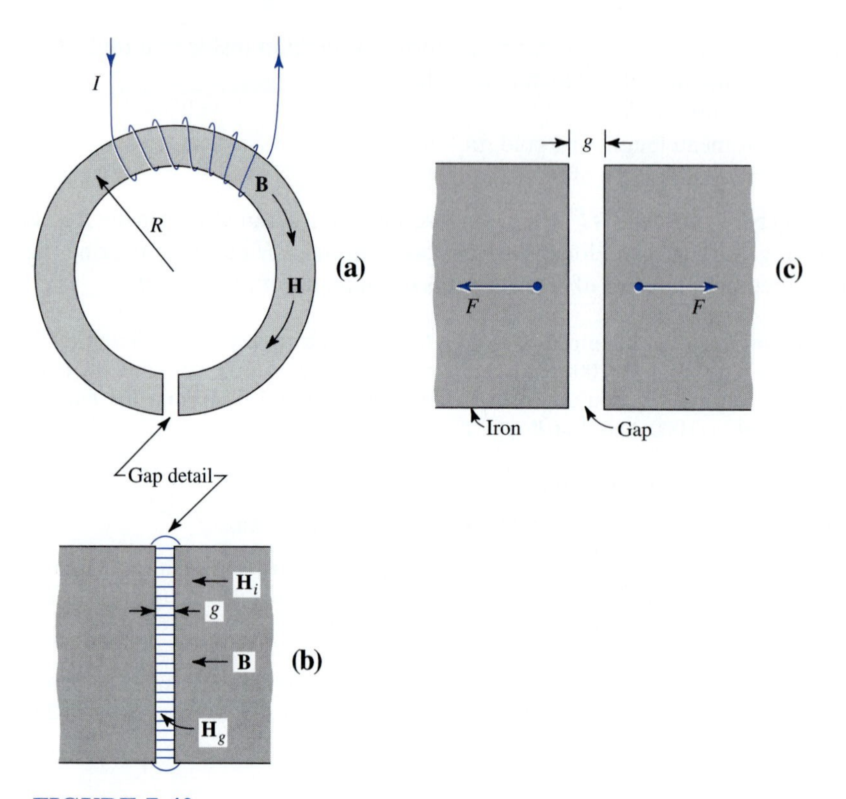

FIGURE 7-42
(*a*) Iron ring with air gap. (*b*) Gap detail. (*c*) Force *F* required to hold gap open.

Example 7-16. Iron ring with air gap. Let the iron ring of Fig. 7-42*a* have a cross-sectional area $A = 1000 \text{ mm}^2$, an air gap of width $g = 2$ mm, and a mean length $L = 2\pi R = 600$ mm, including the air gap. Find the number of ampere-turns required to produce a flux density $B = 1$ T.

Solution. We have

$$NI = \oint \mathbf{H} \cdot d\mathbf{L} = H_i(L - g) + H_g g \tag{4}$$

where $H_i = H$ field in iron
$H_g = H$ field in gap.

From a BH curve for the iron, $H_i = 1000 \text{ A m}^{-1}$, and from (2) we know H_g in terms of H_i. Hence (4) becomes

$$NI = H_i[(L - g) + \mu_r g] \tag{5}$$

where $\mu = 795$ is the relative permeability of the iron ring at $B = 1$ T. Therefore,

$$NI = 1000[(0.6 - 0.002) + 759 \times 0.002] = 2188 \text{ A turns} \quad \textit{Ans.}$$

The introduction of the *narrow air gap makes it necessary to increase the ampere-turns from* 600 *to* 2188 to maintain the flux density at 1 T.

The problem can also be solved by calculating the total reluctance of the magnetic circuit. Thus, from (4) we have

$$NI = \frac{\mu A}{\mu A} H_i(L-g) + \frac{\mu_0 A}{\mu_0 A} H_g g \tag{6}$$

and

$$NI = BA(\mathcal{R}_i + \mathcal{R}_g) \tag{7}$$

where $\mathcal{R}_i = (L-g)/\mu A$ = reluctance of iron part of circuit
$\mathcal{R}_g = g/\mu_0 A$ = reluctance of air gap.

Problem 7-25-1. Iron ring with gap. An iron ring has a uniform cross-sectional area of 150 mm ^{2} and a mean radius of 200 mm. The ring is continuous except for an air gap 1 mm wide. Find the number of ampere-turns required on the ring to produce a flux density $B = 0.5$ T in the air gap. Neglect fringing. When $B = 0.5$ T in the iron, $\mu_r = 250$. *Ans.* 2.4 kA-turns.

7-26 MAGNETIC GAP FORCE

Referring to Fig. 7-42, the effect of the magnetic field is to exert forces which tend to close the air gap. That is, the magnetic poles of opposite polarity at the sides of the gap are attracted to each other. Forces produced by magnetic fields find application in numerous electromechanical devices. In this section an expression for the force between magnetic pole pieces is developed.

The density of energy stored in a magnetic field is

$$w_m = \frac{1}{2}\frac{B^2}{\mu} \qquad (\text{J m}^{-3}) \tag{1}$$

If the gap is small, we may assume a uniform field in the air gap. The total energy W_m stored in the gap is then

$$W_m = w_m A g = \frac{B^2 A g}{2\mu_0} \qquad (\text{J}) \tag{2}$$

where A = area of gap, m^2, and g = width of gap, m.

Suppose now that the iron ring in Fig. 7-42*a* is perfectly flexible, so that the gap must be held open by a force F as in Fig. 7-42*c*. If the force is increased so as to increase the gap by an infinitesimal amount dg while at the same time the current through the coil is increased to maintain the flux density B constant, the energy stored in the gap is increased by the infinitesimal amount

$$dW_m = \frac{B^2 A}{2\mu_0} dg \tag{3}$$

Equation (3) has the dimensions of energy. But energy may also be expressed as force times distance, which in this case is $F\,dg$, where F is the attractive force between the poles. It is equal in magnitude to the force required to hold them apart. Thus,

$$F\,dg = \frac{B^2 A}{2\mu_0} dg$$

or

$$\text{Gap force} = F = \frac{B^2 A}{2\mu_0} \quad \text{(N)} \tag{4}$$

where

F = attractive force, N
B = flux density, T
A = area of gap, m^2
μ_0 = permeability of air = 400π nH m^{-1}

Dividing by the gap area A yields the pressure P. That is,

$$\text{Gap pressure} = P = \frac{F}{A} = \frac{B^2}{2\mu_0} \quad (\text{N m}^{-2}) \tag{5}$$

Problem 7-26-1. Iron ring. Gap force. An iron ring magnet of 0.02-m^2 cross-sectional area and 300-mm radius has a 1-mm air gap and a winding of 1200 turns. If the current through the coil is 6 A, what is the force tending to close the gap? Take $\mu_r = 1000$ for the iron and neglect fringing. *Ans.* 78.3 kN.

7-27 PERMANENT MAGNET WITH GAP

Suppose that a closed iron ring is magnetized to saturation with a uniform toroidal coil wound on the ring. When the coil is removed, the flux density in the iron is equal to the retentivity (see Fig. 7-43). If, however, the system has an air gap,

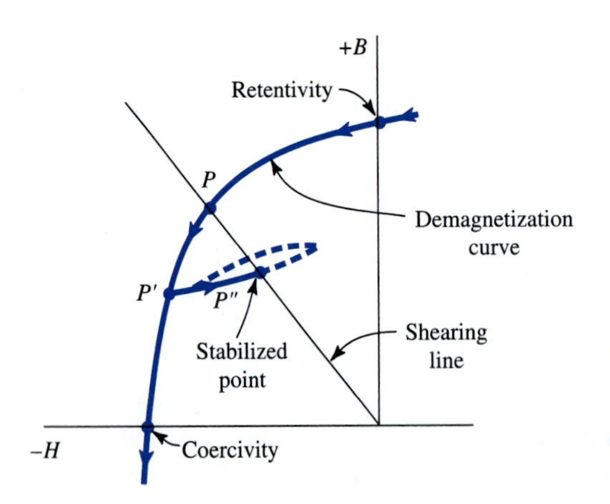

FIGURE 7-43
Demagnetization curve for permanent magnet showing the shearing line.

as in Fig. 7-42, the flux density has a smaller value as given by a point P which lies somewhere on the demagnetization curve (Fig. 7-43).

Further information is needed to locate this point. This may be obtained as follows. The line integral of $\mathbf{H}$ once around a magnetic circuit is $\oint \mathbf{H} \cdot d\mathbf{L} = NI$. Since $NI = 0$,

$$\oint \mathbf{H} \cdot d\mathbf{L} = H_i(L - g) + H_g g = 0$$

or

$$H_i(L - g) = -H_g g \tag{1}$$

where $H_i = H$ field in the iron
$L = 2\pi R =$ total length of magnetic circuit (including gap)
$g =$ width of gap
$H_g = H$ field in gap

Thus H_i and H_g are in opposite directions, as indicated in Fig. 7-44. If leakage is neglected, B is uniform around the circuit. Multiplying (1) by μ_0 and solving for the ratio B/H_i, or the permeability of the iron, we obtain

$$\frac{B}{H_i} = -\mu_0 \frac{L - g}{g} \tag{2}$$

This ratio of the flux density B to the field H_i in the iron gives the slope of a line called the *shearing line,* as shown in Fig. 7-43. The intersection of this line with the demagnetization curve determines the position of the iron on the magnetization curve (point P). This location is a function of the ratio of the iron path length $(L - g)$ to the gap length g.

In most permanent-magnet applications, if it is desired that B remain relatively constant, a moderate demagnetizing field is applied to the iron, moving the position of the iron to P' (Fig. 7-43). On removing the field, the iron moves to the point P'' on the shearing line. The ring magnet is now said to be *stabilized,* and when fields less than about the difference of H between points P' and P'' are applied to the ring and then removed, the iron will always return to approximately point P''. Under these conditions the iron moves along a minor hysteresis loop, as suggested by the dashed lines in Fig. 7-43.

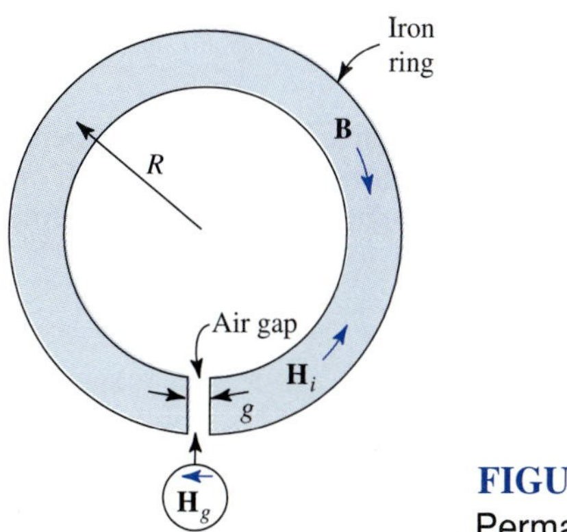

FIGURE 7-44
Permanently magnetized ring with air gap.

Example 7-17. Lifting electromagnet ("pumping iron"). An electromagnet with U-shaped iron yoke just lifts an iron bar (Fig. 7-45). If $\mu_r = 1800$ for the yoke and bar and the coil ampere-turns $NI = 1$ kA, what is the weight of the bar? The yoke and bar length is 1m with a cross-sectional area of 0.1m^2. A 1-mm-thick copper sheet between yoke and bar prevents iron-to-iron contact.

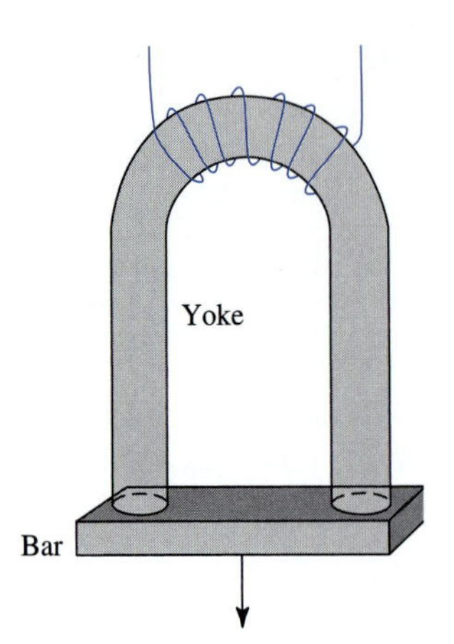

FIGURE 7-45
U-shaped electromagnet.

Solution

$$NI = BA(\mathcal{R}_i + \mathcal{R}_g)$$

$$\mathcal{R}_i = \frac{L-g}{\mu A} = \frac{1-0.002}{(1800 \times 4\pi \times 10^{-7})(0.1)} = 4412 \text{ H}^{-1}$$

Note: Copper sheet 1 mm thick gives a gap at both poles so magnetic circuit gap $g = 2$ mm.

$$\mathcal{R}_g = \frac{g}{\mu_0 A} = \frac{0.002}{4\pi \times 10^{-7} \times 0.1} = 15\,920 \text{ H}^{-1}$$

$$B = \frac{NI}{(\mathcal{R}_i + \mathcal{R}_g)A} = \frac{10^3}{(4412 + 15\,920)(0.1)} = 0.49 \text{ T}$$

$$\text{Weight} = 2F \times 0.102 = \frac{2B^2 A}{2\mu_0} 0.102 = 1.95 \text{ tonnes} \quad \textit{Ans.}$$

Note: At earth's surface, weight (kg) = force (N) × 0.102

Example 7-18. Tapered poles. (*a*) If the yoke contact area of Fig. 7-45 is tapered to half, what weight bar can be just lifted, assuming no change in total flux? (*b*) If (*a*) gives an increased weight, how much can the lifting weight be increased by tapering the yoke contact area to still smaller values?

Solution

$$B = \frac{\psi}{A}$$

so for constant ψ, halving A doubles B.
Thus, relative $B^2 A = 2^2 \times (1/2) = 2$ so weight doubles to 3.9 tonnes *Ans.* (*a*)

Weight can be increased until fringing at taper prevents sufficient increase in B. Thus, if B increases by only $\sqrt{2}$ when A is halved, there is no increase in weight lifted. *Ans.* (*b*).

Problem 7-27-1. Ring with gap. An iron ring 300 mm in diameter and 10 cm^2 in cross-section has a 3-mm air gap. If an mmf of 2000 A-turns is applied, find the flux density B in the gap. Neglect fringing. The iron characteristics are the same as in Fig. 7-26. A trial-and-error solution is required since H_i and μ_r are unknowns and are related in a nonlinear way. Guess a value of H_i and note corresponding μ_r from Fig. 7-26. Use these values to calculate NI. If it is not 2000, try again. If first value is too high and second one too low, interpolate. *Ans.* 327 mT.

Problem 7-27-2. Gap force. A C-shaped iron electromagnet, shown in cross-section in Fig. 7-46, is designed to withstand a gap-closing force equivalent to the weight of a mass of 10 tonnes (10^4 kg). What is the maximum allowable current for which the force will not exceed this value? The magnet coil has 10,000 turns. Take μ_r (for iron) = 400. *Ans.* 377 A.

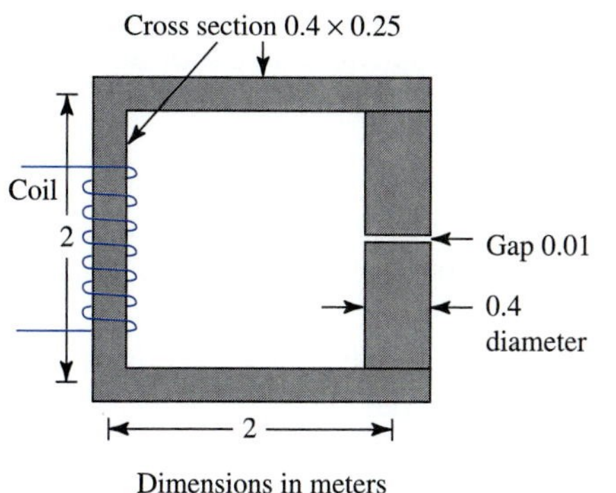

FIGURE 7-46
C-shaped electromagnet.

7-28 ALTERNATING-CURRENT BEHAVIOR OF FERROMAGNETIC MATERIALS

The permeability of iron is not a constant. In spite of this, the permeability of the iron in an iron-cored coil carrying alternating current may be taken as a constant for certain applications, but its value, in this case, needs further explanation.

Where μ is not a constant, the inductance L of a coil of N turns is given by

$$L = N\frac{d\psi_m}{dI} \tag{1}$$

For a toroidal type of coil, $d\psi_m = A\,dB$, and $dI = l\,dH/N$, where A equals the area and l equals the length of the coil. Therefore (1) becomes

$$L = \frac{N^2 A}{l}\frac{dB}{dH} \tag{2}$$

In (2) dB/dH has the dimensions of permeability. It is equal to the slope of the hysteresis curve. Thus, at point P (Fig. 7-47), dB/dH is greater than the ordinary

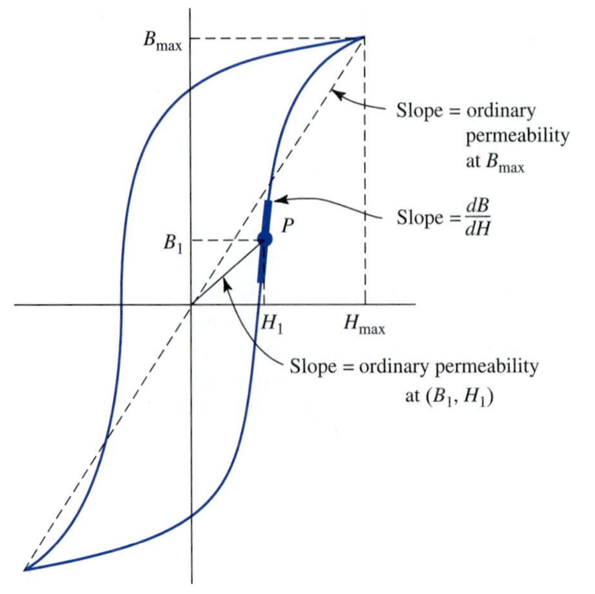

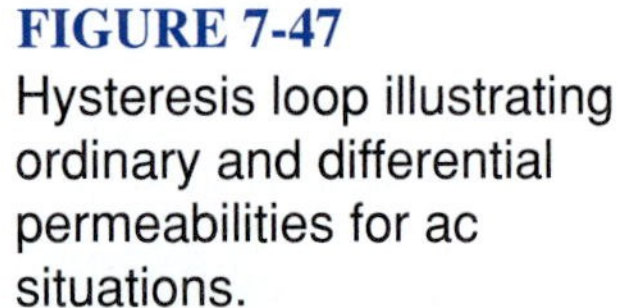

FIGURE 7-47
Hysteresis loop illustrating ordinary and differential permeabilities for ac situations.

permeability, B_1/H_1, which is equal to the slope of the line from the origin to the point P. Since dB/dH involves infinitesimals, it is sometimes called the *infinitesimal* or *differential permeability.*

If alternating current is applied to an iron-cored coil so that the magnetization of the iron moves around a hysteresis loop (Fig. 7-47) once per cycle, the slope dB/dH varies over a wide range and the instantaneous value of the inductance will, from (2), vary over a corresponding range. Under these conditions it is often convenient to consider the average inductance (over one cycle) as obtained from (2), using the average value of the slope dB/dH. This is equal to the *ordinary permeability* or the permeability at the maximum value of B attained in the cycle (see Fig. 7-47). Thus

$$L_{av} = \frac{N^2 A}{l}\left(\frac{dB}{dH}\right)_{av} = \frac{N^2 A}{l}\mu \tag{3}$$

where $\mu = B_{max}/H_{max}$ is the ordinary permeability at B_{max}.

The above discussion is for the case where the only current through the coil is an alternating one. If in addition to a small alternating current there is a relatively large steady, or direct, current through the coil, the situation is as suggested in Fig. 7-48. The magnetic condition of the iron then follows a minor hysteresis loop as indicated. In this case the average value of the slope dB/dH is given by the line passing through the tips of the minor hysteresis loop and is called the ***incremental permeability*** μ_{inc}. Referring to Fig. 7-48,

$$\mu_{inc} = \left(\frac{dB}{dH}\right)_{av} = \frac{\Delta B}{\Delta H} \tag{4}$$

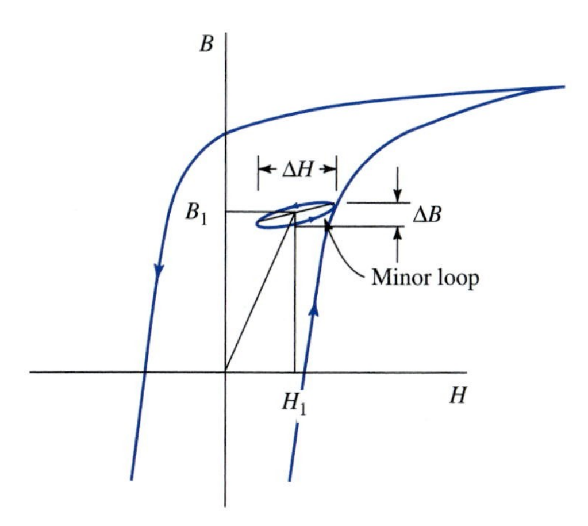

FIGURE 7-48
Minor hysteresis loop illustrating incremental permeability.

For a point at the center of the minor loop in Fig. 7-48 the incremental permeability is much less than the ordinary permeability B_1/H_1.

7-29 EDDY CURRENTS

When large conducting specimens are subjected to transformer or motional induction, currents tend to be induced in the specimen. They flow in closed paths in the specimen and are called *eddy currents.* In accordance with Lenz's law, the eddy current tends to oppose the change in field inducing it.

Eddy currents result in joule heating in the conducting specimen. The energy loss due to eddy currents in the ferromagnetic cores of ac devices is in addition to the energy lost in the magnetization process (proportional to the area of the hysteresis loop). In order to reduce the eddy currents in iron-cored ac devices, the core is commonly made of thin sheets or laminations of iron insulated electrically from each other. Thus, the eddy currents are confined to individual sheets, and the power loss is reduced. Each sheet is continuous in the direction of the magnetic flux through the core, but because of its thinness it has a relatively large reluctance. By stacking a sufficient number of sheets in parallel the total reluctance of the magnetic circuit can be reduced to the desired value. To reduce eddy currents to a minimum, iron wires are sometimes used in place of sheets, while at radio frequencies powdered-iron or ferrite cores are commonly employed.

PROBLEMS

Answers are given in Appendix E to problems followed by the @ symbol.

7-9-1. Aluminum plate capacitor for radio transmitter. A capacitor in the final amplifier of a radio transmitter has a stack of 8 aluminum plates 2 mm thick and 200 mm square as one electrode. The other electrode consists of seven plates of the same dimensions interleaved with the first stack so that there is a uniform spacing of 8 mm between all plates. (*a*) Find the capacitance. (*b*) Find the dc breakdown voltage. Neglect fringing. The edges of the plates are rounded. @

7-9-2. Breakdown for sphere. What is the maximum voltage to which a 200-mm-diameter sphere can be charged if it is situated (*a*) in air or (*b*) in mineral oil? (*c*) Compare these voltage with those for a 400-mm sphere.

7-9-3. Breakdown voltage. Two metal spheres mounted side by side with a spacing d between their edges are connected across a high-voltage transmission line. To prevent voltage surges exceeding 20 kV on the line, what spacing d should be used if the spheres are 100 mm in diameter. @

7-10-4. Energy storage. If a small flashlight battery can store $VIt = 1.5 \times 10^{-3} \times 10^7 = 1.5 \times 10^4$ J, how many 100-mF, 5-V capacitors are required to store the same amount of energy? These capacitors are available as a battery substitute to power CMOS circuits during power outages.

7-10-5. Capacitor energy. A parallel-plate capacitor of plate area A, plate separation d, and dielectric permittivity ε has a plate surface charge density ρ_s. Starting with A, d, and ρ_s, show how $\frac{1}{2}QV$, where Q is the charge on one plate and V the voltage between the plates, can lead to an energy density $\frac{1}{2}\varepsilon E^2$ in the dielectric.

7-10-6. Energy stored around sphere. (*a*) What is the electrostatic energy of a charged metal sphere of radius R and charge Q? The sphere is in air. Note that the energy is actually stored in the region outside the sphere as though the sphere is one electrode of a capacitor with the other electrode a sphere at infinity. (*b*) Within what radius from the sphere is half the energy stored? @

7-10-7. Capacitance of earth. What is the capacitance of the earth?

7-10-8. Energy in capacitor. A typical 12-V automobile battery can store 1 kW-h of energy. How large (in volume) must a capacitor be to store an equal amount of energy? Consult a radio parts catalog for capacitance, voltage rating, and dimensions of capacitors such as electrolytic types. Under what conditions will a battery be preferable and under what conditions will a capacitor (or capacitor bank) be preferable? @

7-10-9. Electrostatic energy in a thundercloud. Electrostatically, a typical thundercloud may be represented by a capacitor model with horizontal plates 10 km^2 in area separated by a vertical distance of 1 km. The upper plate has a positive charge of 200 C and the lower plate an equal negative charge. (*a*) Find the electrostatic energy stored in the cloud. (*b*) What is the potential difference V between the top and bottom of the cloud? (*c*) What is the average electric field E in the cloud? (*d*) How close is this value to the dielectric strength of dry air?

7-12-3. Compass deflection. If the earth's horizontal field component is 40 μT north, find the magnetic moment of a short bar magnet 0.5 m to the west of the compass which deflects the compass needle 45°. @

7-12-4. Bar magnet. Translational force. A bar magnet in a uniform magnetic field is acted on only by a torque, there being no translational force on the magnet. In a nonuniform field, there is a net translational force. This is the force that draws iron objects to a magnet. Find the maximum value of this force on a uniformly magnetized bar magnet 6 mm long with a magnetic moment of 2 A m^2 situated 100 mm from one pole of a very long slender bar magnet having a pole strength of 600 A m.

7-15-3. Energy in bar magnet. A bar magnet of 800 A m^2 magnetic moment has length of 100 mm and a diameter of 15 mm. Find (*a*) the energy density at the center of the bar and (*b*) the total energy in the bar. @

7-15-4. Energy between current sheets. Two large, parallel flat conducting sheets with vertical 20-mm separation carry a sheet current density of 1,000 A/m in opposite directions. Find (*a*) the energy density between the sheets and (*b*) the total energy between the sheets over an area of 2 m^2.

7-15-5. Coaxial line with helical inner conductor. (*a*) To raise the inductance of a coaxial line the inner conductor may be wound as a helix. If the inner conductor is a fine-wire helix of radius a and the outer conductor a thin-walled tube of radius b, find the inductance per unit length. The helix has N turns per meter. (*b*) If the helical inner conductor is wound on a nonconducting ferrite core of radius a with relative permeability μ_r, find the line inductance. @

CHAPTER 8

WAVEGUIDES, RESONATORS, AND FIBER OPTICS

8-1 INTRODUCTION

The waves on transmission lines that we have discussed earlier are transverse electromagnetic (TEM) waves, with electric and magnetic fields entirely transverse to the direction of propagation. In this chapter we continue the discussion of transmission systems with emphasis on waves of higher-order modes, that is, having components of **E** or **H** in the direction of propagation. Transmission systems which can convey electromagnetic waves *only* in higher-order modes are usually called *waveguides* or simply *guides*.

The infinite-parallel-plane transmission line is discussed first, leading to the hollow rectangular and cylindrical waveguides and their field configurations, cutoff wavelengths, and attenuation. Waves propagating along the outside of single conductors with and without dielectric coatings, and waves guided in dielectric sheets, rods, and fibers are considered next. Finally, highly resonant systems with waves trapped inside enclosures or cavities are treated.

8-2 CIRCUITS, LINES, AND GUIDES: A COMPARISON

At low frequencies, a concept of currents, voltages, and lumped circuit elements is practical. Thus, for the simple circuit of Fig. 8-1*a* on page 448, consisting of a generator *G* and resistor *R*, circuit theory involving lumped elements can be used.

At somewhat higher frequencies these ideas can be extended satisfactorily to lines of considerable length provided that the velocity of propagation and the distributed constants of the line are considered. Thus, the behavior of the transmission

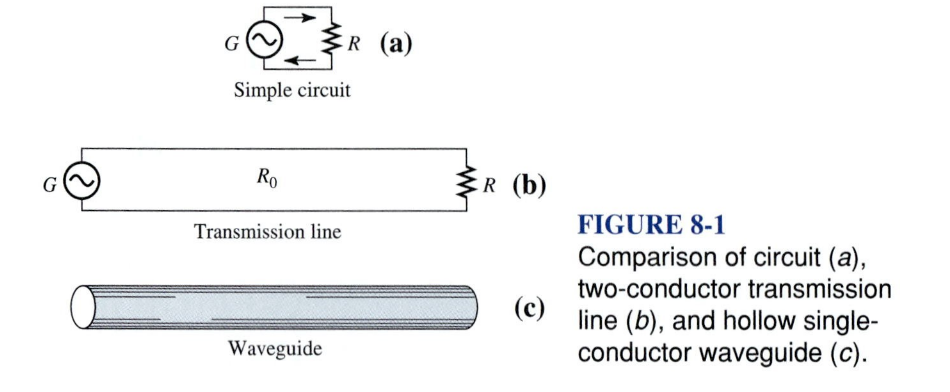

FIGURE 8-1
Comparison of circuit (*a*), two-conductor transmission line (*b*), and hollow single-conductor waveguide (*c*).

line of Fig. 8-1*b* can be handled by an extension of circuit theory involving distributed elements.

Another type of transmission system, shown in Fig. 8-1*c*, consists of a hollow cylindrical or rectangular pipe or tube of metal. Suppose we ask: Can such a pipe convey electromagnetic energy? If our experience were restricted to simple circuits or transmission lines, as in Fig. 8-1*a* and *b*, our answer would be *no,* since there is only a single conductor and no return circuit for the current. However, with an awareness of optics, our answer would be *yes*, since light will pass through a straight metal pipe and light consists of electromagnetic waves of extremely high frequency (10^{16} Hz).

The complete answer is *yes* and *no,* depending on the frequency. Carrying our reasoning further, we might deduce that if the metal pipe will not transmit low frequencies but will transmit extremely high frequencies, there must be some intermediate frequency at which there is a transition from one condition to the other. In the following sections on waveguides we shall find that this transition, or low-frequency *cutoff*, occurs when the wavelength is of the same order of magnitude as the diameter of the pipe.

In explaining the transmission of electromagnetic energy through the pipe of Fig. 8-1*c* it is found that the circuit theory which worked for lumped circuits and two-conductor transmission lines is inadequate. For the hollow metal pipe or tube, it is necessary to direct our attention to the empty space inside the tube and to the electric and magnetic fields **E** and **H** inside the tube. From the field-theory point of view we realize that the energy is actually conveyed through the empty space inside the tube and that the currents or voltages are merely associated effects.

8-3 TE MODE WAVE IN THE INFINITE-PARALLEL-PLANE TRANSMISSION LINE OR GUIDE

As an introduction to waveguides, let us consider an infinite-parallel-plane transmission line, as in Fig. 8-2. This is a two-conductor line which is capable of guiding energy in a *transverse electromagnetic* (TEM) mode with **E** in the *z* direction. However, at sufficiently high frequencies it can also transmit higher-order modes, and

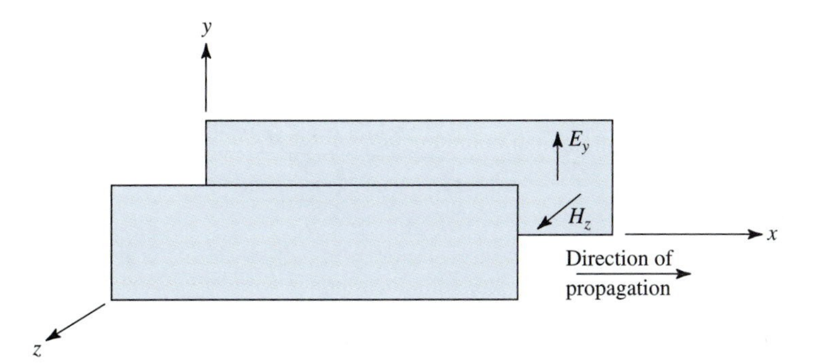

FIGURE 8-2
Transmission system consisting of two conducting planes parallel to the *xy* plane. The planes are assumed to be infinite in extent (infinite-parallel-plane transmission line). This line can transmit a TEM mode with **E** in the *z* direction and also TE modes with **E** in the *y* direction as suggested.

this type of transmission between the parallel planes serves as a good starting point for our discussion of higher-order modes.

Consider the higher-order mode where the electric field is everywhere in the y direction, with transmission in the x direction; i.e., the electric field has only an E_y component. Since E_y is transverse to the direction of transmission, this mode is designated a *transverse electric* (TE) mode. Although **E** is everywhere transverse, **H** has longitudinal, as well as transverse, components. Assuming perfectly conducting sheets, boundary conditions require that E_y vanish *at* the sheets. However, E_y need not be zero at points between the sheets. It is possible to determine the properties of a TE wave of the type under discussion by regarding it as being made up of two plane TEM waves reflected obliquely back and forth between the sheets.

First, however, consider the situation that exists when two plane TEM waves of the same frequency traveling in free space intersect at an angle, as suggested in Fig. 8-3 (on next page). It is assumed that the waves are linearly polarized with **E** normal to the page. Wavefronts, or surfaces of constant phase, are indicated for the two waves.

The solid lines (marked "Max") show where the field is a maximum with **E** directed out from the page. These lines may be regarded as representing the crests of the waves. The dashed lines (marked "Min") show where the field is a minimum, i.e., where **E** is maximum but directed into the page. These lines may be regarded as representing the troughs of the waves. Wherever the crest of one wave coincides with the trough of the other wave, there is cancellation, and the resultant **E** at that point is zero. Wherever crest coincides with crest or trough with trough, there is reinforcement, and the resultant **E** at that point doubles. Referring to Fig. 8-3, it is therefore apparent that at all points along the dash-dot lines the field is always zero, while along the line indicated by dash–double dot the field will be reinforced and will have a maximum value.

Since **E** is zero along the dash-dot lines, boundary conditions will be satisfied at plane, perfectly conducting sheets inserted along these lines normal to the page. The

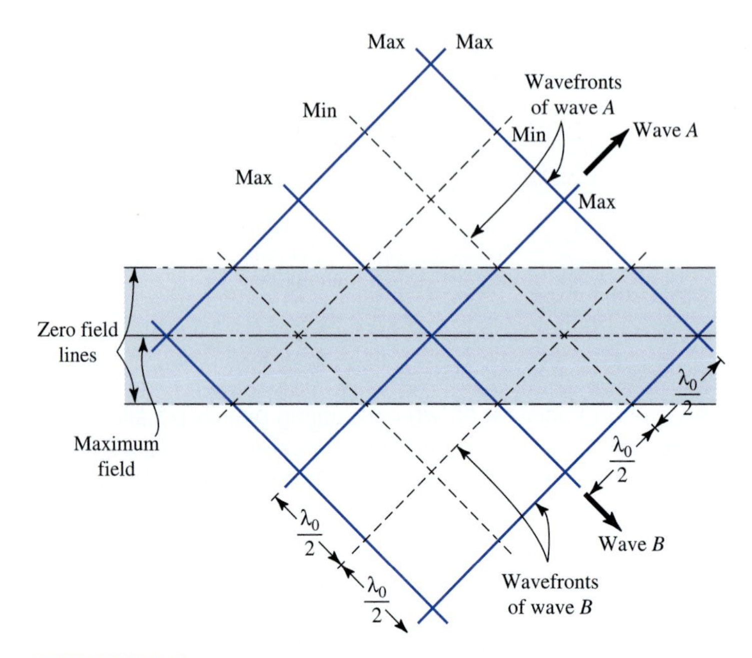

FIGURE 8-3
Two plane TEM waves traveling in free space in different directions result in maximum field along middle horizontal line (dash double–dot) and zero field along the two dash-dot lines.

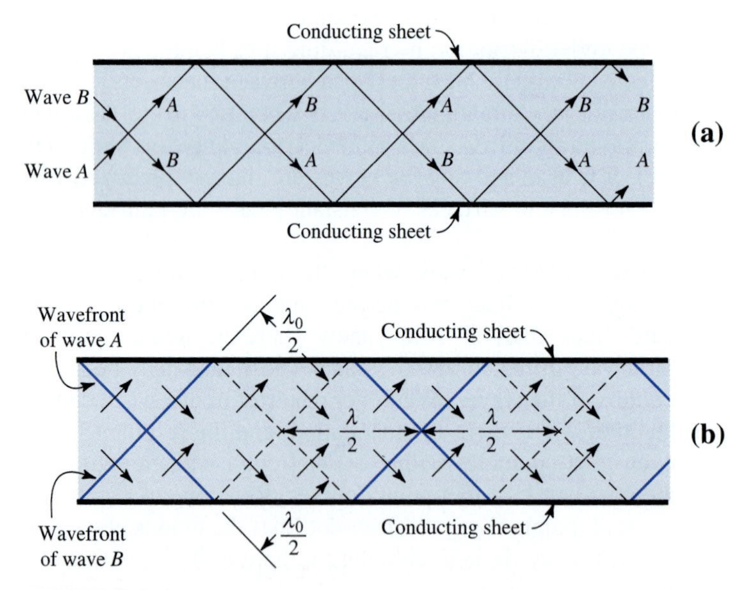

FIGURE 8-4
(*a*) Wave paths and (*b*) wave fronts between infinite parallel conducting sheets acting as a waveguide for a higher-order (TE) mode.

waves, however, will now be reflected at the sheets with an angle of reflection equal to the angle of incidence, and waves incident from the outside will not penetrate to the region between the sheets. But if two plane waves (A and B) are launched between the sheets from the left end, they will travel to the right via multiple reflections between the sheets, as suggested by the wave paths in Fig. 8-4*a*. The wavefronts (normal to the wave paths) for these waves are as indicated in Fig. 8-4*b*. Here the field between the sheets is the same as in Fig. 8-3, with solid lines indicating that **E** is outward (a maximum) and dashed lines that **E** is inward (a minimum). At the sheets the resultant **E** is always zero.

Although the two component waves we have been considering are plane TEM-mode waves, the *resultant wave* belongs to a higher-order TE mode. It is an important property of the TE-mode wave that it will not be transmitted unless the wavelength is sufficiently short. The critical wavelength at which transmission is no longer possible is called the *cutoff wavelength.* It is possible by a very simple analysis, which will now be given, to calculate the cutoff wavelength as a function of the sheet spacing.

Referring to Fig. 8-5, let the TE wave be resolved into two component TEM waves traveling in the x' and x'' directions. These directions make an angle θ with the conducting sheets (and the x axis). The electric field is in the y direction (normal to the page). The spacing between the sheets is b. From Fig. 8-5 we note that E_y' of the x' wave and E_y'' of the x'' wave cancel at a point such as A at the conducting sheet and reinforce at point B midway between the sheets provided that the distance

$$CB = BD = C'B = \frac{\lambda_0}{4} \tag{1}$$

where λ_0 is the wavelength of the TEM wave in unbounded space filled with the same medium as between the sheets.

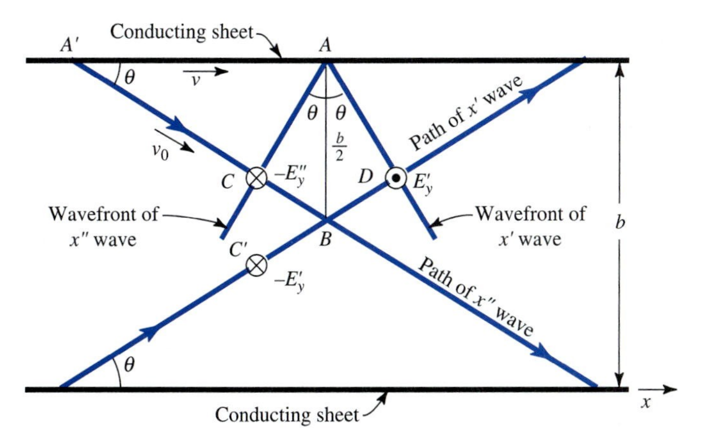

FIGURE 8-5
Component waves between infinite-parallel-plane conducting sheets acting as a waveguide. We note that as the phase front of the component wave travels from A' to C the same wavefront sweeps along the conducting sheet a greater distance ($A'A$) so that $v > v_0$.

Thus, if E_y'' is into the page (negative) at the point C and E_y' is out of the page (positive) at the point D, the two waves will cancel at A. They will also reinforce at B since by the time the field $-E_y''$ moves from C to B the field $-E_y'$ will have moved from C' to B. More generally we can write

$$CB = \frac{n\lambda_0}{4} \tag{2}$$

where n is an integer (1, 2, 3, ...).†

It follows that

$$AB \sin\theta = \frac{b}{2}\sin\theta = \frac{n\lambda_0}{4} \tag{3}$$

or

$$\lambda_0 = \frac{2b}{n}\sin\theta \tag{4}$$

where λ_0 = wavelength, m
b = spacing of conducting sheets, m
n = 1, 2, 3, ...
θ = angle between component wave direction and conducting sheets

According to (4), we note that for a given sheet separation b, the longest wavelength that can be transmitted in a higher-order mode occurs when $\theta = 90°$. This wavelength is the cutoff wavelength λ_{oc} of the higher-order mode. Thus, for $\theta = 90°$,

$$\lambda_{oc} = \frac{2b}{n} \tag{5}$$

Each value of n corresponds to a particular higher-order mode. When $n = 1$, we find that

$$\boxed{\lambda_{oc} = 2b \qquad \textit{Cutoff wavelength}} \tag{6}$$

This is the longest wavelength which can be transmitted between the sheets in a higher-order mode. That is, the spacing b must be at least $\lambda/2$ for a higher-order mode to be transmitted.

When $n = 1$, the wave is said to be the lowest of the higher-order types. When $n = 2$, we have the next higher-order mode and for this case

$$\lambda_{oc} = b \tag{7}$$

Thus, the spacing b must be at least 1λ for the $n = 2$ mode to be transmitted. For $n = 3$, $\lambda_{oc} = \left(\frac{2}{3}\right)b$, etc.

†For n even, the field halfway between the sheets is zero, with maximum fields either side of the centerline.

Introducing (5) in (4) yields $\sin\theta = \lambda_0/\lambda_{oc}$, or

$$\theta = \sin^{-1}\frac{\lambda_0}{\lambda_{oc}} \tag{8}$$

Hence, at cutoff for any mode ($\lambda_0 = \lambda_{oc}$) the angle $\theta = 90°$. Under these conditions the component waves for this mode are reflected back and forth between the sheets, as in Fig. 8-6*a*, and do not progress in the x direction. Hence there is a standing wave between the sheets, and no energy is propagated. If the wavelength λ_0 is slightly less than λ_{oc}, θ is less than 90° and the wave progresses in the x direction although making many reflections from the sheets, as in Fig. 8-6*b*. As the wavelength is further reduced, θ becomes less, as in Fig. 8-6*c*, until at very short wavelengths the transmission for this mode approaches the conditions in an unbounded medium.

It is apparent from Fig. 8-5 that a constant-phase point of the TE wave moves in the x direction with a velocity v that is greater than that of the component waves. The phase velocity v_0 of the component TEM waves is the same as for a wave in an unbounded medium of the same kind as fills the space between the conducting sheets. That is,

$$v_0 = \frac{1}{\sqrt{\mu\varepsilon}} \qquad (\text{m s}^{-1}) \tag{9}$$

where μ = permeability of medium, H m^{-1}, and ε = permittivity of medium, F m^{-1}. From Fig. 8-5 it follows that

$$\frac{v_0}{v} = \frac{A'C}{A'A} = \cos\theta \tag{10}$$

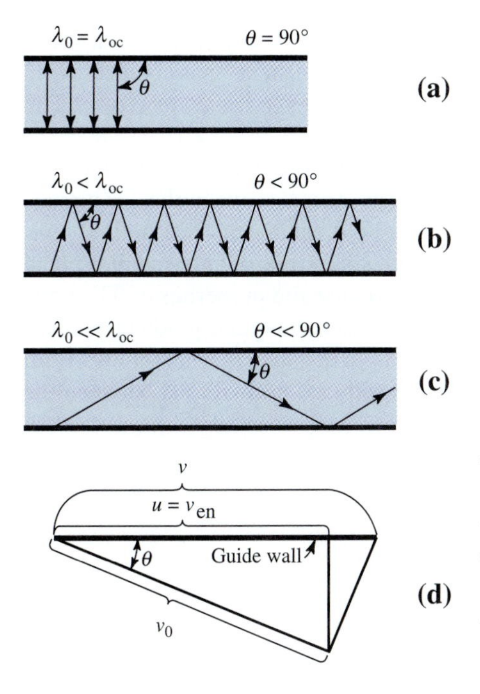

FIGURE 8-6
(*a, b, c*) As the wavelength λ_0 becomes less, the component waves make fewer reflections. (*d*) Triangle showing magnitude of phase velocity v, group velocity u, and energy velocity v_{en}, in the guide relative to phase velocity v_0 of the component wave (equal to phase velocity of wave in an unbounded medium).

or
$$v = \frac{v_0}{\cos\theta} = \frac{1}{\sqrt{\mu\varepsilon}\cos\theta} \qquad (\text{m s}^{-1}) \qquad \textbf{\textit{Phase velocity}} \qquad (11)$$

According to (11), the phase velocity v of a TE wave approaches an infinite value as the wavelength is increased toward the cutoff value. On the other hand, v approaches the phase velocity v_0 in an unbounded medium as the wavelength becomes very short. Thus, the phase velocity of a higher-order mode wave in the guide formed by the sheets is always equal to or greater than the velocity in an unbounded medium. The energy, however, is propagated with the velocity of the zigzag component wave. Thus $v_{en} = v_0 \cos\theta$. Accordingly, the energy velocity v_{en} is always equal to or less than the velocity in an unbounded medium.† When, for instance, the wavelength approaches cutoff, the phase velocity becomes infinite while the energy velocity approaches zero. This is another way of saying that the wave degenerates into a standing wave and does not propagate energy at the cutoff wavelength or longer wavelengths. The relative magnitudes of the various velocities are shown by the triangle in Fig. 8-6*d*.

Since the wavelength is proportional to the phase velocity, the wavelength λ of the higher-order mode in the guide is given in terms of the wavelength λ_0 in an unbounded medium by

$$\lambda = \frac{\lambda_0}{\cos\theta} \qquad (12)$$

The phase velocity and group (or energy) velocity in the guide as a function of θ are shown in Fig. 8-7. As θ approaches 90°, the phase velocity becomes infinite while the energy velocity goes to zero. The velocities are expressed in terms of the phase velocity v_0 of the wave in an unbounded medium. The situation here is analogous to the action of water waves at a breakwater. Thus, as suggested in Fig. 8-8, a water plume moves along the breakwater where a wave crest (constant-phase point) strikes the breakwater. The velocity v of the plume is greater than the wave velocity v_0. The plume velocity can become infinite if θ becomes 90°.

The infinite-parallel-plane transmission line we have been considering is an idealization and not a type to be applied in practice. Actual waveguides for higher-order modes usually take the form of a single hollow conductor. The hollow rectangular guide is a common form. The above analysis for the infinite-parallel-plane transmission line is of practical value, however, because the properties of TE-mode waves, such as are discussed above, are the same in a rectangular guide of width b as between two infinite parallel planes separated by a distance b. This follows from the fact that if infinitely conducting sheets are introduced normal to **E** between the

†The waveguide behaves like a lossless dispersive medium. It follows that

$$u = v_{en} = \frac{v_0^2}{v}$$

where u = group velocity
v_{en} = energy velocity
v_0 = phase velocity in unbounded medium
v = phase velocity in guide

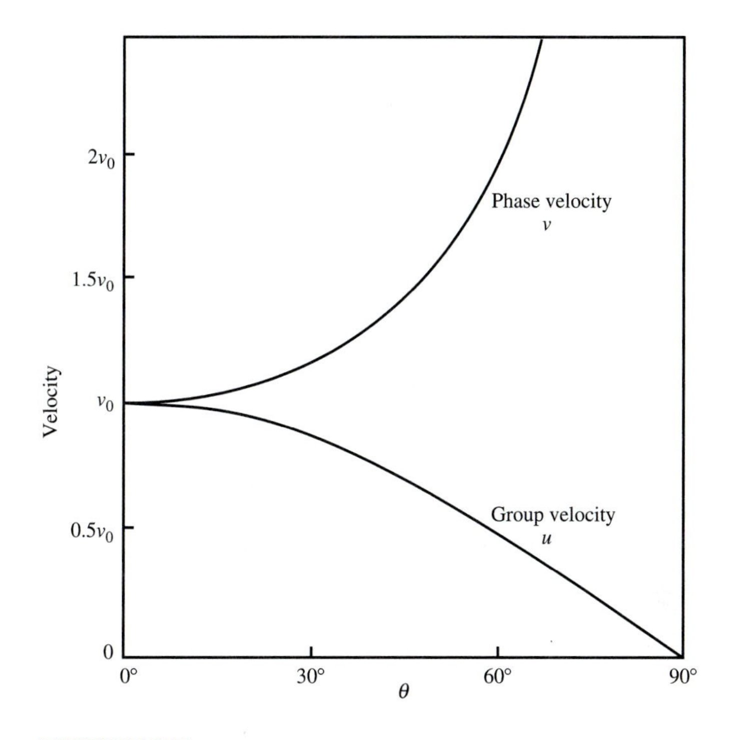

FIGURE 8-7
Phase and group velocity as a function of wave angle θ. The ordinate gives v and u in terms of the velocity v_0 for a wave in an unbounded medium of the same type that fills the waveguide.

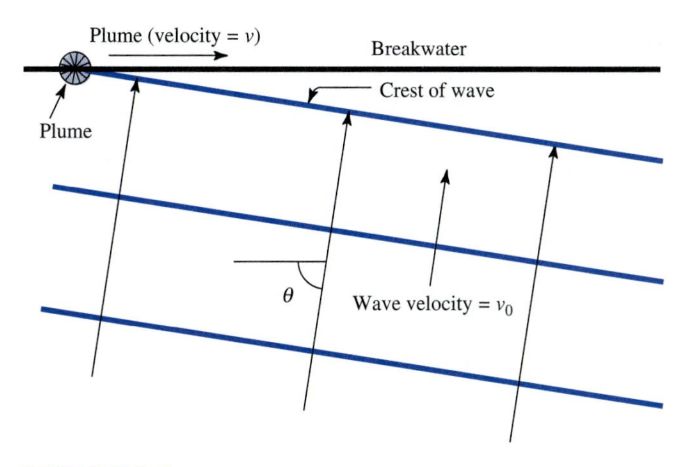

FIGURE 8-8
A plume of water moves along a breakwater with a phase velocity v that is greater than the wave velocity v_0.

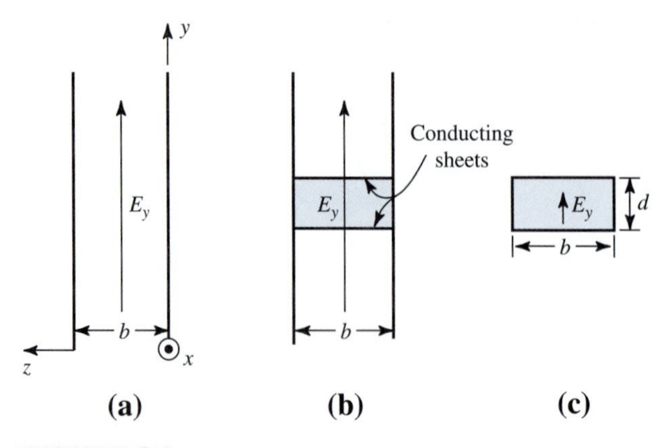

FIGURE 8-9
(a) Infinite-parallel-plane transmission line acting as a waveguide for TE wave. **E** is in y direction with wave in x direction (out of page). The guide consists of two parallel conducting sheets separated by a distance b. Additional sheets introduced normal to E_y as in (b) result in the hollow rectangular waveguide (c).

parallel planes, the field is not disturbed. Thus, if a TE-mode wave with electric field in the y direction is traveling in the x direction, as indicated in Fig. 8-9a, the introduction of sheets normal to E_y, as in Fig. 8-9b, does not disturb the field. The conducting sheets now form a complete enclosure of rectangular shape. Proceeding a step further, let the sheets beyond the rectangular enclosure be removed, leaving the hollow rectangular waveguide shown in Fig. 8-9c. The cutoff wavelengths for the TE modes as given by (5) for the infinite-parallel-plane line also apply for this rectangular guide of width b. For the type of TE modes thus far considered (E_y component only), the dimension d (Fig. 8-9c) is not critical.

Although the above simple analysis yields information about cutoff wave length, phase velocity, etc., it gives little information concerning the field configuration and fails to consider more complex higher-order modes of wave transmission in which, for example, **E** is transverse but has *both* y and z components. To obtain complete information concerning the waves in a hollow waveguide, we need to solve the wave equation subject to the boundary conditions for the guide. This is done for the hollow rectangular guide in the next section.

8-4 THE HOLLOW RECTANGULAR WAVEGUIDE†

In Sec. 8-3 certain properties of an infinite-parallel-plane transmission line and of a hollow rectangular guide were obtained by considering that the higher-mode wave

†A more detailed development is given in the Fourth Edition of this book. See also: Lord Rayleigh, On the Passage of Electric Waves through Tubes, *Phil. Mag.*, **43**: 125–132 (February 1897); L. J. Chu and W. L. Barrow, Electromagnetic Waves in Hollow Metal Tubes of Rectangular Cross Section, *Proc. IRE*, **26**: 1520–1555 (December 1938).

consists of two plane TEM component waves and then, applying the boundary condition, that the tangential component of the resultant **E** must vanish at the perfectly conducting walls of the guide. This method could be extended to provide more complete information about the waves in a hollow waveguide. However, in this section we shall use a more general approach, which involves the solution of the *wave equation* subject to the above-mentioned boundary condition for the tangential component of **E**.

In this method we start with Maxwell's equations and develop a wave equation in rectangular coordinates (Fig. 8-10). This choice of coordinates is made in order that the boundary conditions for the rectangular guide can be easily applied later. The restrictions are then introduced of harmonic variation with respect to time and a wave traveling in the x direction (direction of guide). Next a choice is made of the type of higher-order mode of transmission to be analyzed. Thus, we may consider a transverse electric (TE) wave for which $E_x = 0$ or a transverse magnetic (TM) wave for which $H_x = 0$. If, for example, we select the TE type, we know that there must be an H_x component, since a higher-mode wave always has a longitudinal field component and E_x being zero means that H_x must have a value. It is then convenient to write the remaining field components in terms of H_x. Next a solution of a scalar-wave equation in H_x is obtained that fits the boundary conditions of the rectangular guide. This solution is substituted back into the equations for the other field components (E_y, E_z, H_y, and H_z). In this way we end up with equations giving the variation of each field component with respect to space and time as in the following examples.

Example 8-1. TE$_{10}$ MODE. For this mode $m = 1$ and $n = 0$, and we have, as mentioned above, only three components E_y, H_x, and H_z that are not zero. The six field components for the TE$_{10}$ mode are then

$$E_x = 0 \quad \text{TE mode requirement}$$

$$E_y = \frac{\gamma Z_{yz} H_0}{k^2} \frac{\pi}{z_1} \sin \frac{\pi z}{z_1} e^{-\gamma x}$$

$$E_z = 0$$

$$H_x = H_0 \cos \frac{\pi z}{z_1} e^{-\gamma x}$$

$$H_y = 0$$

$$H_z = \frac{\gamma H_0}{k^2} \frac{\pi}{z_1} \sin \frac{\pi z}{z_1} e^{-\gamma x}$$

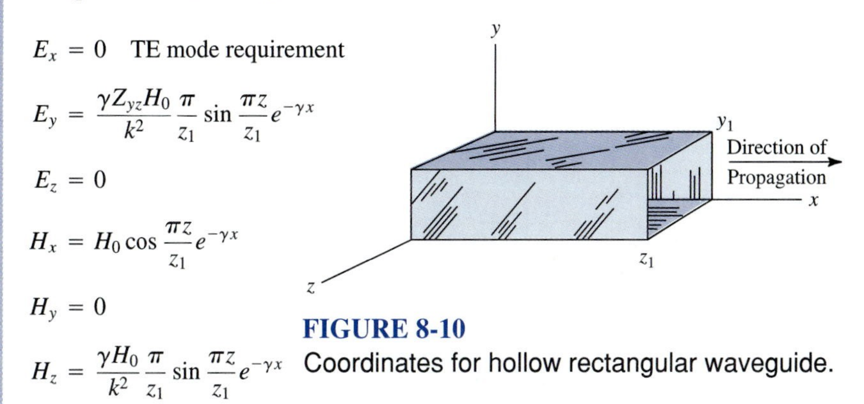

FIGURE 8-10
Coordinates for hollow rectangular waveguide.

The variation of these components as a function of z is portrayed in Fig. 8-11*a*. There is no variation with respect to y. This mode has the longest cutoff wavelength of any higher-order mode, and hence, the lowest frequency of transmission in a hollow rectangular waveguide must be in the TE$_{10}$ mode. In Fig. 8-12*a* the field configuration of the TE$_{10}$ mode is illustrated for a guide cross-section and in Fig. 8-12*b* for a longitudinal section of the guide (top view).

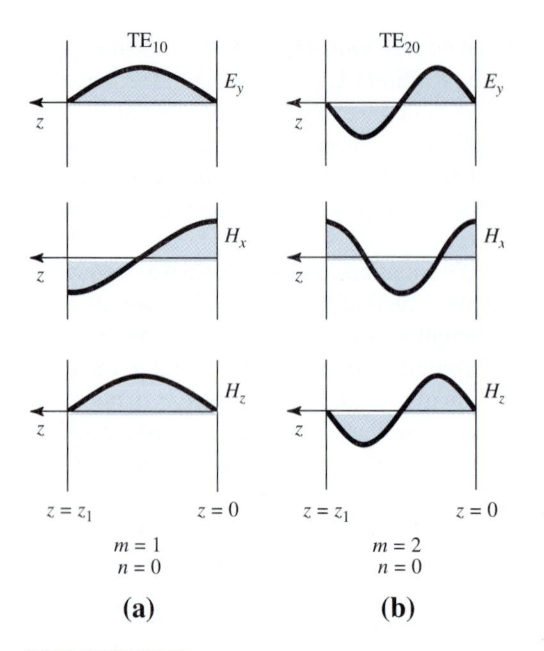

FIGURE 8-11
Variation of field components for TE$_{10}$ and TE$_{20}$ modes in a hollow rectangular waveguide. (Wave traveling out of page.)

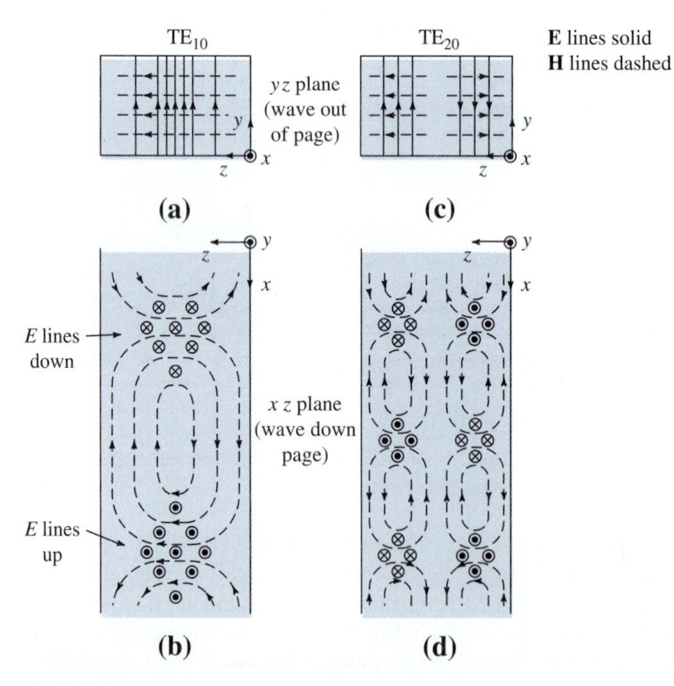

FIGURE 8-12
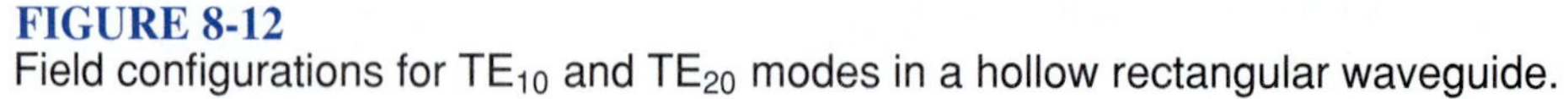
Field configurations for TE$_{10}$ and TE$_{20}$ modes in a hollow rectangular waveguide.

Example 8-2. TE_{20} MODE.† The variation of the field components as a function of z for the TE_{20} mode ($m = 2$, $n = 0$) is shown in Fig. 8-11*b*. The field configuration for a TE_{20} mode is shown in cross-section in Fig. 8-12*c* and in longitudinal section (top view) in Fig. 8-12*d*.

Example 8-3. TE_{11} MODE. For this mode $m = 1, n = 1$, and the field components are given by

$$E_x = 0 \qquad \text{TE mode requirement} \qquad H_x = H_0 \cos\frac{\pi y}{y_1}\cos\frac{\pi z}{z_1}e^{-\gamma x}$$

$$E_y = \frac{\gamma Z_{yz} H_0}{k^2}\frac{\pi}{z_1}\cos\frac{\pi y}{y_1}\sin\frac{\pi z}{z_1}e^{-\gamma x} \qquad H_y = \frac{\gamma H_0}{k^2}\frac{\pi}{y_1}\sin\frac{\pi y}{y_1}\cos\frac{\pi z}{z_1}e^{-\gamma x}$$

$$E_z = -\frac{\gamma Z_{yz} H_0}{k^2}\frac{\pi}{y_1}\sin\frac{\pi y}{y_1}\cos\frac{\pi z}{z_1}e^{-\gamma x} \qquad H_z = \frac{\gamma H_0}{k^2}\frac{\pi}{z_1}\cos\frac{\pi y}{y_1}\sin\frac{\pi z}{z_1}e^{-\gamma x}$$

For this mode, five field components have a value, only E_x being everywhere and always zero. The variation of the five field components with respect to z and y is shown in Fig. 8-13 (on next page). It is assumed that the guide has a square cross-section ($y_1 = z_1$) The field configuration for the TE_{11} mode in a square guide is illustrated in cross-section (end view) in Fig. 8-14*a* and in longitudinal section (side view) in Fig. 8-14*b*.

The solution we have obtained tells us what modes are possible in the hollow rectangular waveguide. However, the particular mode or modes that are actually present in any case depend on the guide dimensions, the method of exciting the guide, and the irregularities or discontinuities in the guide. The resultant field in the guide is equal to the sum of the fields of all modes present.

Returning now to a consideration of the general significance of the solution, we have

$$\left(\frac{n\pi}{y_1}\right)^2 + \left(\frac{m\pi}{z_1}\right)^2 = k^2 \tag{1}$$

Also k^2 is given by

$$k^2 = \gamma^2 - j\omega\mu(\sigma + j\omega\varepsilon) \tag{2}$$

Assuming a lossless dielectric medium in the guide, we can put $\sigma = 0$. Then equating (1) and (2) and solving for γ yields

$$\boxed{\gamma = \sqrt{\left(\frac{n\pi}{y_1}\right)^2 + \left(\frac{m\pi}{z_1}\right)^2 - \omega^2\mu\varepsilon}} \tag{3}$$

†The significance of the subscripts mn in TE_{mn} is as follows:

m = number of half-cycle variations of field in z direction

n = number of half-cycle variations of field in y direction

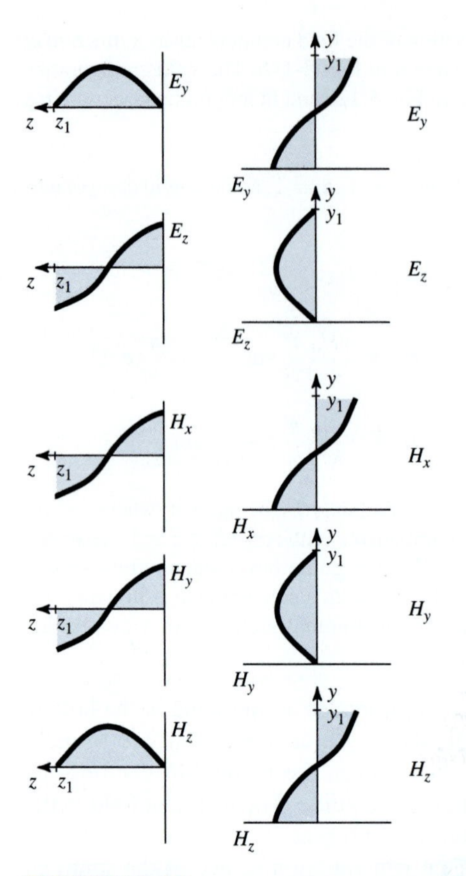

FIGURE 8-13
Variation of field components for TE_{11} mode in a square waveguide. (Wave traveling out of page.)

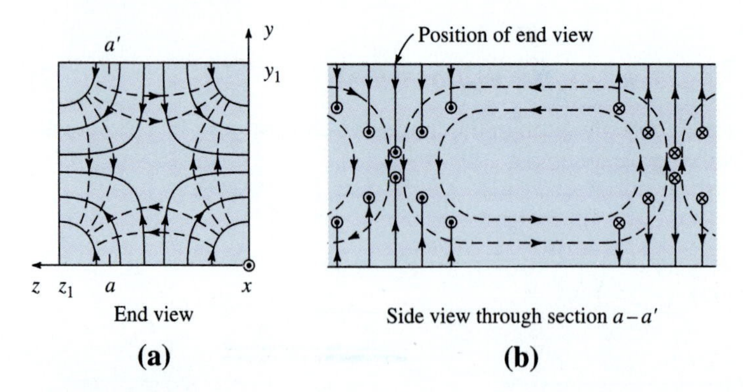

FIGURE 8-14
Field configurations for TE_{11} mode in a square waveguide. **E** lines are solid, and **H** lines are dashed.

At sufficiently low frequencies the last term in (3) is smaller than the sum of the first two terms under the square-root sign. It follows that for this condition γ is real and therefore that the wave is attenuated. Under this condition it is said that the wave (or mode) is not propagated.

At sufficiently high frequencies the last term in (3) is larger than the sum of the first two terms under the square-root sign. Under this condition γ is imaginary, and therefore the wave is propagated without attenuation.

At some intermediate frequency the right-hand side of (3) is zero, and hence $\gamma = 0$. This frequency is called the ***cutoff frequency*** for the mode under consideration. At frequencies higher than cutoff, this mode propagates without attenuation, while at frequencies lower than cutoff, the mode is attenuated.

To summarize the three cases:

1. At low frequencies, ω small, γ real, guide opaque, wave does not propagate.
2. At an intermediate frequency, ω intermediate, $\gamma = 0$, transition condition (cutoff).
3. At high frequencies, ω large, γ imaginary, guide transparent, wave propagates.

Referring to (3), it is to be noted that $\sqrt{\omega^2\mu\varepsilon}$ is equal to the phase constant β_0 for a wave traveling in an unbounded medium of the same dielectric material that fills the guide. Thus we can write

$$\gamma = \sqrt{k^2 - \beta_0^2} \qquad (\mathrm{m}^{-1}) \tag{4}$$

where $\beta_0 = \sqrt{\omega^2\mu\varepsilon} = 2\pi/\lambda_0 =$ phase constant in unbounded medium
$\lambda_0 =$ wavelength in unbounded medium
$k = \sqrt{(n\pi/y_1)^2 + (m\pi/z_1)^2}$

Thus, at frequencies higher than cutoff, $\beta_0 > k$, and

$$\gamma = \sqrt{k^2 - \beta_0^2} = j\beta \tag{5}$$

where $\beta = 2\pi/\lambda = \sqrt{\beta_0^2 - k^2} =$ phase constant in guide, rad m^{-1}, and $\lambda =$ wavelength in guide, m.

At sufficiently high frequencies ($\beta_0 \gg k$) we note that the phase constant β in the guide approaches the phase constant β_0 in an unbounded medium. On the other hand, at frequencies less than cutoff $\beta_0 < k$, and

$$\gamma = \sqrt{k^2 - \beta_0^2} = \alpha \tag{6}$$

where $\alpha =$ attenuation constant.

At sufficiently low frequencies ($\beta_0 \ll k$) we note that the attenuation constant α approaches a constant value k.

At the cutoff frequency, $\beta_0 = k$ and $\gamma = 0$. Thus, at cutoff

$$\omega^2\mu\varepsilon = \left(\frac{n\pi}{y_1}\right)^2 + \left(\frac{m\pi}{z_1}\right)^2 \tag{7}$$

It follows that the *cutoff frequency* is

$$f_c = \frac{1}{2\sqrt{\mu\varepsilon}}\sqrt{\left(\frac{n}{y_1}\right)^2 + \left(\frac{m}{z_1}\right)^2} \quad \text{(Hz)} \qquad \textbf{Cutoff frequency} \tag{8}$$

and the *cutoff wavelength* is

$$\lambda_{oc} = \frac{2\pi}{\sqrt{(n\pi/y_1)^2 + (m\pi/z_1)^2}} = \frac{2}{\sqrt{(n/y_1)^2 + (m/z_1)^2}} \quad \text{(m)} \qquad \textbf{Cutoff wavelength} \tag{9}$$

where λ_{oc} is the wavelength in an unbounded medium at the cutoff frequency (or, more concisely, the *cutoff wavelength*).†

Equations (8) *and* (9) *give the cutoff frequency and cutoff wavelength for any* TE_{mn} *mode in a hollow rectangular guide*. For instance, the cutoff wavelength of a TE_{10} mode is

$$\lambda_{oc} = 2z_1 \tag{10}$$

This is identical with the value found in Sec. 8-3 since $z_1 = b$.

At frequencies above cutoff ($\beta_0 > k$)

$$\beta = \sqrt{\beta_0^2 - k^2} = \sqrt{\omega^2\mu\varepsilon - \left(\frac{n\pi}{y_1}\right)^2 - \left(\frac{m\pi}{z_1}\right)^2} \tag{11}$$

It follows that the *phase velocity* v_p *in the guide* is equal to

$$v_p = \frac{\omega}{\beta} = \frac{v_0}{\sqrt{1 - (n\lambda_0/2y_1)^2 - (m\lambda_0/2z_1)^2}} \quad (\text{m s}^{-1}) \tag{12}$$

or

$$v_p = \frac{v_0}{\sqrt{1 - (\lambda_0/\lambda_{oc})^2}} \qquad \textbf{Phase velocity} \tag{13}$$

where $v_0 = 1/\sqrt{\mu\varepsilon}$ = phase velocity in unbounded medium (= 300 Mm s^{-1} for air)
λ_0 = wavelength in unbounded medium
λ_{oc} = cutoff wavelength

The ratio v_p/v_0 as a function of the wavelength λ_0 is shown in Fig. 8-15 for several TE modes in a hollow waveguide of square cross-section ($y_1 = z_1$).

†Note that $k = 2\pi/\lambda_0$. If this value of k is introduced, (5) can be used to relate λ, λ_0, and λ_{oc} when $\lambda_0 < \lambda_{oc}$.

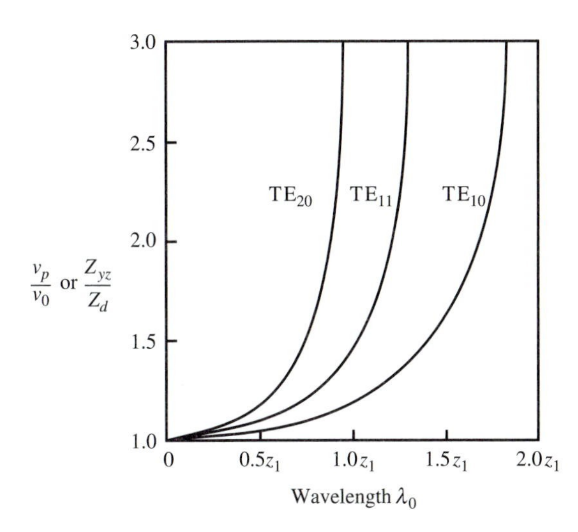

FIGURE 8-15
Relative phase velocity v_p/v_0 or relative transverse impedance Z_{yz}/Z_d as a function of the wavelength λ_0 for TE modes in a hollow square guide (height y_1 equal to width z_1).

In the above analysis there is no attenuation whatsoever at frequencies above cutoff. This results from the assumption of perfectly conducting guide walls and a lossless dielectric medium filling the guide. However, if the walls are not perfectly conducting or the medium is not lossless, or both, there is attenuation.†

If the guide is filled with air, the dielectric loss is usually negligible compared with losses in the guide walls, so that the attenuation at frequencies greater than cutoff is mainly determined by the conductivity of the guide walls. The fact that the guide walls are not perfectly conducting means that the tangential component E_t of the electric field is not zero at the walls but has a finite value. However, for walls made of a good conductor, such as copper, E_t will generally be so small that the above analysis (based on $E_t = 0$) is not affected to any appreciable extent. However, as a result of the finite wall conductivity, α is not zero. Thus, in most practical problems where the wall conductivity is high (but not infinite), the field configuration in the guide, the wavelength λ, the phase constant β, the phase velocity v, etc. can all be calculated with high accuracy on the assumption that the walls have infinite conductivity, as done earlier in this section. The small (but not zero) attenuation may then be calculated separately to find the power lost per unit area in the guide wall, it being assumed that the **H**-field distribution is the same as with perfectly conducting walls.

Finally, let us determine the value of the *transverse-wave impedance* Z_{yz} for TE modes in a rectangular hollow guide. Thus,

$$Z_{yz} = \frac{j\omega\mu}{\gamma} \tag{14}$$

†That γ may have both a real and an imaginary part at frequencies greater than cutoff can be shown by solving (2) for γ under these conditions, with σ not equal to zero.

At frequencies higher than cutoff $\gamma = j\beta$; so

$$Z_{yz} = \frac{\omega\mu}{\beta} = \frac{Z_d}{\sqrt{1-(\lambda_0/\lambda_{oc})^2}} \quad (\Omega) \qquad \textbf{\textit{Transverse-wave impedance}} \qquad (15)$$

where Z_d = intrinsic impedance of dielectric medium filling guide
$= \sqrt{\mu/\varepsilon} = 377\ \Omega$ for air
λ_0 = wavelength in unbounded medium
λ_{oc} = cutoff wavelength

The ratio of Z_{yz} (transverse-wave impedance) to Z_d (intrinsic impedance) as a function of the wavelength λ_0 is shown in Fig. 8-15 for several TE modes in a hollow waveguide of square cross-section ($y_1 = z_1$).

Thus far only TE-mode waves have been considered. To find the field relations for transverse magnetic (TM) mode waves, we proceed as before in this section except that where TE appears we substitute TM and where E_x appears we substitute H_x, and vice versa. In the TM wave $H_x = 0$, and the longitudinal field component is E_x. This analysis will not be carried through here. However, it may be mentioned that (9) for the cutoff wavelength applies to both TE and TM waves, as does (12) for the phase velocity, but this is not the case with (15) for the transverse impedance. The notation for any TM mode, in general, is TM_{mn}, where m and n are integers (1, 2, 3, ...). It is to be noted that neither m nor n may be equal to zero for TM waves. Thus, the lowest-frequency TM wave that will be transmitted by a rectangular waveguide is the TM_{11} mode.

We have seen that each mode of transmission in a waveguide has a particular cutoff wavelength, velocity, and impedance. When the frequency is high enough to permit transmission in more than one mode, the resultant field is the sum of the fields of the individual mode fields in the guide.

For example, suppose that a rectangular waveguide, as shown in cross-section in Fig. 8-16*a*, is excited in the TE_{10} mode. The variation of E_y across the guide is sinusoidal, as shown in Fig. 8-16*b*. Suppose now that z_1 exceeds 1λ, so that the TE_{20} mode can also be transmitted.† If only the TE_{10} mode is excited, no TE_{20} will

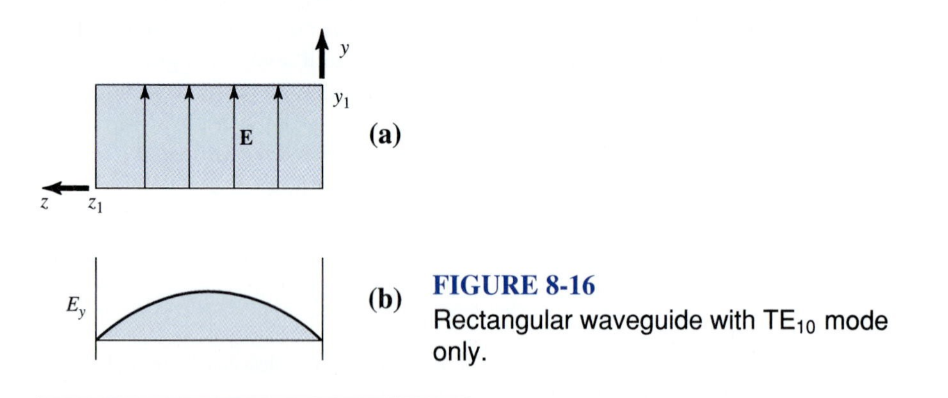

FIGURE 8-16
Rectangular waveguide with TE_{10} mode only.

†And $y_1 < \lambda_0/2$, so that no TE_{01} mode (**E** in z direction) is transmitted.

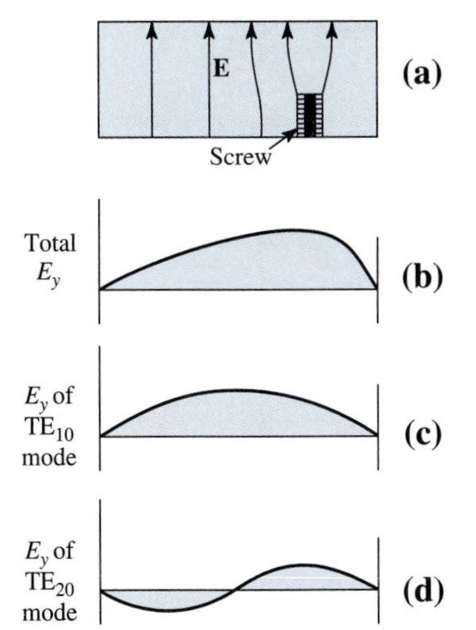

FIGURE 8-17
Rectangular waveguide with TE_{20} mode induced from mode TE_{10} mode by asymetrically placed projection (screw).

appear, provided that the guide is perfectly regular. However, in practice certain asymmetries and irregularities will be present, and these will tend to convert some of the TE_{10}-mode energy into TE_{20}-mode energy. Thus, if an asymmetrically located screw projects into the guide as in Fig. 8-17*a*, the total E_y field will tend to become asymmetrical, as suggested in Fig. 8-17*b*. The total field may be resolved into TE_{10} and TE_{20} components as shown in Fig. 8-17*c* and *d*. If both TE_{10} and TE_{20} modes can be transmitted, the field in the guide beyond the screw location will have energy in both modes.

In effect the screw is a receiving antenna that extracts energy from the incident TE_{10}-mode wave and reradiates it so as to excite the TE_{20} mode. However, if the frequency is decreased so that only the TE_{10} wave can be transmitted, the asymmetric field (Fig. 8-17*b*) will exist only in the vicinity of the screw and farther down the guide the field will be entirely in the TE_{10} mode. To avoid the problems of multiple-mode transmission, a waveguide is usually operated so that only one mode is capable of transmission.† For instance, to ensure transmission only in the TE_{10} mode, z_1 must be less than 1λ and y_1 less than $\lambda/2$. But to allow transmission in the TE_{10} mode, z_1 must exceed $\lambda/2$. Hence z_1 must be between $\lambda/2$ and 1λ, and a value of 0.7λ is often used since this is well below 1λ and yet enough more than $\lambda/2$ so that the velocity and transverse impedance values are not too-critical functions of frequency.

We recall that at cutoff ($z_1 = \lambda/2$), the velocity and impedance approach infinite values. The height y_1 may be as small as desired without preventing transmission of the TE_{10} mode. Too small a value of y_1, however, increases attenuation (because of power lost in the guide walls) and also reduces the power-handling capabilities of the guide. It is often the practice to make $y_1 = z_1/2$. Many TE_{mn} and TM_{mn} modes of a rectangular guide for which $y_1 = z_1/2$ are shown in Fig. 8-18. The slant scale

†The lowest-frequency mode that a guide can transmit is called the *dominant mode*.

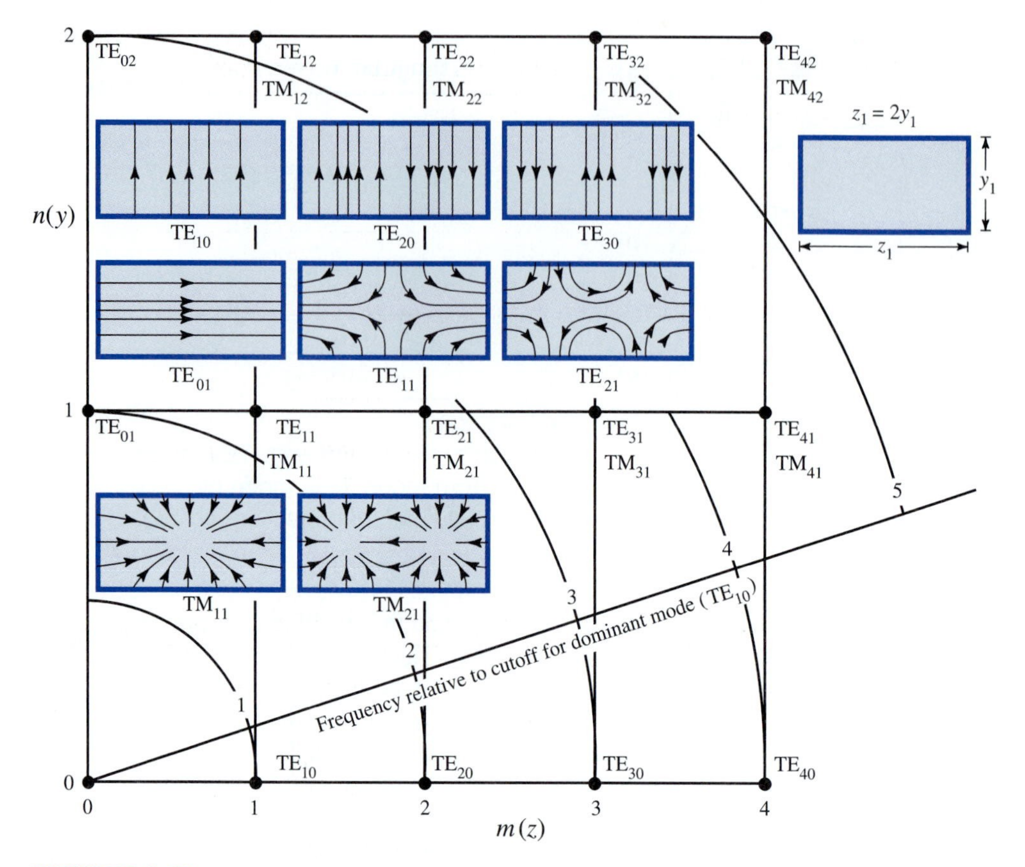

FIGURE 8-18
Possible TE and TM modes in hollow rectangular waveguide as a function of frequency. At 3 times the cutoff frequency for the TE_{10} mode, there are seven modes which will pass (see text) and one mode (TE_{30}) at cutoff. Guide cross-sections with electric field configurations are also shown for the first eight modes. Magnetic field lines are not shown but would be perpendicular to the electric field lines.

gives the frequency relative to the cutoff frequency for the dominant mode (TE_{10}). Thus, if the frequency is three times this value, we note from Fig. 8-18 that the following modes will be transmitted: TE_{10}, TE_{01}, TE_{20}, TE_{11}, TM_{11}, TE_{21}, and TM_{21}. An additional mode (TE_{30}) is at its cutoff frequency. Electric field configurations are also shown in Fig. 8-18 for the first eight modes in a rectangular waveguide.†

The relations derived in this section for a hollow rectangular waveguide (see Fig. 8-10) are summarized in Table 8-1.

†Field configurations for the first 36 modes are given by C. S. Lee, S. W. Lee, and S. L. Chuang, Plot of Modal Field Distribution in Rectangular and Circular Waveguides, *IEEE Trans. Microwave Theory and Techniques,* **MTT-33**: 271–274, March 1985.

TABLE 8-1
Relations for TE_{mn} modes in hollow rectangular waveguides†

Name of relation	Relation
Cutoff frequency	$f_c = \dfrac{1}{2\sqrt{\mu\varepsilon}}\sqrt{\left(\dfrac{n}{y_1}\right)^2 + \left(\dfrac{m}{z_1}\right)^2}$ (Hz)
Cutoff wavelength	$\lambda_{oc} = \dfrac{2}{\sqrt{(n/y_1)^2 + (m/z_1)^2}}$ (m)
Wavelength in guide	$\lambda_g = \dfrac{\lambda_0}{\sqrt{1-(\lambda_0/\lambda_{oc})^2}}$ (m)
Phase velocity	$v_p = \dfrac{v_0}{\sqrt{1-(n\lambda_0/2y_1)^2 - (m\lambda_0/2z_1)^2}} = \dfrac{v_0}{\sqrt{1-(\lambda_0/\lambda_{oc})^2}} = \dfrac{v_0}{\sqrt{1-(f_c/f)^2}}$ (m s^{-1}) where $v_0 = 1/\sqrt{\mu\varepsilon}$
Transverse-wave impedance	$Z_{yz} = \dfrac{Z_d}{\sqrt{1-(n\lambda_0/2y_1)^2 - (m\lambda_0/2z_1)^2}} = \dfrac{Z_d}{\sqrt{1-(\lambda_0/\lambda_{oc})^2}} = \dfrac{Z_d}{\sqrt{1-(f_c/f)^2}}$ (Ω) where $Z_d = \sqrt{\mu/\varepsilon}$

† All the relations also apply to TM_{mn} modes except for the transverse-wave impedance relation. The velocity and impedance relations involving $(\lambda_0/\lambda_{oc})^2$ apply not only to rectangular guides but also to TE modes in hollow single-conductor guides of any shape.

The significance of the subscripts mn in TE_{mn} or TM_{mn} is as follows:

m = number of half-cycle variations of field in z direction
n = number of half-cycle variations of field in y direction

Example 8-4. Rectangular waveguide. A rectangular air-filled waveguide has a cross-section of 45 × 90 mm. Find (*a*) cutoff wavelength λ_{oc} for the dominant mode and (*b*) the relative phase velocity (v/c) in the guide at 1.6 times the cutoff frequency, (*c*) the cutoff wavelength if the guide is filled with dielectric of relative permittivity $\varepsilon_r = 1.7$, and (*d*) the relative phase velocity (v/c) with dielectric at 1.6 times the cutoff frequency.

Solution

$$\lambda_{oc} = 2 \times 90 = 180 \text{ mm} \quad \textit{Ans.} \quad (a)$$

$$v = \frac{c}{\sqrt{1-(\lambda_0/\lambda_{oc})^2}} = \frac{c}{\sqrt{1-(f_c/f)^2}} = \frac{c}{\sqrt{1-(1/1.6)^2}} = 1.28c \quad \textit{Ans.} \quad (b)$$

$$\lambda_{oc} = 2\sqrt{1.7} \times 90 = 235 \text{ mm} \quad Ans. \quad (c)$$

$$v = \frac{1}{\sqrt{1.7}} \frac{c}{\sqrt{1 - (1/1.6)^2}} = 0.98c \quad Ans. \quad (d)$$

Problem 8-4-1. Square waveguide. What modes are passed at frequencies below 3.75 GHz for a square waveguide 100 mm on a side? *Ans.* TE_{10}, TE_{01}, TE_{02}, TE_{20}, TE_{11}, TM_{11}, TE_{12}, TM_{12}, TE_{21}, and TM_{21}.

Problem 8-4-2. Rectangular waveguides. An air-filled waveguide has a cross-section which is 50 mm wide by 40 mm high. Find (*a*) the modes (both TE and TM) which this guide will transmit with $\lambda > 37.5$ mm and (*b*) the relative phase velocity (v/c) in the guide for each of the modes passed if $\lambda = 0.6\lambda_{oc}$. *Ans.* (*a*) TE_{10}, TE_{20}, TE_{01}, TE_{11}, TM_{11}, TE_{02}, TE_{21}, TM_{21}, (*b*) 1.25.

8-5 THE HOLLOW CYLINDRICAL WAVEGUIDE†

Consider the problem of describing wave propagation in a hollow (circular) cylindrical waveguide of radius r_0. This problem is most easily handled with a cylindrical coordinate system, as shown in Fig. 8-19. The procedure is similar to that used in the preceding section for the rectangular waveguide. We assume time-harmonic variation, perfectly conducting walls, and a lossless interior medium ($\sigma = 0$) containing no charge ($\rho = 0$).

The solutions involve Bessel functions and their roots. The roots (or eigenvalues) for the TM modes written k_{nr} (unprimed) correspond to zero values of the Bessel function $J_n(kr)$, whereas the roots for the TE modes written k'_{nr} (primed) correspond to zero values of the derivative (with respect to r) of the Bessel function. The relations for TE and TM modes are displayed in Fig. 8-20 and their numerical values

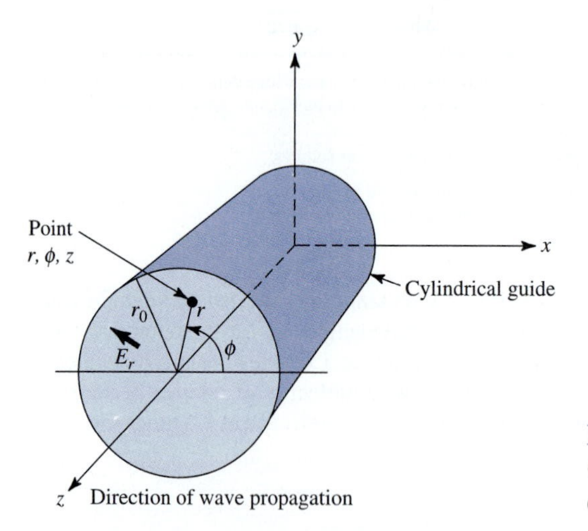

FIGURE 8-19
Coordinates for hollow cylindrical waveguide.

†A more detailed development is given in the Fourth Edition of this book. See also: W. L. Barrow, Transmission of Electromagnetic Waves in Hollow Tubes of Metal, *Proc. IRE,* **24**: 1298–1328, October 1936; G. C. Southworth, Some Fundamental Experiments with Wave Guides, ibid., **25**: 807–822, July 1937.

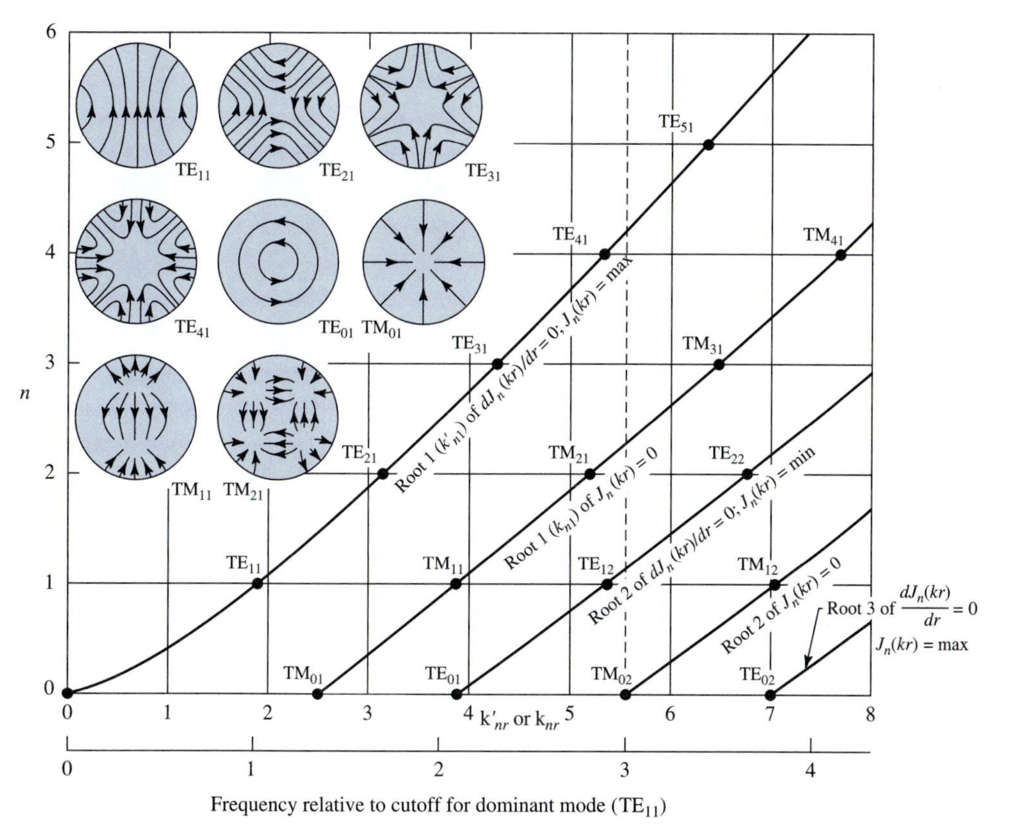

FIGURE 8-20
Possible TE and TM modes in a hollow cylindrical waveguide as a function of frequency. (*After C-T Tai.*) At 3 times the cutoff frequency for the TE_{11} mode, there are nine modes which will pass and one mode (TM_{02}) at cutoff. Electric field configurations are also shown for the first eight modes. Magnetic field lines are not shown but would be perpendicular to the electric field lines.

are listed in Table 8-2. Electric field configurations are also shown in Fig. 8-20 for the first eight modes in a cylindrical waveguide.†

Let us now examine the conditions necessary for propagation inside a cylindrical waveguide. Solving

$$\frac{k'_{nr}}{r_o} = \sqrt{\gamma^2 + \omega^2 \mu \varepsilon}$$

†Field configurations for the first 30 modes are given by C. S. Lee, S. W. Lee, and S. L. Chuang, Plot of Modal Field Distribution in Rectangular and Circular Waveguides, *IEEE Trans. Microwave Theory and Techniques,* **MTT-33**: 271–274, March 1985.

†C-T Tai, On the Nomenclature of TE_{01} Modes in a Cylindrical Guide, *Proc. IRE,* **49**: 1442–1444, September 1961.

TABLE 8-2
Cylindrical waveguide modes

Mode designation†	Eigenvalues k'_{nr}	Eigenvalues k_{nr}	Cutoff wavelength, λ_{oc}
TM_{01}		2.405	$2.61r_0$
TE_{01} (low loss)	3.832		$1.64r_0$
TM_{02}		5.520	$1.14r_0$
TE_{02}	7.016		$0.89r_0$
TE_{11} (dominant)	1.840		$3.41r_0$
TM_{11}		3.832	$1.64r_0$
TE_{12}	5.330		$1.18r_0$
TM_{12}		7.016	$0.89r_0$
TE_{21}	3.054		$2.06r_0$
TM_{21}		5.135	$1.22r_0$
TE_{22}	6.706		$0.94r_0$
TE_{31}	4.201		$1.49r_0$
TM_{31}		6.379	$0.98r_0$
TE_{41}	5.318		$1.18r_0$
TM_{41}		7.588	$0.83r_0$
TE_{51}	6.416		$0.98r_0$

†The subscripts nr as in TE_{nr} or k_{nr} have the following significance:

n = nth-order Bessel function
r = order of root of nth-order Bessel function

for the propagation constant γ, we obtain

$$\gamma = \pm\sqrt{\left(\frac{k'_{nr}}{r_0}\right)^2 - \omega^2\mu\varepsilon} = \alpha + j\beta \qquad \textbf{\textit{Propagation constant}} \qquad (1)$$

where k'_{nr} = rth root of the derivative of the nth order Bessel function (see Table 8-2 and Fig. 8-20).

As in the rectangular guide, there are three conditions:

1. At low frequencies, ω small, γ real, guide opaque (wave does not propagate).
2. At an intermediate frequency, ω intermediate, $\gamma = 0$, transition condition (cutoff).
3. At high frequencies, ω large, γ imaginary, guide transparent (wave propagates).

Putting $\gamma = 0$ in (1), we find for the *cutoff frequency* and *cutoff wavelength*

$$f_c = \frac{1}{2\pi\sqrt{\mu\varepsilon}}\frac{k'_{nr}}{r_0} \qquad \text{(Hz)} \qquad (2)$$

$$\lambda_{oc} = \frac{2\pi r_0}{k'_{nr}} \qquad \text{(m)} \qquad \textbf{\textit{Cutoff wavelength}} \qquad (3)$$

For the TE_{11} mode $\lambda_{oc} = 2\pi r_0/1.84 = 3.41 r_0$. Thus, the cutoff wavelength for the TE_{11} mode corresponds to a wavelength 3.41 times the radius of the guide. The cutoff wavelengths for various modes in a cylindrical guide are listed in Table 8-2.

At frequencies above cutoff

$$\beta = \sqrt{\omega^2 \mu \varepsilon - \left(\frac{k'_{nr}}{r_0}\right)^2} \qquad (\text{rad m}^{-1}) \tag{4}$$

From (4) and (3) we get for the *wavelength in the guide* (in z direction)

$$\lambda_g = \frac{\lambda_0}{\sqrt{1 - (\lambda_0/\lambda_{oc})^2}} \qquad (\text{m}) \qquad \textbf{\textit{Wavelength in guide}} \tag{5}$$

where λ_0 = wavelength in unbounded medium of same type that fills guide, m
λ_{oc} = cutoff wavelength, m.

For the *phase velocity in the guide* ($v_p = f\lambda_g$) we obtain

$$v_p = \frac{\omega}{\beta} = \frac{v_0}{\sqrt{1 - (\lambda_0/\lambda_{oc})^2}} \qquad (\text{m s}^{-1}) \qquad \textbf{\textit{Phase velocity}} \tag{6}$$

where $v_0 = 1/\sqrt{\mu\varepsilon}$.

Example 8-5. Cylindrical waveguide. An air-filled cylindrical waveguide has a diameter of 120 mm. Find (*a*) cutoff wavelength λ_{oc} for the TE_{01} (low-loss) mode and (*b*) the modes passed at wavelengths longer than $0.85\lambda_{oc}$ (TE_{01}).

Solution

$$\lambda_{oc}(TE_{01}) = \frac{2\pi 60}{3.832} = 98.4 \text{ mm} \quad \textit{Ans.} \quad (a)$$

From Fig. 8-20: TE_{01}, TE_{31}, TE_{11}, TE_{21}, TM_{01}, TM_{11} *Ans.* (*b*)

Problem 8-5-1. Cylindrical waveguide. An air-filled cylindrical waveguide has a diameter of 90 mm. Find (*a*) the TE and TM modes which the guide will transmit at wavelengths greater than 60 cm and (*b*) the relative velocity (v/c) in the guide of each of these modes at a frequency 1.1 times the mode cutoff frequency. *Ans.* TE_{11}, TM_{01}, TE_{21}, TM_{11}, TE_{01}, TE_{31}; (*b*) 2.4.

Problem 8-5-2. Tunnel. (*a*) A communications service using mobile units wants communication to be maintained even when its radio-equipped automobiles and trucks are in a vehicular tunnel. If the smallest tunnel diameter encountered is 5 m, what is the lowest frequency which can be employed? (*b*) What can be installed in the tunnel to permit communication at lower frequencies? No frequency converters are permitted. *Hint:* The tunnel is a waveguide. What can be installed so it can transmit TEM waves? *Ans.* (*a*) 35.2 MHz; (*b*) Run an insulated wire through the tunnel (like a streetcar trolley) converting the tunnel to an asymmetrical coaxial line and making it possible to transmit TEM modes (cutoff = 0 Hz). By extending the wire some distance from each end of the tunnel, the extensions would function as antennas coupled to the line.

Equations (5) and (6) are identical to those derived earlier for the rectangular waveguide (see Table 8-1). They also apply to waves in hollow guides of any cross-section.

Problem 8-5-3. Cylindrical waveguide. If the frequency is 3 times that required to pass the dominant mode (TE_{11}), what modes will be passed? *Ans.* TE_{11}, TM_{01}, TE_{21}, TM_{11}, TE_{01}, TE_{31}, TM_{21}, TE_{12}, and TE_{41}. An additional mode (TM_{02}) is at its cutoff frequency.

8-6 HOLLOW WAVEGUIDES OF OTHER CROSS-SECTION

In earlier sections we considered rectangular and cylindrical waveguides. These are only two of an infinite variety of forms in which single-conductor hollow waveguides may be made. For example, the waveguide could have an elliptical† cross-section, as in Fig. 8-21*d*, or a reentrant‡ cross-section, as in Fig. 8-21*f*.

All these forms and many others may be regarded as derivable from the rectangular type (Fig. 8-21*a*). Thus the square cross-section (Fig. 8-21*b*) is a special case of the rectangular guide. By bending out the walls, the square guide can be transformed to the circular shape (Fig. 8-21*c*). By flattening the circular guide the elliptical form of Fig. 8-21*d* is obtained. On the other hand, by bending the top and bottom surfaces of the rectangular waveguide inward the form shown in Fig. 8-21*e* is obtained. A still further modification is the reentrant form with central ridge in Fig. 8-21*f*. The value of regarding these as related forms is that often certain properties of a guide of a particular shape can be interpolated approximately from the known properties of waveguides of closely related shape.

For example, the longest wavelength that the square guide (Fig. 8-21*b*) will transmit is equal to $2b$. This is for the TE_{10} mode. This information can be used to predict with fair accuracy the longest wavelength that a circular guide can transmit. Thus, if the cross-sectional area of the square guide is taken equal to the area of the circular guide,

$$b^2 = \pi\left(\frac{d}{2}\right)^2 \tag{1}$$

where d is the diameter of the circular guide. Now $\lambda_{oc} = 2b$ for the square waveguide or $b = \lambda_{oc}/2$, and so we have

$$\left(\frac{\lambda_{oc}}{2}\right)^2 = \pi\left(\frac{d}{2}\right)^2 \tag{2}$$

or

$$\lambda_{oc} = \sqrt{\pi}d = 1.77d = 3.54r \tag{3}$$

as the cutoff wavelength for a circular waveguide of diameter d or radius r. This approximate value is only 4 percent greater than the exact value of $3.41r$ (see Table 8-2).

Parameters of rectangular and cylindrical waveguides are summarized in Table 8-3.

---

†L. J. Chu, Electromagnetic Waves in Hollow Elliptic Pipes of Metal, *J. Appl. Phys.*, **9**, September 1938.
‡S. B. Cohn, Properties of Ridge Wave Guide, *Proc. IRE,* **35**: 783–789, August 1947.

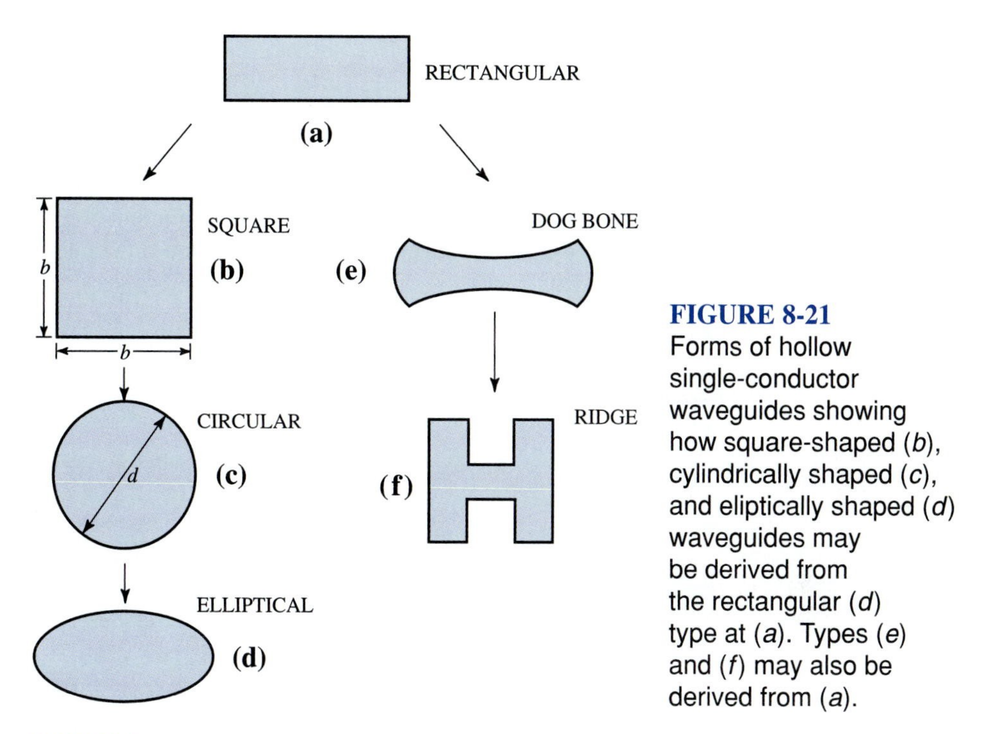

FIGURE 8-21
Forms of hollow single-conductor waveguides showing how square-shaped (*b*), cylindrically shaped (*c*), and eliptically shaped (*d*) waveguides may be derived from the rectangular (*d*) type at (*a*). Types (*e*) and (*f*) may also be derived from (*a*).

TABLE 8-3
Rectangular and circular cylindrical waveguide parameters

Symbol	Name	Equation (rectangular or cylindrical)	
α†	Attenuation constant:		
	Below cutoff frequency	$\alpha = \dfrac{2\pi}{\lambda_0}\sqrt{\left(\dfrac{\lambda_0}{\lambda_{oc}}\right)^2 - 1}$ (Np m^{-1})	
	Above cutoff frequency	$\alpha = \dfrac{\text{Re}\, Z_c \int \lvert H_{t1}\rvert^2\, dl}{2\, \text{Re}\, Z_{yz} \iint \lvert H_{t2}\rvert^2\, ds}$ (Np m^{-1})	
β†	Phase constant	$\beta = \sqrt{\left(\dfrac{2\pi}{\lambda_0}\right)^2 - k^2}$ (rad m^{-1})	
v_p	Phase velocity	$v_p = \dfrac{\omega}{\beta} = \dfrac{v_0}{\sqrt{1 - (\lambda_0/\lambda_{oc})^2}}$ (m s^{-1})	
		Rectangular	**Cylindrical**
Z_{yz}, $Z_{r\phi}$	Transverse-wave impedance	$Z_{yz} = \dfrac{Z_d}{\sqrt{1 - (\lambda_0/\lambda_{oc})^2}}$ TE$_{mn}$ mode only	$Z_{r\phi} = \dfrac{Z_d}{\sqrt{1 - (\lambda_0/\lambda_{oc})^2}}$ (Ω) TE$_{nr}$ mode
λ_{oc}	Cutoff wavelength	$\lambda_{oc} = \dfrac{2}{\sqrt{(n/y_1)^2 + (m/z_1)^2}}$ TE$_{mn}$ or TM$_{mn}$ mode	$\lambda_{oc} = \dfrac{2\pi r_0}{k'_{nr}}$ or $\dfrac{2\pi r_0}{k_{nr}}$ (m) TE$_{nr}$ mode (k'_{nr}), TM$_{nr}$ mode (k_{nr})

† γ = propagation constant = $\alpha + j\beta$.

8-7 WAVEGUIDE DEVICES

Several basic waveguide devices are discussed in this section, namely, *terminations, power dividers,* and *guide-to-line transitions.*

A matched ***termination*** for a rectangular or circular waveguide is shown in cross-section in Fig. 8-22*a*. A card of resistance material is placed transversely in the guide $\lambda/4$ from a metal plate capping the end of the guide. The situation here is similar to that discussed in Sec. 4-7 for the terminated wave. For zero reflection it

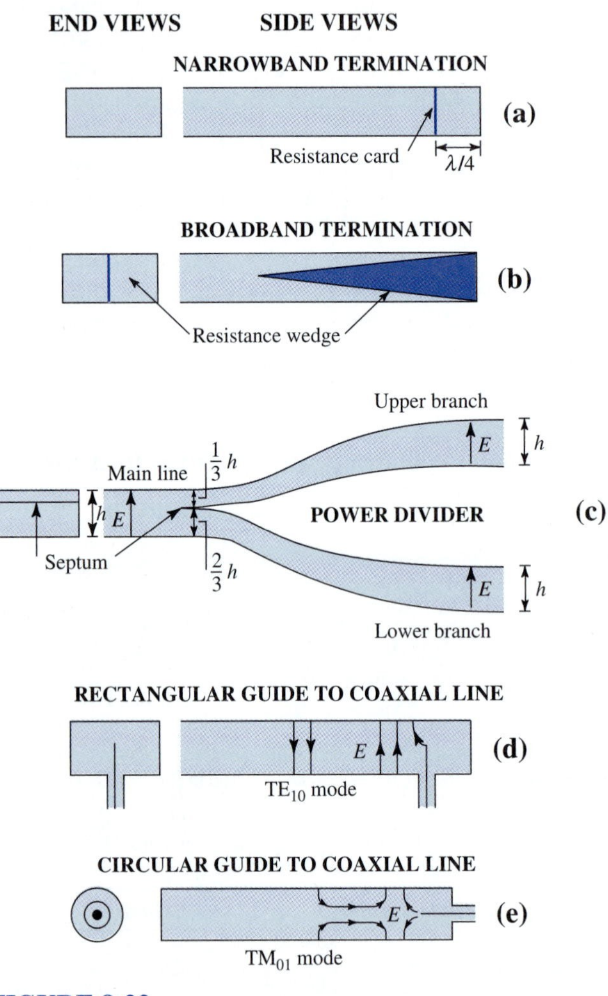

FIGURE 8-22
(*a* and *b*) Waveguide terminations, (*c*) power divider, and (*d* and *e*) waveguide-to-coaxial-line transitions. The figures at the left are end-view cross-sections while those at the right are side-view longitudinal sections.

is necessary only that the card have a resistance per square equal to the transverse-wave impedance of the guide. The termination of Fig. 8-22*a* will be matched at the design frequency for which the card-to-end-plate distance is $\lambda/4$, but not at adjacent frequencies. This termination is a narrowband device. To provide a broadband termination, a wedge of resistance material can be used, as suggested in Fig. 8-22*b*. The length of the wedge should be of the order of a wavelength or more.

In applications where waveguides feed two or more antennas, a ***power divider*** may be required to divide the power in a predetermined ratio. Figure 8-22*c* shows a power divider for a rectangular waveguide (with TE_{10} mode) delivering twice the power to the lower branch as compared to the upper. The division is achieved by inserting a thin septum which divides the guide cross-sectional area in the ratio 2:1. Note that the septum is perpendicular to the direction of the electric field vector **E** so that it does not disturb the field configuration in the guide. The height of the guide to the right of the septum in each branch is increased gradually over a distance of several wavelengths back to the standard height h. If both branches are connected into nonreflecting loads or antennas, there will be no reflection from this power divider and the device can be used over a broad band of frequencies.

At some point in most waveguide systems it is necessary to convert to TEM-mode transmission on a coaxial line, or vice versa. Two ***waveguide-to-coaxial-line transitions*** are shown in Fig. 8-22*d* and *e*. The one in Fig. 8-22*d* provides a transition from a TE_{10} mode in a rectangular guide to a coaxial line, while the one in Fig. 8-22*e* provides a transition from a TM_{01} mode in a circular waveguide to a coaxial line.

8-8 WAVES TRAVELING PARALLEL TO A PLANE BOUNDARY

Consider the plane boundary between two media shown in Fig. 8-23, assuming that medium 1 is air and medium 2 is a perfect conductor. From the boundary condition that the tangential component of the electric field vanishes at the surface of a perfect conductor, the electric field of a TEM wave traveling parallel to the boundary must be exactly normal to the boundary, as portrayed in the figure. However, if medium 2

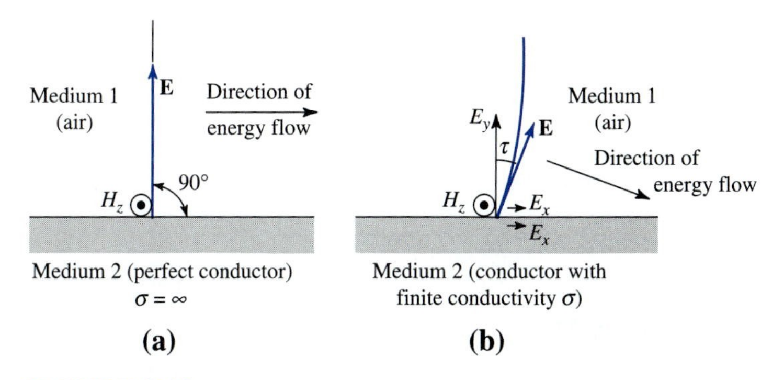

FIGURE 8-23
TEM wave traveling to right (*a*) along surface of perfectly conducting medium and (*b*) along surface of medium with finite conductivity.

has a finite conductivity σ, there will be a tangential electric field E_x at the boundary, and, as a result, the electric field of a wave traveling along the boundary has a *forward tilt,* as suggested in Fig. 8-23*b*. From the continuity relation for tangential electric fields, the field on both sides of the boundary is E_x.

The direction and magnitude of the power flow per unit area are given by the Poynting vector. The average value of the Poynting vector is

$$\mathbf{S}_{\text{av}} = \tfrac{1}{2}\text{Re}\,\mathbf{E} \times \mathbf{H}^* \qquad (\text{W m}^{-2}) \tag{1}$$

At the surface of the conducting medium (Fig. 8-23*b*) the power into the conductor is in the negative y direction, and from (1) its average value per unit area is

$$S_y = -\tfrac{1}{2}\text{Re}\,E_x H_z^* [= (S_{\text{av}})_y] \tag{2}$$

The space relation of E_x, H_z (or H_z^*), and S_y is shown in Fig. 8-24*a*. But

$$\frac{E_x}{H_z} = Z_c \tag{3}$$

where Z_c is the intrinsic impedance of the conducting medium, so that (2) can be written

$$S_y = -\tfrac{1}{2}H_z H_z^* \text{Re}\,Z_c = -\tfrac{1}{2}H_{z0}^2 \text{Re}\,Z_c \tag{4}$$

where $H_z = H_{z0}e^{-j\xi - \gamma x}$
$H_z^* = H_{z0}e^{j\xi - \gamma x}$ = complex conjugate of H_z
ξ = phase lag of H_z with respect to E_x

At the surface of the conducting medium (Fig. 8-23*b*), the power per unit area flowing parallel to the surface (x direction) is

$$S_x = \tfrac{1}{2}\text{Re}\,E_y H_z^* \tag{5}$$

The space relation of E_y, H_z (or H_z^*), and S_x is illustrated by Fig. 8-24*b*. But

$$\frac{E_y}{H_z} = Z_d \tag{6}$$

where Z_d is the intrinsic impedance of the dielectric medium (air). It follows that

$$S_x = \tfrac{1}{2}H_{z0}^2 \text{Re}\,Z_d \tag{7}$$

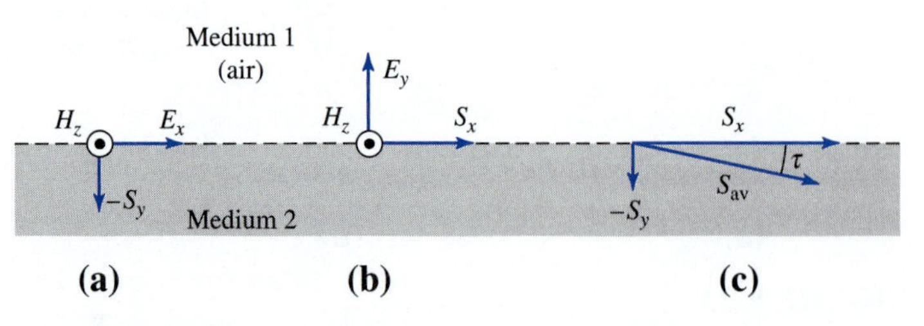

FIGURE 8-24
Fields and Poynting vector at surface of a conducting medium with wave traveling parallel to surface.

The total average Poynting vector is then

$$\mathbf{S}_{av} = \hat{\mathbf{x}}S_x + \hat{\mathbf{y}}S_y = \frac{H_{z0}^2}{2}(\hat{\mathbf{x}}\text{Re } Z_d - \hat{\mathbf{y}}\text{Re } Z_c) \quad (8)$$

The relation of $\mathbf{S}_{av}$ to its x and y components is illustrated in Fig. 8-24c. It is to be noted that the average power flow (per unit area) is not parallel to the surface but inward at an angle τ. This angle is also the same as the angle of forward tilt of the average electric field (see Fig. 8-23b). If medium 2 were perfectly conducting, τ would be zero.

It is of interest to evaluate the tilt angle τ for a couple of practical situations. This is done in the following examples.

Example 8-6. Tilt angle over copper. Find the forward tilt angle τ for a vertically polarized 3-GHz wave traveling in air along a sheet of copper.

Solution. From (8) the tilt angle τ is given by

$$\tau = \tan^{-1}\frac{\text{Re } Z_c}{\text{Re } Z_d} \quad (9)$$

At 3 GHz, we have for copper that $\text{Re } Z_c = 14.4\ \text{m}\Omega$. The intrinsic impedance of air is independent of frequency ($\text{Re } Z_d = 377\ \Omega$). Thus

$$\tau = \tan^{-1}\frac{1.44 \times 10^{-2}}{377} = 0.0022°$$

Although τ is not zero in the above example, it is very small, so that **E** is nearly normal to the copper surface and **S** nearly parallel to it. This small value of tilt is typical at most air-conductor boundaries but accounts for the power flow into the conducting medium. If the conductivity of medium 2 is very low, or if it is a dielectric medium, τ may amount to a few degrees. Thus, the forward tilt of a vertically polarized radio wave propagating along poor ground is sufficient to produce an appreciable horizontal electric field component. In the Beverage, or wave, antenna this horizontal component is utilized to induce emfs along a horizontal wire oriented parallel to the direction of transmission of the wave.

In contrast to Example 8-6, in which medium 2 is copper, the following example considers the case of freshwater as medium 2.

Example 8-7. Tilt angle over water. Find the forward tilt angle τ for a vertically polarized 3-GHz wave traveling in air along the surface of a smooth freshwater lake.

Solution. At 3 GHz the conduction current in freshwater is negligible compared with the displacement current, so that the lake may be regarded as a dielectric medium of relative permittivity $\varepsilon_r = 80$. Thus,

$$\tau = \tan^{-1}\frac{1}{\sqrt{80}} = 6.4°$$

In this case the forward tilt of 6.4° is sufficient to be readily detected by a direct measurement of the direction of the electric field.

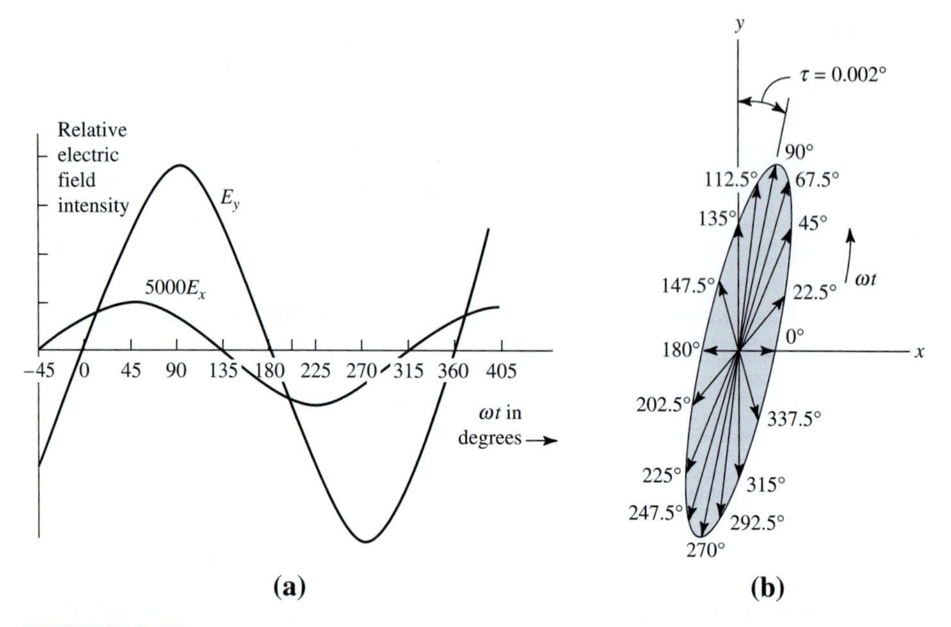

FIGURE 8-25
(a) Magnitude variation with time of E_y and E_x components of **E** in air at the surface of a copper region for a 3000-MHz TEM wave traveling parallel to the surface. (b) Resultant values of **E** (space vector) at 22.5° intervals over one cycle, illustrating elliptical cross-field at the surface of the copper region. The wave is traveling to the right. Note that although **E** has x and y components (cross-field), it is linearly polarized (in y-z plane).

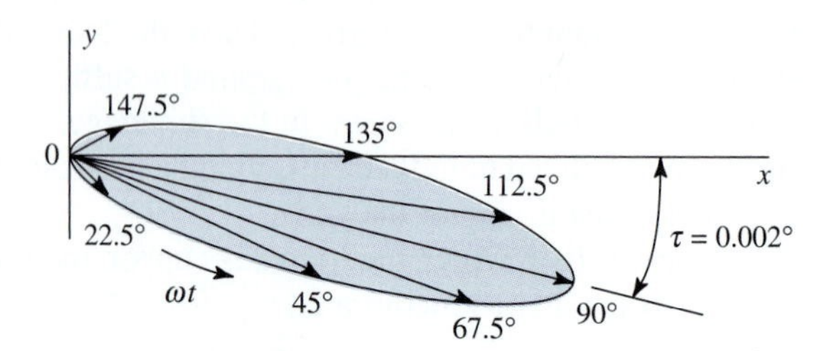

FIGURE 8-26
Power flow downward into a conducting medium (copper) from wave traveling along surface (to right) shown by Poynting vector at 22.5° intervals over one-half cycle. The ordinate values are magnified 5000 times compared with the abscissa values.

The angle τ discussed above is an average value. In general, the instantaneous direction of the electric field varies as a function of time. In the case of a wave in air traveling along a copper sheet, E_y and E_x are in phase octature (45° phase difference), so that at one instant of time the total field $\mathbf{E}$ may be in the x direction and 1/8 period later it will be in the y direction. (See Fig. 8-25*a*.) With time, the locus of the tip of $\mathbf{E}$ describes a cross-field ellipse as portrayed in Fig. 8-25 for a 3-GHz wave in air traveling along a copper sheet (in the x direction) as in Example 8-6. The ellipse is not to scale, the abscissa values being magnified 5000 times. The positions of $\mathbf{E}$ for various values of ωt are indicated. The variation of the instantaneous Poynting vector for this case is shown in Fig. 8-26. Here the ordinate values are magnified 5000 times. It is to be noted that the tip of the Poynting vector travels around the ellipse twice per cycle.

Example 8-8. Surface-wave power. A plane 2-GHz wave traveling in air parallel to a flat ground plane with $\mathbf{H}$ parallel to the ground plane loses 3 μW/m^2 into the ground plane (= S into ground plane). If the constants for the ground plane are $\sigma = 10^7$ ℧/m, $\mu_r = \varepsilon_r = 1$, find (*a*) H, (*b*) E (in air), (*c*) S (parallel to ground plane), and (*d*) ratio S (parallel)/S (into ground plane).

Solution

$$S \text{ (into ground plane)} = H^2(\text{rms}) \operatorname{Re} Z_c$$

$$\operatorname{Re} Z_c = \left[\frac{\mu_0 2\pi f}{2\sigma}\right]^{1/2} = \left[\frac{(4\pi \times 10^{-7})(2\pi)(2 \times 10^9)}{2 \times 10^7}\right]^{1/2}$$

$$= 2\pi(2 \times 10^{-5})^{1/2} = 0.0281\ \Omega$$

$$H^2 = \frac{S}{0.0281} = \frac{3 \times 10^{-6}}{0.0281} = 106.8 \times 10^{-6}$$

$$H = 10.33 \times 10^{-3} \text{ A/m} = 10.33 \text{ mA/m (rms)} \quad \textit{Ans.} \quad (a)$$

$$E = 377\, H = 3.89 \text{ V/m} \quad \textit{Ans.} \quad (b)$$

$$S \text{ (parallel to ground plane)} = E^2/Z = H^2\, Z = 0.0403 \text{ W/m}^2$$

$$= 40.3 \text{ mW/m}^2 \quad \textit{Ans.} \quad (c)$$

$$\text{Ratio:} \quad \frac{S \text{ (parallel to ground plane)}}{S \text{ (into ground plane)}} = \frac{40.3 \times 10^{-3}}{3 \times 10^{-6}} = 13\,433 \quad \textit{Ans.} \quad (d)$$

Problem 8-8-1. Surface wave powers. A 100-MHz wave is traveling parallel to a copper sheet ($|Z_c| = 3.7 \times 10^{-3}\ \Omega$) with E (= 100 V/m rms) perpendicular to the sheet. Find (*a*) Poynting vector (W/m^2) parallel to sheet and (*b*) Poynting vector into the sheet. *Ans.* (*a*) 26.5 W/m^2; (*b*) 182 μW/m^2

Problem 8-8-2. Surface wave powers. A 100-MHz wave is traveling parallel to a conducting sheet for which $|Z_c| = 0.02\ \Omega$. If E is perpendicular to the sheet and equal to 150 V/m (rms), find (*a*) watts per square meter traveling parallel to sheet and (*b*) watts per square meter into sheet. *Ans.* (*a*) 60 W/m^2; (*b*) 2.24 mW/m^2.

Whereas copper has a complex intrinsic impedance, freshwater, at the frequency considered in Example 8-5, has a real intrinsic impedance. It follows that the E_x and E_y components of the total field **E** are in time phase so that the cross-field ellipse in this case collapses to a straight line (linear cross-field) with a forward tilt of 6.4°.

8-9 OPEN WAVEGUIDES

In the previous section we have seen that a wave traveling along an air-conductor or air-dielectric boundary has a longitudinal (E_x) component of the electric field, resulting in a forward tilt of the total electric field. Hence, the Poynting vector is not entirely parallel to the boundary but has a component directed from the air into the adjacent medium, as suggested in Fig. 8-24*c*. This tends to keep the energy in the wave from spreading out and to concentrate it near the surface, resulting in a ***bound wave*** or ***surface wave.*** The phase velocity of such a bound wave is always less than the velocity in free space. Although the field of this guided wave extends to infinity, such a large proportion of the energy may be confined within a few wavelengths of the surface that the surface can be regarded as an open type of waveguide. It should be noted, however, that even though the forward-tilt effect is present along all finitely conducting surfaces, the bound wave may be of negligible importance without a launching device of relatively large dimensions (several wavelengths across) to initiate the wave. If the surface is perfectly smooth and perfectly conducting, the tangential component of the electric field vanishes, there is no forward tilt of the electric field, and there is no tendency whatever for the wave to be bound to the surface.

In 1899 Arnold Sommerfeld† showed that a wave could be guided along a round wire of finite conductivity. Jonathan Zenneck‡ pointed out that for similar reasons a wave traveling along the earth's surface would tend to be guided by the surface.

The guiding action of a flat conducting surface can be enhanced by adding corrugations or a dielectric coating or slab. If the dielectric slab is sufficiently thick, it can act alone as an effective *nonmetallic guide.* The characteristics of a number of these open waveguides are discussed in the following sections.

Consider a perfectly conducting flat surface of infinite extent with transverse conducting corrugations, as in Fig. 8-27*a*. The corrugations have many teeth per wavelength ($s \ll \lambda$). The slots between the teeth can support a TEM wave traveling down into the slots with field E_x. Thus, each slot acts like a short-circuited section of a parallel-plane two-conductor transmission line of length d. Assuming lossless materials, the impedance Z presented to a wave traveling vertically downward into

†A. Sommerfeld, Fortpflanzung elektrodynamischer Wellen an einem zylindrischen Leiter, *Ann. Phys. u. Chem.*, **67**: 233, December 1899.

‡J. Zenneck, Uber die Fortpflanzung ebener elektromagnetischer Wellen langs einer ebenen Leiterflache und ihre Beziehung zur drahtlosen Telegraphie, *Ann. Phys.*, **23**(4): 846–866, September 20, 1907.

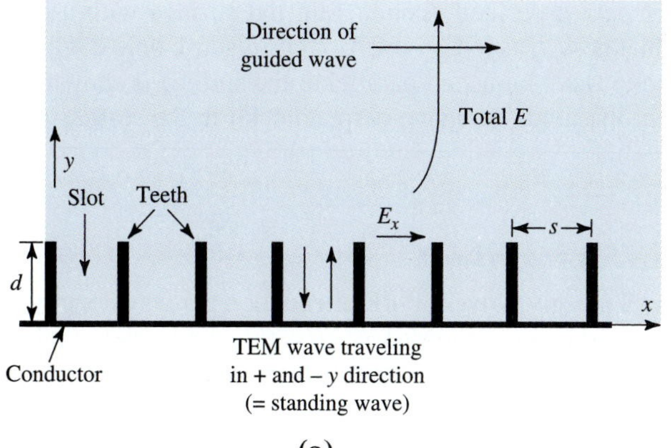

(a)

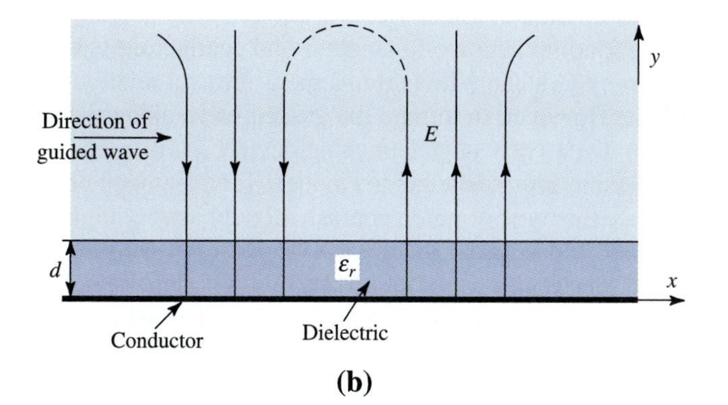

(b)

FIGURE 8-27
(*a*) Wave guided by corrugated surface. (*b*) TM_0 wave guided by conductor with dielectric coating of thickness *d*.

the slotted (or corrugated) surface is a pure reactance, or

$$Z = jZ_d \tan \frac{2\pi\sqrt{\varepsilon_r}d}{\lambda_0} \qquad (\Omega) \tag{1}$$

where Z_d = intrinsic impedance of medium filling the slots, Ω
ε_r = relative permittivity of the medium, dimensionless
λ_0 = free-space wavelength, m
d = depth of the slots, m

The slots may be regarded as storing energy from the passing wave. When $2\pi\sqrt{\varepsilon_r}d/\lambda_0 < 90°$, the surface is inductively reactive. When $d = \lambda/4$ and $\varepsilon_r = 1$, $Z = \infty$. When $d = \lambda/2$ and $\varepsilon_r = 1$, $Z = 0$ and the corrugated surface acts like a conducting sheet.

Consider next a perfectly conducting flat surface with a coating of lossless dielectric of thickness d, as in Fig. 8-27*b*.† The electric field configuration for a TM_0 (dominant) mode wave launched parallel to the surface is shown. For a sufficiently thick coating d, the fields attenuate perpendicular to the surface (in the y direction) as $e^{-\alpha y}$, where

$$\alpha = \frac{2\pi}{\lambda_0}\sqrt{\varepsilon_r - 1} \qquad (\text{Np m}^{-1}) \tag{2}$$

For $\varepsilon_r = 2$ this gives over 50-dB attenuation per wavelength. Thus, the field is effectively confined close to the surface. Assuming a lossless dielectric, the attenuation in the x direction will be due entirely to energy lost by radiation into the region above the dielectric coating.

In the preceding paragraphs we have discussed guiding by open flat structures of infinite extent in the z direction (perpendicular to the page). Guiding can also occur along round wires of metal, of dielectric, or a combination of both. As an example, let us consider the guiding action of a single round conducting wire with dielectric coating. As shown by Goubau,‡ this arrangement forms a relatively efficient open type of waveguide. However, to initiate the guided wave along the wire with good efficiency requires a relatively large launching device, its function being to excite a mode, closely related in form to the guided mode, over a diameter of perhaps several wavelengths. Hence this type of guide is practical only at very high frequencies.

A dielectric-coated single-wire waveguide with typical dimensions is illustrated in Fig. 8-28. The dielectric coat consists of a layer of enamel of relative permittivity $\varepsilon_r = 3$ having a thickness of only 0.0005λ. The wire diameter is 0.02λ. The configuration of the electric field lines in the launcher and along the wire guide is suggested in the figure. The mode on the wire is a TM type, but it is like a plane TEM wave to a considerable distance from the wire.

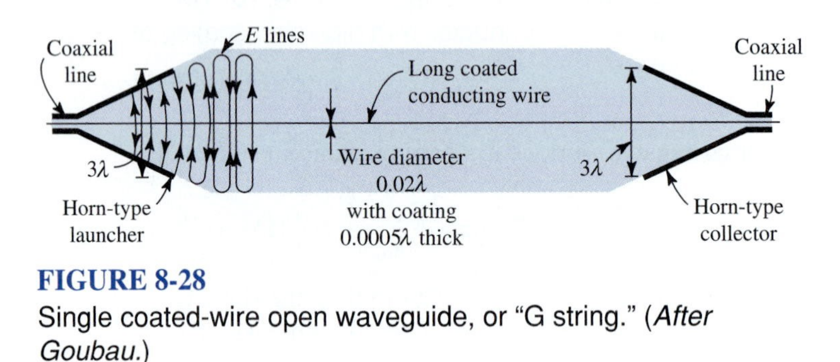

FIGURE 8-28
Single coated-wire open waveguide, or "G string." (*After Goubau.*)

†S. A. Schelkunoff, Anatomy of "Surface Waves," *IRE Trans. Antennas and Propagation,* **AP-7**: S133–139, December 1959; R. F. Harrington, "Time-Harmonic Electromagnetic Fields," p. 168, McGraw-Hill, New York, 1961.

‡G. Goubau, Surface Waves and Their Application to Transmission Lines, *J. Appl. Phys.,* **21**: 1119–1128, November 1950.

Wires wound in the form of long helixes are also effective single-conductor open-type waveguides. Helix diameters as large as 0.4λ can be used successfully.

For efficient transmission of energy by a guiding system, the attenuation should be small. With two-conductor transmission lines this requires that the series resistance R and the shunt conductance G be small. The conductor separation must also be small compared to the wavelength in order that radiation losses be negligible. Under these conditions the fields vary as $1/r^2$, where r is the distance perpendicular to the line and the power density varies as $1/r^4$. Thus, most of the power flow is close to the line. Waves carried in a single hollow conducting waveguide will be unattenuated if the guide walls are perfectly conducting and the material filling the guide is lossless. Perfectly conducting walls also prevent any radiation from the guide. With open guides losses due to radiation tend to become significant, and modes which confine the power flow close to the guiding surface are desirable for efficient transmission. High attenuation of the fields perpendicular to guides is also important to reduce coupling or cross talk between adjacent transmission systems.

8-10 DIELECTRIC SHEET WAVEGUIDES

The guides we have been discussing are totally or partially metallic. Let us consider now guides which are entirely of dielectric material such as an infinite dielectric sheet of thickness d as suggested in Fig. 8-29. The sheet extends infinitely far in the x and z directions, and its permittivity ε_1 is greater than the permittivity ε_2 of the medium above and below.

In Chap. 4 we discussed wave propagation transversely through dielectric sheets. Now let us consider wave propagation longitudinally in the x direction (to right) due to a TEM wave launched into the sheet from the left. This wave can be largely confined inside the sheet by multiple reflections provided its angle of incidence θ_i with respect to the upper and lower surfaces is more than the critical angle θ_{ic}. Under this condition, the wave will be totally internally reflected and

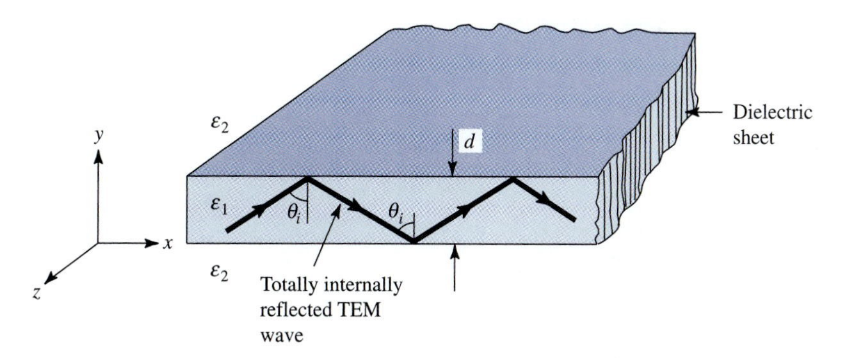

FIGURE 8-29
Section of dielectric sheet waveguide of infinite extent (in x and z directions) and of thickness d (in y direction). Zigzag line shows path of TEM wave propagating inside the sheet (in x direction) by total internal reflection.

will propagate in the x direction via a zigzag path in a manner like that between two infinite conducting sheets, one on the upper and the other on the lower surface. However, there are important differences between the parallel conducting sheet waveguide of Sec. 8-3 and the dielectric sheet waveguide we are now considering. For perfectly conducting sheets, the tangential electric field must be zero at the sheets ($E_x = E_z = 0$ at $y = 0$ and $y = d$) and zero outside. For the dielectric sheet, the field is not zero at the surfaces and it extends outside, theoretically to infinity. But this external field may attenuate very rapidly away from the sheet, indicating that it is tightly bound to the sheet.

At first glance one might suppose that any wave for which $\theta_i > \theta_{ic}$ will propagate in the dielectric sheet. However, because of interference, waves will actually propagate only at certain angles. Referring to Fig. 8-30, consider two rays, 1 and 2, belonging to the same TEM wave and incident at an angle $\theta_i > \theta_{ic}$. This requires that

$$\theta_i > \theta_{ic} = \sin^{-1}\sqrt{\frac{\varepsilon_2}{\varepsilon_1}} \tag{1}$$

where ε_1 = permittivity of sheet, F m^{-1}, and ε_2 = permittivity of medium above and below sheet, F m^{-1}, with $\varepsilon_1 > \varepsilon_2$.

The ray paths are shown by solid lines and the constant-phase fronts by dashed lines. The necessary condition for wave propagation in the sheet is that the phase length a of ray 2 and the phase length b of ray 1, including phase shifts on reflection, be equal or equal plus or minus an integral number of 2π radians.† Thus, in symbols, the requirement is

$$\frac{2\pi}{\lambda_0}\eta_1(b - a) + \phi = 2\pi n \tag{2}$$

where η_1 = index of refraction of medium 1 $(= \sqrt{\varepsilon_{1r}})$
ϕ = phase shift on reflection from surface
λ_0 = free space wavelength, m
n = integer (= 0, 1, 2, 3, ...)

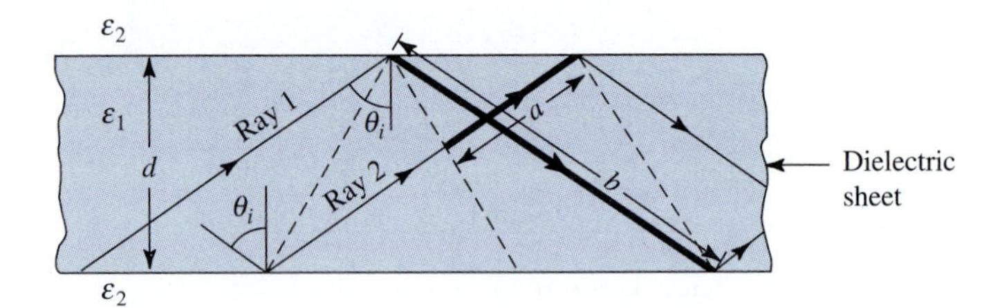

FIGURE 8-30
Geometry of wave reflection in dielectric sheet waveguide for permitted angles of reflection.

---

†Dietrich Marcuse, "Theory of Dielectric Optical Waveguides," Academic Press, New York, 1974.

or

$$\frac{2\pi\eta_1 d}{\lambda_0}\left[\frac{1}{\cos\theta_i} - \sin\theta_i\left(\tan\theta_i - \frac{1}{\tan\theta_i}\right)\right] + \phi = 2\pi n \tag{3}$$

which can be reduced to

$$\frac{4\pi\eta_1 d\cos\theta}{\lambda_0} + \phi = 2\pi n \tag{4}$$

Restricting our attention to waves with **E** perpendicular to the plane of incidence (**E** in z direction, out of page in Fig. 8-30), we have from (4-14-7) that the reflection coefficient for $\theta_i > \theta_{ic}$ is

$$\rho_\perp = \frac{\cos\theta_i - j\sqrt{\sin^2\theta_i - (\varepsilon_2/\varepsilon_1)}}{\cos\theta_i + j\sqrt{\sin^2\theta_i - (\varepsilon_2/\varepsilon_1)}} = 1\angle\phi \tag{5}$$

where $\phi = -2\tan^{-1}(\sqrt{\sin^2\theta_i - (\varepsilon_2/\varepsilon_1)}/\cos\theta_i)$.

Thus, the reflection coefficient $\rho_\perp$ has unit magnitude and a phase shift ϕ. Introducing (5) into (4),

$$\frac{4\pi\eta_1 d\cos\theta_i}{\lambda_0} - 2\pi n = 2\tan^{-1}\frac{\sqrt{\sin^2\theta_i - (\varepsilon_2/\varepsilon_1)}}{\cos\theta_i} \tag{6}$$

or

$$\tan\left(\frac{2\pi\eta_1 d\cos\theta_i}{\lambda_0} - \pi n\right) = \frac{\sqrt{\eta_1^2\sin^2\theta_i - \eta_2^2}}{\eta_1\cos\theta_i} \tag{7}$$

where η_1 = index of refraction of sheet = $\sqrt{\varepsilon_{1r}}$

η_2 = index of refraction of medium above and below sheet = $\sqrt{\varepsilon_{2r}}$

d = thickness of sheet, m

θ_i = angle of incidence, rad or deg

λ_0 = free space wavelength, m

n = integer (= 0, 1, 2, 3, ...)

To illustrate the significance of (7), consider the following example.

Example 8-9. Dielectric sheet waveguide. A sheet of dielectric has a thickness $d = 10$ mm and index of refraction $\eta_1 = 1.5$. The medium above and below is air ($\eta_2 = 1$). The electric field is parallel to the sheet (in z direction in Fig. 8-29 with wave progressing in x direction). Find the angles of incidence θ_i of permitted propagation if the wavelength $\lambda_0 = 10$ mm.

Solution. We note that the field **E**, although parallel to the sheet, is perpendicular to the plane of incidence. From (1) the critical angle

$$\theta_{ic} = \sin^{-1}\sqrt{\frac{\varepsilon_2}{\varepsilon_1}} = \sin^{-1}\frac{\eta_2}{\eta_1} = \sin^{-1}\frac{1}{1.5} = 41.8^\circ \tag{8}$$

Evaluating the left-hand side of (7) for angles of incidence θ_i greater than 41.8° yields the solid curves of Fig. 8-31. Evaluating the right-hand side of (7) in the same way yields the dashed curve. The three θ_i values, where the dashed and solid curves

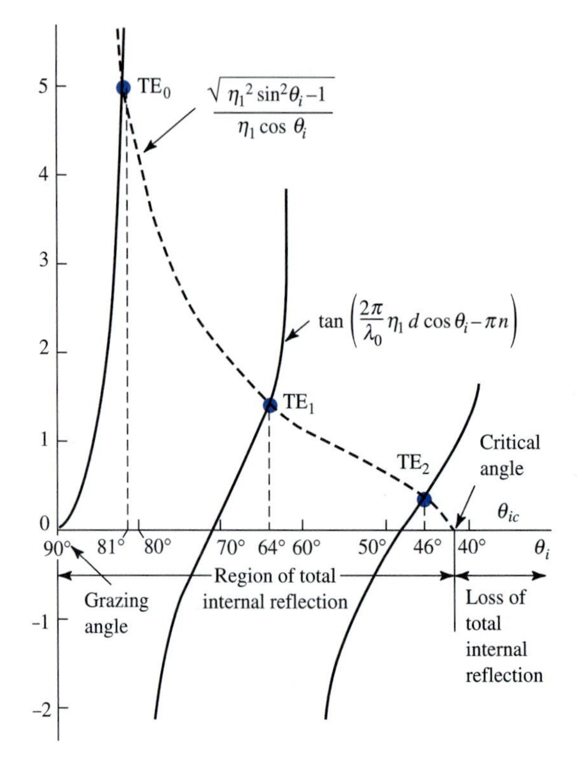

FIGURE 8-31
Solution of (7) for permitted angles of incidence θ_i in dielectric sheet (Example 8-9). The three θ_i values (81°, 64°, and 46°) where the curves intersect are the three solutions, or eigenvalues.

intersect, are the three solutions, or eigenvalues, for which equation (7) is satisfied, corresponding to $\theta_i = 46°, 64°$, and 81°. (*Ans.*) These permitted angles correspond to three transverse electric modes in the sheet: TE$_2$, TE$_1$, and TE$_0$. All may be present simultaneously, but if the thickness d is decreased or the wavelength λ increased (or both), fewer solutions (eigenvalues) or modes will be possible. However, one solution (for the TE$_0$ mode) will always exist so that, at least in theory, waves can propagate to zero frequency.

Problem 8-10-1. Coated surface wave cutoff. A perfectly conducting flat sheet of large extent has a dielectric coating ($\varepsilon_r = 3$) of thickness $d = 5$ mm. Find (*a*) the cutoff frequency for the TM (dominant) mode and (*b*) its attenuation per unit distance. *Ans.* (*a*) $f_c = 0$, (*b*) $8.9/\lambda_0$ Np/m.

The variation of the field E_z across the sheet (in y direction) is shown in Fig. 8-32 for the TE$_0$ mode. The maximum field is at the center of the sheet with a spillover (evanescent) field above and below which attenuates rapidly with distance above and below the sheet.

In addition to the transverse electric (TE) modes we have discussed, other modes are also possible, such as transverse magnetic (TM) modes with **H** transverse (in z direction).

If the infinitely wide sheet (of thickness d) is reduced in width to some value w, it forms a dielectric strip waveguide or transmission line, as in Fig. 8-33a, with

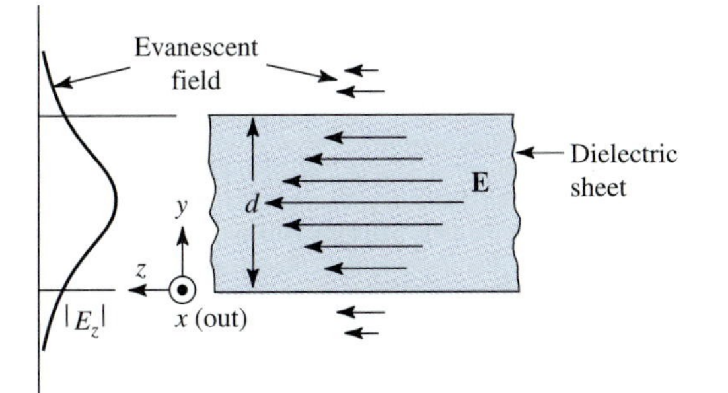

FIGURE 8-32
End view of sheet (from x direction) with graph (at left) and arrows (at right) suggesting variation of E_z with respect to y for TE_0 mode wave propagating out of page. Note that there is a spillover (evanescent) field above and below the sheet.

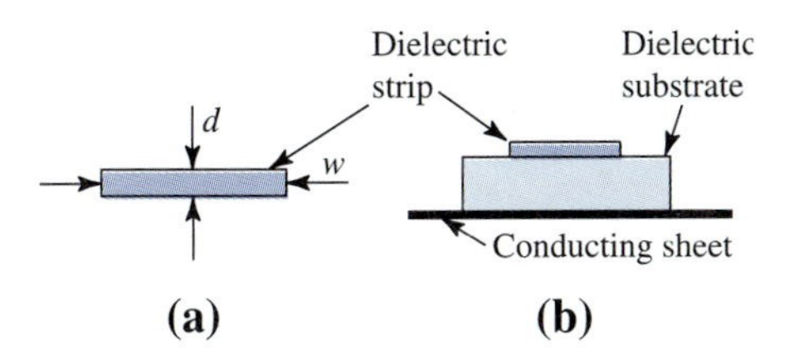

FIGURE 8-33
(*a*) End view of dielectric strip guide of thickness d and width w. (*b*) Dielectric strip on dielectric substrate of lower index of refraction as employed for light transmission in integrated circuits.

properties like those discussed above for the infinite sheet. Dielectric strips are often used as light guides in integrated circuits, the strip being mounted on a dielectric substrate of smaller index of refraction than that of the strip, as suggested in Fig. 8-33*b*.

8-11 DIELECTRIC FIBER AND ROD WAVEGUIDES: FIBER OPTICS

With Sec. 8-10 on the dielectric sheet waveguide as background, let us consider next the dielectric cylinder waveguide, which is usually referred to as a *fiber* or a *rod* depending on its diameter. Although the basic principles of the cylindrical dielectric guide are similar to those for the sheet, there are significant differences.

At or near optical wavelengths the dielectric cylinder guide can be physically small or threadlike in diameter. Such guides, called ***optical fibers,*** consist typically of a transparent core fiber of glass of index of refraction η_1 surrounded by a transparent glass sheath, or cladding, of slightly lower index η_2, with both enclosed in an opaque protective jacket. See Fig. 8-34. A typical optical fiber is as fine as a human hair. With the great bandwidths available at infrared wavelengths, it is possible for a single such fiber to carry 20 million telephone channels or 20 thousand TV channels, or some combination of both, with an attenuation of only $\frac{1}{4}$ dB/km.†

Figure 8-35 shows a cross-section through the axis of an optical-fiber core of index of refraction η_1 with cladding of index η_2. A ray entering the core from an external medium of index η_0 at an angle θ_e will make an angle θ_t with respect to the axis inside the core. The relation between the angles, as given by Snell's law, is

$$\sin\theta_t = \frac{\eta_0}{\eta_1}\sin\theta_e \tag{1}$$

The ray continuing in the core will be incident on the core-cladding boundary at an angle θ_i. If $\theta_i > \theta_{ic}$, where θ_{ic} is the critical angle, the ray will be totally internally reflected and continue to propagate inside the core.‡ From (1) and

$$\sin\theta_{ic} = \frac{\eta_2}{\eta_1} \tag{2}$$

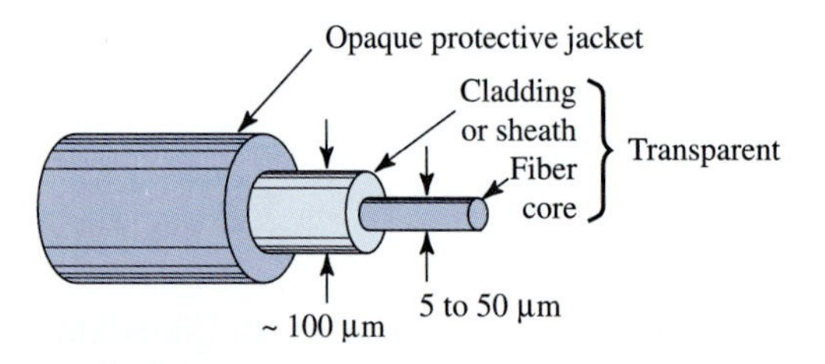

FIGURE 8-34
Optical-fiber guide with transparent core, transparent cladding, or sheath, and opaque protective jacket. Typical dimensions are indicated.

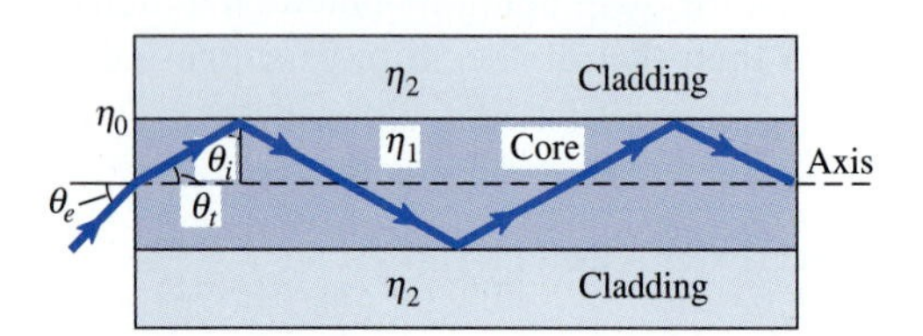

FIGURE 8-35
Wave entering fiber-optic core on axis will be trapped and will propagate by total internal reflection if the entry angle θ_e is less than a critical value.

†G. G. Kudriavtzev, L. E. Varakin, and P. S. Kurakov, Trans-Soviet Fibre-optic Communication Line (TSL) and Global Digital Telecommunications Ring, *Telecommunications Journal,* **57**, 678–688, October 1990.

‡N. S. Kapany, "Fiber Optics," Academic Press, New York, 1967.

we have

$$\sin\theta_e = \frac{\eta_1}{\eta_0}\sin\theta_t = \frac{\eta_1}{\eta_0}\sin(90^\circ - \theta_{ic}) = \frac{\eta_1}{\eta_0}\cos\theta_{ic} \tag{3}$$

or

$$\sin\theta_e = \frac{\sqrt{\eta_1^2 - \eta_2^2}}{\eta_0} \tag{4}$$

where θ_e = entrance angle on axis of core, rad or deg
η_1 = index of refraction of core, dimensionless
η_2 = index of refraction of cladding, dimensionless
η_0 = index of refraction of external medium, dimensionless

For air as the external medium ($\eta_0 = 1$), (4) reduces to

$$\sin\theta_e = \sqrt{\eta_1^2 - \eta_2^2} \tag{5}$$

For core index $\eta_1 = 1.5$, cladding index $\eta_2 = 1.485$ and external medium of air, $\theta_e = 12.2^\circ$. Thus, any ray entering the end of the fiber core on axis with $\theta_e < 12.2^\circ$ will be trapped inside (totally internally reflected) and propagate by multiple reflections down the inside of the core. Although many modes may propagate in a fiber under suitable conditions, there is one mode (as in a dielectric sheet) for which no cutoff exists.

Thus, if

$$\lambda_0 > \frac{2\pi a\sqrt{\eta_1^2 - \eta_2^2}}{k_{01}} = \frac{2\pi a\eta_1\cos\theta_{ic}}{k_{01}} \quad \text{(m)} \tag{6}$$

where λ_0 = free space wavelength, m
a = core radius, m
η_1 = core index of refraction
η_2 = cladding index of refraction
k_{01} = 2.405 = first root of zero-order Bessel function (J_0) (see Table 8-2)
θ_{ic} = critical angle of incidence at core-cladding boundary

then only one mode propagates and the fiber is a single-mode guide.

If the index of refraction of the core fiber decreases continuously as a function of radius, it is possible to change the path from an angular zigzag as in Fig. 8-35 to a smooth undulating curve which does not reach the core boundary, as suggested in Fig. 8-36. Under these conditions the wave propagates as though in an unbounded optical medium.

A typical optical-fiber communication link is illustrated in Fig. 8-37 with a laser or light-emitting diode (LED) as the transmitter and a phototransistor or other photosensitive device as the receiver.

In typical optical fibers, lowest attenuation occurs in the 1000- to 2000-nm range. This is in the infrared region. The light wavelengths to which the human eye is sensitive are nominally from 400 nm (violet light) to 700 nm (red light). See the electromagnetic spectrum chart in Table 1-1.

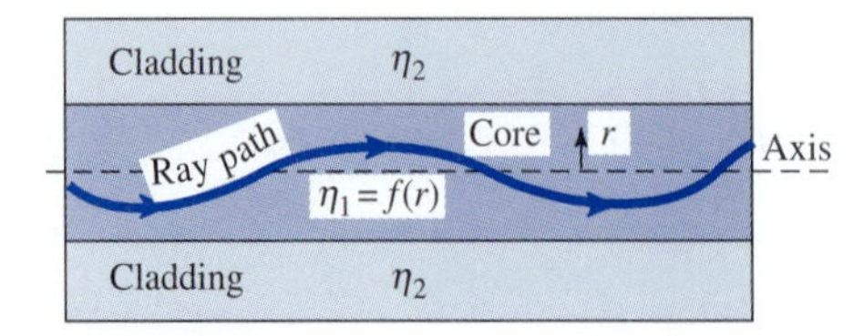

FIGURE 8-36
By gradually decreasing the index of refraction η_1 of the optical-fiber core as a function of radius r, the ray path becomes a smooth undulating curve which does not reach the core-cladding boundary so that the wave propagates as though in an unbounded medium.

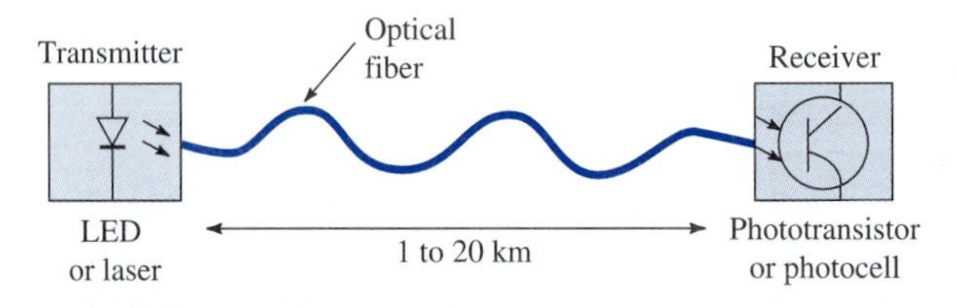

FIGURE 8-37
Optical-fiber communication link with laser or light-emitting diode (LED) transmitter and phototransistor or other photosensitive device as receiver.

Optical fibers usually have core diameters of 5 to 50 μm, which are many light or infrared wavelengths in diameter. Losses due to radiation are small. As the diameter is decreased or wavelength increased, radiation losses increase. For moderate indexes of refraction ($\sim$1.5) rods or fibers greater than 1λ in diameter are predominantly guides (most energy inside and radiation small), while rods or fibers less than 1λ in diameter have most of the energy traveling along outside the dielectric with radiation becoming important. Thus, if the dielectric cylinder or rod is gradually tapered down from a diameter of more than 1λ to less than 1λ, the guiding action will shift from energy mostly inside to mostly outside accompanied by radiation in the direction of the rod so that it behaves as an end-fire antenna.† Because tapered dielectric rod antennas for centimeter wavelengths have often been made of polystyrene, they are called *polyrod* antennas. (See Fig. 5-35*a*.)

A detailed analysis of wave propagation in dielectric cylinders is a complex topic and beyond the scope of this introductory treatment. (See footnotes on page 488.)

Example 8-10. Infrared fiber guide. A fiber guide has a core index of refraction $\eta_1 = 1.55$ and cladding index $\eta_2 = 1.52$. For $\lambda = 1$ μm, find the maximum angle θ_e at which rays will enter the fiber guide and be trapped.

†In practice, bends, nick, irregularities, and discontinuities also induce radiation.

Solution

$$\sin\theta_e = \sqrt{\eta_1^2 - \eta_2^2} \qquad \theta_e = \sin^{-1}\sqrt{1.55^2 - 1.52^2} = 17.7^\circ \quad \textit{Ans.}$$

Problem 8-11-1. Infrared fiber guide. A fiber guide has a core of index 1.53 and cladding of index 1.51. For $\lambda = 1\ \mu$m find the maximum angle θ_e at which rays will enter the fiber and be trapped. *Ans.* 14.3°.

Problem 8-11-2. Infrared fiber core diameter. (*a*) Find the core diameter required for a single mode of propagation in a fiber guide at the infrared wavelength of 1.1 μm if the core index is 1.54 and the cladding index is 1.535. (*b*) Find the maximum entrance angle θ_e. *Ans.* (*a*) 6.8 μm, (*b*) 7.13°.

Example 8-11. Fiber guide attenuation. An infrared fiber guide operating at $\lambda_0 = 1.1\ \mu$m has core constants $\sigma = 0$, $\mu_r = 1$ and $\eta = 1.5\sqrt{1 - j10^{-10}}$. Find the attenuation in dB/km.

Solution. From Table 8-2

$$\alpha = \operatorname{Re} j\frac{2\pi}{\lambda_0}\sqrt{\varepsilon_r}$$

where $\varepsilon_r = \eta^2$. Thus

$$\alpha = \frac{2\pi}{\lambda_0}\operatorname{Re} j1.5\sqrt{1 - j10^{-10}} = \frac{2\pi}{\lambda_0}\operatorname{Re} j1.5\left(1 - j\frac{1}{2}10^{-10}\right)$$

$$= \frac{2\pi}{\lambda_0}1.5\operatorname{Re}\left(j + \frac{1}{2}10^{-10}\right) = \frac{2\pi}{\lambda_0}\frac{1.5}{2}10^{-10} = \frac{\pi 1.5}{1.1}10^{-4}$$

$$= 4.28 \times 10^{-4}\ \text{Np m}^{-1} = 4.28 \times 8.69 \times 10^{-4}\ \text{dB m}^{-1}$$

$$= 3.72\ \text{dB km}^{-1} \quad \textit{Ans.}$$

Problem 8-11-3. Fiber guide attenuation. If $\lambda_0 = 2\ \mu$m and $\eta = 0.73\sqrt{1 - j10^{-11}}$, find the attenuation α in dB/km. *Ans.* 0.1 dB/km.

8-12 CAVITY RESONATORS

The purpose of transmission lines and waveguides is to transmit electromagnetic energy efficiently from one point to another. A *resonator*, on the other hand, is an energy storage device. As such it is equivalent to a resonant circuit element. At low frequencies, a parallel-connected capacitor and inductor, as in Fig. 8-38*a*, form a resonant circuit. To make this combination resonate at shorter wavelengths, the inductance and capacitance can be reduced, as in Fig. 8-38*b*. Parallel straps reduce the inductance still further, as in Fig. 8-38*c*. The limiting case is the completely enclosed rectangular box, or *cavity resonator*, shown in Fig. 8-38*d*. In this cavity resonator the maximum voltage is developed between points 1 and 2 at the center of the top and bottom plates.

Resonators can also be constructed using sections of open- or short-circuited transmission lines, as in Fig. 8-39. The type at (*a*) uses a two-conductor transmission

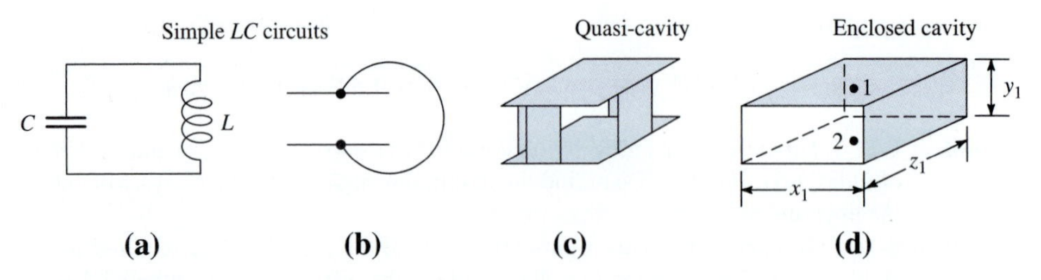

FIGURE 8-38
Evolution of (enclosed) cavity resonator from simple *LC* circuit.

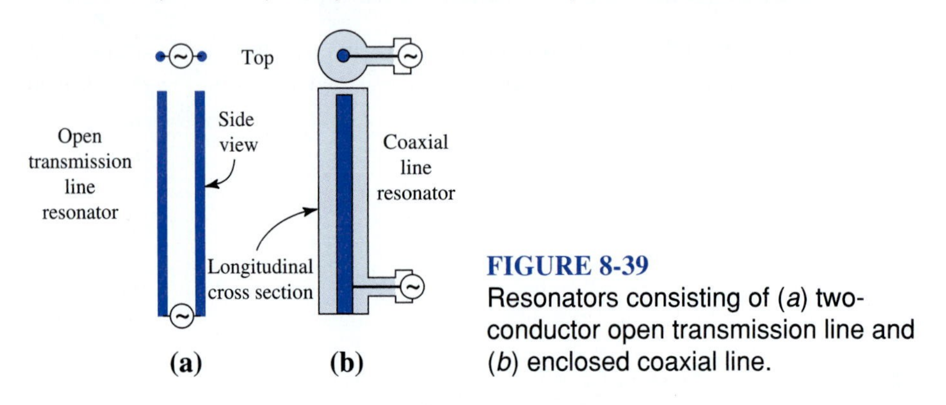

FIGURE 8-39
Resonators consisting of (*a*) two-conductor open transmission line and (*b*) enclosed coaxial line.

line, while that at (*b*) uses a coaxial line. The disadvantage of the two-conductor (open) type is that there can be a small but significant loss due to radiation. In the resonators of Fig. 8-39 the fields are in the TEM mode, whereas in the cavity resonator of Fig. 8-38*d* the fields must be in higher-order modes.

The basic principle of a cavity resonator was described in connection with the pure standing wave of Sec. 4-3. Here the energy oscillates back and forth from entirely electric to entirely magnetic twice per cycle. Let us now consider the case of a rectangular cavity resonator in more detail and determine the *resonant frequency* and Q. It is convenient to begin by recalling the situation for a TE_{m0}-mode wave in a hollow rectangular waveguide. Referring to Fig. 8-40, let a TE_{m0}-mode wave traveling in the $-x$ direction be incident on a conducting plate across the guide at $x = 0$, producing a pure standing wave in the guide. This standing wave is the resultant of two traveling waves of equal amplitude traveling in the negative x direction (incident wave) and in the positive x direction (reflected wave). The fields of these traveling waves (with time shown explicitly) are given by

$$E_y = \frac{j\beta Z_{yz}}{k_z} H_0 \sin k_z z \, e^{j(\omega t \pm \beta x)} \tag{1}$$

$$H_x = H_0 \cos k_z z \, e^{j(\omega t \pm \beta x)} \tag{2}$$

$$H_z = \frac{j\beta}{k_z} H_0 \sin k_z z \, e^{j(\omega t \pm \beta x)} \tag{3}$$

where $k_z = m\pi/z_1$.

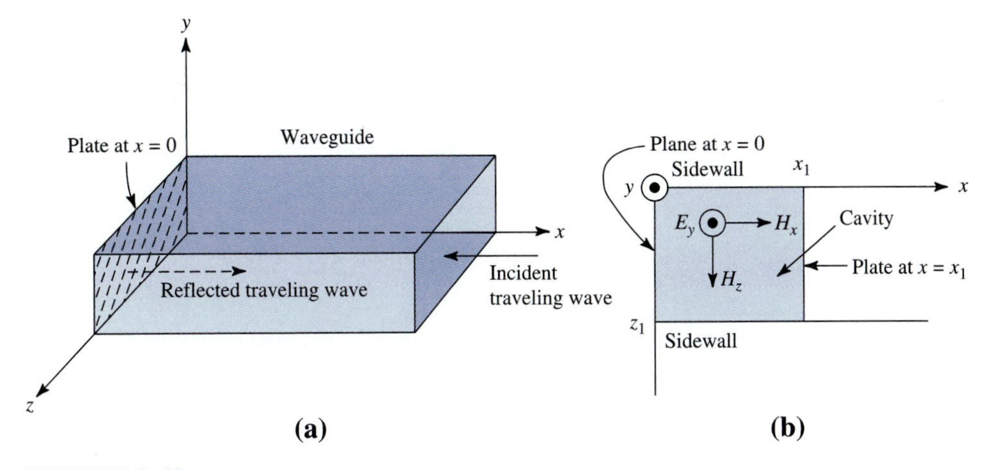

FIGURE 8-40
(*a*) Perspective view of rectangular waveguide closed with plate at $x = 0$. (*b*) Cross-sectional top view with additional plate at $x = x_1$ trapping wave inside the cavity.

The plus sign in the exponent refers to a wave traveling in the $-x$ direction and the minus sign to one traveling in the $+x$ direction. Adding the fields of the two traveling waves to obtain the fields of the standing wave, we obtain

$$E_y = \frac{-j\beta Z_{yz}}{k_z} H_0 \sin k_z z (e^{j\beta x} - e^{-j\beta x}) e^{j\omega t} \tag{4}$$

$$= \frac{2\beta Z_{yz}}{k_z} H_0 \sin k_z z \sin \beta x e^{j\omega t} \tag{5}$$

Inserting another conducting plate across the guide at $x = x_1$ requires that $\beta = k_x = l\pi/x_1$, where l is an integer as will be explained. Noting that the transverse wave impedance $Z_{yz} = \omega\mu/\beta = \omega\mu/k_x$, we get

$$E_y = \frac{2\omega\mu}{k_z} H_0 \sin k_x x \sin k_z z e^{j\omega t} \tag{6}$$

Proceeding in like manner for the magnetic field components, we get

$$H_x = -2H_0 \sin k_x x \cos k_z z e^{j[\omega t + (\pi/2)]} \tag{7}$$

$$H_z = \frac{2k_x}{k_z} H_0 \cos k_x x \sin k_z z e^{j[\omega t + (\pi/2)]} \tag{8}$$

With conducting plates across the waveguide at $x = 0$ and $x = x_1$ the wave is trapped in the rectangular enclosure forming a cavity resonator. We note that the electric and magnetic fields are in time-phase quadrature ($\pi/2$ in exponent for H_x and H_z but not for E_y), as is typical of a standing wave.

The mode of a TE wave in a rectangular cavity is, in general, designated as a TE_{lmn} mode, where l refers to (half-cycle) variations of the fields in the x

direction, m in the z direction, and n in the y direction. Since we assumed $n = 0$ in the above discussion, the designation appropriate to our example would be TE_{lm0}. Now $k^2 = k_z^2 = \gamma^2 + \omega^2\mu\varepsilon$, but $\gamma^2 = -\beta^2$ ($\alpha = 0$) and $\beta = k_x$. Thus,

$$k_z^2 = -k_x^2 + \omega^2\mu\varepsilon = -k_x^2 + (2\pi f)^2\frac{1}{(f\lambda)^2}$$

So

$$\boxed{\lambda = \frac{2}{\sqrt{(l/x_1)^2 + (m/z_1)^2}} \qquad \textbf{\textit{Resonant wavelength}}} \tag{9}$$

where λ is the *resonant wavelength* and x_1 and z_1 are the resonator dimensions. (See Fig. 8-40*b*.)

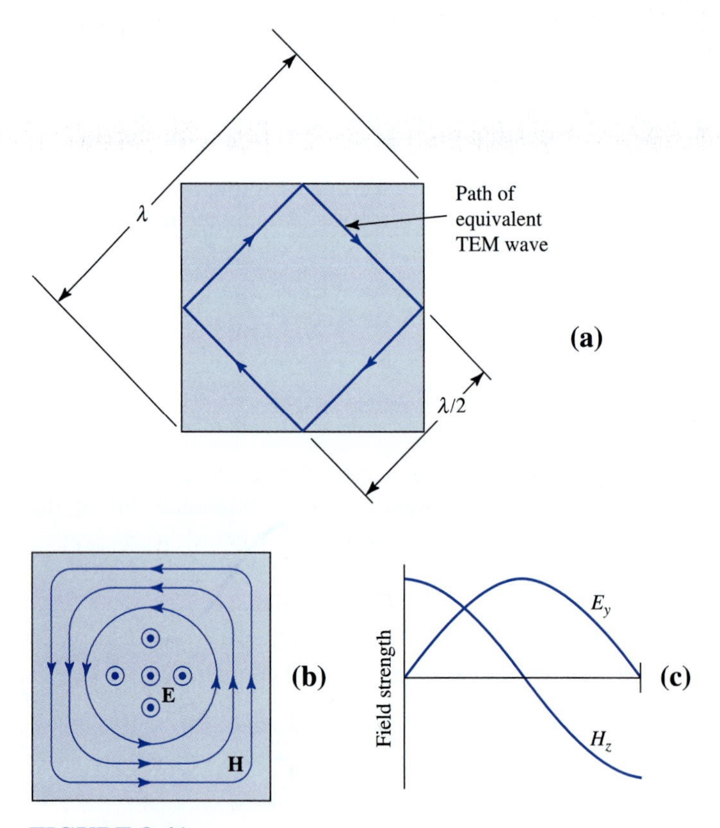

FIGURE 8-41
(*a*) The resonant wavelength of a square cavity is equal to the diagonal distance. The path of the equivalent TEM wave is also shown. (*b*) Electric and magnetic field configuration in square cavity resonator with TE_{110} mode. (*c*) Variation of E_y and H_z field components across centerline of cavity.

For example, the resonant wavelength for a TE_{110} mode in a square-box resonator ($x_1 = z_1$) is given by

$$\lambda = \frac{2}{\sqrt{2/x_1^2}} = 1.41x_1 \qquad \text{(m)} \tag{10}$$

Thus, the resonant wavelength is equal to the diagonal of the square box, as suggested in Fig. 8-41*a*. The resonant frequency is given by $f = c/\lambda = (3 \times 10^8)/\lambda$ Hz. The electric and magnetic field configuration in the resonator is as indicated in Fig. 8-41*b*. There is no variation in the y direction ($n = 0$). The wave inside the cavity is equivalent to a TEM-mode wave reflected at 45° angles, as in Fig. 8-41*a*, assuming that the box is infinitely long in the y direction (perpendicular to the page in Fig. 8-41*a*). In principle the y dimension is noncritical for the mode being considered ($n = 0$), but in practice too large a value of y could permit modes with field variations in the y direction ($n \neq 0$).

At one instant of time ($t = 0$) the energy is all in electric form, as suggested in Fig. 8-42*a*, with accumulation of positive and negative charges on the top and bottom surfaces of the cavity, as in Fig. 8-42*b*. One-quarter cycle later ($t = T/4$) the energy is all in magnetic form (Fig. 8-42*c*) with electric currents flowing down the sidewalls, as in Fig. 8-42*d*.

To find the Q of the cavity resonator, we note that by definition

$$Q = 2\pi \frac{\text{total energy stored}}{\text{decrease in energy in 1 cycle}}$$

The energy situation as a function of time is presented in Fig. 8-43. To get the total energy stored we can integrate the electric energy density w_e ($= \frac{1}{2}\varepsilon_0 E_y^2$) over the

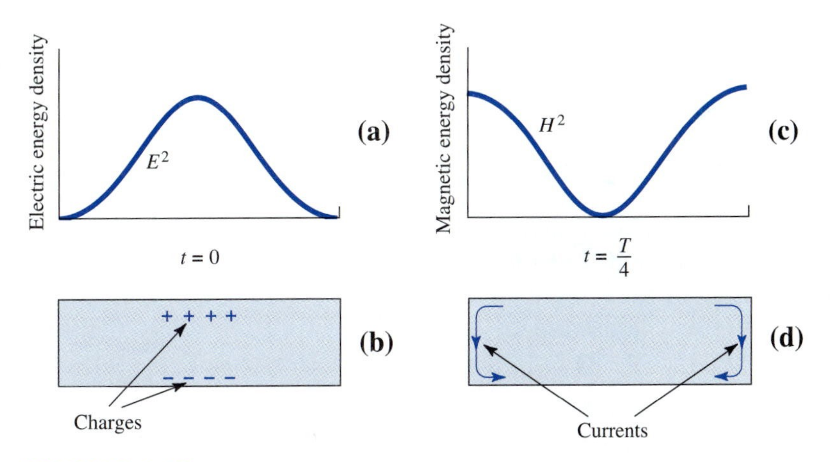

FIGURE 8-42
Electric and magnetic energy along centerline of cavity resonator. At time $t = 0$, all the energy is electric, (*a*) and (*b*), while one-quarter period later ($t = T/4$) all the energy is magnetic, (*c*) and (*d*).

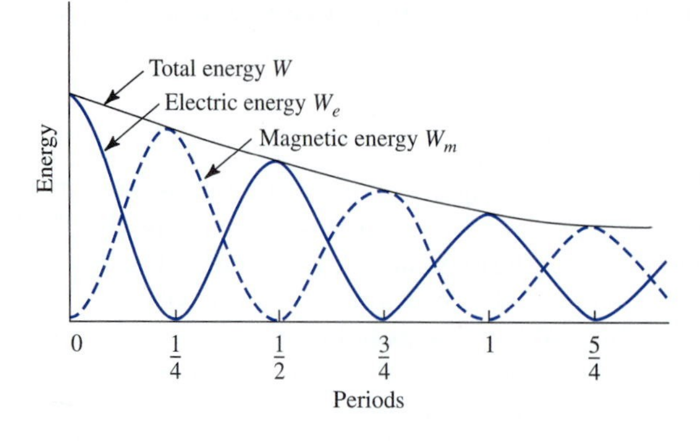

FIGURE 8-43
Decrease of stored energy with time in a resonator. The energy oscillates from all electric energy at one instant to all magnetic energy $\frac{1}{4}$ period later.

interior volume of the cavity when E_y is a maximum, and to get the decrease in energy we can integrate the average power into the walls of the cavity and multiply this by the period T $(= 1/f)$. It is assumed that the medium filling the cavity is air or vacuum. Thus, we have

$$Q = \frac{2\pi W}{T(-dW/dt)} = \frac{2\pi \iiint w_e \, dv}{T\frac{1}{2}\mathrm{Re}\, Z_c \iint |H_t|^2 \, ds} \tag{11}$$

where H_t is the magnetic field component tangent to the cavity wall.

Noting (6) and performing the integration for the total energy stored, we obtain

$$W = W_e = \frac{1}{2}\varepsilon_0 \left(\frac{2\omega\mu_0 H_0}{k_z}\right)^2 \int_0^{x_1}\int_0^{y_1}\int_0^{z_1} \sin^2 k_z x \sin^2 k_z z \, dx\, dy\, dz$$

$$= \frac{1}{8}\varepsilon_0 x_1 y_1 z_1 \left(\frac{2\omega\mu_0 H_0}{k_z}\right)^2 = \frac{\mu_0 H_0^2 x_1 y_1 z_1}{2}\left[\left(\frac{l z_1}{m x_1}\right)^2 + 1\right] \quad \text{(J)} \tag{12}$$

The factor $-dW/dt$ in (11) is the power lost in the walls. It is convenient to calculate this as the power lost in three pairs of faces: two perpendicular to the x direction, two perpendicular to the y direction (top and bottom), and two perpendicular to the z direction. Thus, the power lost in the faces perpendicular to the three coordinate directions is given by

$$P_x = 2 \times \frac{1}{2}\mathrm{Re}\, Z_c \int_0^{y_1}\int_0^{z_1} H_z^2 \, dy\, dz = 2H_0^2 y_1 z_1 \mathrm{Re}\, Z_c \left(\frac{l z_1}{m x_1}\right)^2 \tag{13}$$

$$P_y = H_0^2 x_1 z_1 \mathrm{Re}\, Z_c \left[\left(\frac{l z_1}{m x_1}\right)^2 + 1\right] \tag{14}$$

$$P_z = 2H_0^2 x_1 y_1 \mathrm{Re}\, Z_c \tag{15}$$

where Z_c is the intrinsic impedance of the conducting material forming the walls of the cavity.

Substituting these results in (11), we have for the Q of a rectangular cavity with a TE_{lm0} mode

$$Q = \frac{\mu_0 \pi f}{\operatorname{Re} Z_c} x_1 y_1 z_1 \frac{(lz_1/mx_1)^2 + 1}{2(lz_1/mx_1)^2 y_1 z_1 + [(lz_1/mx_1)^2 + 1]x_1 z_1 + 2x_1 y_1} \tag{16}$$

For a square cavity ($x_1 = z_1$) and a TE_{110} mode

$$Q = \frac{2\mu_0 \pi f}{\operatorname{Re} Z_c} \frac{x_1 y_1 z_1}{2(y_1 z_1 + x_1 z_1 + x_1 y_1)}$$

$$= \frac{2\mu_0 \pi f}{\operatorname{Re} Z_c} \frac{\text{volume of cavity}}{\text{interior surface area of cavity}} \tag{17}$$

or

$$Q = \frac{2}{\delta} \frac{\text{volume of cavity}}{\text{interior surface area of cavity}} \qquad \textbf{\textit{Q, TE}}_{\textbf{\textit{110}}} \textbf{\textit{ mode}} \tag{18}$$

where $\delta = 2\operatorname{Re} Z_c/\omega\mu_0 = 1/e$ depth of penetration.

For copper $\delta = 6.6 \times 10^{-2}/\sqrt{f}$. If $x_1 = z_1 = 100$ mm and $y_1 = 50$ mm, we have that the resonant wavelength $\lambda = 141$ mm and $Q = 17\,500$ (dimensionless).

Example 8-12. Gold-plated square cavity. An air-filled cavity resonator operates in the TE_{110} mode. The cavity is square, $x_1 = z_1 = 80$ mm with height $y_1 = 35$ mm. The cavity is copper with gold plating inside. Find (*a*) resonant wavelength, (*b*) resonant frequency and (*c*) Q.

Solution. From (8-12-9)

$$\text{Resonant } \lambda = \frac{2}{\sqrt{(1/x_1)^2 + (1/z_1)^2}} = \frac{\sqrt{2}}{\sqrt{(1/80)^2}} = 113 \text{ mm} \qquad \textit{Ans.} \quad (a)$$

$$f = \frac{c}{\lambda} = \frac{3 \times 10^8}{113 \times 10^{-3}} = 2.65 \text{ GHz} \qquad \textit{Ans.} \quad (b)$$

From (8-12-18),

$$Q = \frac{2}{\delta} \frac{\text{volume}}{\text{area}}$$

where

$$\delta = \frac{2\operatorname{Re} Z_c}{\omega\mu_0}; \qquad Z_c = \sqrt{\frac{\omega\mu_0}{\sigma}} \angle 45^\circ$$

σ (gold) $= 4.1 \times 10^7$ ℧/m

$$\delta = \frac{\sqrt{2}}{\omega\mu_0}\sqrt{\frac{\omega\mu_0}{\sigma}} = \frac{\sqrt{2}}{\sqrt{\omega\mu_0\sigma}} = 1/e \text{ depth} \qquad \text{(m)}$$

$$= \frac{\sqrt{2}}{\sqrt{(2\pi \times 2.65 \times 10^9)(4\pi \times 10^{-7})(4.1 \times 10^7)}} = 1.53 \times 10^{-6} \qquad \text{(m)}$$

Therefore

$$Q = \frac{2}{\delta}\frac{\text{volume}}{\text{area}} = \frac{2}{\delta}\frac{x_1^2 y_1}{4x_1 y_1 + 2x_1^2} = \frac{2}{\delta}\frac{x_1 y_1}{4y_1 + 2x_1} = \frac{2 \times 10^6}{1.53}\frac{80 \times 35 \times 10^{-3}}{4 \times 35 + 2 \times 80}$$

$$= 12\,200 \quad \textit{Ans.} \quad (c)$$

Problem 8-12-1. Cylindrical cavity. A cylindrical cavity resonator of height equal to its radius operates at $\lambda_0 = 10$ mm in the TM_{02} cylindrical guide mode. The cavity is made of copper. Find: (*a*) the cavity diameter and (*b*) the Q. *Ans.* (*a*) 17.5 mm; (*b*) 11 400.

Problem 8-12-2. Silver-plated cylindrical cavity. A cavity resonator is constructed of a short section of cylindrical tubing of diameter d closed at both ends by flat plates separated by $d/2$. The cavity is made of brass and is silver-plated inside. (*a*) What is the required value of d if the cavity is to operate in its dominant mode at 10 mm wavelength? (*b*) What is Q? *Ans.* (*a*) 7.7 mm; (*b*) 5176.

Methods for coupling of cavity resonators are illustrated in Fig. 8-44 for a square cavity with TE_{110} mode. Couplings to coaxial lines are shown in Fig. 8-44*a*. These involve an electric probe at the center of the cavity (**E** = max) and/or a current loop at the wall (**H** = max). Coupling to a hollow rectangular waveguide can be via a hole in the cavity wall, as in Fig. 8-44*b*. Coupling to an electron beam can be accomplished with holes in the top and bottom surfaces of the cavity, as in

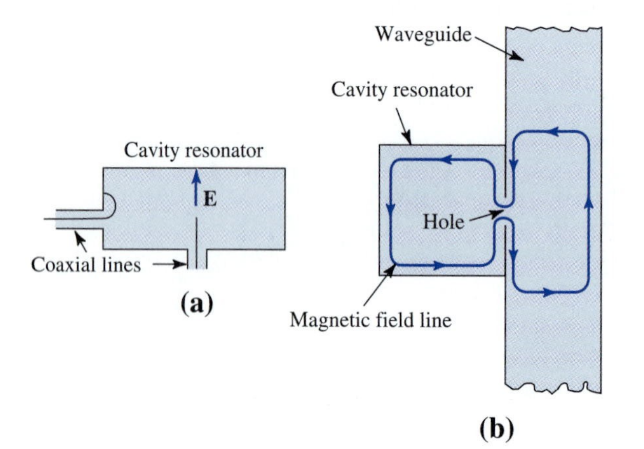

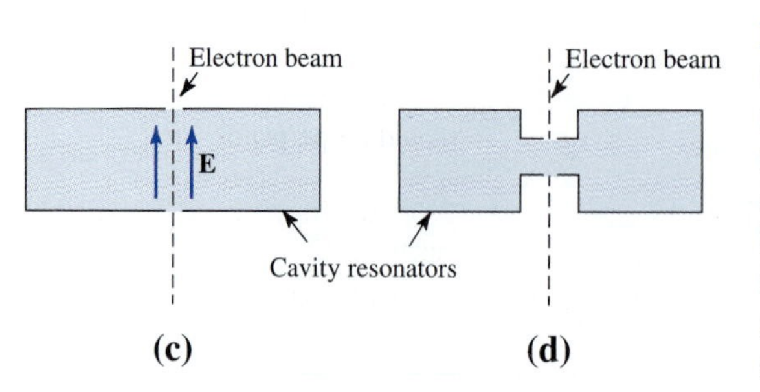

FIGURE 8-44
Cross-sections through rectangular cavity resonators showing couplings (*a*) to coaxial lines, (*b*) to a hollow rectangular waveguide, and (*c*) and (*d*) to electron beams.

Fig. 8-44*c*. Here the electrons move parallel to the electric field where it is a maximum. To reduce the transit time for the electrons, the cavity may be modified as in Fig. 8-44*d*.

PROJECT

Project P8-1. Waveguides. With transmitting antenna at point 1, receiving antenna at point 2, and both of same polarization, adjust for maximum signal. Then move receiving antenna to the side until signal is weaker (point 3) and perform the following tests:

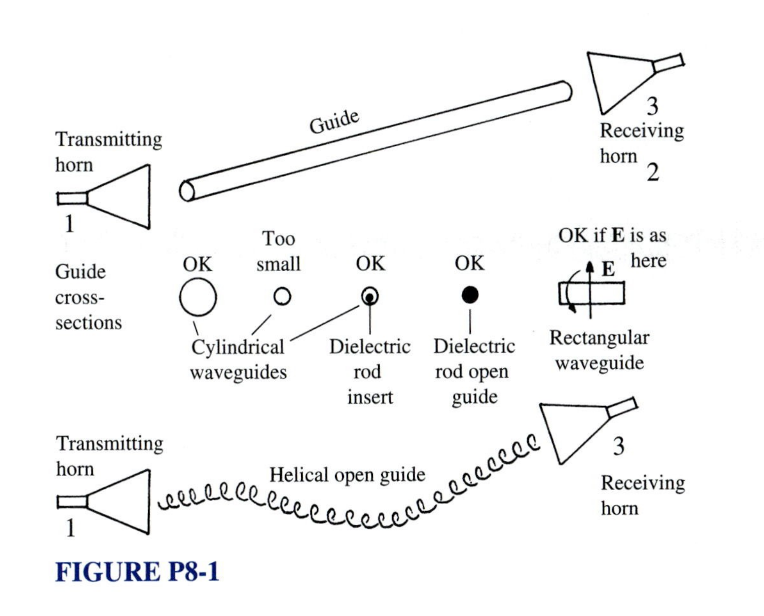

FIGURE P8-1

1. Place a circular metal pipe or tube between the antennas as suggested and note increase in signal, provided the inside diameter of the tube is large enough to pass the TE_{11} mode.
2. Repeat with smaller diameter tubes until the signal cuts off.
3. Slide a long dielectric rod into the tube and note if signal is restored. Try different diameter dielectric rods.
4. Place a dielectric rod between the antennas and note its effect as an open guide. Try rods of different diameter.
5. Place a rectangular metal guide (large enough to pass the TE_{01} mode) between the antennas. Note that the electric field vector **E** should be perpendicular to the narrow dimension of the guide cross-section. Rotate the guide 90° and note signal cutoff. Note that the cylindrical and dielectric guides work with either linear or circular polarization while the rectangular guide requires linear polarization.
6. With a long helix or "slinky" extending between antennas, observe its action as a flexible open guide. Try different diameter helixes.

PROBLEMS

Answers are given in Appendix E to problems followed by the @ symbol.

8-4-3. Windows. A metal-clad building has window openings 1.5 m by 0.5 m. What is the longest wavelength which can readily penetrate into the interior of the building?

8-4-4. Rectangular waveguide. A rectangular air-filled waveguide has a cross-section of 80×40 mm. Find: (*a*) Cut-off wavelength for dominant mode; (*b*) How many modes are passed at 2.5 times cut-off frequency? @

8-4-5. Cylindrical waveguide. A cylindrical air-filled waveguide has a diameter of 50 mm. (*a*) What is the cut-off frequency for the dominant mode? (*b*) How many modes are passed at twice the cut-off frequency of the dominant mode?

8-8-3. Surface wave power. A plane 3-GHz wave in air is traveling parallel to the boundary of a conducting medium with **H** parallel to the boundary. The constants for the conducting medium are $\sigma = 10^{-7}$ ℧/m and $\mu_r = \varepsilon_r = 1$. If the traveling-wave rms electric field $E = 75$ mV/m, find the average power per unit area lost in the conducting medium. @

8-8-4. Tilt angle. A vertically polarized 1-GHz wave is traveling in air along a horizontal aluminum ground plane. Find the tilt angle.

8-9-1. Surface wave. A surface wave is guided by a dielectric ($\varepsilon_r = 4$) layer 7 mm thick on a flat perfectly conducting ground plane. For the TM_0 (dominant) mode, find (*a*) cut-off frequency and (*b*) attenuation in decibels per meter.

8-11-4. Low-loss fiber guide. For a fiber guide operating at 15 THz with an attenuation of 0.05 dB/km, what index value is required? $\sigma = 0$, $\mu_r = 1$. @

8-11-5. Optical fiber ray angle. Rays entering the core of an optical fiber guide with core index 1.70 and cladding index 1.65 must be less than what angle θ_e to be trapped and propagate?

8-11-6. Fiberguide attenuation. If $\lambda_0 = 1.8$ μm, $\eta = 0.9\sqrt{1 - j10^{-11}}$, find the attenuation in dB/km. @

8-12-3. Square cavity. An air-filled square cavity resonator has dimensions 50×50 mm by 20 mm in height. For the TE_{110} mode, find (*a*) resonant frequency and (*b*) Q if walls are silver plated.

8-12-4. Cylindrical cavity. An air-filled cylindrical cavity resonator has a height 0.9 of the radius r. For the TM_{02} cylindrical guide mode at $\lambda_0 = 25$ mm, find (*a*) r for resonance, (*b*) Q if the wall conductivity $\sigma = 50$ M℧/m, and (*c*) Q if the walls are gold plated.

8-12-5. Square cavity. An air-filled cavity resonator operates in the TE_{110} mode. The cavity is square, $x_1 = z_1 = 50$ mm, with height $y_1 = 35$ mm. The cavity is made of copper. Find (*a*) resonant frequency, (*b*) resonant wavelength, and (*c*) Q. @

8-12-6. Cavity dimensions. (*a*) For a resonant frequency of 15 GHz, find x_1 and y_1. (*b*) If gold-plated, find Q.

8-12-7. Resonant frequency of cavity. What is the resonant frequency of a TE_{110} mode wave in a rectangular cavity with dimensions $x_1 = z_1 = 40$ mm? @

CHAPTER 9

Good planets are hard to find. We have a good one and we need good planetary stewardship to preserve it.

BIOELECTROMAGNETICS

9-1 INTRODUCTION

The bodies of humans and all animals are actuated by a complex network of noiseless, lossless coaxial transmission lines, or axons, controlled by a parallel-processing computer, or brain. This chapter includes sections on the axon, retinal optic fibers, the heart dipole field, cardiac pacemakers, defibrillators, electric fields and pulse signals of fish, magnetic navigation of birds, bone healing, and finally radiation hazards and the related environmental and health issues.

9-2 THE AXON: AN ACTIVE, LOSSLESS, SHIELDED, NOISELESS TRANSMISSION LINE

The nervous systems of animals consist of many *neurons* (nerve cells) each having an active transmission line, or *axon,* with input and output terminals. At the input end, structures called *dendrites* interface with specialized transducers sensitive to heat, pressure, or other stimuli. The dendrites are connected to a central cell body (soma), and when the algebraic sum of the excitations it receives from the dendrites exceeds a certain threshold value, it fires a signal down the axon to the terminal region, activating a motor unit (muscle) or another axon. A neuron with axon 1 m long is shown in Fig. 9-1.

The axon is an *active* transmission line with emf inputs all along it, and this results in zero attenuation of the signal. The other transmission lines we have considered are *passive,* having no energy inputs except at the input terminals.

Many neurons may be connected in series by structures called *synapses,* in which the output dendrites of one neuron connect with the input dendrites of the next neuron. The velocity of signal propagation along a particular axon transmission

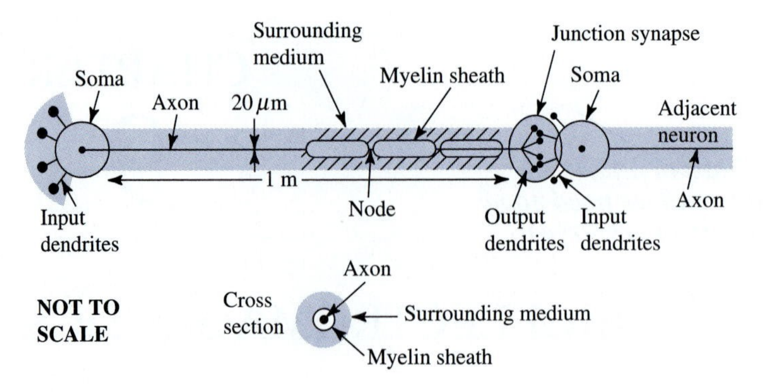

FIGURE 9-1
Idealized diagram of a typical neuron, as found in the sciatic (leg) nerve of a large mammal, with connection shown to an adjacent neuron. The axon acts as the inner conductor, the myelin sheath as its insulation, and the surrounding medium as the outer conductor of a coaxial transmission line. It is a ***noiseless, lossless*** transmission line. A bundle (cable) of thousands of such axons or nerve fibers forms the sciatic nerve.

line is constant, but different axons may have different velocities. Axons of larger diameter (20 μm) may have signal velocities of 100 m s^{-1}.

The axon is enclosed in a myelin sheath, which is electrically passive and acts as an insulator. At millimeter intervals along the axon the sheath may be interrupted at nodes exposing the axon to the surrounding medium. By diffusion of ions from the surrounding medium through the outer membrane of the axon, emfs are applied between the inner axon (as one conductor) and the surrounding medium (as the second conductor) like the voltage across a coaxial transmission line. This voltage produces a current via the axon and surrounding medium through the next node, triggering emfs there and so on down the line.

The first definitive theory of axon behavior was published by A. L. Hodgkin and A. F. Huxley in 1952. Their work, based on research on the properties of a giant axon of the squid, earned them a Nobel Prize.

The equivalent circuit of an axon transmission line according to Hodgkin and Huxley's theory is shown in Fig. 9-2. The line has series resistance and shunt conductance and capacitance. There is no series inductance, but some models include a shunt inductance. In addition, shunt emfs are applied through variable conductances which act like switching elements. Normally the diffusion of potassium (K) ions and miscellaneous leakage (L) ions keeps the inner axon negative by about 100 mV. But on excitation, the diffusion of sodium (Na) ions swings the potential positive for the period of the impulse, which is typically a few tenths of a millisecond. The recovery of the axon to its normal negative potential after the passage of the impulse is accomplished in less than 1 ms. Since the full impulse voltage is received at the terminals, the axon transmission line has *zero attenuation.* It is also a *"noiseless" line* in that it either transmits a full impulse or none at all. There is no intermediate condition.

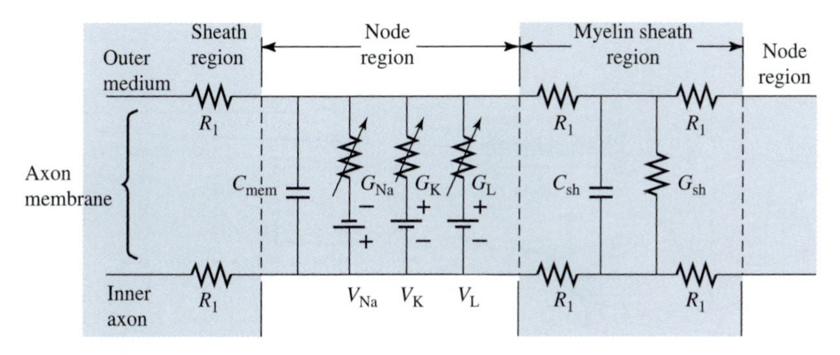

FIGURE 9-2
Equivalent circuit of an axon transmission line, divided into node and sheath regions. The axon has only a membrane separating it from the surrounding medium in the node region but it has a myelin enclosure in the sheath region.

Although this brief discussion is greatly oversimplified, it provides some insights into the remarkable properties of the active coaxial transmission lines present in great numbers in all animals.

Example 9-1. Axon action. If there are 7 neurons in series between the brain and leg of an animal and each neuron is 8 cm long, find (*a*) time for brain pulse to arrive at leg. Take velocity $= 75$ m s^{-1}. (*b*) If the brain pulse is 1 mV, what is the pulse voltage at the leg?

Solution. Time $t = 7 \times 0.08/75 = 7.5$ ms. *Ans.* (*a*).
Since attenuation $= 0$, $V = 1$ mV. *Ans.* (*b*).

Problem 9-2-1. Axon voltage. A coaxial axon transmission line is insulated by a tubular membrane of 100-μm wall thickness. The membrane maintains a low sodium ion concentration inside at a 100-mV difference. What is the electric field in the membrane insulation between the inner and outer conductors? *Ans.* 1000 V/m.

9-3 RETINAL OPTIC FIBERS

The retina of a human eye contains a bundle of more than 100 million optical fibers, each acting as both a light waveguide and also a photon detector (receptor).† There are two classes of these fibers: the *cones,* occupying the central area of the retina, and the more numerous *rods* in the outer surrounding regions. Almost all of the cones are individually connected by nerve transmission lines (axons) to the brain, where signal processing and image formation occurs, and it is with the cones that fine details

†All electromagnetic waves are transmitted in quantum units, called *photons*, of energy equal to hf, where h = Planck's constant ($= 6.63 \times 10^{-34}$ J s) and f = frequency (Hz).

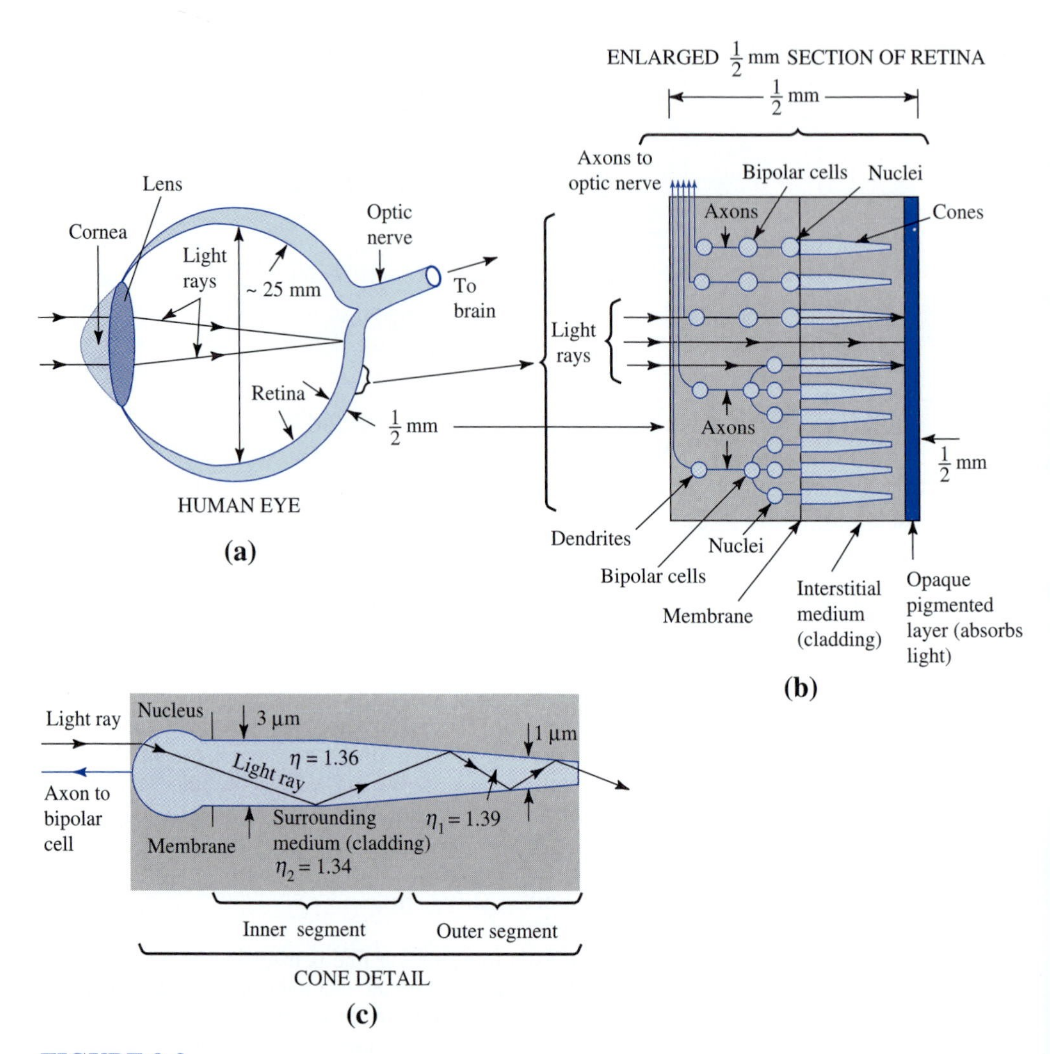

FIGURE 9-3

(*a*) Section of human eye, (*b*) enlarged section of retina, and (*c*) details of cone. The retina contains over 100 million optical fibers (rods and cones) which act as both polyrod antennas and photon detection devices. Normal vision wavelengths cover the range from 400 to 700 nm, which is approximately half the diameter of the outer segments of the rods and cones. The diagrams are simplified and somewhat schematic. Note that the nerve (axon) connectors are *in front of* the rods and cones. This does not interfere with transmission of light to the rods and cones because the connections are transparent.

(such as the book printing you are reading) can be distinguished. The rods provide much poorer resolution of image details but can give better vision at low light levels because of their higher sensitivity and the fact that many rods may be connected in parallel to a single axon-brain line. The rods also provide peripheral vision.

Figure 9-3*a* is a cross-section of a human eye showing lens, retina, and optic nerve to the brain. Figure 9-3*b* is an enlarged view of a section of the retina, which is a transparent medium containing rods, cones, cells, and dendrites. It has an opaque backing called the *pigmented layer.* Figure 9-3*c* is a still more enlarged view of a single cone. The narrow ends of the rods and cones, called *outer segments,* are of the order of 1 μm in diameter and about 20 times as long. The outer segments have an index of refraction η_1 of about 1.39 with the cladding or surrounding (interstitial) medium index η_2 a few percent less. These index values are very close to those employed in typical commercial optical fibers ($\eta_1 = 1.46$, $\eta_2 = 1.44$), but the diameter of the outer segments is less (1.5 to 2λ), so that the outer segments apparently have more radiating or receiving action than a standard commercial optical fiber.

The nucleus of a cone or rod acts as a lens, concentrating light into the interior, where it travels through both segments by total internal reflection. Any light photons not absorbed in the outer segment pass out the far end and impinge on the opaque pigmented layer. In human beings the pigmented layer absorbs light, preventing any reflection, but in night-hunting animals, such as cats, the pigmented layer is replaced by a tapetum which is highly reflective so that light not absorbed on its way to the tapetum is reflected back to the rods and cones. This gives cats a 6-dB dark vision advantage over humans.

It is to be noted that although the index of refraction of the rods and cones varies somewhat with position, it is always greater than that of the cladding or surrounding medium, which is a necessary condition for total internal reflection.

When photons are absorbed by molecules in the outer segment, a current is initiated which flows back to the bipolar cell, which fires the impulse that travels to the brain via the axons and dendrites. Thus, a rod or cone might be described as similar to an end-fire (polyrod) antenna (with unity front-to-back ratio) which is also equipped with sensitive detectors that convert the light photon frequencies (10^{15} Hz) to near dc impulses for transmission to and processing by the brain. Hence, the human retina may be said to have an array of more than 100 million polyrod antennas.

Example 9-2. Cone of retina. (*a*) At what diameter (in μm) does the retinal cone of Fig. 9-4 become a single-mode guide for $\lambda > 550$ nm? This wavelength is at the center of the visual spectrum (400 to 700 nm). (*b*) What is the cone diameter in λ?

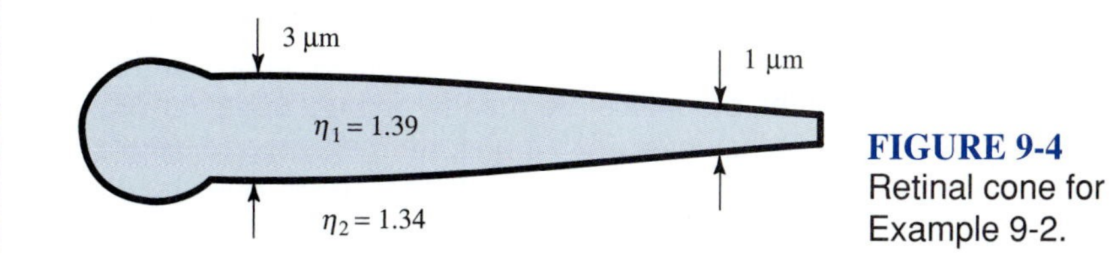

FIGURE 9-4
Retinal cone for Example 9-2.

Solution. From (8-11-6)

$$\text{Diameter } d = 2a = \frac{\lambda_0 \times 2.405}{\pi\sqrt{\eta_1^2 - \eta_2^2}} = \frac{550 \times 10^{-9} \times 2.405}{\pi\sqrt{1.39^2 - 1.34^2}} = 1.14\ \mu\text{m} \quad Ans. \quad (a)$$

$$d = \frac{1.14 \times 10^{-6}}{550 \times 10^{-9}} = 2.07\lambda \quad Ans. \quad (b)$$

Problem 9-3-1. Retinal cone as guide. If a 2-μm-diameter retinal cone has an index of refraction $\eta_1 = 1.36$, with a cladding index $\eta_2 = 1.34$, what is the lower limit for which the cone is a single-mode guide? *Ans.* 607 nm.

9-4 HEART DIPOLE FIELD

The heart of all mammals contracts or beats in response to an electric potential difference across it which reaches a maximum value just prior to the start of the blood-pumping contraction. The potentials measured on the skin of the animal at this instant of maximum voltage have a distribution like that from a dipole aligned with the heart. The similarity of the field measured on a human chest (Fig. 9-5*a*) to the field of a dipole in an isotropic medium (Fig. 9-5*b*) is apparent. (See also

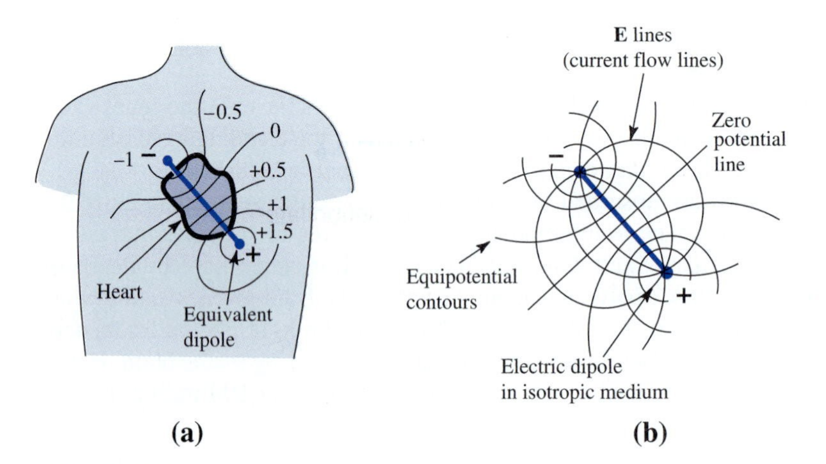

FIGURE 9-5
(*a*) Equipotential contours of electric feed from the heart measured on a human chest just prior to heart contraction with approximate position of the equivalent heart dipole. The field shown is a somewhat simplified version of an actual measured field. (*b*) Equipotential and field lines of electric dipole in an isotropic medium. Note that the heart dipole is within an imperfect (conducting) dielectric medium and the potentials are measured on a surface displaced from it. The dipole in (*b*), however, is in a uniform medium and the map is in a plane parallel to and coincident with the dipole. Thus, one should not expect the two maps to be identical. The equipontential contours in (*a*) are in millivolts. [(*a*) *is after B. Taccardi, Circ. Res., 1963.*]

Fig. 7-1, page 381). The electric fields of animals are of great diagnostic value. For example, a different (or abnormal) heart position would be evident from a field map. An advantage of this mapping technique is that it is harmless, in contrast to X-ray or certain other techniques for acquiring similar information.

9-5 DEFIBRILLATORS AND PACEMAKERS†

The heart is a bioelectrically controlled blood pump. During a heart attack the action of the heart muscle deteriorates from a regular contraction to a convulsive quiver called *fibrillation.* To restore normal heart action to a human subject, a capacitor may be discharged through electrodes, called paddles, placed across the chest as in Fig. 9-6.

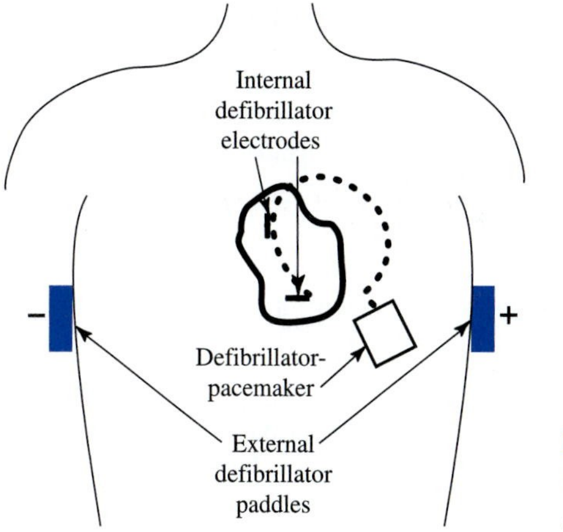

FIGURE 9-6
Placement of defibrillator electrodes and of defibrillator-pacemaker.

Example 9-3. External defibrillator. For a capacitor energy $\mathscr{E} = 400$ J discharged in a time $t = 3$ ms as the defibrillation pulse, find (*a*) pulse current, I, and (*b*) pulse voltage, V.

Solution. From Example 9-4, the chest resistance R between the paddles is 50 Ω. Thus,

$$\text{Energy, } \mathscr{E} = I^2 Rt = VIt$$

And the pulse current

$$I = \sqrt{\frac{\mathscr{E}}{Rt}} = \sqrt{\frac{400}{50 \times 3 \times 10^{-3}}} = 51.6 \text{ A} \quad \textit{Ans.} \quad (a)$$

and pulse voltage

$$V = \frac{\mathscr{E}}{It} = \frac{400}{51.6 \times 3 \times 10^{-3}} = 2.58 \text{ kV} \quad \textit{Ans.} \quad (b)$$

---

†"Design of Cardiac Pacemakers," J. G. Webster, ed., IEEE Press, New York, 1995.

Example 9-4. Chest resistance. For a chest width $w = 250$ mm, find the resistance between two defibrillator paddles placed under the armpits across the chest.

Solution. Draw a field map as in Fig. 9-7. From Table 2-1 the conductivity $\sigma = 0.2\ \mho\ \text{m}^{-1}$. Thus,

$$R = \frac{n}{N} R_0 = \frac{n}{N} \frac{1}{\sigma d}$$

where n = number of cells in series = 3
N = number of cells in parallel = 2
$\sigma = 0.2\ \mho\ \text{m}^{-1}$
d = chest dimension (perpendicular to page)
= 150 mm

Thus, $R = \dfrac{3}{2} \dfrac{1}{0.2 \times 0.15} = 50\ \Omega$ *Ans.*

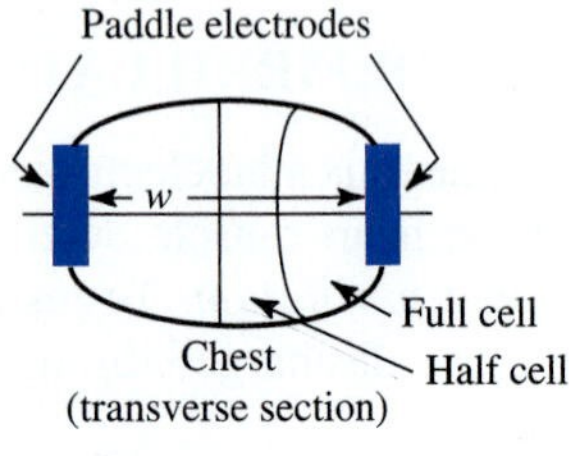

FIGURE 9-7
Field map for chest resistance.

With the development of small implantable (or internal) defibrillators having electrodes placed directly in the heart as in Fig. 9-6, the required pulse energy is much reduced. The great advantage of an implanted defibrillator is that the pulse is delivered immediately when the defibrillator unit detects the onset of fibrillation. There is no wait for a medical team to arrive with a paddle-type defibrillator.

Example 9-5. Internal defibrillator. An implanted defibrillator with electrodes in the heart, as in Fig. 9-6, delivers a 25-J, 5-A, 10-ms pulse. Find the circuit resistance R.

Solution. $R = \dfrac{\mathscr{E}}{I^2 t} = \dfrac{25}{5^2 \times 10 \times 10^{-3}} = 100\ \Omega$ *Ans.*

Although the path between electrodes is much less than with external paddles, the electrodes are tiny compared to the paddles, thus increasing the path resistance.

Problem 9-5-1. Defibrillator voltage. What is the pulse voltage for the defibrillator of Example 9-5? *Ans.* 500 V.

Implantable heart pacemakers are widely used to maintain normal heart action. Whereas a defibrillator may fire one big pulse, a pacemaker fires small pulses at heart beat rates of 50 to 100 per minute. Each pulse may have only one-millionth the energy of a defibrillator pulse.

Example 9-6. Pacemaker. A pacemaker delivers 5-V, 10-mA, 0.5-ms pulses. Find: (*a*) pulse energy and (*b*) path resistance.

Solution.

$$\text{Energy } \mathscr{E} = VIt = 5 \times 0.01 \times 0.5 \times 10^{-3} = 25\ \mu\text{J}$$
$$= 25 \times 10^{-6}\text{ J} = 25\ \mu\text{J} \quad \textit{Ans.} \quad (a)$$

$$\text{Resistance } R = V/I = 500\ \Omega \quad \textit{Ans.} \quad (b)$$

Problem 9-5-2. Pacemaker battery life. A pacemaker delivers 5-V, 10-mA, 0.5-ms pulses 60 times per minute. How many years will a 1-A-h battery last assuming no deterioration? *Ans.* 23 years.

Many implantable devices combine a defibrillator and a pacemaker in one 0.1-liter unit (see Fig. 9-6). Both may remain on standby until the onset of fibrillation when the defibrillator fires. The pacemaker may then activate for a time to help restore normal heart action. Or in other situations where a pacemaker is needed to maintain regular heart action, the pacemaker may be on continuously with the defibrillator on standby.

Problem 9-5-3. Pacemaker shielding and radiation. Pacemaker-defibrillators may be adversely affected by strong magnetic fields (of stereo speakers), by magnetic wands used at airports, by welding machines, by motors of cordless power tools, by cellular phones, and by wireless remote-control units for doors and toys. Conversely, the pacemaker-defibrillator can trigger security alarms. Explain. See Fig. 10-4 (page 529).

9-6 BIOLOGICAL FIELDS

Walk across a rug on a dry day, touch a metal object and zap! You are a *triboelectric* (friction) generator. If the spark you drew is only a millimeter long, you developed a potential of 5000 V. The friction of your feet on the rug drew electrons to your body and the electrons then jumped from your finger to the metal object. Although startled, you were not injured because the amount of electric charge was small. Much less voltage can be lethal if the charge available is large enough.

Frictionally generated charges attract dust and lint to photographic films and circuit boards. They can damage microprocessor chips. In the presence of combustibles they can start a fire or cause an explosion. But they also have many useful applications.

Out-of-doors on a clear day we are in an electrostatic field of about 200 V/m. Under a thundercloud the field may be 20,000 V/m. (See "Hey man, hit the ditch," Example 9-13). And rain or shine, we are in a magnetic field of about 1 gauss (1 G). The thundercloud may alert us when our hair stands on end. But without a compass or other device, we are unaware of the earth's magnetic field. Not so for a shark which detects it by internal voltages developed as it moves through the field. Bees, some bacteria, many birds, and other animals also have a magnetic sense via internal magnetic bodies of an iron oxide, magnetite (Fe_3O_4), which act as a compass.

Many aquatic forms both generate electric fields and are sensitive to them. Some fish signal via electric currents in the same way that scuba divers communicate. Some fish use continuous waves of the order of 700 Hz while others use 2-ms pulses at a repetition rate of 50 Hz.

Perhaps the most spectacular aquatic form is the electric eel that can deliver 500-V shocks for stunning prey or warding off predators. (For more information on electric species see *IEEE Spectrum,* March 1996, p. 22.)

Example 9-7. Electric eel. A freshwater eel, Fig. 9-8*a*, develops 500 V between electrodes on its body spaced 750 mm apart. If the eel's internal battery resistance is 15 Ω, find: (*a*) resulting current and (*b*) power developed. Eel's radius = 40 mm, water conductivity $\sigma = 0.01\ \mho\ \text{m}^{-1}$.

Solution. Draw field map, Fig. 9-8*b*. Assume most current flow within volume of map (to right in water, to left in eel). Note that this is a 3-dimensional situation. The resistance of the water path is given by the resistance of the nine annular sections in series. From map, $r_2 = 160$ mm so the resistance of the shaded annulus is given by

$$R = \frac{r_2 - r_1}{\sigma\pi(r_2^2 - r_1^2)} = \frac{0.12}{0.01\pi(0.16^2 - 0.04^2)} = 159\ \Omega$$

Taking 159 Ω as the average value

$$R\ \text{(water path)} = 159 \times 9 = 1432\ \Omega$$

$$R\ \text{(internal)} = 15\ \Omega$$

$$R\ \text{(total)} = 1432 + 15 = 1447\ \Omega$$

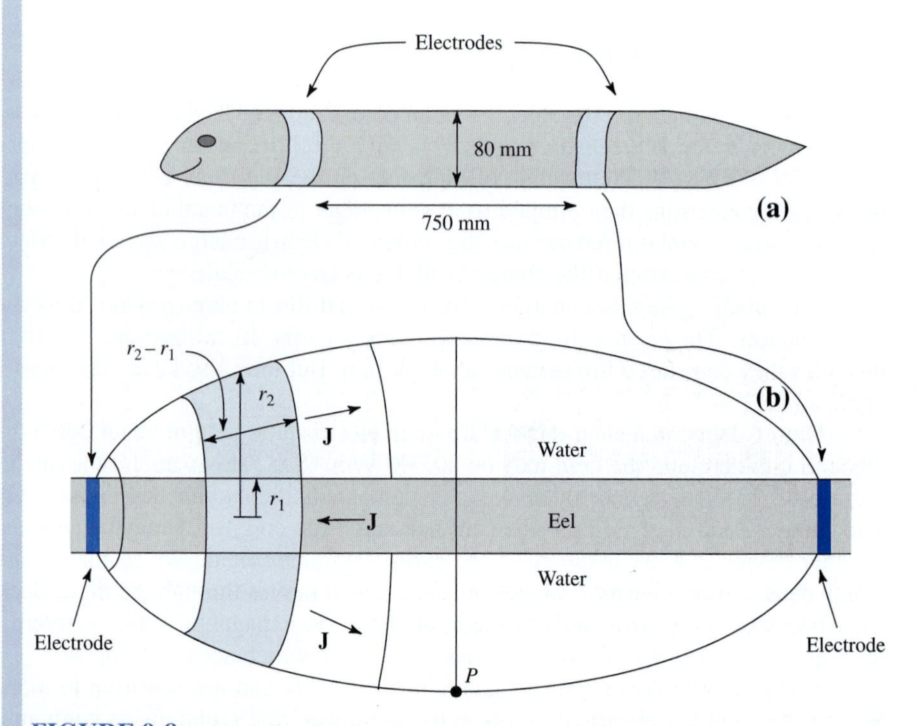

FIGURE 9-8

(*a*) Freshwater electric eel. (*b*) Field map between eel's electrodes.

Thus,

$$I = \frac{V}{R} = \frac{500}{1447} = 346 \text{ mA} \quad Ans. \quad (a)$$

$$P = \frac{V^2}{R} = \frac{500^2}{1447} = 173 \text{ W} \quad Ans. \quad (b)$$

Note: Of the 173 W generated, 99 percent is delivered externally and only 1 percent is dissipated internally in the eel, as in a well-designed generating station.

Problem 9-6-1. Electric field of eel. What is the electric field at point *P* in Fig. 9-8*b*? *Ans.* 343 V/m.

Problem 9-6-2. Eel current. If a 500-V eel delivers a 25-J shock in 0.1 s, what is the current? *Ans.* 500 mA.

Another aspect of biological fields is that bone is *piezoelectric,* that is, a pressure applied produces a potential difference while a potential difference applied produces a mechanical stress. In simplistic terms, exercise stresses a bone, producing an electric potential difference, which, in turn, promotes calcium deposit, strengthening the bone. Astronauts in orbit exercise to maintain bone strength. Persons with a broken arm in a cast can't exercise it, but if a potential difference is applied across the arm, it can produce an electric field at the bone, promoting calcium deposit and more rapid healing.

Example 9-8. Field in bone. If 50 V is applied by insulated "capacitor plates" across a 75-mm-diameter arm, as in Fig. 9-9, what is the electric field E in the bone? ε_{ri} (insulation) = 1.5, ε_{rb} (bone) = 2, ε_{rt} (tissue) = 4. Bone diameter = 25 mm. Note that since the plates are insulated, there is no current flow. The insulation thickness = $\frac{1}{2}$ mm.

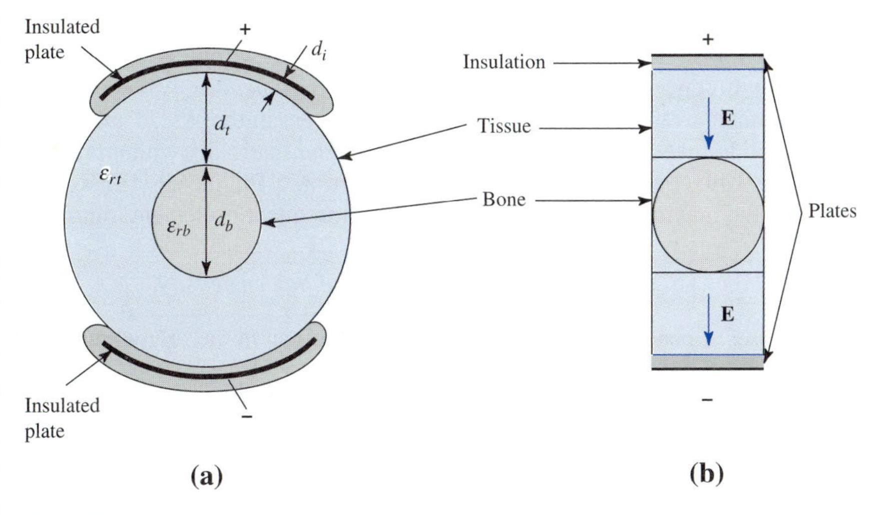

FIGURE 9-9
Capacitor plates across arm for application of electric field to bone.

Solution. Consider that the field at the bone is essentially uniform so that the problem can be treated approximately as a parallel-plate capacitor with different dielectrics as in Fig. 2-22.

Using subscript i for insulation, t for tissue and b for bone we have

$$V = 2E_i d_i + 2E_t d_t + E_b d_b$$

and since D (normal) is continuous

$$D = \varepsilon_0 \varepsilon_{ri} E_i = \varepsilon_0 \varepsilon_{rt} E_t = \varepsilon_0 \varepsilon_{rb} E_b$$

Introducing numerical values gives

$$\text{Field in bone} = E_b = 970 \text{ V m}^{-1} \quad \textit{Ans.}$$

Problem 9-6-3. Voltage applied for bone therapy. If the permittivity ε_r of the insulation $= 3$ and the field in the bone is 700 V/m, what is the applied voltage for the configuration of Fig. 9-9? *Ans.* 36 V.

Problem 9-6-4. Shark velocity. If a shark can detect an electric field of 1 μV/m, what is the minimum velocity at which it can detect the earth's 1-G magnetic field? *Ans.* 10 mm/s.

9-7 ELECTROMAGNETIC HAZARDS AND THE ENVIRONMENT

For eons, lightning bolts were mankind's only electromagnetic hazard. But with the electrical-electronic revolution of the last century many new ones have emerged involving power lines at frequencies of 50 and 60 Hz and radio transmitters at kilohertz to gigahertz frequencies.

Heating with radio frequencies is used for medical diathermy treatments, in the melting of plastics for injection molding, and for cooking food in microwave ovens. In these applications, the radio frequency power is presumed to be used in a controlled manner. The question of hazard also arises from the unintentional exposure to radiation from high-power radio, FM, TV, radar, and wireless transmitters.

It is of concern to humans that radio frequency *heating* can occur internally without much awareness because our heat sensors are in the skin. Thus, *safe power density* guidelines are needed to avoid being cooked internally without realizing it.

The 1991 Institute of Electrical and Electronic Engineers (IEEE) guideline specifies the maximum safe power density level in uncontrolled environments as

2 W/m^2	***IEEE safe level of power density***

The amount of heating this power can produce is given in the following example.

Example 9-9. Heating at 2 W/m^2. An electromagnetic wave with power density 2 W m^{-2} is incident on a 1-square-meter slab of absorbing material 1 cm in thickness (Fig. 9-10). Assuming a perfect match by the slab, find: (*a*) time to raise slab temperature 1°C. (*b*) What is the equivalent volt per meter level?

Solution. The volume of the slab = 10 liters. We assume the slab is thermally equivalent to water, for which 1 kg-cal heat energy raises the temperature of 1 liter 1°C.

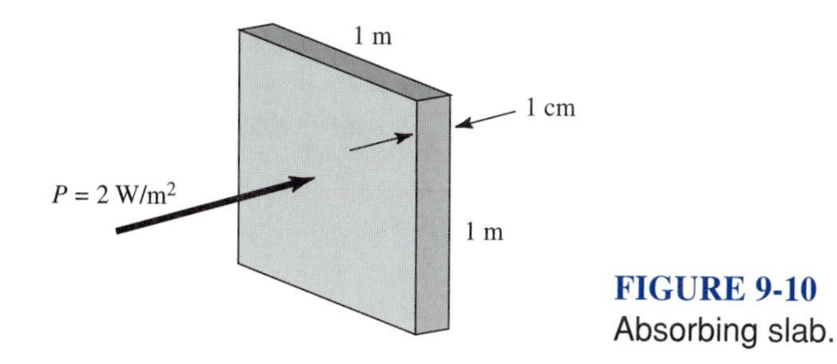

FIGURE 9-10
Absorbing slab.

The energy required to raise the slab temperature 1°C is given by

$$\mathscr{E} = \text{mass} \times \text{specific heat} \times \Delta T = 10 \times 1 \times 1 = 10 \text{ kg-cal}$$

Since 1 kg-cal = 4.2 kJ,

$$\text{Time required} = \frac{\text{energy}}{\text{power}} = \frac{42\,000}{2 \times 1} = 21\,000 \text{ s} = 5.8 \text{ hrs} \quad \textit{Ans.} \quad (a)$$

$$E = \sqrt{PZ} = \sqrt{2 \times 377} = 27.5 \text{ V m}^{-1} \quad \textit{Ans.} \quad (b)$$

Problem 9-7-1. Melting plastics. A $\lambda = 3$-m oscillator is used to melt plastic for injection molding. If $\varepsilon_r = 2.9$ and PF $= 0.04$ at $\lambda = 3$ m, what applied electric field is required to bring 1 kg of plastic to the melting point of 90°C in 1 min? Take $\sigma = 0$ and the starting $T = 23$°C (room temperature). Take the specific heat of the plastic as 0.6 cal/g. *Ans.* 2.1 kV/m. (Additional heat is required to soften or liquify the plastic.)

In the above example and problem, power density and field strengths have been translated into a temperature increase. This gives another way of measuring the effect of radio-frequency fields. The next example involves the baking of a potato.

Example 9-10. Baked Idaho potato in microwave oven. Power factor (PF). A homogeneous 200-ml, 200-g Idaho potato has constants $\varepsilon_{rd} = 65 - j15$ at 2.45 GHz. If the oven applies an electric field $E = 30$ kV m^{-1} at 2.45 GHz, how long will it take to bring the potato temperature from 23°C (room temperature) to 100°C and hold it there long enough to convert 25 percent of its water content to steam? This may be considered sufficient for baking the potato. Assume that the potato is equivalent to an equal volume of water for which the specific heat is 1 cal/g and the heat of vaporization is 550 cal/g. $\sigma = 0$; 1 gcal = 4.2 J.

The relative permittivity ε_{rd} is a complex quantity. Putting $\mathbf{J} = \sigma\mathbf{E}$ and $\varepsilon_{rd} = \varepsilon' - j\varepsilon''$ is Maxwell's equation, we have

$$\nabla \times \mathbf{H} = j\omega\varepsilon'\mathbf{E} + (\sigma + \omega\varepsilon'')\mathbf{E}$$

where $\sigma' = \sigma + \omega\varepsilon''$ is an equivalent conductivity and

$$\mathbf{J}_{\text{total}} = (\sigma + \omega\varepsilon'')\mathbf{E} + j\omega'\varepsilon'\mathbf{E} = \sigma'\mathbf{E} + j\omega\varepsilon'\mathbf{E}$$

with a conduction current density $\sigma'\mathbf{E}$ and a displacement current density $\omega\varepsilon'\mathbf{E}$ in time-phase quadrature. The ratio

$$\frac{\sigma'}{\omega\varepsilon'} = \tan\delta = \textbf{\textit{loss tangent}}$$

and

$$90° - \delta = \theta$$
$$\cos\theta = \textbf{\textit{power factor}} \text{ (PF)}$$

For small δ, PF $\approx \tan\delta$.

For small dc conductivity ($\sigma \approx 0$), $\sigma' = \omega\varepsilon''$ so the power factor for the potato is approximately

$$\text{PF} = \frac{\omega\varepsilon''}{\omega\varepsilon'} = \frac{15}{65} = 0.23$$

Solution. The energy required for baking is

$$\begin{aligned}\mathscr{E} &= (\text{mass} \times \text{specific heat} \times \Delta T) + (\text{mass} \times \text{heat of vaporization})\\ &= 200 \times 1 \times (100 - 23) + (200 \times 0.25)550 = 42\,900 \text{ g-cal}\end{aligned}$$

and

$$\mathscr{E} = 42\,900 \times 4.2 = 180 \text{ kJ}$$

The equivalent conductivity

$$\sigma' = \omega\varepsilon'' = 2\pi \times 2.45 \times 10^9 \times 15 \times 8.85 \times 10^{-12} = 2.0 \text{ ℧/m}$$

From the oven geometry of Fig. 9-11, we have

For E through the potato: $V = E_0 h + E_d h$

For E alongside the potato: $V = 2hE$ so $2E = E_0 + E_d$

The normal components of D are continuous so

$$\varepsilon_0 E_0 = \varepsilon_d E_d$$
$$E_0 = \varepsilon_{rd} E_d$$

and

$$E_d = \frac{2E}{\varepsilon_{rd} + 1} = \frac{2 \times 3 \times 10^4}{65 + 1} = 909 \text{ V m}^{-1}$$

$$\begin{aligned}\text{Baking time} &= \frac{\text{energy}}{\text{power}} = \frac{\mathscr{E}}{(\sigma' E_d^2) \times \text{potato volume}}\\ &= \frac{180 \times 10^3}{(2 \times 909^2)(200 \times 10^{-6})} = 545 \text{ s} = 9 \text{ min} \quad \textit{Ans.}\end{aligned}$$

Experienced cooks make slits or holes in the potato to allow steam to escape and prevent the potato from exploding.

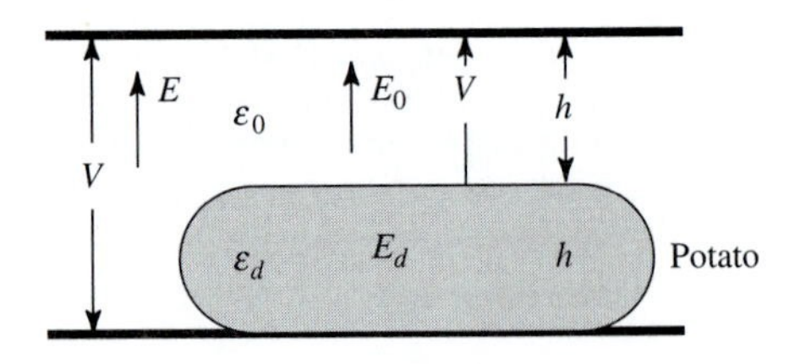

FIGURE 9-11
Microwave oven.

Problem 9-7-2. Potato. In Example 9-10, (*a*) what is the value of E_0? (*b*) If $h = 4$ cm, what is V? *Ans.* (*a*) 59 kV; (*b*) 2.4 kV.

A microwave oven is a resonator and, if well constructed, has negligible radiation leakage. Although an electric field of 30 kV m^{-1} is considered safe inside the oven, such a level in the open is hazardous. An electromagnetic wave of this field strength has a power density

$$P = \frac{E^2}{Z} = \frac{(3 \times 10^4)^2}{377} = 2.4 \times 10^6 \text{ W m}^{-2}$$

This exceeds the IEEE safe level of 2 W m^{-2} by

$$\frac{2.4 \times 10^6}{2} = 1.2 \times 10^6 \qquad \text{or} \qquad 61 \text{ dB!}$$

Many high-power radio and radar antennas have such megawatt per square meter power density levels in their vicinity and warnings to keep away are posted.

This power density level applied to the square-meter 10-liter slab of Fig. 9-10 reduces the time for a 1°C temperature rise from 5.8 hr to

$$\frac{21\,000}{1.2 \times 10^6} = 17.5 \times 10^{-3} \text{ s} = 18 \text{ ms}$$

And if the 30 kV m^{-1} field of the microwave oven was not attenuated to 909 V m^{-1} in the potato, the baking time would be reduced from 9 min to $\frac{1}{2}$ s as given by

$$540 \times \left(\frac{909}{3 \times 10^4}\right)^2 = 0.5 \text{ s}$$

The difference of the field E_d in the potato and the applied field E involves a mismatch. Thus, if a person stands in an applied field E, the internal field may be less. However, if the wavelength is a multiple of the person's height, the person could resonate like a $\lambda/4$ or $\lambda/2$ antenna and develop higher internal fields.

In addition to the safe-level guideline for radio-frequency waves, there are even more stringent requirements for controlling unintentional radiation from electronic equipment which could interfere with other systems. A requirement of the U.S. Federal Communications Commission (FCC) is that unintentional radiation from electronic equipment be less than

100 μV/m at a distance of 3 m	***FCC level***

To comply with this electromagnetic interference (EMI) rule, electronic equipment manufacturers test for spurious emission from their units in an electromagnetic compatibility (EMC) chamber, such as shown in Figs. 10-14 and 10-15 (pages 540 and 541). A field strength of 100 μV m^{-1} is equivalent to a power density level of

$$P = \frac{E^2}{Z} = \frac{(100 \times 10^{-6})^2}{377} = 2.7 \times 10^{-11} \text{ W m}^{-2} = 27 \text{ pW m}^{-2}$$

This is 109 dB less than the IEEE safe level of 2 W m^{-2}.

TABLE 9-1
Table of power density, field strength, and time to heat 10 liters 1°C

Category	Power density	dB	E	Time to heat 10 liters 1°C
Megawatt transmitter	2.4×10^6 W m^{-2}	+61	30 kV m^{-1}	18 ms
IEEE safe level for humans	2 W m^{-2}	0	27.5 V m^{-1}	5.8 hrs
FCC level for equipment	27 pW m^{-2}	−109	100 μV m^{-1}	53×10^6 years (53 million years) (hypothetical)

The above field and power density levels are summarized in Table 9-1.

PROJECTS

Project 9-1. EM survey unit. Build the simple unit in Fig. 9-12 and observe rf emission from a few hertz to a few gigahertz. For example, observe the strong broadband radiation from TV and computer screens and from fluorescent lamps. Note where AM broadcast signals and TV station 60-Hz carriers are strongest and weakest in houses and buildings. Observe continuous emissions from digital clocks and scanners and pulses from garage door openers, wireless keys and from cellular phones when you push the "Send" button. Trace 60-Hz wiring. The unit can do this and much more, making you aware of the electromagnetically polluted world in which we live. On FM signals the unit detects only the spurious AM that accompanies the signal. None of the parameters of the circuit are critical.

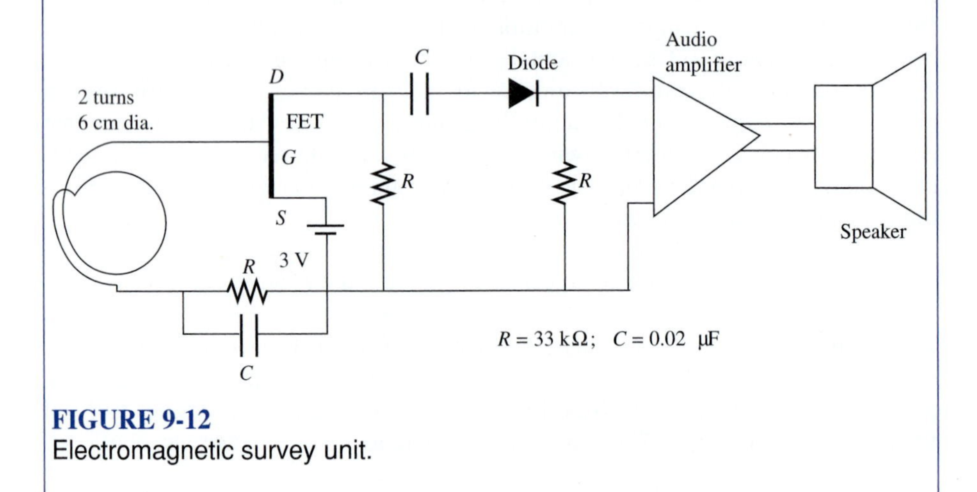

FIGURE 9-12
Electromagnetic survey unit.

Project 9-2. Storm fields and thundercloud potentials. Very high electric potentials develop under thunderclouds resulting in lightning strikes. To observe the fluctuation of thundercloud potentials, John Kraus assembled a very simple monitoring device (Fig. 9-13*a*). A field-effect transistor (FET) inside a plastic pipe has gate connected to two branching wires as an electrostatic antenna, like a bug's antenna, with drain connected to a voltage supply and recorder. The pipe is in an open area with the recorder in the Kraus home.

An actual record is shown in Fig. 9-13*b*. The potential rises quickly until there is a single lightning discharge, then it builds up again to a multiple discharge. The discharges

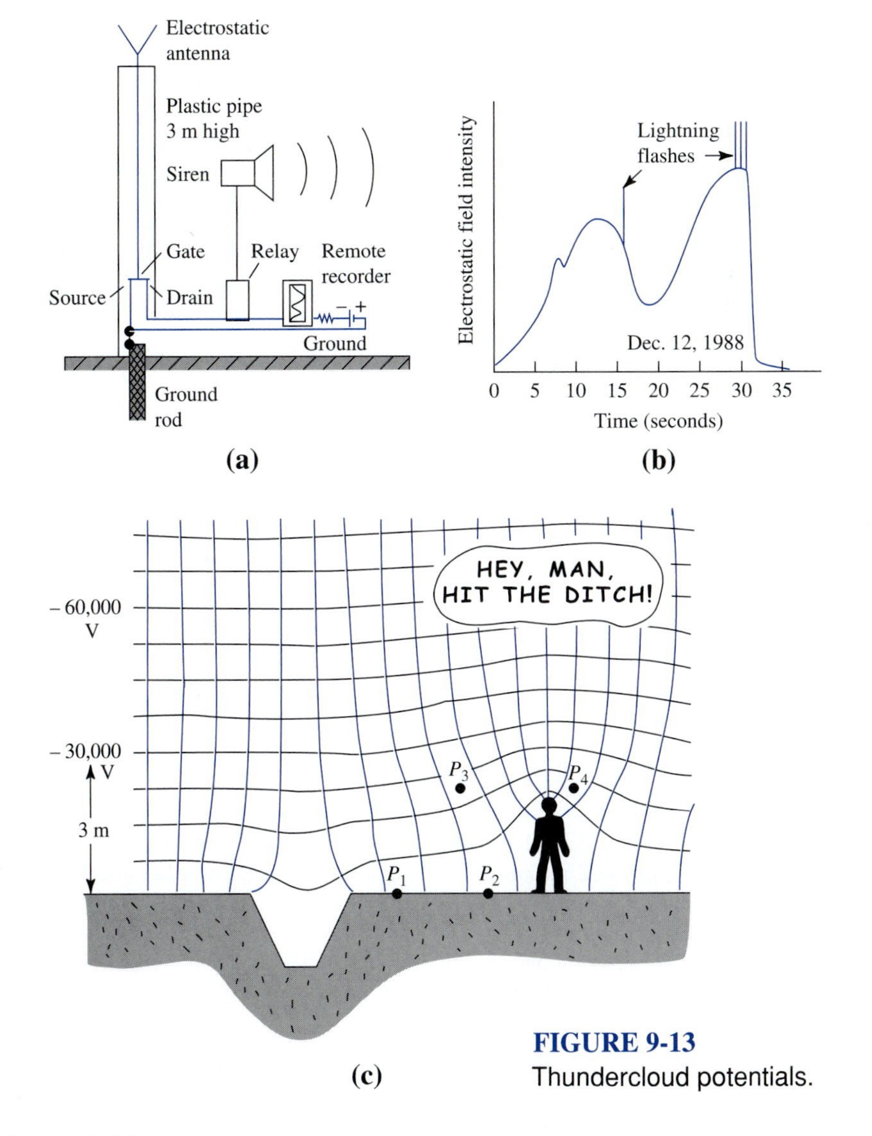

FIGURE 9-13
Thundercloud potentials.

(*Project 9-2 is continued on the next page.*)

or lightning strikes are gigantic "sparks" that produce radio waves. These travel long distances and make loud "cracks" or "pops" in an AM receiver. The strikes at the time of the observations of (*b*) were heard at the instant the spikes appeared on the record.

The fast rise to dangerous potentials does not give the man in Fig. 9-13*c* much time to "hit the ditch" where the electric field is 1000 times less than at his head.

With relay and siren, the system can warn when to seek shelter.

Problem P9-2-1. Storm fields. What are the electric fields at points P_1, P_2, P_3 and P_4 in Fig. 9-13*c*? *Ans.* 9, 7, 11, and 15 kV m^{-1}.

Solar Power to Food

It is solar power, converted by photosynthesis into edible plants, which supplies the world's food for humans and the animals that humans eat. Without solar-power-photosynthesis of green leaves, life as we know it would not be possible on the earth.

Example 9-11. From solar power to cole slaw. Alice Kraus, wife of the senior author, is an avid gardener who grows big Savoy cabbages. A plant takes about 10 weeks to mature and during that time the leaves of a cabbage plant with 5-kg head cover an average area of about 0.4 m^2. Assuming an equivalent of 32 six-hour days of "full sunshine" at an average elevation angle of 50°, what is the conversion efficiency from sun to body energy? Raw cabbage (no mayonnaise) gives 250 calories per kilogram. 1 kg-cal = 4186 J. We assume that the energy required for maintaining the electrochemical gradient in the roots necessary for the acquisition of water and minerals is negligible since the soil was moist. Take sun's power at normal incidence = 1 kW m^{-2}. The "solar constant" or sun's power above the atmosphere = 1353 W m^{-2}. Allowing for atmosphere absorption, haze, jet trails and scattered clouds, we take the incident power at the ground = 1 kW m^{-2}. (1 kg-cal = amount of heat to raise 1 kg or 1 liter of water 1°C.)

Solution.

$$\text{Solar energy} = \text{incident power (W m}^{-2}) \times \sin(\text{elevation angle}) \times \text{time (s)} \times \text{area (m}^2)$$

$$= 1000 \times \sin 50° \times 32 \times 6 \times 3600 \times 0.4 = 210 \times 10^6 \text{ J}$$

$$\text{Body energy} = 5 \times 250 \times 4186 = 5.2 \times 10^6 \text{ J}$$

$$\text{Efficiency} = \frac{\text{body energy}}{\text{solar energy}} = \frac{5.2}{210} = 0.025 \quad \text{or} \quad 2.5\% \quad \textit{Ans.}$$

This is a typical value for solar energy to body energy conversion efficiency. When humans eat animal meat, the conversion efficiency is much lower because of the two-step conversion: plant to animal — animal to human.

If the leaves and roots of the cabbage plant were eaten, the efficiency would be about 20 percent higher (or 3.0 percent) but these parts are customarily discarded.

Photovoltaic solar panels convert solar energy to electrical energy with somewhat higher efficiencies of 5 to 10 percent.

Problem E9-11-1. Cole slaw to the top of the Washington Monument. The monument is as high as a 55-story building (3 m per story) or 1000 steps. (*a*) For a body weight of 70 kg, how much slaw equivalent is required to walk to the top? Assume a typical body-work-output to body-energy-intake efficiency of 10 percent. (*b*) How many kilowatt-hours of incident solar energy are required via slaw to the top of the monument? (*c*) What is the overall efficiency? *Ans.* (*a*) 1000 g slaw (a *big* dish) equivalent to 50-g milk chocolate (one bar); (*b*) 11.7 kW-hr or 42×10^6 J; (*c*) 0.26 percent.

Project 9-3. Solar energy to body energy. Repeat the experiment of Example 9-11 for a cabbage plant and/or other vegetables and plant foods. What is the maximum possible efficiency of the photosynthesis process and other thermal-powered processes?

PROBLEMS

Answers are given in Appendix E to problems followed by the @ symbol.

9-7-3. Hamburger in microwave oven. A $\lambda = 15$ cm oscillator is used to cook 2 hamburger patties (100 g per patty). If the hamburger constants are $\varepsilon_r = 69 - j18$ at $\lambda = 15$ cm, what value of E applied across the patties is required to cook them in 5 min? Take $\sigma = 0$. Use same assumptions as for the potato (Example 9-10).

9-7-4. Wiener in microwave oven. A $\lambda = 15$ cm oscillator is used to cook an eight-pack of wieners (0.45 kg). If $\varepsilon_r = 75 - j20$ at $\lambda = 15$ cm, how long will it take a field of 25 kV/m (rms) applied across the wieners to cook them? Take $\sigma = 0$. Use same assumptions as for the potato (Example 9-10). @

9-7-5. Oven design. Design a microwave oven operating at 2450 MHz with 750-W radio-frequency power and an oven volume of 200 mm by 400 mm by 400 mm. (*a*) What maximum rate of temperature rise will be possible for 3 kg of dielectric material with complex permittivity $\varepsilon_r = 4 + j1$ if its specific heat is 5 J/g/°C? (*b*) If the oven has a window covered by a metal screen, give the screen specifications, material, wire size, and hole size required to keep radiation loss through the screen to less than 2 W/m^2. Also is it important that the screen wires be bonded at all contact points? (*c*) Describe the standing-wave mode or modes in the oven and what means (such as fan or paddles) can be used to "stir" the standing-wave pattern. (*d*) How can the edges of the oven door be sealed to prevent radio-frequency leakage? See Problem P4-2-2, page 239.

9-7-6. Magnetic field under high-voltage line. A typical 765-kV, 60-Hz, 3-phase power transmission line has three conductors at equal heights of 12 m, spaced 16 m apart as in Fig. P9-7-6. With 4000 A per phase, what is (*a*) the magnetic field at ground level directly under an outside wire at point *A* and (*b*) the field at ground level at point *B* 20 m outside? (*c*) How do these values compare with the presumed safe level of 2 mG? See Example 2-8 (Fig. 2-11). @

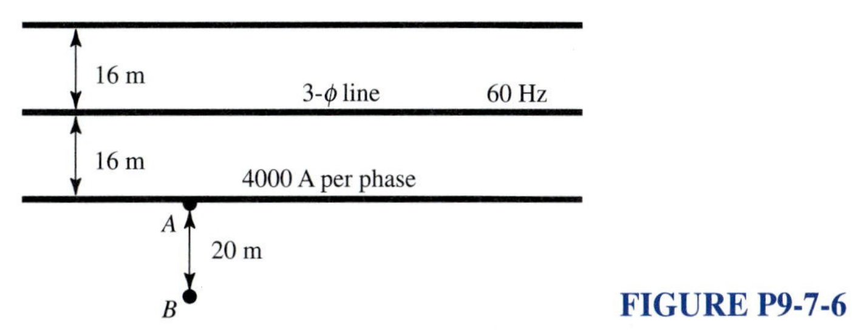

FIGURE P9-7-6

9-7-7. Birds on high-voltage lines. On high-voltage transmission lines, birds often sit on the ground wire (highest wire of the line) but not on the energized wires. Why? The answer is more involved than saying they may receive a shock because they are not connected to ground. The phenomenon involved is electrostatic.

9-7-8. Shielding from lightning. Explain why the danger of injury from lightning is greater standing in the open, under a power line, or under an isolated tree than it is in a depression, in a metal enclosed vehicle, or in a rod-protected building. Note that all six situations require different reasons. Note that if the depression is an arroyo creek bed, it's important to get out of it before the flash flood that follows the storm.

Solar Power Problems

One of the world's large power companies has shut down a nuclear plant. To compensate for the loss of power generation, the company has worked with customers to save on power with more efficient refrigerators, air-conditioners, and lighting systems. The company is also providing customers with solar power panels for the roofs of their homes and offices, resulting in a *distributed power generating network.* Inverters are used to convert the panel dc to 60 Hz. If a typical 3-kW home panel generates more power than the home is using, the rest is fed into the power grid. At night the power flows the other way from grid to home.

In 2010 the world's electric power consumption is projected to be 20 TW (20×10^{12} W) with 10 percent of that in the United States. With a projected 10 billion world population in 2010, this is an average of 2 kW per person worldwide with 5 kW per person in the United States.

9-SP-1. Solar panel area. The power from the sun at normal incidence at the earth's surface is approximately 1 kW/m^2. How much solar panel area would be required to generate the 20 TW at 10 percent efficiency? @

9-SP-2. Personal power generation. For a projected 10 billion world population by 2010, how big a solar panel would be required at 10 percent efficiency for 2 kW per person. Assume ideal weather and a 12-h day for solar power with battery or superconducting magnetic energy storage for night. @

9-SP-3. Regional power generation. For continuous 24-hr power generation consider solar power "farms" of hundreds of square kilometers each, distributed around the world and interconnected by superconducting transmission lines.

9-SP-4. Consider lofting solar panels into low earth orbit (LEO), converting the dc to microwaves, beaming the power to earth-based antennas, and finally converting it to 60 Hz for the power grid. The advantage of this scheme is that the panels are above the earth's atmosphere and not affected by clouds, rain, or snow. They may also be in full sunshine 24 hr a day. Note that with each step, dc to rf, rf antenna-to-antenna, and rf to 60 Hz, there is a loss due to the efficiency factor. Also, very importantly, there must be a fail-safe system to shut off the microwave beam to the earth if it strays away from the earth antenna.

9-SP-5. Consider putting solar panels on the moon and beaming the power from there. The distance is much greater but much bigger antenna arrays can be deployed.

REFERENCES

Becker, Robert O. and G. Selden, "The Body Electric," William Morrow, New York, 1974.
Uman, M. A., "The Lightning Discharge," Academic Press, New York, 1987.
Webster, J. G., "Design of Cardiac Pacemakers," IEEE Press, New York, 1995.
"Dynamics of Cardiac Arrhythmias," *Physics Today,* August 1996.
"The Cellular Phone Scare," *IEEE Spectrum,* June 1993.
"Toward an Artificial Eye," *IEEE Spectrum,* May 1996.
"The Strange Senses of Other Species," *IEEE Spectrum,* March 1996.

CHAPTER 10

ELECTROMAGNETIC EFFECTS IN HIGH-SPEED DIGITAL SYSTEMS

by Prof. Samuel H. Russ
Mississippi State University

Whether one expects to design computer circuitry or not, this chapter provides some excellent insights into important real-world problems that involve not only the principles of electromagnetics but also the restraints of economics and government regulations.

10-1 INTRODUCTION

Modern computer systems are getting faster. Not too long ago, a 33-MHz personal computer (PC) was considered really fast. Now it is considered obsolete! One consequence of this is that computers are becoming harder to design.

What hampers the design of computers? You can probably think of some fairly obvious things like memory density, the speed of light, and power dissipation. But there are also some more subtle effects that can give computer designers, or even innocent bystanders, some real headaches. Let's look at some everyday situations.

Scene I: A Strangely Behaving Reset Signal on a PC Card

Frank got back a prototype of a new card that goes inside a PC. After two or three frustrating hours in the lab, he discovers that the reset signal is being tripped. It's as if the card is resetting itself. Further investigation discloses that the reset signal goes high when the address bus coming onto the card changes from 7 to 8.

What is happening?

It turns out that when the address bus changes, the chip on the card that drives the address bus draws a lot of current from the power and ground connections of the card. This current is being used to change the voltage on the address lines. When the current changes, the inductance of the power and ground connections forms a voltage $V = \mathcal{L}\, dI/dt$. This voltage changes the ground voltage seen by the chip. Since the gates that drive the reset signals and the address signals share the same chip, the reset driver puts out a voltage V instead of zero volts and the board is reset.

Scene II: CD Player on Airplane Interferes with the Navigation System

The pilot of the 757 jet began to sweat. All of the navigation displays on his console showed strange readings. The plane was flying from Atlanta to Baltimore, but the display indicated he was over Texas. He began to worry that the autopilot would begin to alter course, or that he really was way off course, or that the problem might get worse. Just then, everything returned to normal.

After talking with the flight crew, he learned that a passenger had turned on his CD player the moment the display went bad. A flight attendant borrowed the player. Sure enough, every time the portable CD player was on, and was located between rows 27 and 34, the navigation display was wrong.

What happened?

The CD was transmitting at the same frequency used by navigation beacons in the area. The CD player put out much less energy than the beacons, but was much closer to the airplane's navigation antenna, located above row 29.

These stories only touch on two specific problems. Besides electromagnetic interference, computer designs can fail because of cross talk, ringing, reflections, and other effects.

10-2 TWO VIEWPOINTS: LUMPED OR DISTRIBUTED

Many problems can be traced to two fundamental physical properties.

First, systems can no longer be considered "lumped." The time it takes a high-speed computer signal to change from 0 to 1 can be less than the time it takes the signal to move down the line. Thus, the wire that conducts the signal can have different voltages along its length; it isn't a short circuit any more!

Second, inductance and capacitance cause unwanted degradation of signals and unwanted coupling.

A designer needs a thorough understanding of these concepts.

10-3 DISTRIBUTED SYSTEMS

Speed and Distance

We often act as if light and electricity are infinitely fast. However, we know that electricity has a finite speed, and that can cause the behavior of simple circuits to change profoundly.

Consider, for example, a data bus driver on the bus that connects the card slots. (A "driver" is a high-power, fast logic gate used to drive long or heavily loaded lines.)

The driver may be able to switch from 0 to 1 (from 0 volts to 5 V in CMOS logic, 0 to 3.7V in TTL†) very rapidly. For example, the 74F family (a high-speed CMOS family) can switch in about 1.5 ns. The line that is being driven may be very long, for example 30 cm.

The velocity of propagation is inversely proportional to the square root of the effective relative permittivity of the line. Thus,

$$\text{Velocity} = \frac{c}{\sqrt{\varepsilon_{\text{eff}}}} \tag{1}$$

where c = velocity of light and ε_{eff} = effective relative permittivity.

It is often convenient in these discussions to use units of picoseconds per cm and talk in terms of propagation delay. Since the speed of light is equivalent to 33 picoseconds per centimeter, the propagation delay D along a line surrounded by a material of dielectric constant ε_r is

$$D = 33.3\sqrt{\varepsilon_{\text{eff}}} \qquad (\text{ps cm}^{-1}) \tag{2}$$

For example, for a fiberglass circuit board, with $\varepsilon_{\text{eff}} = 4.5$, the propagation delay is 71 ps cm^{-1}.

Rise Time and Length; Lumped versus Distributed Circuits

A typical logic gate drives its output to logic 1 or logic 0 and stays there most of the time. When a gate changes, or "switches," it moves between logic 0 and logic 1 or vice versa. The time that it takes to do this is called the *rise time* T_r. It is also called the *fall time* T_f, but for the purposes of this discussion we will consider the fall time equal to the rise time, and call it rise time, or $T_f = T_r$.

On an oscilloscope, we can see the output voltage ramp up (or down) as the gate is switched. The height of the ramp corresponds to the change in voltage (or "voltage swing" or "ΔV") associated with that logic family. (For example, ΔV is 5 V in CMOS and 3.7 V in TTL.) The length of the ramp corresponds to the rise time. The rise time is traditionally measured from where the voltage exceeds 10 percent of the logic swing to where it exceeds 90 percent, as illustrated in Fig. 10-1.

If a signal is high (or low) for a short time, the waveform is called a *pulse.* The duration of a pulse is measured from where the rising edge crosses one-half of a logic swing to where the falling edge crosses one-half the swing and is called the *pulse width,* as shown in Fig. 10-1.

A digital signal edge looks like a unit step, except that the rise (or fall) occurs over some nonzero amount of time. Since it takes time to change, and since the corresponding voltage wave travels at a finite speed, you can think of the rising edge as having a length, as defined by the following equation:

$$\text{Length of rising edge} = L = \frac{T_r}{D} \qquad (\text{cm}) \tag{3}$$

where T_r = rise time, ps
D = propagation delay, ps cm^{-1}

†CMOS = complementary metal-oxide semiconductor; TTL = transistor-transistor logic.

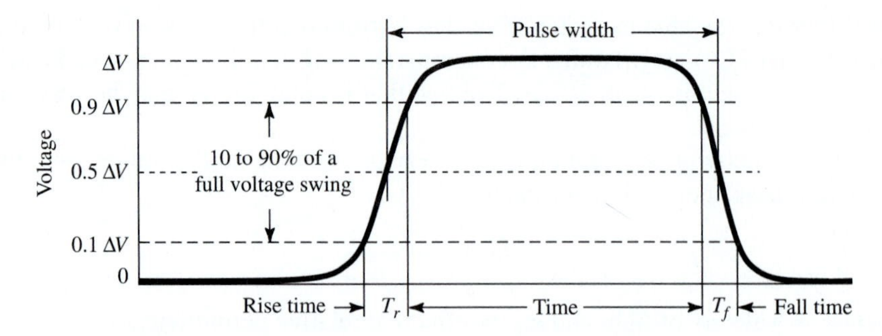

FIGURE 10-1
A rising and falling edge, showing the definition of rise time, fall time, and pulse width.

For example, if a gate has a rise time of 1.5 ns and $\varepsilon_{\text{eff}} = 4.5$, then

$$L = \frac{T_r}{D} = \frac{1.5 \text{ ns}}{71 \text{ ps cm}^{-1}} = \frac{1500 \text{ ps}}{71 \text{ ps cm}^{-1}} = 21.1 \text{ cm} \tag{4}$$

Note the units conversion of nanoseconds to picoseconds. Also note that the terms "wire," "line," and "trace" are used interchangeably to describe the electrical conductor along which a signal travels.

If the wire connected to the driver is 30 cm long, the voltage at the end is still zero even after reaching almost 5 V at the input end. The line can no longer be considered a short circuit. At any moment in time, there can be different voltages along the wire. This is illustrated in Fig. 10-2.

We can look at the circuit in two ways: (1) by "lumped" analysis, in which we treat wires as entities along which the voltage is constant and electromagnetic waves as infinitely fast, and (2) by "distributed" analysis, in which we treat wires as entities along which the voltage can be different and electromagnetic waves as having a finite speed.

When do we consider a line to be "long"? That is, when is a line too long to be considered a single, "lumped" line? The exact break point is arbitrary, but occurs approximately when the line is longer than 1/6 the length of a rising edge. Recalling

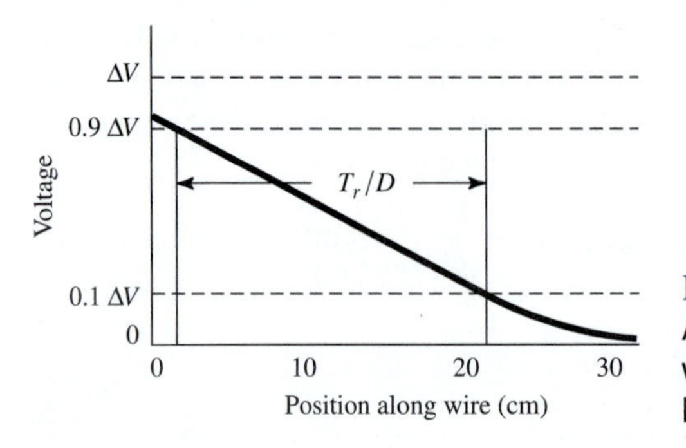

FIGURE 10-2
A fast-rising edge along a long wire about 1.5 ns after the gate has switched.

that the rising edge has a length of T_r/D, where T_r is the rise time and D is the propagation delay on the line, we consider a line to be lumped when its length L_{lump} is bounded by

$$L_{\text{lump}} < \frac{T_r}{6D} = \frac{T_r}{200\sqrt{\varepsilon_r}} \qquad \text{(cm)} \tag{5}$$

where T_r = rise time, ps

Thus, this break point is a function of the rise time of the circuit and the dielectric constant of the board material. This is not a hard boundary. But lengths well below are clearly lumped and lengths well above it are clearly distributed. In between is a "gray area."

Why is this important? It has to do with the complexity of the circuit and, more importantly, the complexity of analyzing and characterizing the circuit. A lumped circuit is like ones in an introductory class, with perfect resistors and capacitors. Even when real capacitors turn out to have series inductance and resistance, the model is still fairly simple. A distributed circuit, however, has time-domain effects. Simply put, when a gate driving a distributed circuit changes voltage, the voltage change is completed before all of the gates "know" that the voltage changed. For example, as will be seen later, a gate can launch too much current into the line and cause reflections.

The bad news is that distributed circuits are harder to understand and analyze. (That is why students start by looking at lumped circuits.) Even worse news is found by looking at the equation above.

The rise time of the logic gates governs whether a circuit is distributed. For example, consider designing a circuit board inside a PC. If you use gates with a 10-ns rise time, any line over 24 cm is distributed and few lines would be considered distributed. However, if you use gates with a 2-ns rise time, any line over 5 cm is considered distributed, which would be most of the lines on the board. As gates get faster, circuits can no longer be treated as lumped, that is, passive elements connected by perfect conductors.

Example 10-1. ECL (Emitter-Coupled Logic) technology on a fiberglass circuit board. Consider a fiberglass circuit board with an $\varepsilon_{\text{eff}} = 2.8$ and an ECL technology with $T_r = 0.5$ ns. (*a*) What is the speed of propagation on the board, in m s^{-1} and ps cm^{-1}? (*b*) How long is a rising edge? (*c*) How long can a trace be and still be considered lumped? (*d*) If you were designing a board that was 20 cm by 20 cm, which kind of analysis would you use—lumped or distributed?

Solution

$$\text{Speed} = \frac{c}{\sqrt{\varepsilon_r}} = \frac{300 \times 10^6}{\sqrt{2.8}} = 179 \text{ Mm s}^{-1}$$

and

$$D = 33.36\sqrt{\varepsilon_r} = 33.36\sqrt{2.8} = 55.8 \text{ ps cm}^{-1} \quad \textit{Ans.} \quad (a)$$

$$\text{Length of rising edge} = \frac{T_r}{D} = \frac{0.5\text{ ns}}{55.8\text{ ps cm}^{-1}} = \frac{500}{55.8} = 9.0\text{ cm} \quad \textit{Ans.} \quad (b)$$

$$\text{Lumped} < \frac{T_r}{6D} = \frac{T_r}{200\sqrt{\varepsilon_r}} = \frac{500}{200\sqrt{2.8}} = 1.49\text{ cm} \quad \textit{Ans.} \quad (c)$$

Since many traces will be longer than 1.49 cm, you should use distributed analysis. *Ans.* (*d*)

Problem 10-3-1. Determining parameters for a CMOS technology on alumina. Consider an alumina circuit board with an ε_{eff} of 8.0 and a CMOS technology with $T_r = 1.5$ ns. What is the speed of propagation on the board (*a*) in m/s and (*b*) ps/cm? (*c*) How long is a rising edge? (*d*) How long can a trace be and still be considered lumped? *Ans.* (*a*) 106 Mm/s; (*b*) 94.4 ps/cm; (*c*) 15.9 cm; (*d*) as long as 2.6 cm.

Knee Frequency

Another important issue has to do with the frequency content of digital signals. From Fourier analysis, a signal can be treated as the sum of pure sine waves at different frequencies. For example, a square wave is the sum of sine waves having frequencies at odd multiples of the square wave frequency.

What about digital signals? The highest frequencies in the frequency domain correspond to the steepest ramps in the time domain. As rise time drops (that is, as gates get faster) the highest frequencies contained in the signal go up.

There is an easy way to remember the formula for the highest frequency found in a digital signal. Consider a logic gate that has an output that rises from 0 to 1 in T_r ns. The waveform of this rise looks a lot like half of a complete sine wave. If you imagine the gate rising from 0 to 1 and immediately going back to 0, the output would form a complete sine wave. What is its period? Twice the rise time. Thus, the highest frequency contained in a digital signal, which we will call the *knee frequency,* is found by $f_{\text{knee}} = 1/(2T_r)$. This is illustrated in Fig. 10-3.

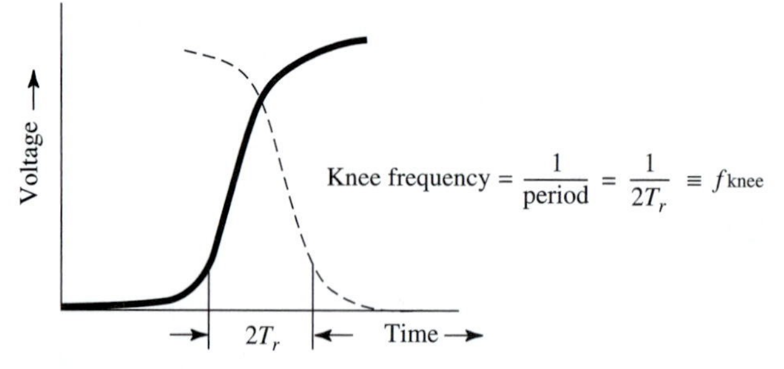

FIGURE 10-3
A way to remember the definition of knee frequency f_{knee}.

Looking back at our PC design example, the 10-ns gate puts out energy up to 50 MHz. The 2-ns gate puts out energy up to 250 MHz. One way to think about knee frequency is this. If the circuit board can handle frequencies up to the knee frequency, it will likely conduct all signals with little or no degradation; otherwise, expect problems.

There are two extremely important points that one needs to understand to design digital systems that work.

First, notice that modern logic gates, with rise times on the order of 1 ns or even fractions of 1 ns, have frequency components up to hundreds of megahertz (MHz) and sometimes even a few gigahertz (GHz). This is traditionally the domain of microwave designers. Nothing in this domain acts like a wire any more—everything is an inductor, capacitor, resistor, or some combination of the three. And many components, such as wires and leads, act like antennas.

Second, notice that "clock speed" has not been mentioned anywhere in this discussion. High frequency effects are *not* a direct function of clock speed. They are a function of rise time! They only seem to be a function of clock speed because as clock speeds go up, faster gates are needed to keep pace.

Problem 10-3-2. ECL technology on a fiberglass circuit board. Consider the same ECL technology with $T_r = 0.5$ ns. What is the knee frequency f_{knee}? *Ans.* $1/(2 \times 0.5 \text{ ns}) = 1/(1 \text{ ns}) = 1.0$ GHz

Review of Transmission Line Theory

Chapter 3 covered the basic theory of transmission lines. One reason why transmission line theory is so important is that circuit board traces act like transmission lines if they are long enough. In fact, they act like lossless transmission lines at many frequencies, which simplifies their analysis considerably. (Note that on a chip, where wiring resistance is much higher because wires are so small, transmission lines are lossy.)

There are several concepts in transmission lines that are important here.

First, they have a characteristic impedance, as given by (3-2-10) and (3-2-11). Another derivation for digital transmission lines may help to understand the concept.

Consider a transmission line with capacitance per unit length $= C_0$ and inductance per unit length $= \mathcal{L}_0$. From (3-2-4), a voltage pulse travels with a velocity

$$v = \frac{1}{\sqrt{\mathcal{L}_0 C_0}} \qquad (\text{m s}^{-1}) \tag{6}$$

Since charge

$$Q = CV \qquad (\text{C}) \tag{7}$$

the amount of charge on a length l of line is given by

$$\Delta Q = C_0 l \, \Delta V \qquad (\text{C}) \tag{8}$$

where C_0 = capacitance per unit length, C m^{-1}
l = line length, m
ΔV = pulse voltage, V

The amount of time t it takes the voltage pulse to traverse the distance l is then

$$t = \frac{l}{v} = l\sqrt{\mathcal{L}_0 C_0} \qquad \text{(s)} \tag{9}$$

Since current is charge divided by time, the current required to charge a length l of line is

$$I = \frac{\Delta Q}{t} = \frac{C_0 l \Delta V}{l\sqrt{\mathcal{L}_0 C_0}} = \Delta V\sqrt{\frac{C_0}{\mathcal{L}_0}} \qquad \text{(A)} \tag{10}$$

The characteristic impedance is then

$$Z_0 = \frac{\Delta V}{I} = \frac{\Delta V}{\Delta V\sqrt{\frac{C_0}{\mathcal{L}_0}}} = \sqrt{\frac{\mathcal{L}_0}{C_0}} \qquad (\Omega) \tag{11}$$

In other words, the characteristic impedance determines the amount of current a pulse voltage ΔV produces as it starts down the line. But when the pulse reaches the load, the load impedance determines the current which will be different if the line is not matched.

The transmission line source "sees" different impedances at different times. For example, the source sees an impedance of Z_0 to ground during the time the line charges up and then sees an impedance equal to the load impedance once the line has charged up (assuming the line is lossless).

Digital signals look like unit step inputs on the line. Concepts like the VSWR require steady-state signals and are not applicable.

The definitions and formulas for the transmission and reflection coefficients at the source and load are listed in Table 3-3 (page 142).

An additional coefficient, the *acceptance coefficient at the source* A_s, is the ratio of the voltage pulse introduced onto the line to the voltage pulse produced by the source. Since the source and line form a voltage divider, the coefficient is given by

$$A_s = \frac{Z_0}{Z_0 + Z_s} \tag{12}$$

The following example follows the saga of a pulse bouncing back-and-forth on a transmission line as was illustrated graphically in Fig. 3-30 of Example 3-23.

Example 10-2. The saga of a pulse reflected on a CMOS-driven transmission line. Consider a CMOS driver and receiver connected by a 50-Ω transmission line. The driver has an output impedance of 20 Ω. The receiver has an input impedance of 1 MΩ. The rise time of the driver is 2.0 ns and the rising edge takes 5.0 ns to travel from one end of the line to the other. The voltage swing at the source is from 0 to 5 V. (*a*) Find the reflection and transmission coefficients at the source and load and the acceptance coefficient at the source. (*b*) Use the coefficients to estimate the voltage seen at the load as a function of time.

Solution. From (12) and Table 3-1 (page 130),

$$A_s = \frac{50}{50 + 20} = 0.714$$

$$\text{Reflection coefficient at source} = \rho_s = \frac{20 - 50}{20 + 50} = -0.428 \quad \textit{Ans.} \quad (a)$$

$$\text{Transmission coefficient at source} = \tau_s = -0.428 + 1 = 0.571 \quad \textit{Ans.} \quad (a)$$

$$\text{Reflection coefficient at load} = \rho_L = \frac{10^6 - 50}{10^6 + 50} \approx 1 \quad \textit{Ans.} \quad (a)$$

$$\text{Transmission coefficient at load} = \tau_L = 1 + 1 = 2 \quad \textit{Ans.} \quad (a)$$

The 5-V rise passes through the voltage divider and enters the line. Since the line is lossless, it arrives at the end intact. An amount equal to the pulse voltage times the load transmission coefficient emerges at the load, and amount equal to the pulse voltage times the load reflection coefficient is introduced back onto the line at the load. The reflected pulse arrives intact at the source, and again there is a transmitted amount at the source and a reflection introduced back onto the line. This process repeats until the reflections are essentially zero. (See Fig. 3-30, page 164). Referring to Fig. 10-4, the load starts at 0 volts and the source has a rising edge 2 ns long from 0 to 5 V at $t = 0$.

A pulse equal to $\Delta V A_s = 5 \times 0.714 = 3.57$ V is introduced onto the line. It reaches the load 5 ns later.

At the load 5 ns after launching the wave, the 3.57 V pulse arrives. A pulse of $3.57\tau_L = 7.14$ V emerges and a pulse of $3.57\rho_L = 3.57$ V is reflected onto the line. The load voltage rises to $0 + 7.14 = 7.14$ V (point A on Fig. 10-4).

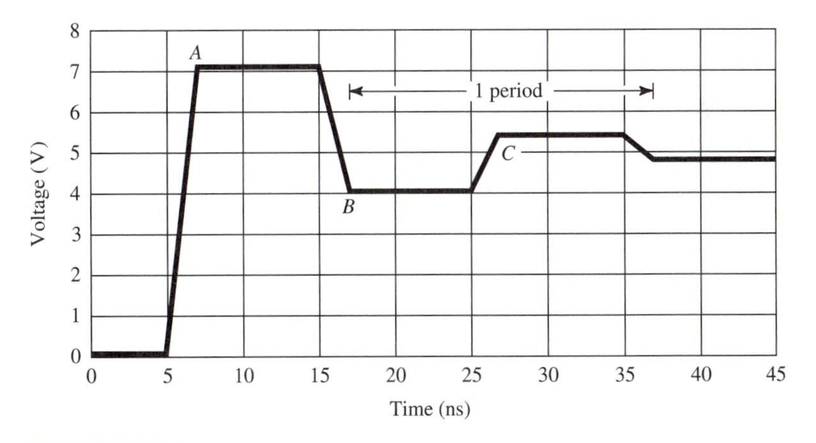

FIGURE 10-4
Oscillogram of load voltage versus time in CMOS Example 10-2 with oscillations of 20-ns period or 50 MHz. If this CMOS circuit is not well shielded, the bouncing pulse could radiate an interfering signal at 50 MHz. The CD player on the jet aircraft (Scene II in the Introduction) may have had such a pulse.

At the source 10 ns after launching the wave, the 3.57 V pulse arrives. A pulse of $3.57\rho_s = -1.53$ V is reflected back onto the line.

At the load 15 ns after launching the wave, the -1.53-V pulse arrives. A pulse of $-1.53V_L = -3.06$ V emerges and a pulse of $-1.53\rho_L = -1.53$ V is reflected onto the line. The load voltage drops to $7.14 - 3.06 = 4.08$ V (point *B* on Fig. 10-4).

At the source 20 ns after launching the wave, the -1.53-V pulse arrives. A pulse of $-1.53\rho_s = 0.66$ V is reflected back onto the line.

At the load 25 ns after launching the wave, the 0.66-V pulse arrives. A pulse of $0.66V_L = 1.32$ V emerges and a pulse of $0.66\rho_L = 0.66$ V is reflected onto the line. The load voltage rises to $4.08 + 1.32 = 5.40$ V (point *C* on Fig. 10-4).

This process repeats, and eventually the load voltage settles to 5.0 V, as shown in Fig. 10-4. *Ans.* (*b*).

Problem 10-3-3. Line length. If the relative effective permittivity of the 50-Ω transmission line of Example 10-2 is 16, how long is the line? *Ans.* 37.5 cm.

Example 10-2 illustrates several important points:

1. Reflections on a transmission line can oscillate. This occurs when the source impedance is less than the characteristic impedance or the load impedance is greater than the characteristic impedance. These oscillations are very similar to ringing. The difference between ringing and oscillatory reflections is that ringing occurs on short lines and reflections on long lines.
2. CMOS has strong output drivers, which means low source impedance, and sensitive gate inputs, which means high load impedance. Thus, it is especially prone to oscillation.
3. Notice that after the second reflection, the load voltage drops to about 4.08 V. This is about 0.92/5 = 18 percent of a full voltage swing below 5 V, which is dangerously low. Reflections can cause the voltage at the load (the gate input) to appear less than a true logic 1. The gate input may appear to go to logic 1, temporarily drop to logic 0, and return to logic 1. If the gate is "edge-triggered" (by a clock input, for example) then the logic may not operate properly at all.

Reflections in the Presence of Capacitance

Unwanted capacitance is very common on circuit boards. In the context of traces, it occurs in two different places. First, the gate input usually has a capacitance, often in the range of tens of picofarads. Second, it can appear in the middle of a trace (and therefore in the middle of the transmission line).

In the first case, its effects can be taken into account (approximately) by adjusting the line's Z_0 and one-way propagation delay, T_0.

Because of the load capacitance, the capacitance of the entire structure has been increased from $C_0 l$ (where C_0 is the capacitance per length and l is the length) to $C_0 l + C_{\text{load}}$ (where C_{load} is the capacitance of the load). Thus, the ratio $\mathcal{R}$ of the new capacitance to the old is given by

$$\mathcal{R} = \frac{C_0 l + C_{\text{load}}}{C_0 l} = \frac{C_{\text{load}}}{C_0 l} + 1 \tag{13}$$

The effective Z_0, $Z_{0\text{eff}}$, is

$$Z_{0\text{eff}} = \sqrt{\frac{\mathcal{L}_0}{C_0 \mathcal{R}}} = \frac{Z_0}{\sqrt{\mathcal{R}}} \tag{14}$$

The effective one-way transition time, $T_{0\text{eff}}$, is

$$T_{0\text{eff}} = \frac{l}{v} = l\sqrt{\mathcal{L}_0 C_0 \mathcal{R}} = T_0\sqrt{\mathcal{R}} \tag{15}$$

For example, in the line from Example 10-2, it is possible to estimate the effects of added load capacitance.

Example 10-3. Added load capacitance. The line from Example 10-2 is 30 cm long and has a C_0 of 3.33 pF/cm. The gate adds 40 pF of load capacitance. (*a*) Find the ratio for the load capacitance and the line. (*b*) Use $\mathcal{R}$ to find the effective impedance Z_0 and one-way transmission time T_0.

Solution

$$\mathcal{R} = \frac{C_{\text{load}}}{C_0 l} + 1 = \frac{40}{3.33 \times 30} + 1 = 1.4 \quad \textit{Ans.} \quad (a)$$

$$Z_{0\text{eff}} = \frac{Z_0}{\sqrt{\mathcal{R}}} = \frac{50}{\sqrt{1.4}} = 42.25\ \Omega \quad \textit{Ans.} \quad (b)$$

$$T_{0\text{eff}} = T_0\sqrt{\mathcal{R}} = 5\sqrt{1.4} = 5.92\text{ ns} \quad \textit{Ans.} \quad (b)$$

Problem 10-3-4. Estimating reflections. Rework Example 10-2 using the new Z_0 and T_0 from Example 10-3. Find the maximum undershoot (the lowest voltage after the rising edge) and determine when it will occur. Is the maximum undershoot serious? *Ans.* 4.37 V at 17.8 ns.

Another place where capacitance can cause problems is when it occurs in the middle of a transmission line. For example, on a PC's backplane, each card slot adds a tiny capacitance to the bus signals. The effect of the capacitance is to split the transmission line into two separate transmission lines.

Notice the difference from the previous case. When the capacitance is at the end of the line, it simply contributes to the reflections already present. When the capacitance is in the middle of the line, it causes degradation both "downstream" as the signal continues to propagate and "upstream" back to the source.

A rising edge moving down a line that hits the capacitance is partially reflected and partially transmitted. The magnitude of the reflection onto the first line and the signal transmission onto the second line depends on the frequency content of the pulse (that is, the pulse rise time) and the amount of capacitance. The circuit is illustrated in Fig. 10-5.

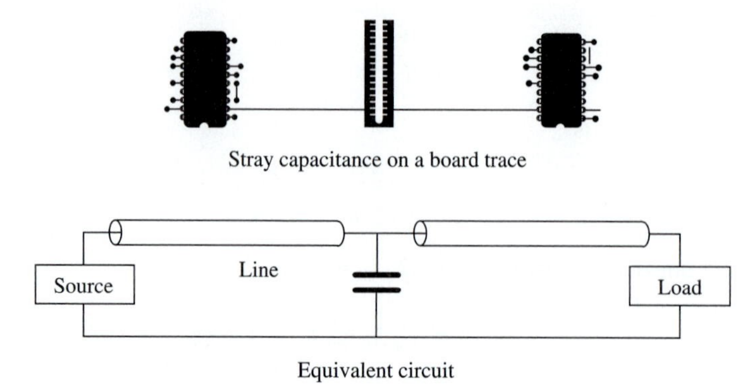

FIGURE 10-5
A line interrupted by a capacitor.

To determine the reflection coefficient of the capacitor, consider first that the capacitance is in parallel with the rest of the transmission line (the part to the right of the capacitor). Thus, the load impedance Z_L is that of the capacitor ($1/j\omega C$) in parallel with the characteristic impedance Z_0, as given by

$$Z_L = \frac{\dfrac{Z_0}{j\omega C}}{Z_0 + \dfrac{1}{j\omega C}} = \frac{Z_0}{j\omega C Z_0 + 1} \quad (\Omega) \tag{16}$$

Second, from the reflection coefficient formula we have

$$\rho_L = \frac{\dfrac{Z_0}{j\omega C Z_0 + 1} - Z_0}{\dfrac{Z_0}{j\omega C Z_0 + 1} + Z_0} = \frac{Z_0 - j\omega C Z_0^2 - Z_0}{Z_0 - j\omega C Z_0^2 + Z_0} = \frac{-j\omega C Z_0}{2 + j\omega C Z_0} \tag{17}$$

What does this formula tell us? Consider two different conditions.

First, if ω is so small that $\omega C Z_0 \ll 2$, then the reflection coefficient is approximately $-j\omega(CZ_0/2)$ and the reflection is approximately the derivative of the incoming wave multiplied by $-CZ_0/2$. Thus, a rising edge reflects a small negative pulse if the knee frequency is small enough. What will reduce the reflected pulse? From the formula, we see that lowering C or lowering Z_0 will reduce the reflection.

Second, if ω is so large that $\omega C Z_0 \gg 2$ then the reflection coefficient is approximately -1, meaning a negative total reflection. Note that in this case the transmission coefficient is approximately $1 + (-1) = 0$. There is a total reflection and the signal takes much longer to pass through the capacitor.

How is the break point found? Recall that the rising edge has a frequency response up to the knee frequency, which is $1/2T_r$. Since

$$\omega = 2\pi f_{\text{knee}} = \frac{2\pi}{2T_r} = \frac{\pi}{T_r} \tag{18}$$

the break point occurs where

$$\frac{\pi}{T_r} C Z_0 = 2 \tag{19}$$

or, alternatively, the break point between the two reflection conditions is determined by the rise time according to

$$T_r = \pi \frac{C Z_0}{2} \tag{20}$$

Thus, when the rise time is smaller than the break point, reflection is near-total and transmission is dramatically slowed down. When the rise time is larger than the break point, a rising edge reflects a negative pulse.

Terminations

We have seen that improperly terminated transmission lines create reflections. So how can the line be terminated to eliminate them? Four schemes that have been developed are illustrated in Fig. 10-6.

There are two obvious termination schemes. If a series resistor is placed between the source and the line with a value of Z_0, then any reflections from the load back to the source are eliminated at the source. This is called series source termination or source termination. If a resistor with a value of Z_0 is placed in parallel with a high-impedance load (such as a CMOS input), then the load produces almost no reflections.

What are the differences between the two?

First, consider the source acceptance coefficient A_s in the case of *source termination.* From (12) we can see that the coefficient is always 0.5. Thus, a rising (or falling) edge introduced into the line initially only has half of the desired voltage change. A 5-V rise, for example, produces a 2.5-V rise on the line. Thus, any input

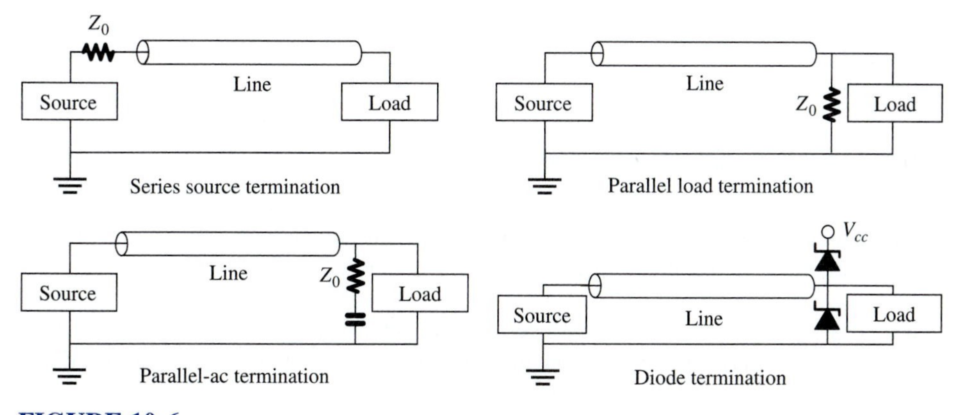

FIGURE 10-6
Some termination schemes.

connected to any part of the line except the load (the right end) will see a partial rise for some time followed by the remaining rise later. In other words, it works correctly only at the load. *Load termination* does not have this problem.

Second, consider the steady-state current required by the load in the case of *load termination.* The current required is approximately V_{cc}/Z_0. For example, a 50-Ω Z_0 and a 5-V V_{cc} requires 0.1 A, or 100 mA. A typical digital output is limited to 10 or 20 mA, so the current requirement is too large. In effect, parallel termination is not feasible in CMOS or TTL.

Parallel-AC termination is an attempt to solve the second problem. The series capacitor eliminates the need for dc current, making it a usable termination for CMOS and TTL. The difficulty is sizing the capacitor. It has to be large enough to offer low impedance, so that the impedance of the load is approximately Z_0, but small enough not to slow down the circuit too much.

Both parallel and parallel ac termination have an alternate version. Instead of placing a single resistor (or resistor-capacitor combination) between the load end of the line and ground, one can place a pair of resistors, one from the line to ground and another from the line to V_{cc}. For example, two resistors of size $2Z_0$ have a parallel impedance of Z_0, and so will terminate the line. The advantage is that the dc current requirement is halved.

Finally, *diode termination* uses zener diodes to clamp the line (hold voltage between limits) and damp out reflections. A zener diode has the property of clamping voltage transients below −0.7 V and above 5.2 V. (The exact voltages actually depend on the diode. These are typical values.) Thus, reflections, which look like transients outside that range, are "clamped," or eliminated. A voltage transient above 5.2 V, for example, is held at 5.2 V by the diode, with the excess current carried away by the power connection. It has the advantage of working for any Z_0 and not requiring a precision resistor value. It is also suitable if unexpected reflection problems appear.

10-4 INDUCTANCE AND CAPACITANCE

Inductance and capacitance limit the speed of circuit boards and chips. A knowledge of circuit board design is useful for understanding where inductances and capacitances occur.

How Circuit Boards Are Made

A circuit board consists of integrated circuits (chips) mounted on a dielectric-substrate-coated ground plane with copper lines or "traces" connecting them. Circuit boards are made in a special process, which is illustrated in Fig. 10-7.

1. Pieces of fiberglass covered with copper are "photoetched." Combinations of optical processing, special light-sensitive chemicals, and special acids are used to form the actual wires on the front and back of the fiberglass.
2. Alternating layers are stacked up. Pieces of fiberglass that have been etched are stacked with pieces of a special material between them. The special material

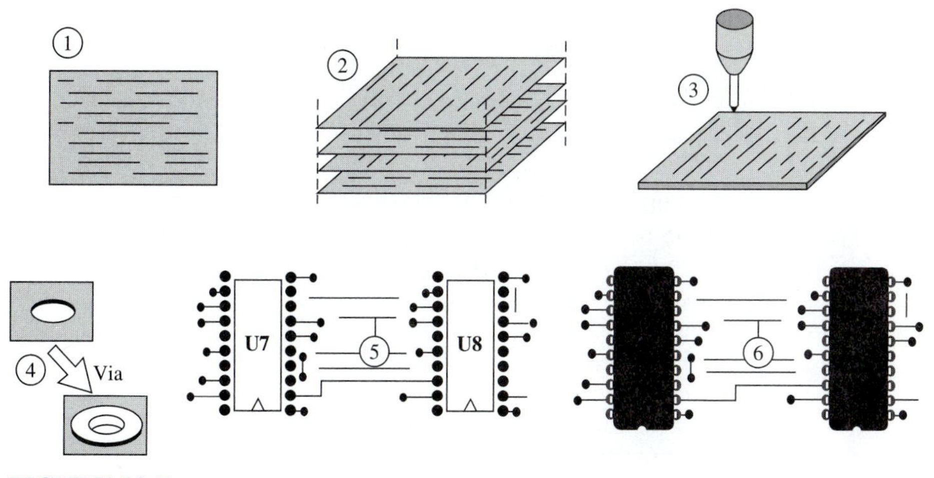

FIGURE 10-7
The process of making circuit boards.

becomes soft under heat and pressure, and so the fiberglass boards are literally glued together.

3. Holes are drilled. The outside layers (top and bottom) are etched.
4. If the board designer wants a hole to be a conductor, it is lined with copper. This is how signals travel between layers of copper etching. These copper-filled holes are called "vias" and look like metal dots on the top and bottom of the circuit board.
5. The top of the board is silk-screened (the same process that's used to make T-shirts). The silk-screening is the white labeling on the top of most boards. The labeling tells where the chips should be mounted.
6. Parts are placed on the board and soldered in place.

Thus, a circuit board consists of layers of copper wiring separated by fiberglass and epoxy insulation. Most of the traces are very thin. Since digital signals are relatively low current, even tiny copper traces have negligible resistive losses.

However, traces that carry power (and ground) are called upon to carry much more current; 10 or 20 A at 5 V is not uncommon. This much current is handled with low losses by using an entire sheet or ground plane of copper.

Figure 10-8 shows the two types of lines used in circuit boards: microstrip and strip line. A "microstrip" configuration is one where a conductor lies on top of a substrate on a ground plane. Traces on the top and bottom of a circuit board are microstrip. A "strip line" configuration is one where a conductor or trace lies between two ground planes. Traces in the middle of the circuit board are strip line.

The differences between them are important. Since strip line conductors are surrounded by insulating material, they have more capacitance per length than microstrip. More capacitance means that the line is "slower." However, strip lines are

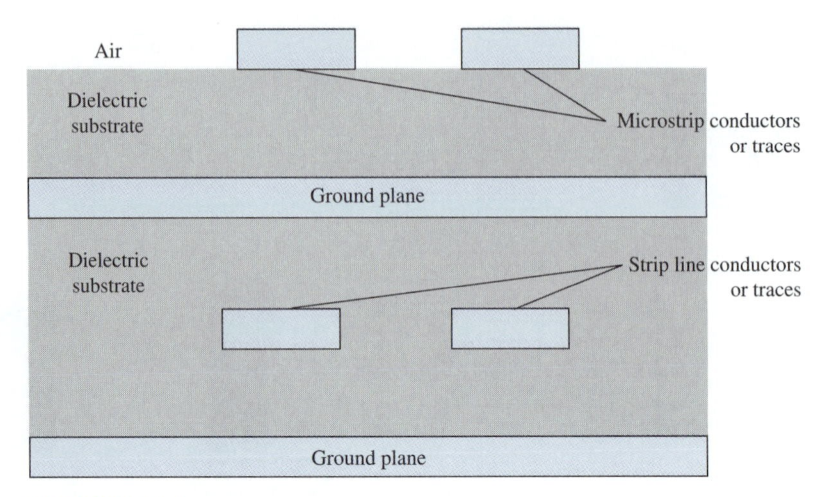

FIGURE 10-8
Configurations for microstrip, above, and strip line, below.

confined inside conductive material, and so are better shielded. Microstrip conductors, however, are faster but more liable to radiate.

Where are the inductors and capacitors? The trace separated from a ground plane by a dielectric substrate forms a capacitor. The circuit is a loop which forms an inductor. Thus, each trace acts as if it has a capacitor in parallel to ground and an inductor in series, as shown in Fig. 10-9.

The same structures exist not only between conductors and ground planes but also between nearby conductors. Pairs of conductors have capacitive and inductive coupling, as shown in Figs. 10-10 and 10-11.

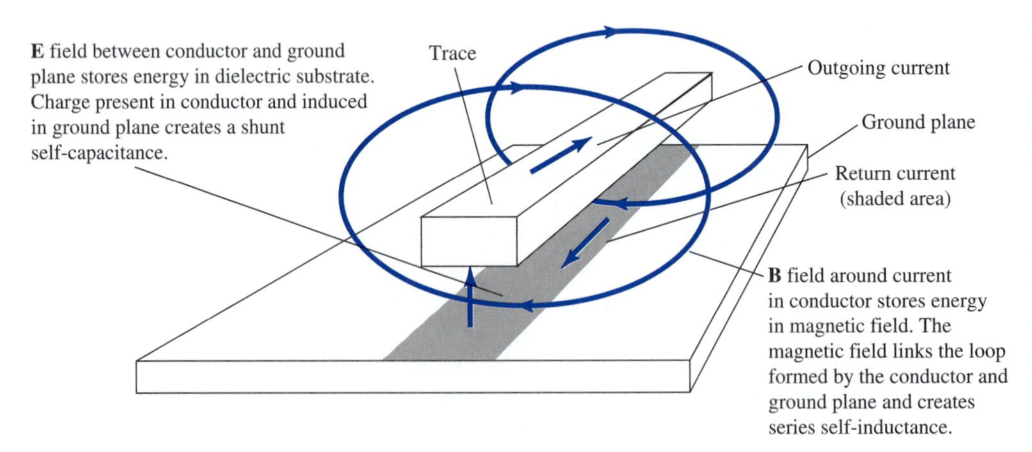

FIGURE 10-9
Self-inductance and capacitance of a single trace.

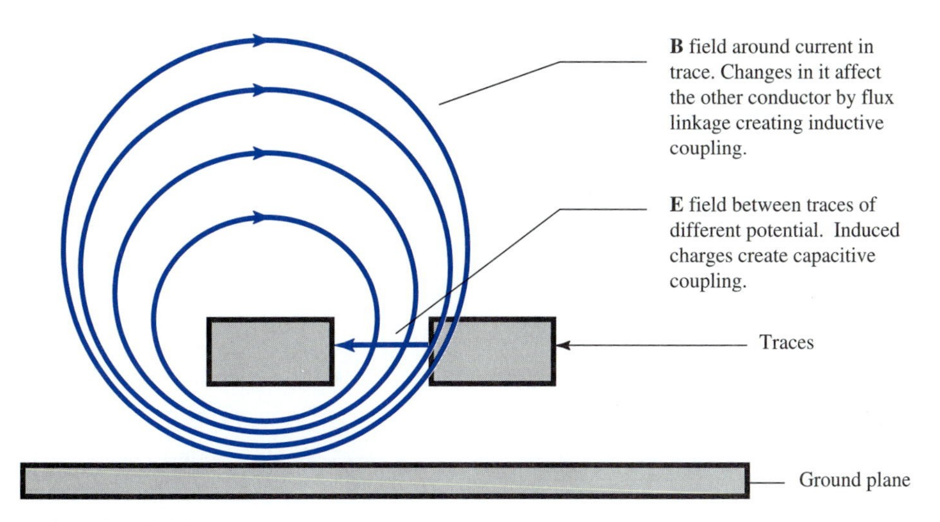

FIGURE 10-10
Mutual inductance and capacitance between two traces (schematic representation).

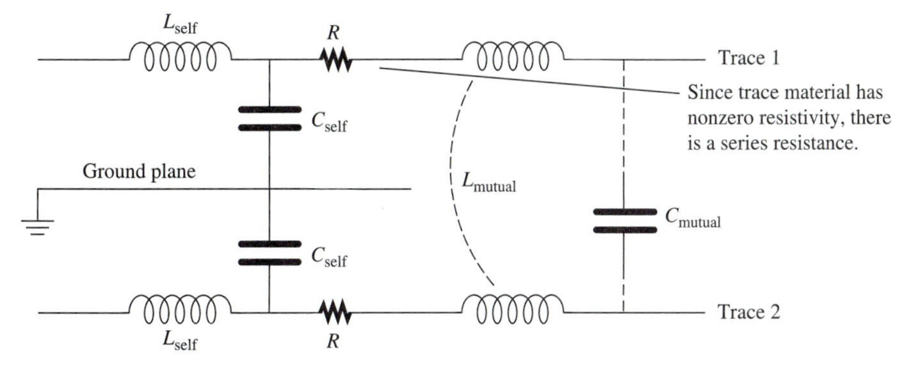

FIGURE 10-11
Equivalent circuit of two traces showing combination of self and mutual inductance and capacitance.

Cross Talk

The voltage V_1 induced in one microstrip line from the current I_2 in an adjacent one is proportional to the mutual inductance M coupling the lines. Thus,

$$V_1 = M\frac{dI_2}{dt} \qquad \text{(V)} \tag{1}$$

This voltage may be described as “cross talk” producing undesirable interference.

For the two adjacent microstrip lines of Fig. 10-12, this mutual effect on line 2 is proportional to the magnetic field **H** from line 1 at the ground plane under line

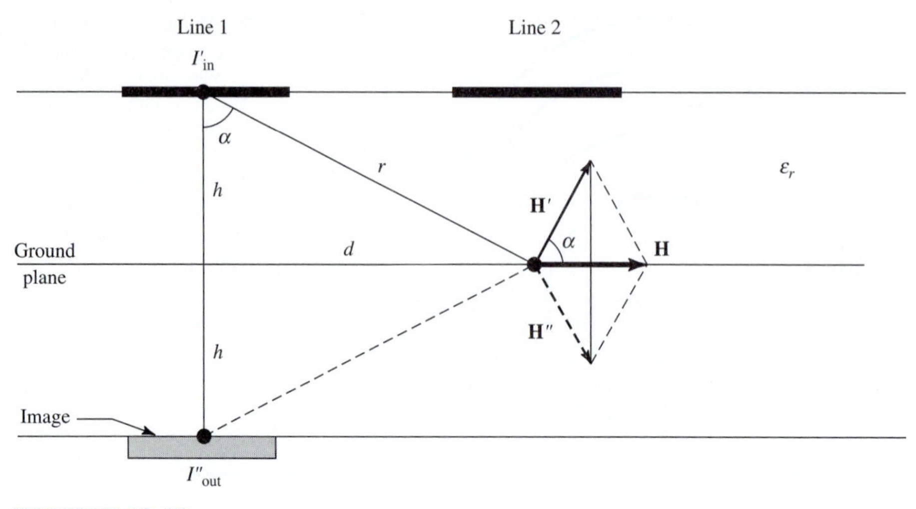

FIGURE 10-12
Construction for finding the cross talk field under the trace of an adjacent strip line.

2. **H** is the vector sum of the fields **H**′ from the actual line 1 and **H**″ from its image in the ground plane. Assuming all of the current on the strip is concentrated at its center point, we have from (2-12-2) that

$$H = \frac{2I'}{2\pi r}\cos\alpha = \frac{I'}{\pi r}\frac{h}{r} = \frac{I'}{\pi}\frac{h}{r^2} = \frac{I'}{\pi}\frac{h}{h^2+d^2} = \frac{I'}{\pi h}\frac{1}{1+\left(\dfrac{d}{h}\right)^2} \quad (\text{A m}^{-1}) \tag{2}$$

Thus, the cross talk is proportional to the last factor of (2), or

$$\text{Cross talk} \propto \frac{1}{1+\left(\dfrac{d}{h}\right)^2} \tag{3}$$

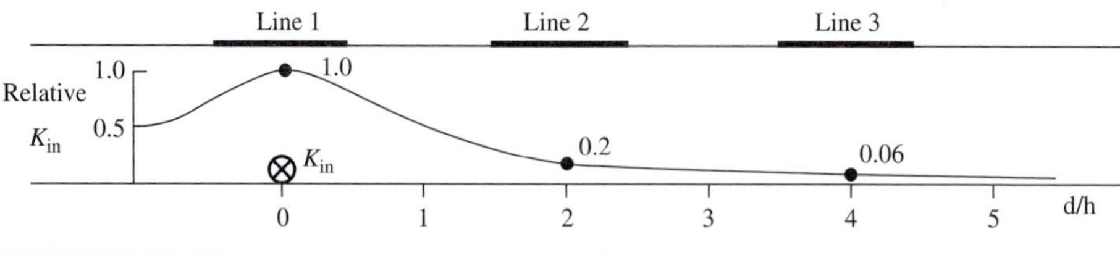

FIGURE 10-13
Relative sheet current density **K** on the ground plane due to current in line 1. From (3), the relative current under line 1 is 1.0, under line 2 is 0.2, and under line 3 is 0.06. Thus, the cross talk at line 3 is 0.06/0.2 = 0.3 or 30 percent of that at line 2. Compare with Fig. 2-47 (page 96).

a useful approximation. The relative variation of this factor at the ground plane is shown in Fig. 10-13, being equal to unity under line 1 ($d = 0$). The sheet current **K** (A m^{-1}) on the ground plane is numerically equal to **H**. Thus, $|\mathbf{K}| = |\mathbf{H}|$, but **K** is into the page and **H** is to the right.

If cross talk from line 1 to data signals on line 2 is objectionable, the data signals can be moved to line 3 and line 2 used as a dc bus or removed.

Problem 10-4-1. Cross talk reduction. To reduce the cross talk between the two traces of Fig. 10-12 from 30 percent to 5 percent, by what factor must the spacing be increased? *Ans.* 2.85.

10-5 ELECTROMAGNETIC INTERFERENCE

Governments regulate the amount of radio-frequency interference that equipment can produce so that computers and televisions, for example, do not interfere with each other.

Several terms are important in dealing with this subject. *EMI,* or *electromagnetic interference,* is the production of unwanted RF energy. *EMC,* or *electromagnetic compatibility,* is the ability of two or more devices to work together without interfering with one another. *Susceptibility* is a measure of a device's sensitivity to external interference. Because a good receiving antenna is a good transmitting antenna, devices with high susceptibility are usually devices that generate EMI.

Suppression, or source suppression, is the preferred way of reducing EMI. This means that the source of rf energy has somehow been turned off or reduced. *Containment,* which is confining the energy inside some enclosure, is the next best strategy.

Energy that is dispersed through radio waves in air is *radiated emission.* The opposite type, conducted through power lines and cables, is *conducted emission.*

Government regulation is monitored through *regulatory agencies.* In the United States, the regulatory agency is the Federal Communications Commission, or FCC. The regulation dealing with computer-generated EMI not only specifies the amount of EMI that is acceptable but also the testing methods used to measure EMI. Testing and approval is usually done by private companies authorized to perform testing.

For there to be an EMI problem there must be: (1) a source of rf power, such as a digital device, (2) coupling via a "transmission line" to (3) a radiating structure or "antenna." Two examples illustrate the problem:

1. **Desktop PC.** A new desktop personal computer is placed on a turntable in an electromagnetic compatibility (EMC) chamber which is a shielded room lined with absorbing pyramids. See Figs. 10-14 and 10-15. Three meters away there is a broadband antenna connected to a spectrum analyzer. As the PC rotates on the turntable, a technician scans across the frequency spectrum. The tests reveal an unwanted emission at one-half the clock speed of the Pentium microprocessor, and it appears to be worse when the back of the PC faces the antenna. The PC fails its test. A small probe, called a "sniffer," is used to localize the source of emission. Not only does the Pentium radiate but other devices connected to it

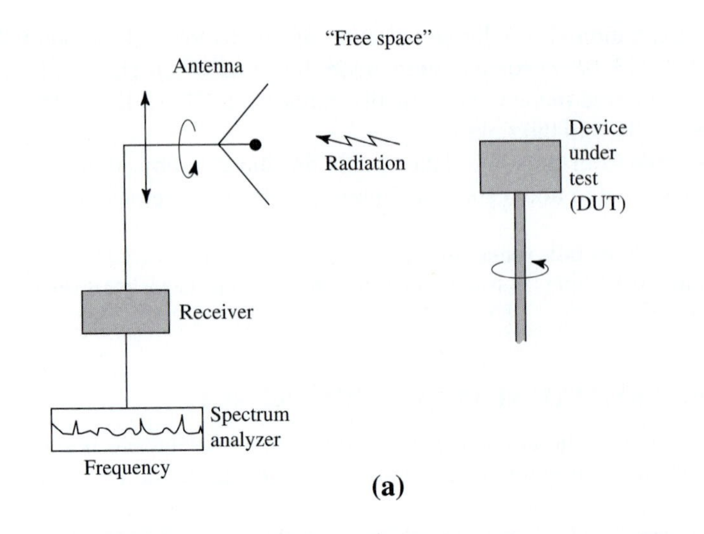

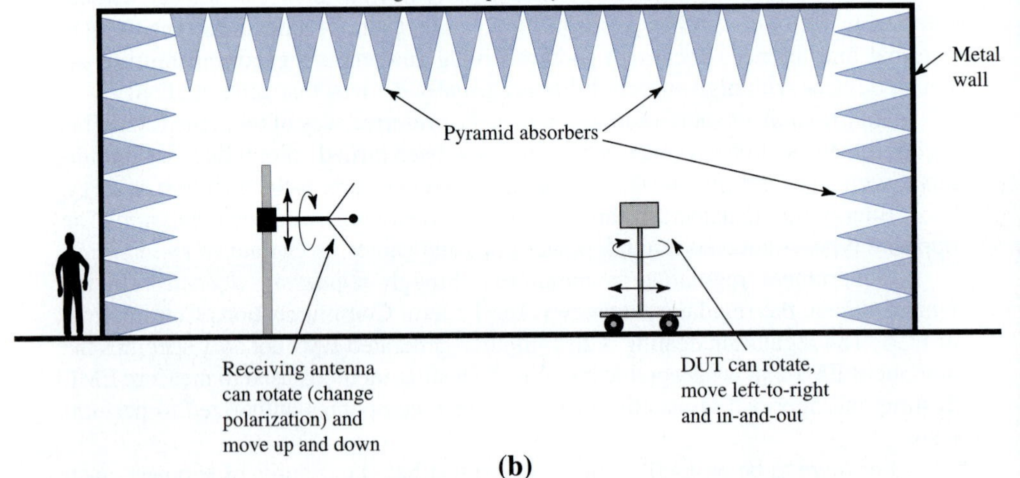

FIGURE 10-14
Typical electromagnetic compliance (EMC) chamber with broadband pyramid absorbers to simulate an ideal test environment for the device under test (DUT).

radiate. A graphics card operating at the Pentium clock speed also radiates and it is on the back side of the PC from which radiation is strongest. The problem is resolved by rearranging the positions of the cards in the PC and adding more screws to the cover lid of the metal card box to prevent leakage.

2. **CD player on jet aircraft (page 522).** The jet landed safely (at the right airport) after the compact disk (CD) player was turned off. The CD player and others like

FIGURE 10-15
EMC chamber at AT&T (now Lucent Technologies), Columbus, Ohio. The device under test on the turntable (in the foreground) is a central office switching unit. The engineer is checking the broadband biconical receiving antenna (in horizontal position) which can be rotated to change polarization and moved up and down by remote control with shielded doors closed.

it were subjected to tests in an EMC and found to have radiation at the frequency of the air navigation system. The players were several months old and the aluminum case had oxidized, forming an insulating layer. Aluminum oxide, or alumina, is a good insulator and radiation was leaking out. Although emitting little energy, the CD player was at a seat next to a window and only a couple of meters from the jet's navigation antenna. On the other hand, the navigation beacon could be 100 km away, giving the CD player a $(10^5/2)^2 = 25 \times 10^8$ or 94 dB advantage as an interfering source. A better box design was required, and also the flight attendant now tells you to turn off your CD player at takeoff and landing.

Although there is much more to high-speed digital system design, this brief chapter and its apocryphal stories highlight a few of the very important electromagnetic aspects.

For an encyclopedic treatment see "The Art of Electronics," 2d ed., by Paul Horowitz and Winfield Hill, Oxford University Press, 1989. A shorter and very practical treatment is given in "High Speed Digital Design, A Handbook of Black Magic," by Howard W. Johnson and Martin Graham, Prentice Hall, Englewood Cliffs, N.J., 1993.

See this book's Web site: www.elmag5.com for more information.

PROBLEMS

Answers are given in Appendix E to problems followed by the @ symbol.

10-3-5. Trace parameters. Consider a CMOS part with a rise time of 2.5 ns and ECL part with a rise time of 0.8 ns. (*a*) How long is a rising edge for each part on FR4 ($\varepsilon_r = 3$)? On alumina ($\varepsilon_r = 9$)? (*Hint:* Remember that the speed of a wave is equal to $c/\sqrt{\varepsilon_r}$.) (*b*) How long can a circuit board trace be and still be considered "lumped"? (*c*) What is f_{knee}? @

10-3-6. Microstrip line behavior. Consider a CMOS logic gate with an output impedance of 12 Ω and an input impedance of 4 MΩ in parallel with a capacitance of 12 pF. The circuit board material is FR-4, with an ε_r of 5. (*a*) Assuming a 30-cm 50-Ω microstrip line with propagation delay of 58.4 ps/cm is used, find T_{pd}, T_0, C_0, and then the loaded Z_0. What are the source and load reflection coefficients? The source acceptance and load transmission coefficients? (*Hint:* Use the loaded Z_0 to find the reflection, acceptance, and transmission coefficients.) (*b*) Recall that on a line with this sort of termination, the voltage will overshoot first, and then undershoot. Estimate the voltage overshoot and undershoot seen at the gate input when it switches from 1 to 0 (5 V to 0 V). Is the undershoot significant? That is, is it above 10 percent of the total voltage swing? @

10-3-7. Line termination. Compare source termination to parallel ac termination for power consumption, overall effectiveness in reducing reflections, and limitations on use. (That is, are there situations where you cannot use one of the termination strategies?)

10-3-8. Parallel termination. Consider a 50-Ω transmission line terminated with a 50-Ω resistor in parallel with a CMOS logic gate. The CMOS gate is the equivalent of a 1-MΩ resistor in parallel with a 15-pF capacitor. The load impedance is the parallel combination of the 50-Ω resistance, the 1-MΩ resistance, and the 15-pF capacitance. (*a*) Given that the 1-MΩ resistance is negligible, what is the reflection coefficient off of the load? (*b*) Is the termination perfect? Why or why not?

10-3-9. Strip line. Consider the circuit board trace geometry in Fig. P10-3-9 for connecting signals. (*a*) If the geometry of a trace is as shown, and you are using FR-4 fiberglass ($\varepsilon_{\text{eff}} = 2.8$), how much capacitance does a 50-mm trace add? That is, what is the capacitance between the trace and the ground planes, given that the traces are 50 mm long? (*b*) For $d = 0.5$ mm, a cross talk voltage of 0.8 V was found on the second conductor when the first conductor switched. What value of d would reduce the cross talk voltage to 0.3 V?

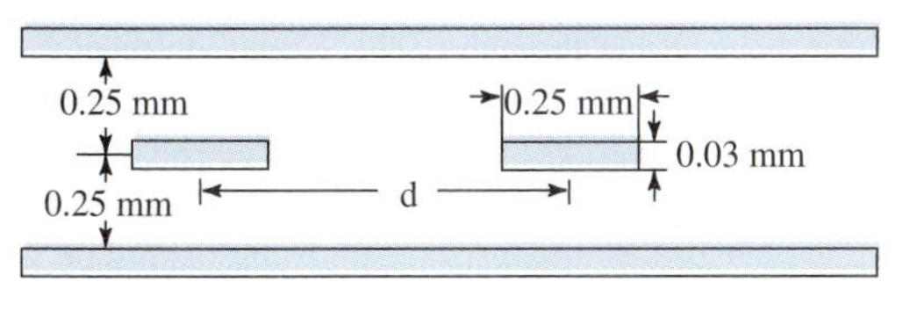

FIGURE P10-3-9

10-3-10. Reflected voltage pulses. Recall that the size of the voltage pulse reflected from a capacitor that is interrupting a transmission line is $-CZ_0/2$ times the derivative of the incoming wave. For a digital signal, the incoming wave is a square wave as illustrated in Fig. 10-1 with a rise time of T_r and voltage swing of ΔV. (*a*) Find an expression for the voltage reflection divided by ΔV in terms of C, Z_0, and T_r. Start by writing an expression for the peak value of the derivative of the square wave in Fig. 10-1. (*b*) What three things can be done to reduce the value of the reflection? (*c*) For a transmission line with a Z_0 of 50 Ω carrying a signal with a 1.5-ns T_r, what is the maximum allowable C that keeps the size of the reflection below 0.1 ΔV? @

10-3-11. Line with capacitor. Consider a line that is interrupted by a capacitor (Fig. P10-3-11). $Z_0 = 50\ \Omega$, $T_r = 4$ ns, $C = 10$ pF. What is the size of the reflected voltage pulse from the capacitor measured at point TP1? See Problem 10-3-10*a*.

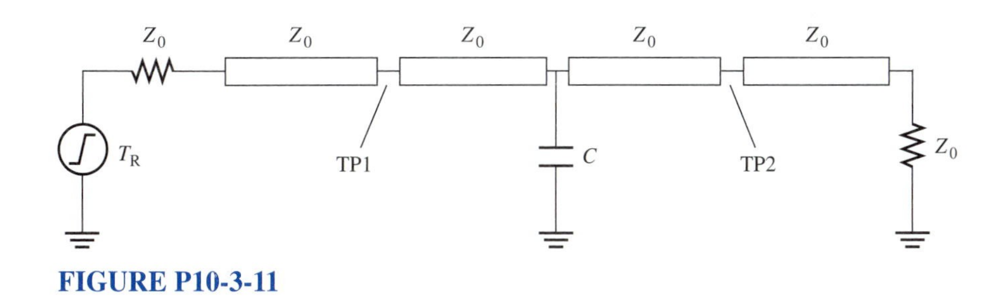

FIGURE P10-3-11

10-4-2. Return current paths. Consider the diagram of return current density shown in Fig. 10-13. The return current is underneath the line because it is attempting to minimize impedance and, at ac, this means minimizing inductance. Any deviation of the return current away from the signal conductor will increase the loop area and, therefore, the series inductance of the signal. (*a*) Is it acceptable for the signal conductor to run alongside the edge of the ground plane? (*b*) What would happen if the ground plane were interrupted under the conductor? What would happen to the return current, and what effect would this have? (*c*) What d/h keeps the cross talk constant below the generally acceptable level of 0.1? (*d*) Using the d/h ratio from (*c*), how far apart should two lines be that are located 0.25 mm above a ground plane? @

10-4-3. Logic circuit. A logic gate switching to a logic 0 looks like a resistor to ground. The trace connecting the gate output to a gate input has a capacitance to ground, as does the gate input. (The trace capacitance will be ignored.) This circuit is shown in Fig. P10-4-3. Assume that the two gates are on circuit boards that are separated by a 1-m-long cable. (For example, the gate on the left could be inside a PC and the gate on the right could be inside a printer.) (*a*) How big is the area of the loop if the cable connecting the PC and printer does have a ground signal? Assume the logic signal and ground signal are 2.5 mm apart along the cable. (*b*) What path does the current take if there is no ground signal on the cable connecting the PC and printer? How big is the area of the loop formed by that circuit? (*c*) What might be the consequence of such a large loop area? Think in terms of neighboring equipment and loop antennas. (*d*) Why do signal-carrying cables almost always have a ground signal on them?

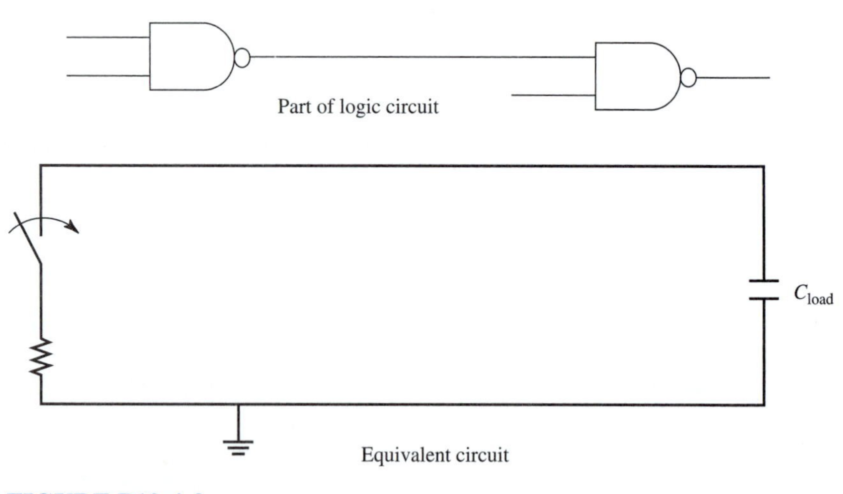

FIGURE P10-4-3

10-5-1. EMI/EMC layout rules. The following are some board layout rules advocated by the leading microprocessor manufacturer. What electromagnetic principles are the basis for the creation of each rule?

(*a*) Route high speed, high-duty-cycle signal traces (strobes, least significant address lines) over a continuous reference plane. For example, when routing on layers adjacent to the power plane, do not route these traces over breaks in this plane. Routing traces over power plane voids increases both common- and differential-mode radiation.

(*b*) Follow sound interconnect principles (like parallel, series, or dual-ended termination) to minimize high-speed-signal ringing.

(*c*) Position system clocks as far as possible from signals that go to cables.

(*d*) Keep chassis seams as far away from the processor as practical. The example below shows that seams become leaky at $\frac{1}{20}$ wavelength. Therefore, even seams 2 inches long leak 300-MHz frequency EMI.

10-5-2. Ferrite beads on cabling. A ferrite bead can be placed around a signal cable in an attempt to eliminate emissions from unwanted return current paths. Any return current that travels on a path outside the bead must change the magnetic flux in the bead, and so sees a much higher impedance. (*a*) An engineer is considering using a ferrite bead on a cable that conducts signals that have a 2.4-ns rise time. What is the maximum frequency content of the signal? What, then, is the lower limit on the frequencies that the bead may attenuate? (*b*) Computer monitor cables often have cylindrical structures on them. What do you think is housed in the structure and why? Why are computer monitors susceptible to unexpected current return paths? (*Hint:* Do both the computer and monitors have power cords with ground prongs on them?) (*c*) Would you expect to find the cylindrical housing nearer the computer or monitor? Why? @

CHAPTER 11

NUMERICAL METHODS

11-1 INTRODUCTION

This chapter discusses several useful numerical methods for solving electromagnetic problems. The repetitive Laplace or finite difference method with iterative solution serves as an introduction. This is followed by integral equation and moment method solutions. The finite difference–time domain technique, finite element, and other methods conclude the chapter.

All electrical science and engineering is based on the action of electric fields (**E** and **D**) and magnetic fields (**B** and **H**). These fields, in turn, are produced by electric charges (Q) and currents (I) in electromagnetic devices of all kinds. To understand how the devices work, we must be able to evaluate the fields in and around them. Thus, fields require a space, or three-dimensional, concept, inviting us to make maps or pictures of them showing field lines and equipotential surfaces. The maps give us information about the field intensities and potential differences; also about the energy, charge, and current densities everywhere on the map. Where squares are small, fields are strong; where squares are large, fields are weak. We learn to think of the field as an aggregation of field cells connected in series and parallel, whose totals give us integrated (or circuit) values of capacitance, resistance, and inductance. Furthermore, a map for a given configuration can be applied to many situations. See Appendix B.

All field maps, whether drawn graphically without mathematics or calculated with mathematics, are solutions of Laplace's (La-plah-s) equation.† In this chapter we proceed further to more formal, rigorous mathematical methods of finding field

†Marquis Pierre Simon de Laplace, French astronomer and mathematician (1749–1827).

configurations for a variety of problems. In the first examples, space is free of charge ($\rho = 0$), and we seek solutions to Laplace's equation ($\nabla^2 V = 0$). With space charge, a solution to Poisson's equation ($\nabla^2 V = -\rho/\varepsilon$) is obtained. Because of the dependence of the field configuration on its boundaries, these solutions are often called ***boundary value problems.***

When you have obtained a solution of Laplace's or Poisson's equation which also satisfies the boundary conditions, it is ***unique.*** No other field configuration is possible. This is important because it means that you have the answer. You don't need to look further.

The solution of a boundary value problem is usually facilitated if it is set up in a coordinate system in which the boundaries can be specified in a simple manner. For instance, a problem involving a rectangular object may be most readily handled with rectangular coordinates, a cylindrical object by cylindrical coordinates, a spherical object by spherical coordinates, etc. A restriction on the formal mathematical method is that the boundaries in many practical problems cannot be simply expressed in any coordinate system, and often in such cases we resort to other methods using graphical, experimental, analog-computer, and numerical techniques. In this chapter we discuss several useful numerical methods.

Laplace's and Poisson's equations are fundamental to science and engineering, wherever fields exist. They are differential equations. In using them, we learn about this important branch of mathematics while using the equation as a tool for finding a solution to an important practical application. Let us now proceed to Laplace's equation and ways for solving it.

11-2 LAPLACE'S EQUATION IN RECTANGULAR COORDINATES; SEPARATION OF VARIABLES

Expanding Laplace's equation ($\nabla^2 V = 0$) in rectangular coordinates (x, y, z), we have

$$\frac{\partial^2 V}{\partial x^2} + \frac{\partial^2 V}{\partial y^2} + \frac{\partial^2 V}{\partial z^2} = 0 \qquad \textit{Laplace's equation} \tag{1}$$

This differential equation is the most general way of expressing the variation of the potential V with respect to x, y, and z. It is a partial differential equation of the second order (no higher than second derivative) and first degree (no higher than the first power).† Equation (1) is the most general way of expressing the variation of V with respect to position (x, y, z), but it is not specific about the potential distribution of a particular problem. For this we must first obtain a solution or solutions and then pick one which satisfies the boundary conditions.

Let us solve (1) by means of the method of *separation of variables* in which we assume that V can be expressed as the product of three functions X, Y, Z, or

†More explicitly, the *degree of a differential equation* is the highest power of the highest-order term.

$$V = XYZ \tag{2}$$

where X = function of x only
Y = function of y only
Z = function of z only

Substituting (2) into (1), we get

$$YZ\frac{d^2X}{dx^2} + XZ\frac{d^2Y}{dy^2} + XY\frac{d^2Z}{dz^2} = 0 \tag{3}$$

Dividing by XYZ separates the variables giving

$$\frac{1}{X}\frac{d^2X}{dx^2} + \frac{1}{Y}\frac{d^2Y}{dy^2} + \frac{1}{Z}\frac{d^2Z}{dz^2} = 0 \tag{4}$$

Since the sum of the three terms is a constant ($= 0$) and each variable is independent, each term must equal a constant. Hence, we may write

$$\frac{1}{X}\frac{d^2X}{dx^2} = a_1^2 \tag{5}$$

or

$$\frac{d^2X}{dx^2} = a_1^2 X \tag{6}$$

and similarly

$$\frac{d^2Y}{dy^2} = a_2^2 Y \tag{7}$$

and

$$\frac{d^2Z}{dz^2} = a_3^2 Z \tag{8}$$

where

$$a_1^2 + a_2^2 + a_3^2 = 0 \tag{9}$$

The problem now is to find a solution for each of the three variables separately (hence the name "separation of variables").

A solution of (6) is

$$X = C_1 e^{a_1 x} + C_2 e^{-a_1 x} \tag{10}$$

where C_1 and C_2 are arbitrary constants that must be evaluated from the boundary conditions. Either term in (10) is a solution, or the sum is a solution, as may be verified by substituting the solution in (6).

It follows that a general solution of (1) is

$$V = (C_1 e^{a_1 x} + C_2 e^{-a_1 x})(C_3 e^{a_2 y} + C_4 e^{-a_2 y})(C_5 e^{a_3 z} + C_6 e^{-a_3 z}) \tag{11}$$

where C_1 and C_2, etc., are constants. The solutions may take the form of exponential, trigonometric, or hyperbolic functions.

Practical examples help illustrate the powerfulness and universality of Laplace's equation, so let's consider a simple example: the parallel-plate capacitor.

11-3 EXAMPLE 11-1: THE PARALLEL-PLATE CAPACITOR

Consider a parallel-plate capacitor as shown in cross-section in Fig. 11-1*a*. The plates are infinite in extent and are separated by a distance x_1. The left-hand plate is at zero potential and the right-hand plate at potential V_1. Let us apply Laplace's equation to find the potential distribution between the plates.

There is no variation in potential in the y and z directions, so that the problem is one-dimensional, and Laplace's equation ($\nabla^2 V = 0$) reduces to

$$\boxed{\frac{d^2V}{dx^2} = 0} \tag{1}$$

For the second derivative of V with respect to x to be zero, the first derivative must be equal to a constant. Thus, we have

$$\frac{dV}{dx} = C_1 \tag{2}$$

or

$$dV = C_1\,dx \tag{3}$$

Integrating (3), we write

$$\int dV = C_1 \int dx \qquad \text{or} \qquad V = C_1 x + C_2 \tag{4}$$

Equation (4) is a solution. For it to be appropriate for our problem (Fig. 11-1*a*), we now need to determine the values of the constants C_1 and C_2 satisfying the boundary conditions (B.C.) which are two in number. They are

B.C.1: $\quad V = 0 \quad$ at $x = 0$

B.C.2: $\quad V = V_1 \quad$ at $x = x_1$

From **B.C. 1,** (4) becomes

$$0 = 0 + C_2 \tag{5}$$

Hence, $C_2 = 0$. From **B.C. 2,** (4) becomes

$$V_1 = C_1 x_1 \tag{6}$$

so that

$$C_1 = \frac{V_1}{x_1} \tag{7}$$

Introducing the values for C_1 and C_2 from (7) and (5) into (4), we have

$$V = \frac{V_1}{x_1}x \qquad \text{(V)} \tag{8}$$

which is a solution of Laplace's equation that satisfies the boundary conditions and, therefore, gives the potential variation between the capacitor plates. It is a linear function of x, as illustrated in Fig. 11-1*b*.

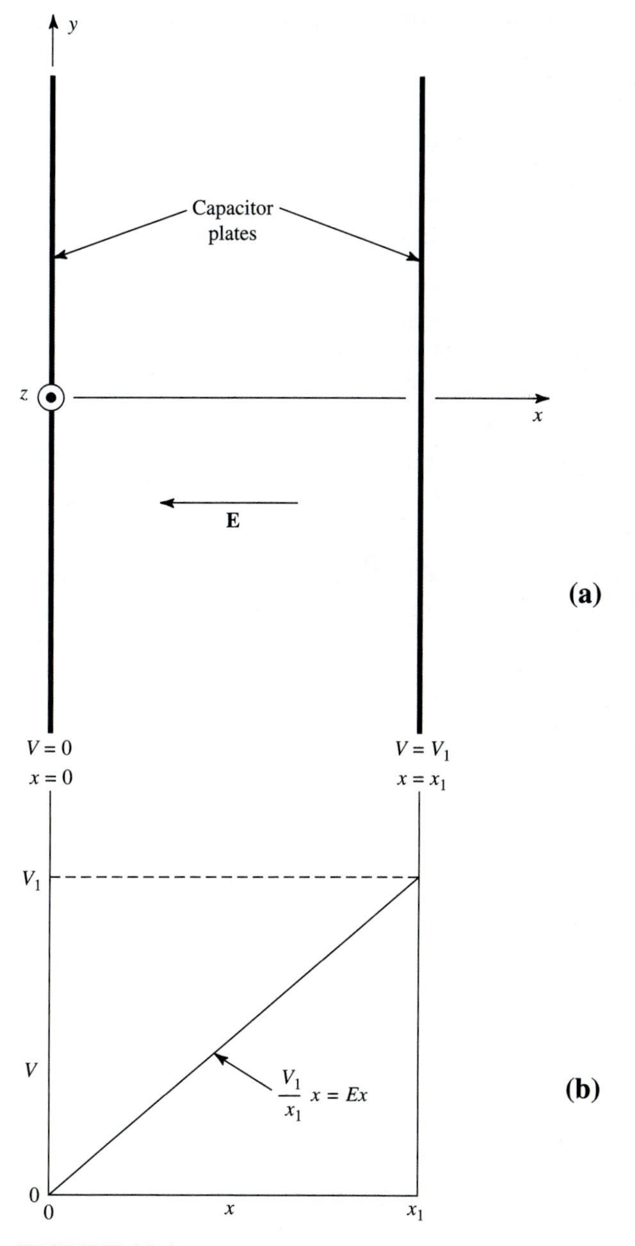

FIGURE 11-1
(*a*) Parallel-plate capacitor and (*b*) potential variation between the plates.

We might have anticipated this result since we would expect a uniform electric field between the parallel plates of the capacitor with magnitude

$$E = \frac{V_1}{x_1} \qquad (\text{V m}^{-1}) \tag{9}$$

with the potential at any point given by

$$V = Ex \qquad (\text{V}) \tag{10}$$

the same as in (8). The constant C_1 is thus equal to the electric field between the plates. Ordinarily we would not have resorted to a formal solution of Laplace's equation for this simple problem, but its application here serves to illustrate the procedures involved in applying Laplace's equation while using a minimum of mathematics.

11-4 REPETITIVE LAPLACE SOLUTION OR FINITE DIFFERENCE METHOD

In Sec. 11-3 we obtained an exact solution of Laplace's equation satisfying the boundary conditions for a parallel-plate capacitor using only pure, formal mathematics. However, we were dealing with a case of very simple geometry. If one of the plates has a bend, dent, or bump, it becomes more difficult to solve in this way so we might resort to a graphical method. Or we might use a numerical method where we solve Laplace's equation by repetitive approximations over smaller and smaller regions until we obtain a solution of desired accuracy. In this section, we discuss how this is done,† and then we apply the method in two examples in the sections which follow.

The basic step is to find a solution of Laplace's equation at a point by noting the slope of the second derivative of the potential V in orthogonal directions around the point. In rectangular coordinates Laplace's equation is given by

$$\frac{\partial^2 V}{\partial x^2} + \frac{\partial^2 V}{\partial y^2} + \frac{\partial^2 V}{\partial z^2} = 0 \tag{1}$$

If the potential variation is independent of z, the problem reduces to a two-dimensional one and (1) simplifies to

$$\frac{\partial^2 V}{\partial x^2} + \frac{\partial^2 V}{\partial y^2} = 0 \tag{2}$$

The first term in (2) is the second partial derivative of V with respect to x, that is, the rate of change of the rate of change of V with respect to x.‡ Similarly, the second term is the rate of change of the rate of change of V with respect to y. The sum of these two terms must be zero.

---

†J. B. Scarborough, "Numerical Mathematical Analysis," 6th ed., pp. 391–422, Johns Hopkins Press, Baltimore, 1966.

‡Or since $\partial V/\partial x = -E$, $\partial^2 V/\partial x^2$ is equivalent to the negative of the rate of change of E with respect to x.

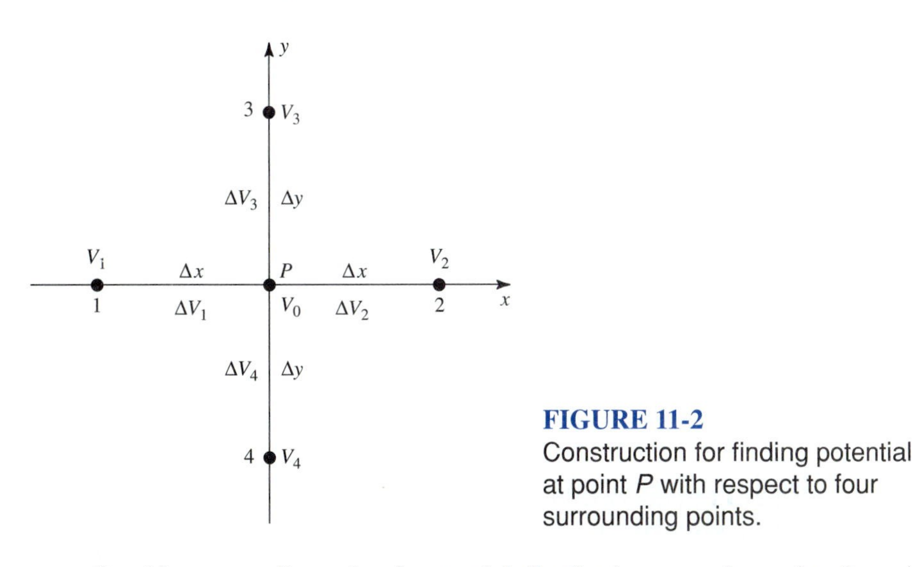

FIGURE 11-2
Construction for finding potential at point *P* with respect to four surrounding points.

Consider a two-dimensional potential distribution around a point *P*, as in Fig. 11-2. Let the potential at *P* be equal to V_0 and at the four surrounding points V_1, V_2, V_3, and V_4, as shown. Now $(V_2 - V_0)/\Delta x$ is the slope of *V* between points 2 and *P*. This is approximately equal to $\partial V/\partial x$ (becomes exact as $\Delta x \rightarrow 0$). Also $(V_0 - V_1)/\Delta x$ is the slope of *V* between points *P* and 1. The difference of these slopes (per distance increment Δx) is approximately equal to $\partial^2 V/\partial x^2$. Hence as discussed in the preceding paragraph, Laplace's equation requires that the difference in slopes of *V* in the *x* direction and the difference of the slopes in the *y* direction must be equal and opposite in sign. Thus, we have

$$\frac{\partial(\partial V/\partial x)}{\partial x} = -\frac{\partial(\partial V/\partial y)}{\partial y} \tag{3}$$

or

$$\frac{[(V_2 - V_0)/\Delta x] - [(V_0 - V_1)/\Delta x]}{\Delta x} \cong -\frac{[(V_3 - V_0)/\Delta y] - [(V_0 - V_4)/\Delta y]}{\Delta y} \tag{4}$$

Letting $\Delta x = \Delta y$, we have

$$V_1 + V_2 + V_3 + V_4 - 4V_0 \approx 0 \tag{5}$$

and

$$V_0 \approx \tfrac{1}{4}(V_1 + V_2 + V_3 + V_4) \tag{6}$$

If we know the potential at points 1, 2, 3, and 4, then according to Laplace's equation, the potential at the point *P* is as given by (6). In other words, the physical significance of Laplace's equation is simply that ***the potential at a point must be the average of the potential at four surrounding points.***† What could be simpler than that?

†In a three-dimensional problem we would find the average of the potential of six surrounding points (four as shown in Fig. 11-2 plus one a distance Δz from *P* into the page and one a distance Δz from *P* out of the page; $\Delta z = \Delta x = \Delta y$) and the potential at *P* would be given by

$$V_0 = \tfrac{1}{6}(V_1 + V_2 + V_3 + V_4 + V_5 + V_6)$$

11-5 EXAMPLE 11-2: THE INFINITE SQUARE TROUGH WITH LID BY REPETITIVE LAPLACE

To illustrate the repetitive method, consider the infinitely long square trough of sheet metal shown in cross-section in Fig. 11-3*a*. The sides and bottom of the trough are at zero potential. The lid, separated by small gaps from the trough, is at a potential of 40 V. Let us use the point-by-point method to obtain the potential distribution. First,

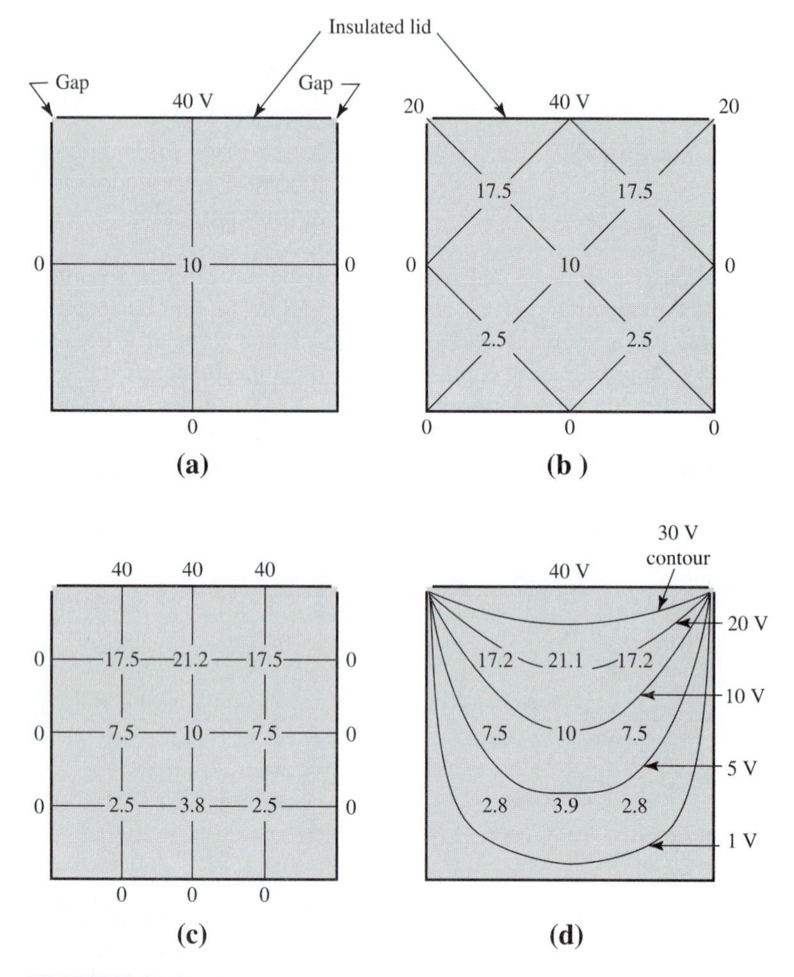

FIGURE 11-3
Application of point-by-point method or repetitive Laplace to determine potential at nine points inside an infinite square trough with insulated lid at a potential of 40 V. The map at (*d*) is a solution of Laplace's equation which satisfies the boundary conditions.

we find the potential at the center of the trough from (11-4-6) as

$$\frac{40 + 0 + 0 + 0}{4} = 10 \text{ V}$$

See Fig. 11-3*a*. Next, we find the potential at the center of the four quadrants of the trough. For this we rotate the xy axes 45° and take the potential at the cover-trough gap as 20 V (average of 40 V and 0). Again, from (11-4-6) we have

$$\frac{20 + 40 + 0 + 10}{4} = 17.5 \text{ V}$$

for the upper left and right quadrants, as in Fig. 11-3*b*, and

$$\frac{0 + 10 + 0 + 0}{4} = 2.5 \text{ V}$$

for the lower left and right quadrants. Next, we find V at four more points, with voltages as shown in Fig. 11-3*c* (xy axes returned to usual orientation).

The procedure is now repeated starting at the upper left, obtaining

$$\frac{40 + 21.2 + 7.5 + 0}{4} = 17.2 \text{ V}$$

This new value of V is now used to recalculate the potential at the adjacent point to the right as

$$\frac{40 + 17.2 + 10 + 17.2}{4} = 21.1 \text{ V}$$

See Fig. 11-3*d*.† All points are recalculated in this way, proceeding left to right and top to bottom over and over again until the values no longer change. This gives the most accurate solution to Laplace's equation we can get by this method with the number of points used. A still more accurate solution can be obtained by using a larger number of points (subdividing the area of the trough into a finer grid). As the distance between points (Δx or Δy) approaches zero, the solution can be made to approach an exact value. This method is well adapted for very accurate calculations using digital computers. However, manual calculations with relatively coarse grids can often give quick solutions of sufficient accuracy for many purposes. Thus, a determination, as above, of the potential at nine points in this trough example is sufficient for drawing the potential contours as in Fig. 11-3*d*.

Referring to the first step (determination of V at the center of the trough), let us reexamine the problem. Using the center point and four neighboring points (Fig. 11-3*a*, shown again in Fig. 11-4*a*), we get the slopes of the potential in the x and y directions as shown in Fig. 11-4*b* and *c*. (We assume $\Delta x = \Delta y = 10$ units). Thus, $\Delta V/\Delta x = 10/10 = \pm 1$, and $\Delta V/\Delta y = 10/10 = 1$ or $30/10 = 3$. Laplace's equation is satisfied since the difference in slopes in the x direction equals the

†In practice it is convenient to tabulate the voltage obtained in each iteration in column format at each grid point so that a running record is kept. When the voltages stabilize, the iteration can stop.

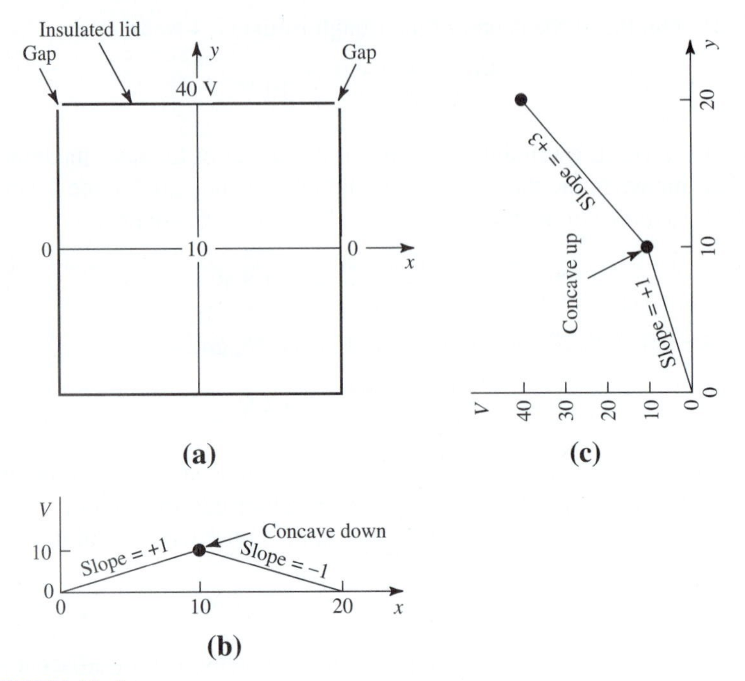

FIGURE 11-4
(*a*) Infinite square trough with insulated lid. (*b*) Slope or gradient of potential in *x* direction. (*c*) Slope or gradient of potential in *y* direction.

negative of the difference of the slopes in the *y* direction.† Thus,

$$\text{Slope difference } (x \text{ direction})\text{: } +1 - (-1) = 2$$

$$\text{Slope difference } (y \text{ direction})\text{: } +1 - (+3) = -2$$

$$\text{Sum:} \quad 2 - 2 = 0$$

We note in this problem that if the slope change is concave downward in one coordinate, it must be concave upward in the other coordinate (see Fig. 11-4*b* and *c*).

11-6 EXAMPLE 11-3: INFINITE SQUARE TROUGH WITH DIFFERENT POTENTIALS ON ALL FOUR SIDES

The method discussed in the preceding section is a repetitive or iterative technique. With a coarse grid (few points), evaluation by the iterative technique is easily done by hand with the aid of a pocket calculator to obtain the averages. With a finer grid (more points), as in Fig. 11-5*a*, one needs a computer. See REPLA program in Appendix C.

†Note that in the one-dimensional case the difference in slopes in one coordinate direction must be zero; that is, *V* must vary linearly with distance as in the parallel-plate-capacitor example (Sec. 11-3). In the three-dimensional case all one can say is that the *sum* of the slope differences in the three orthogonal coordinate directions must be zero.

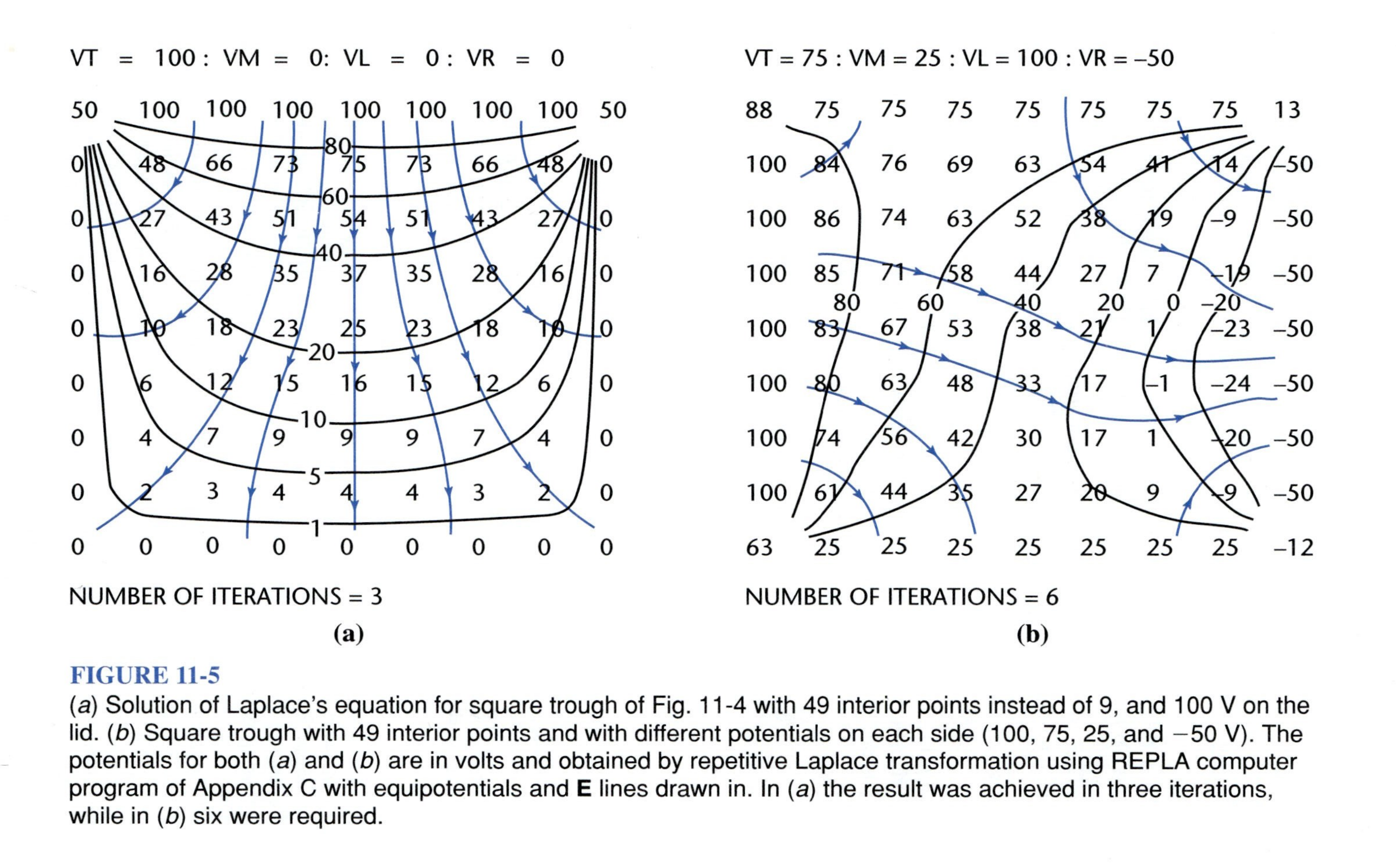

FIGURE 11-5

(*a*) Solution of Laplace's equation for square trough of Fig. 11-4 with 49 interior points instead of 9, and 100 V on the lid. (*b*) Square trough with 49 interior points and with different potentials on each side (100, 75, 25, and −50 V). The potentials for both (*a*) and (*b*) are in volts and obtained by repetitive Laplace transformation using REPLA computer program of Appendix C with equipotentials and **E** lines drawn in. In (*a*) the result was achieved in three iterations, while in (*b*) six were required.

Another trough example is shown in Fig. 11-5*b* with different potentials on all four sides.

Although the geometry of both Examples 11-2 and 11-3 could be handled analytically, they are good examples for the repetitive Laplace method.

11-7 LINE CHARGE DISTRIBUTION: THE INTEGRAL EQUATION AND THE MOMENT METHOD (MM)

From (2-3-14) we have for a line of charge density ρ_L, as in Fig. 11-6, that the potential at point P is given by

$$V = \frac{1}{4\pi\varepsilon_0}\int_0^L \frac{\rho_L(x)}{r}\,dx \qquad (1)$$

where $\rho_L(x)$ = charge per unit length of line as a function of x, C m^{-1}.

If $\rho_L(x)$ is known as a function of x, then (1) can be integrated. However, if $\rho_L(x)$ is not known, (1) represents an *integral equation* with the problem being to find $\rho_L(x)$. The following example uses a ***moment method*** **(MM)** for finding $\rho_L(x)$.

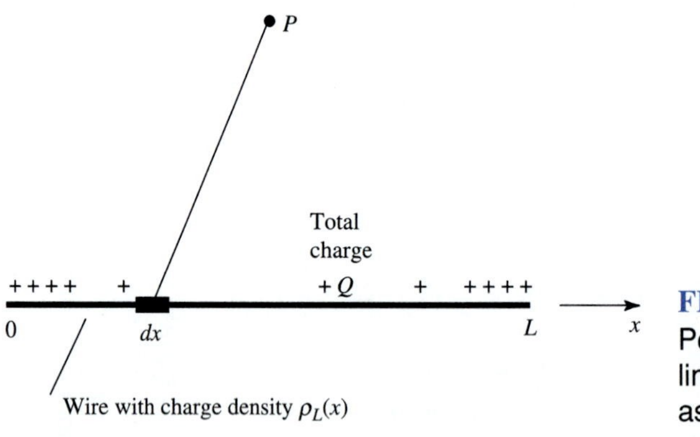

FIGURE 11-6
Potential at P due to a linear charge distribution as given by (1).

Example 11-4. Charge distribution on conducting wire. Let the line of charge be an isolated conducting wire of length L on which a total charge $+Q$ has been placed. Since like charges repel, we anticipate that the charge density will be greater at the ends of the rod than at the center. The problem is to determine the charge distribution using an incremental numerical technique or ***moment method*** **(MM).**

Solution. As a first step or approximation, let us mark off the wire into four sections as in Fig. 11-7. The total charge on each section is Q_1, Q_2, Q_3, and Q_4 as indicated. From symmetry we can set $Q_3 = Q_2$ and $Q_4 = Q_1$. Our problem, then, is to find the ratio of Q_1 to Q_2 (same as Q_4 to Q_3). Now let us assume that all of the charge of one section is situated at its center and calculate the potential from all sections at the observation or test points P_{12} and P_{23}. From symmetry the potential at P_{34} will be the same as at P_{12}. The situation may now be regarded as one with four points of charge (source points) in empty space with the potential to be determined at the observation or test points.

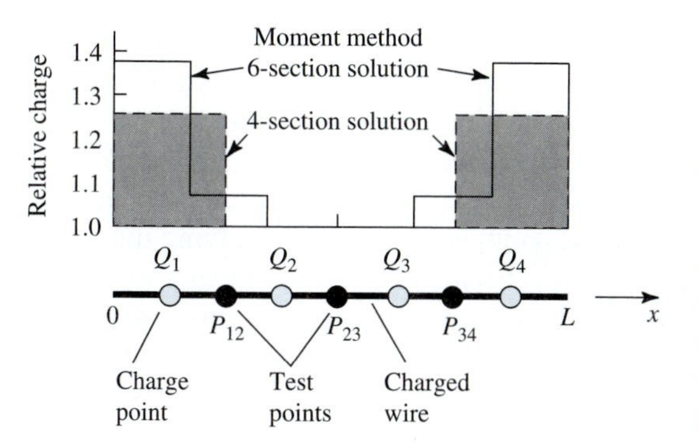

FIGURE 11-7
Charged wire divided into four sections for calculation of actual (nonuniform) charge distribution using the moment method (MM).

Thus, from (1) the potential at P_{12} is given by

$$V\,(\text{at } P_{12}) = \frac{1}{4\pi\varepsilon_0}\sum_{m=1}^{m=4}\frac{Q_m}{r} = \frac{1}{4\pi\varepsilon_0}\left[\frac{Q_1}{0.125L}+\frac{Q_2}{0.125L}+\frac{Q_3}{0.375L}+\frac{Q_4}{0.625L}\right] \quad (2)$$

Likewise the potential at P_{23} is given by

$$V\,(\text{at } P_{23}) = \frac{1}{4\pi\varepsilon_0}\left[\frac{Q_1}{0.375L}+\frac{Q_2}{0.125L}+\frac{Q_3}{0.125L}+\frac{Q_4}{0.375L}\right] \quad (3)$$

A constraint or *boundary condition* is that (even though the charge density varies along the wire) the potential is constant (the wire is a conductor). Therefore, V (at P_{12}) $= V$ (at P_{23}) so that we can equate (2) and (3) from which

$$Q_1 = 1.25Q_2 \quad (4)$$

Thus, the charge (or average charge density) for the outer sections is 25 percent greater than for the inner sections, and we can write the relative charge values as

$$1.25{:}1.00{:}1.00{:}1.25 \quad (5)$$

This is a four-section approximation (see Fig. 11-7). Dividing the wire into six sections, we obtain three simultaneous equations [each of the form of (2) or (3)] with three unknowns. Solving for the unknowns, we obtain the relative charge ratios

$$1.37{:}1.06{:}1.00{:}1.00{:}1.06{:}1.37 \quad (6)$$

In this better approximation we observe that the end sections have 37 percent greater charge than the center ones (see Fig. 11-7). Dividing into still more sections, we obtain a more accurate charge distribution, and we find that the ratio of the end section charge to that of the center section increases even more. Calculations involving many sections are most conveniently done with a computer because of the many steps involved.

An implicit assumption we have made in our solution is that the wire is infinitesimally thin. As the wire is divided into more and more sections, the charge ratio increases without limit (solution does not converge). To make the solution converge, we need to be more realistic and consider the finite thickness of the wire. The only change needed to obtain the distribution for finite-diameter wires is to move the location of the charge points to the circumference of the wire while keeping the test points at the center line of the wire. Thus, with the moment method as above, we can investigate the effect of wire diameter on the charge distribution, but, for simplicity, we did not consider the thickness in our example. In solving this very practical charge distribution problem, we have been introduced to the moment method which finds wide application for many scientific and engineering problems.

Computer Program CHARGED PLATES (Appendix C) uses the moment method to calculate the charge distribution across the plates of two-conductor strip transmission line. See also Problem 11-7-2.

11-8 THE GENERALIZED MULTIPOLE TECHNIQUE (GMT)

This technique was used by Nicolas B. Piller of the Swiss Federal Institute of Technology at Zürich, Switzerland, to generate the field maps of microstrip transmission lines appearing in Chap. 3 (Figs. 3-7, 3-8, and 3-9).

GMT is also used for the simulation of scattering problems. Here the scatterer is decomposed into homogeneous domains represented by an expansion fulfilling Maxwell's equations inside the domain. The boundaries between domains are discretized with D points and 6 boundary conditions for both electric and magnetic fields. This results in a set of equations with N unknowns and $6D$ equations where N equals the number of expansion parameters.

Interactive examples of electromagnetic scattering from objects using GMT can be accessed via the book's web site, www.elmag5.com.

Reference: N. B. Piller and O. J. F. Martin, Extension of the Generalized Multipole Technique to 3D anisotropic scatters, *Optics Letters,* **23**: 579, 1998.

11-9 FINITE DIFFERENCE–TIME DOMAIN (FD-TD) TECHNIQUE†

The repetitive Laplace or finite difference approach used in Sec. 11-4 to solve electrostatic problems can be extended to dynamic problems through a technique called *finite difference–time domain* or FD-TD. Like the repetitive Laplace approach, the FD-TD technique imposes a rectangular grid over the region of interest and solves a discretized version of field equations at the nodes of the grid. However, since dynamic problems involve time-changing electric and magnetic fields, Maxwell's curl equations from Ampére's and Faraday's laws must be solved at each grid point.

In 1966, K. S. Yee‡ introduced these equations amenable to solution by digital computer using a grid such as that shown in Fig. 11-8. Careful examination of this grid (known as "Yee's mesh") shows that the solution points for the electric field are spatially offset from the magnetic-field solution points. This allows the electric or magnetic field at each node to be computed using the surrounding four magnetic or electric field values.

†A. Taflove, Computational Electrodynamics: FD-TD, Artech House, 1995.

‡K. S. Yee, Numerical Solution of Initial Boundary Value Problems Involving Maxwell's Equations, *IEEE Trans. Antennas and Prop.,* **14,** 302, 1966.

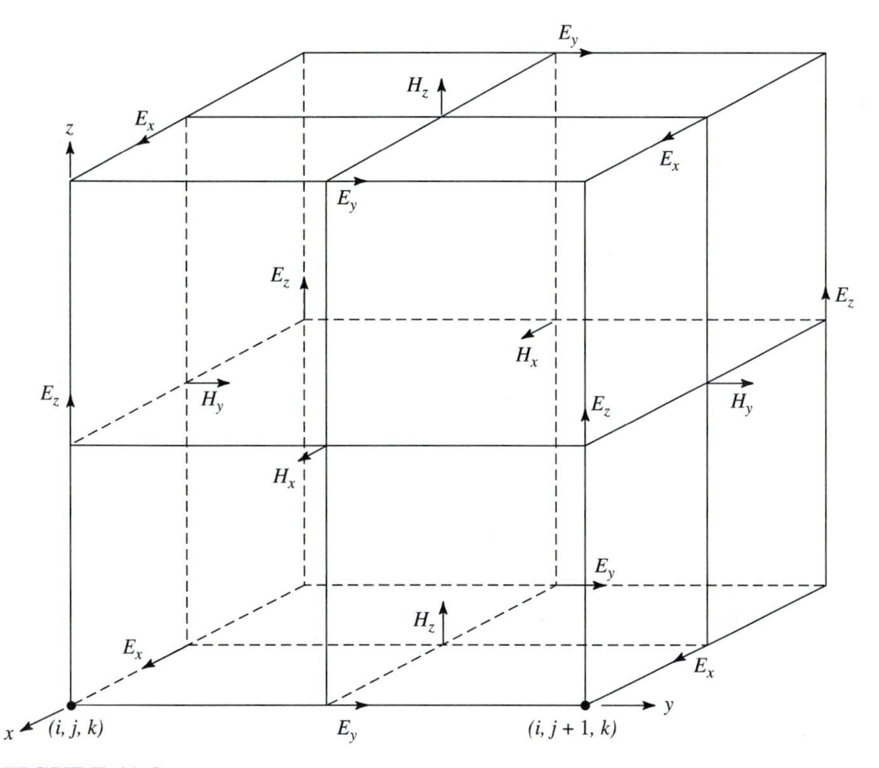

FIGURE 11-8
Solution grid for FD-TD (Yee's mesh).

In addition to discretizing E and H in space, in FD-TD the temporal changes in the fields are computed at discrete time intervals. For spatial grid separations of Δx, Δy, and Δz and time increments of Δt, the fields may be written as

$$F(x, y, z, t) = F(i\Delta x, j\,\Delta y, k\,\Delta z, n\,\Delta t) = F^n(i, j, k) \qquad (1)$$

From the grid of Fig. 11-8, the x-components of the electric and magnetic fields at the solutions points in a charge-free region are

$$\begin{aligned} E_x^{n+1}(i + \tfrac{1}{2}, j, k) &= E_x^n(i + \tfrac{1}{2}, j, k) \\ &+ (\Delta t/\varepsilon)\{[H_z^{n+1/2}(i + \tfrac{1}{2}, j + \tfrac{1}{2}, k) - H_z^{n+1/2}(i + \tfrac{1}{2}, j - \tfrac{1}{2}, k)]/\Delta y \\ &+ [H_y^{n+1/2}(i + \tfrac{1}{2}, j, k - \tfrac{1}{2}) - H_y^{n+1/2}(i + \tfrac{1}{2}, j, k + \tfrac{1}{2})]/\Delta z\} \end{aligned} \qquad (2)$$

and

$$\begin{aligned} H_x^{n+1/2}(i, j + \tfrac{1}{2}, k + \tfrac{1}{2}) &= H_x^{n-1/2}(i, j + \tfrac{1}{2}, k + \tfrac{1}{2}) \\ &+ (\Delta t/\mu)\{[E_y^n(i, j + \tfrac{1}{2}, k + 1) - E_y^n(i, j + \tfrac{1}{2}, k)]/\Delta z \\ &+ [E_z^n(i, j, k + \tfrac{1}{2}) - E_z^n(i, j + 1, k + \tfrac{1}{2})]/\Delta y\} \end{aligned} \qquad (3)$$

with similar expressions for the y- and z- components. While these expressions appear complex, examination of the indices and superscripts reveals that the values for E and H at each time step are calculated simply by updating the values from the previous time step by the curl of the complementary field.

As in the electrostatic case, these equations define the fields in terms of the fields at other locations (and, in this case, at previous times). To arrive at a complete solution, boundary conditions as well as initial conditions must be applied. To understand the application of boundary conditions, consider the microstrip transmission line shown in Fig. 11-9. At $z = 0$, the presence of the ground plane requires that the tangential E and normal H fields must be zero at this plane. Thus for all time increments

$$\begin{aligned} E_x^n(i + \tfrac{1}{2}, j, 0) &= 0 \\ E_y^n(i, j + \tfrac{1}{2}, 0) &= 0 \\ H_z^n(i + \tfrac{1}{2}, j + \tfrac{1}{2}, 0) &= 0 \end{aligned} \tag{4}$$

Additionally, the total fields at the edges of the solution region must be made zero in order to avoid reflections at these artificial boundaries.

The computational starting point for an FD-TD solution is defined by the initial condition, achieved by applying a sinusoidal or impulsive propagating wave beginning at $n = 0$. At each increment of n, the solutions for E and H evolve as the wave propagates. The computational time step Δt must be made sufficiently small to accurately model the propagation delay between different locations. Thus

$$\frac{\Delta r}{\Delta t} > v \tag{5}$$

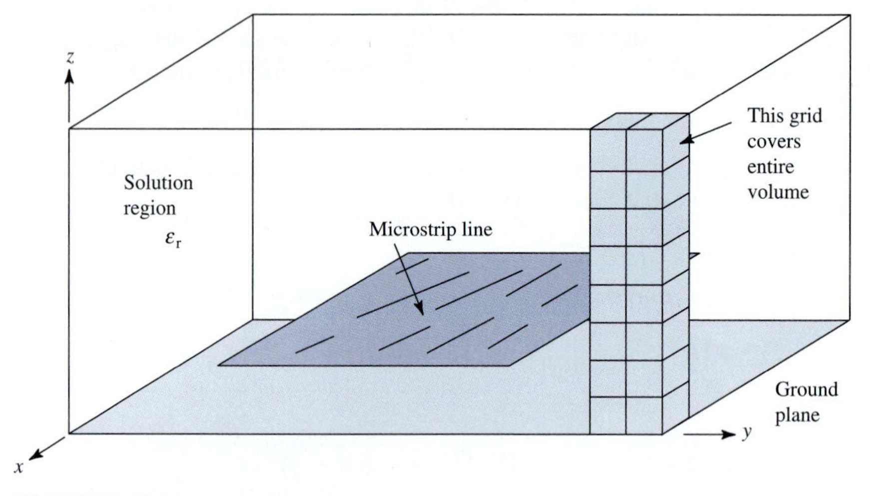

FIGURE 11-9
FD-TD technique applied to microstrip line.

where $\Delta r = (\Delta x^2 + \Delta y^2 + \Delta z^2)^{1/2}$ is the spatial distance between solution points and v is the velocity of propagation within the medium. The solution must therefore be calculated using a time increment which is shorter than the time it takes the wave to propagate between spatial solution points.

To ensure that abrupt field changes do not occur between spatial sample locations, Δr is normally chosen to be no greater than $\lambda/10$. Using the FD-TD approach to find the fields on a microstrip line operating at a frequency of 5 GHz ($\lambda = 60$ mm) in a material with $\varepsilon_r = 4$ ($\lambda = 30$ mm) requires a spatial grid size no larger than 3 mm and a time increment of

$$\Delta t < \frac{\Delta r}{v} = \frac{3 \text{ mm}}{(3 \times 10^8 \text{ m/s})(\frac{1}{2})} = 2 \times 10^{-11} \text{ s} \tag{6}$$

As an example of the required sample density, consider that the number of spatial solution points required to determine the fields on a microstrip transmission line with height of $5\,\Delta z$, width of $20\,\Delta y$, and length of $100\,\Delta x$ over a region of height $50\,\Delta z$, width $200\,\Delta y$, and length of $1000\,\Delta x$ is 10 million. If the line has a relative permittivity of 4 and is operated at a frequency to 5 GHz, the solution must be computed every 20 ps for the time period of interest.

Despite the significant computational burden of the FD-TD technique, this approach is being used to solve a wide range of dynamic electromagnetic problems, including the design of microwave and millimeter-wave integrated circuits (MMICs), the design of subpicosecond photonic devices, and interconnections between high-speed digital circuits in multichip modules (MCMs).

11-10 FINITE ELEMENT METHOD (FEM)†

For problems involving irregular boundaries and nonhomogenous material properties, the finite difference technique described in Sec. 11-4 is difficult to apply. Under such conditions, the finite element method (FEM), a technique originally developed for structural analysis, may be modified and applied to electromagnetic problems.

The finite element method may be implemented in four steps:

1. Discretization of the solution region into elements (usually triangular)
2. Generation of equations for the fields or potentials at each element
3. Integration or assembly of all elements
4. Solution of the resulting system of equations

†Peterson, A. F., Ray, S. L., and Mittra, R., *Computational Methods for Electromagnetics,* IEEE Press, Piscataway, NJ, 1998.

Silvester, P. and Ferrari, R., *Finite Elements for Electrical Engineers,* Cambridge University Press, Cambridge, 1983.

Zienkiewicz, O. and Taylor, R., *The Finite Element Methods,* McGraw Hill, New York, 1989.

Miller, E. K., "A Selective Survey of Computational Electromagnetics," *IEEE Transactions on Antennas and Propagation,* Vol. 36, No. 9, Sept. 1988.

In most FEM implementations, the new solution region is divided into triangular elements. These elements are constructed so that the points of some of the triangles (called "nodes") fall along the boundary of the solution region. Within each element, the parameter of interest (for example, the potential in electrostatic problems) is approximated by a polynomial. A key difference between the FEM and finite difference techniques is that the potential is estimated not only at the nodes, but throughout the element by an interpolation or "shape" function. Typical shape functions have values which vary from one at the node to zero at neighboring nodes.

Using the potential values at each node and the shape function, the value for the potential $\mathbf{V}$ of an element may be written as

$$\mathbf{V}_{\text{element}} = \sum_{\text{i}=1}^{3} f_i(x, y)\mathbf{V}_{\text{node i}} \tag{1}$$

where $f_i(x, y)$ is the shape function as applied to each node and $\mathbf{V}_{\text{node i}}$ is the value of the potential at node i.

Once the series of equations for the potentials at each element is in hand, a matrix may be constructed from all of the coefficients that describes the coupling between the nodes. This "stiffness" matrix is then solved by inversion or iterative techniques to yield the potential throughout the solution region.

An understanding of the strengths and weaknesses of numerical techniques such as FEM can be gained by applying this method to the microstrip transmission line of Ex. 11-5.

Example 11-5. Microstrip transmission line potentials via finite element method. Determine the potential distribution of a microstrip line with dielectric substrate $\varepsilon_r = 10$ using the finite element method.

Solution. To apply FEM to this problem, we first segment the region in and around the line into triangular elements, as shown in Fig. 11-10. Each triangular element and intersection (node) is then numbered, and two arrays are generated. The first of these arrays relates the nodes to the elements (so the computer knows which node numbers belong to each element), and the second contains the x and y coordinates of each node. Next we specify the potential at each of the fixed nodes (10 V on the top plate and 0 V

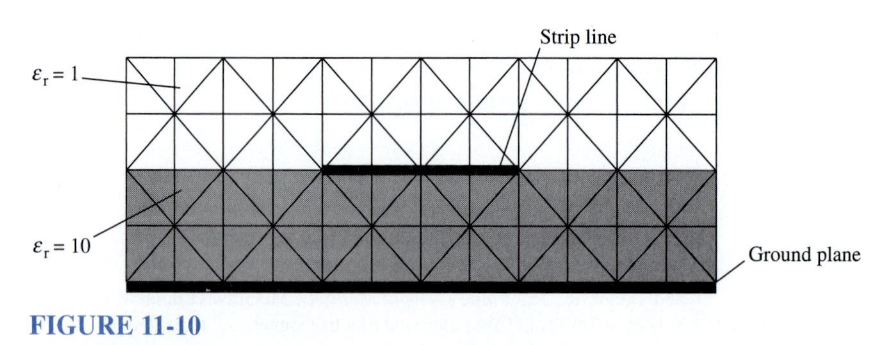

FIGURE 11-10
Discretized solution region for microstrip line.

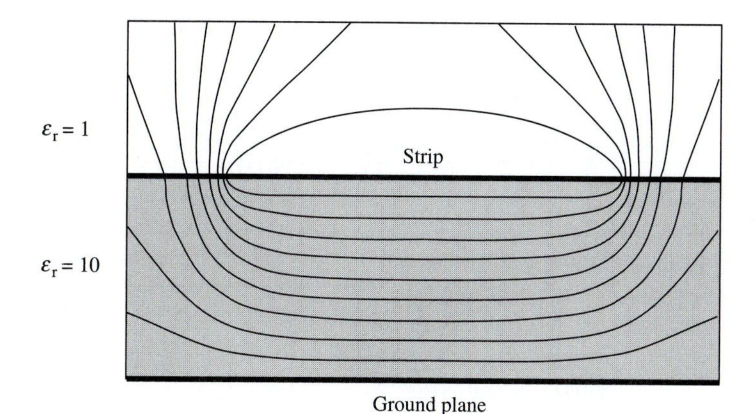

FIGURE 11-11
Equipotential lines for microstrip transmission line computed using the finite element method.

on the ground plane) and the permittivity of each element (all of the shaded elements have $\varepsilon_r = 10$, and all other elements have $\varepsilon_r = 1$).

After the node and element arrays have been set up, the finite element method is used to solve Laplace's equation over the region around the microstrip line. The results of this computation are shown by the equipotential contours in Figure 11-11. Notice that the FEM solution provides a reasonable estimate of the potential distribution, showing the refraction of the potential lines at the air-dielectric interface. However, like all numerical techniques, FEM introduces a quantization of the potential contours, since the solution is obtained over a finite number of elements. Thus, the user of numerical techniques should carefully examine the results and apply an understanding of fields and potentials gained through other techniques such as field mapping.

Compare Fig. 11-11 and Fig. 3-9 which also has electric field lines that divide the map into curvilinear squares and curvilinear rectangles.

11-11 CONTINUOUS WAVE (CW) REFLECTIONS AND FEM

As an extension of the Bouncing Pulses on the $\lambda/4$ transformer of Example 3-23 (Sec. 3-6), let us consider a case of a continuous wave incident on a thick dielectric slab. For the pulse we tracked only amplitude but for the continuous wave we must track *both amplitude and phase.*

Example 11-6. Continuous wave on dielectric slab. A plane 200 MHz wave is incident normally on a dielectric slab of thickness 3.14 m and $\varepsilon_r = 4$, $\mu_r = 1$, $\sigma = 0$. Calculate the VSWR in the three regions (*a*) to the left of the slab, (*b*) within the slab, (*c*) to the right of the slab and (*d*) draw the standing-wave pattern.

Solution. Taking the incident wave field strength equal to 1 V m^{-1} and its phase at the slab boundary as $\angle 0°$, we have from (4-7-9) and (4-7-12) that $-\frac{1}{3}$ is reflected and $\frac{2}{3}$ is transmitted into the slab. At the right-hand boundary $\frac{1}{3} \times \frac{2}{3} = \frac{2}{9}$ $(= 0.222)$ is reflected and $\frac{4}{3} \times \frac{2}{3} = \frac{8}{9}$ $(= 0.889)$ is transmitted.

In crossing the slab the wave travels $2 \times (3.14/1.5) = 4.19\ \lambda$ so at the right boundary the phase equals $0.19 \times 360° = 68°$.

The top of Fig. 11-12 tracks the progress of the wave as it bounces back-and-forth between the two boundaries. Adding the waves in air at the left of the left-hand boundary, we have $0.333 \angle 180° + 0.296 \angle 137° + 0.033 \angle 274° + 0.004 \angle 50° = 0.57 \angle 162°$. Therefore, to the left of the boundary the VSWR $= (1 + 0.57)/(1 - 0.57) = 3.65$. *Ans.* (*a*).

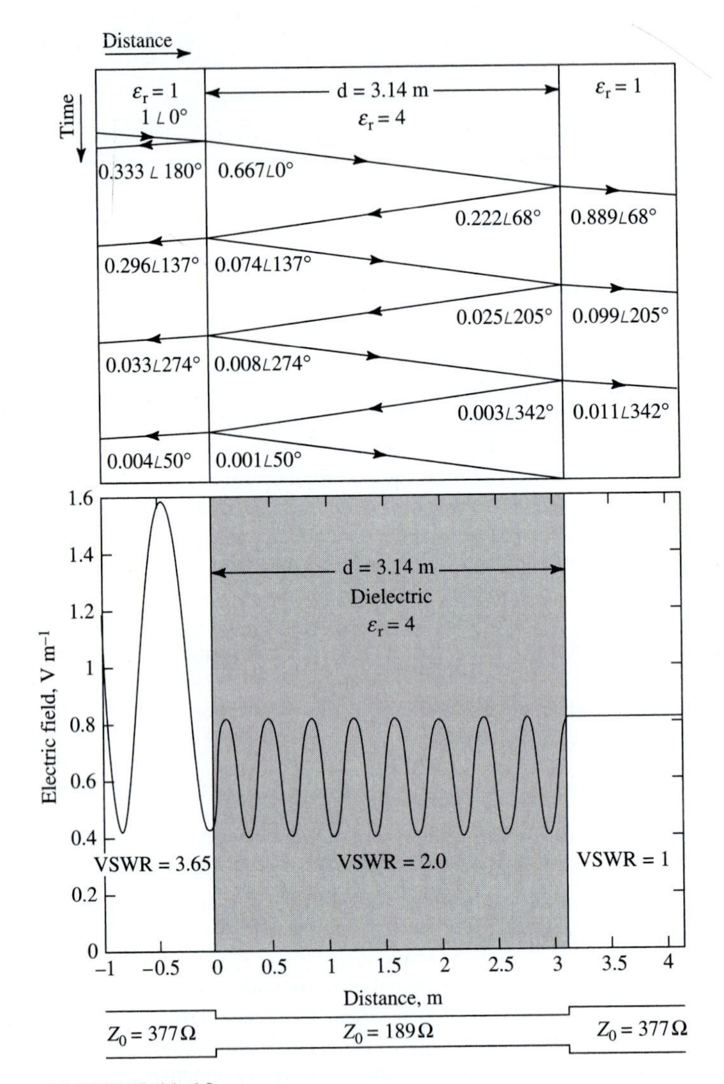

FIGURE 11-12
Continuous wave traveling through dielectric slab. *Top:* Magnitude and phase of wave at multiple reflections. *Center:* Standing-wave pattern as drawn by FEM program of Slone and Lee. *Bottom:* Equivalent transmission line.

Proceeding in like manner in the dielectric the VSWR = 2.00, *Ans.* (*b*), and to the right of the dielectric the VSWR = 1.00, *Ans.* (*c*).

Rodney Slone and Robert Lee of the Ohio State University have developed a finite element method (FEM) program which produces the standing-wave pattern shown in Fig. 11-12.† *Ans.* (*d*).

They explain the finite element method and apply it to several simple cases including the one of Example 11-5. The program has several hundred lines and two subroutines and can solve and plot the results for much more complex geometries. Thus, it can handle arbitrary values of ε_r, μ_r, and σ for multiple or continuously varying parameters.

Problem 11-11-1. VSWRs. Find the three VSWRs of Example 11-6 if $d = 3$ m. *Ans.* 1, 2, 1.

PROBLEMS

Answers are given in Appendix E to problems followed by the @ symbol.

11-6-1. Square trough. A square trough, as in Fig. 11-4*a*, has a potential of 100 V on the lid or top, 0 V on the bottom and 50 V on the two sides. Use the Repetitive Laplace program to obtain the potential distribution inside the trough. (*a*) Draw potential contours at 75, 50, and 25 V. (*b*) If the trough is 10 × 10 cm, what is the maximum electric field inside?

11-7-1. Charged wire. Using the Moment Method (Sec. 11-7), determine the charge density on the end sections of a wire, as in Fig. 11-7, relative to its center when it is divided into (*a*) 8 sections and (*b*) 12 sections. (*c*) Compare with the values in Example 11-3 for 4 and 6 sections. @

11-7-2. Charged plates. Using the Charged Plates program obtain the charge distribution on the upper and lower plates of a two-conductor line for width-to-height (W/H) ratios of (*a*) 10, (*b*) 5, and (*c*) 2.5. (See Fig. P11-7-2.)

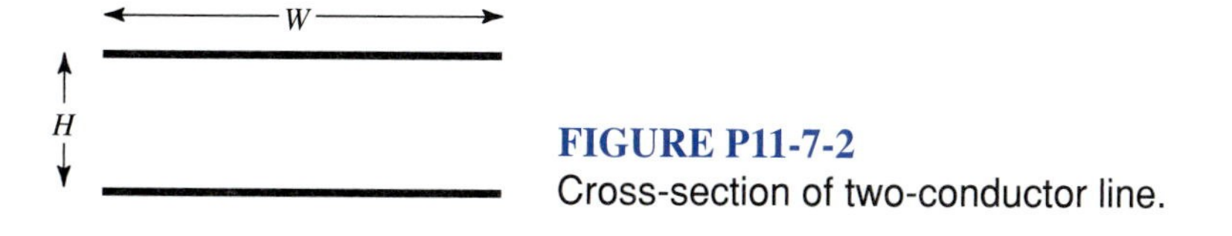

FIGURE P11-7-2
Cross-section of two-conductor line.

11-7-3. Potential between plates with projection. Using the Post program determine the potential distribution when a vertical strip of 3 cm width is introduced into the previously uniform field between two flat plates separated by 10 cm. (See Fig. P11-7-3.) (*a*) For a potential difference of 100 V between the plates, draw equipotentials for 20, 40, 60, and 80 V. (*b*) What is the electric field at point *P*, 5 mm above the strip relative to the uniform field? Compare with Fig. 9-13*c* ("Hey man, hit the ditch"). @

†R. D. Slone and R. Lee, "Finite Element Method for Electromagnetic Modeling," Ohio State University ElectroScience Laboratory, personal communication, 1998.

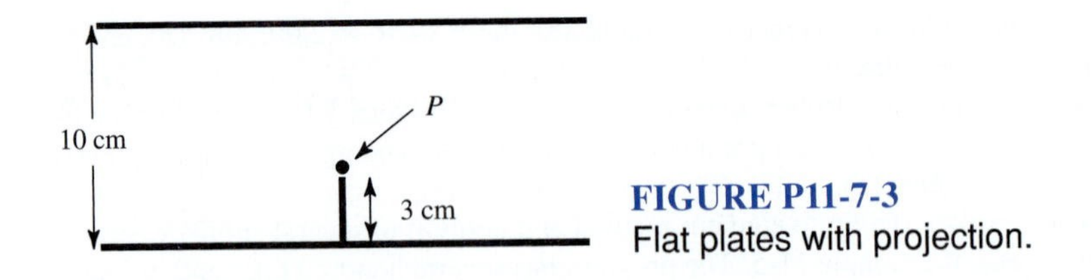

FIGURE P11-7-3
Flat plates with projection.

11-7-4. Square pipe. A hollow square pipe has a potential of 100 V applied. What is (*a*) the potential distribution and (*b*) the electric field distribution inside the pipe? @

11-11-2. VSWRs. Find the three VSWRs of Example 11-6 if $d = 2.38$ m and $\varepsilon_r = 3$. $\mu_r = 1$, with $\sigma = 0$ and $f = 200$ MHz as in the example.

APPENDIX A

UNITS, CONSTANTS, AND OTHER USEFUL RELATIONS

Multiples and submultiples of the basic units are designated by the prefixes listed in Table 1-3 (Chap. 1). Note that in the SI system, multiples and submultiples are in steps of 10^3 or 10^{-3}. Quantities larger than 10^{18} or smaller than 10^{-18} are usually designated by the exponential form. For a discussion of dimensions and units see Chap. 1, Secs. 2, 3, and 4.

There are six columns in Table A-1 as follows:

1. *Name of the dimension or quantity*
2. Common *symbol* used
3. *Description* gives dimension in terms of the fundamental dimensions (mass, length, time, electric current) or other secondary dimensions
4. *SI unit and abbreviation*
5. *Equivalent unit*
6. *Fundamental dimension* in symbols M (mass), L (length), T (time), and I (electric current)

Table A-2 lists trigonometric, hyperbolic, logarithmic, and other relations. Table A-3 lists glyphs.

TABLE A-1
Fundamental, mechanical, electrical, and magnetic units

Name of dimension or quantity	Symbol	Description	SI unit and abbreviation	Equivalent unit	Dimension
Fundamental units					
Current	I	$\frac{\text{charge}}{\text{time}}$	ampere (A)	6.25×10^{18} electron charges per second $= \frac{C}{\text{s}}$	I
Length	L, l, ℓ		meter (m)	1,000 mm = 100 cm	L
Mass	M, m		kilogram (kg)	1,000 g	M
Time	T, t		second (s)	$\frac{1}{60}$ min $= \frac{1}{3600}$ h $= \frac{1}{86\,400}$ day	T
Mechanical units					
Acceleration	a	$\frac{\text{velocity}}{\text{time}} = \frac{\text{length}}{\text{time}^2}$	$\frac{\text{meter}}{\text{second}^2}$ (m s^{-2})	$\frac{\text{N}}{\text{kg}}$	$\frac{L}{T^2}$
Area	A, a, s	length^2	meter^2 (m^2)		L^2
Energy or work	W	force × length = power × time	joule (J)	N m = W s = V C = 10^7 ergs = 10^8 dynes mm	$\frac{ML^2}{T^2}$
Energy density	w	$\frac{\text{energy}}{\text{volume}}$	$\frac{\text{joule}}{\text{meter}^3}$ (J m^{-3})	$\frac{\frac{1}{100}\text{ erg}}{\text{mm}^3}$	$\frac{M}{LT^2}$
Force	**F**	mass × acceleration	newton (N)	$\frac{\text{kg m}}{\text{s}^2} = \frac{\text{J}}{\text{m}}$ $= 10^5$ dynes	$\frac{ML}{T^2}$
Frequency	f	$\frac{\text{cycles}}{\text{time}}$	hertz (Hz)	$\frac{\text{cycles}}{\text{s}}$	$\frac{1}{T}$
Impedance	Z	$\frac{\text{force}}{\text{mass} \times \text{velocity}}$	$\frac{\text{newton-second}}{\text{kilogram-meter}}$	$\frac{\text{N s}}{\text{kg m}}$	$\frac{1}{T}$
Length	L, l, ℓ		meter (m)	1,000 mm = 100 cm	L
Mass	M, m		kilogram (kg)	1,000 g	M

TABLE A-1
(Continued)

Name of dimension or quantity	Symbol	Description	SI unit and abbreviation	Equivalent unit	Dimension
Mechanical units *(continued)*					
Moment (torque)		force × length	newton-meter (N m)	$\frac{\text{kg m}^2}{\text{s}^2} = \text{J}$	$\frac{ML^2}{T^2}$
Momentum	$m\mathbf{v}$	mass × velocity = force × time $= \frac{\text{energy}}{\text{velocity}}$	newton-second (N s)	$\frac{\text{kg m}}{\text{s}} = \frac{\text{J s}}{\text{m}}$	$\frac{ML}{T}$
Period	T	$\frac{1}{\text{frequency}}$	second (s)		T
Power	P	$\frac{\text{force} \times \text{length}}{\text{time}} = \frac{\text{energy}}{\text{time}}$	watt (W)	$\frac{\text{J}}{\text{s}} = \frac{\text{N m}}{\text{s}} = \frac{\text{kg m}^2}{\text{s}^3}$	$\frac{ML^2}{T^3}$
Time	T, t		second (s)	$\frac{1}{60}$ min $= \frac{1}{3600}$ h $= \frac{1}{86{,}400}$ day	T
Velocity (velocity of light in vacuum = 300 Mm s^{-1})	$\mathbf{v}$	$\frac{\text{length}}{\text{time}}$	$\frac{\text{meter}}{\text{second}}$ (m s^{-1})		$\frac{L}{T}$
Volume	v	length^3	meter^3 (m^3)		L^3
Electrical units					
Admittance	Y	$\frac{1}{\text{impedance}}$	mho (℧)	$\frac{\text{A}}{\text{V}} = \frac{\text{C}^2}{\text{J s}} = \text{S}\dagger$	$\frac{I^2T^3}{ML^2}$
Capacitance	C	$\frac{\text{charge}}{\text{potential}}$	farad (F)	$\frac{\text{Q}}{\text{V}} = \frac{\text{C}^2}{\text{J}} = \frac{\text{As}}{\text{V}}$ $= 9 \times 10^{11}$ cm (cgs esu)	$\frac{I^2T^4}{ML^2}$
Charge	Q, q	current × time	coulomb (C)	6.25×10^{18} electron charges = A s $= 3 \times 10^9$ cgs esu = 0.1 cgs emu	IT

TABLE A-1
(Continued)

Name of dimension or quantity	Symbol	Description	SI unit and abbreviation	Equivalent unit	Dimension
		Electrical units (*continued*)			
Charge (volume) density	ρ	$\frac{\text{charge}}{\text{volume}} = \nabla \bullet \mathbf{D}$	$\frac{\text{coulomb}}{\text{meter}^3}$ (C m^{-3})	$\frac{\text{A s}}{\text{m}^3}$	$\frac{IT}{L^3}$
Conductance	G	$\frac{1}{\text{resistance}}$	mho (℧)	$\frac{\text{A}}{\text{V}} = \frac{\text{C}^2}{\text{J s}} = \text{S}\dagger$	$\frac{I^2T^3}{ML^2}$
Conductivity	σ	$\frac{1}{\text{resistivity}}$	$\frac{\text{mho}}{\text{meter}}$ (℧ m^{-1})	$\frac{1}{\Omega\ \text{m}}$	$\frac{I^2T^3}{ML^3}$
Current	I, i	$\frac{\text{charge}}{\text{time}}$	ampere (A)	$\frac{\text{C}}{\text{s}} = 3 \times 10^9$ cgs esu $= 0.1$ cgs emu	I
Current density	J	$\frac{\text{current}}{\text{area}}$	$\frac{\text{ampere}}{\text{meter}^2}$ (A m^{-2})	$\frac{\text{C}}{\text{s m}^2}$	$\frac{I}{L^2}$
Dipole moment	$\mathbf{p}(= q\mathbf{l})$	charge × length	coulomb-meter (C m)	A s m	LIT
Emf	$\mathcal{V}$	$\int \mathbf{E} \bullet d\mathbf{l}$	volt (V)	$\frac{\text{Wb}}{\text{s}} = \frac{\text{J}}{\text{C}}$	$\frac{ML^2}{IT^3}$
Energy density (electric)	w_e	$\frac{\text{energy}}{\text{volume}}$	$\frac{\text{joule}}{\text{meter}^3}$ (J m^{-3})	$\frac{\frac{1}{100}\ \text{erg}}{\text{mm}^3}$	$\frac{M}{LT^2}$
Field intensity	$\mathbf{E}$	$\frac{\text{potential}}{\text{length}} = \frac{\text{force}}{\text{charge}}$	$\frac{\text{volt}}{\text{meter}}$ (V m^{-1})	$\frac{\text{N}}{\text{C}} = \frac{\text{J}}{\text{C m}} = \frac{1}{3} \times 10^{-4}$ cgs esu $= 10^6$ cgs emu	$\frac{ML}{IT^3}$
Flux	ψ	charge $= \iint \mathbf{D} \bullet d\mathbf{s}$	coulomb (C)	A s	IT
Flux density (displacement) (D vector)	$\mathbf{D}$	$\frac{\text{charge}}{\text{area}}$	$\frac{\text{coulomb}}{\text{meter}^2}$ (C m^{-2})	$\frac{\text{A s}}{\text{m}^2} = \frac{A}{\text{m}^2\ \text{s}^{-1}}$	$\frac{IT}{L^2}$
Impedance	Z	$\frac{\text{potential}}{\text{current}}$	ohm (Ω)	$\frac{\text{V}}{\text{A}}$	$\frac{ML^2}{I^2T^3}$
Linear charge density	ρ_L	$\frac{\text{charge}}{\text{length}}$	$\frac{\text{coulomb}}{\text{meter}}$ (C m^{-1})	$\frac{\text{A s}}{\text{m}}$	$\frac{IT}{L}$

TABLE A-1
(Continued)

Name of dimension or quantity	Symbol	Description	SI unit and abbreviation	Equivalent unit	Dimension
		Electrical units (*continued*)			
Permittivity (dielectric constant) (for vacuum, $\varepsilon_0 = 8.85$ pF m^{-1} $\approx 10^{-9}/36\pi$ F m^{-1})	ε	$\frac{\text{capacitance}}{\text{length}}$	$\frac{\text{farad}}{\text{meter}}$ (F m^{-1})	$\frac{\text{C}}{\text{V m}}$	$\frac{I^2T^4}{ML^3}$
Polarization	**P**	$\frac{\text{dipole moment}}{\text{volume}}$	$\frac{\text{coulomb}}{\text{meter}^2}$ (C m^{-2})	$\frac{\text{A s}}{\text{m}^2}$	$\frac{IT}{L^2}$
Potential	V	$\frac{\text{work}}{\text{charge}}$	volt (V)	$\frac{\text{J}}{\text{C}} = \frac{\text{N m}}{\text{C}} = \frac{\text{W s}}{\text{C}} = \frac{\text{W}}{\text{A}} = \frac{\text{Wb}}{\text{s}} = \frac{1}{300}$ cgs esu $= 10^8$ cgs emu	$\frac{ML^2}{IT^3}$
Poynting vector	**S**	$\frac{\text{power}}{\text{area}}$	$\frac{\text{watt}}{\text{meter}^2}$ (W m^{-2})	$\frac{\text{J}}{\text{s m}^2}$	$\frac{M}{T^3}$
Radiation intensity	P	$\frac{\text{power}}{\text{unit solid angle}}$	$\frac{\text{watt}}{\text{steradian}}$ (W sr^{-1})		$\frac{ML^2}{T^3}$
Reactance	X	$\frac{\text{potential}}{\text{current}}$	ohm (Ω)	$\frac{\text{V}}{\text{A}}$	$\frac{ML^2}{I^2T^3}$
Relative permittivity	ε_r	ratio $\frac{\varepsilon}{\varepsilon_0}$			dimensionless
Resistance	R	$\frac{\text{potential}}{\text{current}}$	ohm (Ω)	$\frac{\text{V}}{\text{A}} = \frac{\text{J s}}{\text{C}^2} = \frac{1}{9} \times 10^{-11}$ cgs esu $= 10^{-9}$ cgs emu	$\frac{ML^2}{I^2T^3}$
Resistivity	S	resistance × length $= \frac{1}{\text{conductivity}}$	ohm-meter (Ω m)	$\frac{\text{V m}}{\text{A}}$	$\frac{ML^3}{I^2T^3}$
Sheet-current density	**K**	$\frac{\text{current}}{\text{length}}$	$\frac{\text{ampere}}{\text{meter}}$ (A m^{-1})	$\frac{\text{A}}{\text{m}^2} \times$ m	$\frac{I}{L}$

TABLE A-1
(Continued)

Name of dimension or quantity	Symbol	Description	SI unit and abbreviation	Equivalent unit	Dimension
		Electrical units (*continued*)			
Susceptance	B	$\dfrac{1}{\text{reactance}}$	mho (℧)	$\dfrac{\text{A}}{\text{V}} = \text{S}$	$\dfrac{I^2T^3}{ML^2}$
Wavelength	λ	length	meter (m)		L
		Magnetic units			
Dipole moment (magnetic)	m $(=Q_m\, l)$	pole strength × length = current × area $= \dfrac{\text{torque}}{\text{magnetic flux density}}$	ampere-meter^2 (A m^2)	$\dfrac{\text{C m}^2}{\text{s}}$	IL^2
Energy density (magnetic)	w_m	$\dfrac{\text{energy}}{\text{volume}}$	$\dfrac{\text{joule}}{\text{meter}^3}$ (J m^{-3})	$\dfrac{\frac{1}{100}\text{ erg}}{\text{mm}^3}$	$\dfrac{M}{LT^2}$
Flux (magnetic)	ψ_m	$\iint \mathbf{B} \bullet d\mathbf{s}$	weber (Wb)	$\text{V s} = \dfrac{\text{N m}}{\text{A}}$ $= 10^8\ Mx$‡ (cgs emu)	$\dfrac{ML^2}{IT^2}$
Flux density (B vector)	**B**	$\dfrac{\text{force}}{\text{pole}} = \dfrac{\text{force}}{\text{current moment}}$ $= \dfrac{\text{magnetic flux}}{\text{area}}$	tesla (T) $= \dfrac{\text{weber}}{\text{meter}^2}$ (Wb m^{-2})	$\dfrac{\text{V s}}{\text{m}^2} = \dfrac{\text{N}}{\text{A m}}$ $= 10^4$ G‡ (cgs emu)	$\dfrac{M}{IT^2}$
Flux linkage	Λ	flux × turns	weber-turn (Wb turn)		$\dfrac{ML^2}{IT^2}$
H field (*H* vector)	**H**	$\dfrac{\text{mmf}}{\text{length}}$	$\dfrac{\text{ampere}}{\text{meter}}$ (A m^{-1})	$\dfrac{\text{N}}{\text{Wb}} = \dfrac{\text{W}}{\text{V m}}$ $= 4\pi \times 10^{-3}$ Oe‡ (cgs emu) $= 400\pi$ gammas	$\dfrac{I}{L}$
Inductance	L	$\dfrac{\text{magnetic flux linkage}}{\text{current}}$	henry (H)	$\dfrac{\text{Wb}}{\text{A}} = \dfrac{\text{J}}{\text{A}^2} = \Omega\text{ s}$ $= \frac{1}{9} \times 10^{-11}$ cgs esu $= 10^9$ cm (cgs emu)	$\dfrac{ML^2}{I^2T^2}$
Magnetization (magnetic polarization)	**M**	$\dfrac{\text{magnetic moment}}{\text{volume}}$	$\dfrac{\text{ampere}}{\text{meter}}$ (A m^{-1})	$\dfrac{\text{A m}^2}{\text{m}^3} = \dfrac{\text{A m}}{\text{m}^2}$	$\dfrac{I}{L}$

TABLE A-1
(Continued)

Name of dimension or quantity	Symbol	Description	SI unit and abbreviation	Equivalent unit	Dimension
		Magnetic units *(continued)*			
Mmf	F	$\int \mathbf{H} \bullet d\mathbf{l}$	ampere-turn (A turn)	$\frac{\text{C}}{\text{s}}$	I
Permeability (for vacuum, $\mu_0 = 400\pi$ nH m^{-1})	μ	$\frac{\text{inductance}}{\text{length}}$	$\frac{\text{henry}}{\text{meter}}$ (H m^{-1})	$\frac{\text{Wb}}{\text{A m}} = \frac{\text{V s}}{\text{A m}}$	$\frac{ML}{I^2T^2}$
Permeance	$\mathcal{P}$	$\frac{\text{magnetic flux}}{\text{mmf}} = \frac{1}{\text{reluctance}}$	henry (H)	$\frac{\text{Wb}}{\text{A}}$	$\frac{ML^2}{I^2T^2}$
Pole density	ρ_m	$\frac{\text{pole strength}}{\text{volume}} = \frac{\text{current}}{\text{area}} = \nabla \bullet \mathbf{H} = -\nabla \bullet \mathbf{M}$	$\frac{\text{ampere}}{\text{meter}^2}$ (A m^{-2})		$\frac{I}{L^2}$
Pole strength	Q_m, q_m	current × length $= \iiint \rho_m\, dv$	ampere-meter (A m)	$\frac{\text{C m}}{\text{s}}$	IL
Potential (magnetic) (for **H**)	U	$\int \mathbf{H} \bullet d\mathbf{l}$	ampere (A)	$\frac{\text{J}}{\text{Wb}} = \frac{\text{W}}{\text{V}} = \frac{\text{C}}{\text{s}} = \frac{4\pi}{10}$ Gb§ (cgs emu)	I
Relative permeability	μ_r	ratio $\frac{\mu}{\mu_0}$			Dimensionless
Reluctance	$\mathcal{R}$	$\frac{\text{mmf}}{\text{magnetic flux}} = \frac{1}{\text{permeance}}$	$\frac{1}{\text{henry}}$ (H^{-1})	$\frac{\text{A}}{\text{Wb}}$	$\frac{I^2T^2}{ML^2}$
Vector potential	**A**	current × permeability	$\frac{\text{Weber}}{\text{henry}}$ (Wb m^{-1})	$\frac{\text{H A}}{\text{m}} = \frac{\text{N}}{\text{A}}$	$\frac{ML}{IT^2}$

†S is the SI abbreviation for siemens, used often for mho.
‡Mx, G, and Oe are SI abbreviations for maxwell, gauss, and oersted.
§Gb is the SI abbreviation for gilbert.

TABLE A-2
Trigonometric, hyperbolic, logarithmic, and other relations

Trigonometric relations

$\sin (x \pm y) = \sin x \cos y \pm \cos x \sin y$

$\cos (x \pm y) = \cos x \cos y \mp \sin x \sin y$

$\sin (x + y) + \sin (x - y) = 2 \sin x \cos y$

$\cos (x + y) + \cos (x - y) = 2 \cos x \cos y$

$\sin (x + y) - \sin (x - y) = 2 \cos x \sin y$

$\cos (x + y) - \cos (x - y) = -2 \sin x \sin y$

$\sin 2x = 2 \sin x \cos x$

$\cos 2x = \cos^2 x - \sin^2 x = 2 \cos^2 x - 1 = 1 - 2 \sin^2 x$

$\cos x = 2 \cos^2 \tfrac{1}{2}x - 1 = 1 - 2 \sin^2 \tfrac{1}{2}x$

$\sin x = 2 \sin \tfrac{1}{2}x \cos \tfrac{1}{2}x$

$\sin^2 x + \cos^2 x = 1$

$$\tan (x + y) = \frac{\tan x + \tan y}{1 - \tan x \tan y}$$

$$\tan (x - y) = \frac{\tan x - \tan y}{1 + \tan x \tan y}$$

$$\tan 2x = \frac{2 \tan x}{1 - \tan^2 x}$$

$$\sin x = x - \frac{x^3}{3!} + \frac{x^5}{5!} - \frac{x^7}{7!} + \dots$$

$$\cos x = 1 - \frac{x^2}{2!} + \frac{x^4}{4!} - \frac{x^6}{6!} + \dots$$

$$\tan x = x + \frac{x^3}{3} + \frac{2x^5}{15} + \frac{17x^7}{315} + \frac{62x^9}{2835} + \dots$$

Hyperbolic relations

$$\sinh x = \frac{e^x - e^{-x}}{2} = x + \frac{x^3}{3!} + \frac{x^5}{5!} + \frac{x^7}{7!} + \dots$$

$$\cosh x = \frac{e^x + e^{-x}}{2} = 1 + \frac{x^2}{2!} + \frac{x^4}{4!} + \frac{x^6}{6!} + \dots$$

$$\tanh x = \frac{\sinh x}{\cosh x}$$

$$\coth x = \frac{\cosh x}{\sinh x} = \frac{1}{\tanh x}$$

$\sinh (x \pm jy) = \sinh x \cos y \pm j \cosh x \sin y$

$\cosh (x \pm jy) = \cosh x \cos y \pm j \sinh x \sin y$

$$\left.\begin{aligned} \cosh (jx) &= \tfrac{1}{2}(e^{+jx} + e^{-jx}) = \cos x \\ \sinh (jx) &= \tfrac{1}{2}(e^{+jx} - e^{-jx}) = j \sin x \end{aligned}\right\} \text{de Moivre's theorem}$$

TABLE A-2
(continued)

Hyperbolic relations *(continued)*

$$e^{\pm jx} = \cos x \pm j \sin x$$

$$e^{\pm jx} = 1 \pm jx - \frac{x^2}{2!} \mp j\frac{x^3}{3!} + \frac{x^4}{4!} \pm j\frac{x^5}{5} - \dots$$

$$e^x = \cosh x + \sinh x$$

$$e^{-x} = \cosh x - \sinh x$$

$$e^x = 1 + x + \frac{x^2}{2!} + \frac{x^3}{3!} + \frac{x^4}{4!} + \dots$$

$$\cosh x = \cos jx$$

$$j \sinh x = \sin jx$$

$$\tanh (x \pm jy) = \frac{\sinh 2x}{\cosh 2x + \cos 2y} \pm j\frac{\sin 2y}{\cosh 2x + \cos 2y}$$

$$\coth (x \pm jy) = \frac{\sinh 2x}{\cosh 2x - \cos 2y} \pm j\frac{\sin 2y}{\cosh 2x - \cos 2y}$$

Logarithmic relations

$\log_{10} x = \log x$ common logarithm

$\log_e x = \ln x$ natural logarithm

$$\log_{10} x = 0.4343 \log_e x = 0.4343 \ln x$$

$$\ln x = \log_e x = 2.3026 \log_{10} x$$

$$e = 2.71828$$

$$\text{dB} = 10 \log (\text{power ratio}) = 20 \log (\text{voltage ratio})$$

$$1 \text{ Np (voltage attenuation)} = \frac{1}{e} = 0.368(\text{voltage}) = -8.68 \text{ dB}$$

Approximation formulas for small quantities

(δ is a small quantity compared with unity.)

$$(1 \pm \delta)^2 = 1 \pm 2\delta$$

$$(1 \pm \delta)^n = 1 \pm n\delta$$

$$\sqrt{1 + \delta} = 1 + \tfrac{1}{2}\delta$$

$$\frac{1}{\sqrt{1 + \delta}} = 1 - \tfrac{1}{2}\delta$$

$$e^\delta = 1 + \delta$$

$$\ln(1 + \delta) = \delta$$

$J_n(\delta) = \dfrac{\delta^n}{n!2^n}$ for $|\delta| \ll 1$

where J_n is Bessel function of order n. Thus

$$J_1(\delta) = \frac{\delta}{2}$$

TABLE A-2
(continued)

Series

Binomial:

$$(x+y)^n = x^n + nx^{n-1}y + \frac{n(n-1)}{2!}x^{n-2}y^2 + \frac{n(n-1)(n-2)}{3!}x^{(n-3)}y^3 + \ldots$$

Taylor's:

$$f(x+y) = f(x) + \frac{df(x)}{dx}\frac{y}{1} + \frac{d^2f(x)}{dx^2}\frac{y^2}{2!} + \frac{d^3f(x)}{dx^3}\frac{y^3}{3!} + \ldots$$

Solution of quadratic equation

If $ax^2 + bx + c = 0$, then

$$x = \frac{-b \pm \sqrt{b^2 - 4ac}}{2a}$$

Gradient, divergence, and curl in rectangular, cylindrical, and spherical coordinates

Rectangular coordinates

$$\nabla f = \hat{\mathbf{x}}\frac{\partial f}{\partial x} + \hat{\mathbf{y}}\frac{\partial f}{\partial y} + \hat{\mathbf{z}}\frac{\partial f}{\partial z}$$

$$\nabla \bullet \mathbf{A} = \frac{\partial A_x}{\partial x} + \frac{\partial A_y}{\partial y} + \frac{\partial A_z}{\partial z}$$

$$\nabla \times \mathbf{A} = \hat{\mathbf{x}}\left(\frac{\partial A_z}{\partial y} - \frac{\partial A_y}{\partial z}\right) + \hat{\mathbf{y}}\left(\frac{\partial A_x}{\partial z} - \frac{\partial A_z}{\partial x}\right) + \hat{\mathbf{z}}\left(\frac{\partial A_y}{\partial x} - \frac{\partial A_x}{\partial y}\right) = \begin{vmatrix} \hat{\mathbf{x}} & \hat{\mathbf{y}} & \hat{\mathbf{z}} \\ \frac{\partial}{\partial x} & \frac{\partial}{\partial y} & \frac{\partial}{\partial z} \\ A_x & A_y & A_z \end{vmatrix}$$

Cylindrical coordinates

$$\nabla f = \hat{\mathbf{r}}\frac{\partial f}{\partial r} + \hat{\boldsymbol{\phi}}\frac{1}{r}\frac{\partial f}{\partial \phi} + \hat{\mathbf{z}}\frac{\partial f}{\partial z}$$

$$\nabla \bullet \mathbf{A} = \frac{1}{r}\frac{\partial}{\partial r}rA_r + \frac{1}{r}\frac{\partial A_\phi}{\partial \phi} + \frac{\partial A_z}{\partial z}$$

$$\nabla \times \mathbf{A} = \hat{\mathbf{r}}\left(\frac{1}{r}\frac{\partial A_z}{\partial \phi} - \frac{\partial A_\phi}{\partial z}\right) + \hat{\boldsymbol{\phi}}\left(\frac{\partial A_r}{\partial z} - \frac{\partial A_z}{\partial r}\right) + \hat{\mathbf{z}}\frac{1}{r}\left(\frac{\partial}{\partial r}rA_\phi - \frac{\partial A_r}{\partial \phi}\right) = \begin{vmatrix} \hat{\mathbf{r}}\frac{1}{r} & \hat{\boldsymbol{\phi}} & \hat{\mathbf{z}}\frac{1}{r} \\ \frac{\partial}{\partial r} & \frac{\partial}{\partial \phi} & \frac{\partial}{\partial z} \\ A_r & rA_\phi & A_z \end{vmatrix}$$

Spherical coordinates

$$\nabla f = \hat{\mathbf{r}}\frac{\partial f}{\partial r} + \hat{\boldsymbol{\theta}}\frac{1}{r}\frac{\partial f}{\partial \theta} + \hat{\boldsymbol{\phi}}\frac{1}{r\sin\theta}\frac{\partial f}{\partial \phi}$$

$$\nabla \bullet \mathbf{A} = \frac{1}{r^2}\frac{\partial}{\partial r}r^2A_r + \frac{1}{r\sin\theta}\frac{\partial}{\partial \theta}(A_\theta \sin\theta) + \frac{1}{r\sin\theta}\frac{\partial A_\phi}{\partial \phi}$$

$$\nabla \times \mathbf{A} = \hat{\mathbf{r}}\frac{1}{r\sin\theta}\left(\frac{\partial}{\partial \theta}(A_\phi \sin\theta) - \frac{\partial A_\theta}{\partial \phi}\right) + \hat{\boldsymbol{\theta}}\frac{1}{r}\left(\frac{1}{\sin\theta}\frac{\partial A_r}{\partial \phi} - \frac{\partial}{\partial r}rA_\phi\right) + \hat{\boldsymbol{\phi}}\frac{1}{r}\left(\frac{\partial}{\partial r}rA_\theta - \frac{\partial A_r}{\partial \theta}\right)$$

TABLE A-3

Glyphs (nonalphabetic pictograph symbols)

It is not usually realized that someone had to invent each of these symbols. For example, the first known use of the equals sign was by Robert Record in 1557, who used it to avoid the need for repeating the words "is equal to."

Symbol	Definition
$=$	Equal to
$\sim$ or $\approx$	Approximately equal to
$\cong$	Nearly equal to
$\equiv$	Identical with or by definition
$\neq$	Not equal to
$\propto$	Proportional to
%	Percent
$\rightarrow$	Approaches
$<$	Less than
$>$	Greater than
$\leq$	Less than or equal to
$\geq$	Greater than or equal to
$\ll$	Much less than
$\gg$	Much greater than
∞	Infinity
$\therefore$	Therefore
!	Factorial
$\sqrt{\ }$	Square root
$\lvert\ \rvert$	Absolute value
$\sum$	Summation sign
$\int$	Integral sign
$\oint$	Line integral around a closed path
$\iint$ or $\int_S$	Surface integral
$\oiint$ or $\oint_S$	Surface integral completely enclosing a volume
$\iiint$ or $\int_V$	Volume integral

APPENDIX B

FIELD MAPS, LAPLACE'S EQUATION, FULL VECTOR NOTATION

FIELD MAPPING

A field map is an electromagnetic "road map." As we have seen, it provides an overall physical picture and important insights about electric and magnetic field distributions.

Field maps can be produced by many different methods. The map of the microstrip transmission line of Fig. B-1 was drawn graphically in 30 minutes. A

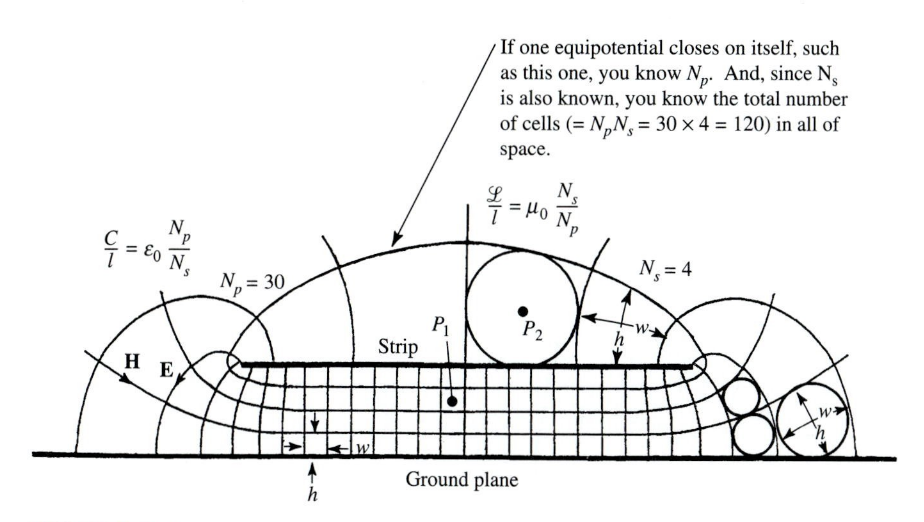

FIGURE B-1
Cross-section of microstrip transmission line mapped into field cells with test circles. ($w = h$ for all cells.) $\varepsilon_r = 1$ everywhere.

graphical map is very useful as a check on solutions by other methods. The technique of graphical mapping is explained in all of the earlier editions of this book.

Field maps by computational methods are discussed in Chap. 11. ***All solutions including graphical, are solutions of Laplace's equation. Such a solution is unique; it is the one and only solution.***

Graphical Solution

Field maps become especially useful when divided into curvilinear squares. A *curvilinear square* is a four-sided area (with field and equipotentials intersecting at right angles) that tends to yield true squares as the area is subdivided into smaller areas. Thus, the large area in Fig. B-2 is a *curvilinear square* while the small ones are true squares.

The microstrip transmission line of Fig. B-1 is mapped into curvilinear squares or field cells. A simple way to determine whether a "square" is a true curvilinear square is to draw a circle inside. All four sides should touch the circle indicating that the median dimensions of the square are equal. Test circles have been drawn in the "squares" of the fringing field of the microstrip line of Fig. B-1. The median distances w and h should be equal.

The field map of Fig. B-1 applies not only to the static electric and magnetic cases but also to the transverse electric and magnetic fields of an electromagnetic wave traveling on the transmission line provided the line is uniform and essentially lossless.

The maps of Figs. B-1 and B-2 are two-dimensional. The field configuration is the same for all planes perpendicular to a uniform transmission line. All such 2-dimensional geometries can be mapped into curvilinear squares.

Returning to the roadmap qualities of a field map, the map of the microstrip transmission line of Fig. B-1 summarizes everything we have discussed. From this

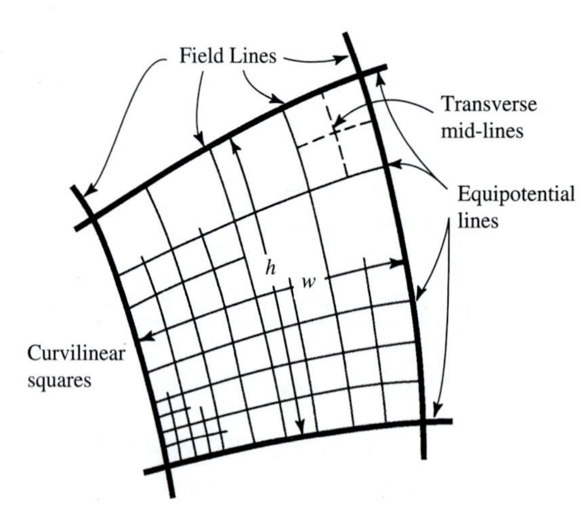

FIGURE B-2
Curvilinear square partially subdivided into smaller curvilinear squares. ($w = h$ for all squares.)

one map we can evaluate 15 electromagnetic quantities. Thus, with the constants of the medium given, we know the inductance, capacitance, and conductance of the line simply from its geometry (no dimensions need be given). From the map we know that

$$\frac{\mathcal{L}}{l} = \mu\frac{N_s}{N_p} = \mu\frac{4}{30} = 0.133\ \mu\ \mathrm{H\,m^{-1}}$$

$$\frac{C}{l} = \varepsilon\frac{N_p}{N_s} = \varepsilon\frac{30}{4} = 7.5\varepsilon\ \mathrm{F\,m^{-1}}$$

$$\frac{G}{l} = \sigma\frac{N_p}{N_s} = \sigma\frac{30}{4} = 7.5\sigma\ \mho\ \mathrm{m^{-1}}$$

$$Z = \sqrt{\mu/\varepsilon}(N_s/N_p) = 377(4/30) = 50\ \Omega \text{ for } \mu_r = \varepsilon_r = 1$$

where N_p = number of cells in parallel and N_s = number of cells in series.

The map also tells us that there are no more than 120 cells (= 30 × 4) ALL THE WAY TO INFINITY.

If the line voltage V is given, we know the line current $I(= V/Z)$ or vice versa. It is assumed that the line is matched (terminated in Z), a condition discussed in Chap. 3. We also know the line power $P(= V^2/Z)$ and the power per cell, which is equal to P divided by the number of cells (N_pN_s).

Example B-1. Microstrip line. Referring to Fig. B-1, if the width of the strip is 10 cm and $V = 100$ V, find P(line), P(cell) and the power density (P.D.) at the points P_1 and P_2. P.D. is the magnitude of the Poynting vector of Chap. 3.

Solution

$$P(\text{line}) = 100^2/50 = 200\ \mathrm{W} \quad \textit{Ans.}$$

$$P(\text{cell}) = 200/120 = 1.67\ \mathrm{W} \quad \textit{Ans.}$$

The power density at any cell is given by

$$\text{P.D.} = P(\text{cell})/\text{cell area}$$

At point P_1 the cell width is 1/20 the strip width so the power density at point P_1 is

$$\text{P.D.(at } P_1) = 1.67\ \mathrm{W}/(5\times10^{-3})^2\mathrm{m}^2 = 67\ \mathrm{kW\,m^{-2}} \quad \textit{Ans.}$$

Above the strip at P_2 the cell width w (or height h) is 5 times that at P_1, and the power density there is

$$\text{P.D.(at } P_2) = 67/5^2 = 2.7\ \mathrm{kW\ m^{-2}} \quad \textit{Ans.}$$

Problem B-1-1. Fringing power. What is the fraction of the power transmitted in the fringing field of the microstrip transmission line of Fig. B-1? *Ans.* 33 percent.

Just from the geometry alone, we know the variation and relative magnitudes everywhere on the map for the

Electric field, **E**
Magnetic field, **H**
Electric flux density, **D**
Magnetic flux density, **B**

and on the strip and ground plane the relative magnitudes of the

Electric charge density, ρ_s
Sheet current density, **K**

Given the line voltage V and strip width, we know the absolute magnitudes of all of the above quantities plus the electric and magnetic fluxes ψ and ψ_m everywhere on the map.

In the following example all of these quantities are evaluated. **This example is a summary or capsule of much that we have discussed in the book.**

Example B-2. Sixteen quantities of microstrip line of Fig. B-1. If $V = 100$, strip width $= 10$ cm, $\mu_r = \varepsilon_r = 1, \sigma = 0$, find the following sixteen quantities:

(a) $\mathcal{L}/l$, C/l, G/l, Z(line), and I.
(b) E, H, D, and B at P_1.
(c) ρ_s and K on the bottom side of the strip above P_1 or on ground plane below P_1.
(d) ψ and ψ_m per tube.
(e) P(line), P(cell), and power density (P.D.) at P_1 and P_2.

Solution with answers

(a1) $\mathcal{L}/l = \frac{4}{30}\mu_0 = 0.133 \times 4\pi 10^{-7} = 167 \text{ nH m}^{-1}$
(a2) $C/l = \frac{30}{4}\varepsilon_0 = 7.5 \times 8.85 \times 10^{-12} = 66 \text{ pF m}^{-1}$
(a3) $G/l = \frac{4}{30}\sigma = 0 \text{ ℧ m}^{-1}$
(a4) $Z(\text{line}) = 377\left(\frac{4}{30}\right) = 50\ \Omega$
(a5) $I = V/Z = 100/50 = 2$ A
(b1) $E = 100/0.02 = 5 \text{ kV m}^{-1}$ at P_1
(b2) $H = E/Z = 5000/50 = 100 \text{ A m}^{-1}$ at P_1
(b3) $D = \varepsilon_0 E = 8.85 \times 10^{-12} \times 5000 = 44 \text{ nC m}^{-2}$ at P_1
(b4) $B = \mu_0 H = 4\pi \times 10^{-7} \times 100 = 126\ \mu\text{Wb m}^{-2}$ at P_1
(c1) $\rho_s = D = 44 \text{ nC m}^{-2}$
(c2) $K = H = 100 \text{ A m}^{-1}$
(d1) $\psi/l = D \times \text{cell width} = \rho_s \omega = 44 \times 10^{-9} \times 0.02/4 = 220 \text{ pC m}^{-1}$

(d2) $\psi_m/l = B \times \text{cell height (same as width)} = Bh = 126 \times 10^{-6} \times 0.02/4$
$= 630 \text{ nWb m}^{-1} = \text{magnetic flux per tube}$

(e1) $P(\text{line}) = V^2/Z = 100^2/50 = 200 \text{ W}$

(e2) $P(\text{cell}) = 200/(30 \times 4) = 1.67 \text{ W}$

(e3) P.D. $(P_1) = 67 \text{ kW m}^{-2}$, P.D. $(P_2) = 2.7 \text{ kW m}^{-2}$ from previous example.

Problem B-1-2. Fields of microstrip line. Find E, H, D, and B at point P_2 of Fig. B-1. *Ans.* $E = 1 \text{ kV m}^{-1}$; $H = 20 \text{ A m}^{-1}$; $D = 8.8 \text{ nC m}^{-2}$; $B = 25 \text{ μWb m}^{-2} = 25 \text{ μT}$.

FULL VECTOR NOTATION

Referring to Fig. B-3, the field **E** in rectangular coordinates at a point (x, y, z) at a distance r from the origin is given by

$$\mathbf{E} = \underbrace{\frac{Q_1}{4\pi\varepsilon_0} \frac{1}{(x^2 + y^2 + z^2)}}_{\text{Magnitude}} \underbrace{\frac{x\hat{\mathbf{x}} + y\hat{\mathbf{y}} + z\hat{\mathbf{z}}}{(x^2 + y^2 + z^2)^{1/2}}}_{\text{Unit vector}} = \frac{Q_1(x\hat{\mathbf{x}} + y\hat{\mathbf{y}} + z\hat{\mathbf{z}})}{4\pi\varepsilon_0(x^2 + y^2 + z^2)^{3/2}} \quad \text{N C}^{-1} \tag{1}$$

The field **E** at a point $P(x, y, z)$ from a charge Q_1 at (x_1, y_1, z_1) (*not* at the origin) is given by

$$\mathbf{E} = \frac{Q_1}{4\pi\varepsilon_0} \frac{(x - x_1)\hat{\mathbf{x}} + (y - y_1)\hat{\mathbf{y}} + (z - z_1)\hat{\mathbf{z}}}{[(x - x_1)^2 + (y - y_1)^2 + (z - z_1)^2]^{3/2}} \tag{2}$$

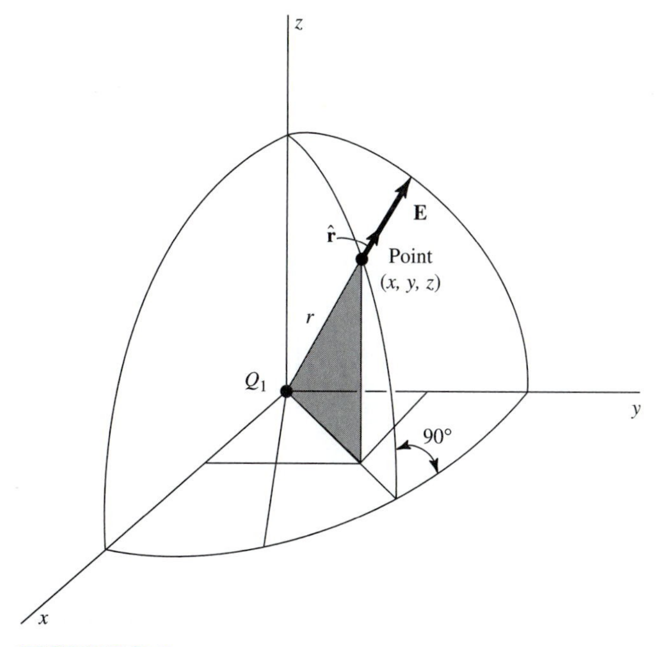

FIGURE B-3

Electric field at a point *(x, y, z)* from charge Q_1 at the origin as given by (3).

or, referring to Fig. B-4, by

$$\mathbf{E} = \frac{Q_1}{4\pi\varepsilon_0 \left|\mathbf{R}_p - \mathbf{R}_1\right|^2} \frac{\mathbf{R}_p - \mathbf{R}_1}{\left|\mathbf{R}_p - \mathbf{R}_1\right|} = \frac{Q_1}{4\pi\varepsilon_0} \frac{(\mathbf{R}_p - \mathbf{R}_1)}{\left|\mathbf{R}_p - \mathbf{R}_1\right|^3} = \frac{Q_1}{4\pi\varepsilon_0} \frac{\mathbf{R}_{1p}}{\left|\mathbf{R}_{1p}\right|^3} \tag{3}$$

If the charge is at the origin ($x_1 = y_1 = z_1 = 0$), (2) reduces to (1).

Example B-3. Electric Field. Find the electric field at point (0,2,3) due to a charge $Q_1 = 10$ nC at (1,0,1). Dimensions in meters, $\varepsilon_r = 1$.

Solution. From Fig. B-4 and (3),

$$\mathbf{R}_p = 0\hat{\mathbf{x}} + 2\hat{\mathbf{y}} + 3\hat{\mathbf{z}}$$

$$\mathbf{R}_1 = 1\hat{\mathbf{x}} + 0\hat{\mathbf{y}} + 1\hat{\mathbf{z}}$$

$$\mathbf{R}_p - \mathbf{R}_1 = -1\hat{\mathbf{x}} + 2\hat{\mathbf{y}} + 2\hat{\mathbf{z}}$$

$$\left|\mathbf{R}_p - \mathbf{R}_1\right| = \sqrt{(-1)^2 + 2^2 + 2^2} = 3$$

Therefore

$$\mathbf{E} = \left|\mathbf{E}\right| \hat{\mathbf{r}} = \frac{9 \times 10^9 \times 10 \times 10^{-9}}{3^2} \frac{(-1\hat{\mathbf{x}} + 2\hat{\mathbf{y}} + 2\hat{\mathbf{z}})}{3} = 10 \frac{-1\hat{\mathbf{x}} + 2\hat{\mathbf{y}} + 2\hat{\mathbf{z}}}{3}$$

$$= -3.33\hat{\mathbf{x}} + 6.67\hat{\mathbf{y}} + 6.67\hat{\mathbf{z}} \quad \text{V m}^{-1} \text{ or N C}^{-1}$$

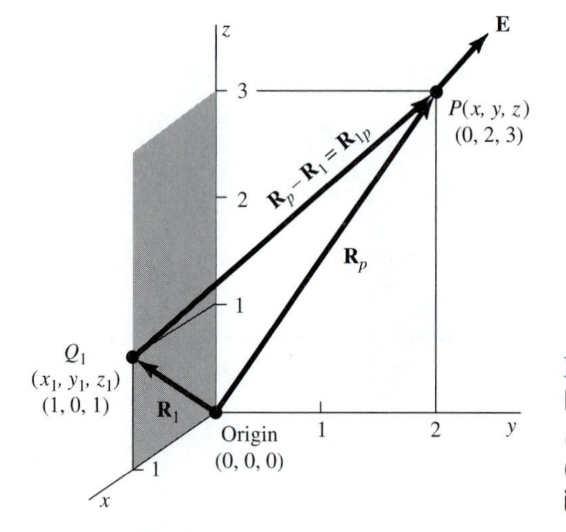

FIGURE B-4
Electric field at point P at (x, y, z) due to charge Q_1 at (x_1, y_1, z_1) (not at the origin) is given by (2) and (3).

Referring to Fig. B-5 (on next page), the field intensity of the charge Q_1 at the point P is $\mathbf{E}_1$ and of the charge Q_2 is $\mathbf{E}_2$. The total field at P due to both charges is the vector sum of $\mathbf{E}_1$ and $\mathbf{E}_2$, as indicated in the figure.

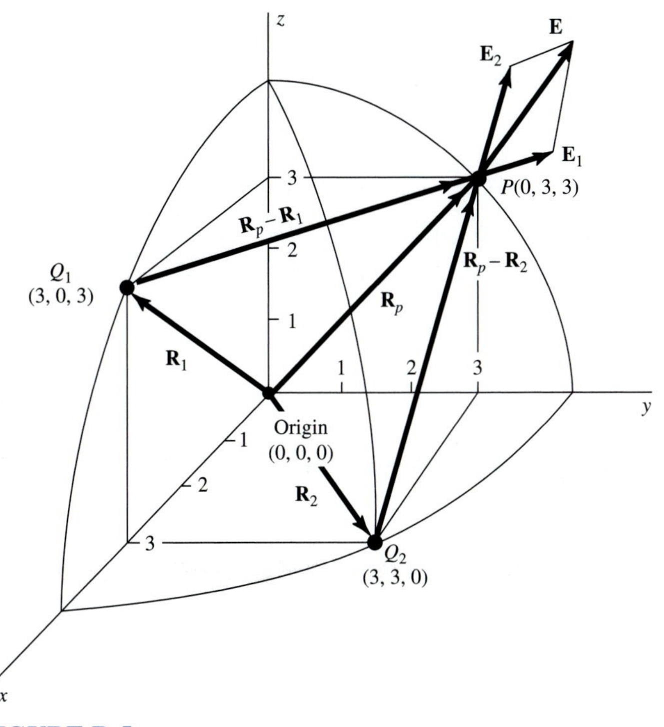

FIGURE B-5
Charge Q_1 at (3, 0, 3) and charge Q_2 at (3, 3, 0) produce field **E** at (0, 3, 3) as given in Example B-4. Dimensions in meters.

Thus,

$$\begin{aligned}\mathbf{E} = \mathbf{E}_1 + \mathbf{E}_2 &= \frac{Q_1}{4\pi\varepsilon_0 \left|\mathbf{R}_p - \mathbf{R}_1\right|^2} \frac{\mathbf{R}_p - \mathbf{R}_1}{\left|\mathbf{R}_p - \mathbf{R}_1\right|} + \frac{Q_2}{4\pi\varepsilon_0 \left|\mathbf{R}_p - \mathbf{R}_2\right|^2} \frac{\mathbf{R}_p - \mathbf{R}_2}{\left|\mathbf{R}_p - \mathbf{R}_2\right|} \\ &= \frac{Q_1}{4\pi\varepsilon_0} \frac{\mathbf{R}_{1p}}{\left|\mathbf{R}_{1p}\right|^3} + \frac{Q_2}{4\pi\varepsilon_0} \frac{\mathbf{R}_{2p}}{\left|\mathbf{R}_{2p}\right|^3}\end{aligned} \tag{4}$$

The field **E** at a point P due to any number n of charges will have n terms like those in (4) with one term for each charge. In more concise notation,

$$\begin{aligned}\mathbf{E} = \mathbf{E}_1 + \mathbf{E}_2 + \cdots + \mathbf{E}_n &= \sum_{m=1}^{m=n} \mathbf{E}_m = \sum_{m=1}^{m=n} \frac{Q_m}{4\pi\varepsilon_0 \left|\mathbf{R}_p - \mathbf{R}_m\right|^2} \frac{\mathbf{R}_p - \mathbf{R}_m}{\left|\mathbf{R}_p - \mathbf{R}_m\right|} \\ &= \sum_{m=1}^{m=n} \frac{Q_m}{4\pi\varepsilon_0} \frac{\mathbf{R}_{mp}}{\left|\mathbf{R}_{mp}\right|^3}\end{aligned} \tag{5}$$

where the summation, indicated by the symbol Σ, includes all integral values of m from $m = 1$ to $m = n$. Thus, if $n = 2$, expression of (5) yields (4).

Example B-4. Electric field of two charges. If Q_1 in Fig. B-5 is 5 nC at point $(x_1, y_1, z_1) = (3, 0, 3)$ and $Q_2 = 8$ nC at $(x_2, y_2, z_2) = (3, 3, 0)$, find the electric field intensity **E** at the point P at $(x, y, z) = (0, 3, 3)$. Dimensions in meters $\varepsilon = \varepsilon_0\ (\varepsilon_r = 1)$.

Solution. We have

$$\mathbf{R}_{1p} = \mathbf{R}_p - \mathbf{R}_1 = 0\hat{\mathbf{x}} + 3\hat{\mathbf{y}} + 3\hat{\mathbf{z}} - (3\hat{\mathbf{x}} + 0\hat{\mathbf{y}} + 3\hat{\mathbf{z}}) = -3\hat{\mathbf{x}} + 3\hat{\mathbf{y}} + 0\hat{\mathbf{z}}$$

$$\left|\mathbf{R}_{1p}\right| = \left|\mathbf{R}_p - \mathbf{R}_1\right| = \sqrt{18}$$

$$\mathbf{R}_{2p} = \mathbf{R}_p - \mathbf{R}_2 = 0\hat{\mathbf{x}} + 3\hat{\mathbf{y}} + 3\hat{\mathbf{z}} - (3\hat{\mathbf{x}} + 3\hat{\mathbf{y}} + 0\hat{\mathbf{z}}) = -3\hat{\mathbf{x}} + 0\hat{\mathbf{y}} + 3\hat{\mathbf{z}}$$

$$\left|R_{2p}\right| = \left|\mathbf{R}_p - \mathbf{R}_2\right| = \sqrt{18}$$

Thus, from (4) or (5),

$$\mathbf{E} = \mathbf{E}_1 + \mathbf{E}_2$$

$$= \frac{9 \times 10^9 \times 5 \times 10^{-9}}{18} \frac{(-3\hat{\mathbf{x}} + 3\hat{\mathbf{y}} + 0\hat{\mathbf{z}})}{\sqrt{18}} + \frac{9 \times 10^9 \times 8 \times 10^{-9}}{18} \frac{(-3\hat{\mathbf{x}} + 0\hat{\mathbf{y}} + 3\hat{\mathbf{z}})}{\sqrt{18}}$$

and

$$\mathbf{E} = \mathbf{E}_1 + \mathbf{E}_2 = (-1.77\hat{\mathbf{x}} + 1.77\hat{\mathbf{y}}) + (-2.83\hat{\mathbf{x}} + 2.83\hat{\mathbf{z}})$$
$$= -4.60\hat{\mathbf{x}} + 1.77\hat{\mathbf{y}} + 2.83\hat{\mathbf{z}} \quad \text{N C}^{-1} \text{ or V m}^{-1}$$

In magnitude and unit vector form

$$\mathbf{E}_1 = 2.50(-0.707\hat{\mathbf{x}} + 0.707\hat{\mathbf{y}})$$
$$\mathbf{E}_2 = 4.00(-0.707\hat{\mathbf{x}} + 0.707\hat{\mathbf{z}})$$
$$\mathbf{E} = 5.68(-0.810\hat{\mathbf{x}} + 0.312\hat{\mathbf{y}} + 0.500\hat{\mathbf{z}}) \quad \text{N C}^{-1} \text{ or V m}^{-1}$$

APPENDIX C

COMPUTER PROGRAMS

With the computer programs now available, computational methods are attractive for solving electromagnetic problems. But they have a fundamental limitation as pointed out by Professor Edward Newman of Ohio State University:

> It is true that numerical techniques can be more accurate than analytical results since they generally involve fewer assumptions. However, a big disadvantage of numerical techniques is that they yield only NUMBERS and not EQUATIONS. Simple equations are useful since we can look at them and gain physical insight into the problem and we can also get design information. Thus, numerical techniques will never replace good (accurate but simple) analytical results.

And nothing will replace experimental measurements for actual results.

Of course, computers are wonderful tools for performing tedious mathematical calculations and displaying results that would take hours by hand. For this reason, 13 computer programs are listed here which are available on the book's Web site at **www.elmag5.com.** We invite you to download, use, and customize these programs.

Program	Function
1. ZX	Calculates impedance on a transmission line
2. VSWR	Evaluates VSWR on a lossless transmission line
3. Bouncing pulses	Tracks progress of a dc pulse on a transmission line
4. Traveling waves	Provides time-lapse plots of a wave on a line
5. Ground bounce	Calculates field strength vs. height on ground-plane range
6. ARRAYPATGAIN	Calculates gain and draws field pattern for an array of sources
7. REPLA	Uses repetitive Laplace to find potentials of a charge distribution
8. Charged plates	Calculates charge distribution, C/ℓ, $\mathcal{L}/\ell$, and Z of a stripline
9. Post	Calculates potentials for an initially uniform field disturbed by the introduction of a post or a fence
10. Lossy line	Calculates voltage along a lossy terminated line. See cover and Fig. 4-11.3.
11. V-LEVEL	Calculates potential values around an arbitrary array of point charges
12. SMITH CHART	Shows your location on a Smith chart as you adjust stub length and distance from a load.
13. QWT	Calculates single and double $\lambda/4$ transformers to match a line.

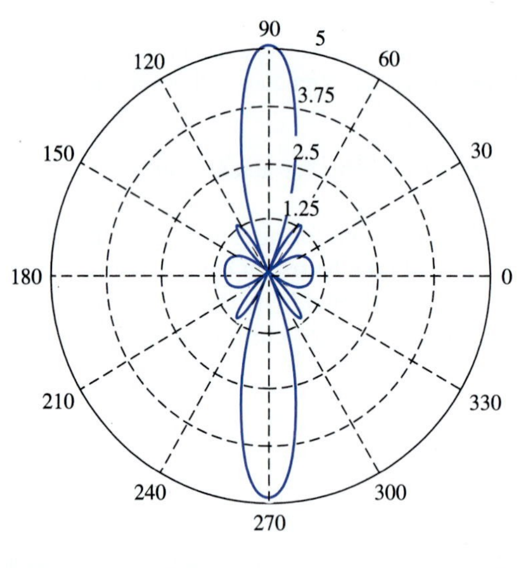

Number of Sources 5

Spacing (wavelengths) 0.5

Phasing (deg) 0

Multiplying Factor 1

FIGURE C-1
User interface and display for broadside array with ARRAYPATGAIN.

Appendix

D

PROJECT EQUIPMENT

10-GHz Gunn oscillators are widely used as doppler radars in automatic door openers and in security systems. These units are available from:

SHF microwave Parts Co.
7102 W 500 S.
LaPorte, IN 46350
www.shfmicro.com
Fax 1-219-789-4552

See also the book's Web site at: **www.elmag5.com.**

APPENDIX E

ANSWERS

Answers to end-of-chapter problems followed by the @ symbol.

CHAPTER 1

1-4-4. (*a*) $10^{15}s^2\ kg^{-1}m^{-1}$; (*b*) $10^{-12}\ s^3\ kg^{-1}m^{-1}$; (*c*) $10^{-3}m^{-1}$.

1-5-3. All balanced except for P (power).

1-6-14. (*a*) 24; (*b*) $\hat{\mathbf{x}}(-4) + \hat{\mathbf{y}}(2) + \hat{\mathbf{z}}(-12)$; (*c*) 28.1°.

1-6-16. 0.17 m.

1-6-18. (*a*) 0.81 m^2; (*b*) 0.38 sr.

1-6-20. −1000 V.

1-6-22. (*a*) 34.8°; (*b*) 4.9 units; (*c*) $\hat{\mathbf{x}}(0.08) + \hat{\mathbf{y}}(0.84) + \hat{\mathbf{z}}(-0.5)$; (*d*) 38.2 $units^2$; (*e*) $\hat{\mathbf{x}}(0.52) + \hat{\mathbf{y}}(0.37) + \hat{\mathbf{z}}(0.77)$.

1-6-24. (*a*) 156.9°; (*b*) 11.9 units; (*c*) $\hat{\mathbf{x}}(-0.55) + \hat{\mathbf{y}}(-0.14) + \hat{\mathbf{z}}(-0.83)$; (*d*) 14.5 $units^2$; (*e*) $\hat{\mathbf{x}}(0.33) + \hat{\mathbf{y}}(0.86) + \hat{\mathbf{z}}(-0.38)$.

1-6-26. (*a*) 15.5 units; (*b*) 9.08 $units^2$; (*c*) 19.9°.

1-6-28. (*a*) 50.7; (*b*) 3.67; (*c*) −0.5.

1-6-30. (*a*) 67; (*b*) 100.

1-7-5. (*a*) (4.47, −63.4°, −1); (*b*) (4.58, 102.6°, −63.4°).

1-7-7. (*a*) 3.2 units; (*b*) 64.3°; (*c*) 3.37 units; (*d*) 5.05 $units^2$.

1-7-9. (*a*) (–3.75, 2.17, 2.5); (*b*) (4.3, −3.01°, 2.5).

1-7-11. (*a*) $\hat{\mathbf{x}}(4.24) + \hat{\mathbf{y}}(11.31) + \hat{\mathbf{z}}(6)$; (*b*) 38.98; (*c*) $\hat{\mathbf{x}}(-26.85) + \hat{\mathbf{y}}(16.94) + \hat{\mathbf{z}}(-12.97)$.

1-7-13. (*a*) (3.61, 56.3°, 5); (*b*) 6.38; (*c*) −0.55; (*d*) 6.

1-7-15. (*a*) $\hat{\mathbf{x}}(6.02) + \hat{\mathbf{y}}(11.73) + \hat{\mathbf{z}}(7)$; (*b*) 54.42; (*c*) 17.3°; (*d*) $\hat{\mathbf{x}}(-4.92) + \hat{\mathbf{y}}(10.08) + \hat{\mathbf{z}}(-12.66)$; (*e*) $\hat{\mathbf{x}}(-0.29) + \hat{\mathbf{y}}(0.59) + \hat{\mathbf{z}}(-0.75)$; (*f*) 16.95 $units^2$.

1-7-17. (*a*) $\hat{\mathbf{x}}(5.99) + \hat{\mathbf{y}}(5.88) + \hat{\mathbf{z}}(3.24)$; (*b*) 6.5; (*c*) 78.9°; (*d*) $\hat{\mathbf{x}}(-0.27) + \hat{\mathbf{y}}(16.19) + \hat{\mathbf{z}}(-28.87)$; (*e*) $\hat{\mathbf{x}}(-0.01) + \hat{\mathbf{y}}(0.49) + \hat{\mathbf{z}}(-0.87)$; (*f*) 33.1 $units^2$.

1-7-19. $\hat{\mathbf{x}}(-1.74) + \hat{\mathbf{y}}(6.74) + \hat{\mathbf{z}}(2.12)$.

1-7-21. $-1/6$.
1-7-23. 2.7.
1-7-25. (*a*) 251.5 units^2; (*b*) 258.2 units^3.
1-7-27. 12.

CHAPTER 2

2-2-4. $\hat{\mathbf{x}}(3.87) + \hat{\mathbf{y}}(-1.29) + \hat{\mathbf{z}}(1.29)$ nN.
2-2-6. (*a*) $\hat{\mathbf{x}}(2.71) + \hat{\mathbf{y}}(-3.25) + \hat{\mathbf{z}}(2.71)$ V/m; (*b*) $\hat{\mathbf{y}}(-5)$ V/m; (*c*) $\hat{\mathbf{x}}(-2.71) + \hat{\mathbf{y}}(-3.25) + \hat{\mathbf{z}}(-2.71)$ V/m.
2-2-8. $\hat{\mathbf{x}}(-3.3) + \hat{\mathbf{y}}(-64.9) + \hat{\mathbf{z}}(-33.5)$ V/m.
2-2-10. 749.3 V/m.
2-3-6. (*a*) 1.08 m; (*b*) 0.27 m.
2-3-8. $\hat{\mathbf{r}}(\phi z^{1/2} r^{-2}) + \hat{\boldsymbol{\phi}}(-z^{1/2} r^{-2}) + \hat{\mathbf{z}}(-\frac{1}{2}\phi z^{-1/2} r^{-1})$ V/m.
2-3-10. $6e^{-2x}$V/m.
2-3-12. 3.3 V.
2-3-14. 29.5 V/m.
2-6-4. 16.7 pC/m.
2-7-2. Yes.
2-7-4. -0.11 C/m^3.
2-7-6. (*a*) 3294 C; (*b*) 3294 C.
2-8-2. $\hat{\mathbf{x}}(15\varepsilon_0) + \hat{\mathbf{y}}(9\varepsilon_0)$.
2-9-5. (*a*) 1.06 pF; (*b*) 12.7 pF.
2-9-7. (*a*) E is halved; (*b*) E is doubled; (*c*) in (*a*) work is done *by* capacitor, while in (*b*) work is done *on* capacitor.
2-9-9. (*a*) 2.26×10^{11} J; (*b*) 2.26×10^9 V; (*c*) 2.26×10^6 V/m.
2-11-8. Silver wire is 7 percent longer.
2-11-10. Electron drift velocity = 39 mm/s. Hole drift velocity = 19 mm/s.
2-11-11. 54 f℧/m.
2-11-12. $5.42/\sigma\ \Omega$.
2-11-14. 4.6 ℧/m.
2-12-10. (*a*) 0 nT; (*b*) 0.14 nT; (*c*) 0.20 nT.
2-12-12. (*a*)$Ir/2\pi r_0^2$; (*b*) $I/2\pi r$.
2-12-14. -13.5 V.
2-15-5. $\hat{\mathbf{z}}$.
2-15-6. $\hat{\mathbf{z}}(2)$.
2-15-7. $\hat{\mathbf{r}}\left(\frac{\cos\theta}{r}\right) + \hat{\boldsymbol{\theta}}\left(\frac{-\sin\theta}{r}\right) + \hat{\boldsymbol{\phi}}(10)$.

CHAPTER 3

3-2-3. (*a*) α 10 percent above exact value, β 10 percent below exact value.
(*b*) α, β within 10^{-7} of exact value.
(*c*) α, β within 10^{-10} of exact value.

3-2-5. (*a*) Line is not distortionless; (*b*) 1.5 mΩ/m.
3-2-7. (*a*) 0.086; (*b*) 0.027; (*c*) 8.6×10^{-3}.
3-2-9. (*a*) 0.44 dB/km; (*b*) 0.44 dB/km; (c) 0.44 dB/km.
3-3-12. (*a*) 250 μW/m^2; (*b*) 0.27 V/m; (*c*) 0.93 mA/m.
3-3-14. (*a*) 65.8 Ω; (*b*) 32.9 Ω; (*c*) 19.0 Ω.
3-3-16. (*a*) 107.4 Ω; (*b*) 53.7 Ω; (*c*) 31.0 Ω.
3-3-18. Coaxial: 13.7 kW, 27.4 kW, 47.5 kW for $\varepsilon_r = 1, 4, 12$;
Single conductor: 8.4 kW, 16.8 kW, 29.1 kW for $\varepsilon_r = 1, 4, 12$.
3-4-16. $15.7 + j\,9.5$ Ω.
3-4-18. (*a*) $0.21\angle 153.6°$; (*b*) 1.55; (*c*) $55.8\angle -17.8°$ Ω; (*d*) 0.213 λ;
(*e*) 116 Ω.
3-4-20. (*a*) 5.8; (*b*) 2; (*c*) 2.16; (*d*) 1.08.
3-4-23. (*a*) 0; (*b*) 0.44; (*c*) 0.44; (*d*) 0.001; (*e*) 1.78.

CHAPTER 4

4-2-4. (*a*) 377 Ω; (*b*) 274 mm.
4-2-5. 1.6 at f_1 and 1.34 at f_2.
4-2-6. (*a*) 2.65 nA/m; (*b*) 1.19 nA/m; (*c*) 709 μA/m.
4-2-8. $103.5 + j42.9$ Ω.
4-5-2. 1.62 mm.
4-5-3. (*a*) $0.82\angle -13°$; (*b*) 0.033; (*c*) 0.033; (*d*) 4.9 mm; (*e*) 9 dB;
(*f*) $0.15\angle -130°$.
4-5-6. (*a*) $|\rho_v| \approx 0.01$; (*b*) $|\rho_v| \approx 0.06$.
4-6-7. 28 μW.
4-6-9. 17 mW.
4-7-11. (*a*) 3.5; (*b*) 2.4 mm.
4-7-13. 13 dB.
4-7-15. 25.9 dB.
4-7-17. 5.4 dB.
4-7-18. 27.5 mm.
4-7-19. −9.9 dB.
4-7-21. 3.4 m.
4-7-23. (*a*) 36.5 V/m; (*b*) 136.9 mA/m; (*c*) 1.06×10^8 m/s; (*d*) 266.6 Ω.
4-8-3. 4.
4-9-3. 75 Mm/s.
4-10-2. (*a*) 8.67×10^7 m/s; (*b*) 2.07 W/m^2; (*c*) 108.8 Ω; (*d*) 137.9 mA/m.
4-10-4. 10.7 μW.
4-10-6. (*a*) 8.74 kW/m^2; (*b*) 3.95×10^{26} W; (*c*) 1.8 kV/m; (*d*) 200 s.
4-11-1. 6.2.
4-12-2. 144 μW.
4-13-3. (*a*) ∞; (*b*) 0°; (*c*) 22.6 mV/m (rms); (*d*) 1.35 μW/m^2.
4-13-4. (*a*) 1; (*b*) 16 mV/m (rms); (*c*) 42.4 μA/m (rms); (*d*) 1.36 μW/m^2; (*e*) LCP.
4-13-5. (*a*) 1; (*b*) 0.5; (*c*) 0; (*d*) 0.5; (*e*) 0.72.

4-13-7.

Point	MF	Point	MF
1	0.5	6	0
2	0.5	7	0.69
3	1.0	8	0.85
4	0.5	9	0
5	0.85		

4-14-4. (*a*) 30°; (*b*) 12.9°; (*c*) total internal reflection.
4-14-5. −0.43.
4-16-5. 42.5 dB.
4-16-6. (*a*) −25.7 dB; (*b*) −19.9 dB; (*c*) −6.4 dB.

CHAPTER 5

5-2-5. 2 mW.
5-2-6. 30 million.
5-2-7. *(a, b, c)* 5.1 versus 3.8, 6 versus 4.6, and 7.1 versus 6.1.
5-2-8. (*a*) 1540 W/m^2; (*b*) 4.3×10^{26} W; (*c*) 1.3×10^{13} years; (*d*) 760 V/m.
Note: The answer to (*c*) is unrealistic because the sun will "shut down" long before all its mass is converted into energy. Its useful life for us on the earth is more like a few billion years.
5-2-11. (*a*) 2.05×10^4 (43.1 dB); (*b*) 1.03×10^4 (40.1 dB).
5-3-11. $E = \cos(54^\circ \cos\phi - 52^\circ)\cos(108^\circ \cos\phi + 90^\circ)$, ϕ measured ccw from N.
5-5-2. (*a*) $2.86 \times 10^{-2}\angle -9^\circ$ V/m; (*b*) $8.77 \times 10^{-2}\angle 81^\circ$ V/m; (*c*) $2.39 \times 10^{-4}\angle 81^\circ$A/m.
5-5-4. (*a*) General: 282 mV/m, quasi-stationary: 121 mV/m.
(*b*) General: 242 mV/m, quasi-stationary: 61 mV/m.
(*c*) General: 784 μA/m, quasi-stationary: 338 μA/m.
5-5-6. (*a*) 2.74 μW; (*b*) 3.0 A.
5-6-2. (*a*) 23.2 Ω.
5-7-2. (*a*) 11 dBi; (*b*) 14 dBi approx.
5-7-2.1. The four collinear pairs in parallel present a feed point impedance at P_2 of 750 Ω. This can be matched to a 500-Ω line with a λ/4 section of impedance $Z = \sqrt{750 \times 500} = 612\ \Omega$. At a frequency of 10 MHz the λ/4 section length = 7.5 m.
5-7-2.2. $d_1 = 4.3$ m, $d_2 = 5.6$ m.
5-7-3. (*a*) 29 dBi approx.; (*b*) at least 960 m.
5-7-4. (*a*) 10 V/m, (*b*) 2 mV/m. The answers for (*a*) and (*b*) bracket the extremes that might be expected. We have neglected the effects of multipath progation via multiple ionosphere-to-earth reflections causing alternate enhancement and cancellation (fading) of signals.
5-7-5. (*a*) 1.38; (*b*) 1.36; (*c*) 1.30.
5-9-9. 3.8 m.
5-9-10. (*a*) 28.8 cm; (*b*) 27.6 cm; (*c*) 26.4, 26.4, 25.8, and 24.0 cm; (*d*) 30 cm; (*e*) 19.1 cm; (*f*) 465 to 535 MHz; (*g*) 10 dBi.
5-9-12. (*a*) 10; (*b*) 6 cm; (*c*) 1.05 (95 percent pure RCP).
5-9-13. 0.19 m^2.
5-9-14. 17.5 cm.

5-9-16. 200d.
5-10-1. 225 km.
5-10-2. (a) 368 μV/m; (b) 50 dB.
5-10-3. (a) 11 dB; (b) 11 dB; (c) 18 dB.
5-10-5. (a) 12.4 W; (b) 257 s.
5-11-7. 1000 LY. The data rate is slow but the first trans-Atlantic cable wasn't much faster. To put 1000 light-years (LY) in perspective, recall that the circumference of the earth is 1/7 light-second, the distance to Jupiter 45 light-minutes, to the planet Pluto 5 light-hours, and to the nearest stars 4 light-years. A distance of 1000 LY is 250 times farther, yet it is only 1 percent of the distance across our galaxy and there are billions and billions of other galaxies, some at distances of more than 10 billion light-years. If the ETC message could be deciphered and understood and the earth sent a reply, it would be at least 2000 years before we might expect an answer. But would we have the vision and patience to wait 2000 years? Even though it is technologically possible, dialogues over such distances may never occur. We will just listen and wonder.
5-11-8. 2×10^{38} W (or 200 trillion trillion trillion watts).
5-11-9. 960 km line-of-sight.
5-11-12. 108.4 dB.
5-11-13. (a) 8.16×10^5 W; (b) 1.0 m; (c) Yes ($G/T = 7.4$ dB K^{-1}); (d) 3.2 dB.
5-11-14. 10.8 dB.
5-11-15. 113.5 dB.
5-11-18. (a) 152 kbits/s; (b) 11 bits/s.
5-11-19. (a) 12 cm; (b) 5; (c) 11 dB.
5-11-20. (a) With blockage by car, signal is below noise but without blockage SNR = 61 dB.
(b) SNR = 73 dB with or without blockage by car.
5-12-4. 45 m/s (= 162 km/h = 101 mi/h).
5-12-5. 160 mW.
5-12-6. 1.2 kHz.
5-12-10. 5.6×10^{-28} m^2 .
5-12-11. 1.25×10^{16} W peak power.
5-12-13. 66 ns.
5-12-15. RCS of plate = 223,000 m^2; RCS of sphere = 4 m^2.
5-12-16. (a) 3.1×10^{-15} W; (b) 9.0×10^{-17} W; (c) Signal-to-noise = S/N = 31 dB, Signal-to-clutter = S/C = 16 dB.
5-14-5. 0.5×2.6 m.
5-16-3. 620 mW.

CHAPTER 6

6-3-2. 258 Mm/s.
6-3-4. 3.2 mm.
6-3-7. 8 mm.
6-4-4. 7.1 G (or 7 times the earth's field).
6-4-6. (a) $F = 0$; (b) $F = 0$; (c) $|\mathbf{F}| = 2 \times 10^{-23}$ N.

6-5-4. (*a*) 1 Mm/s; (*b*) 500 mm.
6-5-6. 181 A.

CHAPTER 7

7-9-1. (*a*) 620 pF; (*b*) 24 kV.
7-9-3. ~7 mm.
7-10-6. (*a*) $Q^2/8\pi\varepsilon_0 R$; (*b*) $2R$.
7-10-8. 7 m^3. Capacitors better for quick release of energy.
7-12-3. 25 A-m^2.
7-15-3. (*a*) 1.29 GJ/m^3; (*b*) 22.8 kJ.
7-15-5. (*a*) $\mathcal{L}/l = \mu_0\left\{\left(N^2\pi a^2/l^2\right) + \left[\left(\frac{1}{2\pi}\right)\ln(b/a)\right]\right\}$

(*b*) $\mathcal{L}/l = \mu_0\left\{\left(\mu_r N^2\pi a^2/l^2\right) + \left[\left(\frac{1}{2\pi}\right)\ln(b/a)\right]\right\}$

CHAPTER 8

8-4-4. (*a*) 160 mm; (*b*) 5.
8-8-3. 1.35 nW/m^2.
8-11-4. $1.8\sqrt{1 - j10^{-11}}$
8-11-6. 0.14 dB/km.
8-12-5. (*a*) 4.23 GHz; (*b*) 71 mm; (*c*) 14,300.
8-12-7. 5.3 GHz.

CHAPTER 9

9-7-4. 16 min.
9-7-6. (*a*) 650 mG; (*b*) 300 mG; (*c*) much above.
9-SP-1. 1.9×10^5 km^2 (approximately equal to area of South Dakota or 1/800 land area of world).
9-SP-2. 40 m^2.

CHAPTER 10

10-3-5. (*a*) CMOS: 0.433 m (FR-4), 0.250 m (alumina)
ECL: 0.138 m (FR-4), 0.080 m (alumina);
(*b*) CMOS: 7.22 cm (FR-4), 4.17 cm (alumina)
ECL: 2.30 cm (FR-4), 1.33 cm (alumina);
(*c*) CMOS: 200 MHz, ECL: 625 MHz.
10-3-6. (*a*) $T_0 = 1.75$ ns, $C_0 = 116.8$ pF/m, $Z_0 = 43.2\ \Omega$, $\rho_s = -0.565$, $\rho_L = 1$, $A_s = 0.783$, $T_L = 2$;
(*b*) Voltage dips to -2.83 V at T_0' (loaded T_0) and rises to 1.59 V at $3T_0'$. The latter is 31.8 percent of ΔV and is therefore very serious.
10-3-10. (*a*) $V_{\text{refl}}/\Delta V = -CZ_0/2T_r$; (*b*) lower C, lower Z_o, raise T_r; (*c*) $C \leq 6$ pF.

10-4-2. (*a*) No; (*b*) apparent series inductance would increase; (*c*) $d/h \geq 3$; (*d*) 0.75 mm.

10-5-2. (*a*) 208 MHz;

(*b*) The “bead” is a cylinder of ferrite which reduces unwanted return currents on power cables;

(*c*) Near the computer to minimize coupling of unwanted signals onto cable.

CHAPTER 11

11-7-1. (*a*) 1.425 : 1.094 : 1.013 : 1.00 : 1.00 : 1.013 : 1.094 : 1.425.

11-7-3. (*b*) $E = 4.5$ times uniform field.

11-7-4. (*a*) 100 V everywhere; (*b*) 0 V/m everywhere.

INDEX[†]

[†]ff = also following pages

rad radian
rad^2 square radian = steradian = sr
S, S Poynting vector, W m^{-2}
S flux density, W m^{-2} Hz^{-1}
S, s distance, m; also surface area, m^2
s second (of time)
sr steradian = square radian = rad^2
T tesla = Wb m^{-2}
T tera = 10^{12} (prefix)
T torque, N m
t time, s
U radiation intensity, W sr^{-1}
U magnetostatic potential, A
V volt
V voltage (also emf), V
$\mathcal{V}$ emf (electromotive force), V
v velocity, m s^{-1}
v volume, m^3
W watt
Wb weber = 10^4 gauss
w energy density, J m^{-3}
X reactance, Ω
X reactance/unit length, Ω m^{-1}
$\hat{\mathbf{x}}$ unit vector in x direction
x coordinate direction
Y admittance, $\mho$
Y admittance/unit length, $\mho$ m^{-1}
$\hat{\mathbf{y}}$ unit vector in y direction
y coordinate direction
Z impedance, Ω
Z impedance/unit length, Ω m^{-1}
Z_c intrinsic impedance, conductor, Ω per square
Z_d intrinsic impedance, dielectric, Ω per square
Z_L load impedance, Ω
Z_{yz} transverse impedance, rectangular waveguide, Ω
$Z_{r\phi}$ transverse impedance, cylindrical waveguide, Ω
Z_0 intrinsic impedance, space, Ω per square
Z_0 characteristic impedance, transmission line, Ω
$\hat{\mathbf{z}}$ unit vector in z direction
z coordinate direction, also red shift
α (alpha) angle, deg or rad
α attentuation constant, Np m^{-1}
β (beta) angle, deg or rad; also phase constant = $2\pi/\lambda$
γ (gamma) angle, deg or rad
δ (delta) angle, deg or rad
δ depth of penetration
ϵ (epsilon) permittivity (dielectric constant), F m^{-1}
ϵ_{ap} aperture efficiency
ϵ_M beam efficiency
ϵ_m stray factor
ϵ_r relative permittivity
ϵ_0 permittivity of vacuum, F m^{-1}
η (eta) index of refraction
θ (theta) angle, deg or rad
$\hat{\boldsymbol{\theta}}$ (theta) unit vector in θ direction
κ (kappa) constant
Λ flux linkage, Wb turn
λ (lambda) wavelength, m
λ_0 free-space wavelength, m
μ (mu) permeability, H m^{-1}
μ_r relative permeability
μ_0 permeability of vacuum, H m^{-1}
ν (nu)
ξ (xi)
π (pi) = 3.1416
ρ (rho) electric charge density, C m^{-3}; also mass density, kg m^{-3}
ρ reflection coefficient, dimensionless
ρ_s surface charge density, C m^{-2}
ρ_c linear charge density, C m^{-1}
σ (sigma) conductivity, $\mho$ m^{-1}
σ radar cross section, m^2
τ (tau) tilt angle polarization ellipse, deg or rad
τ transmission coefficient
ϕ (phi) angle, deg or rad
$\hat{\boldsymbol{\phi}}$ (phi) unit vector in ϕ direction
χ (chi) susceptibility, dimensionless
ψ (psi) angle, deg or rad
ψ_m magnetic flux, Wb
Ω (capital omega) ohm
Ω (capital omega) solid angle, sr or deg^2
Ω_A beam area, rad^2 or deg^2
Ω_M main beam area, rad^2 or deg^2
Ω_m minor lobe area, rad^2 or deg^2
$\mho$ (upsidedown capital omega) mho ($\mho = 1/\Omega$ = **S**, siemens)
ω (omega) angular frequency ($= 2\pi f$), rad s^{-1}